# Introduction to Sociology

Elgin Communiy College Edition

11th Edition

Henry L. Tischler

**CENGAGE Learning·**

Australia • Brazil • Japan • Korea • Mexico • Singapore • Spain • United Kingdom • United States

# CENGAGE Learning

**Introduction to Sociology: Elgin Communiy College Edition, 11th Edition**

Senior Manager, Student Engagement:
Linda deStefano
Janey Moeller

Manager, Student Engagement:
Julie Dierig

Marketing Manager:
Rachael Kloos

Manager, Production Editorial:
Kim Fry

Manager, Intellectual Property Project Manager:
Brian Methe

Senior Manager, Production and Manufacturing:
Donna M. Brown

Manager, Production:
Terri Daley

Cengage Advantage Books: Introduction to Sociology, 11th Edition
Henry L. Tischler

© 2014, 2011 Cengage Learning. All Rights Reserved.
Library of Congress Control Number: 2012955604

ALL RIGHTS RESERVED. No part of this work covered by the copyright herein may be reproduced, transmitted, stored or used in any form or by any means graphic, electronic, or mechanical, including but not limited to photocopying, recording, scanning, digitizing, taping, Web distribution, information networks, or information storage and retrieval systems, except as permitted under Section 107 or 108 of the 1976 United States Copyright Act, without the prior written permission of the publisher.

> For product information and technology assistance, contact us at
> **Cengage Learning Customer & Sales Support, 1-800-354-9706**
> For permission to use material from this text or product,
> submit all requests online at **cengage.com/permissions**
> Further permissions questions can be emailed to
> **permissionrequest@cengage.com**

This book contains select works from existing Cengage Learning resources and was produced by Cengage Learning Custom Solutions for collegiate use. As such, those adopting and/or contributing to this work are responsible for editorial content accuracy, continuity and completeness.

**Compilation © 2014 Cengage Learning**

ISBN-13: 978-1-305-04588-0
ISBN-10: 1-305-04588-2

**WCN: 01-100-101**

**Cengage Learning**
5191 Natorp Boulevard
Mason, Ohio 45040
USA

Cengage Learning is a leading provider of customized learning solutions with office locations around the globe, including Singapore, the United Kingdom, Australia, Mexico, Brazil, and Japan. Locate your local office at:
**international.cengage.com/region.**

Cengage Learning products are represented in Canada by Nelson Education, Ltd.
For your lifelong learning solutions, visit **www.cengage.com/custom.**
Visit our corporate website at **www.cengage.com.**

Printed in the United States of America

# About ECC – Mission, Vision, Values, and Goals

## Mission Statement

To improve people's lives through learning.

## Vision Statement

We pursue our mission by focusing all our efforts on making Elgin Community College one of the best centers of learning in the United States. In recognition of our role as a comprehensive community college, we will strive to create high-quality learning opportunities that respond to the needs of the residents of our district.

## Shared Values

*Excellence* — All college functions and services must strive for the highest level of excellence to successfully achieve our mission. Learning activities must be of the highest quality to help students achieve their goals.

*Freedom of Inquiry* — We believe a learning community is most engaging and viable when a spirit of free inquiry exists, allowing everyone the freedoms to explore new and diverse ideas and to express their interests and attitudes.

*Equity* — We believe that everyone must have an equal opportunity to grow through learning and pledge to provide all who take part in our learning activities with the opportunities and support needed for success.

*Ethical Practices* — As we strive to develop our learning community, we will maintain at all times the highest level of honesty, communication, cooperation, and credibility in all relationships.

*Accountability* — As a public institution, we believe we must assume responsibility for all our decisions and actions, and we must also be open and honest in all our affairs and always ensure that we are making the best use of our resources.

*Respect for Diversity* — All constituencies are important to achieving our vision. Therefore, we must respect the unique and diverse perspectives each person offers and embrace those differences as the means for developing the strongest learning community possible. We promote individual growth and positive sense of self-worth for all members of the college community.

*Community Engagement* — As an active and involved part of our community, we must play an integral role in developing, advancing, and serving the local community.

## Institutional Goals

### Goal 1:
*Foster a learning-centered environment.*

### Goal 2:
*Promote student goal completion.*

### Goal 3:
*Promote a climate of collaboration, equity, and inclusion among all college constituencies.*

### Goal 4:
*Promote greater transparency, efficiency, and accountability in college processes and systems.*

### Goal 5:
*Strengthen educational and workforce partnerships to create a more responsive and sustainable community.*

### Goal 6:
*Enhance Elgin Community College as an employer of choice.*

*What I know about society could fill a book.
What I don't would fill the world.*

*Dedicated to my fellow travelers in the journey of life—
Linda, Melissa, and Ben.*

# Contents in Brief

Preface xv
About the Author xxiii
A Word to the Student xxv

## PART ONE — The Study of Society 1

1. The Sociological Perspective 1
2. Doing Sociology: Research Methods 26

## PART TWO — The Individual in Society 49

3. Culture 49
4. Socialization and Development 73
5. Social Interaction 100
6. Social Groups and Organizations 123
7. Deviant Behavior and Social Control 142

## PART THREE — Social Inequality 180

8. Social Class in the United States 180
9. Global Stratification 202
10. Racial and Ethnic Minorities 227
11. Gender Stratification 258

## PART FOUR — Institutions and Social Issues 278

12. Marriage and Changing Family Arrangements 278
13. Religion 306
14. Education 334
15. Political and Economic Systems 355
16. Health and Aging 378

Glossary 402
References 413
Index 435
Practice Tests 449
Practice Test Answers 509

# Contents

Preface  xv
About the Author  xxiii
A Word to the Student  xxv

## PART ONE
### The Study of Society  1

**CHAPTER 1 — The Sociological Perspective  1**

Sociology as a Point of View  3
   The Sociological Imagination  4
   Is Sociology Common Sense?  4
   Sociology and Science  6
   Sociology as a Social Science  6
The Development of Sociology  9
   Auguste Comte (1798–1857)  9
   Harriet Martineau (1802–1876)  10
   Herbert Spencer (1820–1903)  11
   Karl Marx (1818–1883)  12
   Émile Durkheim (1858–1917)  13
   Max Weber (1864–1920)  14
   The Development of Sociology in the United States  16
Theoretical Perspectives  18
   Functionalism  18
   Conflict Theory  18
   The Interactionist Perspective  19
   Contemporary Sociology  20
   Theory and Research  21
Summary  21
   *How Sociologists Do It* If You Are Thinking About Sociology as a Career, Read This  5
   *Day-to-Day Sociology* How Long Do Marriages Last?  7
   *How Sociologists Do It* What Is the Difference Between Sociology and Journalism?  9
   *Thinking About Social Issues* Are College Students at High Risk for Suicide?  15
   *Thinking About Social Issues* Social Interaction in the Internet Age  20

**CHAPTER 2 — Doing Sociology: Research Methods  26**

The Research Process  28
   Define the Problem  28
   Review Previous Research  29
   Develop One or More Hypotheses  30
   Determine the Research Design  31
   Define the Sample and Collect Data  35
   Analyze the Data and Draw Conclusions  37
   Prepare the Research Report  41
Objectivity In Sociological Research  41
Ethical Issues In Sociological Research  42
Summary  44
   *How Sociologists Do It* A Fashion Model Becomes an Observing Participant  34
   *How Sociologists Do It* How to Spot a Bogus Poll  38
   *How Sociologists Do It* How to Read a Table  40
   *How Sociologists Do It* Facebook, the Internet, and New Ethical Concerns  42
   *Thinking About Social Issues* Famous Research Studies You Cannot Do Today  43

## PART TWO
### The Individual in Society  49

**CHAPTER 3 — Culture  49**

The Concept Of Culture  50
   Culture and Biology  51
   Culture Shock  51
   Ethnocentrism and Cultural Relativism  52
Components Of Culture  54
   Material Culture  54
   Nonmaterial Culture  54
   The Origin of Language  57
   Language and Culture  58
The Symbolic Nature Of Culture  59
   Symbols and Culture  59
Culture And Adaptation  61
   What Produces Cultural Change?  61
   Cultural Lag  63
   Animals and Culture  63
Subcultures  64
   Types of Subcultures  65
Universals Of Culture  65
   The Division of Labor  65
   Marriage, the Family, and the Incest Taboo  66
   Rites of Passage  66
   Ideology  66
Culture And Individual Choice  67
Summary  70
   *Our Diverse World* Marriage to a Perfect Stranger  53
   *Our Diverse World* The United States and Europe—Two Different Worldviews  58
   *Day-to-Day Sociology* Symbols in Cyberspace  60
   *Thinking About Social Issues* Technology Changes Culture  62
   *How Sociologists Do It* Social Science in a War Zone  67
   *How Sociologists Do It* The Conflict Between Being a Researcher and Being a Human Being  68

CONTENTS

## CHAPTER 4 Socialization and Development 73

### Becoming A Person: Biology And Culture 74
- Nature Versus Nurture: A False Debate 75
- Sociobiology 75
- Deprivation and Development 77
- The Concept of Self 79
- Dimensions of Human Development 80

### Theories Of Development 82
- Charles Horton Cooley (1864–1929) 82
- George Herbert Mead (1863–1931) 82
- Sigmund Freud (1856–1939) 83
- Erik H. Erikson (1902–1994) 83

### Early Socialization In American Society 84
- The Family 85
- The School 85
- Peer Groups 87
- Television, Movies, and Video Games 89

### Adult Socialization 91
- Marriage and Responsibility 92
- Parenthood 92
- Career Development: Vocation and Identity 92

### Aging And Society 94

### Summary 94
- *Thinking About Social Issues* Can Socialization Make a Boy into a Girl? 76
- *Day-to-Day Sociology* Does Day Care Create Unruly Brats? 86
- *Our Diverse World* Win Friends and Lose Your Future: The Costs of Not "Acting White" 88
- *Day-to-Day Sociology* Television Made You the Designated Driver 90
- *Our Diverse World* Could You Be a Success at a Japanese Company? 93

## CHAPTER 5 Social Interaction 100

### Understanding Social Interaction 102
- Contexts 102
- Norms 103
- Ethnomethodology 103
- Dramaturgy 105

### Types Of Social Interaction 106
- Nonverbal Behavior 106
- Exchange 107
- Cooperation 108
- Conflict 108
- Competition 108

### Elements Of Social Interaction 109
- Statuses 109
- Roles 110
- Role Sets 112
- Role Strain 112
- Role Conflict 113
- Role Playing 114

### Collective Behavior 114
- Fads and Fashions 115
- Rumors 116
- Public Opinion 117
- Mass Hysteria and Panic 117

### Summary 118
- *Our Diverse World* Cross-Cultural Social Interaction Quiz 104
- *Day-to-Day Sociology* Can You Spot a Liar? 105
- *Day-to-Day Sociology* Laugh and the World Laughs with You 109
- *How Sociologists Do It* Southerners Are Really Friendly Until You Disrespect Them 113
- *Thinking About Social Issues* Don't Call Me. I'll Text You 114

## CHAPTER 6 Social Groups and Organizations 123

### The Nature of Groups 124
- Primary and Secondary Groups 125

### Functions of Groups 126
- Defining Boundaries 126
- Choosing Leaders 126
- Making Decisions 127
- Setting Goals 127
- Assigning Tasks 127
- Controlling Members' Behavior 127

### Reference Groups 129
- Small Groups 129
- Large Groups: Associations 131
- *Gemeinschaft* and *Gesellschaft* 132
- Mechanical and Organic Solidarity 133

### Bureaucracy 134
- Weber's Model of Bureaucracy: An Ideal Type 134
- Bureaucracy Today: The Reality 135
- The Iron Law of Oligarchy 135

### Institutions and Social Organization 136
- Social Institutions 136
- Social Organization 137

### Summary 137
- *How Sociologists Do It* Can One Bad Apple Spoil the Whole Group? 128
- *Thinking About Social Issues* Do You Really Know How Much Your Friends Drink? 130
- *Day-to-Day Sociology* The Strength of the Informal Structure in Job Hunting 132
- *Day-to-Day Sociology* Group Think Versus Crowdsourcing 136

## CHAPTER 7 Deviant Behavior and Social Control 142

### Defining Normal and Deviant Behavior 143
- Making Moral Judgments 144
- The Functions of Deviance 144
- The Dysfunctions of Deviance 144

### Mechanisms Of Social Control 145
- Internal Means of Control 145
- External Means of Control: Sanctions 146

### Theories of Crime and Deviance 147
- Biological Theories of Deviance 147

Psychological Theories of Deviance 148
Sociological Theories of Deviance 150
## The Importance of Law 154
The Emergence of Laws 155
## Crime in the United States 156
Crime Statistics 157
## Kinds of Crime in the United States 159
Juvenile Crime 159
Violent Crime 161
Property Crime 161
White-Collar Crime 162
Victimless Crime 164
Victims of Crime 165
## Criminal Justice in the United States 165
The Police 165
The Courts 166
Prisons 166
A Shortage of Prisons 171
Women in Prison 172
The Funnel Effect 172
Truth in Sentencing 173
## Summary 174
*How Sociologists Do It* It's the Little Things That Matter in Preventing Crime 155
*How Sociologists Do It* How Accurate Is Forensic Science? 160
*Thinking About Social Issues* How Bad Is the Crime Problem? 162
*Thinking About Social Issues* Are Peaceful Pot Smokers Being Sent to Prison? 164
*Our Diverse World* A Bad Country in Which to Be a Criminal 168
*Thinking About Social Issues* The Continuing Debate Over Capital Punishment: Does It Deter Murderers? 169

## PART THREE
# Social Inequality 180

CHAPTER
## 8 Social Class in the United States 180
### The American Class Structure 181
The Upper Class 181
The Upper-Middle Class 182
The Middle-Middle Class 182
The Lower-Middle Class 182
The Lower Class 183
Income Distribution 183
### Poverty 185
The Feminization of Poverty 185
How Do We Count the Poor? 186
Myths About the Poor 187
Government Assistance Programs 188
The Changing Face of Poverty 190
### Consequences of Social Stratification 192
### Why Does Social Inequality Exist? 194
The Functionalist Theory 194
Conflict Theory 195

Modern Conflict Theory 197
The Need for Synthesis 197
### Summary 197
*Day-to-Day Sociology* Would You Be Happier If You Were Richer? 184
*Thinking About Social Issues* Does the Income Gap Between the Rich and Poor Matter? 189
*Our Diverse World* Rich Countries with Poor Children 191
*Thinking About Social Issues* Who Smokes? 192

CHAPTER
## 9 Global Stratification 202
### Stratification Systems 204
The Caste System 205
The Estate System 205
The Class System 206
### Population Dynamics 206
Fertility 207
Mortality 208
Migration 209
### Theories of Population 210
Malthus's Theory of Population Growth 210
Marx's Theory of Population Growth 210
Demographic Transition Theory 211
A Second Demographic Transition 212
### Global Diversity 213
World Health Trends 213
The Health of Infants and Children in Developing Countries 214
HIV/AIDS 215
Population Trends 216
### Summary 222
*Our Diverse World* How Countries Differ—Japan and Nigeria 204
*Our Diverse World* Life Chances of an Adolescent Girl in Liberia 213
*Thinking About Social Issues* Where Are the Baby Girls? 219
*Thinking About Social Issues* What If the Population Problem Is Not Enough People? 221

CHAPTER
## 10 Racial and Ethnic Minorities 227
### The Concept of Race 229
Genetic Definitions 229
Legal Definitions 230
Social Definitions 230
### The Concept of Ethnic Group 233
### The Concept of Minority 233
### Problems In Race and Ethnic Relations 234
Prejudice 234
Discrimination 235
Institutional Prejudice and Discrimination 236
### Patterns of Racial and Ethnic Relations 236
Assimilation 236
Pluralism 237
Subjugation 238
Segregation 239
Expulsion 239
Annihilation 239

## CONTENTS

Racial and Ethnic Immigration To The United States  240
    Immigration Today Compared with the Past  242
    Illegal Immigration  243

America's Ethnic Composition Today  244
    White Anglo-Saxon Protestants  245
    African Americans  245
    Hispanics (Latinos)  246
    Asian Americans  249
    Native Americans  251
    A Diverse Society  252

Summary  252
    *Thinking About Social Issues* How Have Public Attitudes About Racial Intermarriage Changed?  232
    *Our Diverse World* How Many Minorities Are There?  234
    *Our Diverse World* Whites: The New Minority  243
    *Thinking About Social Issues* Are You Hispanic, Latino, or Neither?  247

### CHAPTER 11  Gender Stratification  258

Are the Sexes Separate and Unequal?  259
    Historical Views  259
    Religious Views  260
    Biological Views  262
    Gender and Sex  265
    Sociological View: Cross-Cultural Evidence  266

What Produces Gender Inequality?  267
    The Functionalist Viewpoint  267
    The Conflict Theory Viewpoint  267

Gender-Role Socialization  268
    Childhood Socialization  268
    Adolescent Socialization  269

Gender Inequality And Work  270
    Job Discrimination  270

Summary  273
    *Thinking About Social Issues* Let Women Vote and You Will Get Masculine Women and Effeminate Men  261
    *Our Diverse World* Why Do Women Live Longer Than Men?  265
    *Day-to-Day Sociology* Speaking, Writing, or Blogging—Nowhere to Hide Gender  270
    *How Sociologists Do It* What Happened to the Men?  272
    *Our Diverse World* Who Is a Better Boss?  273

### PART FOUR

Institutions and Social Issues  278

### CHAPTER 12  Marriage and Changing Family Arrangements  278

The Nature Of Family Life  280
    Functions of the Family  281
    Family Structures  282

Defining Marriage  282
    Romantic Love  282
    Marriage Rules  283
    Marital Residence  283
    Mate Selection  284

The Transformation Of The Family  287
    The Decline of the Traditional Family  288
    Changes in the Marriage Rate  288
    Childless Couples  291
    Changes in Household Size  291
    Women in the Labor Force  292
    Family Violence  292
    Divorce  293
    Divorce Laws  294
    Child Custody Laws  296
    Remarriage and Stepfamilies  297

Family Diversity  298
    The Growing Single Population  298
    Single-Parent Families  299
    Gay and Lesbian Couples  300

What Does The Future Hold?  301

Summary  302
    *Day-to-Day Sociology* Marriage and Divorce Quiz  287
    *How Sociologists Do It* Study, Graduate, and Be Married  290
    *How Sociologists Do It* Do 50 Percent of All Marriages Really End in Divorce?  295
    *Thinking About Social Issues* Reluctant to Marry—The Men Who Want to Stay Single  298

### CHAPTER 13  Religion  306

The Nature Of Religion  307
    The Elements of Religion  308

Magic  310

Major Types Of Religions  310
    Supernaturalism  311
    Animism  311
    Theism  312
    Monotheism  312
    Abstract Ideals  312

A Sociological Approach To Religion  312
    The Functionalist Perspective  312
    The Conflict Theory Perspective  315

Organization Of Religious Life  316
    The Universal Church  316
    The Ecclesia  316
    The Denomination  318
    The Sect  318
    Millenarian Movements  318

Aspects Of American Religion  319
    Religious Diversity  319
    Widespread Belief  320
    Secularism  321
    Ecumenism  323

Major Religions In The United States  323
    Protestantism  324
    Catholicism  324
    Judaism  326
    Islam  327
    Social Aspects of Religious Affiliation  329

Summary 329
- *Our Diverse World* Who Is God? 309
- *Our Diverse World* The Worst Offenders of Religious Freedom 317
- *Day-to-Day Sociology* Today's Cult Might Be Tomorrow's Mainstream Religion 319
- *How Sociologists Do It* Is Your Professor an Atheist? 321
- *Thinking About Social Issues* The Rise of No Religious Affiliation 322
- *Day-to-Day Sociology* Changing Religion Early and Often 325

## CHAPTER 14 Education 334

Education: A Functionalist View 335
- Socialization 335
- Cultural Transmission 336
- Academic Skills 336
- Innovation 339
- Child Care 339
- Postponing Job Hunting 339

The Conflict Theory View 340
- Social Control 340
- Screening and Allocation: Tracking 341
- The Credentialized Society 342

Issues In American Education 344
- Unequal Access to Education 344
- Students Who Speak English as a Second Language 345
- High-School Dropouts 346
- Violence in the Schools 346
- Home Schooling 347
- Standardized Testing 349
- The Gifted 349

Summary 351
- *Thinking About Social Issues* Is Education the Great Equalizer? 338
- *Our Diverse World* Illiteracy Is Common Throughout the World 341
- *Day-to-Day Sociology* Is a College Degree Worth the Trouble? 343

## CHAPTER 15 Political and Economic Systems 355

Politics, Power, And Authority 357
- Power 357
- Political Authority 357

Government And The State 358
- Functions of the State 358

Types Of States 359
- Autocracy 359
- Totalitarianism 359
- Democracy 360

Functionalist And Conflict Theory Views Of The State 360

The Economy And The State 361
- Capitalism 361
- The Marxist Response to Capitalism 363
- Socialism 364
- The Capitalist View of Socialism 364
- Democratic Socialism 364

Political Change 365
- Institutionalized Political Change 365
- Rebellions 365
- Revolutions 366

The American Political System 367
- The Two-Party System 367
- Voting Behavior 368
- African Americans as a Political Force 370
- Hispanics as a Political Force 371
- The Role of the Media 371
- Special-Interest Groups 372

Summary 374
- *Day-to-Day Sociology* Eat Your Fresh Fruit and Vegetables or Pay a Fine 362
- *Our Diverse World* Does Suicide Terrorism Make Sense? 366
- *Thinking About Social Issues* I Know It's Not True, But I'm Not Voting for Him Anyway 368

## CHAPTER 16 Health and Aging 378

The Experience of Illness 379
- Health Care in the United States 380
- Gender and Health 380
- Race and Health 382
- Social Class and Health 383
- Age and Health 383
- Education and Health 384
- Women in Medicine 386

Contemporary Health Care Issues 386
- Acquired Immunodeficiency Syndrome 387
- Health Insurance 387
- Preventing Illness 388

The Aging Population 390
- Composition of the Older Population 391
- Aging and the Sex Ratio 392
- Aging and Racial Minorities 392
- Aging and Marital Status 393
- Aging and Wealth 393
- Global Aging 394
- Future Trends 395

Summary 397
- *Our Diverse World* Women Live Longer than Men Throughout the World 381
- *Our Diverse World* Why Isn't Life Expectancy in the United States Higher? 384
- *Day-to-Day Sociology* Marijuana: A Benign Drug or a Health Problem? 385
- *How Sociologists Do It* Can Your Friends Make You Fat? 390
- *Thinking About Social Issues* The Discovery of a Disease 392
- *Our Diverse World* Stereotypes About the Elderly 395
- *Our Diverse World* Global Aging Quiz 396

Glossary 402
References 413
Index 435
Practice Tests 449
Practice Test Answers 509

# Features Contents

## How Sociologists Do It

If You Are Thinking About Sociology as a Career, Read This  5
What Is the Difference Between Sociology and Journalism?  9
A Fashion Model Becomes an Observing Participant  34
How to Spot a Bogus Poll  38
How to Read a Table  40
Facebook, the Internet, and New Ethical Concerns  42
Social Science in a War Zone  67
The Conflict Between Being a Researcher and Being a Human Being  68
Southerners Are Really Friendly Until You Disrespect Them  113
Can One Bad Apple Spoil the Whole Group?  128
It's the Little Things That Matter in Preventing Crime  155
How Accurate Is Forensic Science?  160
What Happened to the Men?  272
Study, Graduate, and Be Married  290
Do 50 Percent of All Marriages Really End in Divorce?  295
Is Your Professor an Atheist?  321
Can Your Friends Make You Fat?  390

## Thinking About Social Issues

Are College Students at High Risk for Suicide?  15
Social Interaction in the Internet Age  20
Famous Research Studies You Cannot Do Today  43
Technology Changes Culture  62
Can Socialization Make a Boy into a Girl?  76
Don't Call Me. I'll Text You  114
Do You Really Know How Much Your Friends Drink?  130
How Bad Is the Crime Problem?  162
Are Peaceful Pot Smokers Being Sent to Prison?  164
The Continuing Debate over Capital Punishment: Does It Deter Murderers?  169
Does the Income Gap Between the Rich and Poor Matter?  189
Who Smokes?  192
Where Are the Baby Girls?  219
What If the Population Problem Is Not Enough People?  221
How Have Public Attitudes About Racial Intermarriage Changed?  232
Are You Hispanic, Latino, or Neither?  247
Let Women Vote and You Will Get Masculine Women and Effeminate Men  261
Reluctant to Marry—The Men Who Want to Stay Single  298
The Rise of No Religious Affiliation  322
Is Education the Great Equalizer?  338
I Know It's Not True, But I'm Not Voting for Him Anyway  368
The Discovery of a Disease  392

## Day-To-Day Sociology

How Long Do Marriages Last?  7
Symbols in Cyberspace  60
Does Day Care Create Unruly Brats?  86
Television Made You the Designated Driver  90
Can You Spot a Liar?  105
Laugh and the World Laughs with You  109
The Strength of the Informal Structure in Job Hunting  132
Group Think Versus Crowdsourcing  136
Would You Be Happier If You Were Richer?  184
Speaking, Writing, or Blogging—Nowhere to Hide Gender  270
Marriage and Divorce Quiz  287
Today's Cult Might Be Tomorrow's Mainstream Religion  319
Changing Religion Early and Often  325
Is a College Degree Worth the Trouble?  343
Eat Your Fresh Fruit and Vegetables or Pay a Fine  362
Marijuana: A Benign Drug or a Health Problem?  385

xiii

## Our Diverse World

Marriage to a Perfect Stranger   53

The United States and Europe—Two Different Worldviews   58

Win Friends and Lose Your Future: The Costs of Not "Acting White"   88

Could You Be a Success at a Japanese Company?   93

Cross-Cultural Social Interaction Quiz   104

A Bad Country in Which to Be a Criminal   168

Rich Countries with Poor Children   191

How Countries Differ—Japan and Nigeria   204

Life Chances of an Adolescent Girl in Liberia   213

How Many Minorities Are There?   234

Whites: The New Minority   243

Why Do Women Live Longer Than Men?   265

Who Is a Better Boss?   273

Who Is God?   309

The Worst Offenders of Religious Freedom   317

Illiteracy Is Common Throughout the World   341

Does Suicide Terrorism Make Sense?   366

Women Live Longer than Men Throughout the World   381

Why Isn't Life Expectancy in the United States Higher?   384

Stereotypes About the Elderly   395

Global Aging Quiz   396

# Preface

As a freshman at Temple University, my first experience with a college textbook was in my sociology course. I dutifully read the assigned chapter during my first week of class hoping to become familiar with the subject matter of this required course. The only problem was that I had no idea what the author was saying. The writing level was advanced, the style dense, and the book downright threatening, without photos or illustrations. After several hours of reading, I felt frustrated and stupid, and I knew no more about sociology than when I started.

If this was what college was going to be like, I was not going to make it, I thought. I remember admitting reluctantly that I was probably not what guidance counselors in that day referred to as "college material." I could picture myself dropping out after the first semester and looking for a job selling furniture or driving a cab. My family would be disappointed, but my father was a factory worker, and there was no family history of college attendance to live up to. I continued to struggle with the book and earned a D on the mid-term exam. After much effort, I managed to finish the course with a C, and a burning disinterest in the field of sociology. I did not take another sociology course for two years, and when I did it was "Marriage and the Family," considered the easiest course on campus.

I often wonder how I came from this inauspicious beginning to become a sociology professor, let alone the author of a widely used introductory sociology textbook. Then again, maybe it is not all that unusual, because that experience continues to have an effect on me each day. Those 15 weeks helped to develop my view that little is to be gained by presenting knowledge in an incomprehensible or unnecessarily complicated way, or by making yourself unapproachable. Pompous instructors and intimidating books are a disservice to education. Learning should be an exciting, challenging, and eye-opening experience, not a threatening one.

One of the real benefits of writing eleven editions of this textbook is that I have periodically examined every concept and theory presented in an introductory course. In doing so, I have approached the subject matter through a new set of eyes and have consistently tried to find better ways of presenting the material. As instructors, we rarely venture into each other's classrooms and hardly ever do we receive honest, highly detailed, and constructive criticism of how well we are transmitting the subject matter. In the writing of a textbook, we receive this type of information, and we can radically restructure or simply fine tune our presentation. It is quite an education for those of us who have devoted our careers to teaching sociology.

## ● STUDENT-ORIENTED EDITION

Before revising this edition of *Introduction to Sociology*, we surveyed dozens of instructors to find out what they wanted in a textbook and what would assist them in the teaching of sociology, as well as satisfy student needs. This revised text reflects their significant input. In the surveys for this and past editions, we learned that both students and instructors continue to be concerned about the cost of textbooks. Introductory textbooks have become very attractive and expensive during the last decade, as publishers have added hundreds of color photos to the typical volume. This trend has caused the price of textbooks to increase, making them a substantial purchase for the typical student. A textbook, after all, is meant to be comprehensive, up-to-date, and to serve as an important supplement to a course. It makes no sense to make a book so colorful, and therefore so expensive, that students often forgo purchasing it.

To give students the best value for the dollar, we use a soft cover rather than a hard cover. In this way, students will be getting far greater value because nothing of educational content is sacrificed to produce this saving.

We are not, however, content to merely provide a better value. We also want to provide a better book. We, therefore, include a full, built-in study guide with this book that is as extensive, if not more so, than those typically sold separately. In this way, students will be able to purchase the combined textbook and study guide for considerably less than the price of a typical textbook. In fact, the price for our textbook/study guide combination will most likely be lower than the used copy price of a typical hardcover introductory sociology textbook.

## ● PRESENTATION

At the end of my sophomore year, I was on academic probation. I went to the university counseling center for advice. A well-meaning counselor asked me what I wanted to do in the future. I told him I wanted to be a professor. To his credit, he did not laugh or encourage me to think of something more in keeping with my

1.91 GPA. I might not have been a good student, but I was fascinated by what college had to offer. Where else could you be exposed to so much about a world that is so interesting? Belatedly, I began to realize that a great deal of what is interesting falls into the field of sociology.

My goal in this book is to demonstrate the vitality, interest, and utility associated with the study of sociology. Examining society and trying to understand how it works is an exciting and absorbing process. I have not set out to make sociologists of my readers (although if that happens, I will be delighted), but rather to show how sociology applies to many areas of life and how it is used in day-to-day activities. In meeting this objective, I have focused on two basic ideas: that sociology is a rigorous scientific discipline and that a basic knowledge of sociology is essential for understanding social interaction in many different settings, whether they be work or social. In order to understand society, we need to understand how it shapes people and how people in turn shape society.

Each chapter progresses from a specific to a general analysis of society. Each part introduces increasingly more comprehensive factors necessary for a broad-based understanding of social organization.

The material is presented through consistently applied learning aids. Each chapter begins with a chapter outline. Then, a thought-provoking opening vignette offers a real-life story of the concepts being covered. Key terms are presented in boldfaced type in the text. Key concepts are presented in italicized type in the text. A chapter summary concludes each chapter. An integrated study guide follows each chapter. A full glossary is in the back of the book for further reference.

Great care has been taken to structure the book in such a way as to permit flexibility in the presentation of the material. Each chapter is self-contained and, therefore, may be taught in any order.

It has taken nearly two years to produce this revision. Every aspect of this book has been updated and a great deal has been changed. The information is as current and up-to-date as possible and there are hundreds of 2000 through 2012 references throughout the book.

## A COMPARATIVE AND CROSS-CULTURAL PERSPECTIVE

Sociology is a highly organized discipline shaped by several theoretical perspectives or schools of thought. It is not merely the study of social problems or the random voicing of opinions. In this book, no single perspective is given greater emphasis; a balanced presentation of both functionalist theory and conflict theory is supplemented whenever possible by the symbolic interactionist viewpoint.

The book has received a great deal of praise for being cross-cultural in approach and for bringing in examples from a wide variety of societies. Sociology is concerned with the interactions of people wherever and whenever they occur. It would be shortsighted, therefore, to concentrate on only our own society. Often, in fact, the best way to appreciate our own situation is through comparison with other societies. We use our cross-cultural focus as a basis for comparison and contrast with U.S. society.

## FEATURES

### Opening Vignettes

Each chapter begins with a lively vignette that introduces students to the subject matter of the chapter. Many of these are from real-life events to which students can relate, such as the scientific validity of the fear that newborn infant theft from hospitals is a problem (Chapter 1), whether the NCAA is honest in broadcasting ads about basketball player graduation rates (Chapter 2), sociologist Peter Moskos and his socialization to being a police officer (Chapter 4), world famous violinist Joshua Bell playing in the Washington, D.C., subway for donations (Chapter 5), gender portrayals of characters in film (Chapter 11), and why modern medicine has become so expensive (Chapter 16). Others deal with unusual circumstances that remind students that there is a wide range of events to which sociology applies. Examples include the norm violation of attempting to order tea with sugar in Tokyo (Chapter 3), the eccentric soprano Florence Foster Jenkins (Chapter 7), and whites who claim to be black (Chapter 10).

### Day-to-Day Sociology

These boxed features examine a trend or interesting sociological research that has a connection to student's lives. The instructor will be able to discuss these with an eye toward showing the relevance of sociology to everyday life. Included in these are such topics as "Symbols in Cyberspace," "How Long Do Marriages Last?" "Does Day Care Create Unruly Brats?" "Television Made You the Designated Driver," "Can You Spot a Liar?" "Laugh and the World Laughs with You," "The Strength of the Informal Structure in Job Hunting," "Group Think Versus Crowdsourcing," "Would You Be Happier If You Were Richer?" "Speaking, Writing, or Blogging—Nowhere to Hide Gender," "Marriage and Divorce Quiz," "Today's Cult Might Be Tomorrow's Mainstream Religion," "Changing Religion Early and Often," "Is a College Degree Worth the Trouble?" "Eat Your Fresh Fruit and Vegetables or Pay a Fine," and "Marijuana: A Benign Drug or a Health Problem?"

### How Sociologists Do It

Social research is an important part of sociology. In this section, we present a variety of studies and information that helps expand our knowledge of the social world.

Included are "If You Are Thinking About Sociology as a Career, Read This," "What Is the Difference Between Sociology and Journalism?" "How to Spot a Bogus Poll," "How to Read a Table," "Facebook, the Internet, and New Ethical Concerns," "A Fashion Model Becomes an Observing Participant," "Social Science in a War Zone," "The Conflict Between Being a Researcher and Being a Human Being," "Southerners Are Really Friendly Until You Disrespect Them," "Can One Bad Apple Spoil the Whole Group?" "It's the Little Things That Matter in Preventing Crime," "How Accurate Is Forensic Science?" "What Happened to the Men?" "Study, Graduate, and Be Married," "Do 50 Percent of All Marriages Really End in Divorce?" "Is Your Professor an Atheist?" and "Can Your Friends Make You Fat?"

### Our Diverse World

To highlight the cross-cultural nature of this book, many chapters include a box entitled "Our Diverse World." These boxed features encourage students to think about sociological issues in a larger context and explore the global diversity present in the world. The United States with its extensive history of immigration has become one of the most diverse countries in the world. How has this diversity expressed itself? In these boxed features we explore such questions as "Marriage to a Perfect Stranger," "The United States and Europe—Two Different Worldviews," "Win Friends and Lose Your Future: The Costs of Not 'Acting White,'" "Could You Be a Success at a Japanese Company?" "Cross-Cultural Social Interaction Quiz," "A Bad Country in Which to Be a Criminal," "Rich Countries with Poor Children," "How Countries Differ—Japan and Nigeria," "Life Chances of an Adolescent Girl in Liberia," "How Many Minorities Are There?" "Whites: The New Minority," "Why Do Women Live Longer Than Men?" "Who Is a Better Boss?" "Who Is God?" "The Worst Offenders of Religious Freedom," "Illiteracy Is Common Throughout the World," "Does Suicide Terrorism Make Sense?" "Women Live Longer than Men Throughout the World," Why Isn't Life Expectancy in the United States Higher?" "Stereotypes About the Elderly," and "Global Aging Quiz."

### Thinking About Social Issues

These boxes will help students realize that most social events require close analysis and that hastily drawn conclusions are often wrong. The students will see that to be a good sociologist, one must be knowledgeable about disparate positions and must be willing to question the validity of all statements and engage in critical thinking. Included in this section are such issues as "Are College Students at High Risk for Suicide?" "Social Interaction in the Internet Age," "Famous Research Studies You Cannot Do Today," "Technology Changes Culture," "Can Socialization Make a Boy into a Girl?" "Don't Call Me. I'll Text You," "Do You Really Know How Much Your Friends Drink?" "Are Peaceful Pot Smokers Being Sent to Prison?" "The Continuing Debate over Capital Punishment: Does It Deter Murderers?" "Does the Income Gap Between the Rich and Poor Matter?" "Who Smokes?" "Are You Hispanic, Latino, or Neither?" "Where Are the Baby Girls?" "What If the Population Problem Is Not Enough People?" "How Have Public Attitudes About Racial Intermarriage Changed?" "Reluctant to Marry—The Men Who Want to Stay Single," "The Rise of No Religious Affiliation," "Is Education the Great Equalizer?" "Let Women Vote and You Will Get Masculine Women and Effeminate Men," and "I Know It's Not True, But I'm Not Voting for Him Anyway."

There are more than 25 new boxes in this edition. Other boxes that appeared in previous editions have been substantially changed.

## NEW TO THE ELEVENTH EDITION

This edition has been substantially revised to make it as current as possible and to reflect new developments in the field of sociology. The goal has always been to make the material accessible and make it possible for instructors to present an interesting and informative course.

The book is organized into four major parts: The Study of Sociology (Chapters 1 and 2); The Individual in Society (Chapters 3-7); Social Inequality (Chapters 8-11) and; Institutions and Social Issues (Chapters 12-16).

### Chapter 1 The Sociological Perspective

This chapter introduces students to the field and asks them to go beyond popular sociology and investigate society more scientifically than they did before. They get to look at major events, as well as at everyday occurrences, a little differently and start to notice patterns they might have never seen before. After students are equipped with the tools of sociology, they should be able to evaluate critically popular presentations of sociology. They will see that sociology represents both a body of knowledge and a scientific approach to the study of social issues.

The chapter also reviews the development of the field of sociology and the various schools of thought. The classical sociology material is presented in a way that it can be related to contemporary developments. For example, Durkheim's suicide study can be discussed in the context of new material presented on whether college students are at higher or lower risk for suicide than non-college students of the same age.

### Chapter 2 Doing Sociology: Research Methods

In this chapter, students are introduced to the research process and the major research methods—social surveys, secondary analysis, participant observation, and controlled experiments. There is new material on a sociologist who used to be a fashion model and used that experience to study the field as an observing participant. The chapter also discusses ethical considerations in social research and presents material on a sociological study of Harvard students and their Facebook usage that was done without their informed consent.

### Chapter 3 Culture

The chapter presents the view that most definitions of culture emphasize certain features, namely, that culture is shared; it is acquired, not inborn; the elements make up a complex whole; and it is transmitted from one generation to the next. New material is presented on arranged marriage in India as well as a discussion of the differences in worldviews between the United States and Europe. The chapter also includes an updated discussion of the Human Terrain program that the U.S. military implemented because they believed they needed a better cultural understanding of life in Afghanistan.

### Chapter 4 Socialization and Development

In Chapter 4, students are introduced to the process of social interaction that teaches the child the intellectual, physical, and social skills needed to function as a member of society. The goal is to have the students become aware of the pull between nature and nurture. Boxes in this chapter discuss such issues as whether socialization can influence gender behavior, the costs of not "acting white" among black students, and the vastly different social expectations of what it takes to succeed in a Japanese company.

### Chapter 5 Social Interaction

This chapter begins with an examination of the basic types of social interaction, whether verbal or nonverbal. How social interaction affects those involved is also explored. Later in the chapter, the discussion broadens to focus on groups and social interactions within them. There are boxes that ask students to look more closely at the role of lying, as well as laughing, in social interaction. Later, the role of texting is examined.

### Chapter 6 Social Groups and Organizations

A good deal of social interaction occurs in the context of groups. The chapter discusses why groups are important to its members and how people can tell an insider from an outsider. The chapter asks students whether one bad member can spoil the whole group. The chapter also discusses the mistaken conceptions students have about how much their fellow students drink.

Finally, there is a discussion comparing group think to crowdsourcing.

### Chapter 7 Deviant Behavior and Social Control

The chapter explores what determines whether a person's actions are seen as eccentric, creative, or deviant. Moral codes differ widely from one society to another and even within a society. Within a society, groups and subcultures exist whose moral codes differ considerably. The latest data on crime trends in the United States and the rest of the world are also presented. Incarceration rates in the United States are compared to those in other countries, and the death penalty is explored in depth. Many students are attracted to criminal justice careers today. The chapter explores the question of how accurate the field of forensic science actually is.

### Chapter 8 Social Class in the United States

The chapter provides an overview of the core ideas in the discussion of social class and social stratification. The goal is to have students see social class in a broader context of social and lifestyle issues. New material discusses the correlation between smoking and social class. It also asks student whether they would be happier if they were richer and presents information showing income as only one factor in happiness. Other material in the chapter looks at the gap between the rich and the poor and asks whether it has become larger in recent years.

### Chapter 9 Global Stratification

After exploring social class in the United States in the previous chapter, this chapter broadens the discussion and looks at the significance of global stratification. The vast social inequality throughout the world is explored and data are presented from developed and less developed countries. The chapter looks at the substantially different issues confronting countries experiencing explosive population growth compared to those losing population. The chapter also looks at the issues of sex-selection abortions taking place in certain countries and the social problems caused by this practice.

### Chapter 10 Racial and Ethnic Minorities

Racial and ethnic issues have been at the forefront of sociological concerns. The chapter explores the significance of race and ethnicity in society. The chapter asks students to think about how many minorities there are in the United States. There is an examination of the fact that the projections suggest that whites will become a minority in the future. Other material examines the growth of Hispanic and Asian populations in American society. The chapter also explores the change in public attitude toward racial intermarriage.

### Chapter 11 Gender Stratification

The chapter explores the social, psychological, and cultural attributes of masculinity and femininity. Gender issues have been changing dramatically in recent years and the chapter presents data highlighting these changes. The chapter examines gender differences in social interaction including differences in the workplace. It also explores such issues as men withdrawing from certain areas of society as women have advanced. Gender changes in certain professions are also looked at, as well as the differences in male and female performance in higher education. The chapter also examines why women live longer than men.

### Chapter 12 Marriage and Changing Family Arrangements

Many family forms are common today: single-parent families (resulting from either unmarried parenthood or divorce), remarried couples, unmarried couples, stepfamilies, and extended or multigenerational families. The family has been undergoing substantial changes and the chapter discusses these, as well as the rise of cohabitation and unmarried mothers. The average age of marriage has also increased rapidly, producing changes in divorce rates and family size. The chapter also explores divorce and remarriage and asks whether there is truth to the belief that 50 percent of all marriages really do end in divorce.

### Chapter 13 Religion

The chapter examines the system of beliefs, practices, and philosophical values shared by a group of people that is commonly referred to as religion. Current trends in religion, such as the rapid increase in the number of people who claim no religious affiliation, are explored. There is also a discussion about the increasingly common practice of changing religion. The chapter looks at the transition from a cult to a mainstream religion and the level of religiosity among faculty members.

### Chapter 14 Education

This chapter examines the role of education in society. It looks at the issue of whether education is the great equalizer as is commonly thought. International comparisons are present throughout the chapter and a discussion of the extensive level of worldwide illiteracy is presented. Students are also asked to evaluate the value of a college degree and its effect on life chances.

### Chapter 15 Political and Economic Systems

This chapter begins by examining the political institution and realizing that the economy is intimately tied to the political system. It also examines that connection. The chapter explores voting behavior, and African Americans and Hispanics as a political force. The role of the media is examined, as well as the influence of special-interest groups. Attempts to influence healthier behavior by political action are explored. The chapter also looks at the role of suicide terrorism in an international context.

### Chapter 16 Health and Aging

Medicine and health care issues are intertwined with our social, emotional, and cultural life. This chapter examines these interactions. The chapter also explores why the life expectancy in the United States is not higher than it is, given the vast amount of money spent on health care. There is a discussion of whether marijuana is really a benign drug that has medical benefits, as well as a discussion of the influence of friends on your weight. The chapter also looks at the growth in the aging population and the impact it has on other areas of society.

## THE ANCILLARY PACKAGE

The primary objective of a textbook is to provide clear information in a format that promotes learning. In order to assist the instructor in using *Introduction to Sociology*, an extensive ancillary package has been developed to accompany the book.

### Supplements for Instructors

*Sociology CourseMate* This website for *Introduction to Sociology*, Eleventh Edition, brings chapter topics to life with interactive learning, study, and exam preparation tools, including quizzes and flash cards for each chapter's key terms and concepts. The site also provides an e-book version of the text with highlighting and note-taking capabilities. For instructors, this text's CourseMate also includes Engagement Tracker, a first-of-its-kind tool that monitors student engagement in the course. Go to login.cengage.com to access these resources.

*WebTutor™ on Blackboard® and WebCT®* Jump-start your course with customizable, rich, text-specific content within your Course Management System:

- **Jump-start**—Simply load a WebTutor cartridge into your Course Management System.
- **Customizable**—Easily blend, add, edit, reorganize, or delete content.
- **Content:** Rich, text-specific content, media assets, e-book, quizzing, web links, videos, and more.

Whether you want to Web-enable your class or put an entire course online, WebTutor delivers. Visit webtutor.cengage.com to learn more.

*The Wadsworth Sociology Video Library Volume I, II, and III* (featuring BBC Motion Gallery video clips) drives home the relevance of course topics through short, provocative clips of current and historical events. Perfect for enriching lectures and engaging students in discussion, many of the segments on this volume have been gathered from the BBC Motion Gallery. Ask your Cengage Learning representative for a list of contents.

*CourseReader for Sociology* allows you to create a fully customized online reader in minutes. You can access a rich collection of thousands of primary and secondary sources, readings, and audio and video selections from multiple disciplines. Each selection includes a descriptive introduction that puts it into context, and every selection is further supported by both critical-thinking and multiple-choice questions designed to reinforce key points. This easy-to-use solution allows you to select exactly the content you need for your courses, and it is loaded with convenient pedagogical features, such as highlighting, printing, note taking, and downloadable MP3 audio files for each reading. You have the freedom to assign and customize individualized content at an affordable price. CourseReader is the perfect complement to any class.

*Instructor's Resource Manual and Test Bank* In addition to the student study guide and practice tests in the textbook, an Instructor's Manual and Test Bank are also available. This provides for unusual consistency and integration among all elements of the teaching and learning package. Both the new and experienced instructor will find plenty of ideas in this Instructor's Manual, which is closely correlated to the textbook and the student study guide. Each chapter of the manual includes teaching objectives, key terms, lecture suggestions, activities, discussion questions, and formatted handouts for many topics. The Instructor's Manual also contains an annotated list of resources for students for reference or as a handout. Instructors will be able to download the Instructor's Manual from the Wadsworth Sociology website.

Consult your sales representative for access information or information on how to secure the printed version. The Test Bank contains multiple-choice, true/false, and essay questions keyed to each learning objective. These test items are page-referenced to the textbook and include significant numbers of application as well as knowledge questions. Story problems use names drawn from a variety of cultures, reflecting the diversity of U.S. society. Instructors requested that the questions be tied to the practice tests, and we followed that suggestion.

## Supplements for Students

*Built-in Study Guide and Practice Tests* The interactive workbook study guide is fully integrated into the book. Each chapter is followed by a study guide section so students can review the material immediately, without having to search for it elsewhere in the book. This encourages students to see the study guide as an integral part of the learning process.

The study guide provides for ample opportunity to review the material with a variety of styles of review questions. All key terms and key sociologists are reviewed with matching questions. Key concepts are revisited with fill-in questions. Critical Thought Exercises help students contextualize concepts covered in the chapter. Often Web site URLs are provided for students to expand on their exploration of the topic. And a matching question-answer key is provided to allow students immediate review of their answers.

Practice tests are in the back of the book to provide students with additional preparation for testing. Whereas other practice tests are limited to recognition and recall items, these questions lead students to engage in such higher-level cognitive skills as analysis, application, and synthesis. The tests encourage students to think critically and apply the material to their experiences. Again, an answer key is provided to allow students full review and preparation.

All of these tools will be very useful for students preparing for essay exams and research papers. The textbook also includes the important section entitled "How to Get the Most Out of Sociology," which discusses how to use the study guides, practice tests, and lecture material in preparing for exams and getting the most out of the introductory sociology course.

*Sociology CourseMate* This website for *Introduction to Sociology,* Eleventh Edition, brings chapter topics to life with interactive learning, study, and exam preparation tools, including quizzes and flash cards for each chapter's key terms and concepts. The site also provides an e-book version of the text with highlighting and note-taking capabilities. Students can access this new learning tool and all other online resources through cengagebrain.com.

## ACKNOWLEDGMENTS

The textbook and study guide manuscripts have been written after an extensive survey of faculty at a wide variety of institutions. I am grateful for the thoughtful contributions of the following people:

Mark Christian, Miami University, Hamilton

Dr. Durrell W. Dickens, Lee College

Mitra Hoshiar, Los Angeles Pierce College

Marion R. Hughes, Towson University

Kimberly A. Reed, Triton College and Elgin Community College

Tina Rushing, Kilgore College

I also wish to thank the many colleagues and reviewers of previous editions of *Introduction to Sociology* for their many contributions and suggestions. I am grateful for the thoughtful contributions of the following people: Laura Dowd, University of Georgia; Nancy Feather, West Virginia University; Hubert Anthony Kleinpeter, Florida A&M University; Steven Patrick, Boise State University; Craig T. Robertson, University of North Alabama; Laurie Smith, East Texas Baptist University; Patrick Ashton, Indiana University—Purdue University; Froud Stephen Burns, Floyd Junior College; Peter Chroman, College of San Mateo; Mary A. Cook, Vincennes University; William D. Curran II, South Suburban College; Ione Y. Deollos, Ball State University; Stanley Deviney, University of Maryland—Eastern Shore; Brad Elmore, Trinity Valley Community College; Cindy Epperson, St. Louis Community College—Meramac; Larry Frye, St. Petersburg College; Richard Garnett, Marshall University; David A. Gay, University of Central Florida; Daniel T. Gleason, Southern State College; Charlotte K. Gotwald, York College of Pennsylvania; Richard L. Hair, Longview Community College; Selwyn Hollingsworth, University of Alabama; Sharon E. Hoga, Longview Community College; Bill Howard, Lincoln Memorial University; Sidney J. Jackson, Lakewood Community College; Michael C. Kanan, Northern Arizona University; Ed Kick, Middle Tennessee State University; Louis Kontos, Long Island University; Steve Liebowitz, University of Texas, Pan American; Thomas Ralph Peters, Floyd College; David Phillips, Arkansas State University; Kanwal D. Prashar, Rock Valley Community College; Charles A. Pressler, Purdue University, North Central; Stephen Reif, Kilgore College; Richard Rosell, Westchester Community College; Catherine A. Stathakis, Goldey Beacom College; Doris Stevens, McLennan Community College; Gary Stokley, Louisiana Tech University; Elena Stone, Brandeis University; Judith C. Stull, La Salle University; Lorene Taylor, Valencia Community College; Paul Thompson, Polk Community College; Brian S. Vargus, Indiana University—Purdue University Indianapolis; Steven Vassar, Minnesota State University—Mankato; Peter Venturelli, Valparaiso University; J. Russell Willis, Grambling State University; Bobbie Wright, Thomas Nelson Community College; William Egelman, Iona College; Carol Apt, South Carolina State University; Lynda Mae, Western Nevada Community College; Margaret E. Preble, Thomas Nelson Community College; Mark Miller, East Texas Baptist University; and Rebecca Stevens, Mount Union College.

At Montclair State University, I would like to thank the following colleagues for their support of the book: Jay Livingston, Gil Klagman, Benjamin Hadis, Janet Ruane, Laura Kramer, and Peter Freund.

A project of this magnitude becomes a team effort, with many people devoting enormous amounts of time to ensure that the final product is as good as it can possibly be. At Cengage, I would like to thank the following people who ushered this project through its many stages: Seth Dobrin, senior acquisitions editor; Liana Sarkisian, developmental editor; Michelle Clark, senior content project manager; John Chell, media editor; and Nicole Bator, editorial assistant.

It was a privilege to have the support and assistance of these very capable people. I am also grateful to all those students and instructors who have shared with me their thoughts about this book over the years. Please continue to let me know how you feel about this book.

**Henry L. Tischler**
**htischler@framingham.edu**

# About the Author

**HENRY L. TISCHLER** grew up in Philadelphia and received his bachelor's degree from Temple University and his masters and doctorate degrees from Northeastern University. He pursued postdoctoral studies at Harvard University. His first venture into textbook publishing took place while he was still a graduate student in sociology when he wrote the fourth edition of *Race and Ethnic Relations* with Brewton Berry. The success of that book led to his authorship of the eleven editions of *Introduction to Sociology*.

Tischler has been a professor at Framingham State University in Framingham, Massachusetts, for several decades. He has also taught at Northeastern University, Tufts University, and Montclair State University. He continues to teach introductory sociology every year and has been instrumental in encouraging many students to major in the field. His other areas of interest are race and ethnicity, and crime and deviant behavior.

Professor Tischler has been active in making sociology accessible to the general population and has been the host of an author interview program on National Public Radio. He has also written a weekly newspaper column called "Society Today" which dealt with a wide variety of sociological topics.

Tischler and his wife Linda divide their time between Boston and New York City. Linda Tischler is a senior editor at a national magazine. The Tischlers have a daughter Melissa, who is a strategy consultant, and a son Ben, who is an interactive producer for an advertising company.

©2000 Al Hirschfeld. Drawing reproduced by special arrangement with the Margo Feiden Galleries Ltd., New York.

# A Word to the Student — How to Get the Most Out of Sociology

## ● EFFECTIVE STUDY: AN INTRODUCTION

Why should you read this essay? If you think you can get an A in the course, perhaps you can skip it. Maybe you are just not interested in sociology or about learning ways to become a really successful student. Maybe you're just here because an advisor told you that you need a social science course. Maybe you feel, "Hey, a C is good. I'll never need this stuff." If so, you can stop reading now.

But if you want to do well in sociology—thereby becoming a more effective participant in society and social life—and if you want to learn some techniques to help you in other classes, too, this is for you. It's filled with the little things no one ever seems to tell you that improve grades, make for better understanding of classes—and may even make classes enjoyable for you. The choice is yours: to read, or not to read. Be forewarned. These contents may challenge the habits of a lifetime—habits that have gotten you this far but ones that may endanger your future success.

This essay contains ways to help you locate major ideas in your textbook. It contains many techniques that will be of help in reading your other course textbooks. If you learn these techniques early in your college career, you will have a head start on most other college students. You will be able to locate important information, understand lectures better, and probably do better on tests. By understanding the material better, you will not only gain a better understanding of sociology but also find that you are able to enjoy your class more.

## ● THE PROBLEM: PASSIVE READING

Do you believe reading is one-way communication? Do you expect the author's facts will become apparent if you only read hard enough or long enough? (Many students feel this way.) Do you believe the writer has buried critical material in the text somewhere and that you need only find and highlight it to get all that's important? And do you believe that if you can memorize these highlighted details you will do well on tests? If so, then you are probably a passive reader.

The problem with passive reading is that it makes even potentially interesting writing boring. Passive reading reduces a chapter to individual, frequently unrelated facts instead of providing understanding of important concepts. It seldom digs beneath the surface, relying on literal meaning rather than sensing implications. Since most college testing relies on understanding of key concepts rather than simple factual recall, passive reading fails to significantly help students to do well in courses.

> ### Key Features of the Study Guide
>
> *For each chapter you will find the following:*
>
> **Key concepts matching exercise**
> Includes every major term defined in the chapter
> Promotes association of major thinkers with their key ideas or findings
> Provides correct answers
>
> **Key thinkers matching exercise (where relevant)**
> Includes most important theorists or researchers discussed in the text
> Promotes association of major thinkers with their key ideas or findings
> Provides correct answers
>
> **Critical-thinking questions**
> Promotes depth in reflecting on the material
> Encourages creative application of the important concepts to everyday life
> Presented in increasing levels of complexity, abstraction, and difficulty
> Provides help in preparing for essay exams and papers
>
> **Comprehensive practice test**
> Includes questions on all major points in the chapter
> Includes true/false, multiple-choice, and essay questions
> Provides correct answers

## ● THE SOLUTION: ACTIVE READING

Active reading is recognizing that a textbook should provide two-way communication. It involves knowing what aids are available to help understand the text and then using them to find the meaning. It involves prereading and questioning. It includes recording of questions, vocabulary learning, and summarizing. Still, with all these techniques, it frequently takes less time and produces significantly better results than passive reading.

This textbook—especially the Study Guide—is designed to help you become an active reader. For your convenience, the Study Guide material related to each chapter appears right after that chapter.

The corners of the Study Guide pages are edged in color for easy reference. In the Study Guide, you will find a variety of learning aids based on the latest research on study skills. If you get into the habit of using the aids presented here, you can apply similar techniques to your other textbooks and become a more successful learner.

## EFFECTIVE READING: YOUR TEXTBOOK

As an active reader, how should you approach your textbook? Here are some techniques for reading text chapters that you should consider.

1. **Think first about what you know.** Read the title of the chapter, and then ask yourself what experiences you have had that relate to the title. For example, if the title is "Society and Social Interaction," ask yourself, "Have I been a part of a social group? In what ways have I interacted with others in social groups? What do I remember about the experience?" Answers to these questions personalize the chapter by making it relate to your experiences. They provide a background for the chapter, which experts say improves your chances of understanding the reading. They show that you do know something about the chapter so that its content won't be so alien.

2. **Review the learning objectives.** Not all textbooks provide learning objectives as this one does, but, where available, they can be a valuable study aid. Learning objectives are stated in behavioral terms—they tell you what you should be able to do when you finish the chapter. Ask yourself questions about the tasks suggested in each learning objective and then read to find the information needed to accomplish that task. For instance, if a learning objective states, "Explain how variations in the size of groups affect what goes on within them," then you'll want to ask yourself something like, "How do groups vary in size?" and "How does each variation affect interaction within the group?"

3. **Prior to reading the textbook chapter, read the chapter summary as a guide to important terms and ideas.** The summary includes all the points you need to make sure you know. Some items you may not know anything about. This tells you where to spend your reading time. A good rule: Study most what you know least. Wherever it is, the summary is often your best guide to important material.

4. **Pay attention to your chapter outline.** This textbook, like most other introductory college textbooks, has an outline at the beginning of each chapter. If you do nothing else besides reading the summary and going through this outline before reading the chapter, you will be far ahead of most students because you will be aware of what is important. The outline indicates how ideas are organized in the chapter and how those ideas relate to one another. Certain ideas are indented to show that they are subsets or parts of a broader concept or topic. Knowing this can help you organize information as you read.

5. **Question as you read.** Turn your chapter title into a question, then read up to the first heading to find your answer. The answer to your question will be the main idea for the entire chapter. In forming your question, be sure it contains the chapter title. For example, if the chapter title is "Doing Sociology: Research Method," your question might be "What research methods does sociology use?" or "Why do you need to know about research methods to do sociology?" As you go through the chapter, turn each heading into a question, and then read to find the answer. Most experts say that turning chapter headings into questions is a valuable step in focusing reading on important information.

    You might also want to use the learning objectives as questions. You know that these objectives will point you toward the most important material in a section. However, it is also a good idea to form your own questions to get into the habit for books that do not contain learning objectives. A good technique might be to think of your own question, and then check it against the appropriate objective before reading. In any case, use a question and then highlight your answer in the text. This will be the most important content under each heading. Don't expect every word to be important. Focus on finding answers.

### Guidelines for Effective Reading of Your Textbook

1. Think first about what you know.
2. Review the learning objectives.
3. Prior to reading the textbook chapter, read the chapter summary as an index to important terms and ideas.
4. Pay attention to your chapter outline.
5. Question as you read.
6. Pay attention to graphic aids.
7. When in doubt, use clues to find main ideas.
8. Do the exercises in the Study Guide.
9. Review right after reading.

6. **Pay attention to graphic aids.** As you read, note those important vocabulary words appearing in bold type. Find the definitions for these words (in this book, definitions appear in italics right next to key words) and highlight them. These terms will be important to remember. Your Study Guide identifies all these important terms in the section headed "Key Concepts." A "Key Thinkers" section identifies the sociologists and other important thinkers in the chapter worth remembering.

Both the "Key Concepts" and "Key Thinkers" sections are organized as matching exercises. Testing yourself after you read a text chapter (the answer key is at the end of the Study Guide) will let you know whether you recognize the main concepts and researchers.

Pay attention to photos and photo captions. They make reading easier because they provide a visualization of important points in the textbook. If you can visualize what you read, you will ordinarily retain material better than people who don't use this technique. Special boxed sections usually give detailed research information about one or more studies related to a chapter heading. For in-depth knowledge, read these sections, but only after completing the section to which they refer. The main text will provide the background for a better understanding of the research, and the visualization provided by the boxed information will help illuminate the text discussion.

7. **When in doubt, use clues to find main ideas.** It is possible that, even after using the questioning technique, there still are sections where you are uncertain whether you're getting the important information. You have clues both in the text and in the Study Guide to help you through such places. In the text, it helps to know that main ideas in paragraphs occur more frequently at the beginning and end. Watch for repeated words or ideas—these are clues to important information. Check examples; whenever the author uses examples to document something, it is usually an important idea. Be alert for key words (such as "first," "second," "clearly," "however," "although," and so on); these also point to important information. Names of researchers (except for those named only within parentheses) will almost always be important. For those chapters in which important social scientists are discussed, you will find a "Key Thinkers" section in your Study Guide.

8. **Do the exercises in the Study Guide.** The exercises in the Study Guide are designed as both an encouragement and a model of active learning. The exercises are not about mere regurgitation of material. Rather, you are asked to analyze, evaluate, and apply what you read in the text. By completing these exercises, you are following two of the most important principles articulated in this essay: You are actively processing the material, and you are applying it to your own life and relating it to your own experiences. This will increase your learning.

9. **Review right after reading.** Most forgetting takes place in the first day after reading. A review right after reading is your best way to hold text material in your memory. A strong aid in doing this review is your Study Guide. If a brief review is all you have time for, return to the learning objectives at the beginning of the chapter. Can you do the things listed in the objectives? If so, you probably know your material. If not, check the objectives and reread the related chapter sections to get a better understanding.

An even better review technique is to complete—if you haven't already done so—the exercises. Writing makes for a more active review, and if you do the exercises, you will have the information you need from the chapter. If there are blanks in your knowledge, you can check the appropriate section of text and write the information you find in your Study Guide. This technique is especially valuable in classes requiring essay exams or papers, as it gives you a comprehensive understanding of the material as well as a sense of how it can be applied to real-world situations.

For a slightly longer but more complete review, do the "Key Concepts" and "Key Thinkers" matching tests. These will assure you that you have mastered the key vocabulary and know the contributions of the most important researchers mentioned in the chapter. Since a majority of test questions are based on the understanding of vocabulary, research findings, and major theories, this is an important study method.

It is also a good idea to review the "Critical Thought" questions in the Study Guide. One key objective of sociology—indeed, of all college courses—is to help you develop critical-thinking skills. Though basic information may change from year to year as new scientific discoveries are made, the ability to think critically in any field is important. If you get in the habit of going beyond surface knowledge in sociology, you can transfer these skills to other areas. This can be a great benefit not only while you're in school but afterward as well. As with the exercises section, these questions provide the kind of background that is extremely useful for essay exams.

What other methods would an active student use to improve understanding and test scores in sociology? The next several sections present a variety of techniques.

## FUNCTIONING EFFECTIVELY IN CLASS

To function effectively in class, you must of course be there. Even if no one is taking attendance or forcing you to be present, studies show that you have a significantly greater chance of succeeding in your class if you attend regularly. Lecture material is generally important—and it is given only once. If you miss a lecture, in-class discussion, game, or simulation, there is no really effective way to make it up.

> **Guidelines for Effective Functioning in Class**
>
> 1. Begin each class period with a question.
> 2. Ask questions frequently.
> 3. Join in classroom discussion.

Assuming you are present, there are two ways of participating in your sociology class: actively and passively. Passive participation involves sitting there, not contributing, waiting for the instructor to tell you what is important. Passive participation takes little effort and produces less learning. Unless you are actively looking for what is significant, the likelihood of finding the important material or of separating it effectively from what is less meaningful is not great. The passive student runs the risk of taking several pages of unneeded notes or of missing key details altogether.

Active students begin each class period with a question. "What is this class going to be about today?" They find an answer to that question, usually in the first minute, and use this as the key to important material throughout the lecture or other activity. When there is a point they don't understand, they ask questions. Active students know that many other students probably have similar questions but are afraid to ask. Asking questions allows you to help others while helping yourself.

Active students also know that what seems to be a small point today may be critical to understanding a future lecture. Such items also have a way of turning up on tests. If classroom discussion is called for, active students are quick to join in. And the funny thing is, they frequently wind up enjoying their sociology class as they learn.

## ● EFFECTIVE STUDYING

As you study your sociology text and notes, both the method you use and the time picked for study will affect your comprehension. Establishing an effective study routine is important. Without a routine, it is easy to put off study—and put it off, and put it off . . . until it is too late. To be most effective, follow the few simple steps listed below.

> **Guidelines for Effective Studying**
>
> 1. When possible, study at the same time and place each day.
> 2. Study in half-hour blocks with five-minute breaks.
> 3. Review frequently.
> 4. Don't mix study subjects.
> 5. Reward yourself when you're finished.

1. **When possible, study at the same time and place each day.** Doing this makes use of psychological conditioning to improve study results. "Because it is 7:00 P.M. and I am sitting at my bedroom desk, I realize it is time to begin studying sociology."
2. **Study in half-hour blocks with five-minute breaks.** Long periods of study without breaks frequently reduce comprehension to the 40% level. That is inefficient. By using short periods (about 30 minutes) followed by short breaks, you can move that comprehension rate into the 70% range. Note that if 30 minutes end while you are still in the middle of a text section, you should go on to the end of that section before stopping.
3. **For even more efficient study, review frequently.** Take about a minute at the end of each study session to mentally review what you've studied so far. When you start the next study session, spend the first minute or two rehearsing in your mind what you studied in the previous session. This weaves a tight webbing in which to catch new associations. Long-term retention of material is aided by frequent review, about every two weeks. A 10-minute review planned on a regular basis saves on study time for exams and ensures that you will remember needed material.

    Another useful way to review is to try to explain difficult concepts or the chapter learning objectives to someone else. One problem students often have is thinking they know the material while studying and reviewing it by themselves, only to have that knowledge leave them at the time of the exam. Trying to explain something to someone else forces us to be clear about key points and to discover and articulate the relationship among the components of an idea. Ask your friends or family to bear with you as you try to explain the material. After all, they will learn something as well!
4. **Don't mix study subjects.** Do all of your sociology work before moving on to another course. Otherwise, your study can result in confusion of ideas and relationships within materials studied.
5. **Finally, reward yourself for study well done.** Think of something you like to do, and do it when you finish studying for the day. This provides positive reinforcement, which makes for continued good study.

## ● SUCCESSFULLY TAKING TESTS

Of course, tests are important to you as a student. Tests are where you can demonstrate to yourself and to the instructor that you really know the material. The trouble is, few people have learned how to take tests effectively.

And knowing how to take tests effectively makes a significant difference in exam scores. Here are a few tips to improve your test-taking skills.

> **Studying for the Tests**
> 1. Think before you study.
> 2. Begin study a week early.
> 3. Put notes and related chapters together for study.
> 4. Take practice tests.

## Studying for Tests

1. **Think before you study.** All material is not of equal value. What did the instructor emphasize in class? What was covered in a week? A day? A few minutes? Were any chapters emphasized more than others? Which learning objectives did your instructor stress? Review the "Key Thinkers" and "Key Concepts" sections in your Study Guide for important people and terms. Which of these were given more emphasis by your instructor? Use these clues to decide where to spend most of your study time.
2. **Begin studying a week early.** When you start early, if you encounter material you don't know, you have time to find answers. If you see that you know blocks of material already, you have saved yourself time in future study sessions. You also avoid much of the forgetting that occurs with last-minute cramming.
3. **Put notes and related chapters together for study.** Integrate the material as much as possible, perhaps by writing it out in a single, comprehensive format. A related technique is to visualize the material on the pages of the text and in your notes. You may even want to think of a visual metaphor for some of the key ideas. This way you can see and remember the connections between similar subjects or similar treatments of the same subject. Grouping the material will also make your studying much more efficient. As you study, don't stop for unknown material. Study what you know. Once you know it, go back and look at what you don't know yet. There is no need to study again what you already know. Put it aside, and concentrate on the unknown.
4. **Take practice tests.** When you have completed your studying, take the appropriate practice test for each chapter. These tests are grouped together at the back of the book. Tests include true/false and multiple-choice questions, with comprehensive or thematic essays at the end. Each test is divided into sections by major headings in the chapter. Within each section, questions are presented in scrambled order, as they are likely to be on the actual test. Taking the practice test contains a double benefit. First, if you get a good score on this test, you know that you understand the material. Second, the format of the practice test is very similar to that of real tests. For this reason, you should develop confidence in your ability to succeed in course tests from doing well on the practice tests. If your course tests include essay questions, you should, in addition to the practice test essays, use the "Critical Thinking" sections to prepare and practice focused, in-depth answers.

> **Taking the Test**
> 1. Don't come early; don't come late.
> 2. Make sure you understand all the directions before you start answering.
> 3. Read through the test, carefully answering only items you know.
> 4. Now that you've answered what you know, look carefully at the other questions.
> 5. If you finish early, stay to check answers.
> 6. Don't be distracted by other test takers.
> 7. When you get your test back, use it as a learning experience.

## Taking the Test

1. **Don't come early; don't come late.** Early people tend to develop anxieties; late people lose test time. Studies show that people who discuss test material with others just before a test may forget that material on the test. This is another reason that arriving too early puts students in jeopardy. Get there about two or three minutes early. Relax and visualize yourself doing well on the test. After all, if you followed the study guidelines discussed above, you can't help but do well! Be confident; repeat to yourself as you get ready for the test, "I can do it! I will do it." This will set a positive mental tone.
2. **Make sure you understand all the directions before you start answering.** Not following directions is the biggest cause of lost points on tests. Ask about whatever you don't understand. The points you save will be your own.
3. **Read through the test, carefully answering only items you know.** Make sure you read every word and every answer choice as you go. Use a piece of paper or a card to cover the text below the line you are reading. This can help you focus on each line individually—and increase your test score.

    Speed creates a serious problem in testing. The mind is moving so fast that it is easy to overlook key words such as *except, but, best example,* and so on. Frequently, multiple-choice questions will contain two close options, one of which is correct, while the other is partly correct.

Moving too fast without carefully reading all items causes people to make wrong choices in these situations. Slowing your reading speed makes for higher test scores. The mind tends to work subconsciously on questions you've read but left unanswered. As you're doing questions later in the test, you may suddenly have the answer for an earlier question. In such cases, answer the question right away. These sudden insights quickly disappear and may never come again.

4. **Now that you've answered what you know, look carefully at the other questions.** Eliminate alternatives you know are wrong, and then guess. Never leave a blank on a test. You will have only a 25% chance when you guess on a four-item multiple choice question, but you will have a chance. And a chance is better than no chance.

5. **If you finish early, stay to check answers.** Speed causes many people to give answers that a moment's hesitation would show to be wrong. Read over your choices, especially those for questions that caused you trouble. Don't change answers because you suddenly feel one choice is better than others. Studies show that this is usually a bad strategy. However, if you see a mistake or have genuinely remembered new information, change your answer.

6. **Don't be distracted by other test takers.** Some people become very anxious because of the noise and movement of other test takers. This is most apparent when several people begin to leave the room after finishing their tests. Try to sit where you will be least apt to see or interact with other test takers. Usually this means sitting toward the front of the room and close to the wall farthest from the door. Turn your chair slightly toward the wall, if possible. The more you insulate yourself from distractions during the test, the better off you will be.

Don't panic when other students finish their exam before you do. Accuracy is always more important than speed. Work at your own pace and budget your time appropriately. For a timed test, always be aware of the time remaining. This means that if a clock is not visible in the classroom, you need to have your own wristwatch.

Take as much of the available time as you need to do an accurate and complete job. Remember, your grade will be based upon the answers you give, not on whether you were the first—or the last—to turn in your exam.

7. **When you get your test back, use it as a learning experience.** Diagnosing a test after it is returned to you is one of the most effective strategies for improving your performance in a course. What kind of material was on the test: theories, problems, straight facts? Where did the material come from: book, lecture, or both? The same kind of material taken from the same source(s) will almost certainly be on future tests.

Look at each item you got wrong. Why is it wrong? If you know why you made mistakes, you are unlikely to make the same ones in the future. Look at the overall pattern of your errors. Did you make most of your mistakes on material from the lectures? Perhaps you need to improve your note-taking technique. Did your errors occur mostly on material from the readings? Perhaps you need to pay more attention to main idea clues and highlight text material more effectively. Were the questions you got wrong evenly distributed between in-class and reading material? Perhaps you need to learn to study more effectively and/or to take steps to reduce test anxiety.

Following these steps can make for more efficient use of textbooks, better note taking, higher test scores, and better course grades.

## ● A FINAL WORD

As you can see, the key to success lies in becoming an active student. Managing time, asking questions, planning effective approaches to increase test scores, and using all aids available to make reading and studying easier are all elements in becoming an active student. The Study Guide and Practice Tests for this textbook have been specially designed to help you be that active student. Being passive may seem easier, but it is not. Passive students spend relatively similar amounts of time but learn less. Their review time is likely to be inefficient. Their test scores are more frequently lower—and they usually have less fun in their classes.

Active students are more effective than passive ones. The benefit in becoming an active student is that activity is contagious; if you become an active student in sociology, it is hard not to practice the same active learning techniques in English and math as well. Once you start asking questions in your textbook and using your Study Guide, you may find that you start asking questions in class as well. As you acquire a greater understanding of your subject, you may find that you enjoy your class more—as well as learn more and do better on tests. That is the real benefit in becoming an active learner. It is a challenge I strongly encourage you to meet.

# 1 The Sociological Perspective

**Sociology as a Point of View**
   The Sociological Imagination
*How Sociologists Do It:* If You Are Thinking About Sociology as a Career, Read This
   Is Sociology Common Sense?
   Sociology and Science
   Sociology as a Social Science
*Day-to-Day Sociology:* How Long Do Marriages Last?
*How Sociologists Do It:* What Is the Difference Between Sociology and Journalism?

**The Development of Sociology**
   Auguste Comte
   Harriet Martineau
   Herbert Spencer
   Karl Marx
   Émile Durkheim

*Thinking About Social Issues:* Are College Students at High Risk for Suicide?
   Max Weber
   The Development of Sociology in the United States

**Theoretical Perspectives**
   Functionalism
   Conflict Theory
   The Interactionist Perspective
*Thinking About Social Issues:* Social Interaction in the Internet Age
   Contemporary Sociology
   Theory and Research

**Summary**

## LEARNING OBJECTIVES

After studying this chapter, you should be able to do the following:

- Understand the sociological point of view and how it differs from that of journalists and talk-show hosts.
- Compare and contrast sociology with the other major social sciences.
- Describe the early development of sociology from its origins in nineteenth-century Europe.
- Know the contributions of sociology's early pioneers: Comte, Martineau, Spencer, Marx, Durkheim, and Weber.
- Describe the early development of sociology in the United States.
- Understand the functionalist, conflict theory, and interactionist perspectives.
- Realize the relationship between theory and practice.

On a hot summer day, a woman wearing hospital scrubs and a backpack entered the maternity ward at Darnell Army Medical Center in Fort Hood, Texas. She went straight to the maternity ward and grabbed a beautiful two-day-old baby that was not hers and made for the exit. Very quickly she set off an alarm system, which produced a building lockdown. The woman panicked, left the baby, and raced out the door. Cameras captured her identity, and a few days later she was arrested.

The thief put a face on a fear that has grown among new parents, as well as hospitals—the fear that strangers are prowling hospital corridors waiting for the chance to steal a baby. Where has this fear come from? Since 1989 the National Center for Missing and Abducted Children has been promoting this danger with nine editions of a book known as "Guidelines on Prevention of and Response to Infant Abductions." Hospitals have responded with alarm systems, security cameras, footprints and photographs of babies, blood samples, and color-coded staff badges that change regularly.

Is this a real danger? In actual fact, the chance that a nonfamily member will abduct a baby is extremely low. Over a 26-year period, only 267 attempts were made to steal a baby. During that time, 108 million babies were born. That makes the chance of a baby being stolen smaller than the chance of a lightning bolt coming through the window and hitting the baby.

The typical baby snatcher is a woman trying to salvage a romance. She fakes a pregnancy and hopes to convince her boyfriend the stolen baby is his. This misguided idea almost always fails. Parents also worry about a baby being switched with another baby. This event seems to be even less common than baby stealing; it happens only a handful of times a year.

The fear of baby stealing is a direct result of press material from the National Center for Missing and Exploited Children (NCMEC). The NCMEC receives 30 million dollars from the federal government and another 10 million from donations. The CEO of the organization is paid more than one million dollars a year. Others who work there make six-figure salaries (American Institute of Philanthropy, Charity Rating Guide & Watchdog Report, December 2011). All of this necessitates that they justify their work, resulting in an exaggeration of the danger of baby stealing.

Much of the information we read every day and mistake for sociology is actually an attempt by one group or another to influence social policy. Other information mistaken for sociology comes from attempts to sell books or efforts of television producers to present entertaining programs.

Given the constant bombardment of information about social issues, we could come to believe that nearly everyone is engaged in the study of sociology to some extent and that everyone has not only the right but also the ability to put forth valid information about society. This is not the case. Some people have no interest in putting forth true and objective information and are, instead, interested only in convincing us to support their position or point of view. In addition, some "researchers" do not have the training and skills required to disseminate accurate information about sociological topics such as drug abuse, homelessness, divorce rates, high-school dropout rates, and white-collar crime.

Sociology studies the interactions among different social groups.

Sociologists have different goals in mind when they investigate a problem than do journalists or talk-show hosts. A television talk-show host needs to make the program entertaining and maintain high ratings, or the show might be canceled. A journalist is writing for a specific readership, which certainly will limit the choice of topics as well as the manner in which issues are investigated. On the other hand, a sociologist must answer to the scientific community as she or he tries to further our understanding of a topic. This means that the goal is not high ratings but, rather, an accurate and scientific approach to the issue being studied.

In this book, we ask you to go beyond popular sociology and investigate society more scientifically than you have done before. You will learn to look at major events, as well as everyday occurrences, a little differently and start to notice patterns you might never have seen before. After you are equipped with the tools of sociology, you should be able to evaluate critically popular presentations of sociological topics. You will see that sociology represents both a body of knowledge and a scientific approach to the study of social issues.

## ● SOCIOLOGY AS A POINT OF VIEW

**Sociology** is *the scientific study of human society and social interactions.* As sociologists, our main goal is to understand social situations and look for repeating patterns in society. We do not use facts selectively to create a lively talk show, sell newspapers, or support one particular point of view. Instead, sociologists are engaged in a rigorous scientific endeavor, which requires objectivity and detachment.

The main focus of sociology is the group, not the individual. Sociologists attempt to understand the forces that operate throughout society—forces that mold individuals, shape their behavior, and, thus, determine social events.

When you walk into an introductory physics class, you might know very little about the subject and hold few opinions about the various topics within the field. On the other hand, when you enter your introductory sociology class for the first time, you will feel quite familiar with the subject matter. You have the advantage of coming to sociology with a substantial amount of information, which you have gained simply by being a member of society. Ironically, this knowledge also can leave you at a disadvantage because these views have not been gathered in a scientific fashion and might not be accurate.

Over the years and through a variety of experiences, we develop a set of ideas about the world and how it operates. This point of view influences how we look at the world and guides our attempts to understand the actions and reactions of others. Even though we accept the premise that individuals are unique, we tend to categorize or even stereotype people to interpret and predict behavior and events.

Is this personalized approach adequate for bringing about an understanding of ourselves and society? Although it might serve us quite well in our day-to-day lives, a sociologist would answer that it does not give us enough accurate information to develop an understanding of the broader social picture. This picture becomes clear only when we know something about the society in which we live, the social processes that affect us, and the patterns of interaction that characterize our lives.

Let us take the issue of theft. Figure 1-1 shows that we could examine the issue in a variety of ways. If we have our laptop stolen, we would have personal information about the experience. If we know three other people who have been victims of theft, we would know more about theft at a specific case level. Although this information is important, it is not yet sociology and is closer to the personalized, common-sense approach to understanding society. Sociology tries to move beyond that level of understanding.

If we rely on our own experiences, we are like the blind men of Hindu legend trying to describe an elephant: The first man, feeling its trunk, asserts, "It is like a snake"; the second, trying to reach around the beast's leg, argues, "No, it is like a tree"; and the third, feeling its solid side, disagrees, saying, "It is more like a wall." In some way, each person is right, but not one of them is able to understand or describe the whole elephant.

If we were to look for recurring patterns in theft, we would now be doing what sociologists do. A sociologist

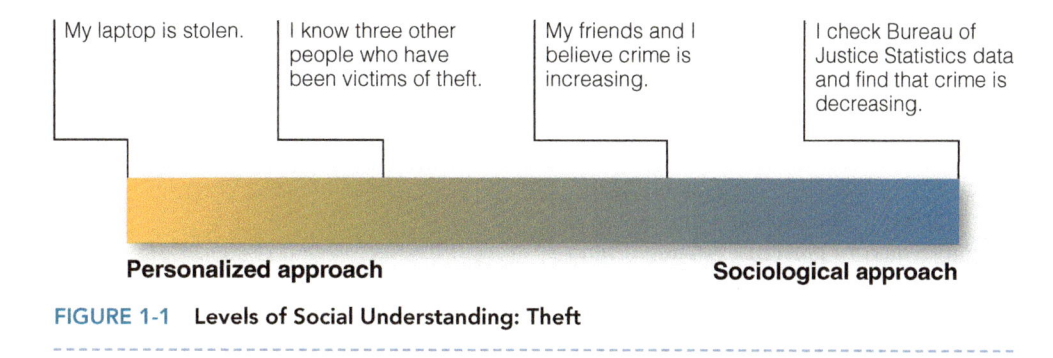

| My laptop is stolen. | I know three other people who have been victims of theft. | My friends and I believe crime is increasing. | I check Bureau of Justice Statistics data and find that crime is decreasing. |

**Personalized approach**           **Sociological approach**

FIGURE 1-1 Levels of Social Understanding: Theft

examining the issue might be interested in the age, socioeconomic level, and ethnic characteristics of the victims of theft. A sociologist might compare these characteristics with the characteristics of victims of other types of property crimes: "Are there differences?" they would ask. "If so, what kinds and why?"

While studying sociology, you will be asked to look at the world a little differently from the way you usually do. Because you will be looking at the world through other people's eyes—using new points of view—you will start to notice things you might never have noticed before. When you look at life in a middle-class suburb, for instance, what do you see? How does your view differ from that of a poor, inner-city resident? How does the suburb appear to a recent immigrant from Mexico, China, or India? How does it appear to a burglar? Finally, what does the sociologist see?

Sociology asks you to broaden your perspective on the world. You will start to see that people act in markedly different ways, not because one person is sane and another is crazy. Rather, it is because they all have different ways of making sense out of what is going on in the world around them. These unique perceptions of reality produce varying lifestyles, which in turn produce different perceptions of reality. To understand other people, we must stop looking at the world from a perspective based solely on our own individual experiences.

## The Sociological Imagination

Although most people interpret social events on the basis of their individual experiences, sociologists step back and view society more as outsiders than as personally involved and possibly biased participants. For example, although we assume that most people in the United States marry for love, sociologists remind us that the decision to marry—or not to marry—is influenced by a variety of social values taught to us since early childhood.

That is, we select our mates based on the social values we internalize from family, peers, neighbors, community leaders, and even our movie heroes. Therefore, we are less likely to marry someone from a different socioeconomic class, from a different race or religion, or from a markedly different educational background. Thus, as we pair off, we follow somewhat predictable patterns. In most cases, the man is older, earns more money, and has a higher occupational status than the woman. These patterns might not be evident to two people who are in love with each other. Indeed, these people might not be aware that anything other than romance has played a role in their choice of a mate.

As sociologists, however, we examine the topic of marriage and begin to discern patterns. We might note that marriage rates vary in different parts of the country, that the average age of marriage is related to educational level, and that social class is related to marital stability.

These patterns (discussed in Chapter 12) show us that forces are at work that influence marriage but might not be evident to the individuals who fall in love and marry.

C. Wright Mills (1959) described the different levels on which social events can be perceived and interpreted. He used the term **sociological imagination** to refer to *the relationship between individual experiences and forces in the larger society that shape our actions.* The sociological imagination is the process of looking at all types of human behavior patterns and finding previously unseen connections among them. We see similarities among individuals with no direct knowledge of one another, and we find that subtle forces mold people's actions. Like a museum patron who draws back from a painting in order to see how the separate strokes and colors form subtly shaded images, sociologists stand back from individual events in order to see why and how they occurred. In so doing, they discover patterns that govern our social existence.

The sociological imagination focuses on every aspect of society and every relationship among individuals. It studies the behavior of crowds at sports events; shifts in styles of dress and popular music; changing patterns of courtship and marriage; the emergence and fading of different lifestyles, political movements, and religious sects; the distribution of income and access to resources and opportunities; decisions made by the Supreme Court, congressional committees, and local zoning boards; and so on. Every detail of social existence is food for sociological thought and relevant to sociological analysis.

The potential for sociology to be used—applied to the solution of real-world problems—is enormous. Proponents of applied sociology believe the work of sociologists can and should be used to help bring about an understanding of, and improvement in, modern society.

The demand for applied sociology is growing, and many sociologists work directly with government agencies or private businesses to apply sociological knowledge to real-world problems. For example, sociologists might investigate such questions as how the building of a dam will affect the residents of the area; how jury make-up affects the outcome of a case; why voters select one candidate over another; how a company can boost employee morale; and how relationships among administrators, doctors, nurses, and patients affect hospital care. The answers to these questions have practical applications. The growing demand for sociological information provides many new career choices for sociologists. (See "How Sociologists Do It: If You Are Thinking About Sociology as a Career, Read This.")

## Is Sociology Common Sense?

Common sense is what people develop through everyday life experiences. In a very real sense, it is the set of expectations about society and people's behavior that

## HOW SOCIOLOGISTS DO IT

### If You Are Thinking About Sociology as a Career, Read This

Speaking from this side of the career-decision hurdle, I can say that being a sociologist has opened many doors for me. It gave me the credentials to teach at the college level and to become an author of a widely used sociology text. It also enabled me to be a newspaper columnist and a talk-show host. Would I recommend this field to anyone else? I would, but not blindly. Realize before you begin that sociology can be an extremely demanding discipline and, at times, an extremely frustrating one.

As in many other fields, the competition for jobs in sociology can be fierce. If you really want this work, do not let the herd stop you. Anyone with motivation, talent, and a determined approach to finding a job will do well. However, be prepared for the long haul: To get ahead in many areas, you will need to spend more than four years in college. Consider your bachelor's degree as just the beginning. Jobs that involve advanced research or teaching at the college level often require a PhD, which means at least four to six years of school beyond the BA.

Now for the job possibilities: As you read through information about careers in sociology, remember that right now your exposure to sociology is limited (you are only on Chapter 1 in your first college sociology text), so do not eliminate any possibilities right at the start. Spend some time thinking about each one as the semester progresses and you learn more about this fascinating discipline.

Most people who go into sociology become teachers. You will need a PhD to teach in college, but often a master's degree will open the door for you at the two-year college or high-school level.

Second in popularity to teaching are nonacademic research jobs in government agencies, private research institutions, and the research departments of private corporations. Researchers perform many functions, including conducting market research, public opinion surveys, and impact assessments. Evaluation research, as the last field is known, has become more popular in recent years because the federal government now requires environmental impact studies on all large-scale federal projects. For example, before a new interstate highway is built, evaluation researchers attempt to determine the effect the highway will have on communities along the proposed route.

This is only one of many opportunities available in government work. Federal, state, and local governments in policymaking and administrative functions also hire sociologists. For example, a sociologist employed by a community hospital provides needed data on the population groups being served and on the health care needs of the community. Further, sociologists working in a prison system can devise plans to deal with the social problems that are inevitable when people are put behind bars. Here are a few additional opportunities in government work: community planner, corrections officer, environmental analyst, equal opportunity specialist, probation officer, rehabilitation counselor, resident director, and social worker.

A growing number of opportunities also exist in corporate America, including market researchers, pollsters, human resource managers, affirmative action coordinators, employee assistance program counselors, labor relations specialists, and public information officers, just to name a few. These jobs are available in nearly every field from advertising to banking, from insurance to publishing. Although your corporate title will not be "sociologist," your educational background will give you the tools you need to do the job and do it well, which, to corporations, is the bottom line.

Whether you choose government or corporate work, you will have the best chance of finding the job you want by specializing in a particular field of sociology while you are still in school. You can become a crime and corrections specialist or become knowledgeable in organizational behavior before you enter the job market. Many demographers, who compile and analyze population data, have specialized in urban sociology or population issues. They may then also be equipped to help a community respond to neighborhood and environmental concerns.

Keep in mind that many positions require a minor or some coursework in fields other than sociology, such as political science, psychology, ecology, law, or business. By combining sociology with one or more of these fields, you will be well prepared for the job market.

What next? Be optimistic and start planning. As the American Sociological Association has observed, few fields are as relevant today and as broadly based as sociology. Yet, ironically, the career potential of this field is just beginning to be tapped. Start planning by reading the "Occupational Outlook Quarterly" (it is available online) published by the U.S. Bureau of Labor Statistics, as well as academic journals, to keep abreast of career trends. Then study hard and choose your specialty. With this preparation, you will be well equipped when the time comes to find a job.

guides our own behavior. Unfortunately, these expectations are not always reliable or accurate because without further investigation, we tend to believe what we want to believe, to see what we want to see, and to accept as fact whatever appears to be logical. Whereas common sense is often vague, oversimplified, and contradictory, sociology as a science attempts to be specific, to qualify its statements, and to prove its assertions.

Upon closer inspection, we find that the proverbial words of wisdom rooted in common sense are often illogical. Why, for example, should you "look before you leap" if "he who hesitates is lost"? How can "absence make the heart grow fonder" when "out of sight, out of mind"? Why should "opposites attract" when "birds of a feather flock together"? The common-sense approach to sociology is one of the dangers the new student encounters. Common sense often makes sense after the fact. It is more useful for describing events than for predicting them. It deludes us into thinking we knew the outcome all along (Hawkins and Hastie, 1990).

One researcher (Teigen, 1986) asked students to evaluate actual proverbs and their opposites. When given the actual proverb, "Fear is stronger than love," most students agreed that it was true, but so did students who were given the reverse statement, "Love is stronger than fear." The same was true for the statements, "Wise men make proverbs and fools repeat them" (actual proverb), and its reversal, "Fools make proverbs and wise men repeat them."

Although common sense gleaned from personal experience might be helpful to us in certain types of interactions, it does not help us understand why and under what conditions these interactions are taking place. Sociologists as scientists attempt to qualify these statements by specifying, for example, under what conditions opposites tend to attract or birds of a feather flock together. Sociology as a science is oriented toward gaining knowledge about why and under what conditions events take place in order to understand human interactions better. (For a discussion of how sociology differs from common sense, see "Day-to-Day Sociology: How Long Do Marriages Last?")

## Sociology and Science

Sociology is commonly described as one of the social sciences. **Science** refers to *a body of systematically arranged knowledge that shows the operation of general laws.* Sociology employs the same general methods of investigation that are used in the natural sciences. Like natural scientists, sociologists use the **scientific method,** *a process by which a body of scientific knowledge is built through observation, experimentation, generalization, and verification.*

The collection of data is an important aspect of the scientific method, but facts alone do not constitute a science. To have any meaning, facts must be ordered in some way, analyzed, generalized, and related to other facts. This is known as theory construction. Theories help organize and interpret facts and relate them to previous findings of other researchers.

Unlike other means of inquiry, science generally limits its investigations to things that can be observed directly or that produce directly observable events. This is known as **empiricism,** *the view that generalizations are valid only if they rely on evidence that can be observed directly or verified through our senses.* For example, theologians might discuss the role of faith in producing true happiness; philosophers might deliberate over what happiness actually encompasses; but sociologists would note, analyze, and predict the consequences of such measurable items as job satisfaction, the relationship between income and education, and the role of social class in the incidence of divorce.

## Sociology as a Social Science

The **social sciences** consist of *all those disciplines that apply scientific methods to the study of human behavior.* Although there is some overlap, each of the social sciences has its own area of investigation. It is helpful to understand each social science and examine how sociology is related to the others.

*Cultural Anthropology* The social science most closely related to sociology is *cultural anthropology*. The two share many theories and concepts and often overlap. The main difference is in the groups they study and the research methods they use. Sociologists tend to study groups and institutions within large, often modern, industrial societies, using research methods that enable them rather quickly to gather specific information about large numbers of people. In contrast, cultural anthropologists often immerse themselves in another society for a long period of time, trying to learn as much as possible about that society and the relationships among its people. Thus, anthropologists tend to focus on the culture of small, preindustrial societies because they are less complex and more manageable using this method of study.

*Psychology* The study of individual behavior and mental processes is part of *psychology*; the field is concerned with such issues as motivation, perception, cognition, creativity, mental disorders, and personality. More than any other social science, psychology uses laboratory experiments.

Psychology and sociology overlap in a subdivision of each field known as *social psychology*—the study of how human behavior is influenced and shaped by various social situations. Social psychologists study such issues as how individuals in a group solve problems and reach a consensus, and what factors might produce

## DAY-TO-DAY SOCIOLOGY

### How Long Do Marriages Last?

A statement that has entered the realm of common sense is that 50 percent of all marriages in the United States end in divorce. As with many simple stereotypes, for certain groups this statement may be true, whereas for others it is not. Here's the crucial difference between sociology and popular wisdom: As sociologists, we don't automatically accept such easy pronouncements as fact. Like scientists—and sociology is, after all, a social *science*—we want proof, and we cultivate a healthy degree of skepticism until we get it. In a case such as this, we would look at research data to determine whether this statement is true. Has it been accurate at a certain point in time but not at another? Does it describe certain marriages and not others?

A look at the data show that most adults in the United States marry only once. The median length of marriage for women is 20.8 years. Twenty-seven states have a marriage length significantly longer than this median. Women in the middle and southern parts of the United States are among those with the longest marriages. Even when we look at second marriages for people who have been divorced, the median length of marriage is 14.5 years.

There are two probable reasons for marriages lasting longer than the popular press has led us to believe. First, there has been an increase in the age at which people marry. People who marry later have more stable marriages. Second, people who marry tend to be better educated than in the past, another factor leading to greater marital stability. Both of these facts combine to produce longer-lasting marriages and lower divorce rates. Another fact that will surprise people if they have been following only media reports is that the divorce rate has been declining over the last decade.

**States with the longest median length of marriage**

Idaho
Indiana
Iowa
Nebraska
Utah

**States with the shortest median length of marriage**

California
Florida
Maryland
Nevada
New York

*Source:* Elliott, Diana B., Tavia Simmons, and Jamie M. Lewis. 2010. "Evaluation of the Marital Events Items on the ACS" (www.census.gov/hhes/socdemo/marriage/data/acs/index.html).

---

nonconformity in a group situation. Generally, however, psychology studies the individual, and sociology studies groups of individuals as well as society's institutions.

The sociologist's perspective on social issues is broader than that of the psychologist, as in the study of alcoholism, for example. The psychologist might view alcoholism as a personal problem that has the potential to destroy an individual's physical and emotional health as well as his or her marriage, career, and friendships. The sociologist, however, would look for patterns in alcoholism. Although each alcoholic makes the decision to take each drink—and each suffers the pain of addiction—the sociologist would remind us to look beyond the personal characteristics and to think about the broader aspects of alcoholism. Sociologists want to know what types of people drink excessively, when they drink, where they drink, and under what conditions they drink. They are also interested in the social costs of chronic drinking—costs in terms of families torn apart, jobs lost, children severely abused and neglected; costs in terms of highway accidents and deaths; costs in terms of drunken quarrels leading to violence and to murder. Noting the rapid rise of chronic alcoholism among women, sociologists might ask what forces are at work to account for these patterns.

*Economics* Economists have developed techniques for measuring such things as prices, supply and demand, money supplies, rates of inflation, and employment. This study of the creation, distribution, and consumption of goods and services is known as *economics*. The economy, however, is just one part of society, and each individual in society decides whether to buy an American car or a Japanese import, whether she or he is able to handle the mortgage payment on a dream house, and so on. Whereas economists study price and availability factors, sociologists are interested in the social factors that influence a person's economic decisions. Does peer pressure result in buying the large flashy car, or does concern about gas mileage lead to the purchase of a fuel-efficient or hybrid vehicle? What social and cultural factors contribute to the differences in the portion of income saved by the average wage earner in different societies? What effect does the unequal allocation of resources have on social interaction? These are examples of the questions sociologists seek to answer.

*History* Although not exactly a social science, history shares certain attributes with sociology. The study of *history* involves looking at the past to learn what happened, when it happened, and why it happened. Sociology also looks at historical events within their social contexts to discover why things happened and, more important, to assess what their social significance was and is. Historians provide a narrative of the sequence of events during a certain period and might use sociological research methods to learn how social forces have shaped historical events. Sociologists examine historical events to see how they influenced later social situations.

Historians focus on individual events—the American Revolution or slavery, for instance—and sociologists generally focus on phenomena such as revolutions or the patterns of dominance and subordination that exist in slavery. They try to understand the common conditions that contribute to revolutions or slavery wherever they occur.

Consider the subject of slavery in the United States. Traditionally, historians might focus on when the first slaves arrived, how many slaves existed in 1700 or 1850, and the conditions under which they lived. Sociologists and modern social historians would use these data to ask many questions: What social and economic forces shaped the institution of slavery in the United States? How did the Industrial Revolution affect slavery? How has the experience of slavery affected the black family? Although history and sociology have been moving toward each other over the past few decades, each discipline still retains a somewhat different focus: sociology on the present, history on the past.

*Political Science* *Political science* is the study of three major areas: political theory, the actual operation of government, and, in recent years, political behavior. This emphasis on political behavior overlaps with sociology. The primary distinction between the two disciplines is that sociology focuses on how the political system affects other institutions in society, whereas political science devotes more attention to the forces that shape political systems and the theories for understanding these forces. However, both disciplines share an interest in why people vote the way they do, why they join political movements, and how the mass media are changing political events.

*Social Work* In the early days of sociology, women were often unable to attend graduate sociology programs and chose social work studies instead, which may explain why the disciplines of sociology and social work are still often confused with each other. Much of the theory and many of the research methods of social work are drawn from sociology and psychology, but social work focuses to a much greater degree on application and problem solving.

The main goal of *social work* is to help people solve their problems, whereas the aim of sociology is to understand why the problems exist. Social workers provide help for individuals and families who have emotional and psychological problems or who experience difficulties that stem from poverty or other ongoing problems rooted in the structure of society. Social workers also organize community groups to tackle local issues such as housing problems and try to influence policymaking bodies and legislation. Sociologists provide many of the theories and ideas used to help others. Although sociology is not social work, it is a useful area of academic concentration for those interested in entering the helping professions.

The work of journalists is also often confused with that of sociologists. It is common for journalists to write articles that examine sociological issues. (For a comparison of the two fields, see "How Sociologists Do It: What Is The Difference Between Sociology and Journalism?")

Sociologists and anthropologists share many theories and concepts. However, sociologists tend to study groups and institutions within large, modern, industrial societies; anthropologists tend to focus on the cultures of small, preindustrial societies.

## HOW SOCIOLOGISTS DO IT

### What Is the Difference Between Sociology and Journalism?

It often seems as if sociologists and journalists are engaged in the same activities. Journalists examine and write about social issues. They interview people. They often conduct polls. They make predictions. They offer recommendations for correcting social problems. If journalists do all this, why would someone need to become trained as a sociologist?

This is a sociology textbook, so needless to say we are going to make the case that there is a difference between sociologists and journalists.

Newspapers, weekly news magazines, and documentaries are designed for the general public, which wants an overview of a topic. One of the fundamental features of these media is the timely coverage of recent events. In recent years, newspapers have been suffering because the latest information can often be found first on the Internet or 24-hour news channels. There are at least three types of journalists: reporters, who actually write stories; editors, who generate ideas for stories and review the copy; and editorial writers, who interpret events or provide other ways of thinking about them. Journalists usually have a college degree in any of a wide variety of areas or an advanced degree from a professional journalism program. Jargon is kept to a minimum, and elaborate explanations must be presented in manageable terms so that the average reader can understand them.

Sociologists study society with the intent of sharing their work with other sociologists. The methods of investigation, previous research, and theories of explanation are all important. Once completed, the other sociologists may respond with alternate explanations (Baker and Dorn, 1993). Sociologists usually publish their writings as articles in scholarly journals, in chapters in books, or as full-length books. These writings are screened by editors and critics hired to evaluate the merits of the work. Sociologists want their colleagues to recognize their work as truly significant.

So, whereas journalists are often thinking about what is currently capturing the public's interest, sociologists' work does not have such a short time frame. Sociologists, however, hope their work will be relevant to contemporary debates or current issues. Essentially, the two fields represent different approaches to social issues. Journalists get a multifaceted overview of an issue, whereas sociologists have the luxury of exploring a topic in depth and contemplating the ramifications of their findings.

## THE DEVELOPMENT OF SOCIOLOGY

It is hardly an accident that sociology emerged as a separate field of study in Europe during the nineteenth century. That was a time of turmoil, a period in which the existing social order was shaken by the growing Industrial Revolution and violent uprisings against established rulers (the American and French revolutions). People were also discovering, through world exploration, how other people lived. At the same time, the church's power to impose its views of right and wrong was also declining. New social classes of industrialists and businesspeople emerged to challenge the rule of the feudal aristocracies.

Tightly knit communities, held together by centuries of tradition and well-defined social relationships, were strained by dramatic changes in society. Factory cities began to replace the rural estates of nobles as the centers for society at large. People with different backgrounds were brought together under the same factory roof to work for wages instead of exchanging their services for land and protection. Families now had to protect themselves, to buy food rather than grow it, and to pay rent for their homes. These new living and working conditions led to the development of an industrial, urban lifestyle, which, in turn, produced new social problems.

Many people were frightened by these changes and wanted to find some way of coping with the new society. The need for a new understanding of society, together with the growing acceptance of the scientific method, led to the emergence of sociology.

### Auguste Comte (1798–1857)

Born in the French city of Montpellier on January 19, 1798, Auguste Comte grew up in the period of great political turmoil that followed the French Revolution of 1789–1799. In August 1817, Comte met Henri Saint-Simon and became his secretary and eventually his close collaborator. Under Saint-Simon's influence, Comte converted from an ardent advocate of liberty and equality to a supporter of an elitist conception of society.

Saint-Simon and Comte rejected the lack of empiricism in the social philosophy of the day. Instead they turned for inspiration to the methods and intellectual

Auguste Comte coined the term sociology. He wanted to develop "a science of man" that would reveal the underlying principles of society, much as the sciences of physics and chemistry explained nature and guided industrial progress.

framework of the natural sciences, which they perceived as having led to the spectacular successes of industrial progress. They set out to develop a "science of man" that would reveal the underlying principles of society much as the sciences of physics and chemistry explained nature and guided industrial progress. During their association, the two men collaborated on a number of essays, most of which contained the seeds of Comte's major ideas. Their alliance came to a bitter end in 1824 when Comte broke with Saint-Simon for both financial and intellectual reasons.

Comte saw this new science, which he named sociology, as the greatest of all sciences. Sociology would include all other sciences and bring them all together into a cohesive whole. However, financial problems, lack of academic recognition, and marital difficulties combined to force Comte into a shell. Eventually, for reasons of "cerebral hygiene," he no longer read any scientific work related to the fields about which he was writing.

Living in isolation at the periphery of the academic world, Comte concentrated his efforts between 1830 and 1842 on writing his major work, *Cours de Philosophie Positive*, the work in which he actually coined the term *sociology*. Comte devoted a great deal of his writing to describing the contributions he expected sociology would make in the future. He was much less concerned with defining sociology's subject matter than with showing how it would improve society.

## Harriet Martineau (1802–1876)

Harriett Martineau was born in Norwich, England. She was the sixth of eight children and was unhappy in her youth. She became deaf at a young age and had no sense of taste or smell. Hearing aids did not exist then, so she used an ear trumpet.

In 1826, Martineau's father died, leaving her with little money. She needed to find a job, but the typical line of work for young women, teaching, was not open to her because she was deaf. Instead, she turned to writing.

In 1837, Martineau published *Theory and Practice of Society in America*, in which she analyzed the customs and lifestyle present in the nineteenth-century United States. Her book was based on traveling throughout the United States and observing day-to-day life in all its forms, from that which took place in prisons, mental hospitals, and factories to family gatherings, slave auctions, and even proceedings of the Supreme Court and Senate. The book helped map out what a sociological work dealt with by examining the impact of immigration, family issues, politics, and religion as well as race and gender issues. In her book, she also compared social stratification systems in Europe with those in the United States.

Martineau's work demonstrated the level of objectivity she thought was necessary for an analysis of society when she noted, "It is hard to tell which is worse, the wide diffusion of things that are not true or the suppression of things that are." Later in her career, she came

Harriet Martineau was an early and significant contributor to the development of sociology. She believed that scholars should not simply offer observations but should also use their research to bring about social reform.

to the conclusion that scholars should not just offer observations but should also use their research to bring about social reform for the benefit of society. She asked her readers to "judge for themselves…how far the people of the United States lived up to" their stated ideals (Hoecker-Drysdale, 1992).

Martineau was outspoken about the treatment of women in the United States. She thought women were treated like slaves. She was a proponent of expanding the education of women so that they did not have to depend only on marriage to live successfully.

Martineau's second important contribution to sociology was translating into English Auguste Comte's six-volume *Positive Philosophy*. Her two-volume edition of this book (1853) introduced the field of sociology to England and influenced people such as Herbert Spencer as well as early American sociologists. Comte himself recommended Martineau's translation to his students instead of his own.

## Herbert Spencer (1820–1903)

Even though Herbert Spencer was largely self-educated, he had an enormous impact on a variety of fields, including sociology. In 1858, he outlined an enormous project for himself that few people have ever attempted. He wanted to demonstrate how the idea of evolution applied to sociology, biology, psychology, and morality. Spencer thought that producing the ten volumes of this work he planned would take 20 years, but in fact, it took the rest of his life. The books Spencer produced, listed below, helped define these fields and shape the future development of these disciplines.

*Principles of Sociology* (3 volumes)
*The Study of Sociology*
*Principles of Biology* (2 volumes)
*Principles of Psychology* (2 volumes)
*The Principles of Ethics* (2 volumes)

Eventually, Spencer became the most famous philosopher of his time. His works were translated into many languages, and he received numerous honors and awards throughout the world.

Spencer believed that society is similar to a living organism. Just as the individual organs of the body are interdependent and make their specialized contributions to the living whole, so, too, are the various segments of society interdependent. Every part of society serves a specialized function necessary to ensure society's survival as a whole.

Spencer became a proponent of a doctrine known as **social Darwinism,** which *applied to society Charles Darwin's notion of survival of the fittest, in which those species of animals best adapted to the environment survived and prospered, whereas those poorly adapted died out.* Spencer reasoned that people who could not

Herbert Spencer helped define what sociology would examine. Spencer also became a proponent of the doctrine known as social Darwinism.

successfully compete in modern society were poorly adapted to their environment and were therefore inferior. Lack of success was viewed as an individual failing, and that failure was in no way related to barriers (such as prejudice or racism) created by society. In this view, to help the poor and needy was to intervene vainly in a natural evolutionary process. As Spencer noted:

> Human society is always in a kind of evolutionary process in which the fittest—which happened to be those who can make lots of money—were chosen to dominate. There were the armies of unfit, the poor, who simply could not compete. And just as nature weeds out the unfit, an enlightened society ought to weed out its unfit and permit them to die off so as not to weaken the racial stock. (Spencer, 1864, p. 444)

Social Darwinism had a significant effect on those who believed in the inequality of races. They now claimed that those who had difficulty succeeding in the white world were really members of inferior races. The fact that they lost out in the competition for status was proof of their poor adaptability to the environment.

The survivors were clearly of superior stock (Berry and Tischler, 1978).

Many people accepted social Darwinism because it served as a justification for their control over society. It enabled them to oppose reforms or social welfare programs, which they viewed as interfering with nature's plan to do away with the unfit. Social Darwinism thus became a justification for the repression and neglect of African Americans following the Civil War. It was also used to justify policies that resulted in the decimation of Native American populations (Parillo, 1997).

During the last years of his life, Spencer was quite isolated and disillusioned. He never married, and many of his close friends died. He imagined he had many illnesses, which doctors were unable to verify or treat.

Spencer's ties to social Darwinism have led many scholars to disregard his original contributions to the discipline of sociology. However, Spencer originally formulated many of the standard concepts and terms still current in sociology, and their use derives directly from his works.

During the nineteenth century, sociology developed rapidly under the influence of three other scholars of highly divergent temperaments and orientations. Despite their differences, however, Karl Marx, Émile Durkheim, and Max Weber were responsible for shaping sociology into a relatively coherent discipline.

### Karl Marx (1818–1883)

Karl Marx is often thought of as a revolutionary proponent of the political and social system seen in countries once labeled communist. It is true that nearly half of the people in the world live under political systems that claim ties to Marxism. The governments, however, have often modified Marx's original ideas to fit their own philosophies.

Marx lived in Europe during the early period of industrialization, when the overwhelming majority of people in such societies were poor. The rural poor moved to cities where employment was available in the factories and workshops of the new industrial economies. Those who owned the factories exploited the masses who worked for them. Even children, some as young as five or six years old, worked 12-hour days, six and seven days a week (Lipsey and Steiner, 1975), and received barely enough money to survive. The rural poor became the urban poor. Meanwhile, the owners achieved great wealth, power, and prestige.

Marx wanted to understand why society produced such inequities, and he looked for a way to improve the human condition. Marx believed the entire history of human societies could be seen as the history of class conflict—the conflict between the bourgeoisie, who owned and controlled the means of production (capitalists), and the proletariat, who made up the mass of workers—the exploiters and the exploited. He believed the capitalists controlled wealth, power, and even ideas in society. They influenced the political, educational, and religious institutions in their society as well. According to Marx, capitalists make and enforce laws that serve their interests and act against the interests of workers.

Marx predicted that capitalist society eventually would be split into two broad classes: the capitalists and the increasingly impoverished workers. Intellectuals like him would show the workers that the capitalist institutions were the source of exploitation and poverty. Gradually, the workers would become unified and organized and then take over control of the economy.

Marx did not think this change would come about peacefully. Violent revolution would be necessary because those in power would not give up power voluntarily. The socialist system Marx envisioned would

Karl Marx's views on class conflict were shaped by the Industrial Revolution. He believed that capitalist societies produced conflict because of the deep divisions between the social classes.

also require what he called "a dictatorship of the proletariat"—a temporary government in which the needs of the workers were protected. Eventually, this would lead to a true socialist society.

The means of production would then be owned and controlled by the people in a workers' socialist state. After the capitalist elements of all societies had been eliminated, the governments would wither away. New societies would develop in which people could work according to their abilities and take according to their needs. The seeds of societal conflict and social change would then come to an end because the factories were no longer privately owned.

In many capitalist societies today, regulatory mechanisms have been introduced to prevent some of the excesses of capitalism. Unions have been integrated into the capitalist economy and the political system, giving workers a legal, legitimate means through which they can benefit from the capitalist system.

Marx was not a sociologist, but his considerable influence on the field can be traced to his contributions to the development of *conflict theory*, which will be discussed more fully in this chapter.

### Émile Durkheim (1858–1917)

It might have been Spencer who wrote the first textbook of sociology, but it was Émile Durkheim who produced the first true sociological study. Durkheim's work moved sociology fully out of the realm of social philosophy and helped chart the discipline's course as a social science.

In 1895, Durkheim published *Rules of the Sociological Method*, in which he described what sociology was and how research should be done. He also founded the first European Department of Sociology at the University of Bordeaux.

Durkheim believed that individuals were exclusively the products of their social environment and that society shapes people in every possible way. To prove his point, Durkheim studied suicide. He believed that if he could take what was perceived to be a totally personal act and show that it is patterned by social factors rather than exclusively by individual mental disturbances, he would provide support for his point of view. People committed suicide because they were members of different social groups that were influenced by a variety of social factors.

Durkheim began with the theory that the industrialization of Western society was undermining the social control and support that communities had historically provided for individuals. The anonymity and impersonality they encountered in these urban areas caused many people to become isolated from both family and friends. Further, in modern societies, people were frequently encouraged to aspire to goals that were difficult to attain. Durkheim believed suicide rates were influenced by group cohesion and societal stability. He believed that low levels of cohesion—which involve more individual choice, more self-reliance, and less adherence to group standards—would mean high rates of suicide.

To test his idea, Durkheim decided to study the suicide rates of Catholic versus Protestant countries. He assumed the suicide rate in Catholic countries would be lower than in Protestant countries because Protestantism emphasized the individual's relationship to God over community ties. The comparison of suicide records in Catholic and Protestant countries in Europe supported his theory by showing the probability of suicide was indeed higher in Protestant countries.

Recognizing the possibility that lower suicide rates among Catholics could be based on factors other than group cohesion, Durkheim proceeded to test other groups. Reasoning that married people would have more group ties than single people, or people with children more than

Émile Durkheim produced the first true sociological study. Durkheim's work helped move sociology out of the realm of social philosophy and into the direction of social science.

people without children, or non–college educated people more than college-educated people (because college tends to break group ties and encourage individualism), or Jews more than non-Jews, Durkheim tested each of these groups, and in each case, his theory held. Then, characteristic of the scientist that he was, Durkheim extended his theory by identifying three types of suicide—egoistic, altruistic, and anomic—that take place under different types of conditions.

Egoistic suicide comes from low group cohesion, an underinvolvement with others. Durkheim argued that loneliness and a commitment to personal beliefs rather than to group values can lead to egoistic suicide. Therefore, he found that single and divorced people had higher suicide rates than did married people and that Protestants, who tend to stress individualism, had higher rates of suicide than did Catholics.

Altruistic suicide derives from a very high level of group cohesion, an overinvolvement with others. The individual is so tied to a certain set of goals that he or she is willing to die for the sake of the community. This type of suicide, as Durkheim noted in his time, still exists in the military as well as in societies based on ancient codes of honor and obedience. Perhaps the best-known historical examples of altruistic suicide come from Japan in the ceremonial rite of *seppuku,* in which a disgraced person rips open his own belly, and in the kamikaze attacks by Japanese pilots toward the end of World War II.

The Japanese pilots, instead of being morose before the bombing missions that would cause their certain deaths, were often reported to be cheerful and serene. One 23-year-old kamikaze, in a letter to his parents, voiced the feelings of thousands of his fellows when he wrote, "I shall be a shield for His Majesty and die cleanly along with my squadron leader and other friends." There were, said Masuo, sixteen members in his squadron, and he added, "May our deaths be as sudden and clean as the shattering of crystal" (Axell, 2002).

Today, we often see examples of altruistic suicide in terrorists, such as those who flew the planes into the World Trade towers and the Middle Eastern suicide bombers. These individuals are willing to sacrifice their lives for their cause as they blow up a building, plane, or restaurant. In addition to destroying the property, the terrorists often want to kill as many people as possible.

Anomic suicide results from a sense of feeling disconnected from society's values. A person might know what goals to strive for but not be able to attain them, or a person might not know what goals to pursue. Durkheim found that times of rapid social change or economic crisis are associated with high rates of anomic suicide.

Durkheim's study was important not only because it proved that the most personal of all acts, suicide, is in fact a product of social forces, but also because it was one of the first examples of a scientifically conducted sociological study. Durkheim systematically posed theories, tested them, and drew conclusions that led to further theories. He also published his results for everyone to see and criticize. (For a discussion of suicide in contemporary society, see "Thinking About Social Issues: Are College Students at High Risk for Suicide?")

Durkheim's interests were not limited to suicide. His mind ranged the entire spectrum of social activities. Two of his other classics include *The Division of Labor in Society* (1893) and *The Elementary Forms of the Religious Life* (1917). In both works he drew on what was known about nonliterate societies as evolutionary precursors of contemporary societies.

Durkheim focused on the forces that hold society together—that is, on the functions of various parts of society. This point of view, often called the *functionalist theory* or *functionalist perspective,* remains one of the dominant approaches to the modern study of society.

## Max Weber (1864–1920)

Max Weber thought of sociology as the study of social action. He differed from the other founders of sociology in a variety of ways. Herbert Spencer thought society was similar to a living organism. Durkheim was concerned with social cohesion in society. Marx believed the conflicts between social classes determined many things in society. In contrast, Max Weber's primary focus was on the individual meanings people attach to the world around them.

In addition, much of Weber's work attempted to clarify, criticize, and modify the works of Marx. Therefore, we shall discuss Weber's ideas as they relate to and contrast with those of Marx. Unlike Marx, who was not only an intellectual striving to understand society but also a revolutionary conspiring to overturn the capitalist social system, Weber was essentially a German academic attempting to understand human behavior.

Weber believed the role of intellectuals was simply to describe and explain truth, whereas Marx believed the scholar should also tell people what to do. Marx believed that ownership of factories resulted in control of wealth, power, and ideas. Weber, however, showed that economic control does not necessarily result in prestige and power. For example, the wealthy president of a chemical company whose toxic wastes have been responsible for the pollution of a local water supply might have little prestige in the community. Moreover, the company's board of directors might deprive the president of any real power. Although Marx maintained that control of production inevitably results in control of ideologies, Weber stated that the opposite may happen: Ideologies sometimes influence the economic system.

## THINKING ABOUT SOCIAL ISSUES

### Are College Students at High Risk for Suicide?

Émile Durkheim would have been interested to learn that college students are much less likely to commit suicide than the same-aged people who are not in college. A study was done of 1,154 public and private four-year schools and the results showed that suicide rates were 40 percent lower than among the same-aged general population. It appears that colleges and university provide a protective environment (see Figure 1-2).

The potential for what Durkheim called egoistic suicide, which results from low social integration, becomes less likely in higher education environments because of the possible attachments a campus provides. A college or university typically provides support and community for its students—whether it be through orientation programs, resident advisor assistance, clubs and interest groups, or counseling. Even though university mental health programs are often understaffed, their very existence represents a type of resource that is less easily available to nonstudents.

Another reason why college and university students are less likely to commit suicide than others in their age group is that although guns are used in 60 percent of all suicides, campuses for the most part forbid the ownership of guns.

This is not to say that suicide on campus is not a serious problem. Advances in medications and treatments have made it possible for students with major depression, bipolar disorder, and even schizophrenia to attend college who would have not been able to do so in the past. For some of these students, the risk is that they may become overwhelmed by the college experience and have serious problems adjusting. For the vast majority of students, however, the higher education environment provides a strong potential for involvement and integration.

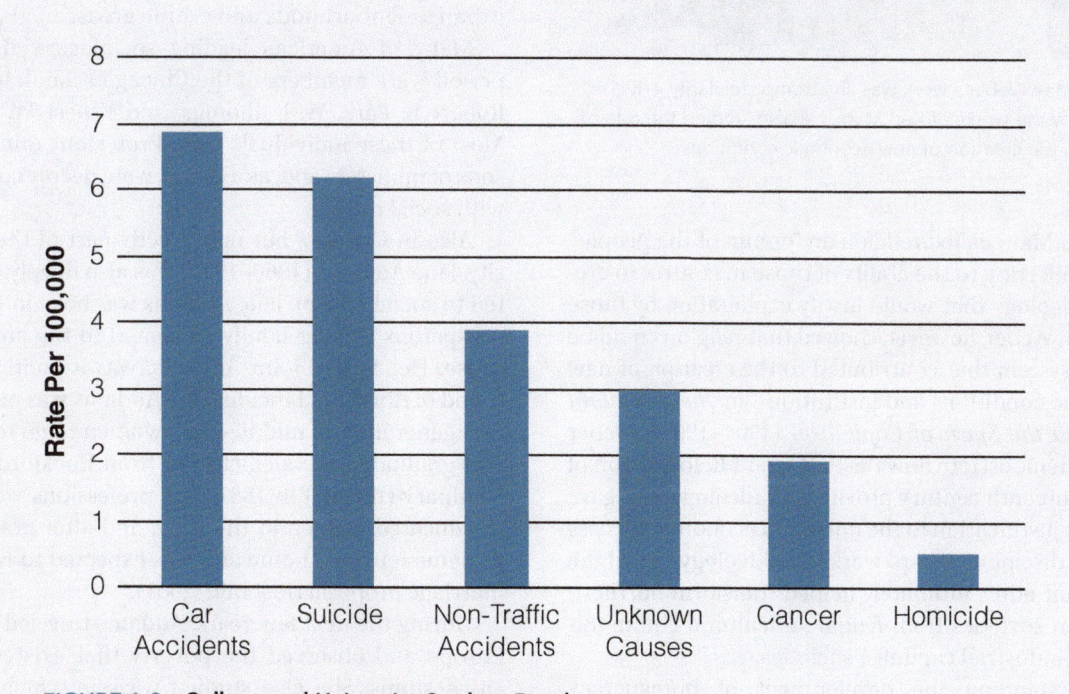

**FIGURE 1-2  College and University Student Deaths**
Source: Turner, James C., and Adrienne Keller, "Leading Causes of Mortality Among American College Students at 4-Year Institutions." Paper presented at the American Public Health Association annual meeting, October 29–November 2, 2011, Washington, D.C.

Much of Max Weber's work was an attempt to clarify, criticize, and modify the works of Karl Marx. He also studied the role of religion in the creation of new economic conditions.

When Marx called religion an "opium of the people," he was referring to the ability of those in control to create an ideology that would justify exploitation by those in power. Weber, however, showed that religion could be a belief system that contributed to the creation of new economic conditions and institutions. In *The Protestant Ethic and the Spirit of Capitalism* (1904–1905), Weber tried to demonstrate how the Protestant Reformation of the seventeenth century provided an ideology that gave religious justification to the pursuit of economic success through disciplined, hard work. This ideology, called the Protestant ethic, ultimately helped transform northern European societies from feudal agricultural communities into industrial capitalist societies.

Understanding the development of bureaucracy interested Weber. Marx saw capitalism as the source of control, exploitation, and alienation of human beings and believed that socialism and communism would ultimately bring an end to this exploitation. Weber believed bureaucracy would characterize both socialist and capitalist societies. He anticipated and feared the domination of individuals by large bureaucratic structures—economic, political, military, educational, and religious. As he foresaw, bureaucracies now rule our modern industrial world in both capitalist and socialist societies. Given the existing situation, it is easy to appreciate Weber's anxiety. As he put it,

> Each man becomes a little cog in the machine and, aware of this, his one preoccupation is whether he can become a bigger cog.... The problem which besets us now is not: how can this evolution be changed?—for that is impossible, but what will become of it? (Quoted in Coser, 1977)

## The Development of Sociology in the United States

Sociology had its roots in Europe and did not become widely recognized in the United States until almost the beginning of the twentieth century. The early growth of American sociology began at the University of Chicago. That setting provided a context in which a large number of scholars and their students could work closely to refine their views of the discipline. It was there that the first graduate department of sociology in the United States was founded in the 1890s. From the 1920s to the 1940s, the so-called Chicago school of sociologists led American sociology in the study of communities, with particular emphasis on urban neighborhoods and ethnic areas.

Many of America's leading sociologists from this period were members of the Chicago school, including Robert E. Park, W. I. Thomas, and Ernest W. Burgess. Most of these individuals were Protestant ministers or sons of ministers and, as a group, were deeply concerned with social reform.

Also in Chicago, but not directly part of the university, Jane Addams (1860–1935) was also deeply committed to social reform. Jane Addams was born in 1860 to a prosperous Quaker family dedicated to the antislavery cause. Her father, John Addams, was a politician and friend of Abraham Lincoln. Jane Addams was part of the first generation of middle-class women to go to college and graduated as valedictorian from Rockford Female Seminary (Illinois) in 1881. Few professions were open to educated women at the time, and after graduation, Addams returned home and was expected to wait for a marriage proposal (Elshtain, 2001).

During the next few years, Addams traveled through Europe and observed the poverty that existed in the cities' slums. She also studied ways in which various organizations attempted to alleviate poverty. During her stay in London, she visited a settlement house run by Oxford University students where they helped the poor. She used this settlement house, called Toynbee Hall, as a model for a program she would later develop in Chicago to assist the poor.

Jane Addams and Ellen Gates Star finally opened the doors to their own version of Toynbee Hall, Hull House, in September 1889. It was designed to serve the immigrant population of Chicago's 19th ward. For 40 years, Hull House successfully served the community by offering a wide variety of clubs and activities.

During this time, Hull House and Jane Addams became known internationally for championing the rights of immigrants and fighting for child labor laws. She also advocated for industrial safety, juvenile courts, labor unions, women's suffrage, and world peace.

Addams wrote extensively about Hull House activities. She published eleven books and numerous articles, and she spoke often at venues throughout the United States and the world. She lived on her inheritance and the proceeds from her writing and speaking engagements because she did not receive a salary from Hull House. She also used her income to underwrite various social causes throughout her life.

In 1907, she published *Newer Ideals of Peace*, from which she became known internationally as a pacifist. This brought her much ridicule when the United States entered World War I. But in time, the public began to embrace her ideals. By 1931, her reputation as a peacemaker was firmly established, and she was awarded the Nobel Peace Prize, shared with Nicholas Murray Butler. After that, people from all over the world began to write her letters and to extoll her work. She received pleas for intervention around the world to help alleviate hunger, poverty, and oppression (Swarthmore College, 2002).

W. E. B. Du Bois (1868–1963) became the first African American to receive a PhD from Harvard in 1896 with his dissertation, *The Suppression of the African Slave-Trade to the United States*. Du Bois then went on to Atlanta University, where he established and was in charge of the sociology program until 1910, when he left to become editor of *The Crisis*, the journal of the National Association for the Advancement of Colored People. By that time, Du Bois had written dozens of articles and books on the history and sociology of African Americans and was the country's leading African American sociologist.

When Du Bois came of age, racism was very much a part of the American landscape on both a popular and academic level. Politicians and writers were openly declaring that blacks belonged to an inferior race that contributed nothing to society. Du Bois believed that doctrines and theories had a powerful effect on social conditions. Slavery and the disenfranchisement of blacks were rooted in the notion of the inferiority of the race. It was important, he felt, to change these beliefs to improve the status of African Americans. Much of his scholarly work was governed by his view that sociological studies of African Americans would have a positive effect on public opinion (Brotz, 1966).

Du Bois argued for the acceptance of African Americans into all areas of society and advocated militant resistance to white racism. He believed that it was not solely the responsibility of blacks, nor was it in their capacity, to alter their collective place in American society, but that it was primarily the responsibility of whites, who held the power to effect such change.

W. E. B. Du Bois was the first African American to receive a PhD from Harvard University. He wrote dozens of articles and books on the history of sociology of African Americans.

In 1903, Du Bois published *The Souls of Black Folk*, a collection of eloquent, well-reasoned essays on race relations. Blending sociology and economics, he described the injustices that had scarred the black experience in the United States. "The problem of the Twentieth Century is the problem of the color line," he declared (Lewis, 2000).

Throughout his life, Du Bois considered himself torn between being a black man and being an American. This conflict led him to feel like an exile in the United States and, eventually, he left and moved to Ghana. As Du Bois noted in his autobiography:

> Had it not been for the race problem early thrust upon me and enveloping me, I should have probably been an unquestioning worshipper at the shrine of the established social order into which I was born. But just that part of this order which seemed to most of my fellows nearest perfection seemed to me most inequitable and wrong; and starting from that critique, I gradually, as the years went by, found other things to question in my environment. (Du Bois, 1968)

Du Bois died in 1963 at the age of 95, one day before the famous march on Washington took place where Martin Luther King Jr. made his "I Have a Dream" speech.

It was ironic that America's preeminent black intellectual died on the eve of this great civil rights gathering, which had gained so much energy from his ideas against segregation. Du Bois had long before concluded that the possibility of racial equality was a receding mirage for people of color. At the time of his death, he was leading the life of a political exile in Ghana.

Talcott Parsons (1902–1979) was the sociologist most responsible for developing theories of structural functionalism in the United States. He presided over the Department of Social Relations at Harvard College from the 1930s until he retired in 1973. Parsons's early research was quite empirical, but he later turned to the philosophical and theoretical side of sociology. In *The Structure of Social Action* (1937), Parsons presented English translations of the writings of European thinkers, most notably Weber and Durkheim. In his best-known work, *The Social System* (1951), Parsons portrayed society as a stable system of well-ordered, interrelated parts. His viewpoint elaborated on Durkheim's perspective.

Robert K. Merton also has been an influential proponent of functionalist theory. In his classic work, *Social Theory and Social Structure* (1968), first published in 1949, Merton spelled out the functionalist view of society. One of his main contributions to sociology was to distinguish between two forms of social functions—manifest functions and latent functions. By **social functions**, Merton meant *those social processes that contribute to the ongoing operation or maintenance of society*. **Manifest functions** are *the intended and recognized consequences of those processes*. For example, one of the manifest functions of going to college is to obtain knowledge, training, and a degree in a specific area. **Latent functions** are *the unintended or not readily recognized consequences of such processes*. Therefore, college can also offer the opportunity of establishing lasting friendships and finding potential marriage partners.

Under the leadership of Parsons and Merton, sociology in the United States moved away from a concern with social reform and adopted a so-called value-free perspective. This perspective, which Max Weber advocated, requires description and explanation rather than prescription; it holds that people should be told what is, not what should be.

## THEORETICAL PERSPECTIVES

How should you begin to think about society? You first need to start with a set of assumptions that offer a framework for interpreting the results of studies. Such assumptions are known as paradigms. For example, "good will triumph over evil" is a paradigm.

**Paradigms** are *models or frameworks for questions that generate and guide research*. Of course, not all paradigms are equally valid, even though at first they seem to be. Sooner or later, some will be found to be rooted in fact, and others will be unusable and finally discarded.

Let us examine the paradigms sociologists are likely to use.

### Functionalism

Functionalism—or structural functionalism, as it is often called—is rooted in the writings of Spencer and Durkheim and the work of such scholars as Parsons and Merton. **Functionalism** *views society as a system of highly interrelated structures or parts that function or operate together harmoniously.*

Functionalists analyze society by asking what each different part contributes to the smooth functioning of the whole. For example, we may assume the education system serves to teach students specific subject matter. However, functionalists might note that it acts as a system to socialize the young so that they can become members of society. The education system serves as a gatekeeper to the rewards society offers to those who follow its rules.

From the functionalist perspective, society appears quite stable and self-regulating. Much like a biological organism, society is normally in a state of equilibrium or balance. Most members of a society share a value system and know what to expect from one another.

Functionalism is a very broad theory in that it attempts to account for the complicated interrelationships of all the elements that make up human societies. In a way, it is impossible to be a sociologist and not be a functionalist because most parts of society serve some stated or unstated purpose. Functionalism is limited in one regard, however: The assumption that societies are normally in balance or harmony makes it difficult for proponents of this view to account for how social change comes about.

If major parts of society fit together smoothly, we can assume that the social system is working well. Conflict is then seen as something that disrupts the essential orderliness of the social structure and produces imbalance between the parts and the whole.

### Conflict Theory

Conflict theory is rooted in the work of Marx and other social critics of the nineteenth century. **Conflict theory** *proposes that each individual or group struggles to attain the maximum benefit. This causes society to change constantly in response to social inequality and social conflict.*

For conflict theorists, social change pushed forward by social conflict is the normal state of affairs. Calm periods are merely temporary stops along the road. Conflict theorists believe social order results from those in power making sure that subordinate groups are loyal to the systems that are the dominant groups' sources of wealth, power, and prestige. The powerful will use coercion, constraint, and even force to control those people who are not voluntarily loyal to the laws and rules those in control have made. When this order cannot be maintained and the subordinate groups rebel, change comes about.

Conflict theorists are concerned with the issue of who benefits from particular social arrangements and how those in power maintain their positions and continue to reap benefits from them. The ruling class is seen as a group that spreads certain values, beliefs, and social arrangements to enhance its power and wealth. The social order then reflects the outcome of a struggle among those with unequal power and resources.

Conflict perspectives are often criticized as concentrating too much on conflict and change and too little on what produces stability in society. They also are criticized for being too ideologically based and using little in the way of research methods or objective statistical evidence. Conflict theorists counter that the complexities of modern social life cannot be reduced to statistical analysis and that doing so has caused sociologists to become detached and removed from the real causes of human problems.

Both functionalist and conflict theories are descriptive and predictive of social life. Each has its strengths and weaknesses, and each emphasizes an important aspect of society and social life.

## The Interactionist Perspective

Functionalism and conflict theory can be thought of as opposite sides of the same coin. Although quite different from one another, they share certain similarities. Both approaches focus on major structural features of entire societies and attempt to give us an understanding of how societies survive and change. Social life, however, also occurs on an intimate scale between individuals. The **interactionist perspective** *focuses on how individuals make sense of—or interpret—the social world in which they participate.* As such, this approach is primarily concerned with human behavior on a person-to-person level. Interactionists criticize functionalists and conflict theorists for implicitly assuming that social processes and social institutions somehow have a life of their own apart from the participants. Interactionists remind us that the educational system, the family, the political system, and, indeed, all of society's institutions are ultimately created, maintained, and changed by people interacting with one another.

The interactionist perspective includes a number of loosely linked approaches. George Herbert Mead devised a *symbolic interactionist* approach that focuses on signs, gestures, shared rules, and written and spoken language. Harold Garfinkel used *ethnomethodology* to show how people create and share their understandings of social life. Erving Goffman took a *dramaturgical* approach in which he saw social life as a form of theater. (We will discuss ethnomethodology and dramaturgy in Chapter 5.) Of these three approaches, the symbolic interactionist approach has received the widest attention and presents us with a well-formulated theory. Table 1-1 compares the functionalist, conflict theory, and interactionist approaches to sociology.

George Herbert Mead (1863–1931) developed an interactionist perspective known as **symbolic interactionism,** which is *concerned with the meanings that people place on their own and one another's behavior.*

Human beings are unique in that most of what they do with one another has meaning beyond the concrete act. According to Mead, people do not act or react automatically but carefully consider and even rehearse what they are going to do. They take into account the other

TABLE 1-1  Major Theoretical Perspectives in Sociology

| Perspective | Scope of Analysis | Point of View | Focus of Analysis |
| --- | --- | --- | --- |
| Structural-Functional | Macro level | The various parts of society are interdependent and functionally related. | The functional and dysfunctional aspects of institutions and society |
|  |  | Social systems are highly stable. |  |
|  |  | Social life is governed by consensus and cooperation. |  |
| Social Conflict | Macro level | Society is a system of accommodations among competing interest groups. | How social inequalities produce conflict |
|  |  | Social systems are unstable and are likely to change rapidly. | Who benefits from particular social arrangements |
|  |  | Social life involves conflict because of differing goals. |  |
| Interactionist | Micro level | Most of what people do has meaning beyond the concrete act. | How people make sense of the world in which they participate |
|  |  | The meanings that people place on their own and on one another's behavior can vary. |  |

## THINKING ABOUT SOCIAL ISSUES: Social Interaction in the Internet Age

The interactionist perspective has received a contemporary update with the enormous increase in the use of social media. Two-thirds of adults (66%) use social media sites such as Facebook, Twitter, or LinkedIn, and the number using such sites in 2011 was twice the number in 2008. Facebook is the most commonly used social media site with more than half of the users visiting every day.

The main reason people use social media sites is to stay in touch with family members and friends. Another common reason for using these sites is to connect with old friends with whom people have lost touch.

It is often suggested that these sites may diminish human relationships and contact, perhaps even increasing social isolation. Research shows that these fears are unfounded. A lack of social ties and community engagement is much more likely to result from traditional factors, such as limited education.

The average Facebook user has 229 Facebook "friends." The friends list is divided as follows:

- 22% people from high school
- 12% extended family
- 10% coworkers
- 9% college friends
- 8% immediate family
- 7% people from voluntary groups
- 2% neighbors

Here is what Facebook users do on a typical day:

- 15% update their own status.
- 22% comment on another's post or status.
- 20% comment on another user's photos.
- 26% "Like" another user's content.
- 10% send another user a private message.

The traditional way in which we think about the interactionist perspective needs to be expanded in light of the influence of social media sites.

*Source:* Brenner, Joanna. March 29, 2012. "Pew Internet: Social Networking" (http://pewinternet.org/Commentary/2012/March/Pew-Internet-Social-Networking-less-detail.aspx); Smith, Aaron. November 15, 2011. "Why Americans Use Social Media" (http://pewinternet.org/Reports/2011/Why-Americans-Use-Social-Media/Main-report.aspx).

---

people involved and the situation in which they find themselves. The expectations and reactions of other people greatly affect each individual's actions. In addition, people give things meaning and act or react on the basis of these meanings. For example, when the flag of the United States is raised, people stand because they see the flag as representing their country.

Because most human activity takes place in social situations—in the presence of other people—we must fit what we as individuals do with what other people in the same situation are doing. We go about our lives assuming that most people share our definitions of basic social situations. This agreement on definitions and meanings is the key to human interactions in general, according to symbolic interactionists. For example, a staff nurse in a mental hospital unlocking a door for an inpatient is doing more than simply enabling the patient to pass from one ward to another. He or she also is communicating a position of social dominance over the patient (within the hospital) and is carrying a powerful symbol of that dominance—the key. The same holds true for a professor asking students to read a chapter or a company vice president informing department heads of new rules.

Such interactions, therefore, although they appear to be simple social actions, also are laden with highly symbolic social meanings. These symbolic meanings are intimately connected with our understanding of what it is to be and to behave as a human being. This includes our sense of self; how we experience others and their views of us; the joys and pains we feel at home, at school, at work, and among friends and colleagues; and so on. (See "Thinking About Social Issues: Social Interaction in the Internet Age" for new forms of symbolic communication.)

Symbolic interaction and its various offshoots have been criticized for paying too little attention to the larger elements of society. Interactionists respond that societies and institutions consist of individuals who interact with one another and do not exist apart from these basic units. They believe that an understanding of the process of social interaction will lead to an understanding of the rest of society. Symbolic interactionism does complement functionalism and conflict theory in important ways and gives us important insights into how people interact.

### Contemporary Sociology

Contemporary sociological theory continues to build on the original ideas proposed in functionalism, conflict theory, and the interactionist perspective. Seeing contemporary sociological theory as either conflict theory or functionalism in the original sense would be difficult. Much of it has been modified to include important aspects of each theory. Even symbolic interactionism has not been wholeheartedly embraced, and aspects of it have instead been absorbed into general sociological writing.

Very little contemporary sociological theory still can be identified as true functionalism. Part of this is because sociologists today no longer try to develop all-inclusive theories and, instead, opt for what Merton (1968) referred to as middle-range theories. **Middle-range theories** *are concerned with explaining specific issues or aspects of society instead of trying to explain how all of society operates.* A middle-range theory might be one that explains why divorce rates rise and fall with certain economic conditions or how crime rates are related to residential patterns.

During the past 30 years, conflict theory has been influenced by a generation of neo-Marxists. These people have helped produce a more complex and sophisticated version of conflict theory that goes beyond the original emphasis on class conflict and instead shows that conflict exists within almost every aspect of society (Gouldner, 1970, 1980; Skocpol, 1979; Starr, 1982, 1992; Tilly, 1978, 1981; Wallerstein, 1974, 1979, 1980, 1991).

Some sociologists have turned to approaches that grew out of developments in Europe after 1960 when approaches known as postmodernism, poststructuralism, or critical theory became popular. Others have turned to the methods of anthropology and have become critical of the previous emphasis on objective or scientific approaches to research.

## Theory and Research

Sociological theory gives meaning to sociological practice. Merely assembling countless descriptions of social facts is not useful for understanding society as a whole. Only when data are collected to answer the specific questions growing out of a specific theory can conclusions be drawn and valid generalizations made. This is the ultimate purpose of all science.

Theory without practice (research to test it) is at best poor philosophy and at worst unscientific, and practice not based on theory is at best trivial and at worst a tremendous waste of time and resources. Therefore, in the next chapter, we shall move from theory to practice—to the methods and techniques of social research.

## ■ SUMMARY

▸ **What is unique about what sociologists do?**
Sociology is the scientific study of human society and social interactions. Sociologists seek an accurate and scientific understanding of society and social life. A great deal of social-issue information comes from sources that have an interest in developing support for a particular point of view

▸ **What is the difference between sociology and the other major social sciences?**
The main focus of sociology is on the group, not on the individual. A sociologist tries to understand the forces that operate throughout the society—forces that mold individuals, shape their behavior, and thus determine social events. The social sciences consist of all those disciplines that apply scientific methods to the study of human behavior. Although the areas of interest do overlap, each of the social sciences has its own area of investigation. Cultural anthropology, psychology, economics, history, political science, and social work all have some things in common with sociology, but each has its own distinct focus, objectives, theories, and methods.

▸ **How did sociology begin?**
The need for a systematic analysis of society, coupled with the acceptance of the scientific method, resulted in the emergence of sociology. Sociology became a separate field of study in Europe during the nineteenth century. It was a time of turmoil and a period of rapid and dramatic social change. Industrialization, political revolution, urbanization, and the growth of a market economy undermined traditional ways of doing things.

▸ **Who were the early pioneers in sociology?**
August Comte developed the new science, which he named sociology. He believed it would bring all the sciences together into a cohesive whole, thereby improving society.

Herbert Spencer believed that every part of society served a specialized function necessary to ensure society's survival as a whole. During the nineteenth century, sociology developed rapidly under the influence of three other scholars of very different orientations. Despite their differences, however, Karl Marx, Émile Durkheim, and Max Weber were responsible for shaping sociology into a relatively coherent discipline.

▸ **Describe the early development of sociology in the United States.**
In the United States, sociology developed in the early twentieth century. Its early growth took place at the University of Chicago, where the first graduate department of sociology in the United States was founded in 1890. The so-called Chicago school of sociology focused on the study of urban neighborhoods and ethnic areas and included many of America's leading sociologists of the period.

### What are the major theoretical approaches in sociology?

Sociologists have developed several perspectives to help them investigate social processes. Functionalism views society as a system of highly interrelated structures that function or operate together harmoniously. Functionalists analyze society by asking what each part contributes to the smooth functioning of the whole. From the functionalist perspective, society appears to be quite stable and self-regulating. Critics have attacked the conservative bias inherent in this assumption.

Conflict theory sees society as constantly changing in response to social inequality and social conflict. For these theorists, social conflict is the normal state of affairs; social order is maintained by coercion. Conflict theorists are concerned with the issue of who benefits from particular social arrangements and how those in power maintain their positions.

The interactionist perspective focuses on how individuals make sense of, or interpret, the social world in which they participate. This perspective consists of a number of loosely linked approaches.

### Media Resources

**CourseMate for *Introduction to Sociology*, Eleventh Edition**

Cengage Learning's Sociology CourseMate brings course concepts to life with interactive learning, study, and exam preparation tools that support the printed textbook. Access an integrated eBook, learning tools including glossaries, flashcards, quizzes, videos, and more in your Sociology CourseMate. Go to CengageBrain.com to register or purchase access.

# CHAPTER ONE STUDY GUIDE

## Key Concepts

*Match each of the following concepts with its definition, illustration, or explanation.*

a. Scientific method
b. Middle-range theory
c. Paradigms
d. Functionalism
e. Conflict theory
f. Interactionist perspective
g. Social cohesion
h. Manifest function
i. Latent function
j. The sociological imagination
k. Egoistic suicide
l. Altruistic suicide
m. Anomic suicide
n. Applied sociology
o. Social Darwinism

____ 1. The degree to which people are bonded to groups and to the society as a whole
____ 2. The belief that inequality in society is the result of a natural selection based on individual capacities and abilities
____ 3. The ability to see the link between personal experiences and social forces
____ 4. Paradigm that proposes that different sectors of a society have different interests and focuses on how groups use resources to secure their own particular interests
____ 5. General views of the world that determine the questions to be asked and the important things to look at in answering them
____ 6. The use of sociological knowledge not just to understand problems in the real world but to solve those problems
____ 7. The paradigm that focuses on how people interpret and attempt to influence the social world
____ 8. Intended outcomes of an institution
____ 9. A process by which a body of scientific knowledge is built through observation, experimentation, generalization, and verification
____ 10. Explanations that focus on specific issues rather than on society as a whole
____ 11. The paradigm that emphasizes how elements of a society work (or do not work) toward accomplishing necessary functions
____ 12. Suicide caused by feelings of normlessness and confusion, the feeling that the rules of the game no longer make sense
____ 13. Unintended, unrecognized, but often useful consequence of an institution
____ 14. Suicide that results from the willingness to sacrifice one's own life for the good of the social group
____ 15. Suicide related to lack of involvement with others

## Key Thinkers

*Match the thinkers with their main idea or contribution.*

a. Auguste Comte
b. Harriet Martineau
c. C. Wright Mills
d. Herbert Spencer
e. Émile Durkheim
f. Karl Marx
g. Max Weber
h. Jane Addams
i. W. E. B. Du Bois
j. Talcott Parsons
k. Robert K. Merton
l. George Herbert Mead

____ 1. Saw society as an organism; applied Darwin's idea of survival of the fittest to explain and justify social conditions of different individuals and groups
____ 2. African American sociologist, early twentieth century; militant opponent of racism and keen observer of its effects (*The Souls of Black Folk*)
____ 3. Theorist whose ideas provide the basis for symbolic interactionism
____ 4. American sociologist; developed concept of the sociological imagination
____ 5. Coined the term *sociology*; emphasized empiricism; thought society was evolving toward perfection
____ 6. American proponent of structural functionalism who saw social systems as complicated but stable interrelations of diverse parts

PART 1   THE STUDY OF SOCIETY

___ 7. Wrote observations of institutions (prisons, factories, and so on); compared American and European class systems
___ 8. Emphasized social solidarity; studied *rates* of behavior in groups rather than individual behavior
___ 9. Advocated middle-range theories and emphasized the distinction between manifest and latent functions of social processes
___ 10. American social reformer; founded Hull House, a settlement house for immigrants in Chicago
___ 11. Viewed social change as resulting from the conflicts between social classes trying to secure their interests; thought that eventually the workers would overthrow the capitalist-run system
___ 12. Thought power, wealth, and status were separate aspects of social class; saw bureaucratization as a dominant trend with far-reaching social consequences; contradicted Marx in arguing that religious ideas influenced economics, specifically that Protestantism brought the rise of capitalism

● **Central Idea Completions**

*Fill in the appropriate concepts and descriptions for each of the following questions.*

1. Recently, there has been much public discussion and some proposed legislation concerning adults who use the Internet to prey on children. Sociologists have also studied this problem. (a) How do the goals of sociologists differ from those of journalists or talk-show hosts in looking at this problem? (b) With regard to Internet predators, how does the information sociologists use differ from the information journalists and talk-show hosts use?

   a. _____

   b. _____

2. Suicide rates among soldiers have risen dramatically for both those in Afghanistan and those who have returned after tours of duty. Using your own knowledge and imagination, identify factors that might promote each of the types of suicide in Durkheim's typology with respect to the U.S. military.

   a. Egoistic suicide _____

   b. Altruistic suicide _____

   c. Anomic suicide _____

3. Tischler says that some people who use sociological data and ideas "have no interest in putting forth objective information." Imagine that there has been a shooting on a campus. How might the interests of the following entities affect the way they treat information?

   a. The local newspaper _____

   b. The school administration _____

   c. The campus police or security _____

   d. The NRA _____

4. What questions and research strategies might each of the major sociological paradigms use in looking at the issue of steroid use among athletes?

   a. Structural functionalism _____

   b. Conflict theory _____

   c. Interactionist perspective _____

5. Think of some institution or organization—a university or elementary school, a court, a church—and list its manifest functions and latent functions.

   a. Manifest functions _____

   b. Latent functions _____

6. Think of certain societies in which inequality, racism, and perceived social injustice are rampant. Contemplate which members of these societies enjoy privilege, opportunity, freedom, and success. Which members don't? How might a modern-day proponent of social Darwinism attempt to explain away such societal imbalances?

## Critical Thought Exercises

1. Women (white, 40–44 years old) with advanced education are much more likely to be married today than were their counterparts in 1980. High-school dropouts (white, 40–44 years old) of both sexes are less likely to be married than were their counterparts in 1980. What might account for these changes?

2. In debates over current political and social issues, which arguments and ideas are most compatible with social Darwinism? Which current ideas are most at odds with social Darwinism?

3. Reread Tischler's insert on sociology as a career. Use the occupational handbooks in your library or on the Internet, as well as flyers from your campus employment service, to conduct a preliminary career inventory of specific occupations for which a major in sociology would be helpful.

## Internet Activities

1. The American Sociological Association website (www.asanet.org/) has a lot of information for sociologists. For prospective sociologists, it also has the ASA booklet on careers in sociology (www.asanet.org/cs/root/leftnav/careers_and_jobs/careers_in_sociology). Visit the site and explore the types of jobs sociologists can do.
2. If the sociological theorists seem too imposing, take a look at the Dead Sociologists' Society (www. http://media.pfeiffer.edu/lridener/dss/). Larry Ridener founded the site after seeing the 1989 Robin Williams movie, *The Dead Poets Society*. Just as the Williams character tried to get his students interested in poetry, Ridener wanted to interest his students in sociological ideas. The site has excellent links to websites with information on a wide variety of sociological topics.
3. For more information on suicide rates in the United States, go to www.suicide.org/suicide-statistics.html. The site has data comparing rates by sex, race, and age. It also shows the rates for each state. Some states have rates that are more than three times the rates of other states. Can you think of social factors that might account for these differences?
4. Where did all the Sophias come from, and what happened to Mildred? Sociology shows us that decisions that seem highly personal and individual (such as the decision to commit suicide) fall into patterns. Another such decision is what to name the baby. You can see these patterns and check your own name at the U.S. Census website (www.ssa.gov/OACT/babynames/). For the same information presented in graphs, visit www.babynamewizard.com/voyager.
5. Blogs. Several sociology blogs appear on the Internet. Sociological Images (http://contexts.org/socimages/) is one of the liveliest, and you'll find links to other blogs. (As the name implies, it always has pictures.)

## Answers to Key Concepts

1. d; 2. e; 3. l; 4. a; 5. f; 6. i; 7. b; 8. o; 9. c; 10. m; 11. j; 12. h; 13. k; 14. n; 15. g

## Answers to Key Thinkers

1. d; 2. i; 3. l; 4. c; 5. a; 6. j; 7. b; 8. e; 9. k; 10. h; 11. f; 12. g

# 2 Doing Sociology: Research Methods

**The Research Process**
   Define the Problem
   Review Previous Research
   Develop One or More Hypotheses
   Determine the Research Design
   *How Sociologists Do It:* A Fashion Model Becomes an Observing Participant
   Define the Sample and Collect Data
   *How Sociologists Do It:* How to Spot a Bogus Poll
   Analyze the Data and Draw Conclusions

   *How Sociologists Do It:* How to Read a Table
   Prepare the Research Report

**Objectivity in Sociological Research**
*How Sociologists Do It:* Facebook, the Internet, and New Ethical Concerns

**Ethical Issues in Sociological Research**
*Thinking About Social Issues:* Famous Research Studies You Cannot Do Today

**Summary**

## LEARNING OBJECTIVES

After studying this chapter, you should be able to do the following:

- Explain the steps in the sociological research process.
- Analyze the strengths and weaknesses of the various research designs.
- Know what independent and dependent variables are.
- Know what sampling is and how to create a representative sample.
- Recognize researcher bias and how it can invalidate a study.
- Explain the strengths and weaknesses of the various measures of central tendency.
- Read and understand the contents of a table.
- Explain the concepts of reliability and validity.
- Understand the problems of objectivity and ethical issues that arise in sociological research.

Do college basketball players perform worse academically than other students? The National Collegiate Athletic Association (NCAA) wanted to counter the common belief that basketball players are not interested in academics and often drop out before graduating. During the last NCAA championship games, an ad ran that showed college athletes running, working out, and engaging in strength training. The voice-over with the ad noted, "African American males who are student athletes are 10 percent more likely to graduate." At the end of the commercial, a female athlete looked into the camera and said, "Still think we're just a bunch of dumb jocks?"

Your first response might be that the common stereotype of poor academic performance by athletes is incorrect. That is what the NCAA wanted you to believe. Was this an honest ad? No. It was a deliberate, and unfortunately common, attempt to manipulate public opinion with dubious research.

The ad used selective information. College athletes do indeed graduate at a higher rate than their nonathlete peers. Most college athletes are involved in golf, swimming, tennis, cross-country running, gymnastics, soccer, lacrosse, crew, rugby, and the like. They are often motivated students who have pursued the sport for a number of years.

The NCAA ad presented these two groups, men's basketball players and college athletes, as similar in graduation rates. This is not true.

In a study at the College Sport Research Institute at the University of North Carolina-Southall, E. Woodrow Eckard and Mark Nagel (2012) compared college basketball players with full-time college students. The basketball players were 20 percent less likely to graduate than the other students. Men's basketball players, especially those in the big-time programs, have very poor graduation rates. Players on March Madness teams were 33 percent less likely to graduate than other students. The ad was misleading because it was shown during the men's basketball tournament.

Every day, we encounter information about social trends. The goal of this chapter is to make you a critical consumer of this information. The more you know about how research should be conducted, the more likely you are to spot gross errors or attempts to manipulate your thinking on the topic. For example, consider the problem of binge drinking. Researchers at the Harvard School of Public Health studied the issue and reported in the *Journal of the American Medical Association* (*JAMA*) that almost half of all college students were binge drinkers, and about 20 percent were frequent binge drinkers. They concluded that binge drinking was widespread on college campuses (Wechsler et al., 1994). Binge drinkers create problems not only for themselves but also for their classmates who might not be drinking. Binge drinking also produces higher rates of assault and unwanted sexual advances.

What is binge drinking? The Harvard researchers defined binge drinking as having five drinks for men and four drinks for women within a few-hour period. It might surprise people to know that binge drinking has been defined very differently in the past compared to how the Harvard researchers defined it. Earlier definitions of binge drinking have involved an extended period of at least two days during which the person repeatedly became intoxicated and ignored his or her usual activities and obligations. Movies such as *Lost Weekend* or *Leaving Las Vegas* show this type of binge drinking.

The definition of binge drinking also differs throughout the world. In Sweden, binge drinking involves drinking a half bottle of hard liquor or two bottles of wine on one occasion. Italians think that even if you have eight drinks a day, you are not bingeing. The English think you have to have eleven or more drinks on one occasion to be a binge drinker.

According to the Harvard definition, if a woman has a pre-dinner drink, a couple of glasses of wine with dinner, and another drink later in the evening, she is a binge drinker. Based on this standard, her blood alcohol level might not be at the level most states define as being drunk. In essence, therefore, many binge drinkers are legally sober.

In the years since the *JAMA* article, repeated studies have shown that binge drinking on college campuses has actually been declining for many years and abstinence has been increasing. Yet, the false impression continues that binge drinking is a growing problem on college campuses (Johnston et al., 2011).

When the media reported on the Harvard School of Public Health study published in a prestigious journal, they thought they were making the public aware of a troubling social issue. The media do not have the time or ability to evaluate carefully the validity of every press release or the accuracy of the information. As you will see in this chapter, the research process requires a number of specific steps to produce a valid study. Only when these steps are followed faithfully can we have any confidence in the results of the study. In this chapter, we examine some of the methods scientists in general—and sociologists in particular—use to collect data to test their ideas.

## ● THE RESEARCH PROCESS

How should you conduct a research study? After reading Chapter 1, "The Sociological Perspective," you know not to approach a study and draw conclusions on the basis of your personal experience and perceptions; rather, you must approach the study scientifically.

To approach a study scientifically, remember that science has two main goals: (1) to describe in detail particular characteristics and events and (2) to propose and test theories that help us understand these characteristics and events.

There is a great deal of similarity between what a detective does in attempting to solve a crime and what a sociologist does in answering a research problem. In the course of their work, both detectives and sociologists must gather and analyze information. For detectives, the object is to identify and locate criminals and collect enough evidence to ensure that their identification is correct. Sociologists develop hypotheses, collect data, and develop theories to help them understand social behavior. Although their specific goals differ, both sociologists and detectives try to answer two general questions: why something happened and under what circumstances it is likely to happen again. That is, sociologists seek to explain and predict.

All research problems require their own special emphasis and approach. The research procedure is usually tailored to the research problem. Nonetheless, the researcher must follow a sequence of steps called the research process when designing a research project. In short, the **research process** involves *defining the problem; reviewing previous research on the topic; developing one or more hypotheses; determining the research design; defining the sample and collecting data; analyzing and interpreting the data; and, finally, preparing the research report.* The sequence of steps in this process and the typical questions asked at each step are illustrated in Table 2-1. Do not become concerned about any unfamiliar terms in this table. We will define them as we examine each of the various steps.

### Define the Problem

"Love leads to marriage." Suppose you were given this statement as a subject for sociological research. How would you proceed to gather data to prove or disprove it? You must begin by defining love, a task that William Shakespeare undertook in one of his plays, *Twelfth Night*, when he asked, "What is love?" Several common phrases focus on love, like Virgil's "Love conquers all" and the Beatles' "All you need is love." You would know at the outset that you have a problem because to this day people are grappling with the question that has been contemplated throughout history, "How do you know when you are in love?"

Concepts of love have varied over time and from one culture to another. Sharon Brehm (1992) noted some of the views of love that have been suggested:

1. Love is insanity.
2. Love is not possible in marriage.
3. Love happens only between people of the same sex.
4. Love should not involve sexual contact.
5. Love is a game.
6. Love is a noble quest.
7. Love is doomed.
8. Love leads to happiness.
9. Love and marriage go together.

You could try a different approach and define love by using a definition a researcher used in another study. For example, Hatfield (1988) defined love as "a state of intense longing for union with another." Now you would have to find some way of determining whether this condition exists, and you also would have to decide whether both people have to be in love for marriage to take place. You might already notice that it can be difficult to achieve the level of precision necessary for a useful research project.

After you defined your terms accurately and provided details to clarify your descriptions, you could begin to test the statement we proposed. Even after arriving at a careful definition of your terms and a detailed description of love, however, you might still have trouble answering the question empirically.

An **empirical question** *can be answered by observing and analyzing the world as it is known.* Examples: How many students in this class have an A average? How many millionaires are there in the United States? Scientists pose empirical questions to collect information, to add to what is already known, and to test hypotheses. To turn the statement about love into

## TABLE 2-1  The Research Process

| Steps in the Process | Typical Questions |
|---|---|
| Define the problem | What is the purpose of the study? |
|  | What information is needed? |
|  | How can we operationalize the terms? |
|  | How will the information be used? |
| Review previous research | What studies have already been done on this topic? |
|  | Do we need additional information before we begin? |
|  | From what perspective should we approach this issue? |
| Develop one or more hypotheses | What are the independent and dependent variables? |
|  | What is the relationship among the variables? |
|  | What types of questions do we need to answer? |
| Determine the research design | Can we use existing data? |
|  | What will we measure or observe? |
|  | What research methods should we use? |
| Define the sample and collect data | Are we interested in a specific population? |
|  | How large should the sample be? |
|  | Who will gather the data? |
|  | How long will it take? |
| Analyze the data and draw conclusions | What statistical techniques will we use? |
|  | Have our hypotheses been proved or disproved? |
|  | Is our information valid and reliable? |
|  | What are the implications of our study? |
| Prepare the research report | Who will read the report? |
|  | What is the reader's level of familiarity with the subject? |
|  | How should we structure the report? |

an empirical question, you must ask how we measure the existence of love.

In trying to define and measure love, one researcher (Rubin, 1970, 1973) used an interesting approach. He prepared a large number of self-descriptive statements that considered various aspects of loving relationships as mentioned by writers, philosophers, and social scientists. After administering these statements to a variety of subjects, he was able to isolate nine items that best reflected feelings of love for another. Three of these items are cited in the following paragraph. In each sentence, the person was to fill in the blank with the name of a particular person and indicate the degree to which the item described the relationship.

The following statements reflect three components of love. The first is attachment-dependency: "If I were lonely, my first thought would be to seek (blank) out." The second component is caring: "If (blank) were feeling bad, my first duty would be to cheer him (or her) up." The final component is intimacy: "I feel that I can confide in (blank) about virtually everything." These three statements show the strong aspect of mutuality in love relationships.

Using Rubin's scale, you can begin to clarify an important component of your research problem. In the language of science, you have operationalized your definition of love. An **operational definition** is *a definition of an abstract concept in terms of the observable features that describe the thing being investigated.* Attachment-dependency, caring, and intimacy can be three features of an operational definition of love and can indicate the presence of love in a research study.

### Review Previous Research

Which questions are the "right" questions? Although there are no inherently correct questions, some are better suited to investigation than are others. To decide what to ask, researchers must first learn as much as possible about the subject. We want to familiarize ourselves with as many of the previous studies on the topic as possible, particularly those closely related to what we want to do. By knowing as much as possible about previous research, we avoid duplicating a previous study and are able to build on contributions others have made to our understanding of the topic.

After reviewing the research, you might discover that the early anthropologist, Ralph Linton, thought love was a form of insanity and that assuming it should lead to marriage was absurd. As he noted:

> All societies recognize that there are occasional violent emotional attachments between persons of the opposite

sex, but our present American culture is practically the only one which has attempted to capitalize these and make them the basis for marriage. Most groups regard them as unfortunate and point out the victims of such attachments as horrible examples.... The percentage of individuals with a capacity for romantic love of the Hollywood type [is] about as large as that of persons able to throw genuine epileptic fits. (Linton, 1936)

Needless to say, Linton would not have thought much of our potential research project.

Helen Fisher (2009) scanned the brains of people in love. She asked them, "What percent of the day and night do you spend thinking about your sweetheart?" For many people in love, thinking about the loved one became an obsession. Fisher found high levels of dopamine activity and low levels of serotonin during these times. The craving for the loved one resembled the experience of addicts who recall the pleasures of their drug use. Fisher also found that people who were deeply in love were similar to people with untreated obsessive-compulsive disorder.

In exploring a question that seems so basic, already we are encountering a good deal of skeptical thinking.

### Develop One or More Hypotheses

Our original statement, "Love leads to marriage," is presented in the form of a hypothesis. A **hypothesis** is *a testable statement about the relationships between two or more empirical variables.* A **variable** is *anything that can change (vary).* The number of highway deaths on Labor Day weekends, the number of divorces that occur each year in the United States, the amount of energy the average American family consumes in the course of a year, the daily temperature in Dallas, the number of marathoners in Boston or in Knoxville, Tennessee—all these are variables. The following are not variables: the distance from Los Angeles to Las Vegas, the altitude of Denver, and the number of marriages in Ohio in 2013. These are fixed, unchangeable facts.

As we review the previous research on the topic of love, we find we can develop additional hypotheses that help us investigate the issue further. For example, our reading might show that a common stereotype people hold is the notion that women are more romantic than men. After all, it appears that women enjoy romantic novels and movies about love more than men do.

But wait a minute. We might begin to suspect that common stereotypes might be all wrong, that they are related to traditional gender-role models. We note that in most traditional societies, the male is the breadwinner, whereas the female depends on him for economic support, status, and financial security. Therefore, it

*In one study, males were more likely than females to agree with the statement, "A person should marry whomever he or she loves regardless of the social position."*

*If you were studying whether love leads to marriage, you would find it difficult to define love.*

would seem that when a man marries, he chooses a companion and, perhaps, a helpmate, whereas a woman chooses a companion as well as a standard of living. This leads us to hypothesize that in traditional societies, men are more likely to marry for love, whereas women are more likely to marry for economic security.

There is support for this hypothesis. One study designed a scale to measure belief in a romantic ideal in marriage. Males were more likely than females to agree with such statements as "A person should marry whomever he loves regardless of social position" and "As long as they love one another, two people should have no difficulty getting along together in marriage." Men were more likely than women to disagree with the statement, "Economic security should be carefully considered before selecting a marriage partner" (Rubin, 1973).

Contrary to popular opinion, men also tend to be more romantic than women (Sprecher and Metts, 1989). This fact should not be that hard to understand when we see that, historically, men, with their control over resources, have had the luxury to be romantic. Women, who often have not been in charge of their economic destiny, have had to think of men as providers more than as lovers. We could hypothesize that as gender-role stereotyping declines in the United States and as more and more families come to depend on the income of both spouses, one of two things could happen: Either the importance of romantic love as a basis for marriage will begin to fade, or it will become stronger as the couple now comes together on the basis of mutual attraction as opposed to economic considerations.

Hypotheses involve statements of causality or association. A **statement of causality** says that *something brings about, influences, or changes something else.* "Love between a man and a woman always produces marriage" is a statement of causality.

A **statement of association,** however, says that *changes in one thing are related to changes in another but that one does not necessarily cause the other.* Therefore, if we propose that "the greater the love relationship between a man and a woman, the more likely it is they will marry," we are making a statement of association. We are noting a connection between love and marriage but also that one does not necessarily cause the other.

Often, hypotheses propose relationships between two kinds of variables. An **independent variable** *causes or changes another variable.* A **dependent variable** *is influenced by the independent variable.* For example, we might propose the following hypothesis: Men who live in cities are more likely to marry young than are men who live in the country. In this hypothesis, the independent variable is the location: Some men live in the city, some live in the country, but presumably, their choice of where to live is not influenced by whether they marry young. The age of marriage is the dependent variable because it is possible that the age of marriage depends on where the men live. If research then showed that the age of marriage (a dependent variable) is indeed younger among urban men than among rural men, the hypothesis probably would be correct. If there were no difference in the age of marriage among urban and rural men—or if it was earlier among rural men—then the hypothesis would not be supported by the data.

Remember that proving a hypothesis false can be scientifically useful; it eliminates unproductive avenues of thought and suggests other, more productive approaches to understanding a problem. Even if research shows that a hypothesis is correct, however, that does not mean the independent variable necessarily produces or causes the dependent variable. For example, even if we can show that love leads to marriage, we still might not know why. In principle at least, it is possible to be in love without getting married. We still do not know what causes people to take the next step.

### Determine the Research Design

After we have developed our hypotheses, we must design a project in which they can be tested. This is a difficult task that frequently causes researchers a great deal of trouble. If a research design is faulty, it might be impossible to conclude whether the hypotheses are true or false, and the whole project will have been a waste of time, resources, and effort.

A research design must provide for the collection of all necessary and sufficient data to test the stated hypotheses. The important word here is *test*. The researcher must not try to prove a point; rather, the goal is to test the validity of the hypotheses. Although it is important to gather as much information as needed, research designs must guard against the collection of unnecessary information, which can lead to a waste of time and money.

When we design our research project, we must also decide which of several research approaches to use. Sociologists use four main methods of research: surveys, participant observation, experiments, and secondary analysis. Each has advantages and limitations. Therefore, the choice of methods depends on the questions the researcher hopes to answer.

*Surveys* A **survey** is *a research method in which a population, or a portion thereof, is questioned to reveal specific facts about itself.* Surveys are used to discover the distribution and interrelationship of certain variables among large numbers of people.

The largest survey in the United States takes place every ten years when the government takes its census. The U.S. Constitution requires this census to determine the apportionment of members to the House of Representatives. In theory, at least, a representative of every family and every unmarried adult responds to a series of questions about his or her circumstances. From these answers, it is possible to construct a picture of the social and economic facts that characterize the American public at one point in time.

*Such a study, which cuts across a population at a given time*, is called a **cross-sectional study.** Surveys, by their nature, usually are cross-sectional. If the same population is surveyed two or more times at certain intervals, a comparison of cross-sectional research can give a picture of changes in variables over time. *Research that investigates a population over a period of time is called* **longitudinal research.**

Another large social survey is conducted by The Nielsen Company to provide television ratings to advertisers and the television industry. In 2010, of the 115,900,000 U.S. television households, only 25,000 total American households participated in the Nielsen daily metered system. Even though this involves only about 100,000 people to represent a country of 295 million people over the age of two, Nielsen claims its results are accurate (Neilsen Wire, 2010).

Survey research usually deals with large numbers of subjects in a relatively short time. One of the shortcomings resulting from this method is that investigators are not able to capture the full richness of feelings, attitudes, and motives underlying people's responses. Some surveys are designed to gather this kind of information through interviewing. An **interview** consists of *a conversation between two (or occasionally more) individuals in which one party attempts to gain information from the other(s) by asking a series of questions.*

It would, of course, be ideal to gather exactly the same kinds of information from each research subject. One way researchers attempt to achieve this is through interviews in which all questions are carefully worked out to get at precisely the information desired (What is your income? How many years of education have you had?). Sometimes, research participants are forced to choose among a limited number of responses to questions (as in multiple-choice tests). This process results in very uniform data easily subjected to statistical analysis.

*A research interview entirely predetermined by a questionnaire (or so-called interview schedule) that is followed rigidly* is called a **structured interview.** Structured interviews tend to produce uniform or replicable data that can be elicited time after time by different interviewers.

The use of this method, however, also can allow useful information to slip into cracks between the predetermined questions. For example, a questionnaire being administered to married individuals might ask about their age, family background, and what role love played in their reasons for getting married. However, if we do not ask about social class or ethnicity, we might not find out that these characteristics are very important for our study. If such questions are not built into the questionnaire from the beginning, it is impossible to recover this lost information later in the process when its importance might become apparent.

One technique that can prevent this kind of information loss is the **semistructured,** or **open-ended, interview,** *in which the investigator asks a list of questions but is free to vary them or even to make up new questions on topics that take on importance in the course of the interview.* Each interview will cover those topics important to the research project but, in addition, will yield additional data somewhat different for each subject. Analyzing such diverse and complex data is difficult, but the results are often rewarding.

Interviewing, although it can produce valuable information, is a complex, time-consuming art. Some research studies try to get similar information by distributing questionnaires directly to the respondents and asking them to complete and return them. This is the way the federal government obtains much of its census data. Although it is, perhaps, the least expensive way of doing social research, it is often difficult to assess the quality of data obtained in this manner. For example, people might not answer honestly or seriously for a variety of reasons. They might not understand the

Participant observation depends for its success on the relationship that develops between the researcher and the research participants.

questions, they might fear the information will be used against them, and so on. But even data gained from personal interviews can be unreliable. In one study, student interviewers were embarrassed to ask preassigned questions on sexual habits, so they left these questions out of the interviews and filled the answers in themselves afterward. In another study, follow-up research found that participants had consistently lied to interviewers.

*Participant Observation* *Researchers entering into a group's activities and observing the group members* are engaged in **participant observation.** Unlike sociologists employing survey research, participant observers do not try to make sure they are studying a carefully chosen sample. Rather, they attempt to know all members of the group being studied to whatever degree possible. This research method is generally used to study relatively small groups over an extended period of time. The goal is to obtain a detailed portrait of the group's day-to-day activities, to observe individual and group behavior, and to interview selected informants.

Participant observation depends for its success on the relationship that develops between the researchers and research participants. The closer and more trusting the relationship, the more information will be revealed to the researcher—especially the kind of personal information often crucial for successful research.

One of the first and most famous studies employing the technique of participant observation was a study of Cornerville, a lower-class Italian neighborhood in Boston. William Foote Whyte moved into the neighborhood and lived for three years with an Italian family. He published his results in a book called *Street Corner Society* (1943). All the information for the book came from his field notes, which described the behavior and attitudes of the people whom he came to know.

Nearly two decades after Whyte's study, Herbert Gans conducted a participant observation study, published as *The Urban Villagers* (1962), of another Italian neighborhood in Boston. The picture Gans drew of the West End was broader than Whyte's study of Cornerville. Gans included descriptions of the family, work experience, education, medical care, relationships with social workers, and other aspects of life in the West End. Although he covered a wider range of activities than Whyte, his observations were not as detailed.

On some occasions, participant observers hide their identities while doing research and join groups under false pretenses. Leon Festinger and his students hid their identities when they studied a religious group preaching the end of the world and the arrival of flying saucers to save the righteous, a group with beliefs similar to the Heaven's Gate group of 1997 (Festinger, Rieken, and Schacter, 1956). However, many sociologists consider this deception unethical. They believe it is better for participant observers to be honest about their intentions and work together with their subjects to create a mutually satisfactory situation. By declaring their positions at the outset, sociologists can then ask appropriate questions, take notes, and carry out research tasks without encountering unnecessary and unethical risks to their study.

There are occasions, however, when a researcher may have unusual access to a group and can become what could be called an observing participant. In this case, the level of disclosure may be less than that needed under full participant observation. (For an example of this type of research, see "How Sociologists Do It: A Fashion Model Becomes an Observing Participant.")

Participant observation is a highly subjective research approach. In fact, some scholars reject it outright because the results often cannot be duplicated by another researcher. This method, however, has the benefit of revealing the social life of a group in far more depth and detail than surveys or interviews alone. The participant observer who is able to establish good rapport with the subjects is likely to uncover information that would never be revealed to a survey taker.

The participant observer is in a difficult position, however. He or she will be torn between the need to become trusted (and, therefore, emotionally involved in the group's life) and the need to remain a somewhat detached observer striving for scientific objectivity.

*Experiments* The most precise research method available to sociologists is the controlled **experiment,** *an investigation in which the variables being studied are controlled and the researcher obtains the results through precise observation and measurement.* Because of their precision, experiments are an attractive means of doing research. Experiments have been used to study patterns of interaction in small groups under a variety of conditions such as stress, fatigue, or limited access to information.

Although experimentation is appropriate for small-group research, most of the issues that interest sociologists cannot be investigated in totally controlled situations. Social events usually cannot be studied in controlled experiments because even if all the variables can be identified, they simply cannot be controlled, so experiments remain the least used research method in sociology.

*Secondary Analysis* **Secondary analysis** is *the process of using data that has been collected by others.* Often, the original investigator gathered the data for a specific

## HOW SOCIOLOGISTS DO IT

### A Fashion Model Becomes an Observing Participant

Ashley Mears spent five years as a fashion model, first in Atlanta for local department stores and eventually in Milan, Tokyo, New York, and Hong Kong. By the time she turned 23, she was already at the older age range for modeling assignments. She decided to pursue a graduate degree at New York University. A modeling agent spotted her, and she began a second stint as a model. Only this time she was also a sociology student. Mears soon realized she had a unique opportunity to do a sociological study of the fashion industry.

From the outside, the fashion industry appears to be a very glamorous field. Yet, it takes a great deal of behind-the-scenes work to produce what is a seemingly effortless natural look. Fashion is not unique in this respect; many fields have an underside that insiders wish to hide from outsiders.

Fashion Week takes place twice a year in New York City. The fashion designers show their wears and retail purchasers are on hand to decide what to feature in the stores. It brings more than $400 million in business into the city each year. The models receive very little of this money. Models get paid on a per-project basis. They have little control over the conditions of their work and are arbitrarily selected and easily dismissed. Top models can earn between $1,000 and $5,000 a show, while others are not paid at all. Some models are even in debt to agencies that brought them from other countries and paid for their visas, plane tickets, and apartments.

There also appears to be an inverse relationship between the prestige of the job and how much the model is paid. A full-day fashion shoot for Vogue may pay only $150, while models for retail circulars and catalogs can make several hundred dollars per hour.

The dilemma for Mears as she did her research was that, on the one hand, she introduced herself as an 18-year-old model, as fashion-shoot bookers told her to do, but on the other she also needed to observe events for her sociology PhD dissertation. By being a working fashion model, she was able to gain valuable industry information. In order to not disrupt her access, she had to limit the information she shared with the people who hired the models. Near the end of her modeling career, as she was approaching 26, she did follow up with clients and models. Eventually she was able to interview other models, fashion designers, agents, and magazine editors.

Mears used what she called "shallow cover" (Fine, 1993) in that she did not keep her role as a researcher secret, but she did not disclose the focus of her research. Her subjects were somewhere between being informed and uninformed. Usually participant observers disclose

---

study. At other times, it was collected merely as part of the process of keeping records. The researcher engaged in secondary analysis might use this same data for a new study and a very different purpose. For example, Émile Durkheim, in his classic study of suicide in France in the 1890s, engaged in a secondary analysis of official records and developed his theories based on that research.

The enormous amount of material the federal government collects is often used for secondary analysis. The U.S. Bureau of the Census has data on income, birth rates, migration, marriage, divorce, and education levels in the United States that are invaluable for doing social research. Other agencies that provide data sociologists use for secondary analysis include the Federal Bureau of Investigation, the Department of Labor, the National Center for Health Statistics, and many others.

An advantage of secondary analysis is that it is useful for collecting or analyzing historical and longitudinal data. It also saves the time and money involved in doing a new study. There are some disadvantages, however. The data can be flawed. You might not know whether the original researchers had biases that skewed the data. Possibly, they were not qualified or knowledgeable enough to collect the data. In addition, the data might not be suitable for your current study. If you are trying to perform a study of economic well-being at different points in history, you might decide to gather that information from certain questions on the U.S. censuses of 1950, 1970, and 2010. But which questions should you use? Should you use those on income, those on poverty, those on net worth, those on satisfaction with the political situation, or some other questions? Different types of questions can produce different results, and your study might not turn out to be valid because of the choices you make.

Therefore, using secondary analysis requires a thorough understanding of the research process and the problems that can arise from a poorly conceived study.

their role to the subjects if at all possible. In Mears's case, she was an observing participant who limited the information about her research. She straddled a fine line between having comfortable and useful conversations with her subjects and feeling like she was spying on her colleagues.

Mears summed up her work by noting that success in high-fashion modeling has a great deal to do with chance and marketing. The field has an abundance of attractive young models who are used and then discarded. Eventually, near the end of her research, Mears was also dismissed from her agency by way of an e-mail that had the heading "Hey Doll." Had she not gone on to become a sociology professor, she would have been like the other ex-models in her study who had to support themselves as restaurant wait staff, yoga instructors, or mall retail clerks.

*Sources:* Mears, Ashley. September 15, 2011. "Poor Models. Seriously." *The New York Times,* P. A35; Copeland, Libby. September 7, 2011. "America's Next Top Sociologist." *Slate* (www.slate.com/articles/double_x/doublex/2011/09/americas_next_top_sociologist.html); Mears, Ashley. 2011. *Pricing Beauty: The Making of a Fashion Model.* Berkeley, Ca.; University of California Press; Fine, Gary Alan. 1993. "Ten Lies of Ethnography: Moral Dilemmas of Field Research." *Journal of Contemporary Ethnography,* 22:267–94.

Former fashion model Ashley Mears, who is now a sociology professor.

Table 2-2 compares the advantages and disadvantages of the various research methods.

### Define the Sample and Collect Data

After determining how the needed information will be collected, the researchers must decide what group will be observed or questioned. Depending on the study, this group might be college students, Texans, or baseball players. *The particular subset of the population chosen for study is known as a* **sample.**

**Sampling** *is a research technique through which investigators study a manageable number of people, known as the sample, selected from a larger population or group.* If the procedures are carried out correctly, the sample can be called a **representative sample,** or *one that shows, in equivalent proportion, the significant variables that characterize the population as a whole.* In other words, the sample will be representative of the larger population, and the findings from the research will tell us something about the larger group. *The failure to achieve a representative sample is called* **sampling error.**

Suppose you wanted to sample the attitudes of the American public on some issue such as military spending or federal aid for abortions. You could not limit your sample to only New Yorkers or Republicans or Catholics or African Americans or home owners. These groups do not represent the nation as a whole, and any findings you came up with would contain a sampling error.

How do researchers make sure their samples are representative? The basic technique is to use a **random sample** *to select subjects so that each individual in the population has an equal chance of being chosen.* For example, if we wanted a random sample of all college students in the United States, we might choose every fifth or tenth or hundredth person from a comprehensive list of all registered college students in this country. Or we might assign each student a number and have a

TABLE 2-2  Advantages and Disadvantages of Various Research Methods

| Research Method | Advantages | Disadvantages |
| --- | --- | --- |
| Social survey | Large numbers of people can be surveyed with questionnaires. | Respondents can give false information or responses they think the researcher wants to hear. |
| | Data can be quantified and comparisons made among groups. | Surveys do not leave room for answers that might not fit the standardized categories. |
| | Measures can be taken at different points. | Response rate might be low. |
| Participant observation | Researcher can observe people in their natural environments. | Findings are open to interpretation and subject to researcher bias. |
| | This observation provides a more in-depth understanding of the people being studied. | The researcher might have an unintended influence on the subjects. |
| | Hypotheses and theories can be developed and changed as the research progresses. | It is time-consuming. |
| | | The results can be difficult to replicate. |
| Experiment | Variables can be isolated and controlled. | The laboratory setting creates an artificial social environment. |
| | A cause-and-effect relationship can be found. | The study has to be limited to a few variables. |
| | Experimentation is easy to replicate. | |
| Secondary analysis | Secondary analysis is useful for collecting or analyzing historical and longitudinal data. | Most behaviors sociologists investigate cannot be studied in a laboratory. |
| | This analysis saves the time and money involved in performing a new study. | The data can be flawed. |
| | | Data might not be suitable for the current study. |

*The representativeness of a sample is more important than its size.*

computer pick a sample randomly. However, there is a possibility that, simply by chance, a small segment of the total college student population would fail to be represented adequately. This might happen with Native American students, for instance, who make up less than 1 percent of college students in the United States. For some research purposes, this might not matter, but if ethnicity is an important aspect of the research, it would be important to make sure that Native American students were included.

*The method to prevent certain groups from being under- or overrepresented in a sample* is to choose a **stratified random sample.** With this technique, the population being studied is first divided into two or more groups (or strata) such as age, sex, or ethnicity. A simple random sample is then taken within each group. Finally, the subsamples are combined (in proportion to their numbers in the population) to form a total sample. In our example of college students, you as the researcher would identify all ethnic groups represented among college students in the United States. Next you would calculate the proportion of the total number of college students represented by each group. Then you would create a random sample separately from each ethnic group. The number chosen from each group should be proportional to its size in the entire college student population. The sample would still be random, but it would be stratified for ethnicity.

For a study to be accurate, it is crucial to choose a sample with care. The most famous example of sampling error occurred in 1936, when *Literary Digest* magazine incorrectly predicted that Alfred E. Landon would win the presidential election. Using telephone directories and automobile registration lists to recruit subjects, *Literary Digest* pollsters sent out more than 10 million straw vote ballots and received 2.3 million completed responses. The survey gave the Republican candidate Landon 55 percent of the vote and Franklin D. Roosevelt only 41 percent. (The remaining 4% went to a third candidate.) Based on this poll, *Literary Digest* confidently predicted Landon's victory. Instead, Landon has become known as the candidate who was buried in a landslide vote for Roosevelt (Squire, 1988).

How could this happen? Two major flaws in the sample accounted for the mistake. First, although the *Literary Digest* sample was large, it was not representative of the nation's voting population because it contained a major sampling error. During the Depression years, only the well-to-do could afford telephones and automobiles, and these people were likely to vote Republican.

The second problem with the study was the response rate. Of those who claimed to have received a *Literary Digest* ballot, 55 percent claimed they would have voted for Roosevelt and 44 percent for Landon. If these people had actually voted in the poll, *Literary Digest* would have predicted the correct winner. As it turned out, there was a low response rate, and those who did respond were generally better educated, wealthier people who could afford cars and telephones and who tended to be Landon supporters (Squire, 1988).

The outcome of the election was not entirely a surprise to everyone. A young pollster named George Gallup forecast the results accurately. He realized that the majority of Americans supported the New Deal policies proposed by the Democrats. Gallup's sample was much smaller but far more representative of the American public than that of *Literary Digest*. This points out that the representativeness of the sample is more important than its size.

An example of a biased sample is the voting method used by the popular television program *American Idol*. Voting is done in two ways: by voice through toll-free numbers, and by text messaging (for which "standard text messaging fees will apply"). Voting by phone is often time-consuming because the lines can be busy. Text messaging is much easier. The two ways of voting cause problems. The audience can use the toll-free lines with slow one-at-a-time votes, or a fast fee-based text messaging system. This voting system produces a highly biased sample because the voting is influenced by the number people who are willing to pay for a text message and the number who are willing to spend the time calling a toll-free number (Renka, 2010). (For a further discussion of deceptive research, see "How Sociologists Do It: How to Spot a Bogus Poll.")

*Researcher Bias* One of the most serious problems in data collection is **researcher bias,** *the tendency for researchers to select data that support, and to ignore data that seem to go against, their hypotheses.* We see this quite often in mass media publications. They might structure their study to produce the results they wish to obtain, or they might publicize only information that supports their viewpoint.

Researcher bias often takes the form of a self-fulfilling prophecy. A researcher who is strongly inclined toward one point of view might communicate that attitude through questions and reactions in such a way that the subject fulfills the researcher's expectations. For example, a researcher who is trying to prove an association between poverty and antisocial behavior might question low-income subjects in a way that would indicate a low regard for their social attitudes. The subjects, perceiving the researcher's bias, might react with hostility and thus fulfill the researcher's expectations.

Researcher bias was behind a study that was presented many years ago by a market researcher, James Vicary. He declared that he had discovered how to make people buy something even if they did not want to do so. He called it subliminal advertising. Vicary claimed that he showed the words *Eat Popcorn* or *Drink Coca-Cola* on a movie theater screen every five seconds as the films played. The words were flashed at three-thousandths of a second, so fast that the audience did not know they had seen the words. He said sales of popcorn and Coke went up dramatically.

People were alarmed. Brainwashing was a concern in those days because some prisoners of war had been indoctrinated by their Chinese captors and had defected to communism. The television networks assured the public they would not use subliminal advertising.

Researchers tried to replicate Vicary's experiment without any success. Eventually, Vicary admitted he made up the research to get publicity and business for his market-research firm (Crossen, 2005).

One of the standard means for dealing with research bias is to use **blind investigators**—*investigators who do not know whether a specific subject belongs to the group of actual cases being investigated or to a comparison group.* For example, in a study on the causes of child abuse, the investigator looking at the children's family backgrounds would not be told which children had been abused and which were in the nonabused comparison group.

Sometimes double-blind investigators are used. **Double-blind investigators** *are kept uninformed not only of the kinds of subjects (case subjects or comparison group subjects) they are studying but also of the hypotheses being tested.* This eliminates any tendency on their part to find cases that support or disprove the research hypothesis.

## Analyze the Data and Draw Conclusions

In its most basic sense, **analysis** is *the process through which large and complicated collections of scientific data are organized so that comparisons can be made and conclusions drawn.* It is not unusual for a sociological research project to result in hundreds of thousands of individual pieces of information. By itself, this vast array of data has no particular meaning. The analyst must find ways to organize such data into useful categories

## HOW SOCIOLOGISTS DO IT

### How to Spot a Bogus Poll

Opinion surveys can look convincing and be completely worthless, but asking four simple questions of any poll can separate the good numbers from the trash.

Politicians use opinion polls as verbal weapons in campaign ads. Journalists use them as props to liven up infotainment shows. Executives are more likely to pay attention to polls when the numbers support their decisions. But this isn't how polls are meant to be used. Opinion polls can be a good way to learn about the views Americans hold on important subjects only if you know how to cut through the contradictions and confusion.

Conducting surveys is difficult, especially when attempting a meaningful survey of public opinion, because opinion is subjective and can change rapidly from day to day. Poll questions sometimes produce conflicting or meaningless results, even when they are carefully written and presented by professional interviewers to scientifically chosen samples. That is why the best pollsters are careful about the order and wording of questions and the way in which data are coded, analyzed, and tabulated.

So, the next time a poker-faced person tries to give you the latest news about how Americans feel, ask some pointed questions of your own, such as the following.

#### Did You Ask the Right People?

Even when you start out with a representative sample, you could end up with a biased one. This is a risk all pollsters take, but some particular methods lend themselves to greater error. For example, people are frequently asked to fill out surveys on weighty subjects such as crime and sexual behavior. Not only do such polls ignore the opinions of those who did not receive the survey, but they also are biased toward those who take the trouble to fill out and return the questionnaire, often at their own expense.

Likewise, many have criticized online polling because Internet users tend to be wealthier, more educated, and more likely to be male than the larger population. Online polling can also easily lead to people voting numerous times, making this one of the least reliable ways to gather information.

Conflicts make news. When journalists are trying to liven up a boring political story, they need angry, well-informed citizens like a fish needs water. This is one reason older men might be quoted more often than other groups. Those aged 50 and older are more likely than younger adults to follow news stories "very closely," according to the Pew Research Center for the People & the Press in Washington, D.C. Men are more likely than women to follow stories about war, business, sports, and politics.

In the past decade, angry white men have dominated media programs designed to give ordinary people a chance to speak out in public. Two-thirds of regular listeners to political talk-radio programs are men. Republicans outnumber Democrats three to one in the talk-radio audience, and 89 percent of listeners are white. Three in five regular listeners

---

so that the relationships that exist can be determined, the hypotheses forming the core of the research can be tested, and new hypotheses can be formulated for further investigation.

One important device to aid in the analysis of data is the table. (See "How Sociologists Do It: How to Read a Table.")

Sociologists often summarize their data by calculating central tendencies or averages. Actually, sociologists use three types of averages: the mean, the median, and the mode. Each type is calculated differently, and each can result in a different figure.

Suppose you are studying a group of ten college students whose verbal SAT scores are as follows:

| 450 | 690 | 280 | 450 | 760 |
| 540 | 520 | 450 | 430 | 530 |

Although you can report the information in this form, a more meaningful presentation would give some indication of the central tendency of the ten scores. The three measures of central tendency, like the three types of averages mentioned earlier, are the mean, median, and mode.

The *mean* is commonly called the *average*. To calculate the mean, you add up all the figures and divide by the number of items. In our example, the SAT scores add up to 5,100. Dividing by 10 gives a mean of 510.

The *median* is the figure that falls midway in a series of numbers; there are as many numbers above it as below it. Because we have ten scores—an even number—in our example, the median is the mean (the average) of the fifth and sixth figures, the two numbers in the middle. To calculate the median, rearrange the data in order from the lowest to highest (or vice versa). In

to political talk radio perceive a liberal bias in the mainstream media, compared with one in five non-listeners.

### What Is the Margin of Error?

No matter how carefully a survey sample is chosen, some margin of error can still exist. If you selected ten sets of 1,000 people, using the same rules, and asked each group the same question, the results would not be identical. The difference between the results is sampling error. Statisticians know that the error is equally likely to be above or below the true mark and that larger samples have smaller margins of error if they are properly drawn. Statisticians are also able to estimate the margin of error, the amount by which the result could be above or below the truth. Sampling error will always exist unless you survey every member of a population. If you do that, you have conducted a census.

Sampling error is one reason two professionally conducted polls can show different results and both be correct. Reputable surveys report a margin of error—usually of 3 or 4 percentage points—at a particular confidence level, typically 95 percent. This means that 5 percent of the time, or 1 time in 20, the poll's results will not be reliable. The other 95 percent of the time, it is accurate within 3 or 4 percentage points.

### Which Came First?

The order in which questions are asked can have a significant effect on the results. Most people want to appear consistent with others and to be consistent in their own minds. When a pollster asks a series of related questions, this desire can lead people to take positions they might not have taken if they were asked only one question. Neither way—asking a series of questions or asking only one—produces an obviously correct response, but the results are different.

One way to handle this problem is to rotate the order of questions. Then the degree of differences due to question order can be described and interpreted, but not everyone heeds such fine distinctions.

### What Was the Question?

"Do you want union officials, in effect, to decide how many municipal employees you, the taxpayer, must support?" Well, do you? This question, taken from an actual survey, is obviously biased. The results might make good propaganda for an antiunion group, but they are bogus as a poll, so before you pass a survey finding on to others, or even believe it yourself, be sure to look at the actual question.

When you are presented with a new survey, as with a used car, it helps to ask a few key questions before you buy. But for all their flaws, surveys are essential to the work of politicians, journalists, and businesspeople.

*Source:* Adapted from Edmondson, Brad. 1996. "How to Spot a Bogus Poll." *American Demographics* 18(10):10, 12–15. Used with permission from Primedia Specialty Group, Inc.

---

our example, you would list the scores as follows: 280, 430, 450, 450, 450, 520, 530, 540, 690, 760. The median is 485, midway between the fifth score (450) and the sixth score (520).

The *mode* is the number that occurs most often in the data. In our example, the mode is 450.

These three measures are used for different reasons, and each has its advantages and disadvantages. The mean is most useful when a narrow range of figures exists because it has the advantage of including all the data. It can be misleading, however, when one or two scores are much higher or lower than the rest. The median deals with this problem by not allowing extreme figures to distort the central tendency. The mode enables researchers to show which number occurs most often. Its disadvantage is that it does not give any idea of the entire range of data. Realizing the problems inherent in each average, sociologists often state the central tendency in more than one form.

Scientists usually are careful in drawing conclusions from their research. One of the purposes of drawing conclusions from data compiled in the course of research is the ability to apply the information gathered to other, similar situations. Thus, problems can develop if there are faults in the research design. For example, the study must show **validity**—that is, *the study must actually test what it was intended to test*. If you want to say one event is the cause of another, you must be able to rule out other explanations to show that your research is valid.

Suppose you conclude that marijuana use leads to heroin use. You must show that it is marijuana use, and

## HOW SOCIOLOGISTS DO IT

### How to Read a Table

Sociologists use statistical tables frequently both to present the findings of their own research and to study the data of others. We will use Table 2-3 to outline the steps to follow in reading and interpreting a table.

1. *Read the title.* The title tells you the subject of the table. Table 2-3 presents data on births to unmarried women in various countries.
2. *Check the source.* At the bottom of a table, you find its source. In this case, the source is *Statistical Abstract of the United States: 2012.* Knowing the source of a table can help you decide whether the information it contains is reliable. It also tells you where to look to find the original data and how recent the information is. In our example, the source is both reliable and recent. If the source were the 1958 abstract, its value would be limited in telling you about births to unmarried mothers in those countries today. Improvements in health care, the education of women, and changes in income, among other factors, might have altered births to single mothers since 1958. Likewise, consider a table of data about AIDS cases in Thailand. If its source were a government agency (which might be trying to alleviate the fears of tourists about the rampant spread of the disease in that country), you might well be skeptical about the reliability of the information in the table.
3. *Look for headnotes.* Many tables contain headnotes directly below the title. These might explain how the data were collected, why certain variables (and not others) were studied, why the data are presented in a particular way, whether some data were collected at different times, and so on. In our table, we do not have headnotes, but the title states that the data are for live births, so stillborn births are not included.
4. *Look for footnotes.* Many tables contain footnotes that explain limitations or unusual circumstances surrounding certain data.
5. *Read the labels or headings for each row and column.* The labels tell you exactly what information is contained in the table. It is essential for you to understand the labels, both the row headings on the left and the column headings at the top. Here, the row headings tell you the names of the countries being compared for births. For each country, a number is given. Note the units used in the table. In this case, the units are percentages. Often, the figures represent percentages or rates. Many population and crime statistics are given in rates per 100,000 people.
6. *Examine the data.* Suppose you want to find the percentage of live births to unmarried women in the United States. First, look down the row at the left until you come to "United States." Then, look across the columns until you come to the number. Reading across, you discover that, on average, 40.6 percent of live births in 2008 were to unmarried women.
7. *Compare the data.* Compare the data in the table both horizontally and vertically. Suppose you want to know which country has the highest percentage of births to unmarried women. Looking down the percentage column, we find that Sweden has the highest percentage, with 54.7 percent. Japan has the lowest percentage of births to unmarried women, with 2.1 percent.
8. *Draw conclusions.* Draw conclusions about the information in the table. After examining the data in the table, you might conclude that in Western Europe (Sweden, France, Denmark), it is common for women to give birth without being married. The attitudes toward births to unmarried women must be markedly different in Japan, given the low percentage.
9. *Pose new questions.* The conclusions you reach might well lead to new questions that could prompt further research. Why, you might want to know, are the percentage of births to unmarried women so much lower in Japan? Are there government policies in Sweden that encourage women to give birth without being married?

**TABLE 2-3** Births to Unmarried Women as a Percentage of All Live Births, 2008

| Country | Percentage |
| --- | --- |
| Sweden | 54.7 |
| France | 52.6 |
| Denmark | 46.2 |
| United Kingdom | 43.7 (2006) |
| United States | 40.6 |
| Netherlands | 41.2 |
| Ireland | 33.2 (2006) |
| Germany | 32.1 |
| Spain | 31.7 |
| Canada | 27.3 (2007) |
| Japan | 2.1 (2006) |

*Source:* U.S. Bureau of the Census. 2011. *Statistical Abstract of the United States: 2012*, 131st ed. Washington, D.C. (www.census.gov/compendia/statab/), Table 1334, Internet Release Date 12-10-2010.

not some other factor, such as peer pressure or emotional problems, that leads to heroin use.

The study must also demonstrate **reliability**—that is, *the findings of the study must be repeatable.* To demonstrate reliability, we must show that research can be replicated—repeated to determine whether initial results can be duplicated. Suppose you conclude from a study that whites living in racially integrated housing projects, who have contact with African Americans in the same projects, have more favorable attitudes toward blacks than do whites living in racially segregated housing projects. If you or other researchers carry out the same study in housing projects in various cities throughout the country and get the same results, the study is reliable.

It is highly unlikely that any single piece of research will provide all the answers to a given question. In fact, good research frequently leads to the discovery of unanticipated information requiring further research. One of the pleasures of research is that ongoing studies keep opening up new perspectives and posing further questions.

### Prepare the Research Report

Research that goes unreported is wasted. Scientific progress is made through the accumulation of research that tests hypotheses and contributes to the ongoing process of understanding our world. Therefore, it is usual for agencies that fund research to insist that scientists agree to share their findings.

Researchers generally publish their findings in scientific journals. If the information is relevant to the public, many popular and semi-scientific publications will report these findings as well. It is especially important for research in sociology and other social sciences to be made available to the public because much of this research has a bearing on social issues and public policies.

Unfortunately, the general public is not always cautious in interpreting research findings. Special-interest groups, politicians, and others who have a cause to plead are often too quick to generalize from specific research results, frequently distorting them beyond recognition. This happens most often when the research focuses on something of national or emotional concern. It is, therefore, important to double-check reports of sociological research, appearing in popular media, with the original research.

## OBJECTIVITY IN SOCIOLOGICAL RESEARCH

One of the goals of sociology specifically and of science generally is to uncover the truth and bring about a fuller understanding of an issue. The sociologist seeks the truth by interviewing subjects, examining data, and piecing records together. Using this information, the researcher makes generalizations about what is likely to occur in the future and proposes and tests new hypotheses. Even though it might seem objective, this search for the truth is still influenced by personal biases and political orientations.

Sometimes a scientist is looking for one thing and discovers something different and unexpected. At other times, a discovery might be made by accident or because the original experiment was done incorrectly. None of this invalidates the truth of what was discovered. For example, in 1895, Wilhelm Roentgen accidentally discovered X-rays, a type of radiation that can pass through material where ordinary light cannot. X-rays revolutionized the fields of physics and medicine, and for this discovery, Roentgen received the Nobel Prize in 1901. Penicillin, too, was an accidental discovery, as were many others that have changed our lives.

If a scientist cheats, exaggerates, or even plagiarizes his or her findings, ultimately, the truth or falsity of the information is what remains. In the courtroom, there is something known as the exclusionary rule, which throws out any evidence that has not been obtained legally. There is no such thing as an exclusionary rule in science, and truth achieved by unfair means is more likely to endure than falsity achieved by fair means. Scientifically discovered truths can always be reconsidered based on new evidence. Scientific research involves a continuous process of building on and refining other people's findings (Dershowitz, 1996).

Max Weber believed that the social scientist should describe and explain what exists rather than prescribe what should be. His goal was a value-free approach to sociology. More and more sociologists today, however, are admitting that completely value-free research might not be possible. In fact, one of the trends in sociology today that could ultimately harm the discipline is that some sociologists who are more interested in social reform than in social research have abandoned all claims to objectivity.

Does this mean that all science—sociology in particular—is hopelessly subjective? Is objectivity in sociological research an impossible goal? There are no simple answers to these questions. The best sociologists can do is strive to become aware of the ways in which these factors influence them and to make such biases explicit when sharing the results of their research. We can think of this as disciplined, or "objective," subjectivity, and it is a reasonable goal for sociological research. (For an example of how difficult it is to be objective, see "How Sociologists Do It: Facebook, the Internet, and New Ethical Concerns.")

## HOW SOCIOLOGISTS DO IT

### Facebook, the Internet, and New Ethical Concerns

When you post information on Facebook, should it be possible for someone to use it for sociological research? Starting with the freshman class of 2006 at what was described as "an anonymous, northeastern American university," a group of sociologists downloaded all the information posted by nearly 1,700 people. They continued to download more information during the entire four years that the class moved through the university. They ended up collecting a snapshot of an entire class over its four years in college. The researchers made good-faith attempts to hide the identity of the institution and to protect the privacy of the data subjects. Very quickly however, others figured out that that the information was from Harvard University.

The researchers had permission for the research from Facebook and the university, but not from the students. Of the freshmen students enrolled at the college, 97.4 percent maintained Facebook profiles and 59.2 percent of these students had updated their profile within 5 days. The first wave of data collection took place in 2006, during the spring of the class's freshman year, and data collection was repeated annually until 2009, when the vast majority of the students had graduated, providing four years of data about this collegiate social network.

The uniqueness of this student information has obvious value for sociologists. The data came directly from Facebook without direct interaction with the subjects or reliance on self-reporting methods, either of which could taint the data collected. The dataset includes demographic, relational, and cultural information on each subject. Most importantly, the dataset represented nearly a complete class of college students. It could be used for future research projects, which had been difficult or impossible before. The sociologists believed the Internet and Facebook provided a new way to do research and made it possible to collect information that previous generations of researchers could never have hoped to obtain.

The Association of Internet Researchers has issued a set of recommendations for engaging in ethical research online that places considerable focus on informed consent and respecting the ethical expectations within the venue under study. As noted, the researchers in the Facebook study did not obtain any informed consent from the subjects within the dataset. They failed to recognize that Facebook users might expect that information shared on Facebook is meant to stay on Facebook.

This study might very well be ushering in new concerns about doing social science using appropriate ethical practices. "Concerns over consent, privacy and anonymity do not disappear simply because subjects participate in online social networks; rather, they become even more important" (Zimmer, 2010).

*Source:* Zimmer, Michael. June 4, 2010. "'But the Data Is Already Public': On the Ethics of Research." In *Facebook Ethics Information Technology.* Springer. Published online: http://collections.lib.uwm.edu/cipr/image/592.pdf.

## ETHICAL ISSUES IN SOCIOLOGICAL RESEARCH

All research projects raise fundamental questions. Whose interests are served by the research? Who will benefit from it? How might people be hurt? To what degree do subjects have the right to be told about the research design, its purposes, and possible applications? Who should have access to and control over research data after a study is completed—the agency that funded the study, the scientists, the subjects? Should research subjects have the right to participate in planning projects? Is it ethical to manipulate people without their knowledge to control research variables? To what degree do researchers owe it to their subjects not to invade their privacy and to keep secret (and, therefore, not report anywhere) things that were told in strict confidence? What obligations do researchers have to the society in which they are working? What commitments do researchers have in supporting or subverting a political order? Should researchers report to legal authorities any illegal behavior discovered in the course of their investigations? Is it ethical to expose subjects to such risk by asking them to participate in a study?

In the 1960s, the federal government began to prescribe regulations for "the protection of human subjects." According to Herbert Gans (1979), these regulations are designed to force scientists to consider one central issue, "how to judge and balance the intellectual and [societal] benefits of scientific research against the actual or possible physical and emotional costs paid by the people who are being studied." Gans discusses three potential dilemmas for the researcher.

The first situation involves the degree of "permissible risk, pain, or harm." Gans writes, "Suppose a study which temporarily induces [severe emotional distress] promises significant benefits. Experimental researchers may justify the study." However, we might wonder "whether the promised benefits can be realized, and whether they justify the potential dangers to the subjects, even if these are volunteers who know what to expect, and when all possible protective measures are taken."

A second dilemma is the extent to which subjects should be deceived in a study. It is now necessary for researchers to obtain the "informed consent," in writing, of the people they study. Questions still arise, however, about whether subjects are informed about the true nature of the study and "whether, once informed, they can freely decline to participate."

A third problem in research studies concerns the "disclosure of confidential or personally harmful information." Is the researcher entitled to delve into personal

## THINKING ABOUT SOCIAL ISSUES: Famous Research Studies You Cannot Do Today

The proper procedures for research have changed considerably over the past few decades. A number of older, famous studies have been referred to hundreds of times in textbooks, other studies, and the classroom that do not, for a variety of reasons, meet contemporary standards of ethical research.

*Zimbardo's Prison Environment.* In 1972, Philip Zimbardo tried to recreate a prison environment with Stanford University undergrads. Some of the students became guards; others played prisoners. The experiment, which was to last two weeks, had to be canceled after six days because some student guards became sadistic, and some student prisoners became distraught and depressed. The subjects had not been properly informed about what might happen in this experiment and had not given informed consent. Today, such protective procedures would have to be followed.

*Tearoom Trade Observation.* In 1970, sociologist Laud Humphreys wrote about his research observing homosexual behavior in public restrooms. After each observation of an older man having sex with a younger male, Humphreys would note the license plate number of the older man's vehicle. Through a friend in the police department, he would trace the addresses of these men. Months later he would go to the homes of these individuals and under false pretenses ask them questions for his study.

*Milgrim's Obedience to Authority.* This was the title of a book by psychologist Stanley Milgram who, in 1974, wanted to see how susceptible people were to pressure from authority figures. He used a fake machine that caused the subjects to think they were giving electrical shocks to people in the next room. He wanted to see whether the subjects would continue to give increasing shocks at the request of the researcher even though the person behind the wall was pleading to be released. The experiment produced great stress in the subjects and today would violate contemporary research ethics.

*Cyril Burt's Twin Studies.* Cyril Burt spent a lifetime trying to prove that intelligence is primarily an inherited characteristic. Between 1943 and 1966, he published numerous studies comparing the intelligence of identical twins who had been raised in different homes. Each study came to the same conclusion, that twins' intelligence test scores were very close. After Burt's death, critics pointed out that the results were just too consistent. Questions also started to arise about whether his research assistants actually existed. On October 24, 1976, Oliver Gillie, the medical correspondent for the *London Sunday Times*, wrote, "Leading scientists are convinced that Burt published false data and invented crucial facts to support his controversial theory that intelligence is largely inherited" (cited in Plucker, 2003).

*The Guatemala and Tuskegee Syphilis Studies.* A U.S. Public Health Service doctor conducted two highly unethical experiments. In the 1940s, Guatemalan prison inmates and mental patients were deliberately infected with syphilis. The U.S. government was interested in studying the then-new drug penicillin to treat the disease. Eventually the Guatemalan subjects were treated, although not always successfully. The Tuskegee study did not involve treatment, since the U.S. government was mainly interested in the long-term effects of syphilis. That experiment involved withholding penicillin from black male sharecroppers. By 1947, it was clear that penicillin was an effective treatment, yet the men were still left untreated. The subjects did not know they were part of these unethical studies and, at least in the Tuskegee case, their conditions deteriorated.

lives? What if the researcher uncovers some information that should be brought to the attention of the authorities? Should confidential information be included in a published study?

Every sociologist must grapple with these questions and find answers that apply to particular situations. However, two general points are worth noting. The first is that social research rarely benefits the research subjects directly. Benefits to subjects tend to be indirect and delayed by many years, as when new government policies are developed to correct problems discovered by researchers. Second, most subjects of sociological research belong to groups with little or no power because they are easier to find and study. It is hardly an accident that poor people are the most studied, and rich people the least. Therefore, research subjects typically have little control over how research findings are used, even though such applications might affect them greatly.

This means that sociologists must accept responsibility for recruiting research subjects who might become vulnerable as a result of their cooperation. It is important for researchers to establish safeguards limiting the use of their findings, protecting the anonymity of their data, and honoring all commitments to confidentiality made in the course of their research.

The ideal relationship between scientist and research participant is characterized by openness and honesty. Deliberately lying to manipulate the participant's perceptions and actions goes directly against this ideal. Yet, often, researchers must choose between deception and abandoning the research. With few, if any, exceptions, social scientists regard deception of research participants as a questionable practice to be avoided if at all possible. It diminishes the respect due to others and violates the expectations of mutual trust on which organized society is based. When the deceiver is a respected scientist, it can have the undesirable effect of modeling deceit as an acceptable practice. Conceivably, it can contribute to the growing climate of cynicism and mistrust bred by widespread use of deception by important public figures. (For some examples of questionable research, see "Thinking About Social Issues: Famous Research Studies You Cannot Do Today.")

It is useful for human beings to understand themselves and the social world in which they live. Sociology has a great contribution to make to this endeavor, both in promoting understanding for its own sake and in providing social planners with scientific information with which well-founded decisions can be made and sound plans for future development adopted. However, sociologists must also shoulder the burden of self-reflection, of understanding the role they play in contemporary social processes while assessing how these social processes affect their findings.

## SUMMARY

▶ **What goals influence social research?**
Science has two important goals: first, to describe in detail particular circumstances or events; and second, to propose and test theories that help us understand these circumstances or events. Like detectives, sociologists want to know "Why did it happen?" and "Under what circumstances is it likely to happen again?"

▶ **What are the steps in the sociological research process?**
The scientific research process involves a sequence of steps that must be followed to produce a valid study. The process begins by defining the problem. Next, the researcher attempts to discover as much as possible about previous studies on the same topic. The researcher must then develop a hypothesis, a testable statement about the relationship between two or more empirical variables. Hypotheses are tested by constructing a research design, a strategy for collecting appropriate data needs to be adopted, and a final report has to be written.

▶ **What are the research designs sociologists use?**
Sociologists use four main research designs. A survey is a research method in which a population is questioned to reveal specific facts about itself. In participant observation, researchers enter into a group's activities and observe the group members. Secondary analysis involves using data that has been collected by others. In an experiment, the variables being studied are controlled and the researcher obtains the results through observation and measurement.

▶ **What are independent and dependent variables?**
An independent variable causes or changes another variable. A dependent variable is influenced by the independent variable.

▶ **What is sampling and how do you create a representative sample?**
The particular subset of the population chosen for study is known as a sample. Sampling is a research technique through which investigators study a manageable number of people selected from a larger population. If the procedures are carried out correctly, the sample will be representative, one that shows, in equivalent proportion, the significant variables that characterize the population as a whole. Failure to achieve a representative sample is known as sampling error.

▶ **What do the concepts of reliability and validity mean?**
A reliable study can be repeated. A valid study measures what it is supposed to measure.

> **What role does objectivity play in social research?**
>
> Sociology, like any other science, is molded by factors that impose values on research. Thus, completely value-free research might not be possible. Nevertheless, objectivity, or a kind of disciplined subjectivity, is a reasonable goal for sociological research.

> **What ethical issues arise in sociological research?**
>
> The central ethical concern in research on human participants is how to judge and balance the intellectual and societal benefits of scientific research against the actual or possible physical and emotional costs to the research participants.

## Media Resources

**CourseMate for *Introduction to Sociology*, Eleventh Edition**

Cengage Learning's Sociology CourseMate brings course concepts to life with interactive learning, study, and exam preparation tools that support the printed textbook. Access an integrated eBook, learning tools including glossaries, flashcards, quizzes, videos, and more in your Sociology CourseMate. Go to CengageBrain.com to register or purchase access.

# CHAPTER TWO STUDY GUIDE

## Key Concepts

*Match each of the following concepts with its definition, illustration, or explanation.*

a. Empirical question
b. Operational definition
c. Hypothesis
d. Variable
e. Independent variable
f. Dependent variable
g. Association
h. Survey
i. Cross-sectional research
j. Longitudinal research
k. Structured interview
l. Open-ended interview
m. Participant observation
n. Secondary analysis
o. Sample
p. Representative sample
q. Random sample
r. Sampling error
s. Stratified random sample
t. Research bias
u. Blind investigators
v. Double-blind investigators
w. Validity
x. Reliability
y. Experiment
z. Analysis

____ 1. A study that observes a population over a period of time
____ 2. Organizing data for the purpose of making comparisons and drawing conclusions
____ 3. Based on, or capable of being based on, observed evidence
____ 4. Research in which the researcher follows a set of questions but can add follow-up questions on his or her own
____ 5. A study that asks short-answer questions of a fairly large number of people
____ 6. The simultaneous change of two variables without one necessarily causing the change in the other
____ 7. The conversion of abstract ideas into specific, observable circumstances or events
____ 8. The relation between what a study is supposed to test and what it actually tests
____ 9. The influence, deliberate or not, a researcher exerts to get the preferred result
____ 10. The population on which a researcher gathers data to assess the entire population
____ 11. A testable statement about the relation between variables
____ 12. The most precise research method because researchers can control which variables come into play
____ 13. The degree to which the results of a study would be repeated in other, similar studies
____ 14. A sample in which each individual in the population has an equal chance of being selected
____ 15. Anything that can change or that can be sorted into more than one category or value
____ 16. A study that looks at a population at a single point in time
____ 17. Research in which the researcher strictly follows a given set of questions
____ 18. The use of available data gathered by another researcher or agency such as the Census Bureau
____ 19. A variable that changes in response to changes in the independent variable
____ 20. The variable that influences another variable without being influenced by that other variable
____ 21. The failure to achieve a representative sample
____ 22. Research in which the researchers mingle with the people they are researching
____ 23. A sample in which the relevant variables are distributed in the same proportions as in the entire population
____ 24. A sample in which subgroups are sampled separately to ensure that no group is disproportionately represented
____ 25. Researchers who do not know which category a subject is in
____ 26. Researchers who are unaware of both the category of the subject and the hypotheses being tested

CHAPTER 2 DOING SOCIOLOGY: RESEARCH METHODS 47

## • Central Idea Completions

*Fill in the appropriate concepts and descriptions for each of the following questions.*

1. A researcher wanted to see whether the amount of time a student spent studying was associated with the student's grades. In this study, what was

    a. The independent variable _____

    b. The dependent variable _____

2. Suppose researchers wanted to investigate cheating in college. Explain the advantages and disadvantages of each of the following four research methods that might be used to assess the extent of cheating and the factors that influence it.

    a. Social survey

      advantages _____

      disadvantages _____

    b. Participant observation

      advantages _____

      disadvantages _____

    c. Experiment

      advantages _____

      disadvantages _____

    d. Secondary analysis

      advantages _____

      disadvantages _____

3. How does research bias take the form of a self-fulfilling prophecy?

4. What precautions might you use to detect a bogus poll?

5. Why do inaccurate statistics, such as Vicary's study on subliminal advertising, take on a life of their own and become widely cited?

## • Critical Thought Exercises

1. Find a poll result in a newspaper (*USA Today* often has them) or popular magazine. From the information given, which of the critical questions for detecting a bogus poll can you answer? If the information is not provided for a question, try to imagine what might have been done to make the poll bogus on that question.

2. Find a statement in an e-mail or letter to the editor, in an advertisement, or in a song, and try to turn it into a hypothesis about the relation between variables. (For example, an old ad stated, "Blondes have more fun.") How would you create an operational definition to turn abstract ideas such as "fun" into something you could measure?

3. There has been much public discussion (and some proposed legislation) about video games and their effects. Groups that express concern about the games claim that they harm children. Companies that make the games say they are harmless or even beneficial. Both sides cite research. What steps would you take to assess the validity of the research claims coming from people on either side of the debate?

4. There are two main sources of data on crime in the United States, the FBI's Uniform Crime Reports (UCR) and the National Crime Victimization Survey (NCVS). The UCR is the total of all crimes reported to the police. The NCVS is based on a sample of people who are interviewed about whether they have been victims of crime. Would you expect the NVCS to be more or less accurate than the UCR? Explain your answer.

5. Plan a research study. What types of biases could enter into your research. What types of preconceived notions might influence the study and its results. How might you eliminate the possibility of bias in your study?

## Internet Activity

The data from the U.S. census are among the most important social science data used throughout the United States. Because the census is taken only every ten years, the data can become outdated. To correct this problem, the U.S. Census Bureau conducts the American Community Survey (ACS). Go to the U.S. Census website (www.census.gov) and examine the bureau's discussion of the methods it uses to collect these data.

The Gallup Poll is one of the oldest and most respected surveys. You can go to that website to see some of those results. You can also find a brief description of their methods, sampling, interview techniques, and question design at http://media.gallup.com/PDF/FAQ/HowArePolls.pdf.

## Answers to Key Concepts

1. j; 2. z; 3. a; 4. l; 5. h; 6. g; 7. b; 8. w; 9. t; 10. o; 11. c; 12. y; 13. x; 14. q; 15. d; 16. i; 17. k; 18. n; 19. f; 20. e; 21. r; 22. m; 23. p; 24. s; 25. u; 26. v

# 3 Culture

## The Concept of Culture
- Culture and Biology
- Culture Shock
- Ethnocentrism and Cultural Relativism

*Our Diverse World:* Marriage to a Perfect Stranger

## Components of Culture
- Material Culture
- Nonmaterial Culture
- The Origin of Language

*Our Diverse World:* The United States and Europe—Two Different Worldviews
- Language and Culture

*Day-to-Day Sociology:* Symbols in Cyberspace

## The Symbolic Nature of Culture
- Symbols and Culture

## Culture and Adaptation
- What Produces Cultural Change?

*Thinking About Social Issues:* Technology Changes Culture
- Cultural Lag
- Animals and Culture

## Subcultures
- Types of Subcultures

## Universals of Culture
- The Division of Labor
- Marriage, the Family, and the Incest Taboo
- Rites of Passage
- Ideology

*How Sociologists Do It:* Social Science in a War Zone

## Culture and Individual Choice

*How Sociologists Do It:* The Conflict Between Being a Researcher and Being a Human Being

## Summary

## LEARNING OBJECTIVES

After studying this chapter, you should be able to do the following:

- Understand how culture makes possible the variation in human societies.
- Distinguish between ethnocentrism and cultural relativism.
- Know the difference between material and nonmaterial culture.
- Understand the importance of language in shaping our perception and classification of the world.
- Discuss whether animals have language.
- Understand the roles of innovation, diffusion, and cultural lag in cultural change.
- Explain what subcultures are.
- Describe cultural universals.

Cultural differences often pop up where you least expect them. Sheena Iyengar was at a restaurant in Tokyo and ordered green tea with sugar. The waiter replied that one does not drink green tea with sugar. Having been raised in the United States and expecting that as a customer her request would be honored, she said she understood, but she liked her tea with sugar. The waiter went to the manager of the restaurant and the two had a lengthy conversation. The waiter returned and said they were sorry, but they did not have sugar. At that point, Sheena changed her order to a cup of coffee. A few minutes later the coffee arrived with two packets of sugar.

This encounter highlights some cultural differences between Japan and the United States. From an American perspective, when a paying customer makes a reasonable request, it is the waiter's job to fulfill it. From the Japanese perspective, having tea with sugar is inappropriate according to their cultural standards, and the waiter was merely trying to prevent her from making a foolish mistake.

Americans are used to the idea that when they make a choice, it is fine to think about what will make them happy. In fact, our founding fathers incorporated the ideas of liberty and the pursuit of happiness into the Constitution and the Bill of Rights. Asian cultures are influenced by an emphasis on duty and fate developed over thousands of years. In these cultures, individuals view their lives more in terms of duties and less in terms of preferences. In 2011, when the Japanese experienced the meltdown of several nuclear reactors, Americans had a hard time understanding how numerous workers at these plants were willing to risk certain illness and death by working inside the plant to shut down various systems. From an American standpoint, you would make a decision that would preserve your life. From a Japanese standpoint, it was your duty to do what was necessary for the collective good.

All human societies have complex ways of life that differ greatly from one to the other. These ways have come to be known as *culture*. In 1871, Edward Tylor gave us the first definition of this concept. Culture, he noted, "is that complex whole which includes knowledge, belief, art, law, morals, custom, and other capabilities and habits acquired by man as a member of society" (Tylor, 1958).

Robert Bierstadt simplified Tylor's definition by stating, "Culture is the complex whole that consists of all the ways we think and do and everything we have as members of society" (Bierstadt, 1974).

Most definitions of culture emphasize certain features, namely, that culture is shared; it is acquired, not inborn; the elements make up a complex whole; and it is transmitted from one generation to the next.

## THE CONCEPT OF CULTURE

We will define **culture** as *all that human beings learn to do, to use, to produce, to know, and to believe as they grow to maturity and live out their lives in the social groups to which they belong.* Culture is basically a blueprint for living in a particular society. In common speech, people often refer to a "cultured person" as someone with an interest in the arts, literature, or music, suggesting that the individual has a highly developed sense of style or aesthetic appreciation of finer things. To sociologists, however, every human being is cultured. All human beings participate in a culture, whether they are Harvard educated and upper class or illiterate and living in a primitive society. Culture is crucial to human existence.

When sociologists speak of culture, they are referring to the general phenomenon that is a characteristic of all human groups. However, when they refer to a culture, they are pointing to the specific culture of a particular group. In other words, all human groups have a culture, but it often varies considerably from one group to the next.

Consider the example of the concept of time, which Westerners accept as entirely natural—time marches on steadily and predictably, with past, present, and future divided into units of precise duration (minutes, hours, days, months, years, and so on). In the Native American culture of the Sioux, however, the concept of time simply does not exist apart from ongoing events: nothing can be early or late—things just happen when they happen. For the Navajo, the future is a meaningless concept—immediate obligations are what count. For natives of the Pacific island of Truk, however, the past has no independent meaning—it is a living part of the present (Hall, 1981). These examples of cultural differences in the

All human groups have a culture, but it often varies considerably from one group to the next.

perception of time point to a basic sociological fact: Each culture must be investigated and understood on its own terms before it is possible to make valid cross-cultural comparisons.

Native Americans often wish to follow the traditional practices and customs of their culture. At the same time, the culture of the larger society urges them to adopt the conventions of mainstream American society.

In every social group, culture is transmitted from one generation to the next. Unlike other creatures, human beings do not pass on many behavioral patterns through their genes. Rather, culture is taught and learned through social interaction.

## Culture and Biology

Human beings, like all other creatures, have basic biological needs. We must eat, sleep, protect ourselves from the environment, reproduce, and nurture our young, or we could not survive as a species. In most other animals, such basic biological needs are met in more or less identical ways by all the members of a species through inherited behavior patterns or instincts. These instincts are specific for a given species as well as universal for all members of that species. Thus, instinctual behaviors, such as the web spinning of specific species of spiders, are constant and do not vary significantly from one individual member of a species to another.

This is not true of humans, whose behaviors are highly variable and changeable, both individually and culturally. It is through culture that human beings acquire the means to meet their needs. In contrast to those little spiders, for instance, human infants cry when hungry or uncomfortable, and the responses to those cries vary from group to group and even from person to person. In some groups, infants are breast-fed; in others, they are fed prepared milk formulas from bottles; and in still others, they are fed according to the mother's preference. Some groups breast-feed children for as long as five or six years, others for no more than ten to twelve months. Some mothers feed their infants on demand—whenever they seem to be hungry; other mothers hold their infants to a rigid feeding schedule. In some groups, infants are picked up and soothed when they seem unhappy or uncomfortable. Other groups believe that infants should be left to cry it out.

In the United States, parents differ in their approaches to feeding and handling their infants, but most are influenced by the practices they have observed among members of their families and their social groups. Such habits, shared by the members of each group, express the group's culture. Group members learn the habits and keep them more or less uniform by social expectations and pressures.

## Culture Shock

Every social group has its own specific culture, its own way of seeing, doing, and making things, its own traditions. Some cultures are quite similar to one another; others are very different. When individuals travel abroad to countries with cultures that are very different from their own, the experience can be quite upsetting. Meals are scheduled at different times of day, strange or even repulsive foods are presented, and the traveler never quite knows what to expect from others or what others in turn might expect. Local customs might seem charming or brutal. Sometimes travelers are unable to adjust easily to a foreign culture; they might become anxious, lose their appetites, or even feel sick. Sociologists use the term **culture shock** to describe *the difficulty people have adjusting to a new culture that differs markedly from their own.*

Jonah Blank experienced culture shock often as he traveled throughout India. One day, he observed three bulls walking in the village:

> [They] ambled lazily by the storefront, leaving three steaming piles of dung in their wake. A few minutes later an old woman waddled along, dropped to her knees, and scooped up the fresh patties with her clapped hands. She slapped them onto an already laden tin plate, and shuffled down the alley.... Around the corner from the manure collector, another old woman hung a string of dried cow patties outside her door for luck. A large mound of dung sat at the step, stuck each day with newly plucked flower stems. Had I been rude enough to tell her that the custom was unhygienic, she would assuredly have laughed at my science. (Blank, 1992)

Culture shock can also be experienced within a person's own society. Picture the army recruit having to adapt to a whole new set of behaviors, rules, and expectations in basic training—a new cultural setting.

## Ethnocentrism and Cultural Relativism

*People often make judgments about other cultures according to the customs and values of their own*, a practice sociologists call **ethnocentrism.** Thus, an American might call a Guatemalan peasant's home filthy because the floor is made of packed dirt or believe that the family organization of the Watusi (of East Africa) is immoral because a husband may have several wives. Ethnocentrism can lead to prejudice and discrimination and often results in the repression or domination of one group by another.

Immigrants, for instance, often encounter hostility when their manners, dress, eating habits, or religious beliefs differ markedly from those of their new neighbors. Because of this hostility and because of their own ethnocentrism, immigrants often establish their own communities in their adopted country. Many Cuban Americans, for example, have settled in Miami where they have built a power base through strength in numbers. In Dade County, which includes Miami, Cuban Americans represent 60 percent of the population.

To avoid ethnocentrism in their own research, sociologists are guided by the concept of **cultural relativism,** *the recognition that social groups and cultures must be studied and understood on their own terms before valid comparisons can be made.* Cultural relativism frequently is taken to mean that social scientists never should judge the relative merits of any group or culture. This is not the case. Cultural relativism is an approach to performing objective cross-cultural research. It does not require researchers to abdicate their personal standards. In fact, good social scientists will take the trouble to spell out exactly what their standards are so that both researchers and readers will be alert to possible bias in their studies.

American Moshe Rubinstein encountered the contrasting values between American and Arab cultures after a traditional Arabic dinner. Rubinstein was presented with a parable by his host, Ahmed:

> "Moshe," Ahmed said as he put his fable in the form of a question, "imagine that you, your mother, your wife, and your child are in a boat, and it capsizes. You can save yourself and only one of the remaining three. Whom will you save?" For a moment, I froze, thoughts raced through my mind. . . . No matter what I might say, it would not be right from someone's point of view, and if I refused to answer, I might be even worse off. I was stuck. So I tried to answer by thinking aloud as I progressed to a conclusion, hoping for salvation before I said what came to my mind as soon as the question was posed, namely, to save the child.
>
> Ahmed was very surprised. I flunked the test. As he saw it, there was only one correct answer. "You see," he said,

As of 2012, King Mswati III of Swaziland had 13 wives and 23 children. His father had 70 wives. Cultural relativism asks us to withhold judgment about these facts based on our culture alone.

> "you can have more than one wife in a lifetime, you can have more than one child, but you have only one mother. You must save your mother!" (Rubinstein, 1975)

This example shows us how the value of individuals such as children, spouses, and mothers can vary greatly from one culture to the next. We can see how what we might consider to be a natural way of thinking is not the case at all in another culture.

Cultural relativism requires behaviors and customs to be viewed and analyzed within the context in which they occur. The packed-dirt floor of the Guatemalan house should be noted in terms of the culture of the Guatemalan peasant, not in terms of suburban America. Researchers, however, might find that dirt floors contribute to the incidents of parasites in young children and might, therefore, judge such construction to be less desirable than wood or tile floors.

King Mswati III of Swaziland has 13 wives and 23 children. His personal wealth is estimated at between $100 and $200 million, while most of Swaziland's 1.4 million people live below the poverty line. More than 25 percent of the adult population is infected with HIV/AIDS (*The Telegraph,* April 27, 2012). According to Swazi culture, the king is expected to marry a woman from every clan. The king's 13 wives is still far fewer than the 70 wives his father had (Bearak, 2008).

The customs of our society make us frown on this type of behavior, but cultural relativism does not ask us to approve of these actions—only to understand them from the cultural context within which they occur. (See "Our Diverse World: Marriage to a Perfect Stranger.")

## OUR DIVERSE WORLD

### Marriage to a Perfect Stranger

Anita Jain recalls that after learning the words *mummy* and *papa,* she learned *shaadi,* the word for marriage in many Indian languages. Even though she is a college-educated journalist living in Brooklyn and is part of the singles dating scene, her father is active in finding her a husband. He spends hours on websites such as shaadi.com, indiamatrimony.com, and punjabimatrimoy.com, posting ads describing his daughter and hoping to find the ideal husband for her. To most Indians of his generation, only two professions are legitimate for his future son-in-law, doctor or engineer.

Many Indians believe that people in the West make the process of finding a mate unnecessarily difficult. They cannot understand how a couple in America can spend time together for years and still not be sure about whether they want to get married. Indian women who study in the United States watch their friends complain about how they have been waiting years for their boyfriends to propose. Indian women know that dating resulting from an arranged marriage introduction leads directly to marriage.

In very traditional Indian arranged marriages, couples may meet only once or twice before their wedding day. Indians living in the United States or in large cities throughout the world might stretch the courtship out for a few months, but not longer.

What about love? People in the West cannot fathom marrying without first falling in love. Ruchika Tulshyan, another Indian woman who went to college in the United States and United Kingdom does not reject the idea of love. Yet, dating the Western way has forced her to question how you really know if you are in love. "Is it love when you're living together for seven years but your boyfriend refuses to propose because 'he's not ready'? Is it love when your family and friends hate him but you're crazy about him? How do you know?"

In an arranged marriage, instead of love being the motivator for marriage, it is commitment. People commit themselves to each other and let the feelings grow stronger throughout the marriage. Love comes after the marriage, and you will have a lifetime to get to know the person and learn to love him or her.

The Indian marriage is less a relationship between two people than it is between two families, especially between the women on both sides and the rest of the husband's family. It is common for the daughter to be part of a joint family that includes the husband's family and even the husband's brothers and their wives.

The arranged marriage system works because there is trust that the parents will do what is best for the daughter and that they have the wisdom to make a good decision. The parents emphasize certain important factors such as similar beliefs, backgrounds, and life ambitions. After that, fate takes care of the rest. In case you are still skeptical, remember that divorce in India is very low at slightly over 1 percent of all marriages.

*Sources:* Jain, Anita. May 21, 2005. "Is Arranged Marriage Really Any Worse Than Craigslist?" *New York Magazine*; Tulshyan, Ruchika. March 29, 2010. "Match Dot Mom and Dad" (www.forbes.com/2010/03/29/arranged-marriage-love-tradition-forbes-woman-well-being-family.html), accessed April 22, 2012.

Arranged marriages account for the overwhelming majority of marriages in India.

## COMPONENTS OF CULTURE

The concept of culture is not easy to understand, perhaps because every aspect of our social lives is an expression of it and because familiarity produces a kind of nearsightedness toward our own culture, making it difficult for us to take an analytical perspective toward our everyday social lives. Sociologists find it helpful to break down culture into separate components: material culture (objects), and nonmaterial culture (rules and shared beliefs) (Hall and Hall, 1990).

### Material Culture

**Material culture** consists of *human technology—all the things human beings make and use, from small, handheld tools to skyscrapers.* Without material culture, our species could not survive long because material culture provides a buffer between humans and their environment. Using it, human beings can protect themselves from environmental stresses, as when they build shelters and wear clothing to protect themselves from the cold or from strong sunlight.

Even more important, humans use material culture to modify and exploit the environment. They build dams and irrigation canals, plant fields and forests, convert coal and oil into energy, and transform ores into versatile metals. Using material culture, our species has learned to cope with the most extreme environments and to survive and even to thrive on all continents and in all climates. Human beings have walked on the floor of the ocean and on the surface of the moon. No other creature can do this; none has our flexibility. Material culture has made human beings the dominant life-form on earth.

### Nonmaterial Culture

Every society also has a **nonmaterial culture,** which consists of *the totality of knowledge, beliefs, values, and rules for appropriate behavior.* The nonmaterial culture is structured by such institutions as the family, religion, education, economy, and government.

Housing is an aspect of material culture that can vary widely, as displayed by comparing this elaborate house in Barcelona with this unusual home in Fire Island, New York.

Whereas material culture is made up of things that have a physical existence (they can be seen, touched, and so on), the elements of nonmaterial culture are the ideas associated with their use. Although engagement rings and birthday flowers have a material existence, they also reflect attitudes, beliefs, and values that are part of American culture, with rules for their appropriate use in specified situations. Norms are central elements of nonmaterial culture.

**Norms** are *the rules of behavior that are agreed upon and shared within a culture and that prescribe limits of acceptable behavior.* They define normal expected behavior and help people achieve predictability in their lives. For example, one of the few truly universal gestures is the kiss.

Anthropologists have speculated that kissing evolved from the time when mothers would pass food, mouth to mouth, to their infants. Think for a moment how the kiss has permeated our lives. Mothers kiss bruises to make them better. Athletes kiss their trophies. The French sculptor Auguste Rodin sculpted a famous one—*The Kiss*.

Yet, most cultures follow unwritten norms concerning kissing in public. In some cultures, cheek kissing is a normal way of greeting another person. In Russia, you actually kiss the cheek. In other places, such as France, Italy, and Latin America, you kiss the air—that is, cheeks touch and lips make the sound of kissing, but the lips do not actually press against the cheek. In Latin America, only one cheek is kissed. In France, each cheek is kissed. In Belgium and Russia, you kiss one cheek, then the other, and back to the first.

In some countries, kissing the hand is the acceptable form of greeting. French etiquette suggests that the man kisses the woman's hand without actually touching it with his lips. Kissing your own fingertips is a European gesture that conveys the message "That's great! That's beautiful." The origin for this gesture probably stems from the custom of ancient Greeks and Romans who, when entering or leaving the temple, threw a kiss to sacred objects. In Mexico, a kissing sound summons a waiter in a restaurant. In the Philippines, street vendors use it to attract the attention of customers.

In the United States, kissing between men and women in public is common. Presidential candidates have even given extended mouth-to-mouth kisses to their spouses during prime-time broadcasts. At the other extreme, in certain Asian countries, such as Japan, kissing is considered an intimate sexual act and is not permissible in public, even as a social greeting (Axtell, 1998).

**Mores** (pronounced more-ays) are *strongly held norms that usually have a moral connotation and are based on the central values of the culture.* Violations of mores produce strong negative reactions, which are often supported by law. Desecration of a church or temple, sexual molestation of a child, rape, murder, incest, and child beating are all violations of American mores.

Not all norms command such absolute conformity. Much of day-to-day life is governed by traditions, or **folkways,** which are *norms that permit a wide degree of individual interpretation as long as certain limits are not overstepped.* People who violate folkways are seen as peculiar or possibly eccentric, but rarely do they elicit strong public response.

For example, a wide range of dress is now acceptable in most theaters and restaurants. Men and women may wear clothes ranging from business attire to jeans, an open-necked shirt, or a sweater. However, extremes in either direction will cause a reaction. Many establishments limit the extent of informal dress; signs might specify that no one with bare feet or without a shirt may enter. On the other hand, a person in extremely formal attire might well attract attention and elicit amused comments in a fast-food restaurant.

Good manners in our culture also show a range of acceptable behavior. A man might or might not open a door or hold a coat for a woman, who might also choose to open a door or hold a coat for a man—all four options are acceptable behavior and cause neither comment nor negative reactions from people.

These two examples illustrate another aspect of folkways: They change with time. Not too long ago, a man was

According to the norms of American culture, a common way for men to greet each other is to shake hands. In Japanese society, bowing is common.

*always* expected to hold a door open for a woman, and a woman was *never* expected to hold a coat for a man.

Folkways also vary from one culture to another. In the United States, for example, it is customary to thank someone for a gift. To fail to do so is to be ungrateful and ill mannered. Subtle cultural differences can make international gift giving, however, a source of anxiety or embarrassment to well-meaning business travelers. For example, if you give a gift on first meeting an Arab businessman, it might be interpreted as a bribe. If you give a clock in China, it is considered bad luck. In fact, the Mandarin word for *clock* is similar to the one for *death*. In Latin America, you will have a problem if you give knives, letter openers, or handkerchiefs. The first two indicate the end of a friendship; the last is associated with sadness.

Norms are specific expectations about social behavior, but it is important to add that they are not absolute. Even though we learn what is expected in our culture, there is room for variation in individual interpretations of these norms that deviate from the ideal norm.

**Ideal norms** are *expectations of what people should do under perfect conditions*. These are the norms we first teach our children. They tend to be simple, making few distinctions and allowing for no exceptions. In reality, however, nothing about human beings is ever that dependable.

**Real norms** are *norms that are expressed with qualifications and allowances for differences in individual behavior*. They specify how people actually behave. They reflect the fact that a person's behavior is guided by norms as well as unique situations.

At Russian universities, for example, although bribery is frowned upon, professors are so poorly paid that, over the years, a system has developed in which they supplement their salaries by taking bribes from their students in return for higher grades. Approximately 50 percent of the students pay bribes to their instructors on a regular basis, and about one-third of all instructors take them, equal to about $150 million U.S. dollars every year. Sociologist Igor Klyamkin, who has studied Russian university corruption, found that students and instructors know what bribes are to be given for what outcomes, and everyone seems to know that some of the money has to go to the deans also (Goble, 2009).

The concepts of ideal and real norms are useful for distinguishing between mores and folkways. For mores, the ideal and the real norms tend to be very close, whereas folkways can be much more loosely connected: Our mores say, "Thou shalt not kill," and really mean it, but we might violate a folkway by neglecting to say thank you, for example, without provoking general outrage. More important, the very fact that a culture legitimizes the difference between ideal and real expectations allows us room to interpret norms to a greater or smaller degree according to our own personal dispositions.

**Values** are *a culture's general orientations toward life—its notions of what is good and bad, what is desirable*

The thumbs-up gesture is appropriate in American society but might be an insult in other countries.

*and undesirable*. For example, each year, the University of California at Los Angeles surveys college students to get an idea of what values are important to them. In 2011, the Higher Education Research Institute (HERI) surveyed several hundred thousand freshman students at 270 of the nation's baccalaureate institutions.

Nearly 28 percent of students characterized themselves as liberal. The number of freshmen who described themselves as politically middle-of-the-road in 2011 was 47.4 percent, roughly the same percentage as in 1970. One in 5 students (20.7%) identified themselves as conservative in 2011.

Liberal causes also gained support. In 2011, more than two-thirds (71.3%) of college freshmen supported the right to legal marriage for same-sex couples, up from 51 percent in 1997. Forty-one percent agreed with the statement, "Students from disadvantaged social backgrounds should be given preferential treatment in college admissions," up from 37.4 percent in 2009 (Higher Education Research Institute, 2011).

Values can also be understood by looking at patterns of behavior. Sociologist Seymour Martin Lipset (1996) believes the United States has unique cultural values that set it apart from the rest of the world. However, these values can produce both positive and negative outcomes. For example, the United States is the most religious, optimistic, productive, well-educated, and individualistic country in the world. It is also one of the most violent crime-ridden and litigious nations, with a wide gap in income distribution and some of the lowest levels of welfare benefits.

How can widely held cultural values produce both good and bad outcomes? Part of the answer lies in how people attempt to fulfill these values. For example, the American emphasis on individualism and individual happiness might produce technological innovation in a wide variety of areas, but it is also responsible for the country's high divorce rate. Americans believe you should be satisfied with your life, be happy with your work, and like your spouse. If that is not the case, you are expected to make changes to correct the situation. If there are marital problems, we want to know why you are staying in the marriage. The Japanese do not automatically assume that if you are dissatisfied with your circumstances you must change them.

The United States is a highly achievement-oriented society. At the same time, it also leads the world in many types of crime. In the United States, a lack of success causes the individual to feel much more dissatisfied than in societies that are less achievement-oriented. Hence, people will try to get ahead by whatever means necessary. For some, this might mean committing crime.

In American society, a disdain for authority stems from the country's revolutionary past. The early founders rejected the control of England and produced a sharp break with the authority of the English powers. Americans do not show the kind of deference to authority that is commonly the case in countries such as Canada or Britain, which have not had the same kind of revolutionary history. (For a further discussion of value differences, see "Our Diverse World: The United States and Europe—Two Different Worldviews.")

### The Origin of Language

Language enables humans to organize the world around them into labeled cognitive categories and use these labels to communicate with one another. Language, therefore, makes possible teaching and sharing the values, norms, and nonmaterial culture we just discussed. It provides the principal means through which culture is transmitted and the foundation on which the complexity of human thought and experience rests.

Language allows humans to transcend the limitations imposed by their environment and biological evolution. It has taken tens of millions of years of biological evolution to produce the human species. On the other hand, in a matter of decades, cultural evolution has made it possible for us to travel to the moon. Biological evolution had to work slowly through genetic changes, but cultural evolution works quickly through the transmission of information from one generation to the next. In terms of knowledge and information, each human generation, because of language, is able to begin where the previous one left off. Each generation does not have to begin anew, as is the case in the animal world.

Over the past 75 years, sociologists and anthropologists have formed a standard view of the interplay between language and culture. The current view is that whereas animals are rigidly controlled by their biology, human behavior is determined by culture and language. Free from biological constraints, human cultures can vary from one another in countless ways.

Human infants are born with nothing more than a few reflexes and an ability to learn. Children learn their culture through their culture's language, socialization, and role models. Some scientists (Pinker, 1995) believe that the human capacity to use language is one of the most distinctive human attributes and that this critical step in cultural development has a biological basis as well as a cultural one.

The study of the genomes of people and chimpanzees has yielded some insight into the origin of language. It appears that language is a relatively recent development, having evolved only in the past 100,000 years. Some believe the emergence of behaviorally modern humans about 50,000 years ago was set off by a major genetic change that made modern language possible.

In 2001, the first human gene involved specifically in language was discovered. Designated as FOXP2, the gene is known to switch on other genes that are important for speech and language. The discovery took place as part of research on three generations of the KE family, half of whom had speech and language disorders. The thinking at first was that the gene merely caused low intelligence and made speech unintelligible. Testing suggested that the disorder was more complex. Some of the family members did score lower than average on IQ tests. Others, however, scored in the normal range but still had a problem with language. The language problems were not just due to motor control. The family members also had trouble understanding sentences or grammar rules. They even had trouble performing tasks that the average four-year-old could perform.

It appears that Neanderthals, an ancient human species that lived in Europe some 45,000 years ago, possessed a critical gene known to facilitate speech, according to DNA evidence retrieved from a cave in

## OUR DIVERSE WORLD

### The United States and Europe—Two Different Worldviews

It would not surprise many people if we pointed out that there are value differences between people in the United States and people in Saudi Arabia, Japan, or China. What about Western Europe? Many U.S. citizens have ancestors who came from Europe. Yet, American values differ from those of Western Europeans in significant ways also.

Americans are more individualistic than Europeans. Nearly 6 in 10 (58%) believe it is important for everyone to be free to pursue their life's goals without interference from the government. In addition, nearly two-thirds of Americans believe we are responsible for our own success. Only about one-quarter of Germans share this view.

Americans are less supportive of antipoverty programs than are people in Britain, France, Germany, and Spain. Only 35 percent of Americans believe it is important for the government to guarantee that nobody is in need. In contrast, at least 6 in 10 in Spain (67%), France (64%), and Germany (62%) and 55 percent in Britain share this view.

Americans are also considerably more religious than Western Europeans. Half of all Americans believe religion is *very* important in their lives. Fewer than a quarter in Spain (22%), Germany (21%), Britain (17%), and France (13%) think so. Moreover, Americans are far more likely to believe it is necessary to believe in God to be moral and have good values. Fifty-three percent of Americans have this view, but only 15 percent of the people in France share this belief.

Finally, about half of Americans (49%) agree with the statement, "Our people are not perfect, but our culture is superior to others." In Britain and France, only about a third or fewer (32% and 27%, respectively) think their culture is better than others.

So, although many of us may have European roots, our values have evolved in a different direction.

**Our People Are Not Perfect but Our Culture Is Superior to Others**

| Country | Agree | Disagree |
|---|---|---|
| United States | 49% | 46% |
| Spain | 44% | 55% |
| Great Britain | 32% | 63% |
| France | 27% | 73% |

**Which Is More Important?**

| Country | Nobody in Need | Freedom to Pursue Life's Goals |
|---|---|---|
| United States | 35% | 58% |
| Spain | 67% | 30% |
| France | 64% | 36% |
| Great Britain | 55% | 38% |

**Importance of Religion**

| Country | Religion Is Very Important |
|---|---|
| United States | 50% |
| Spain | 22% |
| Great Britain | 17% |
| France | 13% |

*Source:* Kohut, Andrew, Richard Wike, Juliana Menasce Horowitz, Jacob Poushter, and Cathy Barker. Updated February 29, 2012. "American Exceptionalism Subsides the American-Western European Values Gap." Pew Research Center, Pew Global Attitudes Project.

---

northern Spain. The thinking is that the FOXP2 gene had remained largely unaltered during the evolution of mammals, but suddenly changed in humans after they split off from the chimpanzee line of descent. The human version of the FOXP2 gene differs at two critical points from the chimpanzee version, suggesting that these changes have something to do with the fact that people can speak and chimps cannot (Wade, 2007).

### Language and Culture

All people are shaped by the **selectivity** of their culture, *a process by which some aspects of the world are viewed as important while others are virtually neglected.* The language of a culture reflects this selectivity in its vocabulary and even its grammar. Therefore, as children learn a language, they are being molded to think and even to experience the world in terms of one particular cultural perspective.

This view of language and culture, known as the **Sapir-Whorf hypothesis,** *argues that the language a person uses determines his or her perception of reality* (Sapir, 1961; Whorf, 1956). This idea caused some alarm among social scientists at first because it implied that people from different cultures never quite experience the same reality. Although more recent research has modified this extreme view, it remains true that different languages classify experiences differently—that language is the lens through which we experience the world. The prominent anthropologist Ruth Benedict (1961) pointed out, "We do not see the lens through which we look."

The category corresponding to one word and one thought in language A might be regarded by language B as two or more categories corresponding to two or more words or thoughts. For example, we have only one word for water, but the Hopi Indians have two words—*pahe* (for water in a natural state) and *keyi* (for water in a container). Yet, the Hopi have only one word to cover every being or thing that flies except birds. Strange as it might seem to us, they call a flying insect, an airplane, and a pilot by the same word. Verbs also are treated differently in different cultures. In English, we have one verb, *to go*. In New Guinea, however, the Manus language has three verbs, depending on direction, distance, and whether the going is up or down.

Language helps define our view of the world and how we see other's actions. The Indians of the North American plains are willing to accept someone's unusual behavior rather than label the person as mentally ill. In fact, in American Indian culture, such people might be considered gifted and, thus, very spiritual. Some American Indian cultures believe people with special needs are considered waken, or holy, and belong to the creator. They are, therefore, treated with great respect.

Many Native American groups do not have the concept of mental illness. They view mental illness as a white person's disease defined by mainstream society as shameful and unnatural. As such, American Indians are likely to see mental health as a mainstream concept that does not really apply to them. Mental and emotional problems are thought to be brought on by biological, social, or cultural violations or taboos such as excessive drinking, for example. Wellness takes place when there is harmony among body, mind, and spirit.

The American Indian conception of respect is also different from that of mainstream culture. It is inappropriate to pry into the innermost thoughts and feelings of another person, as is done by mental health professionals. One lives in harmony with all other beings because it is spiritually necessary to do so. All parts of life are interrelated and thus worthy of respect. To be in a state of conflict with people or to offend them is to be in disharmony and thus in a dangerous and vulnerable state.

In the Northern Plains Indians' view of communication, asking direct questions about mental illness might actually produce the behavior. In American Indian culture, such questioning can allow spirits to enter the person's essence, producing ghost illness (National Institute of Justice, June 2005).

A little bit closer to home, consider the number of words and expressions pertaining to technology that have entered the English language. These include *tweeting*, *texting*, *cyberspace*, *virtual reality*, *hackers*, *phishing*, *spamming*, *morphing*, and *googling*. These words reflect the preoccupation of American culture with technology.

In contrast, many Americans are at a loss for words when they are asked to describe facets of nature such as the varieties of snow, wind, or rain; kinds of forests; rock formations; earth colors and textures; or vegetation zones. Why? These things are not of great importance in urban American culture.

The translation of one language into another frequently presents problems. Direct translations are often impossible because (1) words might have a variety of meanings and (2) many words and ideas are culture-bound. An extreme example of the first type of these translation problems occurred near the end of World War II. After Germany surrendered, the Allies sent Japan a surrender ultimatum. Japan's premier responded that his government would *mokusatsu* the ultimatum to surrender. *Mokusatsu* has two possible meanings in English: "to consider" or "to take notice of." The premier meant that the government would consider the surrender ultimatum.

The English translators, however, used the second interpretation, "to take notice of," and assumed that Japan had rejected the ultimatum. This belief that Japan was unwilling to surrender was a factor in the atomic bombing of Hiroshima and Nagasaki (Samovar, Porter, and Jain, 1981). Most likely the bombing would still have taken place even with the other interpretation, but this example does demonstrate the problems in translating words and ideas from one language into another.

These examples demonstrate the uniqueness of language. No two cultures represent the world in exactly the same manner, and this cultural selectivity, or bias, is expressed in the form and content of a culture's language.

## ● THE SYMBOLIC NATURE OF CULTURE

All human beings respond to the world around them. They might decorate their bodies, make drawings on cave walls or canvases, or mold likenesses in clay. These all act as symbolic representations of their society. All complex behavior is derived from the ability to use symbols for people, events, or places. Without the ability to use symbols to create language, culture could not exist.

### Symbols and Culture

What does it mean to say that culture is symbolic? A **symbol** is *anything that represents something else and carries a particular meaning recognized by members of a culture.* Symbols need not share any quality at all with whatever they represent. Symbols stand for things simply because people agree that they do. Thus, when two or more individuals agree about the things a particular object represents, that object becomes a symbol by virtue of its shared meaning for those individuals. When Betsy Ross sewed the first American flag, she was creating a symbol.

The important point about the meanings of symbols is that they are entirely arbitrary, a matter of cultural

## DAY-TO-DAY SOCIOLOGY

### Symbols in Cyberspace

Communication involves the display of many symbols. When we communicate in e-mail, many of the gestures and facial expressions that help clarify a message are missing. In response, people have developed a host of symbols to help clarify the message. They include:

| | |
|---|---|
| :-) | smile |
| :-( | sad |
| :-0 | wow |
| \-o | bored |
| :-c | bummed out |
| :-X | my lips are sealed |
| LOL | laughing out loud |
| :-\|\| | I am angry |
| }:[ | angry, frustrated |
| :-( | I am sad |
| \|-{ | good grief |
| :*) | drunk |
| :-6 | exhausted, wiped out |
| :( | frown |
| \|-\| | asleep |
| \|^o | snoring |
| :-@ | screaming |
| ~ :-( | steaming mad |
| %-) | dazed or silly |
| %-\ | hung over |
| :-))) | Laughing uncontrollably |
| :˜/ | mixed up |
| %-{ | ironic |
| %-( | confused |
| :-C | astonished |

---

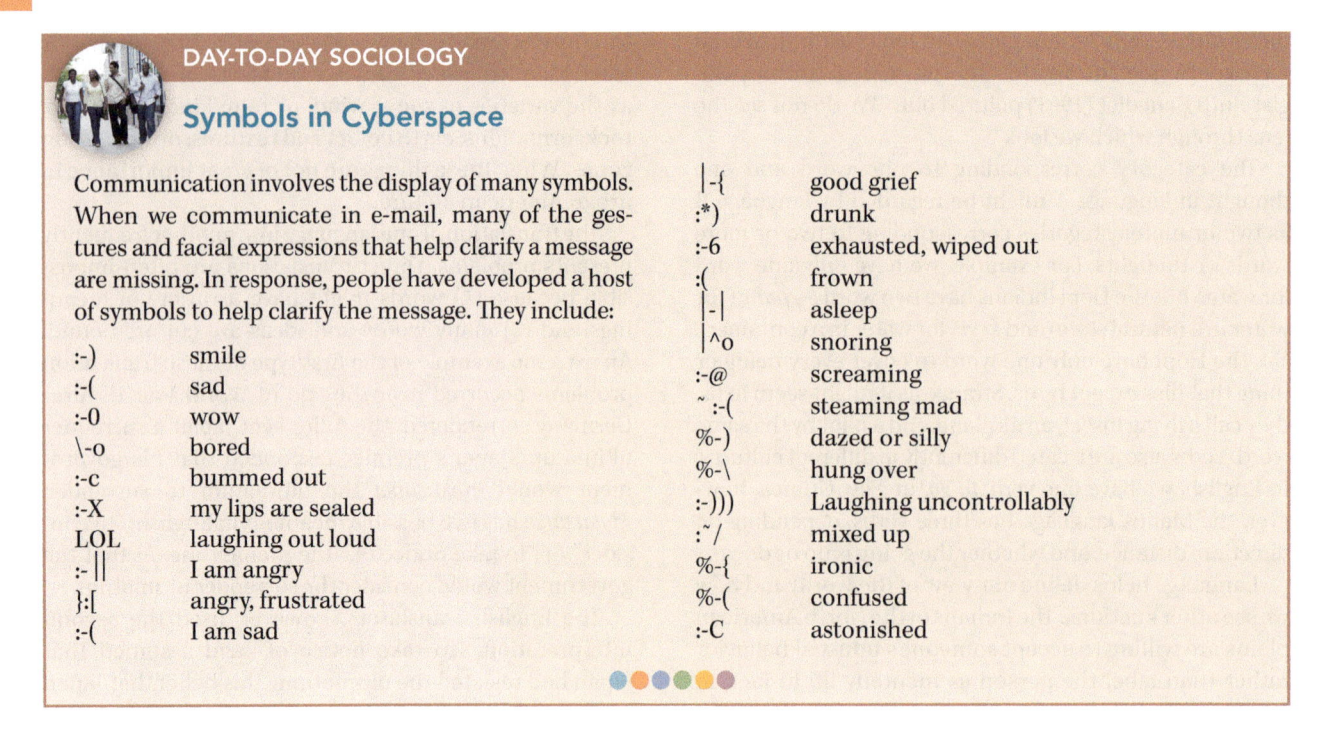

convention. Each culture attaches its own meanings to things. Thus, in the United States, mourners wear black to symbolize their sadness at a funeral. In the Far East, people wear white. In this case, the symbol is different, but the meaning is the same.

On the other hand, the same object can have different meanings in different cultures. Among the Sioux Indians, the swastika (a cross made with ends bent at right angles to its arms) was a religious symbol; in Nazi Germany, its meaning was political.

In recent years, e-mail messages have produced a whole host of symbols for commonly used expressions. (See "Day-to-Day Sociology: Symbols in Cyberspace," for examples of some of these symbols.)

Few travelers would think of going abroad without taking along a dictionary or phrase book to help them communicate with the people in the countries they visit. Although most people are aware that symbolic gestures are the most common form of cross-cultural communication, they do not realize that the language of gestures can be just as different, just as regional, and just as likely to cause misunderstanding as the spoken word can.

After a good meal in Naples, a well-meaning American tourist expressed his appreciation to the waiter by making the "A-OK" gesture with his thumb and forefinger. The waiter was shocked. He headed for the manager. The two seriously discussed calling the police and having the hapless tourist arrested for obscene behavior in a public place.

What had happened? How could such a seemingly innocent and flattering gesture have been so misunderstood? In American culture, everyone from astronauts to

People who live in Venice, Italy, have adjusted how they build their homes and live their lives to fit their watery environment.

display it to ward off evil. In Brazil, women wear it as a sign of good luck. In the United States, baseball players may use it to signal two outs, and in football the referee may use it to indicate a "second down." In Los Angeles, it is a gang symbol (Associated Press, 2005).

Looking at culture from this point of view, we would have to say that all aspects of culture—nonmaterial and material—are symbolic.

## CULTURE AND ADAPTATION

Culture probably has been part of human evolution since the time, some 15 million years ago, when our ancestors first began to live on the ground. As we have stressed throughout this chapter, humans are extraordinarily flexible and adaptable.

**Adaptation** is *the process by which human beings adjust to changes in their environment.* This adaptability, however, is not the result of being biologically fitted to the environment; in fact, human beings are remarkably unspecialized. We do not run very fast, jump very high, climb very well, or swim very far. However, we are specialized in one area: We are culture producing, culture transmitting, and culture dependent. This unique specialization is rooted in the size and structure of the human brain and in our physical ability both to speak and to use tools.

Culture, then, is the primary means by which human beings adapt to the challenges of their environment. Thus, using enormous machines, we strip away layers of the earth to extract minerals and, using other machines, we transport these minerals to yet more machines, where they are converted to a staggering number of different products. Take away all our technology, and American society would cease to exist. Take away all culture, and the human species would perish. Culture is as much a part of us as our skin, muscles, bones, and brains.

When people of one society come in contact with people of another society, cultural diffusion takes place, as evidenced by the KFC in China.

politicians signify that everything is fine by using the sign confidently in public. In France and Belgium, however, it means "You're worth zero"; in Greece and Turkey, it is an insulting or vulgar sexual invitation. In parts of southern Italy, it is an offensive and graphic reference to a part of the anatomy. No wonder the waiter was shocked.

In fact, dozens of gestures take on totally different meanings from one country to another. Is thumbs-up always a positive gesture? It is in the United States and in most of western Europe. When it was displayed by the emperor of Rome, the upright thumb gesture spared the lives of gladiators in the Coliseum. However, do not try it in Sardinia and northern Greece. There the gesture means the insulting phrase, "Up yours."

The same naiveté that can lead Americans into trouble in foreign countries also can cause problems at home. Former President Bush used a "hook 'em, 'horns" salute as a show of support for the University of Texas Longhorns marching band that was playing at his inauguration. The gesture involves raising the right hand with the index and pinky fingers raised. People in Norway were shocked because they know it as a salute to Satan.

The symbol has other meanings throughout the world. In Italy, the gesture means your wife is cheating on you. In the Mediterranean Sea, fishing boats may

### What Produces Cultural Change?

Cultural change takes place at many levels within a society. Some of the radical changes that have taken place often become obvious only in hindsight. When the airplane was invented, few people could visualize the changes it would produce. Not only did it markedly decrease the impact of distance on cultural contact, but it also had enormous impact on areas such as economics and warfare.

It is generally assumed that the number of cultural items in a society (including everything from toothpicks to structures as complex as government agencies) has a direct relation to the rate of social change. A society that has few such items will tend to have few **innovations,** *any new practice or tool that becomes widely accepted in a society.* As the number of cultural items increases, so do the innovations as well as the rate of social change.

For example, an inventory of the cultural items—from tools to religious practices—among the hunting and gathering Shoshone Indians totals a mere 3,000. Modern Americans, who also inhabit the same territory in Nevada and Utah, are part of a culture with items numbering well into the millions. Social change in American society is proceeding rapidly, whereas Shoshone culture, as revealed by archaeological excavations, appears to have changed scarcely at all for thousands of years. (For a discussion of one of the greatest influences on cultural change, see "Thinking About Social Issues: Technology Changes Culture.")

Two simple mechanisms are responsible for cultural evolution: innovation and diffusion. Innovation takes place in several ways, including recombining in new ways elements already available to a society (invention), discovering new concepts, finding new solutions to old problems, and devising and making new material objects.

**Diffusion** is *the movement of cultural traits from one culture to another.* It almost inevitably results when people from one group or society come into contact with another, as when immigrant groups take on the dress or manners of already established groups and in turn contribute new foods or art forms to the dominant culture.

Rarely does a trait diffuse directly from one culture into another. Rather, diffusion is marked by **reformulation,** *in which a trait is modified in some way so that it fits better in its new context.* This process of reformulation can be seen in the transformation of black folk blues into commercial music such as rhythm and blues and rock 'n' roll. Or consider moccasins—today's machine-made, chemically waterproofed, soft-soled cowhide shoes differ from the Native American originals and usually are worn for recreation rather than as part of basic dress, as they originally were. Sociologists would say, therefore, that moccasins are an example of a cultural trait that was reformulated when it diffused from Native American culture to industrial America.

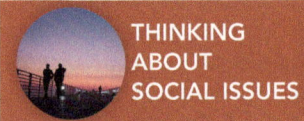

---

## THINKING ABOUT SOCIAL ISSUES

### Technology Changes Culture

People appear to have an innate need to communicate with each other. As they do so, they have the ability to change the world with the information they share. It is small wonder that repressive countries try to stop open communication, because it can bring about cultural and social change.

Today, we are more connected to each other than ever before. We visit social media sites; communicate via e-mail, texting, twitter, and instant messaging; and check news events. It seems as if the goal is not so much to be in touch, as to make sure you are not out of touch.

Changes in how people communicate are not new. With every change, people have worried and lamented that it would affect the world. After Johannes Gutenberg developed the printing press in the fifteenth century, books became available to the masses instead of just the wealthy. People became concerned that not everyone would be able to handle the new knowledge they would encounter. Prior to Gutenberg's invention, the typical book was a masterpiece. With the printing press, the belief emerged that anyone would be able to write a book that was merely entertaining instead of profound and insightful.

In 1845, the short story author Edgar Allan Poe noted, "The enormous multiplication of books . . . is one of the greatest evils of this age since it presents one of the most serious obstacles to the acquisition of correct knowledge by throwing in the reader's way piles of lumber in which he must painfully grope for the scraps of useful lumber" (Clay Shirky, 2010). This statement could very easily apply today. Poe was concerned about the proliferation of books; today we could apply this quote to the proliferation of information on the Internet.

In the mid-1800s, the postal service reduced the cost of sending a letter to 3 cents, a rate that did not change until 1958. Prior to that change, people mailed three letters a year. Afterward, the rate of letters sent increased rapidly until the average person was sending 350 letters a year (CQ Researcher, 2010).

Henry David Thoreau's book *Walden* is essentially about one man's rebellion against the complicated world coming about because of technological change. Mary Shelley's novel *Frankenstein,* published more than 150 years ago, is not just about a monster created in the laboratory, but also about a world where scientists can use technology to create havoc. Ted Kaczynski, the person known as the Unabomber, who sent letter bombs to those he thought were responsible for some of the technological changes in society, was rebelling in a vicious way against a changing world.

There is no real way to hold back the cultural change that comes from advancing technology. Some may rebel and refuse to participate, but the world will change despite their protests.

## Cultural Lag

Although the diverse elements of a culture are interrelated, some can change rapidly while others lag behind. William F. Ogburn (1964) coined the term **cultural lag** to describe *the phenomenon through which new patterns of behavior may emerge, even though they conflict with traditional values.* Ogburn observed that technological change (material culture) is typically faster than change in nonmaterial culture—a culture's norms and values—and technological change often results in cultural lag.

Consequently, stresses and strains among elements of a culture are more or less inevitable. For example, even though the Internet in general and the World Wide Web in particular offer vast educational opportunities, teachers have been slow to incorporate these technologies into the classroom. Traditional school values might be in conflict with use of the Web. Schools often assume that education is best carried out in isolation from the rest of society and that the teacher is the main guide for the students along a path to learning. Education has changed little from one hundred years ago, and we still expect teachers to talk and groups of students to listen.

The Web, however, enables the student to connect to countless sites outside of the classroom and to pursue individual educational goals. The teacher's role and influence is becoming less clear with the introduction of this technology. The teacher, instead of being in charge, must now be ready to collaborate with the student and serve as a partner in the exploration of the resources (Maddux, 1997). Traditional teacher-student roles and values are challenged in the process.

Other instances of cultural lag have considerably greater and more widespread negative effects. Advances in medicine have led to lower infant mortality and greater life expectancy, but there has been no corresponding rapid worldwide acceptance of methods of birth control. The result is a potentially disastrous population explosion in certain parts of the world.

## Animals and Culture

Do animals have culture? Many social scientists would say no. Language often is cited as the major behavioral difference between humans and animals. Humans possess language, whereas it is said animals do not. Language is the crucial ingredient in the ability to transmit culture from one generation to the next. Animals might have traits that can be socially transmitted, but they cannot benefit from the accumulation of knowledge and the ability to improve things over time. In human cultures, things change and improve from one generation to the next. People create things that are useful for survival, and these things evolve and get better, causing human culture to flourish. No single individual created something as complex and useful as a computer, but the history of advances led to its development.

Researchers have had some success teaching sign language to apes. This learning does not always translate into understanding human values and norms.

Others disagree and think animals use language in unique ways that we have overlooked. A number of experiments—the earliest dating back to the mid-1950s—have shown that apes are able to master some of the most fundamental aspects of language. Apes, of course, cannot talk. Their mouths and throats simply are not built to produce speech, and no ape has been able to approximate more than four human words. However, efforts to teach apes to communicate by other means have met with a fair amount of success.

The first and most widely known experiment in ape language research began in 1966, under the direction of Allen and Beatrix Gardner of the University of Nevada, with a chimpanzee named Washoe. This experiment consisted of teaching the chimp American sign language (ASL), the hand-gesture language used by deaf people. Washoe learned more than 200 distinct signs and was able to ask for food, name objects, and make reference to her environment. The Gardners replicated their results with four other chimpanzees.

Another experiment has involved a female gorilla named Koko. Francine Patterson has been working with Koko since 1972. Koko uses approximately 400 signs regularly and another 300 occasionally. She also understands several hundred spoken words (so much so that Patterson has to spell such words as *candy* in her presence). In addition, Koko invents signs or creates sign combinations to describe new things. She tells Patterson when she is happy or sad, refers to past and future events, defines objects, and insults her human companions by calling them such things as "dirty toilet," "nut," and "rotten stink." Once, when Patterson was drilling Koko on body parts, the gorilla signed, "Think eye ear nose boring" (Hawes, 1995).

Koko has taken several IQ tests and has recorded scores just below average for a human child—between 70 and 95 points. However, as Patterson has pointed out, the IQ tests have a cultural bias toward humans, and the gorilla may be more intelligent than the tests indicate. For example, one item instructs the child, "Point to two things that are good to eat." The choices are a block, an apple, a shoe, a flower, and an ice cream sundae. Reflecting her tastes, Koko pointed to the apple and the flower. She likes to eat flowers and has never seen an ice cream sundae. Although this answer is correct for Koko, it is only half right for humans and therefore was scored incorrect.

Some interesting work with apes has been done by Sue Savage-Rumbaugh and her colleagues, who taught a form of computer language to chimpanzees. Using special symbols, they managed to teach apes to name objects and converse with each other. An unexpected turn of events produced some interesting results with a bonobo, or pygmy chimp, named Kanzi. Rumbaugh was having no luck teaching Kanzi's adopted mother Matata to use the keyboard. During the lessons, while the mother was trying in vain to figure out what the experimenters wanted, Kanzi was spontaneously picking up on the tasks while crawling about and generally being more of a distraction than a participant.

When Rumbaugh finally gave up on Matata and turned her attention to Kanzi, assuming that, at the age of two, he might now be old enough to learn, she was shocked to find that he already knew most of what she wanted to teach him and more. Kanzi was far more adept at the language tasks than any previous chimp and developed those abilities further over the years.

Kanzi might have learned so well because he was immature at the time. It suggests the possibility that there might be in chimps, as in humans, a critical period when some special language-learning mechanism is activated. Yet, the problem with the critical learning period approach is that chimps in the wild do not learn a language. Why should a chimp whose ancestors never spoke (and who himself cannot speak) demonstrate a critical period for language learning (Deacon, 1997)?

Steven Pinker, a cognitive scientist at Harvard, believes the various animal language experiments are "exercises in wishful thinking" (Johnson, 1995). He states, "In my mind this kind of research is more analogous to the bears in the Moscow circus who are trained to ride unicycles." Johnson writes: "[Pinker] is not convinced that the chimps have learned anything more sophisticated than how to press the right buttons in order to get the hairless apes on the other side of the console to cough up M & Ms, bananas, and other tidbits of food."

Language and the production and use of tools are central elements of nonmaterial and material culture. So does it make sense to say that culture is limited to human beings? Although scientists disagree in their answers to this question, they do agree that humans have refined culture to a far greater degree than have any other animals and that humans depend on culture for their existence much more completely than do any other creatures.

## ● SUBCULTURES

To function, every social group must have a culture of its own—its own goals, norms, values, and ways of doing things. As Thomas Lasswell (1965) pointed out, such group culture is not just a "partial or miniature" culture. It is a full-blown, complete culture in its own right. Every family, clique, shop, community, ethnic group, and society has its own culture. Hence, every individual participates in a number of different cultures in the course of a day. Meeting the social expectations of various cultures is often a source of considerable stress for individuals in complex, heterogeneous societies such as ours.

Many college students, for example, find that the culture of the campus varies significantly from the culture of their family or neighborhood. At home, they might be criticized for their musical taste, their clothing, their antiestablishment ideas, and for spending too little time with the family. On campus, they might be pressured to open up their minds and experiment a little or to reject old-fashioned values.

Sociologists use the term **subculture** to refer to *the distinctive lifestyles, values, norms, and beliefs of certain segments of the population within a society.* The concept of subculture originated in studies of juvenile delinquency and criminality (Sutherland, 1924), and in some contexts, the *sub* in *subculture* still has the meaning of inferior. However, sociologists increasingly use subculture to refer to the cultures of discrete population segments within a society. The term is primarily applied to the culture of ethnic groups (Italian Americans, Jews, Native Americans, and so on) as well as to social classes (lower or working, middle, upper, and so on). Certain sociologists reserve the term *subculture* for marginal groups—that is, for groups that differ significantly from the so-called dominant culture.

## Types of Subcultures

Several groups have been studied at one time or another by sociologists as examples of subcultures. These can be classified roughly as follows.

*Ethnic Subcultures* Many immigrant groups have maintained their group identities and sustained their traditions even while adjusting to the demands of the wider society. Although originally distinct and separate cultures, they have become American subcultures. America's newest immigrants, Asians from Vietnam, Korea, Japan, the Philippines, Taiwan, India, and Cambodia, have maintained their values by living together in tight-knit communities in New York, Los Angeles, and other large cities, simultaneously encouraging their children to achieve success by American terms.

*Occupational Subcultures* Certain occupations seem to involve people in a distinctive lifestyle even beyond their work. For example, New York's Wall Street is not only the financial capital of the world; it is identified with certain values such as materialism, greed, or power. Construction workers, police, entertainers, and many other occupational groups involve people in distinctive subcultures.

*Religious Subcultures* Certain religious groups, although continuing to participate in the wider society, nevertheless practice lifestyles that set them apart. These include Christian evangelical groups, Mormons, Muslims, Jews, and many religious splinter groups. Sometimes the lifestyle might separate the group from the culture as a whole as well as from the subculture of its immediate community. In a drug-ridden area of Brooklyn, New York, for example, a group of Muslims follows an antidrug creed in a community filled with addicts and dealers. Their religious beliefs set them apart from the general society, and their attitude toward drugs separates them from many other community members.

*Political Subcultures* Small, marginal political groups can so involve their members that their entire way of life is an expression of their political convictions. Often, these so-called left-wing and right-wing groups reject much of what they see in American society but remain engaged in society through their constant efforts to change it to their liking.

*Geographic Subcultures* Large societies often show regional variations in culture. The United States has several geographical areas known for their distinctive subcultures. For instance, the South is known for its leisurely approach to life, its broad dialect, and its hospitality. The North is noted for Yankee ingenuity, commercial cunning, and a crusty standoffishness. California is known for its trendy and ultra-relaxed, or laidback, lifestyle. And New York City stands as much for a driven, elitist, arts and literature–oriented subculture as for a city.

*Social Class Subcultures* Although social classes cut horizontally across geographical, ethnic, and other subdivisions of society, to some degree it is possible to discern cultural differences among the classes. Sociologists have documented that linguistic styles, family and household forms, and values and norms applied to child rearing are patterned in terms of social class subcultures. (See Chapter 8, "Social Class in the United States," for a discussion of social class in the United States.)

*Deviant Subcultures* As mentioned earlier, sociologists first began to study subcultures as a way of explaining juvenile delinquency and criminality. This interest expanded to include the study of a wide variety of groups that are marginal to society in one way or another and whose lifestyles clash with that of the wider society in important ways. Some of the deviant subcultural groups studied by sociologists include prostitutes, strippers, pool hustlers, pickpockets, drug users, and a variety of criminal groups.

## UNIVERSALS OF CULTURE

In spite of their individual and cultural diversity, their many subcultures and countercultures, human beings are members of one species with a common evolutionary heritage. Therefore, people everywhere must confront and resolve certain common, basic problems such as maintaining group organization and overcoming difficulties originating in their social and natural environments. **Cultural universals** are *certain models or patterns that have developed in all cultures to resolve common problems.*

Among those universals that fulfill basic human needs are the division of labor, the incest taboo, marriage, family organization, rites of passage, and ideology. It is important to keep in mind that although these forms are universal, their specific contents are particular to each culture.

### The Division of Labor

Many primates live in social groups in which it is typical for each adult group member to meet most of his or her own needs. The adults find their own food, prepare their own sleeping places, and, with the exceptions of infant care, mutual grooming, and some defense-related activities, generally fend for themselves.

This is not true of human groups. In all societies—from the simplest bands to the most complex industrial nations—groups divide the responsibility for completing necessary tasks among their members. This means that humans constantly must rely on one another; hence, they are the most cooperative of all primates.

## Marriage, the Family, and the Incest Taboo

All human societies regulate sexual behavior. Sexual mores vary enormously from one culture to another, but all cultures apparently share one basic value: sexual relations between parents and their children are to be avoided. (There is evidence that some primates also avoid sexual relations between males and their mothers.) In most societies, it is also wrong for brothers and sisters to have sexual contact (notable exceptions being the brother-sister marriages among royal families in ancient Egypt and Hawaii and among the Incas of Peru). *Sexual relations between family members* is called **incest,** and because in most cultures very strong feelings of horror and revulsion are attached to incest, it is said to be forbidden by taboo. A **taboo** is *the prohibition of a specific action.*

The presence of the incest taboo means that individuals must seek socially acceptable sexual relationships outside their families. All cultures provide definitions of who is or is not an acceptable candidate for sexual contact. They also provide for institutionalized marriages—ritualized means of publicly legitimizing sexual partnerships and the resulting children. Thus, the presence of the incest taboo and the institution of marriage result in the creation of families. Depending on who is allowed to marry—and how many spouses each person is allowed to have—the family will differ from one culture to another.

The basic family unit consisting of husband, wife, and children (called the nuclear family) seems to be a recognized unit in almost every culture, and sexual relations among its members (other than between husband and wife) are almost universally taboo. For one thing, this helps keep sexual jealousy under control. For another, it prevents the confusion of authority relationships within the family. Perhaps most important, the incest taboo ensures that family offspring will marry into other families, thus re-creating in every generation a network of social bonds among families that knits them together into larger, more stable social groupings.

## Rites of Passage

All cultures recognize stages through which individuals pass in the course of their lifetimes. Some of these stages are marked by biological events such as the start of menstruation in girls. However, most of these stages are quite arbitrary and culturally defined. All such stages—whether or not corresponding to biological events—are meaningful only in terms of each group's culture. Rarely do individuals drift from one such stage to another; every culture has established **rites of passage** or *standardized rituals marking major life transitions.*

The most widespread—if not universal—rites of passage are those marking the arrival of puberty (often resulting in the individual's taking on adult status),

Certain patterns of behavior, such as marriage, are found in every culture but take various forms. The marriage ceremony, for example, varies greatly among different cultures.

marriage, and death. Typical rites of passage celebrated in American society include baptisms, bar and bat mitzvahs, confirmations, major birthdays, graduation, wedding showers, bachelor parties, wedding ceremonies, major anniversaries, retirement parties, and funerals and wakes. Such rites accomplish several important functions, including helping the individual achieve a sense of social identity, mapping out the individual's life course, and aiding the individual in making appropriate life plans. Finally, rites of passage provide people with a context in which to share common emotions, particularly with regard to events that are sources of stress and intense feelings such as marriage and death.

## Ideology

A central challenge that every group faces is how to maintain its identity as a social unit. One of the most important ways groups accomplish this is by promoting beliefs and values to which group members are firmly committed. Such **ideologies,** or *strongly held beliefs and values*, are the cement of social structure.

Every culture contains ideologies. Some are religious, referring to things and events beyond the perception of the human senses. Others are more secular—that is, nonreligious and concerned with the everyday world. In the end, all ideologies rest on untestable ideas rooted in the basic values and assumptions of each culture.

## HOW SOCIOLOGISTS DO IT

### Social Science in a War Zone

Southern Afghanistan is a dangerous area where heavily armed American troops move very carefully. A platoon was forming a protective circle around Paula Loyd, a researcher who was interviewing villagers in a local market about the price of cooking fuel. Fuel price appears to be related to whether supply lines have been hijacked by the Taliban.

At the end of one interview, the man she was speaking to thanked her profusely and then lit his jug of fuel on fire and threw it on Loyd. A colleague, Don Ayala, was so outraged by the incident that he shot the man. He was eventually convicted of voluntary manslaughter. Loyd spent two months fighting for her life and died in January 2009.

In February 2007, the army launched the Human Terrain program after general David Petraeus became convinced that the military needed a better cultural understanding of life in Afghanistan. Under the Human Terrain program, social scientists embedded with combat brigades conduct research and provide cultural information to military commanders and staff. United States commanders have been happy with the program and believe it has reduced combat operations by 60 percent.

No sooner had the Human Terrain program been launched than the American Anthropological Association announced that the program violates the group's code of ethics. In February 2008, the Association voted to prohibit this type of research because it is not made public, does not protect the subjects from being forced to participate, and does not guarantee that the subjects will not be harmed. The Network of Concerned Anthropologists sent a letter to Congress in 2010 in which it questioned the effectiveness of the program and called it "dangerous and reckless" and a "waste of taxpayers' money." In addition to Paula Loyd, two other social scientists working for the program have been killed. A bomb killed Nicole Suveges in Afghanistan and Michael Bhatia in Iraq.

When Loyd, Suveges, and Bhatia joined the Human Terrain program, they were well aware of the controversy surrounding it. Yet, they believed that when they were conducting interviews and writing reports about the local population, they were acting as social scientists, not as soldiers. Other researchers are skeptical about whether any useful research can be done in such a hostile environment and whether it is possible for the population to see the embedded social scientists as noncombatants. Critics of the program fear that, no matter how noble the goal, such work for the military could inadvertently cause all researchers to be viewed as intelligence gatherers who care little about the population.

Montgomery McFate, the Human Terrain program's senior social science adviser, has dismissed criticism of scholars working with the military. "I'm frequently accused of militarizing anthropology," she said. "But we're really anthropologizing the military."

*Sources:* Rohde, David. October 5, 2007. "Army Enlists Anthropology in War Zones." *New York Times*, pp. A1, A14; Stockman, Farah. February 12, 2009. "Anthropologist's War Death Reverberates." *Boston Globe;* Hodges, Jim. March 22, 2012. "Cover Stoy: U.S. Army's Human Terrain Experts May Help Defuse Future Conflicts." *Defense News* (www.defensenews.com/article/20120322/C4ISR02/303220015/Cover-Story-U-S-Army-8217-s-Human-Terrain-Experts-May-Help-Defuse-Future-Conflicts), accessed May 1, 2012.

---

Even though ideologies rest on such assumptions, however, their consequences are very real. Ideologies give direction and thrust to our social existence and meaning to our lives. The power of ideologies to mold passion and behavior is well known. History is filled with both horrors and noble deeds people have performed in the name of some ideology: thirteenth-century Crusaders, fifteenth-century Inquisitors, pro–states' rights and pro-union forces in nineteenth-century America, abolitionists, prohibitionists, trade unionists, Nazis and fascists, communists, segregationists, civil rights activists, feminists, consumer activists, environmentalists. These and countless other groups have marched behind their ideological banners, and, in the name of their ideologies, they have changed the world, often in major ways.

(For a discussion of ideological conflicts that can occur during military interventions, see "How Sociologists Do It: Social Science in a War Zone.")

### ● CULTURE AND INDIVIDUAL CHOICE

Very little human behavior is instinctual or biologically programmed. In the course of human evolution, genetic programming gradually was replaced by culture as the source of instructions about what to do, how to do it, and when it should be done. This means that humans have a great deal of individual freedom of action—more than any other creature.

However, as we have seen, individual choices are not entirely free. Simply by being born into a particular society with a particular culture, every human being is presented with a limited number of recognized or socially valued choices. Every society has means of training and social control that are brought to bear on each person, making it difficult for individuals to act or even think in ways that deviate too far from their

> ### HOW SOCIOLOGISTS DO IT
>
> ### The Conflict Between Being a Researcher and Being a Human Being
>
> Sociologists and anthropologists are supposed to understand the importance of cultural relativism and realize that cultures must be studied and understood on their own terms before valid comparisons can be made. A researcher should avoid imposing his or her values on a people or interfering with a culture to such an extent that it moves away from its origins. Is this a realistic goal, or are we being overly idealistic when we think this can be accomplished?
>
> Kenneth Good had to deal with these issues often when he studied the Yanomama. The Yanomama are approximately 10,000 South American Indians who live in 125 villages in southern Venezuela and northern Brazil.
>
>> In terms of their material and technological culture, the Yanomama stand out in their primitiveness. They have no system of numbers ... they have not invented the wheel. They know nothing of the art of metallurgy. Until recently they made fire ... [by] rubbing two sticks together. (Good, 1991)
>
> Traditionally, the Yanomama do not wear clothing, but they paint red designs on their bodies. "Girls and women adorn their faces by inserting slender sticks through holes in the lower lip at either side of the mouth and in the middle, and through the pierced nasal septum." Their lives are characterized by persistent aggression among village members and perpetual warfare with other groups. They engage in club fighting, gang rape, and murder.
>
> Good encountered many events that went against his cultural value system. One day, he saw two groups of tribespeople having a tug of war.
>
>> But instead of pulling on a thick vine, they were pulling on a woman.... Her assailants on one side were three of the wilder teenage [boys]. Trying to pull her away from them were three elderly women. The tug of war went on for 10 minutes or so while I watched, my blood rising as instinct told me to put a stop to it.... "What are they doing to her?" I asked.... "They're going to rape her," came the answer as casually as if she had said, "They're going to have a picnic."
>
>> With a concerted heave, the teenagers pulled her free.... Howling in victory, they ran down the trail, yanking her along. As they ran, they were joined by more shouting teenagers. I followed behind as the stampede bore her into the jungle.
>
>> I stood there, my heart pounding. I had no doubt I could scare these kids away. On the other hand, I was an anthropologist. I wasn't supposed to take sides and make value judgments. This kind of thing went on. If a woman showed up somewhere unattached, chances were she'd be raped. She knew it, they knew it. It was expected behavior. What was I supposed to do, I thought, try to inject my standards of morality? I hadn't come down here to change these people or because I thought I'd love everything they did; I'd come to study them.
>
>> So why was I standing there shaking with anger? Why was I thinking, "Come on, Ken, what's wrong with you? Are you going to stand around with your notebook in your hand and observe a gang rape in the name of anthropological science?"... How could I live with a group of human beings and not be involved with them as a fellow human being?
>
> That afternoon was a turning point for Good. About a month later, there was another woman-dragging episode.
>
>> After half an hour ... one of the other men got fed up with the noise.... "That's enough," he said, picking up his arrows. "This is really annoying. I'm going to stab her; that will put a stop to it." I watched as he walked up to the three of them with his arrows. He was really going to do it.
>
>> When I saw this, I yelled, "Don't do that!" [He] stopped and looked around, surprised. Our eyes met, then he walked back to his hammock and put his arrows down. I knew that I was not the same detached observer I had been before. (Good, 1991)
>
> Good was scheduled to spend 15 months with the Yanomama. After this time passed, however, he did not leave. Instead, he stayed sixteen years and learned to speak their language and hunt fish and gather as they did. He learned what it meant to be a nomad.

culture's norms. To get along in society, people must keep their impulses under some control and express feelings and gratify needs in a socially approved manner at a socially approved time. (See "How Sociologists Do It: The Conflict Between Being a Researcher and Being a Human Being.") This means that humans inevitably feel somewhat dissatisfied, no matter to which group they belong. Coming to terms with this central truth about human existence is one of the great tasks of living.

This Yanomami woman is having her face decorated for an upcoming celebration.

Later, he was adopted into the lineage of the village and given a wife, Yarima, according to Yanomama custom and in keeping with the wishes of the tribal leader. At the time, Yarima was a child and Good was 34 years old. Good formally married her when she was approximately fourteen years old. Yarima had a profound effect on him.

> I'm in love. Unbelievable, intense emotion, almost all the time. In the morning when she gets up to start the day, when I see her come in from the gardens with a basket of plantains, especially when we make love. Sure it's universal, except that being in love in Yanomama culture with a Yanomama girl is different, a different game, different rules.
>
> In my wildest dreams it had never occurred to me to marry an Indian woman in the Amazon jungle. I was from suburban Philadelphia. I had no intention of going native. . . . That was what I had come to, after all these years of struggling to fit into the Yanomama world, to speak their language fluently, to grasp their way of life from the inside. My original purpose—to observe and analyze this people . . . had slowly merged with something far more personal.

Eventually, Kenneth Good brought Yarima out of the Amazon jungle to the United States, where she lived from 1988 to 1993. Yarima had never seen flat cleared land, much less houses and cars. She had never worn clothes or walked in shoes.

They had three children. On a trip back to the Yanomama as part of a National Geographic documentary, Yarima decided to remain. Good was horrified and went back to the United States.

Good's last attempt to persuade his wife to rejoin him in the West nearly succeeded. Lengthy negotiations were held with Yarima's brother, and she eventually agreed to accompany Good to a jungle landing strip where a plane was waiting to fly them out. At the last minute, she decided she could not do it and ran back into the jungle.

Kenneth Good returned to New Jersey, where he lives with their three children. Was Good doing valuable research, or did he impose his values on another culture?

*Sources:* Good, Kenneth, with D. Chanoff. 1991. *Into the Heart.* New York: Simon & Schuster; and an interview with David Chanoff. July 1994; Tsertishvili, Tamar. August 8, 2010. "A Love Story" (http://tamartsertishvili.blogspot.com/2010/08/yarima-and-ken-good.html#!/2010/08/yarima-and-ken-good.html), accessed March 5, 2012.

## SUMMARY

▸ **Why is culture important for society?**

Humans are remarkably unspecialized; culture allows us to adapt quickly and flexibly to the challenges of our environment. Due to a lack of instinctual or biological programming, humans have a great deal of flexibility and choice in their activities. Individual freedom of action is limited, however, by the existing culture.

▸ **How does culture make it possible for human societies to vary so much?**

Culture consists of all that human beings learn to do, to use, to produce, to know, and to believe as they grow to maturity and live out their lives in the social groups to which they belong. All human societies have complex ways of life that differ greatly from one to the other. Each society has its own unique blueprint for living, or culture.

▸ **What is the difference between ethnocentrism and cultural relativism?**

People often make judgments about other cultures according to the customs and values of their own, a practice sociologists call ethnocentrism. To avoid ethnocentrism in their own research, sociologists are guided by the concept of cultural relativism, the recognition that social groups and cultures must be studied and understood on their own terms before valid comparisons can be made.

▸ **What is the difference between material and nonmaterial culture?**

Sociologists view culture as having two major components: material culture and nonmaterial culture. Material culture consists of human technology—all the things human beings make and use, from small, handheld tools to skyscrapers. Every society also has a nonmaterial culture, which consists of the totality of knowledge, beliefs, values, and rules for appropriate behavior.

▸ **How does language shape our perception and classification of the world?**

Language and the production of tools are central elements of culture. Evidence exists that animals engage, or can be taught to engage, in both of these activities. This does not mean that they have culture. Humans have refined culture to a far greater degree than other animals and are far more dependent on it for their existence.

▸ **Do animals have language?**

Language is the crucial ingredient in the ability to transmit culture from one generation to the next. Animals might have traits that can be socially transmitted, but they cannot benefit from the accumulation of knowledge and the ability to improve things over time.

▸ **What are subcultures?**

Sociologists use the term *subculture* to refer to the distinctive lifestyles, values, norms, and beliefs associated with certain segments of the population within a society. Types of subcultures include ethnic, occupational, religious, political, geographical, social class, and deviant subcultures.

▸ **What are cultural universals?**

Cultural universals are certain models or patterns that have developed in all cultures to resolve common problems. Among them are the division of labor, the incest taboo, marriage, family organization, rites of passage, and ideology. Although the forms are universal, the content is unique to each culture.

▸ **Why is the division of labor important?**

People in all societies must confront and resolve certain common, basic problems. By dividing the responsibility for completing necessary tasks among their members, societies create a division of labor.

▸ **What are rites of passage?**

Every culture has established rites of passage, standardized rituals marking major life transitions. Ideologies, or strongly held beliefs and values, are the cement of social structure in that they help a group maintain its identity as a social unit.

## Media Resources

**CourseMate for *Introduction to Sociology*, Eleventh Edition**

Cengage Learning's Sociology CourseMate brings course concepts to life with interactive learning, study, and exam preparation tools that support the printed textbook. Access an integrated eBook, learning tools including glossaries, flashcards, quizzes, videos, and more in your Sociology CourseMate. Go to CengageBrain.com to register or purchase access.

CHAPTER 3 CULTURE   71

# CHAPTER THREE STUDY GUIDE

## Key Concepts

*Match each of the following concepts with its definition, illustration, or explanation.*

| a. Culture | j. Ideal norms | s. Real norms |
|---|---|---|
| b. Culture shock | k. Real norms | t. Ideal norms |
| c. Ethnocentrism | l. Values | u. Selectivity |
| d. Cultural relativism | m. Sapir-Whorf hypothesis | v. Symbol |
| e. Material culture | n. Diffusion | w. Cultural universals |
| f. Nonmaterial culture | o. Reformulation | x. Taboo |
| g. Norms | p. Cultural lag | y. Rites of passage |
| h. Mores | q. Subculture | z. Ideology |
| i. Folkways | r. Rites of passage | |

____ 1. Strongly held norms that usually have important moral implications
____ 2. The conflict between cultural ideas and new patterns of behavior, especially those that arise because of technological innovation
____ 3. Expectations of what people should do under perfect conditions
____ 4. A pattern for living shared by the members of a society
____ 5. Cars, computers, and screwdrivers; anything a human produces
____ 6. Rules, often unwritten, for everyday behavior
____ 7. The movement of cultural traits from one culture to another
____ 8. Standardized rituals marking major life transitions
____ 9. Judging another culture by standards of one's own culture
____ 10. The ideas—knowledge, beliefs, values, and rules for appropriate behavior—shared by members of society
____ 11. The modification of a cultural trait adopted from another culture so that it fits better with one's own
____ 12. Withholding judgment and seeking to understand other societies on their own terms
____ 13. Norms or customs that permit a wide degree of individual interpretation
____ 14. The difficulty people have adjusting to a new culture that differs markedly from their own
____ 15. The distinctive lifestyle or culture of certain segments of the population within a society
____ 16. Shared ideas of what is better
____ 17. Norms that are expressed with qualifications and allowances for differences in individual behavior
____ 18. The language a person uses shapes his or her perception of reality
____ 19. Strongly held but untestable ideas, shared by members of a society, that uphold the basis of that society
____ 20. Something that represents something else and whose meaning is understood by the members of a culture
____ 21. Expectations of what people would do under ideal conditions
____ 22. Rules that allow for differences in individual behavior and describe the way people actually behave
____ 23. Rituals that mark the transition from one stage of life to another
____ 24. The process by which a culture emphasizes some aspects of reality and ignores others
____ 25. A very strongly held prohibition against some specific action
____ 26. Patterns or models that have developed in all cultures to resolve common problems.

## Central Idea Completions

*Fill in the appropriate concepts and descriptions for each of the following questions.*

1. Cell phones, e-mail, twitter, and text messaging are a fairly recent phenomenon. The friction over their use indicates some cultural lag as people try to apply older norms. Do norms for contemporary technology

differ among groups? In some particular situation, what values does cell phone use contradict? What values support the use of this cellular technology? How do you think norms will change in response to cellular technology with its capacity for conversation, Internet access, pictures, text messages, and so on?

   a. Norm difference _____

   b. Values that conflict with cellular technology use _____

   c. Values that support cellular technology use _____

   d. Likely change in norms _____

2. Is there a youth culture in the United States? What differences in values and norms are there between younger people (14–25) and people 26 or more years older?

   a. Differences in norms _____

   b. Differences in values _____

3. What subcultures exist at your school? How do norms and values differ among these subcultures?

4. In recent years, some African cultures have been criticized for practicing female circumcision (the surgical removal of some of the labia or clitoris). Describe both the ethnocentric and cultural relativist reaction to this practice.

   a. Ethnocentrism _____

   b. Cultural relativism _____

## Critical Thought Exercises

1. Tischler discusses the concepts of ideal and real norms. How do the concepts of ideal and real norms apply at your college or university?

2. Discuss the question of whether a culture clash exists between the United States and Middle Eastern societies.

3. In American politics, people often refer to culture wars, and cultural differences between red states and blue states. What values seem to be in dispute in these "wars"? How do these values affect what people actually do?

4. In recent years, America has witnessed an increasing number of "honor-killings" in which relatives have murdered those they suspected of becoming too "Americanized." Read the following material, which describes how these tragedies are growing increasingly prevalent in America: www.cbsnews.com/8301-504083_162-57409395-504083/honor-killing-under-growing-scrutiny-in-the-u.s/. What do you believe could be done to eliminate this tragic phenomenon?

## Internet Activity

The World Values Survey at www.worldvaluessurvey.org/ has a chart showing the location of countries on two broad dimensions, traditional-secular and survival–self-expression. It also tracks changes in happiness in several countries. Compare the United States with China on these dimensions.

## Answers to Key Concepts

1. h;  2. p;  3. j;  4. a;  5. e;  6. g;  7. n;  8. r;  9. c;  10. f;  11. o;  12. d;  13. i;  14. b;  15. q;  16. l;  17. k;  18. m;  19. z;  20. v;  21. t;  22. s;  23. y;  24. u;  25. x;  26. w

# 4 Socialization and Development

**Becoming a Person: Biology and Culture**
 Nature Versus Nurture: A False Debate
 Sociobiology
*Thinking About Social Issues:* Can Socialization Make a Boy into a Girl?
 Deprivation and Development
 The Concept of Self
 Dimensions of Human Development

**Theories of Development**
 Charles Horton Cooley
 George Herbert Mead
 Sigmund Freud
 Erik H. Erikson

**Early Socialization in American Society**
 The Family
 The School
*Day-to-Day Sociology:* Does Day Care Create Unruly Brats?
 Peer Groups

*Our Diverse World:* Win Friends and Lose Your Future: The Costs of Not "Acting White"
 Television, Movies, and Video Games
*Day-to-Day Sociology:* Television Made You the Designated Driver

**Adult Socialization**
 Marriage and Responsibility
 Parenthood
 Career Development: Vocation and Identity
*Our Diverse World:* Could You Be a Success at a Japanese Company?

**Aging and Society**

**Summary**

# LEARNING OBJECTIVES

After studying this chapter, you should be able to do the following:

- Describe socialization.
- Explain primary socialization.
- Discuss how biology and socialization contribute to the formation of the individual.
- Describe how people develop a social identity.
- Know what sociobiology is.
- Explain how extreme social deprivation affects early childhood development.
- Explain the views of Charles Horton Cooley and George Herbert Mead.
- Describe Erik Erikson's model of lifelong socialization.
- Explain how family, schools, peer groups, and the mass media contribute to childhood socialization.
- Know how adult socialization differs from primary socialization.
- Identify where resocialization takes place.

Peter Moskos is a Harvard-trained sociologist who spent more than a year as police officer in Baltimore. The area he was assigned to was 97 percent African American and had a 37 percent poverty rate. The area had high rates of crime, drug abuse, and blight.

At the police academy, Moskos learned to respect the chain of command and that he was at the bottom of that chain. He needed to say "sir" often, salute, and follow orders. He also learned that there was a "right way" and a "wrong way" to do things. Some instructors went so far as to say "This is what they want me to teach you, but you'll see how things are done soon enough." As a sociologist, he knew to refer to these as the formal and informal rules. He learned that officially he could not frisk a suspect for drugs, only for weapons which could be a threat to the officer. If, however, in the process of frisking for a weapon, he discovered drugs, that was a legal search.

According to the Fourth Amendment, police officers must have probable cause for every search and arrest. He learned the way to deal with this is to proceed with a "stop," which requires only "reasonable suspicion." It was important to say, "Do you mind coming over here?" verses "Come here." Moskos learned there was a difference between creative writing and lying in police reports. Creative writing meant using what you had to make your arrest stand up in court. The prosecutor was going to review your report to see whether there was a reasonable suspicion for the stop and a probable cause for the arrest.

By the time his job was over, Moskos understood that police officers see themselves as different from society morally and culturally. The danger of the job created a unique bond among the officers. The police experience had modified their previous years of socialization enough that they now felt most comfortable with others who were also part of that culture. As a police officer, Moskos was willing to risk his life for a fellow officer, and they would do it for him. Later, as a professor at John Jay College of Criminal Justice, he felt he had great colleagues, but none that he would die for. He understood that he had undergone a unique socialization experience (Moskos, 2008).

The process of socialization begins at birth. A baby is helpless. It cannot walk or talk. Somebody has to take care of its every need. How does the baby get the care it needs? It smiles, makes sounds, and does cute things. The baby is developing social skills that are at the heart of what makes us human. Through its ability to get the attention and care of its mother and others, the baby promotes many of the behaviors and emotions that we prize in ourselves and that often distinguish us from animals.

Human babies are dependent for such a long time that humans have had to develop a model of childrearing that is different from that of the apes. Chimpanzee and gorilla mothers are capable of rearing their offspring pretty much through their own powers, but human mothers are not. To raise a child, many people need to be involved. The adults need to cooperate, and the baby needs to respond and interact with them. An elaborate social network needs to develop.

This *process of social interaction that teaches the child the intellectual, physical, and social skills needed to function as a member of society* is called **socialization.** Through socialization experiences, children learn the culture of the society into which they have been born. In the course of this process, each child slowly acquires a **personality**—that is, *the patterns of behavior and ways of thinking and feeling that are distinctive for each individual.* Contrary to popular wisdom, nobody is a born salesperson, criminal, or military officer. These things all are learned and modified as part of the socialization process.

## BECOMING A PERSON: BIOLOGY AND CULTURE

Every human being is born with a set of **genes,** *inherited units of biological material.* Half are inherited from the mother, half from the father. No two people except identical twins have exactly the same genes. Genes are made up of complicated chemical substances, and a full set of genes is found in every body cell. The Human Genome Project has found that humans have about 30,000 genes—barely twice the number of a fruit fly.

Genes influence the chemical processes in our bodies and even control some of these processes completely. For example, such things as blood type, the ability to taste the presence of certain chemicals, and some people's inability to distinguish certain shades of green and red are completely under the control of genes. Most of our body processes, however, are not controlled solely by genes but are the result of the interaction of genes and the environment (physical, social, and cultural). Thus, how tall you are depends on the genes that control the growth of your legs, trunk, neck, and head as well as the amount of protein, vitamins, and minerals you consume in your diet. Genes help determine your blood pressure, but so do the amount of salt in your diet, the frequency with which you exercise, and the amount of stress under which you live.

## Nature Versus Nurture: A False Debate

For more than a century, sociologists, educators, and psychologists have argued about which is more important in determining a person's qualities: inherited characteristics (nature) or socialization experiences (nurture). After Charles Darwin (1809–1882) published *On the Origin of Species* in 1859, human beings were seen to be a species similar to all the others in the animal kingdom. Because most animal behavior seemed to the scholars of that time to be governed by inherited factors, they reasoned that human behavior similarly must be determined by instincts—biologically inherited patterns of complex behavior. Instincts were seen to lie at the base of all aspects of human behavior, and eventually, researchers cataloged more than 10,000 human instincts (Bernard, 1924).

Then, at the turn of the century, a Russian scientist named Ivan Pavlov (1849–1936) made a startling discovery. He found that if a bell were rung just before dogs were fed, eventually they would begin to salivate at the ringing of the bell itself, even when no food was served. The conclusion was inescapable: So-called instinctive behavior could be molded or, as Pavlov (1927) put it, conditioned through a series of repeated experiences linking a desired reaction with a particular object or event. Dogs could be taught to salivate.

Pavlov's work quickly became the foundation on which a new view of human beings was built, one that stressed their infinite capacity to learn and be molded. The American psychologist John B. Watson (1878–1958) taught a little boy to be afraid of a rabbit by startling him with a loud noise every time he was allowed to see the rabbit. What he had done was link a certain reaction (fear) with an object (the rabbit) through the repetition of the experience. Watson eventually claimed that if he were given complete control over the environment of a dozen healthy infants, he could train each one to be whatever he wished—"doctor, lawyer, artist, merchant, even beggar or thief" (Watson, 1925). Among certain psychologists, conditioning became the means through which they explained human behavior.

Sociologists believe humans are unique because they are the only life form able to accumulate knowledge,

Children learn the culture of the society into which they have been born through the socialization process.

improve on it, and pass it on to the next generation. Human society would not be able to advance without this ability. This fact has made it possible for humans to develop societies throughout the world in a wide variety of climates and settings. Genes alone do not make this possible, but the accumulated knowledge from others in the community does (Richerson and Boyd, 2004). (For a discussion of how much of gender is determined by socialization, see "Thinking About Social Issues: Can Socialization Make a Boy into a Girl?")

## Sociobiology

The debate over nature versus nurture took a different turn when Edward O. Wilson published his book *Sociobiology: The New Synthesis* (1975). The discipline of sociobiology, also known as evolutionary psychology or human behavioral ecology, uses biological principles to explain the behavior of all social beings, both animal and human. The study of the biological basis of social behavior in animals had long been accepted by biologists. Wilson, in the twenty-seventh and final chapter of *Sociobiology*, applied the theories of sociobiology to human beings.

According to Wilson (1975), certain behavioral traits are inherited due to the process of natural selection, and they help humans survive. These traits commonly are called human nature and include the division of labor between the sexes, bonding between parents and children, heightened altruism toward closest kin, incest avoidance, suspicion of strangers, dominance orders within groups, male dominance overall, and territorial aggression over limited resources. Although people have free will, our genes make certain behaviors more likely than others. Although cultures can vary, they all begin to lean in certain common directions.

When an especially harsh and prolonged winter leaves an Inuit (Eskimo) family without food supplies,

## THINKING ABOUT SOCIAL ISSUES

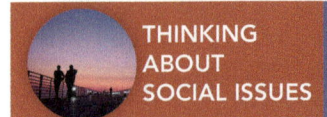

### Can Socialization Make a Boy into a Girl?

In the 1950s and 1960s, social scientists believed that even though male and female children had different genitals, they were born neutral as far as their socialization to gender was concerned. One of the main proponents of this view was Dr. John Money.

In 1963, two healthy twin boys were born. Seven months later, the twin named David Reimer lost his entire penis in a botched circumcision. As a result of the irreparable injury, his parents took him to the renowned Johns Hopkins Hospital in Baltimore. There, Dr. Money persuaded the parents to allow their son to have a surgical sex change. The process involved genital surgery, followed by a 12-year program of social, mental, and hormonal therapy to complete the metamorphosis.

The case was reported in the medical literature as an unqualified success. In 1973, *Time* magazine reported that the case "provides strong support... that conventional patterns of masculine and feminine behavior can be altered. It also casts doubt on the theory that major sex differences, psychological as well as anatomical, are immutably set by the genes at conception." For twenty years, this case was cited in numerous sociology, psychology, and human sexuality textbooks. Researchers on these topics used the case as proof that gender is very much the product of socialization and that if intervention occurs early enough, a child may be raised in a gender that is different from that with which he or she was born. As Money noted, "the gender identity gate is open at birth.... [I]t stays open at least for something over a year." The case also served as a model of how to treat infants with ambiguous genitalia. Many biological boys had operations that turned them into girls, and many girls had procedures that turned them into boys.

The one dissenting voice during this period was Milton Diamond, a young graduate student at the University of Kansas. Diamond was not convinced that there was any reason to believe John Money's theory of psychosexual neutrality in children. He published an article on the topic, and then he contacted Money and offered to do a joint study with him. Money, a respected researcher, saw no reason to pay any attention to the unknown Diamond.

It now appears that Diamond was correct and Money woefully misguided. The view that gender can be changed through socialization and surgery appears to be decidedly wrong. After further investigation, it turned out that the twin David never adjusted to being a girl. He steadfastly refused to grow into a woman and stopped taking the estrogen pills that were prescribed for him at age 12. David also refused to undergo additional surgery that the physicians tried to persuade him he needed to fully become a woman. At 14, he found other physicians who were sympathetic to his plight. He underwent a double mastectomy and a phalloplasty and began a regimen of male hormones. He refused to ever return to Johns Hopkins University where the original diagnosis and surgery had taken place. After David's return to the male gender identity, he felt that his attitudes, behavior, and body were once again united. At 25, David married a woman several years older and adopted her children. He died at an early age.

Many of the people who received these gender-changing operations in early life because of the gender-neutral hypothesis have experienced difficulties. They have formed support groups, and some have undergone surgery to reverse the process. Physicians now reject John Money's views on gender identity and no longer treat infants with ambiguous genitalia the way they did thirty-five years ago.

The current thinking is that people are not gender-neutral at birth and that we are predisposed to be male or female. Gender is part of a complex mixture of nature and nurture.

*Sources:* Colapinto, J. 2000. *As Nature Made Him.* New York: HarperCollins; Diamond, M. and H. K. Sigmundson. March 1997. "Sex Reassignment at Birth: A Long Term Review and Clinical Implications." *Archives of Pediatric and Adolescent Medicine* 151:298–304; Dreifus, Claudia. May 31, 2005. "Declaring with Clarity, When Gender Is Ambiguous." *New York Times*; Money, J. and P. Tucker. 1975. *Sexual Signatures: On Being a Man or Woman.* Boston: Little, Brown; Holden, C. March 1997. "Changing Sex Is Hard to Do." *Science* 275(5307):1745; Colapinto, John. June 3, 2004. "Gender Gap: What Were the Real Reasons Behind David Reimer's Suicide?" Slate.com (www.slate.com/id/2101678/).

---

the family must break camp and quickly find a new site to survive. Frequently, an elderly member of the family, often a grandmother, who might slow down the others and require some of the scarce food, will stay behind and face certain death. Wilson (1975, 1979) saw this as an example of altruism, which ultimately might have a biological component. In sacrificing her own life, the grandmother is improving her kin's chances of survival. She already has made her productive contribution to the family. Now the younger members of the family must survive to ensure the continuation of the family and its genes into future generations.

Wilson concluded that behavior can be explained in terms of the ways in which individuals act to increase the probability that their genes and the genes of their close relatives will be passed on to the next generation. For example, studies of mating strategies in thirty-seven countries show that men and women consistently respond differently to questions about what they look for in their romantic and sexual partners. Men value physical appearance more than women do, and women weigh status and income more than men do. Men's ideal mates are a few years younger than they are on average, and women's a few years older (Buss, 1994). The men are looking for healthy women to bear their children, whereas women are looking for mature, responsible men who will not abandon them.

Until the publication of Wilson's book, the prevailing view in the social sciences had been that there was no biologically based human nature, that human behavior is almost entirely sociocultural in origin, and that, therefore, genes play little or no role in social behavior. Wilson (1994) took the opposite position, proposing that human behavior cannot be understood without biology. Proponents of sociobiology believe that social science one day will be a mere subdivision of biology.

Critics saw sociobiology as not just intellectually flawed but also morally wrong. If human nature is rooted in heredity, they suggested, then some forms of social behavior probably cannot be changed. Tribalism and gender differences could be judged as unavoidable and class differences and war in some manner as natural. People could also argue that some racial or ethnic groups differ irreversibly in personal abilities and emotional attributes. Some people could have inborn mathematical genius, others a bent toward criminal behavior (Wilson, 1994).

The furor over sociobiology was so strong that the American Anthropological Association attempted to pass a resolution to censure sociobiology. Only a passionate speech by noted anthropologist Margaret Mead defeated the motion (Fisher, 1994).

Within Wilson's own department at Harvard University, his chair, Richard Lewontin, and another department member, Stephen Jay Gould, formed the Sociobiology Study Group to publicly denounce Wilson's ideas. Wilson needed bodyguards and was publicly attacked at academic conferences.

Stephen Jay Gould (1976) proposed another, equally plausible scenario, one that discounts the existence of a particular gene programmed for altruism. He perceived the grandmother's sacrifice as an adaptive cultural trait. (It is widely acknowledged that culture is a major adaptive mechanism for humans.) Gould suggested that the elders remain behind because they have been socially conditioned from earliest childhood to the possibility and the appropriateness of this choice. They grew up hearing the songs and stories that praised the elders who stayed behind. Such self-sacrificing elders were the greatest heroes of the clan. Families whose elders rose to such an occasion survived to celebrate the self-sacrifice, but those families without self-sacrificing elders died out.

Wilson made several major concessions to Gould's viewpoint, acknowledging that among human beings, "the intensity and form of altruistic acts are to a large extent culturally determined" and that "human social evolution is obviously more cultural than genetic." He also left the door open to free will, admitting that even though our genetic coding may have a major influence, we still have the ability to choose an appropriate course of action (Wilson, 1978). However, Wilson insisted, "history did not begin 10,000 years ago.... [B]iological history made us what we are no less than culture" (1994).

Gould conceded that human behavior has a biological, or genetic, base. He distinguished, however, between genetic determinism (the sociobiological viewpoint) and genetic potential. What the genes prescribe is not necessarily a particular behavior but rather the capacity for developing certain behaviors. Although the total array of human possibilities is inherited, which of these numerous possibilities a particular person displays depends on his or her experience in the culture.

In later years, two similar research fields grew out of sociobiology: human behavioral ecology and evolutionary psychology. Both disciplines try to understand contemporary human behavior by referring to evolutionary developments. As with sociobiology, both approaches are still the cause of much debate.

Some of the opposition to these fields is related to the nature versus nurture debate. Biology is about nature; culture is about nurture. Critics of evolutionary theories can explain genetically determined behaviors but not behaviors that are culturally learned. As most human behavior is learned, skeptics believe little is to be gained from sociobiology, human behavioral ecology, or evolutionary psychology (Richerson and Boyd, 2004).

The debate over the relative contribution of nature and nurture continues. Probably the best way to think about the influence of genes is to compare them to a recipe, one in which the ingredients are influenced by the environment. Some traits are more sensitive to environmental influence than others. However, just as a winter snowfall is the result of both the temperature and the moisture in the air, so must the human organism and human behavior be understood in terms of both genetic inheritance and the effects of environment. Nurture—that is, the entire socialization experience—is as essential a part of human nature as are our genes. It is from the interplay between genes and the environment that each human being emerges (Fisher, 1994).

## Deprivation and Development

Some unusual events and interesting research have indicated that human infants need more than just food and shelter if they are to function effectively as social creatures.

*Extreme Childhood Deprivation* Only a few recorded cases exist of human beings who have grown up without any real contact with other humans. One such case was found in January 1800, when hunters in Aveyron, in southern France, captured a boy who was running naked through the forest. He seemed to be about 11 years old and apparently had been living alone in the forest for at least five or six years. He appeared to be thoroughly wild and subsequently was exhibited in a cage, from which he managed to escape several times. Finally, he was examined by "experts" who found him to be an incurable "idiot" and placed him in an institute for deaf-mutes.

A young doctor, Jean-Marc Itard, thought differently, however. After close observation, he discovered that the boy was not deaf, mute, or an idiot. Itard believed that the boy's wild behavior, lack of speech, highly developed sense of smell, and poor visual attention span all were the result of having been deprived of human contact. It appeared that the crucial socialization provided by a family had been denied him. Though human, the boy had learned little about how to live with other people. Itard took the boy into his house, named him Victor, and tried to socialize him. He had little success. Although Victor slowly learned to wear clothes, to speak and write a few simple words, and to eat with a knife and fork, he ignored human voices unless they were associated with food and developed no relationships with people other than Itard and the housekeeper who cared for him. He died at the age of 40 (Itard, 1932; Shattuck, 1980).

Another sad case concerns a girl named Anna who grew up in the 1930s and had the misfortune of being born illegitimately to the daughter of an extremely disapproving family. Her mother tried to place Anna with foster parents, was unable to do so, and therefore brought her home. To quiet the family's harsh criticisms, the young mother hid Anna away in a room in the attic, where she could be out of sight and even forgotten by the family. Anna remained there for almost six years, ignored by the whole family, including her mother, who did the very minimum to keep her alive. Finally, Anna was discovered by social workers. The 6-year-old girl was unable to sit up, to walk, or to talk. In fact, she was so withdrawn from human beings that at first she appeared to be deaf, mute, and brain damaged. However, after she was placed in a special school, Anna did learn to communicate somewhat, to walk (awkwardly), to care for herself, and even to play with other children. Unfortunately, she died at the age of 10 (Davis, 1940).

Another case of extreme childhood isolation was that of a girl named Genie who came to the attention of authorities in California in 1970. Genie's nearly blind mother went to the California social welfare offices looking for help for herself, not for Genie. The social worker noticed that the child was small, looked "withered," and had "a halting gait and an unnaturally stooped posture." The worker alerted her supervisor to what she thought was an unreported case of autism in a child estimated to be 6 or 7 years old.

This young boy in India had been brought with wolves before being discovered and placed with a family. He walked on all fours and had no speech ability.

Genie was actually 13½ years old, weighed only 59 pounds, and was only 54 inches tall. She could not focus her eyes beyond 12 feet, could not chew solid food, showed no perception of hot and cold, and could not talk. Her condition was due to her father, who throughout her whole life had confined Genie to a small bedroom, harnessed to an infant's potty seat. Genie was left to sit in the harness, unable to move anything except her fingers and hands, feet and toes, hour after hour, day after day, month after month, year after year (Rhymer, 1993).

When Genie was hospitalized, she was unsocialized, severely malnourished, unable to speak or even to stand upright. After four years in a caring environment, Genie had learned some social skills, was able to take a bus to school, had begun to express some feelings toward others, and had achieved the intellectual development of a 9-year-old. There were still, however, serious problems with her language development that could not be corrected, no matter how involved the instruction (Curtiss, 1977). Today, Genie is in her fifties and lives in an adult care facility.

Another example of a child being deprived of proper care and socialization is Oxana Malaya. Oxana spent much of her childhood between the ages of three and eight living in a kennel with dogs in the back garden of the family home in the Ukraine. On some occasions, she would be allowed in the house with her alcoholic parents, but most of her time was spent with the dogs. With dogs as her constant companions, she growled, barked, and crouched like a wild dog, smelled her food before she ate it, and was found to have acquired extremely acute senses of hearing, smell, and sight.

When her situation came to light in 1991, when she was eight, she could hardly speak. She has acquired some language, but her communication with others is limited. It appears she missed some of the critical periods for normal language development. She lives in a home for the mentally disabled (Feralchildren.com).

These examples of extreme childhood isolation point to the fact that none of the behavior we think of as typically human arises spontaneously. Humans must be taught to stand up, to walk, to talk, even to think. Human infants must develop **social attachments,** that is, they must learn to have *meaningful interactions and affectionate bonds with others.* This seems to be a basic need of all primates, as the research by Harry F. Harlow shows.

In a series of experiments with rhesus monkeys, Harlow and his coworkers demonstrated the importance of body contact in social development (Harlow, 1959; Harlow and Harlow, 1962). In one experiment, infant monkeys were taken from their mothers and placed in cages where they were raised in isolation from other monkeys. Each cage contained two substitute mothers: One was made of hard wire and contained a feeding bottle; the other was covered with soft terry cloth but did not have a bottle. Surprisingly, the baby monkeys spent much more time clinging to the cloth mothers than to the wire mothers, even though they received no food at all from the cloth mothers. Apparently, the need to cling to and to cuddle against even this meager substitute for a real mother was more important to them than being fed.

Other experiments with monkeys have confirmed the importance of social contact in behavior. Monkeys raised in isolation *never* learn how to interact with other monkeys or even how to mate. If placed in a cage with other monkeys, those who were raised in isolation either withdraw or become violent and aggressive, threatening, biting, and scratching the others.

Female monkeys who are raised without affection make wretched mothers themselves. After being impregnated artificially and giving birth, such monkeys either ignore their infants or display a pattern of behavior described by Harlow as "ghastly."

As with all animal studies, we must be very cautious in drawing inferences for human behavior. After all, we are not monkeys. Yet, Harlow's experiments show that without socialization, monkeys do not develop normal social, emotional, sexual, or maternal behavior. Because human beings rely on learning even more than monkeys do, it is likely that the same is true of us.

It is obvious that the human organism needs to acquire culture to be complete; it is very difficult, if not impossible, for children who have been isolated from other people from infancy onward to catch up. They apparently suffer permanent damage, although human beings do seem to be somewhat more adaptable than were the rhesus monkeys Harlow studied.

*Infants in Institutions* Studies of infants and young children in institutions confirm the view that human beings' developmental needs include more than the mere provision of food and shelter. Psychologist Rene Spitz (1945) visited orphanages in Europe and found that in those dormitories, where children were given routine care but otherwise were ignored, they were slow to develop and were withdrawn and sickly. These children's needs for social attachment were not met.

With the fall of the former Soviet Union, people in the West became aware of the conditions in eastern European orphanages. In many of these orphanages, no consistent, responsive caregiving was provided. Children had several caregivers each, which prevented close individual attention from someone who knew the child. The children received little contact due to understaffing and uninformed or disinterested caregivers. Often, the rooms had no toys or other objects for mental stimulation. Because malnutrition was common in eastern European culture, many times the malnourished staff would consume food meant for the children. Frequently, children were not spoken to or called by their names and were left in bed for hours without any attention. There was very little attempt to provide physical, mental, or emotional stimulation for these children.

Children from these experiences usually displayed an **attachment disorder**—they were *unable to trust people and to form relationships with others.* Many people who adopted children from these settings and thought that "love was all the children needed" to overcome these early experiences discovered that extensive treatment was necessary for these children to ever become normally functioning adults (Doolittle, 1995).

It appears clear that human infants need more than just food and shelter if they are to grow and develop properly. Every human infant needs frequent contact with others who demonstrate affection, who respond to attempts to interact, and who themselves initiate interactions with the child. Infants also need contact with people who find ways to interest them in their surroundings and who teach them the physical and social skills and knowledge they need to function. In addition, to develop normally, children need to be taught the culture of their society—to be socialized into the world of social relations and symbols that are the foundation of the human experience.

## The Concept of Self

Every individual comes to possess a social identity by occupying **statuses**—*culturally and socially defined positions*—in the course of his or her socialization. This social identity changes as the person moves through the various stages of childhood and adulthood recognized by the society. New statuses are occupied; old ones are abandoned.

Picture a teenage girl who volunteers in a community hospital. She leaves that position to attend college, joins a sorority, becomes a premedical major, and graduates. She goes to medical school, completes a residency, becomes engaged, and then enters a program for specialized training in surgery. Perhaps she marries; possibly she has a child. All along the way, she is moving through different social identities, often assuming several at once. When, many years later, she returns to the hospital where she was a teenage volunteer, she will have an entirely new social identity: adult woman, surgeon, wife (perhaps), mother (possibly).

This description of the developing girl is the view from the outside, the way other members of the society experience her social transitions, or what sociologists would call changes in her social identity. A **social identity** is *the total of all the statuses that define an individual.* But what of the girl herself? How does this human being who is growing and developing physically, emotionally, intellectually, and socially experience these changes? Is there something constant about a person's experience that allows one to say, "I am that changing person—changing, but yet somehow the same individual?" In other words, do all human beings have personal identities separate from their social identities?

Most social scientists believe that the answer is yes. *This changing yet enduring personal identity* is called the **self.** The self develops when the individual becomes aware of his or her feelings, thoughts, and behaviors as separate and distinct from those of other people. This usually happens at a young age, when children begin to realize that they have their own history, habits, and needs and begin to imagine how these might appear to others. By adulthood, the concept of self is developed fully.

Most researchers would agree that the concept of self includes (1) an awareness of the existence, appearance, and boundaries of one's own body (you are walking among the other members of the crowd, dressed appropriately for the occasion, and trying to avoid bumping into people as you chat); (2) the ability to refer to one's own being by using language and other symbols ("Hi, as you can see from my name tag, I'm Harry Hernandez from Gonzales, Texas"); (3) knowledge of one's personal history ("My folks own a small farm there, and since I was a small boy I've wanted to study farm management"); (4) knowledge of one's needs and skills ("I'm good with the day-to-day requirements of farming, but I need the intellectual stimulation of doing large-scale planning"); (5) the ability to organize one's knowledge and beliefs ("Let me tell you about planning crop rotation"); (6) the ability to organize one's experiences ("I know what I like and what I don't like"); and (7) the ability to take a step back and look at one's being as others do, to evaluate the impressions one is creating, and to understand the feelings and attitudes one stimulates in others ("It might seem unusual that a farmer like me would want to come to a party for the opening of a new art gallery. Well, as far back as I can remember, I always enjoyed art,

If a child's need for meaningful interaction with another is not met, the development of a social identity will be delayed.

and I happen to know someone who works here") (see Cooley, 1909; Erikson, 1964; Gardner, 1978; Mead, 1935).

## Dimensions of Human Development

Clearly, the development of the self is a complicated process. It involves many interacting factors, including the acquisition of language and the ability to use symbols. Three dimensions of human development are tied to the emergence of the self: cognitive development, moral development, and gender identity.

*Cognitive Development* For centuries, most people assumed that a child's mind works in exactly the same way as an adult's mind. The child was thought of as a miniature adult who simply is lacking information about the world. Swiss philosopher and psychologist Jean Piaget (1896–1980) was instrumental in changing that view through his studies of the development of intelligence in children. His work has been significant to sociologists because the processes of thought are central to the development of identity and, consequently, to the ability to function in society.

Piaget found that children move through a series of predictable stages on their way to logical thought and that some never attain the most advanced stages. From birth to age 2, during the sensorimotor stage, the infant relies on touch and the manipulation of objects for information about the world, slowly learning about cause and effect. At about age 2, the child begins to learn that words can be symbols for objects. In this, the preoperational stage of development, the child cannot see the world from another person's point of view.

The *operational stage* is next and lasts from age 7 to about age 12. During this period, the child begins to think with some logic and can understand and work with numbers, volume, shapes, and spatial relationships. With the onset of adolescence, the child progresses to the most advanced stage of thinking—*formal, logical thought*. People at this stage are capable of abstract, logical thought and can develop ideas about things that have no concrete reference such as infinity, death, freedom, and justice. In addition, they are able to anticipate possible consequences of their acts and decisions. Achieving this stage is crucial to developing an identity and an ability to enter into mature interpersonal relationships (Piaget and Inhelder, 1969).

*Moral Development* Every society has a **moral order**—that is, *a shared view of right and wrong*. Without moral order, a society soon would fall apart. People would not know what to expect from themselves and one another, and social relationships would be impossible to maintain. Therefore, the process of socialization must include instruction about the moral order of an individual's society.

The research by Lawrence Kohlberg (1969) suggests that not every person is capable of thinking about morality in the same way. Just as our sense of self and our ability to think logically develop in stages, our moral thinking develops in a progression of steps as well.

To illustrate this, Kohlberg asked children from a number of different societies (including Turkey, Mexico, China, and the United States) to resolve moral dilemmas such as the following: A man's wife is dying of cancer. A rare drug might save her, but it costs $2,000. The man tries to raise the money but can come up with only $1,000. He asks the druggist to sell him the drug for $1,000, and the druggist refuses. The desperate husband then breaks into the druggist's store to steal the drug. Should he have done so? Why or why not?

Kohlberg was more interested in the reasoning behind the child's judgment than in the answer itself. Based on his analysis of this reasoning, he concluded that changes in moral thinking progress step by step through six qualitatively distinct stages (although most people never go beyond stages 3 or 4):

Stage 1. Orientation toward punishment.
Stage 2. Orientation toward reward.
Stage 3. Orientation toward possible disapproval by others.
Stage 4. Orientation toward formal laws and fear of personal dishonor.
Stage 5. Orientation toward peer values and democracy.
Stage 6. Orientation toward one's own set of values.

Kohlberg found that although these stages of moral development correspond roughly to other aspects of the developing self, most people never progress to stages 5 and 6. In fact, Kohlberg subsequently dropped stage 6 from his scheme because it met with widespread criticism that he could not refute. Critics felt that stage 6 was elitist and culturally biased. Kohlberg himself could find no evidence that any of his long-term subjects ever reached this stage (Muson, 1979).

At times, people even regress from a higher stage to a lower one. For example, when Kohlberg analyzed the explanations that Nazi war criminals of World War II gave for their participation in the systematic murder of millions of people who happened to possess certain religious (Jewish), ethnic (gypsies), or psychological (mentally retarded) traits, he found that none of the reasons was above stage 3 and most were at stage 1—"I did what I was told to do; otherwise, I'd have been punished" (Kohlberg, 1967). However, many of these war criminals had been very responsible and successful people in their

During the first years of life, a child needs to develop a sense of trust through supportive relationships with adults.

prewar lives and presumably in those times had reached higher stages of moral development.

*Gender Identity* One of the most important elements of our sense of self is our gender identity. Certain aspects of gender identity are rooted in biology. Males tend to be larger and stronger than females, but females tend to have better endurance than males. Females also become pregnant and give birth to infants and (usually) can nurse infants with their own milk. However, gender identity is mostly a matter of cultural definition. There is nothing inherently male or female about a teacher, a pilot, a carpenter, or a pianist other than what our culture tells us. As we shall see in Chapter 11, "Gender Stratification," gender identity and sex roles are far more a matter of nurture than of nature.

## ● THEORIES OF DEVELOPMENT

Among the scholars who have devised theories of development, Charles Horton Cooley, George Herbert Mead, Sigmund Freud, and Erik Erikson stand out because of the contributions they have made to the way sociologists today think about socialization. Cooley and Mead saw the individual and society as partners. They were symbolic interactionists (see Chapter 1, "The Sociological Perspective") and, as such, believed that the individual develops a self solely through social relationships—that is, through interaction with others. They believed that all our behaviors, our attitudes, even our ideas of self arise from our interactions with other people. Hence, they were pure environmentalists in that they believed that social forces rather than genetic factors shape the individual.

Freud, on the other hand, tended to picture the individual and society as enemies. He saw the individual as constantly having to yield reluctantly to the greater power of society, to keep internal urges (especially sexual and aggressive ones) under strict control.

Erikson presented something of a compromise position. He thought of the individual as progressing through a series of stages of development that express internal urges yet are greatly influenced by societal and cultural factors.

### Charles Horton Cooley (1864–1929)

Cooley believed that the self develops through the process of social interaction with others. This process begins early in life and is influenced by such primary groups as the family. Later on, peer groups become very important as we continue to progress as social beings. Cooley (1909) used the phrase **looking-glass self** to describe *the three-stage process through which each of us develops a sense of self.* First, we imagine how our actions appear to others. Second, we imagine how other people judge these actions. Finally, we make some sort of self-judgment based on the presumed judgments of others. In effect, other people become a mirror or looking glass for us.

In Cooley's view, therefore, the self is entirely a social product—that is, a product of social interaction. Each individual acquires a sense of self in the course of being socialized and continues to modify it in each new situation throughout life. Cooley believed that the looking-glass self constructed early in life remains fairly stable and that childhood experiences are very important in determining our sense of self throughout our lives.

One of Cooley's principal contributions to sociology was his observation that although our perceptions are not always correct, what we believe is more important in determining our behavior than is what is real. This same idea was also expressed by sociologist W. I. Thomas (1928) when he noted, "If men define situations as real, they are real in their consequences." If we can understand the ways in which people perceive reality, then we can begin to understand their behavior.

### George Herbert Mead (1863–1931)

Mead was a philosopher and a well-known social psychologist at the University of Chicago. His work led to the development of the school of thought called symbolic interactionism (described in Chapter 1). As a

Although certain aspects of gender identity are rooted in biology, socialization plays a large role in learning the appropriate cultural behavior.

student of Cooley's, Mead (1934) built on Cooley's ideas, tracing the beginning of a person's awareness of self to the relationships between the caregiver (usually the mother) and the child.

According to Mead, the self becomes the sum total of our beliefs and feelings about ourselves. The self is composed of two parts, the I and the me. The **I** *portion of the self wishes to have free expression, to be active and spontaneous.* The I wishes to be free of the control of others and to take the initiative in situations. It is the part of the individual that is unique and distinctive. The **me** *portion of the self is made up of those things learned through the socialization process from family, peers, school, and so on.* The me makes normal social interaction possible, whereas the I prevents it from being mechanical and totally predictable.

Mead used the term **significant others** to refer to *those individuals who are most important in our development, such as parents, friends, and teachers.* As we continue to be socialized, we learn to be aware of the views of the generalized others. These **generalized others** are *the viewpoints, attitudes, and expectations of society as a whole or of a community of people whom we are aware of and who are important to us.* We might believe it is important to go to college, for example, because significant others have instilled this viewpoint in us. While at college, we might be influenced by the views of selected generalized others who represent the community of lawyers that we hope to join one day as we progress with our education.

Mead (1934) believed that the self develops in three stages. The first, or **preparatory stage** is *characterized by the child imitating the behavior of others, which prepares the child for learning social-role expectations.* In the second, or **play stage**, *the child has acquired language and begins not only to imitate behavior but also to formulate role expectations*—playing house, cops and robbers, and so on. In this stage, the play features many discussions among playmates about the way things ought to be. "I'm the boss," a little boy might announce. "The daddy is the boss of the house." "Oh no," his friend might counter, "Mommies are the real bosses." In the third, or **game stage**, *the child learns that there are rules that specify the proper and correct relationship among the players.* For example, a baseball game has rules that apply to the game in general as well as to a series of expectations about how each position should be played. During the game stage, according to Mead, we learn the expectations, positions, and rules of society at large. Throughout life, in whatever position we occupy, we must learn the expectations of the various positions with which we interact, as well as the expectations of the general audience, if our performance is to go smoothly.

Thus, for Mead, the self is rooted in, and begins to take shape through, the social play of children and is well on its way to being formed by the time the child is 8 or 9 years old. Therefore, like Cooley, Mead regarded childhood experience as very important to charting the course of development.

### Sigmund Freud (1856–1939)

Freud was a pioneer in the study of human behavior and the human mind. He was a doctor in Vienna, Austria, who gradually became interested in the problem of understanding mental illness.

In Freud's view, the self has three separately functioning parts: the id, the superego, and the ego. The **id** consists of *the drives and instincts that Freud believed every human being inherits but which, for the most part, remain unconscious.* Of these instincts, two are most important: the aggressive drive and the erotic or sexual drive (called libido). Every feeling derives from these two drives. The **superego** represents *society's norms and moral values as learned primarily from our parents.* The superego is the internal censor. It is not inherited biologically, like the id, but is learned in the course of a person's socialization. The superego keeps trying to put the brakes on the id's impulsive attempts to satisfy its drives. So, for instance, the superego must hold back the id's unending drive for sexual expression (Freud, 1920, 1923). The id and superego, then, are eternally at war with each other. Fortunately, there is a third, functional part of the self called the **ego,** which *tries not only to mediate in the eternal conflict between the id and the superego but also to find socially acceptable ways for the id's drives to be expressed.* Unlike the id, the ego constantly evaluates social realities and looks for ways to adjust to them (Freud, 1920, 1923).

Freud pictured the individual as constantly in conflict: The instinctual drives of the id (essentially sex and aggression) push for expression while the demands of society set certain limits on the behavior patterns that will be tolerated. Even though the individual needs society, society's restrictive norms and values are a source of ongoing discontent (Freud, 1930). Freud's theories suggest that society and the individual are enemies, with the latter yielding to the former reluctantly and only out of compulsion.

### Erik H. Erikson (1902–1994)

In 1950, Erikson, an artist turned psychologist who studied with Freud in Vienna, published an influential book called *Childhood and Society* (1950). In it, he built on Freud's theory of development but added two important elements. First, he stressed that development is a lifelong process and that a person continues to pass through new stages even during adulthood. Second, he paid greater attention to the social and cultural forces operating on the individual at each step along the way.

In Erikson's view, human development is accomplished in eight stages (see Table 4-1). Each stage amounts to a crisis of sorts brought on by two factors:

TABLE 4-1  Erikson's Eight Stages of Human Development

| Stage | Age Period | Characteristic to Be Achieved | Major Hazards to Achievement |
|---|---|---|---|
| Trust vs. mistrust | Birth to 1 year | Sense of trust or security—achieved through parental gratification of needs and affection | Neglect, abuse, or deprivation; inconsistent or inappropriate love in infancy; early or harsh weaning |
| Autonomy vs. shame and doubt | 1–4 years | Sense of autonomy—achieved as child begins to see self as individual apart from his/her parents | Conditions that make the child feel inadequate, evil, or dirty |
| Initiative vs. guilt | 4–5 years | Sense of initiative—achieved as child begins to imitate adult behavior and extends control of the world around him/her | Guilt produced by overly strict discipline and the internalization of rigid ethical standards that interfere with the child's spontaneity |
| Industry vs. inferiority | 6–12 years | Sense of duty and accomplishment—achieved as the child lays aside fantasy and play and begins to undertake tasks and schoolwork | Feelings of inadequacy produced by excessive competition, personal limitations, or other events leading to feelings of inferiority |
| Identity vs. role confusion | Adolescence | Sense of identity—achieved as one clarifies sense of self and what he/she believes in | Sense of role confusion resulting from the failure of the family or society to provide clear role models |
| Intimacy vs. isolation | Young adulthood | Sense of intimacy—the ability to establish close personal relationships with others | Problems with earlier stages that make it difficult to get close to others |
| Generativity vs. stagnation | 30s–50s | Sense of productivity and creativity—resulting from work and parenting activities | Sense of stagnation produced by feeling inadequate as a parent and stifled at work |
| Integrity vs. despair | Old age | Sense of ego integrity—achieved by acceptance of the life one has lived | Feelings of despair and dissatisfaction with one's role as a senior member of society |

*Source:* Adapted from *Childhood and Society*, 2d. ed., by E.H. Erikson. Copyright © 1950, 1963 by W.W Norton & Company, Inc., renewed © 1978, 1991 by Erik H. Erikson. Used by permission of W.W. Norton & Company, Inc. and the Random House Group Limited (Hogarth Press)

biological changes in the developing individual and social expectations and stresses. At each stage, the individual is pulled in opposite directions to resolve the crisis. In normal development, the individual resolves the conflict experienced at each stage somewhere toward the middle of the opposing options. For example, very few people are entirely trusting, and very few trust nobody at all. Most of us are able to trust at least some other people and thereby form enduring relationships while also staying alert to the possibility of being misled.

Erikson's view of development has proved to be useful to sociologists because it seems to apply to many societies. In a later work (1968), he focused on the social and psychological causes of the identity crisis that seems to be so prevalent among American and European youths. Erikson's most valuable contribution to the study of human development has been to show that socialization continues throughout a person's life and does not stop with childhood. People continue to develop after age 30—and after 60 and 70 as well. The task of building the self is lifelong; it can be considered our central task from cradle to grave. We construct the self—our identity—using the materials made available to us by our culture and our society.

## EARLY SOCIALIZATION IN AMERICAN SOCIETY

Children are brought up very differently from one society to another. Each culture has its own childrearing values, attitudes, and practices. No matter how children are raised, however, each society must provide certain minimal necessities to ensure normal development. The infant's physical needs must, of course, be addressed, but more than that is required. Children need speaking social partners. (Some evidence suggests that a child who has received no language stimulation at all

According to Erik Erikson, adolescence is a time when the teenager must develop an identity as well as the ability to establish close personal relationships with others.

in the first five to six years of life will be unable ever to acquire speech [Chomsky, 1975].) They also need physical stimulation; objects that they can manipulate; space and time to explore, to initiate activity, and to be alone; and finally, limits and prohibitions that organize their options and channel development in certain culturally specified directions (Provence, 1972).

Every society provides this basic minimum care in its own culturally prescribed ways. A variety of agents, which also vary from culture to culture, are used to mold the child to fit into the society. Here we consider some of the most important agents of socialization in American society.

## The Family

For young children in most societies—and certainly in American society—the family is the primary world for the first few years of life. Children also have significant early experiences in day-care centers with no family members. The values, norms, ideals, and standards presented in both places are accepted by the child uncritically as correct—indeed, as the only way things could possibly be. Even though later experiences lead children to modify much of what they have learned within the family, it is not unusual for individuals to carry into the social relationships of adult life the role expectations that characterized the family of their childhood. Many of our gender-role expectations are based on the models of female and male behavior we witnessed in our families.

Every family, therefore, socializes its children to its own particular version of the society's culture. In addition, however, each family exists within certain subcultures of the larger society: It belongs to a geographical region, a social class, one or two ethnic groups, and possibly a religious group or other subculture. Families differ with regard to how important these factors are in determining their lifestyles and their childrearing practices. For example, some families are very deeply committed to a racial or ethnic identification such as African American, Hispanic, Chinese, Native American, Italian American, Polish American, or Jewish American. Much of family life might revolve around participation in social and religious events of the community and can include speaking a language other than English.

Evidence also shows that social class and parents' occupations influence how children are raised in the United States. Parents who have white-collar occupations are accustomed to dealing with people and solving problems. As a result, white-collar parents value intellectual curiosity and flexibility. Blue-collar parents have jobs that require specific tasks, obeying orders, and being on time. They are likely to reward obedience to authority, punctuality, and physical or mechanical ability in their children (Kohn and Schooler, 1983).

The past four decades have seen major changes in the structure of the American family. High divorce rates, the dramatic increase in the number of single-parent families, and the common phenomenon of two-worker families have meant that the family as the major source of socialization of children is being challenged. Child-care providers have become a major influence in the lives of many young children. (For a discussion of the effects of day care on the socialization of children, see "Day-to-Day Sociology: Does Day Care Create Unruly Brats?")

## The School

The school is an institution intended to socialize children in selected skills and knowledge. In recent decades, however, the school has been assigned additional tasks. For instance, in poor communities and neighborhoods, school lunch (and breakfast) programs are an important source of balanced nutrition for children. The school must also confront a more basic

Every family socializes its children to its own particular version of the society's culture. The values and worldview are a reflection of the larger society and the particular subculture the family is a part of.

## DAY-TO-DAY SOCIOLOGY

### Does Day Care Create Unruly Brats?

The vast majority of Americans believe it is best for one parent to stay at home solely to raise children while the other parent works. Very few say the ideal situation is for both parents to work full time outside the home. Yet, the reality often does not allow for the ideal arrangement.

Of the 20 million children younger than 5 years old whose mothers were employed, about 63 percent spent time in some type of regular child-care arrangement every week. Nearly 25 percent were cared for in an organized care facility such as a day-care center, nursery, or preschool. Twenty-three percent were cared for by a grandparent, 16 percent received care from their fathers, and 3 percent from siblings.

For more than two decades, the Study of Early Child Care and Youth Development has tracked more than 1,300 children in various child-care arrangements, including staying home with a parent, being cared for by a nanny or a relative, or attending a large day-care center. This study, which is the largest and longest-running study of American child care, found that the longer a child spent in day care, the more aggressive and disobedient the child was in elementary school, and the effect persisted through the sixth grade.

The most troubling aspect of this finding was that it held up regardless of the child's sex or family income or the quality of the day-care center. With more than 3 million American preschoolers attending day care, the increased disruptiveness makes the elementary school classroom harder to manage.

On the positive side, the study also found that time spent in high-quality day-care centers was correlated with higher vocabulary scores through elementary school.

The debate about child care started in the late 1980s, when social scientists questioned the impact of day care. Day-care centers and working parents argued that it was the quality of the care that mattered, not the setting. Jay Belsky, one of the study's principal authors, in 1986 suggested that nonparental child care, even in good-quality day care, could cause developmental problems. The cause of the disruptive behavior later on may be due to preschool peer groups that model disruptive behavior that is allowed to continue because of limited supervision in day-care centers.

*Sources:* Benedict, Carey. March 26, 2007. "Poor Behavior Is Linked to Time in Day Care." *New York Times* (www.nytimes.com/2007/03/26/us/26center.html?scp=18&sq=day%20care&st=cse); Belsky, J. 2007. "Recent Child Care Findings." *Pediatrics for Parents*, 23; Bernstein, Robert. February 28, 2008. "Nearly Half of Preschoolers Receive Child Care from Relatives" (www.census.gov/Press-Release/www/releases/archives/children/011574.html); U.S. Bureau of the Census. 2005. "Survey of Income and Program Participation (SIPP)" (www.census.gov/population/www/socdemo/childcare.html); Belsky, Jay. February 2011. "Child Care and Its Impact on Young Children." In *Encyclopedia on Early Childhood Development*, 2nd rev. ed. United Kingdom: Birkbeck University of London (www.child-encyclopedia.com/documents/BelskyANGxp3-Child_care.pdf), accessed June 3, 2012; Laughlin, Lynda. 2010. "Who's Minding the Kids? Child Care Arrangements: Spring 2005 and Summer 2006." *Current Population Reports*, pp. 70–121. Washington, D.C.: U.S. Bureau of the Census.

Day-care quality depends on the adult-child ratios in the center and the overall size of the group within which the children are cared for.

problem. As an institution, it must resolve the conflicting values of the local community and of the state and regional officials whose job it is to determine what should be taught. For example, in many school systems, parents want to reintroduce school prayer or discussions of religion, even though the Supreme Court has ruled that such actions are not permissible. In other instances, education officials make curriculum changes in the classroom despite the complaints of parents, whose objections are dismissed as perhaps uninformed or tradition bound.

AIDS education is a vivid example of schools deciding what issues should or should not be presented to children. Some parents have objected to teaching young children about sexuality, condoms, and homosexuality despite the health risks that ignorance could pose. Many school boards have taken the position that the schools have a responsibility to provide this information, even when large numbers of parents object.

In coming to grips with their multiple responsibilities, many school systems have established a philosophy of education that encompasses socialization as well as academic instruction. Educators often aim to help students develop to their fullest capacity, not only intellectually but also emotionally, culturally, morally, socially, and physically. By exposing the student to a variety of ideas, the teachers attempt to guide the development of the whole student in areas of interests and abilities unique to each. Students are expected to learn how to analyze these ideas critically and reach their own conclusions. The ultimate goal of the school is to produce a well-integrated person who will become socially responsible. Two questions arise: Is such an ambitious, all-embracing educational philosophy working? And is it an appropriate goal for our schools?

In a way, the school is a model of much of the adult social world. Interpersonal relationships are not based on individuals' love and affection for one another. Rather, they are impersonal and predefined by the society with little regard for each particular individual who enters into them. Children's process of adjustment to the school's social order is a preview of what will be expected as they mature and attempt to negotiate their way into the institutions of adult society (job, political work, organized recreation, and so on). Of all the socializing functions of the school, this preview of the adult world can be the most important. (The role of the school in socialization will be discussed more extensively in Chapter 14, "Education.")

## Peer Groups

**Peers** are *individuals who are social equals*. From early childhood until late adulthood, we encounter a wide variety of peer groups. No one will deny that they play a powerful role in our socialization. Often, their influence is greater than that of any other source of socialization.

Peer groups provide valuable social support for children.

Within the family and the school, children are in socially inferior positions relative to figures of authority (parents, teachers, principals). As long as the child is small and weak, this social inferiority seems natural, but by adolescence, a person is almost fully grown, and arbitrary submission to authority is not so easy to accept. Hence, many adolescents withdraw into the comfort of social groups composed of peers.

Parents might play a major role in teaching basic values and the development of the desire to achieve long-term goals, but peers have the greatest influence in lifestyle issues such as appearance, social activities, and dating. Peer groups also provide valuable social support for adolescents who are moving toward independence from their parents. As a consequence, their peer-group values often run counter to those of the older generation. New group members quickly are socialized to adopt symbols of group membership such as styles of dress, use and consumption of certain material goods, and stylized patterns of behavior. A number of studies have documented the increasing importance of peer-group socialization in the United States. One reason for this is that parents' life experiences and accumulated wisdom might not be very helpful in preparing young people to meet the requirements of life in a society that is changing constantly. Not infrequently, adolescents are better informed than their parents are about such things as sex, drugs, and technology.

Peer-group influence, for many youths, can lead to wasted lives and violence. For many gang members who band together for identity, status, petty criminal activity, and mutual protection, drug abuse is often involved. In many urban high schools, for example, drugs are available, and attempts to emphasize the dangers of drugs fall on deaf ears. (See "Our Diverse World: Win Friends and Lose Your Future: The Costs of Not 'Acting White'" for another aspect of peer-group influence.)

The negative effects of peer pressure are felt on college campuses as well as in poor inner-city neighborhoods. Peer pressure has caused deaths from hazing activities in college fraternities. For example, one month after entering the Massachusetts Institute of Technology, freshman Scott Kruger died after consuming excessive amounts of alcohol as part of a hazing ritual at a fraternity party.

## OUR DIVERSE WORLD: Win Friends and Lose Your Future: The Costs of Not "Acting White"

*Children can't achieve unless we raise their expectations and turn off the television sets and eradicate the slander that says a black youth with a book is acting white.*
—Senator Barack Obama, 2004 Democratic National Convention Keynote Address

It has been about 140 years since the end of slavery, yet black students lag significantly behind white students. The average black 17-year-old reads at the level of a white 13-year-old. Black students score one standard deviation below white students on the SAT test. This racial divide exists even when black students attend schools in affluent neighborhoods.

One way to explain this gap is to point to a black student peer culture that discourages students against "acting white." Factors that would be considered examples of acting white could include enrolling in honors or advanced placement courses, raising your hand in class, playing the violin, wearing clothes from Abercrombie and Fitch or the Gap, or speaking Standard English.

Students in high school can attest to the fact that there is a strong desire to be accepted by valued peer groups. Most of the time being accepted by a peer group does not require any major short-term costs. Yes, it is true that being part of one peer group might mean you will be rejected by another peer group. But usually, you will not end up financially poorer or seriously handicapped in some way because you are part of a sports, music, or other social peer group.

Becoming a member of a peer group means you have to decide what proportion of your time you are going to devote to the group and its values and what proportion of time to other activities and your studies. For some peer groups, studying and doing well overlaps with the values of the group. At other times, these behaviors are in conflict with the group, and members can be ostracized for acting in a way that conflicts with the group values. If a peer group does not value academic achievement, loyalty to the group can result in mediocre or even failing grades.

In many high schools, white students who do well actually have more friends than students who do poorly. Often, for black students, that is not the case. A black student who has a 4.0 average has 1.5 fewer friends than the white student with a 4.0 (Cook and Ludwig, 1997).

Let's look at the Frank W. Ballou High School in Washington, D.C., as an example. The principal of the school decided that it would be a good idea to give outstanding students cash for their achievements. A student with perfect grades in any of the year's four marking periods would receive $100. With four marking periods, over the course of a year a student could get $400. There was a catch. The straight-A students had to receive each check personally at an awards assembly. Soon, the assemblies turned into forums in which the winners were jeered, being called "Nerd!" "Geek!" "Egghead!" and, the harshest of all, "Whitey!" The honor students were tormented for months afterward. With each assembly, fewer students showed up to receive their checks (Susskind, 1998).

In the short term, then, for black students, the price of doing well is not being accepted. The benefit of not doing well is that you will have friends and a more active social life. In the future, however, things are different. When high school is over, the benefit of peer acceptance rapidly disappears, but the cost of not doing well in school quickly starts to become important. Employers and colleges look at mediocre or poor grades as a sign that this person should be passed over for the job or college admission. The short-term benefit of being accepted by the peer group has now turned into a long-term liability that is difficult to erase.

A peer-group socialization process that discourages students from doing well in school because it is considered a sign of acting white is exacting a long-term price that most students can ill afford.

*Sources:* Austen-Smith, David and Roland G. Fryer, Jr. May 2005. "An Economic Analysis of 'Acting White.'" *Quarterly Journal of Economics* 120(2):551–83; Fryer, Jr., Roland G. and Paul Torelli. May 1, 2005. "An Empirical Analysis of 'Acting White.'" Harvard University Society of Fellows and NBER (http://post.economics.harvard.edu/faculty/fryer/papers/fryer_torelli.pdf); Cook, Phillip and Jens Ludwig. 1997. "Weighing the Burden of 'Acting White': Are There Race Differences in Attitudes Towards Education?" *Journal of Public Policy and Analysis*, 16:256–78; Susskind, Ron. 1998. *A Hope in the Unseen*. New York: Broadway Books.

As the authority of the family diminishes under the pressures of social change, peer groups move into the vacuum and substitute their own morality for that of the parents. Peer groups are most effective in molding the behavior of those adolescents whose parents do not provide consistent standards, a principled moral code, guidance, and emotional support. David Elkind (1981) has expressed the view that the power of the peer group is in direct proportion to the extent that the adolescent feels ignored by the parents. In fact, more than a half century ago, sociologist David Riesman, in his classic work *The Lonely Crowd* (1950), already thought the peer group had become the single most powerful molder of many adolescents' behavior and that striving for peer approval had become the dominant concern of an entire American generation—adults as well as adolescents. He coined the term *other directed* to describe those who are overly concerned with finding social approval.

### Television, Movies, and Video Games

It is possible that today Riesman would review his thinking somewhat. Since the late 1960s, the mass media—television, radio, magazines, films, newspapers, and the Internet—have become important agents of socialization in the United States. It is almost impossible in our society to escape from the images and sounds of television or radio; in most homes, especially those with children, the media are constantly visible or audible. With the exception of video games, most mass media communication is one-way, creating an audience that is conditioned to receive passively whatever news, messages, programs, or events are brought to them. Today, 96.7 percent of all households in the United States have television sets, with an average of 2.9 sets per home. Most children become regular watchers of television between ages 3 and 6 (Neilsen Wire, April 2010).

Schoolchildren watch an average of 2 hours and 30 minutes of television a day on school days and an average of 4 hours and 20 minutes on an average weekend. Their favorite programs are situation comedies, cartoons, music videos, sports, game shows, talk shows, and soap operas. One study concluded that by the time most people reach the age of 18, they will have spent more waking time watching television than doing anything else—talking with parents, spending time with friends, or even going to school (U.S. Bureau of the Census, 2002).

Because young children are so impressionable, and because in so many American households the television is used as an unpaid mechanical babysitter, social scientists have become increasingly concerned about the socializing role the mass media play in our society. If children learn from experience, such exposure must certainly have an effect. (See "Day-to-Day Sociology: Television Made You the Designated Driver.") By the time the average American child leaves elementary school, she or he has seen 8,000 killings and 100,000 violent acts portrayed on television.

Media violence can increase aggression because of several related factors. First, it creates more positive attitudes, beliefs, and expectations about solving problems with violence. Youth come to believe that aggression is normal, appropriate, and likely to succeed. It also leads to the belief that aggressive behavior explains how the social world works. It leads to the view that the world is a fairly hostile place and decreases the recognition that there are nonviolent ways to handle conflict. Media violence also produces an emotional desensitization to aggression and violence. Normally, people have a pretty negative emotional reaction to conflict, aggression, and violence (Anderson, 2010).

In South Africa, the apartheid government did not allow television to be viewed for fear of its destabilizing effects. Because of this policy, television was not part of South African life until many years after its introduction in the United States and Canada. Yet, just as in the United States and Canada, the homicide rate rose sharply as the first generation of children who grew up viewing television reached adulthood.

Attempts have been made to warn viewers of violent content in programs. In 1997, the television networks agreed to add content ratings to the existing age-based ratings, using the designation TV-Y for a show appropriate for all children; TV-7 for a show appropriate for children age 7 and up; TV-G for a show suitable for all ages; TV-PG recommending parental guidance (this rating might also include a V for violence, S for sexual situations, L for language, or D for suggestive dialog);

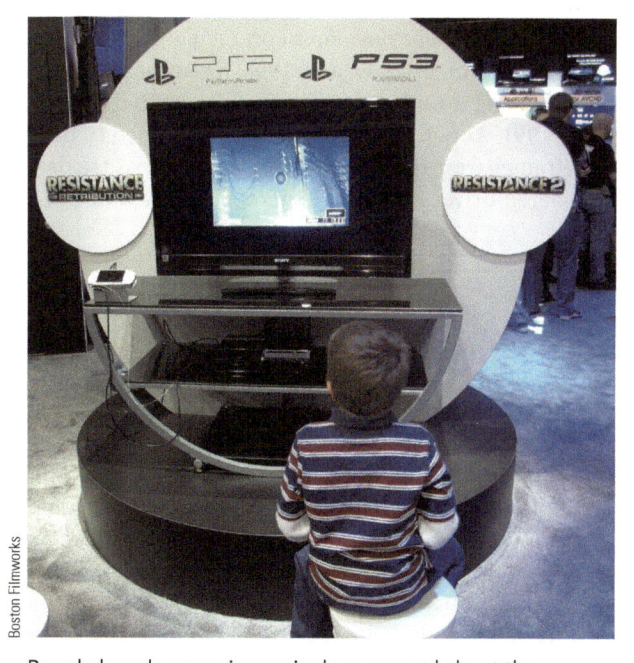

People have become increasingly concerned about the socializing role played by video games and other mass media.

## DAY-TO-DAY SOCIOLOGY

### Television Made You the Designated Driver

As discussed in this chapter, television shows include ample doses of sex, violence, and pathology. But if television contributes to poor behavior, as some suggest, might it also be a vehicle for encouraging good behavior?

In 1988, Jay Winsten, a professor at the Harvard School of Public Health and the director of the school's Center for Health Communication, conceived a plan to use television to introduce a new social concept—the designated driver—to North America. Shows were already dealing with the topic of drinking, Winsten reasoned, so why not add a line of dialog here and there about not driving drunk? With the assistance of then NBC chair Grant Tinker, Winsten met with more than 250 writers, producers, and executives over six months, trying to sell them on his designated-driver idea.

Winsten's idea worked; "designated driver" is now common parlance across all segments of American society and, in 1991, won entry into a Webster's dictionary for the first time. An evaluation of the campaign in 1994 revealed that the designated-driver message had aired on 160 prime-time shows in four seasons and had been the main topic of twenty-five 30-minute or 60-minute episodes. More important, these airings appear to have generated tangible results. In 1989, the year after the designated driver was invented, a Gallup poll found that 67 percent of adults had noted its appearance on network television. What's more, the campaign seems to have influenced adult behavior: Polls conducted by the Roper organization in 1989 and 1991 found significantly increasing awareness and use of designated drivers. By 1991, 37 percent of all U.S. adults claimed to have refrained from drinking at least once to serve as a designated driver, up from 29 percent in 1989. In 1991, 52 percent of adults younger than 30 had served as designated drivers, suggesting that the campaign was having greatest success with its target audience.

In 1988, there were 23,626 drunk-driving fatalities. By 2010, the number of drunk-driving fatalities had declined to 10,228 (National Highway Traffic Safety Administration). Although the Harvard Alcohol Project acknowledges that some of this decline is due to new laws, stricter anti–drunk driving enforcement, and other factors, it claims that many of the 75,000 lives saved by the end of 2007 were saved because of the designated-driver campaign. (The television campaign was only part of the overall campaign; there were strong community-level and public service components as well.)

Of course, making television an explicit vehicle for manipulating behavior has its dangers. My idea of the good might not be yours; if my ideas have access to the airwaves but yours do not, what I'm doing will seem to you like unwanted social engineering. We can all agree that minimizing drunk driving is a good thing—but not everyone agrees on the messages we want to be sending to, say, teenage girls about abstaining from sex versus using condoms, about having an abortion, or about whether interfaith marriages are okay. Television's power to mold viewers' understanding of the world is strong enough that we need to be aware that embedding messages about moral values or social behavior can have potent effects—for good or for ill.

*Sources:* Jane Rosenzweig, "Can TV Improve Us?" *American Prospect* vol. 10:45 (July/August 1999). Reprinted with permission from American Prospect; U.S. Department of Transportation (April 1, 2011) *Traffic Fatalities in 2010 Drop to Lowest Level in Recorded History*. Washington, D.C.: NHTSA National Center for Statistics and Analysis

---

TV-14 for a show that might be unsuitable for children under 14; and TV-MA for a show suitable only for mature audiences. The ratings are designed to work in conjunction with the V-chip, a device in all new television sets produced after 1997. The V-chip allows parents to program their sets to screen out any program rated beyond a desired level (Federal Communications Commission, 2005). Eighty-five percent of the public does not know the V-chip exists.

This might not help the problem. In a study by Mediascope (1996), when boys, especially those aged 10 to 14, saw that a program or movie had a parental-discretion advisory, they found that show more attractive and wanted to watch it. In addition, no evidence was found that antiviolence public service announcements altered adolescents' attitudes toward the appropriateness of using violence to resolve conflict.

Proving cause and effect in sociology is never easy, and the relationship between television and violence might be more complex than is generally acknowledged. Yet, a study (Johnson, 2002) that followed 707 subjects for 17 years uncovered some disturbing findings. According to the study, adolescents who watched more than one hour a day of television—regardless of content—were roughly four times more likely to commit aggressive acts toward other people later in their lives than those who watched less than one hour. Of those who watched more than three hours, 28.8 percent were later involved in assaults, robberies, fights, and other aggressive behavior (Sappenfield, 2002).

A 15-year longitudinal study of 329 youth found that children who watched a great deal of violent TV programs between the ages of 6 and 10 exhibited high levels of aggression as young adults. The findings held true regardless of the child's intellectual capabilities, social status, or type of parenting they received (Huesmann et al., 2003).

As the popularity of video games has increased, and the average child spends about 13 hours a week playing them, new concerns have been raised about the potential socialization aspects of this medium. Many video games have violent themes, and the same concerns about the impact of television on aggressive behavior are now being raised about this entertainment medium. Do games such as Postal 2, Grand Theft Auto III, Manhunt, or Madworld increase a person's aggressive thoughts and behavior? One study (Anderson and Dill, 2000) of 227 college students found that those who had played violent video games in junior high school were more prone to aggressive behavior in college. The thinking is that playing video games provides a way of "practicing aggressive solutions to conflict situations." The concern is that exposure to violent video games is more dangerous than exposure to violent television programs or movies because of the interactive nature of the games. The player is more involved in carrying out the violence than is the passive viewer of a program.

Repeated exposure to video game violence increases the potential for aggressive behavior because (1) it produces more positive attitudes and expectations regarding the use of aggression; (2) it leads to rehearsing more aggressive solutions to problems; (3) it decreases consideration of nonviolent alternatives; and (4) it decreases the likelihood of thinking of conflict, aggression, and violence as unacceptable alternatives (Anderson, 2003).

Clearly, many other factors are involved in the relationship between television, movies, and video games and violence. The relationship between violent acts and antisocial behavior is much more complicated than was originally thought. For both adults and children, the social context, peer influence, values, and attitudes also play an important a role in determining behavior.

## ● ADULT SOCIALIZATION

A person's primary socialization is completed when he or she reaches adulthood. **Primary socialization** means that *individuals have mastered the basic information and skills required of members of a society.* He or she has (1) learned a language and can think logically to some degree, (2) accepted the basic norms and values of the culture, (3) developed the ability to pattern behavior in terms of these norms and values, and (4) assumed a culturally appropriate social identity.

There is still much for a person to learn, however, and many new social identities to explore. Socialization, therefore, continues during the adult years. **Adult socialization** is *the process by which adults learn new statuses and roles.* It differs from primary socialization in two ways. First, adults are much more aware than young people are of the processes through which they are being socialized. In fact, adults deliberately engage in programs such as advanced education or on-the-job training in which socialization is an explicit goal. Second, adults often have more control over how they wish to be socialized and therefore can generate more enthusiasm for the process. Whether going to business school, taking up a new hobby, or signing up for the Peace Corps, adults can decide to channel their energy into making the most effective use of an opportunity to learn new skills or knowledge.

An important aspect of adult socialization is **resocialization,** which involves *exposure to ideas or values that in one way or another conflict with what was learned in childhood.* This is a common experience for college students who leave their homes for the first time and encounter a new environment in which many of their family's cherished beliefs and values are held up to critical examination. Changes in religious and political values are not uncommon during the college years, which often lead to a time of stress for students and their parents.

Erving Goffman (1971) discussed the major resocialization that occurs in **total institutions**—*environments such as prisons or mental hospitals in which the participants are physically and socially isolated from the outside world.* Goffman noted several factors that produce effective resocialization. These include (1) isolation from the outside world, (2) spending all of one's time in the same place with the same people, (3) shedding individual identity by giving up old clothes and possessions for standard uniforms, (4) a clean break with the past, and (5) loss of freedom of action. Under these circumstances, an individual usually changes in a major way along the lines prescribed by those doing the resocialization.

During the 1970s and 1980s, a number of religious cults gained notoriety because they attracted thousands of followers. Hundreds of cults continue to exist today, but we notice them only when they have trouble with authorities. The methods various cults use to indoctrinate their members can be seen as a conscious attempt at resocialization. New members are swept up in the communal spirit of the cult. Group pressure eventually can produce major personality changes in the recruits, and new value systems replace the ones learned previously. Consequently, friends and family members might no longer recognize the person who has been resocialized. The tactics that some religious cults use have been criticized widely. The cults defend their programs as simply a means through which they encourage people to rid themselves of old ideas and replace them with new ones.

It is not unusual for those undergoing resocialization experiences to become confused and depressed and to question whether they have chosen the right course. Some drop out; others eventually stop resisting and accept the values of their instructors.

In the following sections, we will discuss four events in adult socialization: marriage, parenthood, work, and aging.

## Marriage and Responsibility

As Ruth Benedict (1938) noted in a now classic article on socialization in America, "our culture goes to great extremes in emphasizing contrasts between the child and the adult." We think of childhood as a time without cares, a time for play. Adulthood, by contrast, is marked by work and taking up the burden of responsibility. One of the great adult responsibilities in our society is marriage.

Indeed, today's young adults no longer accept uncritically many of the traditional role expectations of marriage. For both men and women, choices loom large. How much of oneself should one devote to a career? How much to self-improvement and personal growth? How much to a spouse? Ours is a time of uncertainty and experimentation. Even so, marriage still retains its primacy as a life choice for adults. Although divorce has become common in many circles, marriage still is treated seriously as a public statement that both partners are committed to each other and to stability and responsibility. (We will discuss marriage and family arrangements in greater detail in Chapter 12, "Marriage and Alternative Family Arrangements.")

Once married, the new partners must define their relationships to each other and in respect to the demands of society. This is not as easy today as it was when these choices largely were determined by tradition. Although friends, parents, and relatives usually are only too ready to instruct the young couple in the shoulds and should nots of married life, such attempts at socialization are often resented by young people who wish to chart their own courses. One choice they must make is whether to become parents.

## Parenthood

When a couple has a child, their responsibilities increase enormously. They must find ways to provide the care and nurturing necessary to the healthy development of their baby while working hard to keep their own relationship intact because the arrival of an infant inevitably is accompanied by stress. This requires a reexamination of the role expectations each partner has of the other, both as a parent and as a spouse.

Of course, most parents anticipate some stresses and try to resolve them before the baby is born. They make financial plans, create a living space, and study baby care. They ask friends and relatives for advice and secure their future babysitting services. However, not all the stresses of parenthood are so obvious. One that is overlooked frequently is the fact that parenthood is itself a new developmental phase.

The psychology of becoming and being a parent is extremely complicated. During the pregnancy, both parents experience intense feelings—some expected, others quite surprising. Some of these feelings might even be upsetting, such as the fear that one will not be an adequate parent or that one might even harm the child. Sometimes, such feelings lead people to reconsider their decision to become parents.

The birth of the child brings forth new feelings in the parents, many of which can be traced to the parents' own experiences as infants. As their child grows and passes through all the stages of development we have described, parents relive their own development. In psychological terms, parenthood can be viewed as a second chance: Adults can bring to bear all they have learned to resolve the conflicts that were not resolved when they were children. For example, it might be possible for some parents to develop a more trusting approach to life while observing their infants grappling with the conflict of basic trust and mistrust (Erikson's first stage).

## Career Development: Vocation and Identity

Taking a job involves more than finding a place to work. It means stepping into a new social context with its own statuses and roles, and it requires a person to be socialized to meet the needs of the situation. These new conditions can even include learning how to dress appropriately. For example, a young management trainee in a major corporation was criticized for wearing his keys on a ring snapped to his belt. "Janitors wear their keys," his supervisor told him. "Executives keep them in their pockets." The keys disappeared from the trainee's belt.

Aspiring climbers of the occupational ladder even might have to adjust their personalities to fit the job. In the 1950s and 1960s, corporations looked for quiet, loyal, tradition-oriented men to fill their management positions—men who would not upset the status quo (Whyte, 1956)—and most certainly not women. Today, especially in high-tech industries, the trend has been toward recruiting men and women who show drive and initiative and a capacity for creative thinking and problem solving.

Some occupations require extensive resocialization. Individuals wishing to become doctors or nurses, for example, must overcome their squeamishness about blood, body wastes, genitals, and the inside of the body. They also must accept the undemocratic fact that they will receive much of their training while caring for poor patients (usually ethnic minorities). Wealthier patients are more likely to receive care from fully trained personnel.

## OUR DIVERSE WORLD: Could You Be a Success at a Japanese Company?

In American society, hard work is seen as admirable, and individual accomplishments are rewarded. In other societies, work success depends on the relationship the individual develops with superiors and others in the work group.

Lifetime employment is a central feature of Japanese business culture. Young people in Japan hope to get a job at a large corporation after graduating from a university. They expect that they will retire from that same company when they reach age 65. This feature holds true for the majority of employees at Japan's major corporations. Strong bonds develop between the company and employees and between coworkers and managers.

Japanese companies hire people with the expectation that Japanese society has socialized them to accept:

1. complete loyalty and dedication to the company, even when it competes with family concerns
2. never criticize their company, coworkers, or managers, and conform
3. be reliable, conservative, and make all decisions in committees

Not only are many American students poorly suited for Japanese work expectations, but so are Japanese students who have studied in the United States. Japanese companies, often look at these American-trained students as being too elite to fit in, too eager to get ahead, and too likely to be poached or to switch employers before long. These students also tend to be too old given the rigid age preferences that graduates be no older than their mid-20s.

Kenta Koga was attending Yale University when he returned to Japan for a summer internship at a big Japanese advertising agency in Tokyo. On one occasion, he accompanied his boss to a meeting discussing trends in technology and social media. Kenta, a computer science major, expressed his views, because his boss was misinformed about certain points. He was careful to preface his comments with "I'm terribly sorry to interrupt," or "My deepest apologies if you already knew this." His boss was still annoyed and said he overstepped his bounds and prevented other people from speaking.

If it is difficult for a Kenta Koga to readjust to Japanese culture, imagine how difficult it would be for a young American new to Japanese management rules. Jeff was a 28-year-old who worked for Motorola Nippon. Jeff believed that he had to prove himself by working hard. He got up at 6:30 a.m. and scheduled four sales visits a day. After nine months in the job, Jeff was outselling the three other sales executives in his unit, all Japanese. After eighteen months, he had outsold all three of them put together. This was, he felt, because of his hard work and persistence. He was out making sales calls 95 percent of the time and rarely bothered his boss. You can imagine how shocked he was when at his annual appraisal meeting he received an "acceptable." This was the lowest mark possible, short of "unacceptable," which would have led to being fired. It was also the worst appraisal in the department.

According to American values, Jeff had performed extremely well. He was living and working in Japan, however, and the cultural expectations were different. It was assumed that as he did well, he would keep his boss informed and invite him to share the successes. Jeff also should have kept his coworkers informed so that they could benefit from his knowledge.

Jeff needed to be more modest and should have sought advice every step of the way. Shared knowledge is vital in Japanese organizations. Adjusting to such an approach is not easy when one is accustomed to American workplace cultural values. We often do not realize how much we are socialized by our culture until we encounter it directly.

*Sources:* Tabuchi, Hiroko. May 29, 2012. "Young and Global Need Not Apply in Japan." *The New York Times*; Venture Japan. "Japanese Business Culture" (www.venturejapan.com/japanese-business-culture.htm), accessed June 3, 2012; Hampden-Turner, C. and F. Trompenaars. 2000. *Building Cross-Cultural Competence.* New Haven, CT: Yale University Press, pp. 175–77.

The armed forces use basic training to socialize recruits to obey orders without hesitating and to accept killing as a necessary part of their work. For many people, such resocialization can be quite confusing and painful. (For an example of career resocialization for an American in Japan, see "Our Diverse World: Could You Be a Success at a Japanese Company?")

For some individuals, career and identity are so intertwined that job loss can lead to personal crisis. This occurs for many people who are downsized or encouraged to retire. For many, losing a job means reevaluation and a new direction. For others, it means spending months looking for a new job and feeling a profound loss of self-identity.

## ● AGING AND SOCIETY

In many societies, such as Japan and China, age brings respect and honor. Older people are turned to for advice, and their opinions are valued because they reflect a full measure of experience. Often, older people are not required to stop their productive work simply because they have reached a certain age. Rather, they work as long as they are able to, and their tasks might be modified to allow them to continue to work virtually until they die. In this way, people maintain their social identities as they grow old—and their feelings of self-esteem as well.

This is not the case in the United States. Most employers expect their employees to retire well before they have reached age 70, and Social Security regulations restrict the amount of nontaxable income that retired people may earn.

Perhaps the biggest concern of the elderly is where they will live and who will take care of them when they get sick. The American nuclear family ordinarily is not prepared to accommodate an aging parent who is sick or whose spouse has recently died. In addition, with the increasing life span, many elderly who might be in their late 70s or early 80s have sons and daughters who themselves can be in their 50s or even 60s. As a result, those older people who have trouble moving around or caring for themselves often have no choice but to live in protected environments.

This means that late in life, many people are forced to acquire another social identity. Sadly, it is not a valued one but rather one of being less valued and less important. This change can be damaging to older people's self-esteem, and it can even hasten them to their graves. The past two decades have seen some attempts at reform to address these issues. Age discrimination in hiring is illegal, and some companies have extended or eliminated arbitrary retirement ages. However, the problem will not be resolved until elderly people achieve a position of respect and value in American culture equal to that of younger adults.

Even though aging is a biological process, becoming old is a social and cultural one. Only society can create a senior citizen. From infancy to old age, both biology and society play important parts in determining how people develop over the course of their lives.

## ■ SUMMARY

### ▸ What is socialization?

The process of social interaction that teaches children the intellectual, physical, and social skills and the cultural knowledge they need to function as members of society is called socialization. In the course of this process, each child acquires a personality, patterns of behavior, and ways of thinking that are distinctive for each individual.

### ▸ What is primary socialization?

Primary socialization means that individuals have mastered the basic information and skills required of members of a society. They have (1) learned a language and can think logically to some degree, (2) accepted the basic norms and values of the culture, (3) developed the ability to pattern behavior in terms of these norms and values, and (4) assumed a culturally appropriate social identity.

### ▸ How does biology and socialization contribute to the formation of the individual?

From infancy to old age, both biology and society play important parts in determining how people develop. Unlike other animal species, human offspring have a long period of dependency. During this time, parents and society work together to make children social beings.

### ▸ How do people develop a social identity?

Every individual comes to possess a social identity by occupying culturally and socially defined positions. In addition, the individual acquires a changing yet enduring personal identity called the self, which develops when the individual becomes aware of his or her feelings, thoughts, and behaviors as distinct from those of other people.

### ▸ What is sociobiology?

The discipline of sociobiology, also known as evolutionary psychology or human behavioral ecology, uses biological principles to explain the behavior of all social beings, both animal and human.

### ▸ How does extreme social deprivation affect early childhood development?

Examples of extreme childhood isolation point to the fact that none of the behavior we think of as typically human arises spontaneously. Humans must be taught to stand up, to walk, to talk, even to think. Human infants must develop social attachments;

they must learn to have meaningful interactions and affectionate bonds with others.

▶ **Explain the views of Charles Horton Cooley and George Herbert Mead.**

Cooley and Mead saw the individual and society as partners. They were symbolic interactionists and, as such, believed that the individual develops a self solely through social relationships—that is, through interaction with others. They believed that all our behaviors, our attitudes, even our ideas of self arise from our interactions with other people. Hence, they were pure environmentalists in that they believed that social forces rather than genetic factors shape the individual.

▶ **Describe Erik Erikson's model of lifelong socialization.**

Erikson believed that human development is accomplished in eight stages. Each stage amounts to a crisis of sorts brought on by two factors: biological changes in the developing individual and social expectations and stresses. At each stage, the individual is pulled in opposite directions to resolve the crisis. In normal development, the individual resolves the conflict experienced at each stage.

▶ **How do the family, schools, peer groups, and the mass media contribute to childhood socialization?**

In American society, the family is the most important socializing influence in early childhood development. As the child grows older and moves into society, other agents of socialization come into play. Schools are increasingly expected to meet a variety of social and emotional needs as well as to pass on knowledge and help children develop skills. From school age to early adulthood, peers powerfully influence lifestyle orientations and, in some cases, values. The mass media present today's children with an enormous amount of information, both for better and for worse.

▶ **How does adult socialization differ from primary socialization?**

Although socialization continues throughout one's life, adult socialization differs from primary socialization in that adults are much more aware of the processes through which they are socialized, and they often have more control over the processes.

▶ **What is resocialization, and where does it take place?**

One important form of adult socialization is resocialization, which involves exposure to ideas or values that conflict with what was learned in childhood. Resocialization occurs in occupational settings, among others, as well as in total institutions such as prisons and mental hospitals, where participants are physically, socially, and psychically isolated from the outside world.

## Media Resources

**CourseMate for *Introduction to Sociology*, Eleventh Edition**

Cengage Learning's Sociology CourseMate brings course concepts to life with interactive learning, study, and exam preparation tools that support the printed textbook. Access an integrated eBook, learning tools including glossaries, flashcards, quizzes, videos, and more in your Sociology CourseMate. Go to CengageBrain.com to register or purchase access.

# CHAPTER FOUR STUDY GUIDE

## Key Concepts

*Match each of the following concepts with its definition, illustration, or explanation.*

a. Sociobiology
b. Nature vs. nurture
c. Attachment disorder
d. Statuses
e. Social identity
f. Self
g. Cognitive development
h. Moral development
i. Looking-glass self
j. The I and the me
k. Significant others
l. Generalized others
m. Superego
n. Primary socialization
o. Adult socialization
p. Resocialization
q. Total institutions
r. Personality

____ 1. The child's assimilation of the basic elements of culture—language, norms, behavior—and adoption of a culturally appropriate identity
____ 2. The combination of statuses that define who a person is in society
____ 3. The use of Darwinian principles of evolution to explain the social behavior of animals and humans
____ 4. Places in which the incarcerated (inmates, patients, etc.) are cut off from the outside world and live under the control and authority of staff members
____ 5. A series of stages of increasingly complex thinking about what is right and wrong in specific situations
____ 6. The inability to form relationships or trust other people, often found in children who grew up having minimal interaction with adults
____ 7. A set of stages in a person's ability to think logically and abstractly about how the world works
____ 8. A sense of who you are based on a three-step process of how you think other people would judge you
____ 9. Term used by Mead to include those individuals who are most important in our development, for example, parents and friends
____ 10. The debate over the relative importance of biological and genetic factors on the one hand and cultural forces on the other in shaping human behavior
____ 11. The viewpoints, attitudes, and expectations of society as a whole or of a community of people of whom we are aware and who are important to us
____ 12. The parts of the self, according to Mead, one more active and spontaneous, the other more a product of socialization
____ 13. According to Freud, the part of the self that represents society's norms and moral values learned primarily from parents
____ 14. The process by which adults learn new statuses and roles
____ 15. Culturally or socially defined positions in a social system
____ 16. Exposure to ideas or values that in one way or another conflict with what was learned in childhood
____ 17. An individual's changing yet enduring personal identity
____ 18. The patterns of behavior and ways of thinking and feeling that are distinctive for each individual

## Key Thinkers

*Match the thinkers with their main idea or contribution.*

a. Harry Harlow
b. Jean Piaget
c. Lawrence Kohlberg
d. Sigmund Freud
e. Stephen Jay Gould
f. Ivan Pavlov
g. Edward O. Wilson
h. George Herbert Mead
i. Erik Erikson
j. Charles Horton Cooley

____ 1. Through experiments with dogs, demonstrated that behavior could be conditioned
____ 2. Proposed a theory of socialization based on the development of the me; saw children's relation to rules as moving through three stages: preparatory, play, and game
____ 3. Coined the term *sociobiology* and was its major advocate as an explanation of human behavior
____ 4. Biologist who criticized sociobiology, offering instead explanations based on culture rather than on genetics and evolution
____ 5. Illustrated the harmful effects of social isolation through his experiments with rhesus monkeys
____ 6. Offered a theory of childhood development based on developmental problems rooted both in biological changes in the individual and in social expectations in the culture
____ 7. Maintained that moral thinking developed through five to six distinctive stages
____ 8. Argued that society's demand for civilized behavior constantly conflicted with the individual's basic instincts of sex and aggression
____ 9. Studied the stages of cognitive development that children go through in learning to think logically about the world
____ 10. Offered a theory of childhood development based on the looking-glass self—a person's sense of other people's evaluations

## Central Idea Completions

*Fill in the appropriate concepts and descriptions for each of the following questions.*

1. Discuss the roles biology and socialization play in the formation of the individual.

   _____
   _____
   _____
   _____
   _____

2. Use the case histories presented in Chapter 4 to discuss how extreme social isolation and deprivation affect a human's early childhood development.

   _____
   _____
   _____
   _____

3. Briefly discuss each of the following sources of influence on the socialization of children.
   a. Family _____
   b. School _____
   c. Peer groups _____
   d. Mass media _____

4. Define and discuss *resocialization*.

5. Explain the key features of the developmental-stage models of Piaget and Erikson.

   a. Piaget

   b. Erikson

## ● Critical Thought Exercises

1. The nature versus nurture debate has been an ongoing controversy in the social sciences. Develop a pro and con list of evidence for both positions, as one might for a debate situation. Discuss each position in the debate, providing examples to support the position.

2. Visit the children's section of your local library. After examining several books aimed at different age levels, discuss how these books might reflect Piaget's work on the developmental stages of childhood. Find examples of books you feel to be ideally suited to appeal to children at different levels of development. Select three titles you would want to read to your own children someday and explain why you think they are examples of valuable children's books.

3. As you think about the material in this chapter, consider the role that day care plays in the lives of our children. How does the time spent in day care shift the locus of childhood socialization? What might be the strengths and weaknesses of a childhood spent in day-care situations?

4. How important are peer groups compared to families as agents of socialization? After rereading the box on "Acting White," reflect on your own high-school experiences and those of people you knew. Did the values and norms of the peer group conflict with those of parents or others? If so, how did these differing worlds affect your reactions, thoughts, and feelings and those of your friends?

5. Think of things you have done that might have been of moral interest, for example, telling a lie or telling an uncomfortable truth, cheating or not cheating on an exam (or in a relationship), revealing or refusing to reveal something told to you in confidence, hazing or refusing to haze a fraternity pledge, and so on. Jot down your main justification for each of these actions. Where would each of these justifications fall on Kohlberg's scale? How consistent is the level of moral reasoning? Then try this same exercise by asking someone else about similar decisions in that person's life.

## Internet Activities

1. Visit http://mentalhelp.net/psyhelp/chap3/chap3h.htm. This website begins with a discussion and critique of Kohlberg's ideas on moral development. It then takes you to pages that allow you to write your own philosophy of life by answering a list of questions. It also shows you how college students years ago answered these questions. Check your answers against theirs. Most important, ask yourself to what extent your answers can be seen as the product of socialization in a particular time and place (America around the turn of the twenty-first century).

2. Watch the following informative and entertaining video on classical conditioning developed by Ivan Pavlov (www.youtube.com/watch?v=hhqumfpxuzl).

## Answers to Key Concepts

1. n;  2. e;  3. a;  4. q;  5. h;  6. c;  7. g;  8. i;  9. k;  10. b;  11. l;  12. j;  13. m;  14. o;  15. d; 16. p;  17. f;  18. r

## Answers to Key Thinkers

1. f;  2. h;  3. g;  4. e;  5. a;  6. i;  7. c;  8. d;  9. b;  10. j

# 5 Social Interaction

## Understanding Social Interaction
- Contexts
- Norms

*Our Diverse World:* Cross-Cultural Social Interaction Quiz
- Ethnomethodology
- Dramaturgy

*Day-to-Day Sociology:* Can You Spot a Liar?

## Types of Social Interaction
- Nonverbal Behavior
- Exchange
- Cooperation
- Conflict
- Competition

*Day-to-Day Sociology:* Laugh and the World Laughs with You

## Elements of Social Interaction
- Statuses
- Roles
- Role Sets

*How Sociologists Do It:* Southerners Are Really Friendly Until You Disrespect Them
- Role Strain
- Role Conflict
- Role Playing

*Thinking About Social Issues:* Don't Call Me. I'll Text You

## Collective Behavior
- Fads and Fashions
- Rumors
- Public Opinion
- Mass Hysteria and Panic

## Summary

## LEARNING OBJECTIVES

After studying this chapter, you should be able to do the following:

- Understand why it is important to understand social interaction.
- Know what the major types of social interaction are.
- Understand the concept of role.
- Explain the role norms play in social interaction.
- Describe the main features of statuses.
- Understand how the context of a situation influences social interaction.
- Know the difference between role strain and role conflict.
- Describe the role collective behavior plays in social interaction.

A young man in jeans and a long-sleeved T-shirt went into the Washington, D.C., subway on a Friday morning. He took his violin out of its case, placed the open case at his feet, threw a few dollars and pocket change in it, and began to play. It was the middle of the morning rush hour, and for the next 43 minutes, he performed six classical pieces. During that time, 1,097 people passed by. Each passerby had a quick decision to make: to stop and listen, to throw some money into the violin case, or to hurry past.

No one knew that the violinist was Joshua Bell, one of the finest classical musicians in the world. He was playing on a 1713 violin made by Antonio Stradivari that he had bought for $3.5 million. The musician did not play popular tunes, but instead played masterpieces that have been played in concert halls.

So, what do you think happened?

In the three-quarters of an hour that Joshua Bell played, seven people stopped what they were doing to hang around and take in the performance for at least a minute. Twenty-seven gave money, most of them as they walked by without listening—for a total of $32. Three days earlier Bell had filled Boston's Symphony Hall, where modest seats went for $100. That means 1,070 people hurried by, oblivious to what they had missed.

Afterward Bell was mystified by the number of people who did not pay attention at all, as if he were invisible. He thought the inattention might be deliberate, because then the people would not have to give any money. People also said they were busy and had other things on their mind.

In 1831, a young French sociologist named Alexis de Tocqueville visited the United States and found himself impressed, as well as dismayed, by the degree to which Americans were driven, to the exclusion of everything else, in the pursuit of wealth. A violinist would not be much of a force in such a situation.

Our ability to know how to respond to a person playing his violin depends on us knowing how to put it in context with our previous experiences. Where a social interaction occurs makes a difference in what it means. There seem to be three elements that help us to understand a social interaction: (1) *the physical setting or place*—Bell was playing in a subway, not a concert hall; (2) *the social environment*—most of the people were by themselves on their way to work; and (3) *the activities surrounding the social interaction*—many were distracted by other demands. Some people were talking on cell phones. Nobody had set out that day to hear a violin concert in the subway. The context of an interaction consists of many elements. Without knowledge of these elements, it is impossible to know the meaning of even the simplest interaction.

After graduating from college, Jason Williams took a job with a company that sent him on business trips to Indonesia and other Asian countries. He enjoyed the opportunity to travel but found some of his experiences confusing. Indonesia is a crowded society and the people are used to invasions of personal space. If Jason was riding on an empty escalator, a stranger would walk down the steps until he was standing on the same step as him. He had similar experiences in other countries also. He would be in an empty elevator and another person would come in and stand uncomfortably close to him. He could not wait to escape from the ride.

What Jason did not realize is that in Asia people gravitate toward each other and are not offended by invasions into their personal space. Such actions make Americans like Jason uncomfortable, but it is appropriate behavior in that part of the world. Without realizing it, Jason was engaging in a pattern of social interaction with these strangers.

Most Americans use four main distance zones in their business and social relations: intimate, personal, social, and public. Intimate distance varies from direct physical contact with another person to a distance of 6 to 18 inches and is used for private activities with another close acquaintance. Personal distance varies from 2 to 4 feet and is the most common spacing that people use in conversation. The third zone, social distance, is employed during business transactions or interactions with a clerk or salesperson. As they work with each other, people tend to use this distance, which is usually between 4 and 12 feet. Public distance, which is used, for instance, by teachers in classrooms or speakers at public gatherings, can be anywhere from 12 to 25 feet or more (Hall, 1969).

Jason tended to think of himself as being in the fourth zone and thought that the people in Asia were progressing to the second or third zone. As Jason discovered, there is no way not to interact with others. Social interaction has no opposite. If we accept the fact that all social interaction has a message value, it follows that no matter how we try, we cannot not send a message. Activity or inactivity, words or silence all contain a message. They influence others, and others respond to these messages. The mere absence of talking or taking notice of each other is no exception because that behavior is a message also.

When sociologists study human behavior, they are interested primarily in how people affect each other through their actions. They look at the overt behaviors that produce responses from others as well as at the subtle cues that can result in unintended consequences. Human social interaction is very flexible and quite unlike that of the social animals.

## UNDERSTANDING SOCIAL INTERACTION

Max Weber (1922) was one of the first sociologists to stress the importance of social interaction in the study of sociology. He argued that the main goal of sociology is to explain what he called **social action,** a term he used to refer to *anything people are conscious of doing because of other people.* Weber claimed that to interpret social actions, we have to put ourselves in the positions of the people we are studying and try to understand their thoughts and motives. The German word Weber used for this is *verstehen,* which can be translated as "sympathetic understanding."

In a public space, people tend to stay anywhere from 12 to 25 feet from each other.

Weber's use of the term *social action* identifies only half of the puzzle because it deals only with one individual taking others into account before acting. A **social interaction** involves *two or more people taking one another into account.* It is the interplay between the actions of these individuals. In this respect, social interaction is a central concept to understanding the nature of social life.

In this chapter, we will explain how sociologists investigate social interaction. We will start with the basic types of social interaction, whether verbal or nonverbal. Next, we will examine how social interaction affects those involved in it. We then will broaden our focus a bit and move on to groups and social interactions within them. Finally, we will look at the large groupings of people that make social life possible and that ultimately make up the social structure. In other words, we will start with social behavior at the most basic level and move outward to ever more complicated levels of social interaction.

Social interaction is a central concept to understanding the nature of social life.

### Contexts

Where a social interaction occurs makes a difference in what it means. Edward T. Hall (1974) identified *three elements that, taken together, define the* **context** *of a social interaction: (1) the physical setting or place, (2) the social environment, and (3) the activities surrounding the interaction—preceding it, happening simultaneously with it, and coming after it.*

Without knowledge of the context, it is impossible to know the meaning of even the simplest interaction. For example, Germans and Americans treat space very differently. Hall (1969) noted that in many ways, the difference between German and American doors gives us a clue about the space perceptions of these two cultures. In Germany, public and private buildings usually have double doors that create a soundproof environment. Germans feel that American doors, in contrast, are flimsy and light, inadequate for providing the privacy that Germans require.

In American offices, doors usually are kept open; in German offices, they are kept closed. In Germany, the closed door does not mean that the individual wants to be left alone or that the people inside are planning something that should not be seen by others. Germans simply think that open doors are sloppy and disorderly. As Hall explained it:

> I was once called in to advise a firm that has operations all over the world. One of the first questions asked was, "How do you get the Germans to keep their doors open?" In this company, the open doors were making the Germans feel exposed and gave the whole operation an unusually relaxed and unbusinesslike air. Closed doors, on the other hand, gave the Americans the feeling that there was a

The location and context in which a social interaction takes place make a difference in what it means.

conspiratorial air about the place and that they were being left out. The point is that whether the door is open or shut, it is not going to mean the same thing in the two countries. (Hall, 1969)

In Japan is a third view on the issue. Most Japanese executives prefer to share offices to ensure that information can flow easily and each person knows what is happening in the other's area of responsibility. The Japanese executive does not want to risk being unaware of events as they are developing. Japanese firms have ceremonial rooms for receiving visitors, but few other work areas afford any privacy (Hall and Hall, 1987). (See "Our Diverse World: Cross-Cultural Social Interaction Quiz.")

## Norms

Human behavior is not random. It is patterned and, for the most part, quite predictable. What makes human beings act predictably in certain situations? For one thing, there is the presence of **norms**, *specific rules of behavior, agreed upon and shared, that prescribe limits of acceptable behavior.*

Norms tell us the things we should both do and not do. In fact, our society's norms are so much a part of us that we often are not aware of them until they are violated. Consider the unfortunate circumstances that happened in Suzanne Berger's life, for example. One day, she bent down to pick up her child only to discover she could not straighten up again. In the process of bending, she had injured her back so severely that for years thereafter she could neither walk nor sit for more than a few minutes. Traveling anywhere meant that she had to take along a mat and immediately lie down. In effect, she had to violate the norms that assume that you will not stretch out on the floor in a department store, a train station, a classroom, or at public events. As she described it:

> Strangers try not to stare.... At airports and train stations, people have thought I was a derelict or crazy or maybe homeless; only the dispossessed lie on floors, children lie on floors, dogs lie on floors... but adults? *What's that woman doing over there?* a security guard said at the airport. *Dunno, leave her alone. Must be drunk.* With friends inside my house, being down here upsets a balance of conviviality, of the *whereness* that grounds a conversation. I am always looking up, as though younger or subservient. Outside, I live down with mother-dirt, grass, the asphalt of the city. Wherever I go, I lie down with my mat. *Hey, lady, what the hell you doing down there?* says a child on a city playground. *You sick? You tired?* (Berger, 1996)

We also have norms that guide us in how we present ourselves to others. We realize that how we dress, how we speak, and the objects we possess relay information about us. In this respect, North Americans are a rather outgoing people. The Japanese have learned that it is a sign of weakness to disclose too much of oneself by overt actions. They are taught very early in life that touching, laughing, crying, or speaking loudly in public are not acceptable ways of interacting.

Not only can the norms for behavior differ considerably from one culture to another, but they also can differ within our own society. Conflicting interpretations of an action can exist among different ethnic groups. Unfamiliarity with such cultural communication can lead to misinterpretations and even unintended insults. For example, African Americans and white Americans might have different interpretations of particular styles of dress or nonverbal behaviors. Among African Americans, listeners are expected to avert their eyes from the speaker and in doing so are showing respect. Among white Americans, looking at the speaker directly is seen as a sign of respect, and looking away can indicate disinterest or boredom.

Let us assume that a white teacher is speaking directly to an African American boy. The youngster might avert his eyes as she is speaking. The teacher might think he is not listening. Worse yet, she might say something like, "Look at me when I speak to you," and the boy will be even more confused. He might have been led to believe that looking at the teacher for an extended time would be regarded as a challenge to her authority.

Sociologists thus need to understand the norms that guide people's behavior because without this knowledge, it is impossible to understand social interaction.

## Ethnomethodology

Many of the social actions we engage in every day are commonplace events. They tend to be taken for granted and rarely are examined or considered. Harold Garfinkel

## OUR DIVERSE WORLD: Cross-Cultural Social Interaction Quiz

### Saudi Arabia

You are visiting Saudi Arabia. You go to the restroom, and a group of Saudi men are washing their feet in the sink. They are doing this because:

1. They have been sweating and their feet stink.
2. Water is scarce in Saudi Arabia.
3. They are preparing to read their prayers.

CORRECT ANSWER: 3. Saudis believe the way to get to Paradise is through prayer, and a key part of prayer is cleanliness.

### Brazil

You walk down a busy street and see a man in a group forming two tubes with his hands and looking through them. What is he doing?

1. Trying to read the license plate on a distant car.
2. Making a joking gesture that he is a spy.
3. Letting the other men know that he has spotted a beautiful woman.

CORRECT ANSWER: 3. This is a common gesture men in Brazil use to let on that they admire an attractive woman.

### Japan

You have spent several days preparing a business plan for a group of Japanese executives. During your presentation, you notice a few of them are sitting with their arms folded and eyes closed. They are:

1. Exhausted from last night's dinner and social events.
2. Listening intently.
3. Letting you know that they hate your ideas.

CORRECT ANSWER: 2. The Japanese believe that when you close your eyes, you can hear better, because you are screening out visual stimuli and focusing only on the words.

### China

You walk along a street and notice that people spit on the sidewalk or blow their nose without a handkerchief. This is considered:

1. An act of personal hygiene that rids the body of waste.
2. An insult to the foreigner who has just walked past.
3. Rude behavior by ignorant people.

CORRECT ANSWER: 1. Even though the Chinese government is trying to get people to stop doing this, it is still considered appropriate, much like washing your hands.

*Sources:* Axtell, Roger E. 2007. *Essential Do's and Taboos: The Complete Guide to International Business and Leisure Travel.* New York: John Wiley and Sons; Axtell, Roger E. 1998. *Gestures: The Do's and Taboos of Body Language around the World.* New York: John Wiley and Sons; Kwintessential (www.kwintessential.co.uk/resources/culture-tests.html), accessed June 2, 2012.

---

(1967) has proposed that studying the commonplace is important. The things we take for granted have a tremendous hold over us because we accept their demands without question or conscious consideration.

**Ethnomethodology** is *the study of the sets of rules or guidelines that individuals use to initiate behavior, respond to behavior, and modify behavior in social settings.* For ethnomethodologists, all social interactions are equally important because they provide information about a society's unwritten rules for social behavior—the shared knowledge that is basic to social life.

Garfinkel asked his students to participate in a number of experiments in which the researcher would violate some of the basic understandings among people. For example, when two people hold a conversation, each assumes that certain things are perfectly clear and obvious and do not need further elaboration. Examine the following conversation and notice what happens when one individual violates some of these expectations:

*Bob:* "That was a very interesting sociology class we had yesterday."

*John:* "How was it interesting?"

*Bob:* "Well, we had a lively discussion about deviant behavior, and everyone seemed to get involved."

*John:* "I'm not certain I know what you mean. How was the discussion lively? How were people involved?"

*Bob:* "You know, they really participated and seemed to get caught up in the discussion."

*John:* "Yes, you said that before, but I want to know what you mean by lively and interesting."

*Bob:* "What's wrong with you? You know what I mean. The class was interesting. I'll see you later."

Bob's response is quite revealing. He is puzzled and does not know whether John is being serious. The normal expectations and understandings around which day-to-day forms of expression occur have been challenged. Still, is it not reasonable to ask for further elaboration of certain statements? Obviously not, when it goes beyond a certain point.

Another example of the confusion brought on by the violation of basic understandings surfaced when Garfinkel asked his students to act like boarders in their own homes. They were to ask whether they could use the phone, take a drink of water, have a snack, and so on. The results were quite dramatic:

> Family members were stupefied. They vigorously sought to make the strange actions intelligible and to restore the situation to normal appearances. Reports were filled with accounts of astonishment, shock, anxiety, embarrassment, and anger and with charges by various family members that the student was mean, inconsiderate, selfish, nasty, or impolite. Family members demanded explanations: "What's the matter?" "What's gotten into you?" "Did you get fired?" "Are you sick?" "What are you being so superior about?" "Why are you mad?" "Are you out of your mind?" "Are you stupid?" One student acutely embarrassed his mother in front of her friends by asking if she minded if he had a little snack from the refrigerator. "Mind if you have a little snack? You've been eating little snacks around here for years without asking me. What's gotten into you?" (Garfinkel, 1972)

Ethnomethodology seeks to make us more aware of the subtle devices we use in creating the realities to which we respond. These realities are often intrinsic in human nature rather than imposed from outside influences. Ethnomethodology addresses questions about the nature of social reality and how we participate in its construction.

## Dramaturgy

People create impressions, and others respond with their own impressions. Erving Goffman (1959, 1963, 1971) concluded that a central feature of human interaction is impression formation—the attempt to present oneself to others in a particular way. Goffman believed that much human interaction can be studied and analyzed on the basis of principles derived from the theater. This approach, known as **dramaturgy**, states that *to create an impression, people play roles, and their performance is judged by others who are alert to any slips that might reveal the actor's true character.* For example, a job applicant at an interview tries to appear composed, self-confident, and capable of handling the position's responsibilities. The interviewer is watching for whether the applicant is really able to work under pressure and perform the necessary functions of the job.

Most interactions require a person to undertake some type of play-acting to present an image that will bring about the desired behavior from others. Dramaturgy sees these interactions as governed by planned behavior designed to enable an individual to present a particular image to others. (For a discussion of how successful we might be when judging the truthfulness of others' image presentation, see "Day-to-Day Sociology: Can You Spot a Liar?")

---

### DAY-TO-DAY SOCIOLOGY

### Can You Spot a Liar?

Can you spot a person who is lying? Do liars always look to the left or down? Do they cover their mouths? Do they fidget or do they hold very still? Do they cross their legs or cross their arms? Do they look you in the eye or do they fail to make eye contact? All of these are things, people claim, that are sure signs that a person is lying.

Charles Bond, a researcher who studied 2,520 adults in 63 countries, found that more than 70 percent thought liars tend to look away. Most also believed that liars squirm, stutter, touch or scratch themselves, or tell longer stories than usual. Every culture seems to have certain liar stereotypes. Yet, there is no reason to believe any of this is true. Bond found that just as there are many kinds of liars, there are many ways to lie, and there is no clear tip-off.

Most people think they are pretty good at spotting liars, yet that is not correct. No more than 5 percent of us seem to be good at spotting liars; the rest of us perform not much better than chance. Even people who you would think are experienced lie catchers, such as judges and customs officials, perform as poorly as the rest of us. We could probably flip a coin and get the same results.

Lying is an important part of social interaction. There are at least 112 English words for lying, including *deception, collusion, fakery,* and *prevarication,*

(*Continued*)

indicating that lying is an important part of social interaction. Can you imagine always telling the truth? You would have to tell people you don't like them, they made stupid choices in clothes, or they have bad breath. Surely, you would leave lots of hurt feelings in your wake. You would also have to reveal things about yourself that you might want to keep private. Everyone would have access to your innermost thoughts and feelings. Think about what you would really say the next time someone asked, "How are you?" Most likely, just saying "Fine" or "OK" would be a lie.

We begin to understand what a lie is at about the age of 3 or 4. That's when we realize that what we might be thinking is different from what others are thinking. Lying means we know something that others do not know. After a while, lying becomes part of our normal social interaction.

How can you become better at spotting liars? There are certain things you should look for. If a person's voice, hand movements, and posture do not seem to fit with what he or she is describing, that might be a clue. Getting agitated about what should be a calm description might be part of a lie. If the person's speech pattern or use of gestures is different than usual, the person might be lying. When people lie, they tend to use distancing language with fewer first-person pronouns and more third-person ones. They might also stall for time as they figure out what they want to say. They might ask for clarification for what might seem like a clear question, or they might repeat the question.

As Mark Twain pointed out, "Everybody lies every day; every hour; awake; asleep; in his dreams; in his joy; in his mourning."

*Sources:* Bond, Jr., Charles F. and Bella M. DePaulo. July 2008. "Individual Differences in Judging Deception: Accuracy and Bias." *Psychological Bulletin* 134(4):477–92; Marantz Henig, Robin. February 5, 2006. "Looking for the Lie." *New York Times*, pp. 47–53, 76, 80.

## TYPES OF SOCIAL INTERACTION

When two individuals are in each other's presence, they inevitably affect each other. They might do so intentionally, as when one person asks the other for change for a dollar, or they might do so unintentionally, as when two people drift toward opposite sides of the elevator in which they are riding. Whether intentional or unintentional, both behaviors represent types of social interaction.

The following sections examine five different types of social interaction: nonverbal behavior, exchange, cooperation, conflict, and competition.

### Nonverbal Behavior

Many researchers have focused our attention on how we communicate with one another by using body movements. This study of body movements, known as *kinesics*, attempts to examine how such things as "slight head nods, yawns, postural shifts, and other nonverbal cues, whether spontaneous or deliberate, affect communication" (Samovar, Porter, and Jain, 1981).

Our culture has taught us a variety of appropriate communication procedures. When they are followed, we feel comfortable with the other person; otherwise, it seems like something is out of place. In the United States, we think of side-by-side conversation as impersonal, to be used when speaking to someone standing next to us at a public event. When we arrange ourselves at a 90-degree angle with another person, we are likely to feel close to the other person and share more personal information. You are likely to do this at a social gathering where you and a friend can speak to each other, but also watch what is going on in the room. The most intimate kind of communication is face-to-face interaction, which gives you much more information about what the other person might be thinking and feeling. You are comfortable doing this with someone you know, but may feel awkward if you are forced to communicate in this manner with a stranger.

All of this becomes even more complex as we move from one culture to the other and try to use communication patterns that may be natural to us, but not to the person from another background. For example, the anthropologist Edward Hall commented:

> [I]t used to puzzle me that a special Arab friend seemed unable to walk and talk at the same time. After years in the United States, he could not bring himself to stroll along, facing forward while talking. Our progress would be arrested while he edged ahead, cutting slightly in front of me and turning sideways so we could see each other. Once in this position, he would stop. His behavior was explained when I learned that for the Arabs to view the other person peripherally is regarded as impolite, and to sit or stand back-to-back is considered very rude. You must be involved when interacting with Arabs who are friends. (Hall, 1969)

The uses of hand and arm movements, eye contact, and norms of nonverbal behavior are markedly different for Arab adults than they are for Americans.

Eye contact is another area where cultural differences are likely to show up. Samovar, Porter, and Jain explain that in the United States, the following has been noted:

1. We tend to look at our communication partner more when we are listening than when we are talking. The search for words frequently finds us, as speakers, looking into space, as if to find the words imprinted somewhere out there.
2. The more rewarding we find the speaker's message to be, the more we will look at him or her.
3. The amount of eye contact we try to establish with other people is determined in part by our perception of their status. . . . When we address someone we regard as having high status, we attempt a modest-to-high degree of eye contact. But when we address a person of low status, we make very little effort to maintain eye contact.
4. We tend to feel discomfort if someone gazes at us for longer than ten seconds at a time.

These notions of eye contact found in the United States differ from those of other societies. In Japan and China, for example, "it is considered rude to look into another person's eyes during conversation." Looking away is a sign of deference in Japan. As mentioned previously, some Japanese may actually close their eyes when thinking deeply and concentrating on an important point (Neuliep, 2008).

Arabs, in contrast, use personal space very differently; they stand very close to the person they are talking to and stare directly into the eyes. Arabs believe that the eyes are a "key to a person's being and that looking deeply into another's eyes allows one to see another's soul."

The proscribed relationships between males and females in a culture also influence eye contact. Asian cultures, for example, consider it "taboo for women to look straight into the eyes of males. Most men, out of respect for this cultural characteristic, do not stare directly at women." French men, on the other hand, accept staring as a cultural norm and often stare at women in public (Axtell, 1998; Samovar, Porter, and Jain, 1981).

There are also cultural differences in the use of hand and arm movements as a means of communication. We all are aware of the different gestures for derision. For some European cultures, it is a closing fist with the thumb protruding between the index and middle fingers. The Russian expresses this same attitude by moving one index finger horizontally across the other.

In the United States, we can indicate that things are okay by making a circle with the thumb and index finger while extending the others. If you make this gesture in Japan, you are signifying "money." And, in Arab countries, if you bare your teeth while making this gesture, you are displaying "extreme hostility."

In the United States, we say good-bye or farewell by waving the hand and arm up and down. If you wave this way in South America, you might discover that the other person is not leaving but moving toward you. That is because in many countries, the gesture we use as a sign of leaving actually means "come."

### Exchange

*When people do something for each other with the express purpose of receiving a reward or return, they are involved in an* **exchange interaction.**

Most employer-employee relationships are exchange relationships. The employee does the job and is rewarded with a salary. The reward in an exchange interaction, however, need not always be material; it can also be based on emotions such as gratitude. For example, if you visit a sick friend, help someone with a heavy package at the supermarket, or help someone solve a problem, you will expect these people to feel grateful to you.

Sociologist Peter Blau (1964) pointed out that exchange is the most basic form of social interaction. He believes social exchange can be observed everywhere after we are sensitized to it.

## Cooperation

A **cooperative interaction** occurs *when people act together to promote common interests or achieve shared goals.* The members of a basketball team pass to one another, block opponents for one another, rebound, and assist one another to achieve a common goal—winning the game. Likewise, family members cooperate to promote their interests as a family—the husband and wife both might hold jobs as well as share in household duties, and the children might help out by mowing the lawn and washing the dishes. College students often cooperate by studying together for tests. (For a discussion of how laughter facilitates cooperation, see "Day-to-Day Sociology: Laugh and the World Laughs with You.")

## Conflict

In a cooperative interaction, people join forces to achieve a common goal. By contrast, people in conflict struggle with one another for some commonly prized object or value. In most conflict relationships, only one person can gain at someone else's expense. Conflicts arise when people or groups have incompatible values or when the rewards or resources available to a society or its members are limited. Thus, conflict usually involves an attempt to gain or use power.

The fact that conflict often leads to unhappiness and violence causes many people to view it negatively. However, conflict appears to be inevitable in human society. A stable society is not a society without conflicts but, rather, one that has developed methods for resolving its conflicts by justly resolving them or brutally suppressing them temporarily. For example, Lewis Coser (1956, 1967) pointed out that conflict can be a positive force in society. The American civil rights movement in the 1950s and 1960s might have seemed threatening and disruptive to many people at the time, but it helped bring about important social changes that led to greater social stability.

Coercion involves the use of power regarded as illegitimate by those on whom it is exerted. The stronger party can impose its will on the weaker, as in the case of a parent using the threat of punishment to impose a curfew on an adolescent. Coercion rests on force or the threat of force, but usually, it operates more subtly.

## Competition

The fifth type of social interaction, **competition,** is *a form of conflict in which individuals or groups confine their conflict within agreed-upon rules.* Competition is a common form of interaction in the modern world—not only on the sports field but also in the marketplace, the educational system, and the political system.

Spontaneous cooperation that arises from the needs of a particular situation is the oldest and most natural form of cooperation.

American presidential elections, for example, are based on competition. Candidates for each party compete throughout the primaries, and eventually, one candidate is selected to represent each of the major parties. The competition grows even more intense as the remaining candidates battle directly against each other to persuade a nation of voters that he or she is the best person for the presidency.

Some types of relationships might span the entire range of focused interactions; an excellent example is marriage. Husbands and wives cooperate in household chores and responsibilities. They also engage in exchange interactions. Married people often discuss their problems with each other—the partner whose role is listener at one time will expect the spouse to provide a sympathetic ear at another time. Married people also experience conflicts in their relationship. A couple might have a limited amount of money set aside, and each might want to use it for a different purpose. Unless they can agree on a third, mutually desirable use for the money, one spouse will gain at the other's expense, and the marriage might suffer. The husband and wife whose marriage is irreversibly damaged might find themselves in direct competition. If they wish to separate or divorce, their conflict will be regulated according to legal and judicial rules.

Through the course of our lifetimes, we constantly are involved in several types of social interaction because we spend most of our time in some kind of group situation. How we behave in these situations is generally determined by two factors—the statuses we occupy and the roles we play—which together constitute the main components of what sociologists call social organization.

## DAY-TO-DAY SOCIOLOGY

### Laugh and the World Laughs with You

Why do people laugh? It is possible that laughing began as a community's shared sign of relief after some passing danger. The group would be signaling, "We can now let our guard down and relax." You will notice that when people laugh, the muscles in their body do, in fact, relax. And we have all heard stories of people who have laughed so hard they fell over or worse. The relaxation might also encourage us to trust those we are sharing the laugh with.

Most laughter is not about something particularly funny. Laughing is a social event. Studies have shown that people are 30 times more likely to laugh when they are with others than when they are alone. Laughter helps strengthen our social bonds with others. When people are in a group that is laughing, they join in so as not to feel left out. The more everyone laughs, the stronger the bonds become. Most of this laughter is not really in response to a formal effort at telling a joke. Most of the time, the laughter is after a fairly innocuous comment such as, "Wow, look what he's wearing" or "Gee, tell me how I can be so lucky, too."

Laughter is contagious, and from the earliest days of television, producers of comedy shows have known that the show seems funnier when it has a laugh track. The discovery was made by accident in 1950 when a show tried to make up for not having a live audience by using recorded laughter. Ever since then, the laugh track has been one of the few unchanging aspects of television.

Laughter can change the behavior of others. It helps make threatening or embarrassing situations more comfortable. Laughter can also be a way of deflecting anger. If the threatening person laughs, it lowers the risk of a confrontation. The person is saying, "This is not as serious as you thought."

Laughter, however, is related to power. In the workplace, the boss or those with more control use humor more than subordinates. In those types of situations, controlling what is considered funny becomes a way of exercising power. Because men often have more power than women, we see women laughing significantly more when the speaker is a man than when it is a woman. In mixed-gender audiences, both men and women laugh more when the speaker is a male than when it is a female. This makes it tough for female comedians.

The more we examine the role of laughter in social situations, the more complicated we will see it is. In some situations, laughter can be an aggressive act, a sign of winning, or a way of controlling the situation. It can be used to boast about a triumph and belittle a competitor.

*Source:* Provine, Robert A. 2000. *Laughter: A Scientific Investigation.* New York: Penguin Books.

## ELEMENTS OF SOCIAL INTERACTION

People do not interact with one another as anonymous beings. They come together in the context of specific environments and with specific purposes. Their interactions involve behaviors associated with defined statuses and particular roles. These statuses and roles help pattern our social interactions and provide predictability.

### Statuses

Statuses are *socially defined positions that people occupy.* Common statuses can pertain to religion, education, ethnicity, and occupation, for example, Protestant, college graduate, African American, and teacher.

Statuses exist independently of the specific people who occupy them (Linton, 1936). For example, our society recognizes the status of politician. Many people occupy that status, including President Barack Obama, Senator Al Franken, Senator Barbara Boxer, and Governor Mario Cuomo. New politicians appear, while others retire or lose popularity or are defeated, but the status, as the culture defines it, remains essentially unchanged. The same is true for all other statuses—both occupational statuses such as doctor, computer analyst, bank teller, police officer, butcher, insurance adjuster, thief, and prostitute, and nonoccupational statuses such as son or daughter, jogger, friend, Little League coach, neighbor, gang leader, and mental patient.

It is important to remember that from a sociological point of view, status does not refer—as it does in common usage—to the idea of prestige, even though different statuses often do contain differing degrees of prestige. In the United States, for example, research has shown that the status of Supreme Court justice has more prestige than that of physician, which in turn has more prestige than that of sociologist (Nakao, Keoko, and Treas, 1993).

People generally occupy more than one status at a time. Consider yourself, for example: You are someone's daughter or son, a full-time or part-time college student, perhaps also a worker, a licensed car driver, a member of a church or synagogue, and so forth. Sometimes, one of the statuses a person occupies seems to dominate the others in patterning that person's life; such a status is called a *master status*. For example, Barack Obama has occupied a number of diverse statuses: husband, father, U.S. senator, and presidential candidate. After January 20, 2009, however, his master status was that of president of the United States because it governed his actions more than did any other status he occupied at the time.

A person's master status will change many times in the course of his or her life cycle. Right now, your master status probably is that of college student. Five years from now, it might be graduate student, artist, lawyer, spouse, or parent.

Figure 5-1 illustrates the different statuses occupied by a 35-year-old woman who is an executive at a major television network. Although she occupies many statuses at once, her master status is that of vice-president for programming.

In some situations, a person's master status can have a negative influence on the person's life. For example, people who have followed what their culture considers a deviant lifestyle might find that their master status is labeled according to their deviant behavior. Those who have been identified as ex-convicts are likely to be so classified no matter what other statuses they occupy. They will be thought of as ex-convict painters, ex-convict machinists, ex-convict writers, and so on. Their master status has a negative effect on their ability to fulfill the roles of the statuses they would like to occupy. Ex-convicts who are good machinists or house painters might find employers unwilling to hire them because of

This woman's ascribed status is female; her achieved status is based on her profession.

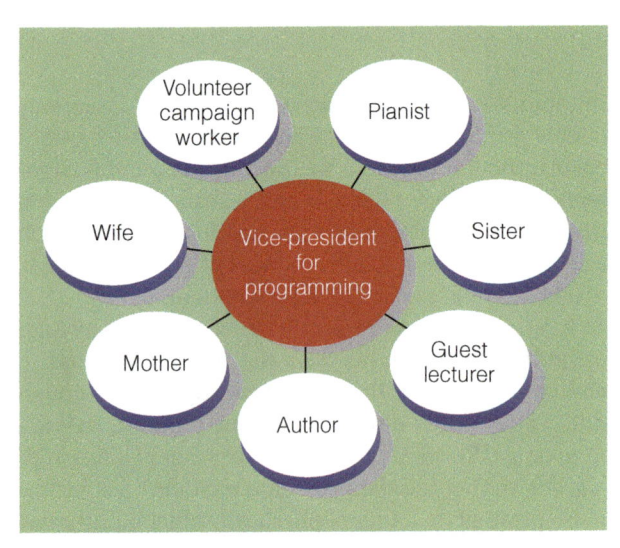

**FIGURE 5-1 Status and Master Status**
Generally, each individual occupies many statuses at one time. The statuses of a female executive at a major television network include author, wife, mother, pianist, and so on. Other statuses could be added to this list. However, one status—vice-president for programming—is most important in patterning this woman's life. Sociologists call such a status a master status.

their police records. Because the label *criminal* can stay with individuals throughout their lives, the criminal justice system is reluctant to label juvenile offenders or to open their records to the courts. Juvenile court files are usually kept secret and often permanently sealed when the person reaches age 18.

Some statuses, called **ascribed statuses,** are *conferred upon us by virtue of birth or other significant factors not controlled by our own actions or decisions; people occupy them regardless of their intentions.* Certain family positions, such as that of daughter or son, are typical ascribed statuses, as are one's gender and ethnic or racial identity. Other statuses, called **achieved statuses,** are *acquired as a result of the individual's actions*—student, professor, garage mechanic, race car driver, artist, prisoner, bus driver, husband, wife, mother, or father.

## Roles

Statuses alone are static—nothing more than social categories into which people are put. Roles bring statuses to life, making them dynamic. As Robert Linton (1936) observed, you occupy a status, but you play a role. **Roles** are *the culturally defined rules for proper behavior that are associated with every status.*

Roles may be thought of as collections of rights and obligations. For example, to be a race car driver, you must become well versed in these rights and obligations because your life might depend on them. Every driver has the right to expect other drivers not to try to pass when the race has been interrupted by a yellow flag because of danger. Turned around, each driver has the obligation not to pass other drivers under yellow-flag conditions. A driver also has a right to expect race committee members to enforce the rules and spectators to stay off the raceway. On the other hand, a driver has an obligation to the owner of the car to try hard to win.

In the case of our television executive, she has the right to expect to be paid on time, to be provided with good-quality scripts and staff support, and to make decisions about the use of her budget. On the other hand, she has the obligation to act in the best interest of the network, to meet schedules, to stay within her budget, and to treat her employees fairly. What is important is that all these rights and obligations are part of the roles associated with the status of vice-president for programming. They exist without regard to the particular individuals whose behavior they guide (see Figure 5-2).

A status can include a number of roles, and each role will be appropriate to a specific social context. For example, as the child of a military officer, Kay Redfield Jamison found that children had to learn the

**FIGURE 5-2** **Status and Roles**
*The status of vice-president for programming at a major television network has several roles attached to it, including attending meetings, making programming decisions, and so on.*

Relatively rapid social change is making less predictable the types of behavior that go along with gender roles.

importance of statuses and roles and the proper behavior to be displayed toward those who occupied those positions.

> [The] Cotillion was where officers' children were supposed to learn the fine points of manners, dancing, white gloves, and other unrealities of life. It also was where children were supposed to learn, as if the preceding fourteen or fifteen years hadn't already made it painfully clear, that generals outrank colonels, who, in turn, outrank majors and captains and lieutenants, and everyone, but everyone, outranks children. Within the ranks of children, boys always outrank girls.
>
> One way of grinding this particularly irritating pecking order into the young girls was to teach them the old and ridiculous art of curtsying. It is hard to imagine that anyone in her right mind would find curtsying an even vaguely tolerable thing to do. But having been given the benefits of a liberal education by a father with strongly nonconforming views and behaviors, it was beyond belief to me that I would seriously be expected to do this. I saw the line of crisply crinolined girls in front of me and watched each of them curtsying neatly. "Sheep," I thought, "Sheep." Then it was my turn. Something inside of me came to a complete boil. It was one too many times watching one too many girls being expected to acquiesce; far more infuriating, it was one too many times watching girls willingly go along with the rites of submission. I refused. A slight matter, perhaps, in any other world, but within the world of military custom and protocol—where symbols and obedience were everything, and where a child's misbehavior could jeopardize a father's promotion—it was a declaration of war. Refusing to obey an adult, however absurd the request, simply wasn't done. Miss Courtnay, our dancing teacher, glared. I refused again. She said she was very sure that Colonel Jamison would be terribly upset by this. I was, I said, very sure that Colonel Jamison couldn't care less. I was wrong. (Jamison, 1995)

## Role Sets

All the roles attached to a single status are known collectively as a **role set.** However, not every role in a particular role set is enacted all the time. An individual's role behaviors depend on the statuses of the other people with whom he or she is interacting. For example, as a college student, you behave one way toward other students and another way toward professors. Similarly, professors behave one way toward other professors, another way toward students, and yet a third way toward deans.

So the role behavior we expect in any given situation depends on the pairs of statuses that the interacting individuals occupy. This means that role behavior really is defined by the rights and obligations that are assigned

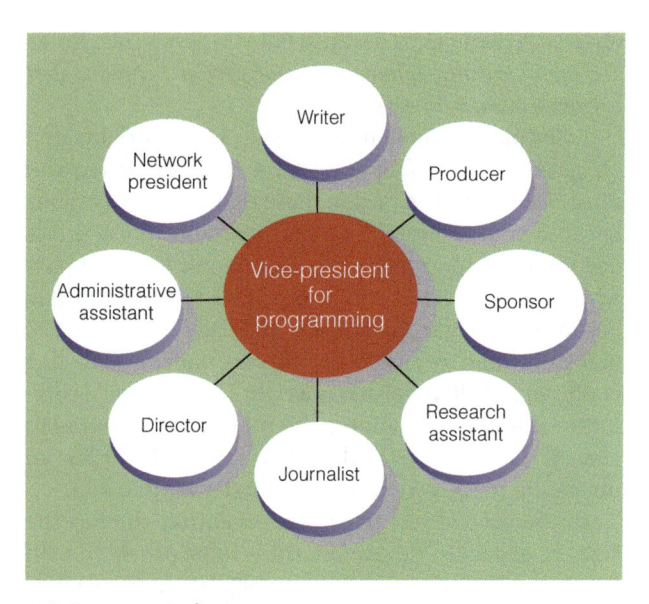

**FIGURE 5-3 Role Sets**
People's role behaviors change according to the statuses of the other people with whom they interact. The female vice-president for programming will adopt somewhat different roles depending on the statuses of the various people with whom she interacts at the station: a writer, a journalist, her assistants, and so on.

to statuses when they are paired with one another (see Figure 5-3).

It would be difficult to describe the wide-ranging, unorganized assortment of role behaviors associated with the status of television vice-president for programming. Sociologists find it more useful to describe the specific behavior expected of a network television vice-president for programming interacting with different people. Such a role set would include the following:

Vice-president for programming/network president
Vice-president for programming/other vice-presidents
Vice-president for programming/script writer
Vice-president for programming/administrative assistant
Vice-president for programming/television star
Vice-president for programming/journalist
Vice-president for programming/producer
Vice-president for programming/sponsor

The vice-president's role behavior in each case would be different, meshing with the role behavior of the individual(s) occupying the other status in each pairing (Merton, 1969).

## Role Strain

Even though most people try to enact their roles as they are expected to, they sometimes find it difficult. *When a single role has conflicting demands attached to*

## HOW SOCIOLOGISTS DO IT

### Southerners Are Really Friendly Until You Disrespect Them

No matter what region of the country you hail from, chances are you've heard a stereotype about the culture and personalities of the people from your area. New Yorkers are pushy. Californians are flaky. New Englanders are standoffish. Texans are arrogant. Oregonians are tree-huggers. But if you tell Southerners that they're more violent than their neighbors to the north, you might just have a fight on your hands.

Still, there's evidence that the characterization might be more than just a regional slur. In their book, *Culture of Honor: The Psychology of Violence in the South,* authors Richard E. Nisbett and Dov Cohen argue that the difference in the two regions' propensity toward violence might stem from culturally acquired beliefs about personal honor that are more embedded in the residents of the South than in those of their northern peers. Southerners fiercely believe that a person's reputation is important and worth defending. As a result, a comment that might be tolerated in Trenton might escalate to lethal violence in Tuscaloosa.

Nisbett and Cohen marshal some impressive evidence to support their hypothesis. They cite the fact that in the South, murder rates are higher than in the North for arguments among friends and acquaintances—where someone's honor is at stake—but not for killings committed in the course of other felonies such as knocking off a convenience store, where there's no relationship with the victim. Per capita income, hot weather, and a history of slavery cannot explain these differences.

The authors also tested to see what kinds of triggers led to violent responses by asking Northerners and Southerners to read vignettes in which a man's honor was challenged. Southern respondents were much more likely to justify a violent response to an insult, saying a guy "wouldn't be much of a man" if he wasn't willing to fight.

Nisbett and Cohen have an interesting theory about the origins of such attitudes. Many of the South's early settlers were Scots-Irish livestock herders. Because it's easy to steal sheep or cows in sparsely populated rural areas, people in herding cultures often cultivate a reputation for being trigger-happy hotheads as a deterrent to theft. This theory gains credence from the fact that Southern white homicide rates are high in poor, rural regions but not in more affluent, densely populated areas. The Scots-Irish code of honor, unfortunately, is less useful when it involves two guys trading insults in a bar rather than two men fighting over a heifer.

*Sources:* Nisbett, Richard E. and Dov Cohen. 1996. *Culture of Honor: The Psychology of Violence in the South.* Boulder, CO: Westview Press; Cohen, Dov, Joseph Vandello, Sylvia Puente, and Adrian Rantilla. 1999. "'When You Call Me That, Smile!' How Norms of Politeness, Interaction Styles, and Aggression Work Together in Southern Culture." *Social Psychology Quarterly* 62(3):257–75.

---

*it, individuals who play that role experience* **role strain** (Goode, 1960).

For example, the captain of a freighter is expected to be sure the ship sails only when it is in safe condition, but the captain also is expected to meet the company's delivery schedule because a day's delay could cost the company thousands of dollars. These two expectations can exert competing pulls on the captain, especially when some defect is reported such as a malfunction in the ship's radar system. The stress of these competing pulls is not due to the captain's personality but, rather, is built into the nature of the role expectations attached to the captain's status. Therefore, sociologists describe the captain's experience of stress as role strain.

(For a special type of role strain created by a culture of honor, see "How Sociologists Do It: Southerners Are Really Friendly Until You Disrespect Them.")

### Role Conflict

*An individual who is occupying more than one status at a time and who is unable to enact the role of one status without violating that of another status* is encountering **role conflict.** Not long ago, pregnancy was considered women's work. An expectant father was expected to get his wife to the hospital on time and to pace the waiting room, anxiously awaiting the nurse's report on the sex of the baby and its health. Today, men are encouraged and even expected to participate fully in the pregnancy and the birth of the child. A role conflict arises, however, in that although the new father is expected to be involved, his involvement is defined along male gender-role lines. He is expected to be helpful, supportive, and essentially a stabilizing force. He is not allowed to indicate that he is frightened, nervous, or possibly angry about the baby. His role as a male, even in modern American society, conflicts with his feelings as a new father (Shapiro, 1987).

As society becomes more complex, individuals occupy increasing numbers of statuses. This increases the chances for role conflict, which is one of the major sources of stress in modern society.

### Role Playing

The roles we play can have a profound influence on both our attitudes and our behavior. Playing a new social role often feels awkward at first, and we might feel we are just acting—pretending to be something that we are not. However, many sociologists feel that the roles a person plays are the person's only true self.

Peter Berger's (1963) explanation of role playing goes further: The roles we play can transform not only our actions but also ourselves. One feels more ardent by kissing, more humble by kneeling, and more angry by shaking one's fist—that is, the kiss not only expresses ardor but creates it. Roles carry with them both certain actions and emotions and attitudes that belong to these actions. The professor putting on an act that pretends to wisdom comes to feel wise. (For a discussion of the roles we play using social media, see "Thinking About Social Issues: Don't Call Me. I'll Text You.")

## ● COLLECTIVE BEHAVIOR

Many of the issues we have discussed in this chapter involve **collective behavior,** *the relatively spontaneous social actions that occur when people respond to unstructured and ambiguous situations.* Collective behavior has the potential for causing the unpredictable, and even the improbable, to happen. Collective actions are capable of unleashing powerful social forces that catch us by surprise and change our lives, at times temporarily but at other times even permanently. It is the more dramatic forms of collective behavior that we tend to remember: riots, mass hysterias, violent uprisings, and panics. Fads, fashion, and rumor, however, also are forms of collective behavior.

With the mass media, television, the Internet, and other systems of communication spreading information instantaneously throughout the entire population,

---

### THINKING ABOUT SOCIAL ISSUES

### Don't Call Me. I'll Text You.

Decades ago parents used to complain about their teenagers spending hours on the phone with their friends. Eventually, some parents relented and had a separate phone line put in for the teens so that the older crowd could also have access to the phone. When people first started to use online types of communication, they were seen as a substitute for face-to-face contact. The assumption was that real direct interaction is best, the phone is second-best, and e-mailing or texting is less ideal, though useful. When people were overscheduled, the "lesser" forms of communication proved very useful.

These ideas would have to be reconsidered today. Communication patterns have changed over the last decades. Younger teens, many of whom never spent a great deal of time talking on the phone, now see it as a relic from an earlier era. Some of these teens, in fact, hate phones and never pick up their voicemail. They take comfort in being in touch with many people, but prefer to keep some of them at a safe distance. Talking on the phone is seen as being too time-consuming. These teens prefer to text, tweet, or look at someone's Facebook wall to get all the information they need.

The newer forms of communication allow us to have greater control over the messages we wish to convey. All the nonverbal cues that would be transmitted in face-to-face interaction are missing from the new forms of communication. The changes in our voice, the movement of our eyes, our breathing and body movement are absent in a text message. Little wonder that some people prefer these newer forms of interaction that allow them to become the person they want to be. We now spend time developing several identities.

People get into elevators and check their phone for messages. A person who may have much in common with the person standing next to him or her will never know that because they are each checking their messages. We may marvel at how people can send 6,000 text messages a month, and yet, it has become clear that this is the new way in which many people wish to communicate.

*Source:* Turkle, Sherry. 2011. "Alone Together: Why We Expect More From Technology and Less from Each Other." New York: Basic Books.

collective behavior shared by large numbers of people who have no direct knowledge of one another has become commonplace. In other words, dispersed forms of collective behavior seem to be universal.

## Fads and Fashions

Fads and fashions are transitory social changes (Vago, 1980), patterns of behavior that are widely dispersed throughout society, but that do not last long enough to become fixed or institutionalized. Yet, it would be foolish to dismiss fads and fashions as unimportant just because they fade relatively quickly. In modern society, fortunes are won and lost trying to predict fashions and fads—in clothing, in entertainment preferences, in eating habits, in choices of investments. Probably the easiest way to distinguish between fads and fashions is to look at their typical patterns of diffusion through society.

**Fads** are *social changes with a very short life span marked by a rapid spread and an abrupt drop in popularity*. This was the fate of the Hula Hoop in the 1950s and the dance known as the "twist" in the 1960s. The rollerskating fad that emerged in 1979 rolled off into the pages of history sometime in the 1980s, as did the Rubik's cube, and Coleco, the company that made Cabbage-Patch dolls, which went bankrupt in 1988. The Tickle Me Elmo stuffed toys were quickly forgotten after the 1996 Christmas season. While you or your parents may remember such past fads as yo-yos and Mr. Potato Head, more recent fads have included the Furby, Teletubbies, and Pokemon. The early 2000s saw the fad of Heelys, the shoes with wheels. Store owners could not keep up with the demand. By 2010, the fad started to fade. Malls, schools, and other public facilities began to ban the shoes for safety reasons. Pediatricians began to complain about the number of broken ankles, arms, and wrists, and dislocated elbows, that were resulting from the shoes.

Some fads may seem particularly absurd. During the Great Depression of the 1930s, when many Americans were having trouble putting food on the table, college students started engaging in the practice of swallowing goldfish. The fad was started by a Harvard freshman who swallowed a single, live fish as fellow students looked on in disgust. Three weeks later a student at Franklin and Marshall College swallowed three fish. The practice quickly escalated and new records were set daily, with 89 being swallowed in one sitting at Clark University. Eventually a pathologist at the U.S. Public Health Service cautioned that goldfish may contain tapeworms that could cause intestinal problems and anemia. The fad disappeared shortly thereafter (Levin, 1993).

On other occasions, what appears to be a fad actually signals a trend and a change in social values. In 1922, newspapers reported the shocking news that smoking was common among female college students. The University of Wisconsin's dean of women said the smoking fad, most popular among women of the "idle, blasé, disappointed class," was already passing. She pointed out that an intelligent woman "cannot see herself rocking a baby or making a pie with a cigarette in her mouth, flicking ashes in the baby's face or dropping them in the pie crust" (Schwarz, 1997).

**Fashions** relate to *the standards of dress or manners in a given society at a certain time*. They spread more slowly and last longer than fads. In his study of fashions in European clothing from the eighteenth to the present century, Alfred A. Kroeber (1963) showed that though minor decorative features come and go rapidly (that is, are faddish), basic silhouettes move through surprisingly predictable cycles that he correlates with degrees of social and political stability. In times of great stress, fashions change erratically; but in peaceful times, the cycle seems to last 100 years he claimed.

Georg Simmel (1957) believed that changes in fashion (such as dress or manners) are introduced or adopted by the upper classes, who seek in this way to keep themselves visibly distinct from the lower classes. Of course, those immediately below them observe these fashions and also adopt them in an attempt to identify

Fashions relate to standards of dress and manners during a particular time.

themselves as upper class. This process repeats itself again and again, with the fashion slowly moving down the class ladder, rung by rung. When the upper classes see that their fashions have become commonplace, they take up new ones, and the process starts all over again.

Blue jeans have shown that this pattern is no longer true today. Jeans started out as sturdy work pants worn by those engaged in physical labor. Young people then started to wear them for play and everyday activities. College students wore them to class. Eventually, fashion designers started to make fancier, higher-priced versions, known as designer jeans, worn by the middle and upper classes. Eventually we came full circle with jeans with rips and signs of wear selling for hundreds of dollars. In this way, the introduction of blue jeans into the fashion scene represents movement in the opposite direction from what Simmel noted.

Of course, the power of the fashion business to shape consumer taste cannot be ignored. Fashion designers, manufacturers, wholesalers, and retailers earn money only when people tire of their old clothes and purchase new ones. Thus, they promote certain colors and widen and narrow lapels to create new looks, which consumers purchase.

Indeed, the study of fads and fashions provides sociologists with recurrent social events through which to study the processes of change. Because they so often involve concrete and quantifiable objects, such as consumer goods, fads and fashions are much easier to study and count than are rumors, another common form of dispersed collective behavior.

## Rumors

A **rumor** is information that is shared informally and spreads quickly through a mass or a crowd. It arises in situations that, for whatever reasons, create ambiguity with regard to their truth or their meaning. Rumors may be true, false, or partially true, but characteristically they are difficult or impossible to verify.

Rumors are generally passed from one person to another through face-to-face contact, but they can be started through television, radio, and the Internet as well. However, when the rumor source is the mass media, the rumor still needs people-to-people contact to enable it to escalate to the point of causing widespread concern (or even panic). Sociologists see rumors as one means through which collectivities try to bring definition and order to situations of uncertainty and confusion. In other words, rumors are "improvised news" (Shibutani, 1966).

Rumors of affairs, rumors of "weapons of mass destruction" and their alleged removal to other countries, and rumors such as "John Kerry is French," "President Obama is a Muslim," "John McCain had an illegitimate black child," "President Obama's health care plan will legalize 'death panels,'"—all of these involve statements whose veracity is in question or that are simply false. Some rumors involve statements whose ambiguous nature makes them potentially appealing to different audiences who may interpret them in particular ways and circulate them. Rumors are often distinguished from gossip, in that a rumor is supposedly about public issues and gossip is about private, trivial things. The emergence of tabloid journalism has blurred this distinction.

Hard-to-believe rumors usually disappear first, but this is not always the case. For 103 years, Procter & Gamble, the maker of familiar household products such as Mr. Clean and Tide laundry detergent, used the symbol of the moon and thirteen stars as a company logo on its products. A rumor started to circulate that the president of the company appeared on the *Phil Donahue Show* in 1994 and announced that he was part of the church of Satan. Supposedly he was giving

Fashion statements are common throughout society. It is difficult to predict which ones will become widespread.

a large portion of the company's profits to support the satanic church. Although there was no evidence to substantiate this rumor, and even though the man in the moon symbol was adopted in 1851, the company finally decided to remove the logo from its products in 1985 (Koenig, 1985).

In 1997, Procter & Gamble was still plagued by the rumor and filed the latest in a series of lawsuits, this time against Amway Corporation and some of its distributors for allegedly spreading rumors that Procter & Gamble is affiliated with the church of Satan. In 2007, a jury awarded P&G $19.25 million after finding that Amway distributors had spread false rumors about Procter and Gamble. Similar rumors were spread about Liz Claiborne and McDonald's.

## Public Opinion

The term *public* refers to a dispersed collectivity of individuals concerned with or engaged in a common problem, interest, focus, or activity. An opinion is a strongly held belief. Thus, **public opinion** refers to *the beliefs held by a dispersed collectivity of individuals about a common problem, interest, focus, or activity.* It is important to recognize that a public that forms around a common concern is not necessarily united in its opinions regarding this concern. For example, Americans concerned about abortion are sharply divided into pro and con camps.

Whenever a public forms, it is a potential source for, or opposition to, whatever its focus is. Hence, it is extremely important for politicians, market analysts, public relations experts, and others who depend on public support to know the range of public opinion on many different topics. Those individuals often are not willing to leave opinions to chance, however. They seek to mold or influence public opinion, usually through the mass media.

Advertisements are attempts to mold public opinion, primarily in the area of consumption. They may create a need where there was none, as they did with fabric softeners, or they may try to convince consumers that one product is better than another when there is actually no difference. *Advertisements of a political nature, seeking to mobilize public support behind one specific party, candidate, or point of view,* technically are called **propaganda** (but usually only by those people in disagreement). For example, radio broadcasts from the former Soviet Union were habitually called "propaganda blasts" by the American press, but similar Voice of America programs were called "news" or "informational broadcasts" by the same American press.

**Opinion leaders** are *socially acknowledged experts whom the public turns to for advice.* The more conflicting sources of information there are on an issue of public concern, the more powerful the position of opinion leaders becomes. The leaders weigh various news sources and then provide an interpretation in what has been called the two-step flow of communication. Those opinion leaders can have a great influence on collective behavior, including voting (Lazarsfeld et al., 1968), patterns of consumption, and the acceptance of new ideas and inventions.

Typically, each social stratum has its own opinion leaders (Katz, 1957). Oprah Winfrey, for example, is an opinion leader in the African American community. The mass media have turned some news anchors into accepted opinion leaders for a broad portion of the American public. Rush Limbaugh has emerged as one of the more influential opinion leaders, as the fortunes of political candidates are determined by his loyal listeners.

When rumor and public opinion grip the public imagination so strongly that facts no longer seem to matter, terrifying forces may be unleashed. Mass hysteria may reign, and panic may set in.

## Mass Hysteria and Panic

A recurring and bizarre example of mass hysteria and panic appears throughout the world. An epidemic of penis theft swept Nigeria between 1975 and 1977. Men could be seen in the streets of Lagos holding on to their genitalia either openly or discreetly with their hand in their pockets. Only by being on guard could one prevent the theft. In a typical incident, someone would suddenly point to an individual and yell, "Thief! My genitals are gone!" The unfortunate "thief" might be assaulted and even killed.

The so-called thievery seems to be an ongoing problem. In April 2001, mobs in Nigeria lynched at least twelve suspected penis thieves. Later that year, there were other deaths in neighboring Benin. One survey found at least thirty-six suspected penis thieves killed at the hands of angry mobs between 1997 and 2003. In 2008, police in Congo arrested thirteen suspected sorcerers accused of using black magic to steal or shrink men's penises. A wave of panic was triggered by the alleged witchcraft (*China Daily*, 2008).

**Mass hysteria** occurs when *large numbers of people are overwhelmed with emotion and frenzied activity or become convinced that they have experienced something for which investigators can find no discernible evidence.* A **panic** is *an uncoordinated group flight from a perceived danger,* as in the public reaction to Orson Welles's 1938 radio broadcast of H. G. Wells's *War of the Worlds*.

According to Irving Janis and his colleagues (1964), people generally do not panic unless four conditions are met. First, people must feel that they are trapped in a life-threatening situation. Second, they must perceive that the threat to their safety is so large that they can do little else but try to escape. Third, they must realize that their escape routes are limited or inaccessible. Fourth, there must be a breakdown in communication between

According to Herbert Blumer, a crowd is a collectivity of people more or less waiting for something to happen. Eventually, something stirs them, and people react without the kind of caution and critical judgments they would normally use.

the front and rear of the crowd. Driven into a frenzy by fear, people at the rear of the crowd make desperate attempts to reach the exit doors, and their actions often completely close off the possibility of escape.

The perception of danger that causes a panic may come from rational as well as irrational sources. A fire in a crowded theater, for example, can cause people to lose control and trample one another in their attempt to escape. This happened when fire broke out at the Beverly Hills Supper Club in Southgate, Kentucky, on May 28, 1977. When employees discovered an out-of-control fire, they warned the 2,500 patrons and tried to usher them out of the building. A panic resulted as people attempted to escape the overcrowded, smoke-filled building. People trampled each other trying to reach the exits, and 165 people died in the process.

Such extreme events are not very common, but they do occur often enough to present a challenge to social scientists, some of whom believe there is a rational core behind what at first glance appears to be wholly irrational behavior (Rosen, 1968). For example, sociologist Kai T. Erikson (1966) looked for the rational core behind the wave of witchcraft trials and hangings that raged through the Massachusetts Bay Colony beginning in 1692. Erikson joins most other scholars in viewing this troublesome episode in American history as an instance of mass hysteria (Brown, 1954). He accounts for it as one of a series of symptoms, suggesting that the colony was in the grip of a serious identity crisis and needed to create real and present evil figures who stood for what the colony was not—thus enabling the colony to define its identity in contrast and build a viable self-image.

Mass hysterias account for some of the more unpleasant episodes in history. Of all social phenomena, they are among the least understood—a serious gap in our knowledge of human behavior.

## ■ SUMMARY

### ▶ Why is it important to understand social interaction?

Humans are symbolic creatures, and everything they do conveys a message to others. Whether we intend it or not, other people take account of our behavior. People do not interact with each other as anonymous beings. People come together in the context of specific environments, with specific purposes and specific social characteristics.

### ▶ What types of social interaction are there?

There are five major forms of social interaction. One way we communicate with one another is through *nonverbal behavior*. When people do something for each other with the express purpose of receiving a reward or return, they are involved in an *exchange interaction*. A *cooperative interaction* occurs when people act together to promote common interests or achieve shared goals. *Conflicts* arise when people or groups have incompatible values or when the rewards or resources available to a society or its members are limited. *Competition* is a form of conflict in which individuals or groups confine their conflict within agreed-upon rules.

### ▶ What are roles?

Roles are the culturally defined rules for proper behavior that are associated with every status. A role is basically a collection of rights and obligations. Statuses and roles help define our social interactions and provide predictability.

### ▶ What role do norms play in social interaction?

Norms are specific rules of behavior, agreed upon and shared, that prescribe limits of acceptable behavior. Norms tell us the things we should both do and not do. In fact, our society's norms are so much a part of us that we often are not aware of them until they are violated.

▶ **What are the main features of statuses?**
Statuses are socially defined positions that people occupy, in a group or society, that help to determine how they interact with one another. Statuses exist independently of the specific people who occupy them.

▶ **How does the context of a situation influence social interaction?**
There are three elements that, taken together, define the *context* of a social interaction: (1) the physical setting or place, (2) the social environment, and (3) the activities surrounding the interaction—preceding it, happening simultaneously with it, and coming after it.

▶ **What is the difference between role strain and role conflict?**
Role conflict occurs when an individual is occupying more than one status at a time and is unable to enact the role of the status without violating that of another status. When a single role has conflicting demands attached to it, individuals who play that role experience role strain.

▶ **What role does collective behavior play in social interaction?**
Because the mass media today spread information quickly among millions of people, collective behavior is often shared by large numbers of people who have no direct knowledge of or contact with one another.

## Media Resources

**CourseMate for *Introduction to Sociology*, Eleventh Edition**
Cengage Learning's Sociology CourseMate brings course concepts to life with interactive learning, study, and exam preparation tools that support the printed textbook. Access an integrated eBook, learning tools including glossaries, flashcards, quizzes, videos, and more in your Sociology CourseMate. Go to CengageBrain.com to register or purchase access.

# CHAPTER FIVE STUDY GUIDE

## Key Concepts

*Match each of the following concepts with its definition, illustration, or explanation.*

- **a.** Social action
- **b.** Social interaction
- **c.** Norms
- **d.** Ethnomethodology
- **e.** Dramaturgy
- **f.** Kinesics
- **g.** Exchange interaction
- **h.** Status
- **i.** Master status
- **j.** Ascribed status
- **k.** Achieved status
- **l.** Roles
- **m.** Role set
- **n.** Role conflict
- **o.** Role strain

____ 1. Specific rules of behavior that are agreed upon and shared and that prescribe limits of acceptable behavior
____ 2. Something done toward another person for the purpose of receiving some reward from that person
____ 3. The study of the sets of rules or guidelines that individuals use to initiate behavior, respond to behavior, and modify behavior
____ 4. Anything people are conscious of doing because of other people
____ 5. The study of how slight nods, yawns, postural shifts, nonverbal cues, and other body movements affect behavior
____ 6. Of the many statuses a person occupies, the one that seems to dominate the others in patterning a person's life
____ 7. A status occupied as a result of an individual's actions
____ 8. Culturally defined rules for proper behavior that are associated with every status
____ 9. Conflicting demands attached to the same role
____ 10. An inability to enact the roles of one status without violating those of another status
____ 11. Use of the framework of theater, performance, and role to understand people's behavior
____ 12. A status conferred on a person because of unchangeable qualities (sex, race, and so on) rather than because of his or her actions
____ 13. Two or more people taking one another into account
____ 14. All the roles associated with a particular status
____ 15. A socially defined position in a social system

## Key Thinkers

*Match the thinkers with their main idea or contribution.*

- **a.** Edward T. Hall
- **b.** Erving Goffman
- **c.** Harold Garfinkel
- **d.** Max Weber

____ 1. Sociological theorist who emphasized sympathetic understanding (*verstehen*) in studying interaction
____ 2. A pioneer in studying the context of social interaction
____ 3. Proposed that it was important to study the commonplace aspects of everyday life
____ 4. Developed an approach that focused on how people try to create a favorable impression of themselves and the manner in which others judge their performances

CHAPTER 5  SOCIAL INTERACTION    121

● **Central Idea Completions**

*Fill in the appropriate concepts and descriptions for each of the following questions.*

1. Differentiate between role strain and role conflict:

   _____
   _____
   _____

   a. What might be two forms of role strain facing a female college professor?

   _____

   b. What are the sources of these role strains?

   _____

   c. How might the individual resolve these strains?

   _____

2. Consider the kinds of interactions in a family.

   a. What activities in a family illustrate exchange behaviors?

   _____
   _____

   b. What activities in a family illustrate conflict behaviors?

   _____
   _____

   c. What activities in a family illustrate competitive behaviors?

   _____
   _____

3. List the four major types of social distance zones outlined in your text.
   (1) _____  (2) _____
   (3) _____  (4) _____

   a. Assuming you were to violate each form of social distance zone, which violation would you define as most serious? _____

   b. What factors might influence an audience witnessing your violation to regard your behavior as serious?

   _____

4. What is the difference between an achieved status and an ascribed status?

   _____

   In which types of situations is it legitimate to think more in terms of ascribed than achieved status?

   _____

5. What is "master status" and what difficulties can it create in social interaction?

   _____

## Critical Thought Exercises

1. Following your text's discussion of the four social distances most Americans use in business and social relations (intimate, personal, social, and public), conduct your own ethnomethodological study by adopting a personal distance incongruent with the setting in which you are interacting. For example, with a close acquaintance, begin a conversation while standing face-to-face. As your conversation unfolds, begin backing away and increase your distance from the customary 6 to 18 inches to 2½ to 4 feet, then to more than 4 feet. What happened to the conversation? What was the reaction of your acquaintance? Hypothesize what might have happened if you had elected to use an intimate personal distance within a public distance setting, for example, at a department or grocery store. How might a salesperson or store clerk have reacted if you had begun to move as close as 6 to 18 inches?

2. Attend a meeting of an international students' organization at your college or university. Observe the public and personal conversational distances engaged in by the students, and try to determine whether these distances vary by the country from which each student comes. After the meeting, approach several of the international students and ask them whether they have noticed any differences in the use of personal space among students in their home country and in the United States. If the opportunity presents itself, inquire whether the students had any initial difficulties adjusting to Americans' use of social space when they first arrived. Finally, ask whether they expect (on the basis of the time they are spending in the United States) to have any adjustment problems when they return to their home country.

3. Make a list of embarrassing incidents. Ask other students or friends to contribute to the list. Analyze the incidents in terms of role, status, and dramaturgy. Are there role requirements that were not being met? Was there role conflict, as when a performance intended for one audience (friends, for example) is seen by someone it was not intended for (a teacher, parents)?

4. Sociologists point out that we are often unaware of the exchange qualities of social interaction, but that we can see them when we become sensitized to them. Look at several ordinary interactions and try to find the exchange component in them. To make the task easier, first look at points of disagreement or conflict, in which one person thought another had not done the right thing. Is the complaint really about the person's failure to live up to the expectations of exchange? Then look at more successful interactions to see how each person contributed to the exchange.

## Internet Activities

1. Visit https://www.orangecoastcollege.edu/academics/library/Pages/Nonverbal-Communication-Research-Guide.aspx This guide will help you start your research on nonverbal communication for your research paper or presentation.
2. Visit www.eyesforlies.com/video.htm#Truth%20About%20Liars. Paul Ekman has been studying nonverbal behavior for many years, particularly fleeting facial expressions that reveal thoughts and emotions. He is especially interested in microexpressions that reveal that someone is lying, and he is a consultant for the TV show *Lie to Me*. This page has videos of Ekman, his work, and the TV show.

## Answers to Key Concepts

1. c;  2. g;  3. d;  4. a;  5. f;  6. i;  7. k;  8. l;  9. o;  10. n;  11. e;  12. j;  13. b;  14. m;  15. h

## Answers to Key Thinkers

1. d;  2. a;  3. c;  4. b

# 6 Social Groups and Organizations

**The Nature of Groups**
    Primary and Secondary Groups

**Functions of Groups**
    Defining Boundaries
    Choosing Leaders
    Making Decisions
    Setting Goals
    Assigning Tasks
    Controlling Members' Behavior
    *How Sociologists Do It:* Can One Bad Apple Spoil the Whole Group?

**Reference Groups**
    *Thinking About Social Issues:* Do You Really Know How Much Your Friends Drink?
    Small Groups
    Large Groups: Associations
    *Gemeinschaft* and *Gesellschaft*

*Day-to-Day Sociology:* The Strength of the Informal Structure in Job Hunting
    Mechanical and Organic Solidarity

**Bureaucracy**
    Weber's Model of Bureaucracy: An Ideal Type
    Bureaucracy Today: The Reality
    The Iron Law of Oligarchy
*Day-to-Day Sociology:* Group Think Versus Crowdsourcing

**Institutions and Social Organization**
    Social Institutions
    Social Organization

**Summary**

Stuart Monk/Shutterstock.com

## LEARNING OBJECTIVES

After studying this chapter, you should be able to do the following:

- Distinguish between primary and secondary groups.
- Explain the functions of groups.
- Understand the role of reference groups.
- Know the influence of group size.
- Understand the characteristics of bureaucracy.
- Know what Michels's concept of "the iron law of oligarchy" is.
- Understand why social institutions are important.

TABLE 6-1    Top 10 Names for 2011

| Rank | Male name | Female name |
| --- | --- | --- |
| 1 | Jacob | Sophia |
| 2 | Mason | Isabella |
| 3 | William | Emma |
| 4 | Jayden | Olivia |
| 5 | Noah | Ava |
| 6 | Michael | Emily |
| 7 | Ethan | Abigail |
| 8 | Alexander | Madison |
| 9 | Aiden | Mia |
| 10 | Daniel | Chloe |

*Source:* Social Security Administration, www.ssa.gov/OACT/babynames/.

TABLE 6-2    Out-of-Favor Baby Names

| Male | Female |
| --- | --- |
| Cecil | Agnes |
| Chester | Bertha |
| Dewey | Bessie |
| Elmer | Beulah |
| Floyd | Gertrude |
| Homer | Myrtle |
| Mack | Pearl |

*Source:* Social Security Administration, www.ssa.gov/OACT/babynames/.

We all feel strongly about our names. We do not like it when people mispronounce our name and we are quick to correct them. Throughout history American immigrants have changed their names to a more Anglicized sounding version. This was the case with Jang Do who came to the United States more than three decades ago. First he changed "Jang" into "John." Then, he added an "e" to his last name. He has been John Doe ever since. He encounters quizzical looks from airport security officials or anyone else who looks at his ID card and sees a Korean male with the American name often used to remain anonymous (Cowan, 2009).

Parents give their children names that they think will help their future life prospects. The popularity of names, however, changes with the times. The parents of today who are naming their little girls Sophia, Isabella, and Emma would be horrified if someone suggested they name their new baby Bertha, Gertrude, or Myrtle. Yet, those were very popular names in the past. (See Tables 6-1 and 6-2.) Names help identify who we are and what groups we belong to. They can be used to show that we are part of a particular ethnic group, social class, or religious group. People have changed their names to avoid being identified with a certain group. Naming a baby, then, is the parents' first act to show that the child is part of a certain group.

## THE NATURE OF GROUPS

A good deal of social interaction occurs in the context of a group. In common speech, the word *group* is often used for almost any occasion when two or more people come together. In sociology, however, we use several terms for various collections of people, not all of which are considered groups.

A **social group** consists of *a number of people who have a common identity, some feeling of unity, and certain common goals and shared norms.* In any social group, the individuals interact with one another according to established statuses and roles. The members develop expectations of proper behavior for persons occupying different positions in the social group. The people have a sense of identity and realize they are different from others who are not members. Social groups have a set of values and norms that might or might not be similar to those of the larger society.

For example, a group of students in a college class can have some common norms that include taboo subjects, open expression of feelings, interrupting or challenging the professor, avoiding conflict, and the length and frequency of contributions. All of these are usually hidden or implicit and new members have to learn them quickly. Violations of the norms involve sanctions (that is, disapproval), which might include comments, disapproving looks, or avoiding the deviant.

Our description of a social group contrasts with our definition of a **social aggregate,** which is made up of *people who temporarily happen to be in physical proximity to each other but share little else.* Consider passengers riding together in one car of a train. They might share a purpose (traveling to Washington, D.C.) but do not interact or even consider their temporary association to have any meaning. It hardly makes sense to call them a group—unless something more happens. If it is a long ride, for instance, and several passengers start a card game, the card players will have formed a social group: They have a purpose, they share certain role expectations, and they attach importance to what they are doing together. Moreover, if the card players continue to meet one another every day (say, on a commuter train), they might begin to feel special in contrast to the rest of the passengers, who are just riders. A social group, unlike an aggregate, does not cease to exist when its members are away from one another.

Members of social groups carry the fact of their membership with them and see the group as a distinct entity with specific requirements for membership. A social group has a purpose and is therefore important to its members, who know how to tell an insider from an outsider. It is a social entity that exists for its members apart from any other social relationships that some of them might share. Members of a group interact according to established norms and traditional statuses and roles. As new members are recruited to the group, they move into these traditional statuses and adopt the expected role behavior—if not gladly, then as a result of group pressure.

Consider, for example, a tenants' group that consists of the people who rent apartments in a building. Most such groups are founded because tenants feel a need for a strong, unified voice in dealing with the landlord on problems with repairs, heat, hot water, and rent increases. Many members of a tenants' group might never have met one another before; others might be related to one another; and some might also belong to other groups such as a neighborhood church, the PTA, a bowling league, or political associations. The group's existence does not depend on these other relationships, nor does it cease to exist when members leave the building to go to work or away on vacation. The group remains, even when some tenants move out of the building and others move in. Newcomers are recruited, told of the group's purpose, and informed of its meetings; they are encouraged to join committees, take leadership responsibilities, and participate in the actions the group has planned. Members who fail to support group actions (such as withholding rent) will be pressured and criticized by the group.

People sometimes are defined as being part of a specific group because they share certain characteristics. If these characteristics are unknown or unimportant to those in the category, it is not a social group. Involvement with other people cannot develop unless one is aware of them. People with similar characteristics do not become a social group unless concrete, dynamic interrelations develop among them (Lewin, 1948). For example, although all left-handed people fit into a group, they are not a social group just because they share this common characteristic. A further interrelationship must also exist. They can, for instance, belong to Left-Handers International of Topeka, Kansas, an organization that champions the accomplishments of left-handers. The group has even designated August 13 as International Left-Handers' Day. Thousands of left-handers belong to this social group.

Even if people are aware of one another, that is still not enough to make them a social group. We can be classified as Democrats, college students, upper class, or suburbanites. Yet, for many of us who fall into these categories, there is no group. We might not be involved with the others in any patterned way that is an outgrowth of that classification. In fact, we personally might not even define ourselves as members of the particular category even if someone else does.

Social groups can be large or small, temporary or long-lasting. Your family is a group, as is your ski club, any association to which you belong, or the clique you hang around with. In fact, it is difficult for you to participate in society without belonging to a number of groups.

In general, social groups, regardless of their nature, have the following characteristics: (1) permanence beyond the meetings of members, that is, even when members are dispersed; (2) means for identifying members; (3) mechanisms for recruiting new members; (4) goals or purposes; (5) social statuses and roles, that is, norms for behavior; and (6) means for controlling members' behavior.

The traits we described are features of many groups. A baseball team, a couple about to be married, a work unit, players in a weekly poker game, members of a family, or a town planning board all can be described as groups. Yet, being a member of a family is significantly different from being a member of a work unit. The family is a primary group, whereas most work units are secondary groups.

## Primary and Secondary Groups

The difference between primary and secondary groups lies in the kinds of relationships their members have with one another. Charles Horton Cooley (1909) defined primary groups as groups that

> are characterized by intimate face-to-face association and cooperation. They are primary in several senses, but chiefly in that they are fundamental in forming the social nature and ideas of the individual. The result of intimate association, psychologically, is a certain fusion of individualities in a common whole, so that one's very self, for many purposes at least, is the common life and purpose of the group. Perhaps the simplest way of describing this wholeness is by saying that it is a "we"; it involves the sort of sympathy and mutual identification of which "we" is the natural expression.

Cooley called primary groups the nursery of human nature because they have the earliest and most fundamental effect on the individual's socialization and development. He identified three basic primary groups: the family, children's play groups, and neighborhood or community groups.

**Primary groups** involve *interaction among members who have an emotional investment in one another and in a situation, who know one another intimately, and who interact as total individuals rather than through specialized roles.* For example, members of a family are emotionally involved with one another and know one another well. In addition, they interact with one another in terms of their total personalities, not just in terms of their social identities or statuses as breadwinner, student, athlete, or community leader.

A **secondary group,** in contrast, is *characterized by much less intimacy among its members. It usually has specific goals, is formally organized, and is impersonal.* Secondary groups tend to be larger than primary groups, and their members do not necessarily interact with all other members. In fact, many members often do not know one another at all; to the extent that they do, they rarely know more about one another than about their respective social identities. Members' feelings about, and behavior toward, one another are patterned mostly by their statuses and roles rather than by personality characteristics. The chair of the General Motors board of directors, for example, is treated respectfully by all General Motors employees, regardless of the chair's gender, age, intelligence, habits of dress, physical fitness, temperament, or qualities as a parent or spouse. In secondary groups, such as political parties, labor unions, and large corporations, people are very much what they do.

Table 6-3 outlines the major differences between primary and secondary groups.

## FUNCTIONS OF GROUPS

To function properly, all groups, both primary and secondary, must (1) define their boundaries, (2) choose leaders, (3) make decisions, (4) set goals, (5) assign tasks, and (6) control members' behavior.

### Defining Boundaries

Group members must have ways of knowing who belongs to their group and who does not. Sometimes devices for marking boundaries are obvious symbols, such as the uniforms worn by athletic teams, lapel pins worn by Rotary Club members, rings worn by Masons, and styles of dress. The idea of the British school tie that, by its pattern and colors, signals exclusive group membership has been adopted by businesses ranging from banking to brewing. Other ways by which group boundaries are marked include the use of gestures (special handshakes) and language (dialect differences often mark people's regional origin and social class). In some societies (including our own), skin color also marks boundaries between groups.

### Choosing Leaders

All groups must grapple with the issue of leadership. A **leader** is *someone who occupies a central role or position of dominance and influence in a group.* In some

TABLE 6-3  Relationships in Primary and Secondary Groups

| | Primary | Secondary |
|---|---|---|
| Physical Conditions | Small number | Large number |
| | Long duration | Shorter duration |
| Social Characteristics | Identification of ends | Disparity of ends |
| | Intrinsic valuation of the relation | Extrinsic valuation of the relation |
| | Intrinsic valuation of other person | Extrinsic valuation of other person |
| | Inclusive knowledge of other person | Specialized and limited knowledge of other person |
| | Feeling of freedom and spontaneity | Feeling of external constraint |
| | Operation of informal controls | Operation of formal controls |
| Sample Relationships | Friend–friend | Clerk–customer |
| | Husband–wife | Announcer–listener |
| | Parent–child | Performer–spectator |
| | Teacher–pupil | Officer–subordinate |
| Sample Groups | Play group | Nation |
| | Family | Clerical hierarchy |
| | Village or neighborhood | Professional association |
| | Work team | Corporation |

*Source:* Davis, K. 1949. *Human Society.* New York: Macmillan.

groups, such as large corporations, leadership is assigned to individuals by those in positions of authority. In other groups, such as adolescent peer groups, individuals move into positions of leadership through the force of personality or through particular skills such as athletic ability, fighting, or debating. In still other groups, including political organizations, leadership is awarded through the democratic process of nominations and voting. Think of the long primary process the presidential candidates must endure to amass enough votes to carry their parties' nominations for the November election.

Leadership need not always be held by the same person within a group. It can shift from one individual to another in response to problems or situations that the group encounters. In a group of factory workers, for instance, leadership can fall to different members, depending on what the group plans to do—complain to the supervisor, head to a bar after work, or organize a picnic for all members and their families.

Politicians and athletic coaches often like to talk about individuals who are natural leaders. Although attempts to account for leadership solely in terms of personality traits have failed again and again, personality factors can determine what kinds of leadership functions a person assumes. Researchers (Bales, 1958; Slater, 1966) have identified two types of leadership roles: (1) **instrumental leadership,** *in which a leader actively proposes tasks and plans to guide the group toward achieving its goals*, and (2) **expressive leadership,** *in which a leader works to keep relations among group members harmonious and morale high*. Both kinds of leadership are crucial to the success of a group.

Sometimes one person fulfills both leadership functions, but when that is not the case, those functions are often distributed among several group members. The individual with knowledge of the terrain who leads a group of airplane crash survivors to safety is providing instrumental leadership. The group members who think of ways to keep the group from giving in to despair are providing expressive leadership. The group needs both kinds of leadership to survive.

### Making Decisions

Closely related to the problem of leadership is the way groups make decisions. In many early hunting and food-gathering societies, important group decisions were reached by consensus—talking about an issue until everybody agreed on what to do (Fried, 1967). Today, occasionally, town councils and other small governing bodies operate in this way. Because this consensus gathering takes a great deal of time and energy, many groups opt for efficiency by taking votes or simply letting one person's decision stand for the group as a whole.

### Setting Goals

As we pointed out before, all groups must have a purpose, a goal, or a set of goals. The goal can be very general, such as spreading peace throughout the world, or it can be very specific, such as playing cards on a train. Group goals can change. For example, the card players might discover that they share a concern about the use of nuclear energy and decide to organize a political action group.

### Assigning Tasks

Establishing boundaries, defining leadership, making decisions, and setting goals are not enough to keep a group going. To endure, a group must do something, if nothing more than ensure that its members continue to make contact with one another. Therefore, it is important for group members to know what needs to be done and who is going to do it.

This assigning of tasks, in itself, can be an important group activity (think of your family discussions about sharing household chores). By taking on group tasks, members not only help the group reach its goals but also show their commitment to one another and to the group as a whole. This leads members to appreciate one another's importance as individuals and the importance of the group in all their lives—a process that injects life and energy into a group.

### Controlling Members' Behavior

If a group cannot control its members' behavior, it will cease to exist. For this reason, failure to conform to group norms is seen as dangerous or threatening, whereas conforming to group norms is rewarded, if only by others' friendly attitudes. Groups not only encourage but often depend for survival on conformity of behavior. A member's failure to conform is met with responses ranging from coolness to criticism or even ejection from the group. Anyone who has tried to introduce changes into the constitution of a club or to ignore long-standing conventions—such as ways of dressing, rituals of greeting, or the assumption of designated responsibilities—probably has experienced group hostility.

Primary groups tend to be more tolerant of members' deviant behavior than are secondary groups. For example, families often will conceal the problems of a member who suffers from chronic alcoholism or drug abuse. Even primary groups, however, must draw the line somewhere, and they will invoke negative sanctions if all else fails to influence the deviant member to show at least a willingness to conform. When primary groups finally do act, their punishments can be far more severe than those of secondary groups. Thus, an intergenerational conflict in a family can result in the commitment of a teenager to an institution or treatment center.

Secondary groups tend to use formal, as opposed to informal, sanctions and are much more likely than primary groups simply to expel, or push out, a member who persists in violating strongly held norms. Corporations fire unsatisfactory employees, the army discharges soldiers who violate regulations, and so on.

Even though primary groups are more tolerant of their members' behavior, people tend to conform more closely to their norms than to those of secondary groups. This is because people value their membership in a primary group, with its strong interpersonal bonds, for its own sake. Secondary group membership is valued mostly for what it will do for the people in the group, not because of any deep emotional ties. Because primary group membership is so desirable, its members are more reluctant to risk expulsion by indulging in behavior that might violate the group's standards or norms than are secondary group members. (For a discussion of the influence of one member on a group, see "How Sociologists Do It: Can One Bad Apple Spoil the Whole Group?")

Although groups must fulfill certain functions to continue to exist, they serve primarily as a point of reference for their members.

## HOW SOCIOLOGISTS DO IT

### Can One Bad Apple Spoil the Whole Group?

Can one person spoil the efforts of a whole group? The common view has been that group pressure can make people conform to group demands and standards. What if the opposite is true? Maybe one person can undermine a group and the members' confidence in their abilities.

A group of researchers (Felt et al., 2006) have found that there are three types of bad apples, people who can spoil a group:

- *The jerk* This is a person who attacks and insults others. He or she tells other people their ideas are not good but offers few alternatives, saying things such as, "Are you kidding me? That's stupid" and "You don't know what you are talking about."
- *The slacker* This is the person who contributes as little as possible to the group. He or she might lean back in a chair, put his or her feet up on a desk, and send text messages during meetings. When the group makes a decision, the slacker might respond, "Whatever, I really don't care."
- *The depressive pessimist* This person is withdrawn and has limited input in the project, viewing the task as unpleasant and boring. He or she doubts the group has the ability to succeed.

The general belief in the workplace or the classroom has been that groups are powerful and should be able to force such individuals to change their behavior. Yet, it turns out that groups with one of these types of bad apples perform 30 to 40 percent worse than groups without a bad apple.

Groups with bad apples tend to argue and fight. They communicate less than other groups and might not share relevant information. One of the real problems is that people in the group often take on the characteristics of the bad apple. For example, if the bad apple is a jerk, the group members will be nasty to each other. If he or she is a slacker, the other people will say, for example, "Let's just get this over with and get out of here." If the bad apple is a depressive pessimist, the other people will start to withdraw and lose interest.

According to the research, the best predictor of how a group will perform is not how good the best member is or the average performance level of the group members but what the worst member is like. Groups descend down to the level of the worst member instead of rising to the level of the best member.

Are you the bad apple? If you are sarcastic and make jokes at other people's expense, you might be. Even if people laugh at your jokes, you might be bringing the group down. If you typically let others do most of the work or show little interest in the task, you might also be a bad apple.

The best way to deal with bad apples is to take them out of the group as quickly as possible. In the workplace, that can mean dismissal. In other settings, it can mean reassignment. If this is not possible, the best way to stop the bad apple is for one person to work to engage all the team members and solicit everyone's opinions. Someone has to work to defuse conflicts and allow everyone to have meaningful input in the project.

*Source:* Felps, Well, Terence R. Mitchell, and Eliza Byington. 2006. "How, When, and Why Bad Apples Spoil the Barrel: Negative Group Members and Dysfunctional Groups." *Research in Organizational Behavior* 27:175–222.

## REFERENCE GROUPS

Groups are more than just bridges between the individual and society as a whole. We spend much of our time in one group or another, and the effect these groups have on us continues even when we are not actually in contact with the other members. The norms and values of groups we belong to or identify with serve as the basis for evaluating our own and others' behavior.

A **reference group** is *a group or social category that an individual uses to help define beliefs, attitudes, and values and to guide behavior.* It provides a comparison point against which people measure themselves and others. A reference group is often a category we identify with rather than a specific group we belong to. For example, a communications major might identify with individuals in the media without having any direct contact with them. In this respect, anticipatory socialization is occurring in that the individual might alter his or her behavior and attitudes toward those he or she perceives to be part of the group he or she plans to join. For example, people who become bankers soon feel themselves part of a group—bankers—and assume ideas and lifestyles that help them identify with that group. They tend to dress in a conservative, bankerish fashion, even buying their clothes in shops that other bankers patronize to make sure they have the "right" clothes from the "right" stores. They join organizations such as country clubs and alumni associations so they can mingle with other bankers and clients. Eventually, the norms and values they adopted when they joined the bankers' group become internalized; they see and judge the world around them as bankers.

We can also distinguish between positive and negative reference groups. Positive reference groups are composed of people we want to emulate. Negative reference groups provide models we do not wish to follow. Therefore, a writer might identify positively with those writers who produce serious fiction but might think of journalists who write for tabloids as a negative reference group.

Even though groups are composed of individuals, individuals are also created to a large degree by the groups they belong to through the process of socialization (see Chapter 4, "Socialization and Development"). Of all the groups, the small group usually has the strongest direct effect on an individual. (See "Thinking About Social Issues: Do You Really Know How Much Your Friends Drink?")

### Small Groups

The term **small group** refers to *many kinds of social groups, such as families, peer groups, and work groups, that actually meet together and contain few enough members so that all members know one another.* The smallest group possible is a **dyad,** which *contains only two members.* An engaged couple is a dyad, as are the pilot and copilot of an aircraft.

George Simmel (1950) was the first sociologist to emphasize the importance of the size of a group on the interaction process. He suggested that small groups have distinctive qualities and patterns of interaction that disappear when the group grows larger. For example, dyads resist change in their group size. On the one hand, the loss of one member destroys the group, leaving the other member alone; on the other hand, a **triad,** or *the addition of a third member*, creates uncertainty because it introduces the possibility of two-against-one alliances.

Often one member in a triad can help resolve quarrels between the other two. When three diplomats are negotiating offshore fishing rights, for example, one member of the triad might offer a concession that will break the deadlock between the other two. If that does not work, the third person might try to analyze the arguments of the other two in an effort to bring about a compromise. The formation of shifting pair-offs within triads also can help stabilize the group. When it appears that one group member is weakening, one of the two paired members will often break the alliance and form a new one with the individual who had been isolated (Hare, 1976). This is often seen among groups of children engaged in games.

In triads in which alliances do not shift and the configuration constantly breaks down into two against one, the group will become unstable and might eventually break up. In George Orwell's novel *1984*, the political organization of the earth was defined by three eternally warring political powers. As one power seemed to be losing, one of the others would come to its aid in a temporary alliance, thereby ensuring worldwide political stability while also making possible endless warfare. No power could risk the total defeat of another because the other surviving power might then become the stronger of the surviving dyad.

As a group grows larger, the number of relationships within it increases, which often leads to the formation of **subgroups,** *splinter groups within the larger group.* Once a group has more than five to seven members, spontaneous conversation becomes difficult for the group as a whole. Then two solutions are available. The group can split into subgroups (as happens informally at parties), or it can adopt a formal means of controlling communication (use of *Robert's Rules of Order*, for instance).

For these reasons, small groups tend to resist the addition of new members. Increasing size threatens the nature of the group. In addition, there can be a fear that new members will resist socialization to group norms and thereby undermine group traditions and values. On the whole, small groups are much more vulnerable than large groups to disruption by new members, and the introduction of new members often leads to shifts in patterns of interaction and group norms.

## THINKING ABOUT SOCIAL ISSUES: Do You Really Know How Much Your Friends Drink?

Do we adjust our behavior to what other people think is appropriate or desirable? Many of you probably said "yes." What if we are wrong, however, in our judgments of what other people are thinking?

Group pressure has a powerful influence over college students' decisions regarding alcohol use, smoking, and exercise. Students make judgments about what they think is typical of their fellow students. Drinking behavior is no exception, and students make drinking decisions based on what they think the other students are doing (see Figure 6-1). Are their judgments correct?

The short answer is "no." Students tend to overestimate how much other students drink. In one study of 76,000 students at 130 colleges and universities, 71 percent of the students overestimated how much other students drank, whereas only 15 percent underestimated the amount. Students also overestimated how comfortable their peers were with heavy drinking. They clearly believed other students were more comfortable with heavy drinking than they themselves were. Male students felt particularly embarrassed by their concern about their friends' heavy drinking and thought they would be mocked if they expressed it.

How could the students be so wrong? We often feel we are different from others even though we may be acting similarly. We may jog even though we hate it. We may claim to like country music even though we prefer classical. We may drink heavily even though we would prefer to limit our intake.

Alcohol use can be greatly misperceived. Suppose you go to a party where heavy drinking is taking place. It appears everyone is having a good time, but you do not know that privately many people disapprove. It is easier for us to see a certain behavior than the absence of it. There are going to be more stories about the events that took place during a night of heavy drinking compared to stories about a night of light drinking. The heavy drinking stories become more memorable and make drinking seem more common.

The problem is that this incorrect belief sets in motion more drinking. First, the students think their fellow students drink more than they do, witnessing such behavior at parties and sports event. The more evidence they see of drinking, the more they think the average student drinks. They do not see the students who are not drinking. They believe they need to drink to be accepted, so they drink more.

Norms help us decide who is part of our group and who is an outsider. Behaving in ways the group considers appropriate is a way of demonstrating to others, and to ourselves, that we belong to the group, even if we do not approve of everything the group does.

*Sources:* Prentice, Deborah A. 2008. "Mobilizing and Weakening Peer Influence as Mechanisms for Changing Behavior." In Prinstein, M. J. and K. A. Dodge (Eds.), *Understanding Peer Influence in Children and Adolescents.* New York: Guilford, pp. 161–80; Perkins, H. Wesley, Michael P. Haines, and Richard Rice. 2005. "Misperceiving the College Drinking Norm and Related Problems: A Nationwide Study of Exposure to Prevention Information, Perceived Norms and Alcohol Misuse." *Journal of Studies of Alcohol,* 66(4):470–78.

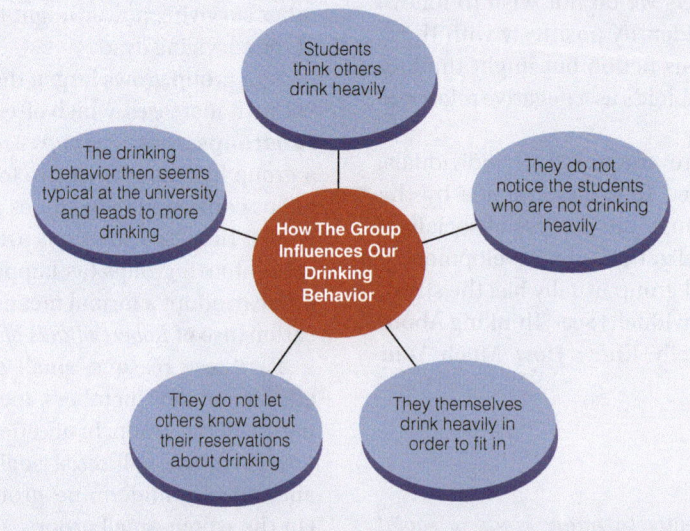

FIGURE 6-1 Do You Really Know How Much Your Friends Drink?

How The Group Influences Our Drinking Behavior:
- Students think others drink heavily
- They do not notice the students who are not drinking heavily
- They themselves drink heavily in order to fit in
- They do not let others know about their reservations about drinking
- The drinking behavior then seems typical at the university and leads to more drinking

Triads are usually unstable groups because the possibility of two-against-one alliances is always present.

## Large Groups: Associations

As patrons or employees of large organizations and governments, we function as part of large groups all the time. Thus, sociologists must study large groups as well as small groups to understand the workings of society. Although all of us probably would be able to identify and describe the various small groups we belong to, we might find it difficult to follow the same process with the large groups that affect us.

Much of the activity of a modern society is carried out through large and formally organized groups. Sociologists refer to these groups as **associations,** which are *purposefully created special-interest groups that have clearly defined goals and official ways of doing things.* Associations include such organizations as government departments and agencies, businesses and factories, labor unions, schools and colleges, fraternal and service groups, hospitals and clinics, and clubs for various hobbies from gardening to collecting antiques. Their goals can be very broad and general—such as helping the poor, healing the sick, or making a profit—or quite specific and limited, such as manufacturing automobile tires or teaching people to speak Chinese. Although an enormous variety of associations exist, they all are characterized by some degree of formal structure with an underlying informal structure.

*Formal Structure* For associations to function, the necessary work is assessed and broken down into manageable tasks that are assigned to specific individuals. In other words, associations are run according to a formal organizational structure that consists of planned, highly institutionalized, and clearly defined statuses and role relationships.

The formal organizational structure of large associations in contemporary society is exemplified best by the organizational structure called bureaucracy. For example, when we consider a college or university, fulfilling its main purpose of educating students requires far more than simply bringing together students and professors. Funds must be raised, buildings constructed, qualified students and professors recruited, programs and classes organized, materials ordered and distributed, grounds kept up, and buildings maintained. Lectures must be given; seminars must be led; and messages need to be typed, copied, and filed. To accomplish all these tasks, the school must create many positions: president, deans, department heads, registrar, public relations staff, groundskeepers, maintenance personnel, purchasing agents, administrative assistants, faculty, and students.

Every member of the school has clearly spelled-out tasks that are organized in relation to one another: Students are taught and evaluated by faculty, faculty members are responsible to department heads or deans, deans to the president, and so on. Underlying these clearly defined assignments are procedures that are never written down but are worked out and understood by those who have to get the job done.

*Informal Structure* Sociologists recognize that formal associations never operate entirely according to their stated rules and procedures. Every association has an informal structure consisting of networks of people who help one another by bending rules and taking procedural shortcuts. No matter how carefully plans are made, no matter how clearly and rationally roles are defined and tasks assigned, every situation and its variants cannot be anticipated. Sooner or later, then, individuals in associations are confronted with situations in which they must improvise and even persuade others to help them do so.

As every student knows, no school ever runs as smoothly as planned. For instance, going by the book—that is, following all the formal rules—often gets students tied up in long lines and red tape. Enterprising students and instructors find shortcuts. A student who wants to change from Section A of Sociology 100 to Section E might find it very difficult or time-consuming to change sections (add and drop classes) officially. However, it might be possible to work out an informal deal—the student stays registered in Section A but attends, and is evaluated in, Section E. The instructor of Section E then turns the grade over to the instructor of Section A, who hands in that grade with all the other Section A grades as if the student had attended Section A all along. The formal rules have been bent, but the major purposes of the school (educating and evaluating students) have been served.

In addition, human beings have their own individual needs even when they are on company time, and these needs are not always met by attending single-mindedly to assigned tasks. To accommodate these needs, people often try to find extra break time for personal business by getting jobs done faster than would be possible if they followed all the formal rules and procedures.

To accomplish these ends, individuals in associations might cover for one another, look the other way at strategic moments, and offer one another useful information about office politics, people, and procedures. Gradually, the reciprocal relationships among members of these informal networks become institutionalized; unwritten laws are established, and a fully functioning informal structure evolves. (For a discussion of applying the informal structure to job hunting, see "Day-to-Day Sociology: The Strength of the Informal Structure in Job Hunting.")

### Gemeinschaft and Gesellschaft

The Chicago sociologists, in their studies of the city, used some of the concepts developed by Ferdinand Tönnies (1865–1936), a German sociologist. In his book, *Gemeinschaft und Gesellschaft*, Tönnies examined the changes in social relations attributable to the transition from rural society (organized around small communities) to urban society (organized around large, impersonal structures).

Tönnies noted that in a **gemeinschaft** (community), *relationships are intimate, cooperative, and personal*. Author Philip Roth (1998), in his book *American Pastoral*, described such a community:

> About one another, we knew who had what kind of lunch in the bag in his locker and who ordered what on his hot dog at Syd's; we knew one another's physical attributes—who walked pigeon-toed and who had breasts, who smelled of hair oil and who oversalivated when he spoke; we knew who among us was belligerent and who was friendly, who was smart and who was dumb; we knew whose mother had an accent and whose father had a mustache, whose mother worked and whose father was dead; somehow we even dimly grasped how every family's different set of circumstances set each family a distinctive difficult human problem.

---

#### DAY-TO-DAY SOCIOLOGY

## The Strength of the Informal Structure in Job Hunting

Here is a common story of a frustrating job-hunting experience. With a great GPA and good references, you send out dozens of résumés to online job postings, newspaper ads, and company recruitment websites and hear ... nothing. Then, you get news that a less-skilled friend whose father's golf buddy works in a top company has landed a plum job. Is the old cliché, "It's not what you know but *who* you know," really true?

In some cases, yes. Mark Granovetter, author of a famous study on the strength of what he called weak ties, found that 56 percent of all professional job applicants found their jobs through personal contacts. Only 16 percent landed jobs through advertisements or employment agencies.

But the nature of those contacts might surprise you. Typically, the person who opened the door was not a close friend or relative but someone who actually did not know the job seeker very well. Granovetter called these relationships "weak ties." These are the bonds that exist between individuals who see one another infrequently and whose relationships are casual rather than intimate. Today, we often characterize these clusters of acquaintances as social networks.

But why would weak ties work better than strong ones? Aren't the people who know you well the ones most likely to have your best interests at heart? That might be, but because of their very closeness, they are likely to be exposed to the same sources of information. People outside that tightly knit circle, however, have networks that reach much further afield. By getting to know them, you essentially tap into their networks, just as they, then, have access to yours.

Additionally, people who know us less well are less likely to pigeonhole us based on our past experiences or skills. This is especially true for job seekers who want to branch out into new areas. When you want to reinvent yourself, the people who know you best can hinder rather than help you. They might be supportive, but they might try to preserve the old identities you are trying to shed.

The best strategy for a job hunter, then, is to try to expand your circle of acquaintances. College buddies, friends of friends, professional associations, social clubs, church groups, and civic organizations all can lead to relationships that qualify as weak-tie networks. (They can also, of course, lead to enduring friendships and personal growth!)

Websites such as Facebook and LinkedIn have been launched to facilitate these kinds of social networks electronically. As people increasingly understand the reciprocal power of social networks, they are more willing to facilitate such relationships. People with contacts in many social networks ultimately do better in the job market.

*Sources:* Granovetter, Mark. 1983. "The Strength of Weak Ties: A Network Theory Revisited." *Sociological Theory* 1: 201–233; Ibarra, Herminia. 2002. *Working Identity: Unconventional Strategies for Reinventing Your Career*. Cambridge: Harvard Business School Press.

In a *gemeinschaft*, the exchange of goods is based on reciprocity and barter, and people look out for the well-being of the group as a whole. Among the Amish, for example, there is such a strong community spirit that if a barn burns down, members of the community quickly come together to rebuild it. In just a matter of days, a new barn will be standing—the work of community members who feel a strong tie and responsibility to another community member who has encountered misfortune.

In a **gesellschaft** (society), *relationships are impersonal and independent.* People look out for their own interests, goods are bought and sold, and formal contracts govern economic exchanges. Everyone is seen as an individual who might compete with others who happen to share a living space.

Tönnies saw *gesellschaft* as the product of mid-nineteenth-century social changes that grew out of industrialization, in which people no longer automatically wanted to help one another or to share freely what they had. There is little sense of identification with others in a *gesellschaft*, in which each individual strives for advantages and regards the accumulation of goods and possessions as more important than the qualities of personal ties. Modern urban society is, in Tönnies's terms, typically a *gesellschaft*, whereas rural areas retain the more intimate qualities of *gemeinschaft*.

In small, rural communities and preliterate societies, the family provided the context in which people lived, worked, were socialized, were cared for when ill or infirm, and practiced their religion. In contrast, modern urban society has produced many secondary groups in which these needs are met. It also offers far more options and choices than did the society of Tönnies's *gemeinschaft*: educational options, career options, lifestyle options, choice of marriage partner, choice of whether to have children, and choice of where to live. In this sense, the person living in today's urban society is freer.

## Mechanical and Organic Solidarity

Tönnies wrote about communities and cities from the standpoint of what we call an ideal type in that no community or city actually could conform to the definitions he presented. They are basic concepts that help us understand the differences between the two. In the same sense, Émile Durkheim devised ideas about mechanical and organic solidarity.

According to Durkheim, every society has a **collective conscience**—*a system of fundamental beliefs and values.* These beliefs and values define for its members the characteristics of the good society, which is one that meets the needs for individuality, for security, for superiority over others, and for any of a host of other values that could become important to the people in that society. **Social solidarity** *emerges from the people's commitment and conformity to the society's collective conscience.*

A **mechanically integrated society** is one in which *a society's collective conscience is strong and there is a great commitment to that collective conscience.* In this type of society, members have common goals and values and a deep and personal involvement with the community. A modern-day example of such a society is that of the Tasaday, a food-gathering community in the Philippines. Theirs is a relatively small, simple society, with little division of labor, no separate social classes, and no permanent leadership or power structure.

In contrast, in an **organically integrated society,** *social solidarity depends on the cooperation of individuals in many positions who perform specialized tasks.* The society can survive only if all the tasks are performed. With organic integration, such as is found in the United States, social relationships are more formal and functionally determined than are the close, personal relationships of mechanically integrated societies.

Although we may take for granted the movement from the *gemeinschaft* to the *gesellschaft*, or

A small town is likely to produce a *gemeinschaft*, in which relationships are intimate, cooperative, and personal; a city is likely to produce a *gesellschaft*, in which relationships are impersonal and people look out for their own interests.

mechanically integrated to organically integrated societies, it is only relatively recently in the course of history that organically integrated societies have become so dominant.

## BUREAUCRACY

In ordinary usage, the term *bureaucracy* suggests a certain rigidity and amount of red tape, but it has a somewhat different meaning to sociologists. Robert K. Merton (1969) defined **bureaucracy** as *"a formal, rationally organized social structure [with] clearly defined patterns of activity in which, ideally, every series of actions is functionally related to the purposes of the organization."*

The German sociologist, Max Weber (1956), provided the first detailed study of the nature and origins of bureaucracy. Although much has changed in society since he developed his theories, Weber's basic description of bureaucracy remains essentially accurate to this day.

### Weber's Model of Bureaucracy: An Ideal Type

Weber viewed bureaucracy as the most efficient—although not necessarily the most desirable—form of social organization for the administration of work. He studied examples of bureaucracy throughout history and noted the elements they had in common. Weber's model of bureaucracy is an **ideal type,** which is *a simplified, exaggerated model of reality used to illustrate a concept.*

When Weber presented his ideal type of bureaucracy, he combined into one the characteristics, found in one form or another, in a variety of organizations. We are unlikely ever to find a bureaucracy that has all the traits presented in Weber's ideal type. However, his presentation can help us understand what is involved in bureaucratic systems. It is also important to recognize that Weber's ideal type is in no way meant to be ideal in the sense that it presents a desired state of affairs. In short, an ideal type is an exaggeration of a situation that simply conveys a set of ideas. Weber outlined six characteristics of bureaucracies:

1. *A clear-cut division of labor.* The activities of a bureaucracy are broken down into clearly defined, limited tasks, which are attached to formally defined positions (statuses) in the organization. This permits a great deal of specialization and a high degree of expertise. For example, a small-town police department might consist of a chief, a lieutenant, a detective, several sergeants, and a dozen officers. The chief issues orders and assigns tasks; the lieutenant is in charge when the chief is not around; the detective does investigative work; the sergeants handle calls at the desk and do the paperwork required for formal booking procedures; and the officers walk or drive through the community, making arrests and responding to emergencies. Each member of the department has a defined status and duty as well as specialized skills appropriate to his or her position.
2. *Hierarchical delegation of power and responsibility.* Each position in the bureaucracy is given sufficient power to enable the individual who occupies it to do assigned work adequately and compel subordinates to follow instructions. Such power must be limited to what is necessary to meet the requirements of the position. For example, a police chief can order an officer to walk a specific beat but cannot insist that the officer join the Lions Club.
3. *Rules and regulations.* The rights and duties attached to various positions are stated clearly in writing and govern the behavior of all individuals who occupy those positions. In this way, all members of the organizational structure know what is expected of them, and each person can be held accountable for his or her behavior. For example, the regulations of a police department might state, "No member of the department shall drink intoxicating liquors while on duty." Such rules

Max Weber viewed bureaucracy as the most efficient—although not necessarily the most desirable—way to organize work. It can lead to the common stereotype of the inflexible bureaucrat. These seven signs in seven languages instruct people not to ask this man any questions.

make the activities of bureaucracies predictable and stable.

4. *Impartiality.* The organization's written rules and regulations apply equally to all its members. No exceptions are made because of social or psychological differences among individuals. Also, people occupy positions in the bureaucracy only because they are assigned according to formal procedures. These positions belong to the organization itself; they cannot become the personal property of those who occupy them. For example, a vice-president of United States Steel Corporation is usually not permitted to pass on that position to his or her children through inheritance.

5. *Employment based on technical qualifications.* People are hired because they have the ability and skills to do the job, not because they have personal contacts within the company. Advancement is based on how well a person does the job. Promotions and job security go to those who are most competent.

6. *Distinction between public and private spheres.* A clear distinction is made between the employees' personal lives and their working lives. It is unusual for employees to be expected to take business calls at home. At the same time, employees' families' lives have no place in the work setting.

Although many bureaucracies strive at the organizational level to attain the goals that Weber proposed, most do not achieve them on the practical level.

## Bureaucracy Today: The Reality

Just as no building is ever identical to its blueprint, no bureaucratic organization fully embodies all the features of Weber's model. One characteristic that most bureaucracies do have in common is a structure that separates those whose responsibilities include overseeing the needs of the entire organization from those whose responsibilities are much narrower and task-oriented. Visualize a modern industrial organization as a pyramid. Management (at the top of the pyramid) plans, organizes, hires, and fires. Workers (in the bottom section) make much smaller decisions limited to carrying out the work assigned to them. A similar division cuts through the hierarchy of the Roman Catholic church. The pope is at the top, followed by cardinals, archbishops, and bishops; the clergy are below. Only bishops can ordain new priests, and they plan the church's worldwide activities. The priests administer parishes, schools, and missions; their tasks are quite narrow and confined.

Although employees of bureaucracies might enjoy the privileges of their positions and guard them jealously, they can be adversely affected by the system in ways they do not recognize. Alienation, adherence to unproductive ritual, and acceptance of incompetence are some of the results of a less-than-ideal bureaucracy.

Robert Michels, a colleague of Weber's, also was concerned about the depersonalizing effect of bureaucracy. His views, formulated at the beginning of the last century, are still pertinent today.

## The Iron Law of Oligarchy

Robert Michels (1911) concluded that the formal organization of bureaucracies inevitably leads to **oligarchy,** *under which organizations that were originally idealistic and democratic eventually come to be dominated by a small self-serving group of people who achieved positions of power and responsibility.* This can occur in large organizations because it becomes physically impossible for everyone to get together every time a decision has to be made. Consequently, a small group is given the responsibility of making decisions. Michels believed that the people in this group would become corrupted by their elite positions and more and more inclined to make decisions to protect their power rather than to represent the will of the group they were supposed to serve.

In effect, Michels was saying that bureaucracy and democracy do not mix. Despite any protestations and promises that they will not become like all the rest, those placed in positions of responsibility and power often come to believe that they are indispensable to, and more knowledgeable than, those they serve. As time goes on, they become further removed from the rank and file.

The iron law of oligarchy suggests that organizations that wish to avoid oligarchy should take a number of precautionary steps. They should make sure that the rank and file remain active in the organization and that the leaders not be granted absolute control of a centralized administration. As long as open lines of communication and shared decision making exist between the leaders and the rank and file, an oligarchy cannot develop easily.

Clearly, the problems of oligarchy, of the bureaucratic depersonalization described by Weber, and of personal alienation all are interrelated. If individuals are deprived of the power to make decisions that affect their lives in many or even most of the areas that are important to them, withdrawal into narrow ritualism and apathy are likely responses. (For another view on the value of groups, see "Day-to-Day Sociology: Group Think Versus Crowdsourcing.")

### DAY-TO-DAY SOCIOLOGY

## Group Think Versus Crowdsourcing

Can a group of diverse people do a better job of solving a problem than a group of experts?

The United States Navy submarine the *Scorpion* had been lost at sea. It was not known exactly where it had gone down, and the Navy had to face the prospect of a search in a huge body of water some 20 miles in diameter and thousands of feet deep.

Some people might think the best way to solve this problem would be to let a group of experts brainstorm about the location of the submarine. All the members of the group could share their views until they came to an agreed-upon answer. During the group discussion, each member of the group would try to convince the others that his or her guess was correct. Unfortunately, with this approach, someone with a strong or dynamic personality could sway the results.

A naval officer came up with a strange plan. He collected a large team of diverse people: "mathematicians, submarine specialists, and salvage men." Rather than having them talk to each other to come up with a single answer, he had each of them make their best individual guess as to where the submarine was. Then he used those individual guesses along with a complex formula to come up with one answer from all the guesses. This gave him the group's collective guess about where the submarine was. Amazingly, the group's guess was a mere 220 yards from where the submarine was found, in spite of the fact that no single person's guess was anywhere near the submarine's actual location.

As individuals, the group members had very little, if any, information. Each person was a specialist in one field and was unlikely to be able to accurately guess where the submarine really was. However, the fact that each person in the group was a specialist meant that they all possessed small pieces of information that could be valuable.

This scenario demonstrates that in some situations, rather than seeking the opinions of experts, it is more valuable to seek the collective opinions of a diverse group of people who will approach problems and decision making from a variety of viewpoints. The next time we think we need to find an expert, we should keep in mind that:

1. Groups can be remarkably intelligent and are often smarter than the smartest people in them.
2. A group can make a good decision if it is diverse, independent, and has limited communication. Too much communication can make the group as a whole less intelligent.
3. The best decisions are a product of disagreement.
4. A fair and neutral person needs to pull the information together to arrive at the answer.

*Sources:* Surowiecki, James. 2004. *The Wisdom of Crowds*. New York: Doubleday; Oinas-Kukkonen, Harri. 2008. "Network Analysis and Crowds of People as Sources of New Organisational Knowledge." In Koohang, A., et al. (Eds), *Knowledge Management: Theoretical Foundation*. Santa Rosa, CA: Informing Science Press, pp. 173–89.

## INSTITUTIONS AND SOCIAL ORGANIZATION

Anyone who has traveled to foreign countries knows that different societies have different ways of doing things. The basic things that get done actually are quite similar—food is produced and distributed; people get married and have children; and children are raised to take on the responsibilities of adulthood. The vehicle for accomplishing the basic needs of any society is the social institution.

### Social Institutions

Sociologists usually speak of five areas of society in which basic needs have to be fulfilled: the family sector, the education sector, the economic sector, the religious sector, and the political sector. For each of these areas, social groups and associations carry out the goals and meet the needs of society.

The behavior of people in these groups and associations is organized or patterned by the relevant **social institutions**—*the ordered social relationships that grow out of the values, norms, statuses, and roles that organize the activities that fulfill society's fundamental needs*. Thus, economic institutions organize the ways in which society produces and distributes the goods and services it needs; educational institutions determine what should be learned and how it should be taught; and so forth.

Of all social institutions, the family is perhaps the most basic. A stable family unit is the main ingredient necessary for the smooth functioning of society. For instance, sexual behavior must be regulated and children must be cared for and raised to fit into society.

Hence, the institution of the family provides a system of continuity from one generation to the next.

Using the family as an example, we can see the difference between the concept of group and the concept of institution. A group is a collection of specific, identifiable people. An institution is a system for organizing standardized patterns of social behavior. In other words, a group consists of people, and an institution consists of actions. For example, when sociologists discuss *a* family (say, the Smith family), they are referring to a particular group of people. When they discuss *the* family, they are referring to the family as an institution—a cluster of statuses, roles, values, and norms that organize the standardized patterns of behavior that we expect to find within family groups. Thus, the family as an American institution typically embodies several master statuses: those of husband, wife, and, possibly, father, mother, and child. It also includes the statuses of son, daughter, brother, and sister. These statuses are organized into well-defined, patterned relationships. Parents have authority over their children, spouses have a sexual relationship with each other (but not with the children), and so on.

Specific family groups, however, might not conform entirely to the ideals of the institution. There are single-parent families, families in which the children appear to be running things, and families in which there is an incestuous parent-child relationship. Although a society's institutions provide what can be thought of as a master plan for human interactions in groups, actual behavior and actual group organization often deviate in varying degrees from this plan.

## Social Organization

If we step back from a mosaic, we see the many multicolored stones composing a single, coordinated pattern or picture. Similarly, if we step back and look at society, we see that the many actions of all its members fall into a pattern or series of interrelated patterns. These consist of social interactions and relationships expressing individual decisions and choices. These choices, however, are not random; rather, they are an outgrowth of a society's social organization. **Social organization** consists of *the relatively stable pattern of social relationships among individuals and groups in society.* These relationships are based on systems of social roles, norms, and shared meanings that provide regularity and predictability in social interaction.

Social organization differs from one society to the next. Thus, Islam allows a man to have up to four wives at once, whereas in American society, with its Judeo-Christian religious tradition, such plural marriage is not an acceptable family form.

Just as statuses and roles exist in ordered relationships to one another, social institutions also exist in patterned relationships with one another in the context of society. All societies have their own patterning for these relationships. For example, a society's economic and political institutions often are closely interrelated. So, too, are the family and religious institutions. Thus, a description of American social organization would indicate the presence of monogamy along with Judeo-Christian values and norms and the institutionalization of economic competition and of democratic political organizations.

A society's social organization tends to be its most stable aspect. The American social organization, however, might not be as static as those of many other societies. American society is experiencing relatively rapid social change because of its complexity and because of the great variety in the types of people who are part of it. This complexity makes life less predictable because new values and norms being introduced from numerous quarters result in changes in social organization. For example, ideas about the behavior associated within female gender roles have changed considerably over the past three decades. Traditionally, married women were expected not to work but to stay home and attend to the rearing of children. Today, the majority of American women are working outside the home, and views on the roles mothers should play in the lives of their children are in flux.

## ■ SUMMARY

▶ **What are social groups?**
Social groups consist of a number of people who have a common identity, some feeling of unity, and certain common goals and shared norms.

▶ **Why are social groups important?**
Sociologists usually speak of five areas of society in which basic needs have to be fulfilled: the family sector, the education sector, the economic sector, the religious sector, and the political sector. For each of these areas, social groups and associations carry out the goals and meet the needs of society.

▶ **What are associations?**
Associations are purposefully created special-interest groups that have clearly defined goals and official ways of doing things.

▶ **What do groups need to do?**
To function properly, all groups must define their boundaries, choose leaders, make decisions, set goals, assign tasks, and control members' behavior.

▶ **What is the difference between primary groups and secondary groups?**
Sociologists distinguish between primary groups, which involve intimacy, informality, and emotional investment in one another, and secondary groups, which have specific goals, formal organization, and much less intimacy.

▶ **What role do reference groups play in social behavior?**

A reference group is a group or social category an individual uses to help define beliefs, attitudes, and values and to guide behavior.

▶ **What are the characteristics of bureaucracy?**

A bureaucracy is a formal, rationally organized social structure with clearly defined patterns of activity that are functionally related to the purposes of the organization.

▶ **What is Michels's concept of "the iron law of oligarchy"?**

Robert Michels concluded that the formal organization of bureaucracies inevitably leads to oligarchy, under which organizations that were originally idealistic and democratic eventually come to be dominated by a small, self-serving group of people who have achieved positions of power and responsibility.

▶ **Why are social institutions important?**

Social institutions consist of the ordered relationships that grow out of the values, norms, statuses, and roles that organize the activities that fulfill society's fundamental needs. Institutions are systems for organizing standardized patterns of social behavior.

▶ **What is the difference between mechanically and organically intergrated societies?**

A mechanically integrated society is one in which a society's collective conscience is strong and there is a great commitment to that collective conscience. In an organically integrated society, social solidarity depends on the cooperation of individuals in many positions who perform specialized tasks.

▶ **What do we mean by social organization?**

Social organization consists of the relatively stable pattern of social relationships among individuals and groups in society. It differs from one society to the next.

## Media Resources

**CourseMate for *Introduction to Sociology*, Eleventh Edition**

Cengage Learning's Sociology CourseMate brings course concepts to life with interactive learning, study, and exam preparation tools that support the printed textbook. Access an integrated eBook, learning tools including glossaries, flashcards, quizzes, videos, and more in your Sociology CourseMate. Go to CengageBrain.com to register or purchase access.

# CHAPTER SIX STUDY GUIDE

## Key Concepts

*Match each of the following concepts with its definition, illustration, or explanation.*

a. Social group
b. Social aggregate
c. Primary group
d. Secondary group
e. Instrumental
f. Expressive
g. Reference group
h. Dyad
i. Triad
j. Subgroup
k. Associations
l. *Gemeinschaft*
m. *Gesellschaft*
n. Collective conscience
o. Mechanical solidarity
p. Organic solidarity
q. Bureaucracy
r. Ideal type
s. Oligarchy
t. Social institutions
u. Social organization

____ 1. A formal, rationally organized social structure divided into offices with specific tasks run on principles of impartiality
____ 2. A group a person uses as a guide to values, beliefs, and behavior
____ 3. A goal-oriented, large, impersonal, more formal group
____ 4. A simplified model used to illustrate a concept
____ 5. The relatively stable pattern of social relationships among individuals and groups in society
____ 6. A splinter group, usually created informally, to enable face-to-face interaction
____ 7. A group of people who know one another well and interact as complete individuals rather than in specialized roles
____ 8. A group of two people
____ 9. A group of three people
____ 10. Social solidarity based on similarity among people and strong commitment to the collective conscience
____ 11. Social solidarity based on difference and the fitting together of specialized tasks
____ 12. Rule by a small group of self-interested people
____ 13. The ordered social relationships that grow out of the values, norms, statuses, and roles that organize the activities that fulfill society's fundamental needs
____ 14. (Community) A group in which relations are intimate, personal, and cooperative
____ 15. People who share goals, norms, and a common identity
____ 16. Purposefully created groups with clearly defined goals and procedures
____ 17. People who have little in common but are in the same place together
____ 18. Durkheim's term for the shared fundamental beliefs and values of a group
____ 19. (Society) A group in which relations are impersonal and independent
____ 20. Focused on accomplishing concrete tasks
____ 21. Concerning feelings and interpersonal relationships

## Key Thinkers

*Match the thinkers with their main idea or contribution.*

a. Charles Horton Cooley
b. Georg Simmel
c. Ferdinand Tönnies
d. Émile Durkheim
e. Max Weber
f. Robert Michels

____ 1. Emphasized the importance of bureaucracy as a social development
____ 2. Proposed the idea that different types of society were held together by different types of solidarity (mechanical and organic)
____ 3. Originated the idea of the iron law of oligarchy
____ 4. Developed the concepts of *gemeinschaft* and *gesellschaft*
____ 5. Sociologist who pioneered the idea that group size affects interaction
____ 6. Developed the concepts of primary and secondary groups

## Central Idea Completions

*Fill in the appropriate concepts and descriptions for each of the following questions.*

1. List and define the six major functions of social groups:

   a. _____

   b. _____

   c. _____

   d. _____

   e. _____

   f. _____

2. What are reference groups? _____

   What functions do reference groups serve? _____

3. Present an example of each of the following:

   a. Aggregate: _____

   b. Social group: _____

   c. Primary group: _____

   d. Secondary group: _____

4. Using the American high school as an organization, pick three specific characteristics of bureaucracy outlined by Weber. Describe how informal structures might subvert the three formal characteristics of your example high school.

   a. _____

   b. _____

   c. _____

5. What is Michels's "iron law of oligarchy"?
   _____

6. What are the characteristics of bureaucracy? (Can you think of any that are not mentioned in the text?)

   a. _____

   b. _____

   c. _____

7. What are the differences between a social group and an institution?

   _____
   _____
   _____
   _____

## Critical Thought Exercises

1. Develop a list of the top three (in terms of importance to your life) primary groups to which you belong and the top three secondary groups. What, if anything, do these six groups share in common? What features are unique to each? How do the concepts of primary group and secondary group help you in your examination of the differences among the six groups? Select one of the three secondary groups and discuss what changes would need to occur for that group to become a primary group. Follow the same procedure for one of the three primary groups becoming a secondary group.

2. Consider the functions that all groups must perform (defining boundaries, setting goals, and so on). Compare how two different groups you are familiar with—one primary, one secondary—carry out these functions, and also how the way in which each group carries out these functions affects how the members of the group feel about the group and about one another.

3. To what extent is education at your college or university bureaucratized? How do the various elements of bureaucracy (division of labor, impartiality, and so on) affect the content and quality of education? How might education be different in a less bureaucratically organized school?

4. Observe a group in operation and try to classify each statement according to whether it is instrumental or expressive. See whether group members seem to specialize more in one area than in the other.

## Internet Activity

Examine the chart at www.vetfriends.com/military_structure/ that depicts the structure of the various branches of the U.S. Armed Forces. Pay particular note to the following: groups, subgroups, structure, organization, and compartmentalization. On which levels are relationships tightest? On which levels do you see bureaucratic difficulties developing? At which levels would you expect relationships to be closest? What steps could be taken to foster cohesion? Would these steps apply equally to all groups?

## Answers to Key Concepts

1. q;  2. g;  3. d;  4. r;  5. u;  6. j;  7. c;  8. h;  9. i;  10. o;  11. p;  12. s;  13. t;  14. l;  15. a; 16. k;  17. b;  18. n;  19. m;  20. e;  21. f

## Answers to Key Thinkers

1. e;  2. d;  3. f;  4. c;  5. b;  6. a

# 7 Deviant Behavior and Social Control

**Defining Normal and Deviant Behavior**
- Making Moral Judgments
- The Functions of Deviance
- The Dysfunctions of Deviance

**Mechanisms of Social Control**
- Internal Means of Control
- External Means of Control: Sanctions

**Theories of Crime and Deviance**
- Biological Theories of Deviance
- Psychological Theories of Deviance
- Sociological Theories of Deviance

**The Importance of Law**
- The Emergence of Laws

*How Sociologists Do It:* It's the Little Things That Matter in Preventing Crime

**Crime in the United States**
- Crime Statistics

*How Sociologists Do It:* How Accurate Is Forensic Science?

**Kinds of Crime in the United States**
- Juvenile Crime
- Violent Crime
- Property Crime
- White-Collar Crime
- Victimless Crime

*Thinking About Social Issues:* Are Peaceful Pot Smokers Being Sent to Prison?
- Victims of Crime

**Criminal Justice in the United States**
- The Police
- The Courts
- Prisons

*Our Diverse World:* A Bad Country in Which to Be a Criminal

*Thinking About Social Issues:* The Continuing Debate over Capital Punishment: Does It Deter Murderers?
- A Shortage of Prisons
- Women in Prison
- The Funnel Effect
- Truth in Sentencing

**Summary**

## LEARNING OBJECTIVES

After studying this chapter, you should be able to do the following:

- Understand deviance as culturally relative.
- Explain the functions and dysfunctions of deviance.
- Distinguish between internal and external means of social control.
- Differentiate among the various types of sanctions.
- Describe and critique biological, psychological, and sociological theories of deviance.
- Discuss the concept of anomie and its role in producing deviance.
- Know how the Uniform Crime Reports and the National Crime Victimization Survey differ as sources of information about crime.
- Describe the major features of the criminal justice system in the United States.

Soprano Florence Foster Jenkins believed she was the goddess of song. Unfortunately, she had no singing talent. She sang wildly out of tune, and her voice was quivering and colorless. She was, however, a wealthy New York socialite and did not hesitate to use her money to let the world know that she should be considered a world-class diva.

Several times a year, she would rent the Ritz Carlton Hotel and give stunningly inept renditions of standard opera arias and songs specifically written for her, which she would mangle with her appalling voice. She would create lavish costumes for these performances, and her pianist, Cosme McMoon, treated her with the utmost respect as he accompanied her with the appropriate music.

Eventually, her reputation produced a following, and tickets to her bizarre performances, which could be purchased only from her directly, were sold out months in advance and were as difficult to get as those for the Metropolitan Opera. Her following continued to build; her final performance took place at Carnegie Hall. When she died a month later, she left the following epitaph on her gravestone: "Some people say I cannot sing, but no one can say that I didn't sing." Florence Foster Jenkins was certainly an eccentric, but was she deviant?

### DEFINING NORMAL AND DEVIANT BEHAVIOR

What determines whether a person's actions are seen as eccentric, creative, or deviant? Why will two men walking hand-in-hand cause raised eyebrows in one place but not in another? Why do Britons waiting to enter a theater stand patiently in line, whereas people from the Middle East jam together at the turnstile? In other words, what makes a given action—men holding hands, cutting into a line—normal in one case but deviant in another?

The answer is culture—more specifically, the norms and values of each culture (see Chapter 3, "Culture"). Together, norms and values make up the **moral code** of a culture, *the symbolic system in terms of which behavior takes on the quality of being "good" or "bad," "right" or "wrong."* Therefore, to decide whether any specific act is normal or deviant, it is necessary to know more than only what a person did. One also must know who the person is (that is, the person's social identity) and the social and cultural contexts of the act. For example, if Florence Foster Jenkins held her recitals on a street corner in a seedy neighborhood instead of on a stage at the Ritz Carlton Hotel, would people still have been as interested in her events? Of course not.

For sociologists, then, **deviant behavior** is *behavior that fails to conform to the rules or norms of the group in question* (Durkheim, 1960/1893). Therefore, when

There is a fine line between eccentric behavior and deviant behavior. Advocates for the homeless often claim that these people are on the street because of a lost job, a low minimum wage, or a lack of affordable housing. Others point to problems that cannot be corrected by economic measures.

we try to assess an act as being normal or deviant, we must identify the group by whose terms the behavior is judged. Moral codes differ widely from one society to another. For that matter, even within a society, groups and subcultures exist whose moral codes differ considerably. Watching television is normal behavior for most Americans, but it would be seen as deviant behavior among the Amish.

## Making Moral Judgments

As we stated, sociologists take a culturally relative view of normalcy and deviance and evaluate behavior according to the values of the culture in which it occurs. Ideally, they do not use their own values to judge the behavior of people from other cultures. Even though social scientists recognize that normal and deviant behavior can vary greatly and that no science can determine what acts are inherently deviant, certain acts are almost universally accepted as being deviant. For example, parent-child incest is severely disapproved of in nearly every society.

Genocide, the willful killing of specific groups of people—as occurred in the Nazi extermination camps during World War II—also is universally considered wrong even if it is sanctioned by a government or an entire society. The Nuremberg trials that were conducted after World War II supported this point. Even though most of the accused individuals tried to claim they were merely following orders when they murdered or arranged for the murder of large numbers of Jews and other groups, many were found guilty. The reasoning was that there is a higher moral order under which certain human actions are wrong regardless of who endorses them. Thus, despite their desire to view events from a culturally relative standpoint, most sociologists find certain actions wrong, no matter what the context.

## The Functions of Deviance

Émile Durkheim (1895) observed that deviant behavior is "an integral part of all healthy societies." Why is this true? The answer, Durkheim suggested, is that in the presence of deviant behavior, a social group becomes united in its response. In other words, opposition to deviant behavior creates opportunities for cooperation essential to the survival of any group. For example, let us look at the response to a scandal in a small town as Durkheim described it:

> [People] stop each other on the street, they visit each other, they seek to come together to talk of the event and to wax indignant in common. From all the similar impressions which are exchanged, from all the temper that gets itself expressed, there emerges a unique temper… which is everybody's without being anybody's in particular. That is the public temper. (1895)

When social life moves along normally, people begin to take one another and the meaning of their social interdependence for granted. A deviant act, however, reawakens their group attachments and loyalties because it represents a threat to the moral order of the group. The deviant act focuses people's attention on the value of the group. Perceiving itself under pressure, the group marshals its forces to protect itself and preserve its existence. Deviance also offers society's members an opportunity to rededicate themselves to their social controls. In some cases, deviant behavior actually helps teach society's rules by providing illustrations of violation. Knowing what is wrong is a step toward understanding what is right.

Deviance, then, might be functional to a group in that it (1) causes the group's members to close ranks, (2) prompts the group to organize to limit future deviant acts, (3) helps clarify for the group what it really does believe in, and (4) teaches normal behavior by providing examples of rule violation. Finally, (5) in some situations, tolerance of deviant behavior acts as a safety valve and actually prevents more serious instances of nonconformity. For example, the Amish, a religious group that does not believe in using such examples of contemporary society as cars, radios, televisions, and fashion-oriented clothing, allows its teenagers a great deal of latitude in their behaviors before they are fully required to follow the dictates of the community. This prevents a confrontation that could result in a major battle of wills.

## The Dysfunctions of Deviance

Deviance, of course, has a number of dysfunctions as well, which is why every society attempts to restrain deviant behavior as much as possible. Included among the dysfunctions of deviant behavior are the following:

1. It is a threat to the social order because it makes social life difficult and unpredictable.
2. It causes confusion about the norms and values of a society. People become confused about what is expected and what is right and wrong. The different social standards compete with one another, causing tension among the different segments of society.
3. It undermines trust. Social relationships are based on the premise that people will behave according to certain rules of conduct. When people's actions become unpredictable, the social order is thrown into disarray.
4. Deviance also diverts valuable resources. To control widespread deviance, vast resources must be shifted from other social needs.

CHAPTER 7 DEVIANT BEHAVIOR AND SOCIAL CONTROL 145

These women are on a public street in the Soho section of New York City. Are their clothes an example of deviant behavior?

In this photo from France at the end of World War II, the woman whose head has been shaven is jeered by the crowd as she is escorted out of town. She had been a Nazi collaborator during the war. Émile Durkheim believed that deviant behavior performs an important function by focusing people's attention on the values of the group. The deviance represents a threat to the group and forces the group to protect itself and preserve its existence.

The forms of dress and the types of behavior that are considered deviant depend on who is doing the judging and what the context might be.

## MECHANISMS OF SOCIAL CONTROL

In any society or social group, it is necessary to have mechanisms of social control or ways of directing or influencing members' behavior to conform to the group's values and norms. Sociologists distinguish between internal and external means of control.

### Internal Means of Control

As we already observed in Chapter 3, "Culture," Chapter 5, "Social Interaction," and Chapter 6, "Social Groups and Organizations," people are socialized to accept the norms and values of their culture, especially in the smaller and more personally important social groups to which they belong, such as the family. The word *accept* is important here. Individuals conform to moral standards not just because they know what they are but also because they have internalized these standards. They experience discomfort, often in the form of guilt, when they violate these norms.

In other words, for a group's moral code to work properly, it must be internalized and become part of each individual's emotional life as well as of his or her thought processes. As this occurs, individuals begin to pass judgment on their own actions. In this way, the moral code of a culture becomes an internal means of control—that is, it operates on the individual even in the absence of reactions by others.

## External Means of Control: Sanctions

**External means of control** consist of *other people's responses to a person's behavior—that is, rewards and punishments.* They include social forces external to the individual that channel behavior toward the culture's norms and values.

**Sanctions** are *rewards and penalties that a group's members use to regulate an individual's behavior.* Thus, all external means of control use sanctions of one kind or another. *Actions that encourage the individual to continue acting in a certain way* are called **positive sanctions**. *Actions that discourage the repetition or continuation of the behavior* are **negative sanctions**.

*Positive and Negative Sanctions* Sanctions take many forms, varying widely from group to group and from society to society. For example, an American audience might clap and whistle enthusiastically to show its appreciation for an excellent artistic or athletic performance, but the same whistling in Europe would be a display of strong disapproval. Or consider the absence of a response. In the United States, a professor would not infer public disapproval because of the absence of applause at the end of a lecture—such applause by students is the rarest of compliments. In many universities in Europe, however, students are expected to applaud after every lecture (if only in a rhythmic, stylized manner). The absence of such applause would be a horrible blow to the professor, a public criticism of the presentation.

Most social sanctions have a symbolic side to them. Such symbolism has a powerful effect on people's self-esteem and sense of identity. Consider the positive feelings experienced by Olympic gold medalists or those elected to Phi Beta Kappa, the national society honoring excellence in undergraduate study. Or imagine the negative experience of being given the silent treatment such as that imposed on cadets who violate the honor code at the military academy at West Point. (To some, this is so painful that they drop out.)

Sanctions often have important material qualities as well as symbolic meanings. Nobel Prize winners receive not only public acclaim but also a hefty check. The threat of loss of employment can accompany public disgrace when an individual's deviant behavior becomes known. In isolated, preliterate societies, social ostracism can be the equivalent of a death sentence.

Both positive and negative sanctions work only to the degree that people can be reasonably sure they actually will occur as a consequence of a given act. In other words, sanctions work on people's expectations. Whenever such expectations are not met, sanctions lose their ability to mold social conformity.

It is important to recognize a crucial difference between positive and negative sanctions. When society applies a positive sanction, it is a sign that social controls are successful—the desired behavior has occurred and is being rewarded. When a negative sanction is applied, it is due to the failure of social controls—the undesired behavior has not been prevented. Therefore, a society that frequently must punish people is failing in its attempts to promote conformity. A school that must expel large numbers of students or a government that frequently must call out troops to quell protests and riots should begin to look for the weaknesses in its own system of internal means of social control to promote conformity.

*Formal and Informal Sanctions* **Formal sanctions** *are applied in a public ritual, as in the awarding of a prize or an announcement of expulsion, and are usually under the direct or indirect control of authorities.* For example, to enforce certain standards of behavior and protect members of society, our society creates laws. Behavior that violates these laws can be punished through formal negative sanctions.

Not all sanctions are formal, however. Many social responses to a person's behavior involve **informal sanctions** or *actions by group members that arise spontaneously with little or no formal direction.* Gossip is an informal sanction that is used universally. Congratulations are offered to people whose behavior has approval. In teenage peer groups, ridicule is a powerful, informal, negative sanction. The anonymity and impersonality of urban living, however, decrease the influence of these controls except when we are with members of our friendship and kinship groups.

*A Typology of Sanctions* Figure 7-1 shows the four main types of social sanctions, produced by combining the two sets of sanctions we have just discussed, informal and formal, positive and negative. Although formal sanctions might appear to be strong influences on behavior, informal sanctions actually have a greater effect on people's self-images and behavior. This is so

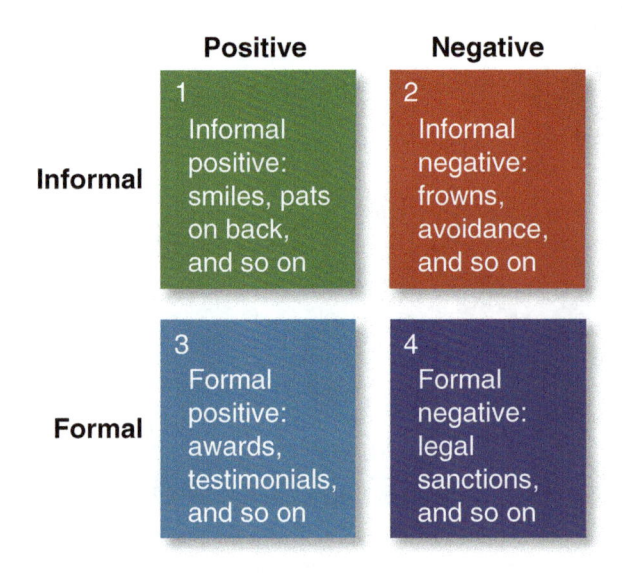

|  | Positive | Negative |
| --- | --- | --- |
| **Informal** | 1 Informal positive: smiles, pats on back, and so on | 2 Informal negative: frowns, avoidance, and so on |
| **Formal** | 3 Formal positive: awards, testimonials, and so on | 4 Formal negative: legal sanctions, and so on |

FIGURE 7-1 Types of Social Sanctions

because informal sanctions usually occur more frequently and come from close, respected associates.

1. **Informal positive sanctions** are *displays people use spontaneously to express their approval of another's behavior.* Smiles, pats on the back, handshakes, congratulations, and hugs are informal positive sanctions.
2. **Informal negative sanctions** are *spontaneous displays of disapproval or displeasure* such as frowns, damaging gossip, or impolite treatment directed toward the violator of a group norm.
3. **Formal positive sanctions** are *public affairs, rituals, or ceremonies that express social approval of a person's behavior.* These occasions are planned and organized. In our society, they include such events as parades that take place after a team wins the World Series or the Super Bowl, the presentation of awards or degrees, and public declarations of respect or appreciation (banquets, for example). Awards of money are a form of formal positive sanctions.
4. **Formal negative sanctions** are *actions that express institutionalized disapproval of a person's behavior.* They usually are applied within the context of a society's formal organizations—schools, corporations, or the legal system, for example—and include expulsion, dismissal, fines, and imprisonment. They flow directly from decisions made by a person or agency of authority, and frequently, specialized agencies or personnel (such as a board of directors, a government agency, or a police force) enforce them.

## THEORIES OF CRIME AND DEVIANCE

Criminal and deviant behavior has been found throughout history. It has been so troublesome and so persistent that much effort has been devoted to understanding its roots. Many dubious ideas and theories have been developed over the ages. For example, a medieval law specified that "if two persons fell under suspicion of crime, the uglier or more deformed was to be regarded as more probably guilty" (Wilson and Herrnstein, 1985). Modern-day approaches to deviant and criminal behavior can be divided into the general categories of biological, psychological, and sociological explanations.

### Biological Theories of Deviance

Nearly all of the early theories of crime tried to provide biological explanations for deviant and criminal behavior. Doing so meant focusing on the importance of inherited factors and downplaying the importance of environmental influences. From this point of view, deviant individuals are born, not made.

Cesare Lombroso (1835–1901) was an Italian doctor who believed that too much emphasis was being put on free will as an explanation for deviant behavior. While trying to discover the anatomic differences between deviant and insane men, he came upon what he believed was an important insight. As he was examining the skull of a criminal, he noticed a series of features, recalling an apish past rather than a human present:

> At the sight of that skull, I seemed to see all of a sudden, lighted up as a vast plain under a flaming sky, the problem of the nature of the criminal—an atavistic being who reproduces in his person the ferocious instincts of primitive humanity and the inferior animals. (Quoted in Taylor et al., 1973)

According to Lombroso's "criminal anthropology," criminals are evolutionary throwbacks whose behavior is more apelike than human. They are driven by their instincts to engage in deviant behavior. These people can be identified by certain physical signs that betray their savage nature. Lombroso spent much of his life studying and dissecting dead prisoners in Italy's jails and concluded that their criminality was associated with an animal-like body type that revealed an inherited primitiveness (Lombroso-Ferrero, 1972). He also believed that certain criminal types could be identified by their head size, facial characteristics (size and shape of the nose, for instance), and even hair color. His writings were met with heated criticism from scholars who pointed out that perfectly normal-looking people have committed violent acts. (Modern social scientists would add that by confining his research to the study of prison inmates, Lombroso used a biased sample, thereby limiting the validity of his investigations.)

Shortly before World War II, anthropologist E. A. Hooten argued that the born criminal was a scientific reality. Hooten believed crime was not the product of social conditions but the outgrowth of "organic inferiority."

> [W]hatever the crime may be, it ordinarily arises from a deteriorated organism.... You may say that this is tantamount to a declaration that the primary cause of penitentiaries of our society are [sic] built upon the shifting sands and quaking bogs of inferior human organisms. (Hooten, 1939)

Hooten went to great lengths to analyze the height, weight, and shape of the body, nose, and ears of criminals. He was convinced that people betrayed their criminal tendencies by the shape of their bodies:

> The nose of the criminal tends to be higher in the root and in the bridge, and more frequently undulating or concave-convex than in our sample of civilians.... [B]ootleggers persistently have broad noses and short faces and flaring jaw angles, while rapists monotonously display narrow foreheads and elongated, pinched noses. (Hooten, 1939)

Following in Hooten's footsteps, William H. Sheldon and his coworkers carried out body measurements of thousands of subjects to determine whether personality traits are associated with particular body types. They found that human shapes could be classified as three particular types: endomorphic (round and soft), ectomorphic (thin and linear), and mesomorphic (ruggedly muscular) (Sheldon and Tucker, 1940). They also claimed that certain psychological orientations are associated with body type. They saw endomorphs as being relaxed creatures of comfort; ectomorphs as being inhibited, secretive, and restrained; and mesomorphs as being assertive, action-oriented, and uncaring of others' feelings (Sheldon and Stevens, 1942).

Sheldon did not take a firm position on whether temperamental dispositions are inherited or are the outcome of society's responses to individuals based on their body types. For example, Americans generally expect heavy people to be good-natured and cheerful, skinny people to be timid, and strongly muscled people to be physically active and inclined toward aggressiveness. Other people may often encourage the person to act along the lines expected. In a study of delinquent boys, Sheldon and his colleagues (1949) found that mesomorphs were more likely to become delinquents than were boys with other body types. Their explanation of this finding emphasized inherited factors, although they acknowledged social variables. The mesomorph is quick to anger and lacks the ectomorph's restraint, they claimed. Therefore, in situations of stress, the mesomorph is more likely to get into trouble, especially if the individual is both poor and not very smart. Sheldon's bias toward a mainly biological explanation of delinquency was strong enough for him to have proposed a eugenic program of selective breeding to weed out those types he considered predisposed toward criminal behavior.

In the mid-1960s, further biological explanations of deviance appeared, linking a chromosomal anomaly in males, known as XYY, with violent and criminal behavior. Typically, males receive a single X chromosome from their mothers and a Y chromosome from their fathers. Occasionally, a child will receive two Y chromosomes from his father. These individuals will look like normal males; however, based on limited observations, a theory developed that these individuals were prone to commit violent crimes. The simplistic logic behind this theory is that because males are more aggressive than females and possess a Y chromosome that females lack, this Y chromosome must be the cause of aggression, and a double dose means double trouble. One group of researchers noted: "It should come as no surprise that an extra Y chromosome can produce an individual with heightened masculinity, evinced by characteristics such as unusual tallness... and powerful aggressive tendencies" (Jarvik, 1972).

Today, the XYY chromosome theory has been discounted. It has been estimated that 96 percent of XYY males lead ordinary lives with no criminal involvement (Chorover, 1979; Suzuki and Knudtson, 1989). A maximum of 1 percent of all XYY males in the United States might spend any time in a prison (Pyeritz et al., 1977). No valid theory of deviant and criminal behavior can be devised around such unconvincing data.

Current biological theorists have gone beyond the simplistic notions of Lombroso, Hooten, Sheldon, and the XYY syndrome. Today, such theories focus on technical advances in genetics, brain functioning, neurology, and biochemistry. Contemporary biological theories are based on the notion that behavior, whether conforming or deviant, results from the interaction of physical and social environments.

The best known of these theories is Sarnoff Mednick's theory of inherited criminal tendencies. Mednick proposed that some genetic factors are passed along from parent to child. Criminal behavior is not directly inherited, nor do the genetic factors directly cause the behavior; rather, one inherits a greater susceptibility to criminality or a predisposition to adapt to normal environments in a criminal way (Mednick, Moffitt, and Stacks, 1987). Mednick believed that certain individuals inherit an autonomous nervous system that is slow to be aroused or to react to stimuli. Such individuals are then slow to learn control of aggressive or antisocial behavior (Mednick, 1977).

More recent research has focused on trying to show that several brain chemical systems might be involved in sensation seeking, impulsivity, negative temperament, and other types of antisocial behavior. Some of this research has focused on neurotransmitters, chemicals in the brain that allow the various regions of the brain to communicate with each other. The neurotransmitter serotonin has been identified with impulsivity and aggression. It is believed that low levels of serotonin produce impulsive and aggressive behavior (Edwards and Kravitz, 1997).

Serotonin levels are related to environmental conditions, however. Studies show that low levels of serotonin are found in individuals who experience high and chronic amounts of stress (Dinan, 1996; Graeff et al., 1996). It also appears that poor parenting lowers serotonin levels and good parenting raises them (Field et al., 1998; Pine et al., 1997).

In recent years, this approach to crime has become known as biocriminology. Today, researchers believe that genes alone do not determine personality traits. The production of behavior is no longer seen as a debate between nature and nurture. Most genes work together with other genes and the environment to produce traits.

Critics of biological theories of criminal behavior claim that such theories present an oversimplified view of genetic and biological influences. They also worry that the research might be used to support the view that there are racial differences in the predisposition to crime (Fishbein, 2001).

## Psychological Theories of Deviance

Psychological explanations of deviance downplay biological factors and emphasize instead the role of parents and early childhood experiences, or behavioral conditioning, in producing deviant behavior. Although such

Homemade meth labs have sprung up throughout the United States. Which theory of deviance would help explain this trend?

explanations stress environmental influences, there is a significant distinction between psychological and sociological explanations of deviance.

Psychological orientations assume that the seeds of deviance are planted in childhood and that adult behavior is a manifestation of early experiences rather than an expression of ongoing social or cultural factors. The deviant individual, therefore, is viewed as a psychologically sick person who has experienced emotional deprivation or damage during childhood.

*Psychoanalytic Theory* Psychoanalytic explanations of deviance are based on the work of Sigmund Freud and his followers. Psychoanalytic theorists believe that the unconscious, the part of us consisting of irrational thoughts and feelings of which we are not aware, causes us to commit deviant acts.

According to Freud, our personality has three parts: the id, our irrational drives and instincts; the superego, our conscience and guide as internalized from our parents and other authority figures; and the ego, the balance among the impulsiveness of the id, the restrictions and demands of the superego, and the requirements of society. Because of the id, all of us have deviant tendencies, though through the socialization process, we learn to control our behavior, driving many of these tendencies into the unconscious. In this way, most of us are able to function effectively according to our society's norms and values.

For some, however, the socialization process is not what it should be. As a result, the individual's behavior is not adequately controlled by either the ego or superego, and the wishes of the id take over. Consider, for example, a situation in which a man has been driving around congested city streets looking for a parking space. Finally, he spots a car that is leaving and pulls up to wait for the space. Just as he is ready to park his car, another car whips in and takes the space. Most of us would react to the situation with anger. We might even roll down the car window and direct some angry gestures and strong language at the offending driver. There have been cases, however, in which the angry driver has pulled out a gun and shot the offender. Instead of simply saying, "I'm so mad I could kill that guy," the offended party acted out the threat. Psychoanalytic theorists might hypothesize that in this case, the id's aggressive drive took over because of an inadequately developed conscience.

Psychoanalytic approaches to deviance have been strongly criticized because the concepts are very abstract and cannot easily be tested. For one thing, the unconscious can be neither seen directly nor measured. Also, such approaches tend to overemphasize innate drives at the same time that they underemphasize social and cultural factors that bring about deviant behavior.

*Behavioral Theories* According to the behavioral view, people adjust and modify their behaviors in response to the rewards and punishments their actions elicit. If we do something that leads to a favorable outcome, we are likely to repeat that action. If our behavior leads to unfavorable consequences, we are not eager to do the same thing again (Bandura, 1969). Those of us who live in a fairly traditional environment are likely to be rewarded for engaging in conformist behavior such as working hard, dressing in a certain manner, or treating our friends in a certain way. We would receive negative sanctions if our friends found out that we had robbed a liquor store.

For some people, however, the situation is reversed. That is, deviant behavior may elicit positive rewards. A 13-year-old who associates with a delinquent gang and is rewarded with praise for shoplifting, stealing, or vandalizing a school is being indoctrinated into a deviant lifestyle. The group might look with contempt at the straight kids who study hard, make career plans, and do not go out during the week. According to this approach, deviant behavior is learned by a series of trials and errors. One learns to be a thief in the same way that one learns to be a sociologist.

*Crime as Individual Choice* James Q. Wilson and Richard Herrnstein (1985) have devised a theory of criminal behavior that is based on an analysis of individual behavior. Sociologists, almost by definition, are suspicious of explanations that emphasize individual behavior because they believe such theories neglect the setting in which crime occurs and the broad social forces that determine levels of crime. However, Wilson and Herrnstein have argued that whatever factors contribute to crime—the state of the economy, the competence of the police, the nurturance of the family, the availability of drugs, the quality of the schools—they must affect the behavior of individuals before they affect crime. They believe that if crime rates rise or fall, it must be due to changes that have occurred in areas that affect individual behavior.

Wilson and Herrnstein contend that individual behavior is the result of rational choice. A person will choose to do one thing as opposed to another because it appears that the consequences of doing it are more desirable than the consequences of doing something else. At any given moment, a person can choose between committing a crime and not committing it.

The consequences of committing the crime consist of rewards and punishments. The consequences of not committing the crime also entail gains and losses. Crime becomes likely if the rewards for committing the crime are significantly greater than those for not committing the crime. The net rewards of crime include not only the likely material gain from the crime but also intangible benefits such as obtaining emotional gratification, receiving the approval of peers, or settling an old score against an enemy. Some of the disadvantages of crime include the pangs of conscience, the disapproval of onlookers, and the retaliation of the victim.

The benefits of not committing a crime include avoiding the risk of being caught and punished and not suffering a loss of reputation or the sense of shame afflicting a person later discovered to have broken the law. All the benefits of not committing a crime lie in the future, whereas many of the benefits of committing a crime are immediate. The consequences of committing a crime gradually lose their ability to control behavior in proportion to how delayed or improbable they are. For example, millions of cigarette smokers ignore the possibility of fatal consequences of smoking because those consequences are distant and uncertain. If smoking one cigarette caused certain death tomorrow, we would expect cigarette smoking to drop dramatically.

### Sociological Theories of Deviance

Sociologists have been interested in the issue of deviant behavior since the pioneering efforts of Émile Durkheim in the late nineteenth century. Indeed, one of the major sociological approaches to understanding this problem derives directly from his work. It is called anomie theory.

*Anomie Theory* Durkheim published *The Division of Labor in Society* in 1893. In it, he argued that deviant behavior can be understood only in relation to the specific moral code it violates: "We must not say that an action shocks the common conscience because it is criminal, but rather that it is criminal because it shocks the common conscience" (1960/1893).

Durkheim recognized that the common conscience, or moral code, has an extremely strong hold on the individual in small, isolated societies where there are few social distinctions among people and everybody more or less performs the same tasks. Such *mechanically integrated societies*, he believed, are organized in terms of shared norms and values. All members are equally committed to the moral code. Therefore, deviant behavior that violates the code is felt by all members of the society to be a personal threat. As society becomes more complex—that is, as work is divided into more numerous and increasingly specialized tasks—social organization is maintained by the interdependence of individuals. In other words, as the division of labor becomes more specialized and differentiated, society becomes more *organically integrated*. It is held together less by moral consensus than by economic interdependence. A shared moral code continues to exist, of course, but it tends to be broader and less powerful in determining individual behavior.

For example, political leaders among the Cheyenne Indians led their people by persuasion and by setting a moral example (Hoebel, 1960). In contrast with the Cheyenne, few modern Americans actually expect exemplary moral behavior from their leaders, despite the public rhetoric calling for it. We express surprise, but not outrage, when less than honorable behavior is revealed about our political leaders. We recognize that political leadership is exercised through formal institutionalized channels, and not through model behavior.

In highly complex, rapidly changing societies such as our own, some individuals come to feel that the moral consensus has weakened. Some people lose their sense of belonging, the feeling of participating in a meaningful social whole. Such individuals feel disoriented, frightened, and alone.

Durkheim used the term **anomie** to refer to *the condition of normlessness in which values and norms have little impact and the culture no longer provides adequate guidelines for behavior.* He found that anomie was a major cause of suicide, as we discussed in Chapter 1, "The Sociological Perspective." Robert Merton built on this concept and developed a general theory of deviance in American society.

*Strain Theory* Robert K. Merton (1938, 1969) believed that American society pushes individuals toward deviance by overemphasizing the importance of monetary success while failing to emphasize the importance of using legitimate means to achieve that success. Those individuals who occupy favorable positions in the social-class structure have many legitimate means at their disposal to achieve success. However, those who occupy unfavorable positions lack such means. Thus, the goal of financial success combined with the unequal access to important environmental resources creates deviance.

As Figure 7-2 shows, Merton identified four types of deviance that emerge from this strain. Each type represents a mode of adaptation on the part of the deviant individual; that is, the form of deviance a person engages

| Mode of adaption | | Culture's goals | Institutionalized means |
|---|---|---|---|
| Conformists | | Accept | Accept |
| Deviants | Innovators | Accept | Reject |
| | Ritualists | Reject | Accept |
| | Retreatists | Reject | Reject |
| | Rebels | Reject/Accept | Reject/Accept |

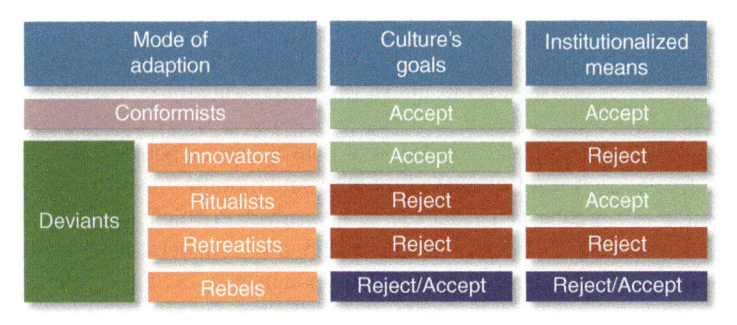

Conformists accept both (a) the goals of the culture and (b) the institutionalized means of achieving them. Deviants reject either or both. Rebels are deviants who might reject the goals of the institutions of the current social order and seek to replace them with new ones that they would then embrace.

**FIGURE 7-2** Merton's Typology of Individual Modes of Adaptation

in depends greatly on the position he or she occupies in the social structure. Specifically, it depends on the availability to the individual of legitimate, institutionalized means for achieving success. Thus, some individuals, called **innovators**, *accept the culturally validated goal of success but find deviant ways of going about reaching it.* Con artists, embezzlers, bank robbers, fraudulent advertisers, drug dealers, corporate criminals, crooked politicians, cops on the take—each is trying to get ahead by using whatever means are available.

**Ritualists** are *individuals who reject or deemphasize the importance of success once they realize they will never achieve it and instead concentrate on following and enforcing rules more precisely than was ever intended.* Because they have a stable job with a predictable income, they remain within the labor force but refuse to take risks that might jeopardize their occupational security. Many ritualists are often tucked away in large institutions such as governmental bureaucracies.

Another group of people also lacks the means to attain success but does not have the institutional security of the ritualists. **Retreatists** are *people who pull back from society altogether and cease to pursue culturally legitimate goals.* They are the drug and alcohol addicts who can no longer function—the panhandlers and street people who live on the fringes of society.

Finally, there are the rebels. **Rebels** *reject both the goals of what to them is an unfair social order and the institutionalized means of achieving them.* Rebels seek to tear down the old social order and build a new one with goals and institutions they can support and accept.

Merton's theory has become quite influential among sociologists. It is useful because it emphasizes external causes of deviant behavior that are within the power of society to correct. The theory's weakness is its inability to account for the presence of certain kinds of deviance that occur among all social strata and within almost all social groups in American society, for example, juvenile alcoholism, drug dependence, and family violence (spouse beating and child abuse).

**Control Theory** In control theory, social ties among people are important in determining their behavior. Instead of asking what causes deviance, control theorists ask what causes conformity. They believe that what causes deviance is the absence of what causes conformity. In their view, conformity is a direct result of control over the individual. Therefore, the absence of social control causes deviance.

According to this theory, people are free to violate norms if they lack intimate attachments with parents, teachers, and peers. These attachments help them establish values linked to a conventional lifestyle. Without these attachments and acceptance of conventional norms, the opinions of other people do not matter, and the individual is free to violate norms without fear of social disapproval. This theory assumes that the disapproval of others plays a major role in preventing deviant acts and crimes.

According to Travis Hirschi (1969), one of the main proponents of control theory, we all have the potential to commit deviant acts. Most of us never commit these acts because of our strong bond to society. Hirschi's view is that there are four ways in which individuals become bonded to society and conventional behavior:

1. *Attachment to others.* People form intimate attachments to parents, teachers, and peers who display conventional attitudes and behavior.
2. *Commitment to conformity.* Individuals invest their time and energies in conventional types of activities, such as getting an education, holding a job, or developing occupational skills. At the same time, people show a commitment to achievement through these activities.
3. *Involvement in conventional activities.* People spend so much time engaged in conventional activities that they have no time to commit or even think about deviant activities.
4. *Belief in the moral validity of social rules.* Individuals have a strong moral belief that they should obey the rules of conventional society.

If these four elements are strongly developed, the individual is likely to display conventional behavior. If these elements are weak, deviant behavior is likely.

More recently, Hirschi and Gottfredson (1993) have also proposed a theory of crime based on one type of control only—self-control. They have suggested that people with high self-control will be less likely during all periods of life to engage in criminal acts. Those with low self-control are more likely to commit crime than those with high self-control. The source of low self-control is ineffective parenting. Parents who do not take an active interest in their children and do not socialize them properly produce children with low self-control. Once established in childhood, the level of social control a person has acquired will guide them throughout the rest of their lives.

*Techniques of Neutralization* Most of us think we act logically and rationally most of the time. To violate the norms and moral values of society, we must have **techniques of neutralization,** *a process that enables us to justify illegal or deviant behavior* (Sykes and Matza, 1957). In the language of control theory, these techniques provide a mechanism by which people can break the ties to the conventional society that would inhibit them from violating the rules.

Techniques of neutralization are learned through the socialization process. According to Sykes and Matza, they can take several forms:

1. *Denial of responsibility*. These individuals argue that they are not responsible for their actions; forces beyond their control drove them to commit the act, such as a troubled family life, poverty, or being drunk at the time of the incident. In any event, the responsibility for what they did lies elsewhere.
2. *Denying the injury*. The individual argues that the action did not really cause any harm. Who really got hurt when the individual illegally copied some computer software and sold it to friends? Who is really hurt in illegal betting on a football game?
3. *Denial of the victim*. The individual sees the victim as someone who deserves what he or she got. The man who made an obscene gesture to us on the highway deserved to be assaulted when we caught up with him at the next traffic light. Some athletes, when accused of sexual assault of a woman, have claimed that the woman consented to sex when she agreed to go to the athlete's hotel room.
4. *Condemnation of the authorities*. The individual justifies the deviant or criminal behavior by claiming that those who are in positions of power or are responsible for enforcing the rules are dishonest and corrupt themselves. Political corruption and police dishonesty leave us with little respect for these authority figures because they are more dishonest than we are.
5. *Appealing to higher principles or authorities*. The individual claims the behavior is justified because he or she is adhering to standards that are more important than abstract laws. Acts of civil disobedience against the government are justified because of the government's misguided policy of supporting a corrupt dictatorship. The behavior might be technically illegal, but the goal justifies the action.

Using these techniques of neutralization, people are able to break the rules without feeling morally unworthy. They might even be able to put themselves on a higher plane specifically because of their willingness to rebel against rules. They are redefining the situation in favor of their actions.

*Cultural Transmission Theory* The cultural transmission theory grows out of the work of Clifford Shaw and Henry McKay, who received their training at the University of Chicago, and relies strongly on the concept of learning. These authors became interested in the patterning of delinquent behavior in that city when they observed that Chicago's high-crime areas remained the same over the decades, even though the ethnic groups living in those areas changed. Further, they found that as members of an ethnic group moved out of the high-crime areas, the rate of juvenile delinquency in that group fell; at the same time, the delinquency rate for the newly arriving ethnic group rose. Shaw and McKay (1931, 1942) discovered that delinquent behavior was taught to newcomers in the context of juvenile peer groups. Also, because such behavior occurred mostly in the context of peer-group activities, youngsters gave up their deviant ways when their families left the high-crime areas.

Edwin H. Sutherland and his student, Donald R. Cressey (1978), built a more general theory of juvenile delinquency on the foundation laid by Shaw and McKay. This theory of differential association is based on the central notion that criminal behavior is learned in the context of intimate groups (see Table 7-1). When criminal behavior is learned, it includes two components:

TABLE 7-1 Sutherland's Principles of Differential Association

1. Deviant behavior is learned.
2. Deviant behavior is learned in interaction with other people in a process of communication.
3. The principal part of the learning of criminal behavior occurs within intimate personal groups.
4. When deviant behavior is learned, the learning includes (a) techniques of committing the act, which are sometimes very complicated or sometimes very simple, and (b) the specific rationalizations, and attitudes.
5. A person becomes deviant because of more definitions favorable to violating the law over definitions unfavorable to violating the law.
6. The process of learning criminal behavior involves all the mechanisms used in any other learning situation.
7. Although criminal behavior is an expression of general needs and values, it is not explained by those general needs and values because noncriminal behavior is an expression of the same needs and values.

*Source:* Adapted from Sutherland, E. H. and D. R. Cressey. 1978. *Principles of Criminology*, 10th ed. Chicago: Lippincott, pp. 80–82.

(1) criminal techniques (such as how to break into houses) and (2) criminal attitudes (rationalizations that justify criminal behavior). People who become criminals are more likely to accept the rationalizations for breaking the law than the arguments for obeying the law. They acquire these attitudes through long-standing interactions with others who hold these views. Thus, among the estimated 70,000 gang members in Los Angeles County, status is often based on criminal activity and drug use. Even arrest and imprisonment are events worthy of respect. A youngster exposed to and immersed in such a value system will identify with it, if only to survive.

In many respects, differential association theory is quite similar to the behavioral theory we discussed earlier. Both emphasize the learning or socialization aspect of deviance. Both also point out that deviant behavior emerges in the same way that conformist behavior emerges; it is merely the result of different experiences and different associations.

**Labeling Theory** Under **labeling theory,** *the focus shifts from the deviant individual to the social process by which a person comes to be labeled as deviant and the consequences of such labeling for the individual.* This view emerged in the 1950s from the writings of Edwin Lemert (1972). Since then, many other sociologists have elaborated on the labeling approach. Labeling theorists note that although we all break rules from time to time, we do not necessarily think of ourselves as deviant, nor are we so labeled by others. However, some individuals, through a series of circumstances, do come to be defined as deviant by others in society. Paradoxically, this labeling process actually helps bring about more deviant behavior.

Being caught in wrongdoing and branded as deviant has important consequences for one's further social participation and self-image. The most important consequence is a drastic change in the individual's public identity. It places the individual in a new status, and he or she might be revealed as a different kind of person than formerly thought to be. Such people might be labeled as thieves, drug addicts, lunatics, or embezzlers and are treated accordingly.

To be labeled as a criminal, one need commit only a single criminal offense. Yet, the word *criminal* carries a number of connotations of other traits characteristic of anyone bearing the label. A man convicted of breaking into a house and thereby labeled criminal is presumed to be a person likely to break into other houses. Police operate on this premise and round up known offenders for questioning after a crime has been committed. In addition, it is assumed that such an individual is likely to commit other kinds of crimes as well because he or she has been shown to be a person without respect for the law. Therefore, apprehension for one deviant act increases the likelihood that this person will be regarded as deviant or undesirable in other respects.

Even if no one else discovers the deviance or endorses the rules against it, the individual who has committed it acts as an enforcer. Such individuals might brand themselves as deviant because of what they did and punish themselves in one way or another for the behavior (Becker, 1963).

At least three factors appear to determine whether a person's behavior will set in motion the process by which he or she will be labeled as deviant: (1) the importance of the norms that are violated, (2) the social identity of the individual who violates them, and (3) the social context of the behavior in question. Let us examine these factors more closely.

1. *The importance of the violated norms.* As we noted in Chapter 3, not all norms are equally important to the people who hold them. The most strongly held norms are mores, and their violation is likely to cause the perpetrator to be labeled deviant. The physical assault of an elderly person is an example. For less strongly held norms, however, much more nonconformity is tolerated, even if the behavior is illegal. For example, running red lights is both illegal and potentially dangerous, but in some American cities, it has become so commonplace that even the police are likely to look the other way rather than pursue violators.
2. *The social identity of the individual.* In all societies, there are those whose wealth or power (or even force of personality) enables them to ward off being labeled deviant despite behavior that violates local values and norms. Such individuals are buffered against public judgment and even legal sanction. A rich or famous person caught shoplifting or even using narcotics has a fair chance of being treated indulgently as an eccentric and let off with a lecture by the local chief of police. Conversely, those marginal or powerless individuals and groups such as welfare recipients or the chronically unemployed, toward whom society has little tolerance for their nonconformity, are quickly labeled deviant when an opportunity presents itself and are much more likely to face criminal charges.
3. *The social context.* The social context within which an action occurs is important. In a certain situation, an action might be considered deviant, whereas in another context it will not. Notice that we say social context, not physical location. The nature of the social context can change even when the physical location remains the same. For example, for most of the year, the New Orleans police manage to control open displays of sexual behavior, even in the famous French Quarter. However, during the week of Mardi Gras, throngs of people freely engage in what at other times of the year would be called lewd and indecent behavior. During Mardi Gras, the social context invokes norms for evaluating behavior that do not so quickly lead to the assignment of a deviant label.

Labeling theory has led sociologists to distinguish between primary and secondary deviance. **Primary deviance** is *the original behavior that leads to the application of the label of deviant to an individual.* **Secondary deviance** is *the behavior people develop as a result of having been labeled as deviant* (Lemert, 1972). For example, a teenager who has experimented with illegal drugs for the first time and is arrested for it might face ostracism by peers, family, and school authorities. Such negative treatment might cause this person to turn more frequently to using illegal drugs and associating with other drug users and sellers, possibly resorting to robberies and muggings to get enough money to buy the drugs. Thus, the primary deviant behavior and the labeling resulting from it lead the teenager to slip into an even more deviant lifestyle. This new lifestyle would be an example of secondary deviance.

A test of labeling theory was done in Florida. Florida law allows judges not to declare people guilty of a felony if they have been sentenced to probation, even though technically they were guilty of a felony offense. For these people, their record does not reflect a felony, and they can lawfully claim they have never been convicted of a felony. The authors looked at 96,000 Florida felony cases and compared those who were declared guilty of a felony with those for whom the judge withheld the label. The researchers found that those who received the formal label of felon were significantly more likely to commit another crime in the next two years than those who did not (Chiricos et al., 2007).

Labeling theory has proved useful. It explains why society will label certain individuals deviant but not others, even when their behavior is similar. There are, however, several drawbacks to labeling theory. First, it does not explain primary deviance. That is, even though we might understand how labeling can contribute to future or secondary acts of deviance, we do not know why the original, or primary, act of deviance occurred. In this respect, labeling theory explains only part of the deviance process. Another problem is that labeling theory ignores the instances when the labeling process might deter a person from engaging in future acts of deviance. It looks at the deviant as a misunderstood individual who really would like to be an accepted, law-abiding citizen. Clearly, this is an overly optimistic view.

It would be unrealistic to expect any single approach to explain deviant behavior fully. In all likelihood, some combination of these various theories is necessary to gain a fuller understanding of the emergence and continuation of deviant behavior. (For a discussion of the broken-windows theory of crime prevention, see "How Sociologists Do It: It's the Little Things That Matter in Preventing Crime.")

## THE IMPORTANCE OF LAW

As discussed earlier in this chapter, some interests are so important to a society that folkways and mores are not adequate to ensure orderly social interaction. Therefore, laws are passed to give the state the power of enforcement. These laws become a formal system of social control, which is exercised when informal forms of control are not effective.

It is important not to confuse a society's moral code with its legal code, nor to confuse deviance with crime. Some legal theorists have argued that the legal code is an expression of the moral code, but this is not necessarily the case. For example, although most states and hundreds of municipalities have enacted some sort of antismoking law, smoking is not a moral offense. Conversely, it is possible to violate American moral sensibilities without breaking the law.

What, then, is the legal code? The **legal code** consists of *the formal rules*, called **laws,** *adopted by a society's political authority*. The code is enforced through formal negative sanctions when rules are broken. Ideally, laws are passed to promote conformity to rules of conduct the authorities believe are necessary for the society to function and that will not be followed if left solely to people's internal controls or informal sanctions. Others argue that laws are passed to benefit or protect specific interest groups with political power rather than society at large (Quinney, 1974; Vago, 1988).

Local police officers are limited to enforcing the law in the communities where they serve.

## The Emergence of Laws

How is it that laws come into society? How do we reach the point at which norms are no longer voluntary and need to be codified and given the power of authority for enforcement? Two major explanatory approaches have been proposed: the consensus approach and the conflict approach.

*Consensus Approach* The **consensus approach** *assumes that laws are merely a formal version of the norms and values of the people.* There is a consensus among the people on these norms and values, and the laws reflect this consensus. For example, people will generally agree that it is wrong to steal from another person. Therefore, laws emerge, formally stating this fact and providing penalties for those caught violating the law.

The consensus approach is basically a functionalist model for explaining a society's legal system. It assumes that social cohesion will produce an orderly adjustment in the laws. As the norms and values in society change,

---

### HOW SOCIOLOGISTS DO IT

### It's the Little Things That Matter in Preventing Crime

The broken-windows theory of preventing crime has been popular for the past 20 years. According to the broken-windows view, if a building has a few broken windows that are not repaired, a series of events are set in motion. Vandals eventually break more windows. The windows are still not repaired. Soon, criminals break into the building and steal anything valuable. Eventually, squatters might move in. Others start to recognize that the street might be dangerous and stay away. Criminals begin to realize that illegal activities can be conducted on the street because no one is watching, and the process of neighborhood deterioration and crime sets in.

The broken-windows view is that quality-of-life problems should be addressed when they are small and within a short period of time. This will show observers that someone cares about the area and that vandals and criminals will not be allowed to take over. The assumption is that if you stop petty crime and low-level antisocial behavior, major crimes will be prevented.

All of this sounds logical, but definitive evidence confirming this view has been lacking since the idea was first presented. Now, a few studies are starting to confirm the broken-windows theory.

The first study was done in Lowell, Massachusetts. Researchers and the police identified thirty-four crime activity hot spots. In half of them, city crews cleaned the streets of trash, fixed broken street lights, secured abandoned buildings, and made arrests for misdemeanor violations. In the other hot spots, policing and services were implemented as before. The results, as seen in Figure 7-3, supported the broken-windows view, as complaints about selling drugs, public drinking, and loitering plunged in the neighborhoods in which cleanup had taken place.

A second experiment took place in the Netherlands. The researchers assumed that if people see one norm or rule violated, such as loose trash or graffiti,

FIGURE 7-3 Change in Complaints After Increased Attention to Quality-of-Life Issues

| Crime | Complaint (%) |
|---|---|
| Selling Drugs | −61.9 |
| Robbery | −41.8 |
| Burglary | −35.5 |
| Nondomestic Assault | −34.2 |
| Graffiti | −23.1 |

*Source:* Braga, Anthony A. and Brenda J. Bond. August 2008. "Policing Crime and Disorder Hot Spots: A Randomized Controlled Trial." *Criminology* 46(3).

people are more likely to violate other norms themselves. In the first part of the experiment, fliers were put on bikes in an alley without graffiti and a sign forbidding it. Trash cans were not available. Two-thirds of the cyclists put the fliers in their pockets. When the fliers were put on bikes in an alley with graffiti, however, 77 percent of the cyclists threw the fliers on the ground. In the second part of the study, envelopes were left partially sticking out of public mailboxes. The envelopes contained the equivalent of a $5 bill. Twenty-three percent of the letters were stolen from mailboxes in dirty areas, but only 13 percent were stolen from mailboxes in clean areas.

The researchers in this two-part study believe that a disorderly environment produces ever more disorder and, eventually, a generalized state of affairs in which crime and illegal activity can flourish.

*Sources:* Kelling, George and Catherine Coles. 1996. *Fixing Broken Windows: Restoring Order and Reducing Crime in Our Communities.* New York: Free Press; Braga, Anthony A. and Brenda J. Bond. August 2008. "Policing Crime and Disorder Hot Spots: A Randomized Controlled Trial." *Criminology* 46(3); Holden, Constance. November 21, 2008. "Study Shows How Degraded Surroundings Can Degrade Behavior." *Science* 322(5905).

so will the laws. Therefore, blue laws, which were enacted in many states during colonial times and prohibited people from working or opening shops on Sunday, have been changed, and now vast shopping malls do an enormous amount of business on Sundays.

*Conflict Approach* The conflict approach to explaining the emergence of laws sees dissension and conflict between various groups as a basic aspect of society. The conflict is resolved when the groups in power achieve control. The **conflict approach** to law *assumes that the elite use their power to enact and enforce laws that support their own economic interests and go against the interests of the lower classes.* As William Chambliss (1973) noted:

> Conventional myths notwithstanding, the history of criminal law is not a history of public opinion or public interest.... On the contrary, the history of criminal law is everywhere the history of legislation and appellate-court decisions which in effect (if not in intent) reflect the interests of the economic elites who control the production and distribution of the major resources of the society.

The conflict approach to law was supported by Richard Quinney (1974) when he noted, "Law serves the powerful over the weak; moreover, law is used by the state ... to promote and protect itself."

Chambliss used the development of vagrancy laws as an example of how the conflict approach to law works. He pointed out that the emergence of such laws paralleled the need of landowners for cheap labor in England during a time when the system of serfdom was breaking down. Later, when cheap labor was no longer needed, vagrancy laws were not enforced. Then, in the sixteenth century, the laws were modified to focus on those who were suspected of being involved in criminal activities and interfering with those engaged in the transportation of goods. Chambliss (1973) noted, "Shifts and changes in the law of vagrancy show a clear pattern of reflecting the interests and needs of the groups who control the economic institutions of the society. The laws change as these institutions change."

## CRIME IN THE UNITED STATES

Crime is behavior that violates a society's legal code. In the United States, what is criminal is specified in written law, primarily state statutes. Federal, state, and local jurisdictions often vary in their definitions of crimes, although they seldom disagree in their definitions of serious crimes.

A distinction is often made between violent crimes and property crimes. A **violent crime** is *an unlawful event such as homicide, rape, and assault that can result in injury to a person.* Aggravated assault, rape, and murder are violent crimes. Robbery is also a violent crime because it involves the use or threat of force against the person.

The United States has one of the highest homicide rates of all industrialized countries in the world. The nation's murder rate has been declining; in 2010, it was 4.8 per 100,000 of population, compared with 10.2 per 100,000 in 1980. (See Figure 7-4.) This lower number is still two to three times that of most European countries. Among the less industrialized countries, more than thirty countries have higher homicide rates than the United States, including South Africa, Russia, El Salvador, Honduras, Columbia, Venezuela, Mexico, and Jamaica.

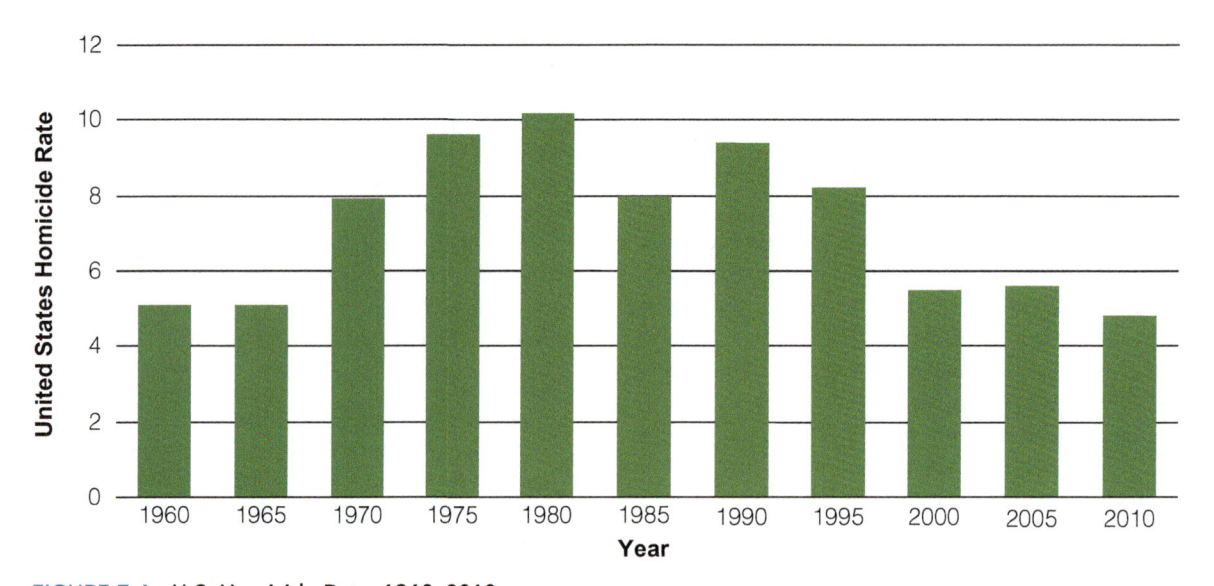

FIGURE 7-4  U.S. Homicide Rate, 1960–2010
Source: Federal Bureau of Investigation. Uniform Crime Reports. Prepared by the National Archive of Criminal Justice Data (www.ucrdatatool.gov/Search/Crime/State/RunCrimeTrendsInOneVar.cfm), accessed June 21, 2012.

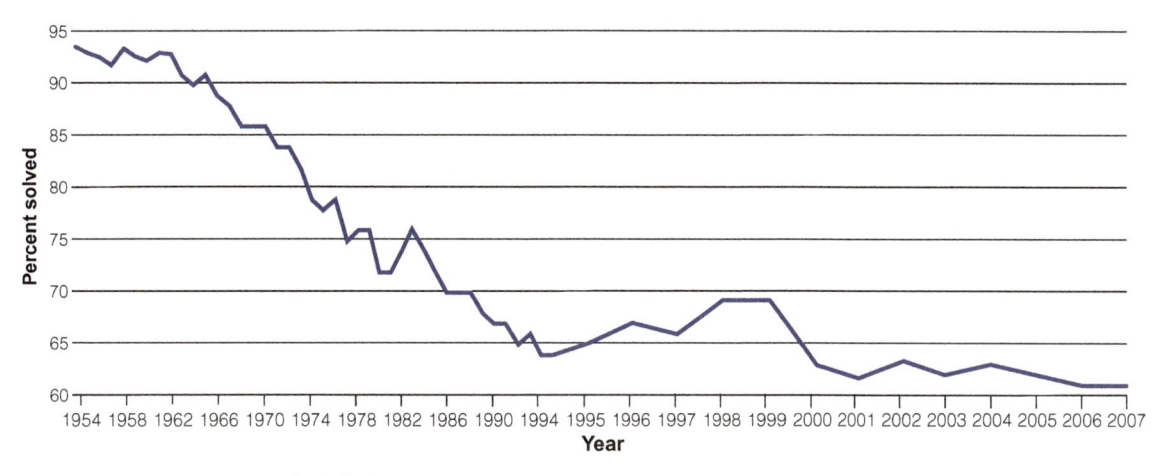

**FIGURE 7-5** U.S. Homicide Solution Rate
*Source: Uniform Crime Reports; FBI Supplement, "Homicide Reports 1976–2007."*

If we look for explanations for this phenomenon, we begin to see that in the United States, homicide has become less of a domestic non-stranger event and more of an event that grows out of other criminal situations. This change has also made it more difficult to solve homicides. When homicide was more likely to be a domestic or intimate relationship event, nearly all were solved. For example, 94 percent of all homicides in 1954 were solved. As of 2010, the solution rate had dropped to 64.8 percent because homicides are increasingly likely to be perpetrated by individuals as they commit a variety of unrelated crimes. (See Figure 7-5 for the homicide solution rates since 1954 more on the relationship between the victim and the perpetrator.)

Homicide is the eighth leading cause of death among blacks and the twentieth leading cause of death among whites in the United States (Anderson and Smith, 2005). When compared with the death toll from other major causes such as heart disease and cancer, the percentage attributed to homicide seems quite modest. There is another way of looking at it, however. Homicide disproportionately involves young victims without any major diseases, making them more like the victims of fatal automobile accidents than those dying from fatal diseases. In any given year, the median age of homicide victims is about 33 (FBI, Supplementary Homicide Reports, 1976–2007). The number of years of life lost to a homicide is usually significantly greater than those lost to a disease. Taken as a whole, the years lost to homicide equal nearly 80 percent of those lost to heart disease and nearly 70 percent of those lost to cancer (Zimring and Hawkins, 1997).

A **property crime** is *an unlawful act that is committed with the intent of gaining property but that does not involve the use or threat of force against an individual*. Larceny, burglary, and motor vehicle theft are examples of property crimes.

Criminal offenses are also classified according to how the criminal justice system handles them. In this respect, most jurisdictions recognize two classes of offenses, felonies and misdemeanors. Felonies are not distinguished from misdemeanors in the same way in all areas, but most states define **felonies** as *offenses punishable by a year or more in state prison*. Although the same act might be classified as a felony in one jurisdiction and as a misdemeanor in another, the most serious crimes are never misdemeanors, and the most minor offenses are never felonies.

## Crime Statistics

It is difficult to know with any certainty how many crimes are committed in the United States each year. Two major approaches are taken in determining the extent of crime. One measure of crime is provided by the Federal Bureau of Investigation (FBI) through its Uniform Crime Reports (UCR). Since 1929, the FBI has been receiving monthly and annual reports from law enforcement agencies throughout the country, currently representing 94 percent of the national population. The UCR consists of reports on eight classifications of crimes: homicide, forcible rape, robbery, aggravated assault, burglary, larceny-theft, motor vehicle theft, and arson. Arrests are reported for twenty-one additional crime categories also. Not included are federal offenses—political corruption, tax evasion, bribery, and violation of environmental protection laws, among others.

The UCR now also includes the National Incident-Based Reporting System (NIBRS). For each crime reported to the police, a variety of data are collected about the incident. Included is information about the offense, characteristics of the victim(s) and offender(s), description of the stolen property, and characteristics of the people arrested in connection with the crime.

Sociologists and critics in other fields note that for a variety of reasons, these statistics are not always reliable. For example, each police department compiles its own figures, and definitions of the same crime vary from place to place. Other factors also affect the accuracy of the crime figures and rates published in these reports. For example, a law enforcement agency or a local government might change its method of reporting crimes so that the new statistics reflect a false increase or decrease in the occurrence of certain crimes. Under some circumstances, UCR data are estimated because some jurisdictions do not participate or report partial data (Bureau of Justice Statistics, September 2002).

Another measure of crime is provided through the National Crime Victimization Survey (NCVS), which began in 1973 to collect information on crimes suffered by individuals and households, whether or not these crimes were reported to the police. The UCR measures reported crimes only. The NCVS collects detailed information on the frequency and nature of the crimes of rape, sexual assault, personal robbery, aggravated and simple assault, household burglary, theft, and motor vehicle theft—both reported and unreported.

The similarity between the categories of crimes in the NCVS and those in the UCR is obvious and intentional. Some crimes are missing from the NCVS that appear in the UCR. Murder, for example, cannot be measured through victim surveys because, obviously, the victim is dead. Arson cannot be measured well through such surveys because the victim might, in fact, have been the criminal. An arson investigator is often needed to determine whether a fire was actually arson. Also, because of problems in questioning the victims, crimes against children younger than age 12 are also excluded.

Whereas the UCR depends on police departments' records of reported crimes, the NCVS attempts to assess the total number of crimes committed. The NCVS obtains its information by asking a nationally representative sample of 87,000 households (about 75,000 people over the age of 12) about their experiences as victims of crime during the previous six months. The households stay in the sample for three years (Bureau of Justice Statistics, April 2009).

Of the crimes that occurred in 2010, the NCVS estimated that slightly more than half of violent victimizations and 39.3 percent of property crimes were reported to the police. Of the violent crimes in 2010, 50 percent of rapes, 57.9 percent of robberies, and 60.1 percent of aggravated assault with injury were brought to the attention of the police. Motor vehicle theft continued to be the property crime most reported (83.4%) to the police (Bureau of Justice Statistics, 2011; see Figure 7-6).

The particular reason most frequently mentioned for not reporting a crime was that it was not important enough. For violent crimes, the reason most often given for not reporting was that it was a private or personal matter. (See Figure 7-7 for information on the likelihood that someone will be arrested for a known crime and Figure 7-8 for the average time served for various types of crimes. Also see "How Sociologists Do It: How Accurate Is Forensic Science?" for a discussion of the techniques used in solving crimes.)

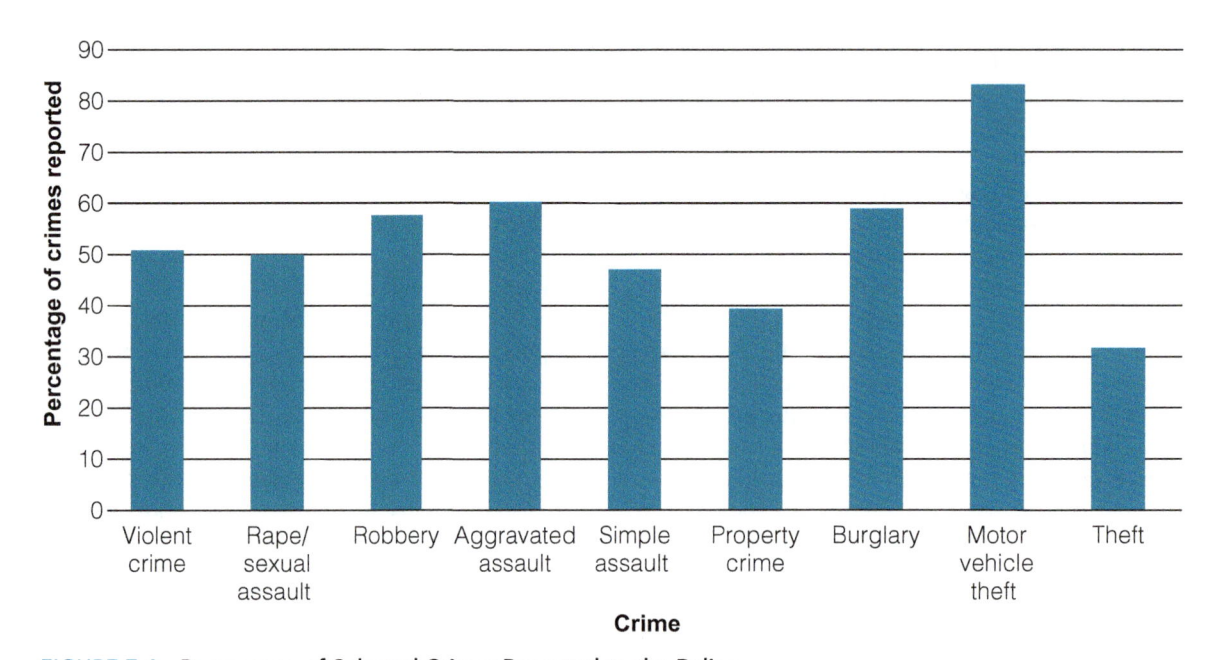

FIGURE 7-6  Percentage of Selected Crimes Reported to the Police
*Source: U.S. Department of Justice, Bureau of Justice Statistics. September 2011. NCJ 235508. BJS Bulletin, Criminal Victimization, 2010 (http://bjs.ojp.usdoj.gov/content/pub/pdf/cv10.pdf), accessed June 21, 2012.*

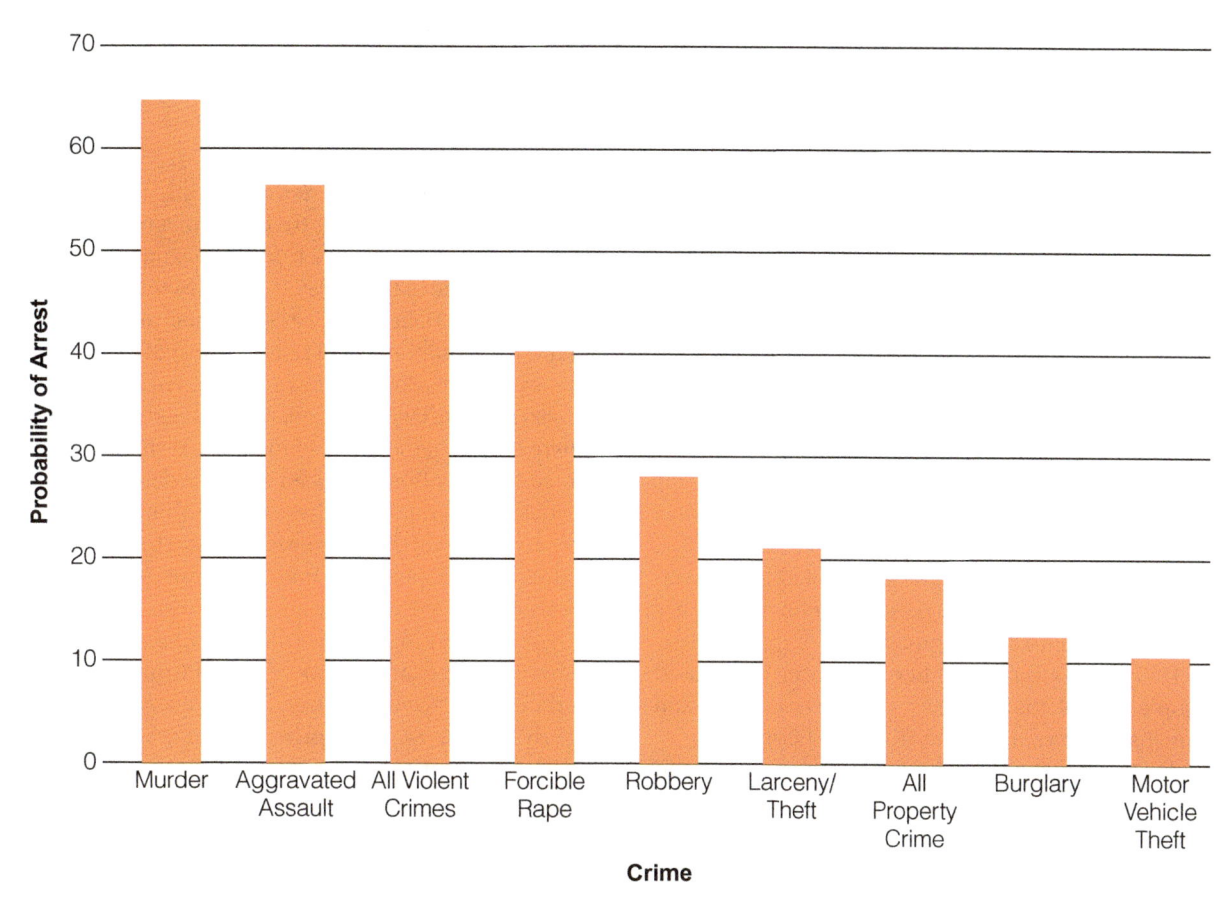

**FIGURE 7-7** Likelihood That Someone Will Be Arrested for a Known Crime
*Source: U.S. Department of Justice, Bureau of Justice Statistics. Crime in the United States, 2010.*

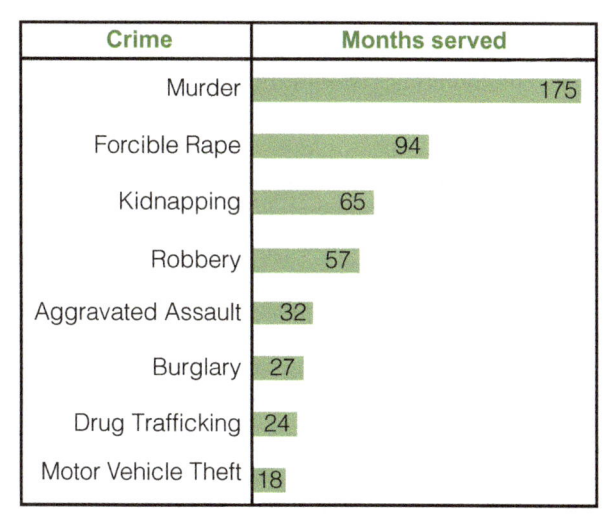

| Crime | Months served |
|---|---|
| Murder | 175 |
| Forcible Rape | 94 |
| Kidnapping | 65 |
| Robbery | 57 |
| Aggravated Assault | 32 |
| Burglary | 27 |
| Drug Trafficking | 24 |
| Motor Vehicle Theft | 18 |

**FIGURE 7-8** Average Time Served for Various Types of Crimes
*Source: U.S. Department of Justice, Bureau of Justice Statistics. May 5, 2011. NCR 0908.csv. Bonczar, Tom. Table 8, "State Prison Releases, 2009: Time Served in Prison, by Offense and Release Type." National Corrections Reporting Program, 2009 (http://bjs.ojp.usdoj.gov/index.cfm?ty=pbdetail&iid=2045), accessed June 21, 2012.*

## KINDS OF CRIME IN THE UNITED STATES

The crime committed can vary considerably in terms of the effect it has on the victim and on the self-definition by the perpetrator of the crime. White-collar crime is as different from street crime as organized crime is from juvenile crime. In the next section, we will examine these differences.

### Juvenile Crime

**Juvenile crime** refers to *the breaking of criminal laws by individuals younger than age 18*. Regardless of the reliability of specific statistics, one thing is clear: Serious crime among our nation's youth is a matter of great concern. Hard-core youthful offenders—perhaps 10 percent of all juvenile criminals—are responsible, by some estimates, for two-thirds of all serious crimes. Although the vast majority of juvenile delinquents commit only minor violations, the juvenile justice system is overwhelmed by these hard-core criminals.

## HOW SOCIOLOGISTS DO IT

### How Accurate Is Forensic Science?

Popular television shows such as *CSI: Crime Scene Investigation* have focused on forensic evidence investigation. The fictional characters in these dramas show the daily operations of crime scene investigators and crime laboratories, including their instrumentation, analytical technologies, and capabilities. Cases are solved in an hour, highly technical analyses are accomplished in minutes, and laboratory and instrumental capabilities are often exaggerated, misrepresented, or entirely fabricated. In courtroom scenes, forensic examiners state their findings are a match between evidence and suspect with unfailing certainty, often demonstrating the technique used to make the determination. The dramas suggest that convictions are quick and no mistakes are made.

Hollywood's version of law and order has caused people to think that forensic evidence is accurate and conclusive. The reality is far from this fiction.

The term *forensic science* encompasses a broad range of disciplines, each with its own distinct practices. The disciplines vary widely with regard to techniques, reliability, and level of error. Some methods are laboratory based, such as DNA analysis, toxicology, and drug analysis. Others are based on interpretation of collected material, such as fingerprints, writing samples, fiber samples, or bite marks. Some activities require the skills and analytical expertise of individuals trained as chemists or biologists. Other activities are conducted by individuals trained in law enforcement.

The biggest problem is the lack of scientific research to support the disciplines. Most have not been studied to determine whether the discipline is valid and reliable, as has been done with DNA evidence. Many of the crime scene activities are subject to interpretation and error.

DNA evidence, which has been used since the 1980s, is the only discipline among the forensic disciplines that consistently produces results that can be relied on when seeking to determine whether a piece of evidence is connected with a particular source. A 1990 congressional report concluded that DNA tests were both reliable and valid. DNA analysis, with its well-defined precision and accuracy, however, applies to only about 10 percent of the cases and may not be relevant to a particular case.

The number of exonerations resulting from the analysis of DNA has grown across the country in recent years, uncovering a disturbing number of wrongful convictions—many for homicides and rapes—and exposing serious limitations in some of the other forensic science approaches commonly used in the United States. According to The Innocence Project, there were 301 postconviction DNA exonerations in the United States from 1989 through November 2012.

Some non-DNA forensic tests do not meet the fundamental requirements of science. Most of these techniques were developed by law enforcement to aid in the investigation of evidence from a particular crime scene. There may be some logic behind these techniques, but no concerted effort has been made to determine the reliability and validity of these tests. Although the precise error rates of these forensic tests are still unknown, it does appear clear that many of these tests, as currently performed, produce erroneous results.

For example, fingerprint identifications were long viewed as an exact means of associating a suspect with a crime scene and were rarely questioned. For nearly a century, fingerprint examiners compared partial fingerprints found at crime scenes to inked fingerprints taken directly from suspects. Recently, however, it has been suggested that fingerprint identifications may not be as reliable as previously assumed. Can someone determine with adequate reliability that the finger that left an imperfect impression at a crime scene is the same finger that left an impression in a file of fingerprints? Baltimore County Circuit Judge Susan M. Souder refused to allow a fingerprint analyst to testify and ruled that the traditional method of fingerprint analysis is "a subjective, untested, unverifiable identification procedure that purports to be infallible."

A 2009 congressional report found that "the forensic science community has had little opportunity to pursue or become proficient in the research that is needed to support what it does…. Most of the studies are… conducted by crime laboratories with little or no participation by the traditional scientific community…. Most disciplines in the profession are hindered by a lack of enforceable standards for interpretation of data."

Television programs may make you think that there is an interesting job opportunity out there using scientific findings to lead to the quick resolution of crimes. The reality, however, is far from what is presented.

*Sources:* National Research Council, Committee on Identifying the Needs of the Forensic Sciences Community. 2009. *Strengthening Forensic Science in the United States: A Path Forward.* Washington, D.C.: National Academies Press (www.nap.edu/catalog.php?record_id=12589#toc), accessed June 25, 2012; *State of Maryland v. Bryan Rose,* No. K06-545 (Circuit Court for Baltimore County).

Serious juvenile offenders are predominantly male, disproportionately minority group members (compared with their proportion in the population), and typically disadvantaged economically. They are likely to exhibit interpersonal difficulties and behavioral problems, both in school and on the job. They are also likely to come from one-parent families or families with a high degree of conflict, instability, and inadequate supervision.

Arrest records for 2009 showed that youths younger than age 18 accounted for 14.1 percent of all arrests. The most common crime that juveniles were arrested for was larceny-theft, followed by burglary (Federal Bureau of Investigation, 2010).

Arrests, however, are only a general indicator of criminal activity. The greater number of arrests among young people might be partly due to their lack of experience in committing crimes and to their involvement in the types of crimes for which apprehension is more likely, for example, theft versus fraud. In addition, because youths often commit crimes in groups, the resolution of a single crime can lead to several arrests. (See Table 7-2 for arrest rates by age.)

Indeed, one of the major differences between juvenile and adult offenders is the importance of gang membership and the tendency of youths to engage in group criminal activities. Gang members are more likely than other young criminals to engage in violent crimes, particularly robbery, rape, assault, and weapons violations. Gangs that deal in the sale of crack cocaine have become especially violent in the past decade.

There is conflicting evidence on whether juveniles tend to progress from less serious to more serious crimes. Many believe violent adult offenders began their criminal careers as perpetrators of violent juvenile crimes. However, many offenses by youths are often dealt with informally.

The juvenile courts—traditionally meant to treat, not punish—have had limited success in coping with such juvenile offenders (Reid, 2012). Strict rules of confidentiality, aimed at protecting juvenile offenders from being labeled as criminals, make it difficult for the police and judges to know the full extent of a youth's criminal record. The result is that violent youthful offenders who have committed numerous crimes often receive little or no punishment.

Defenders of the juvenile courts contend, nonetheless, that there would be even more juvenile crime without them. Others, arguing from learning and labeling perspectives, contend that the system has such a negative effect on children that it actually encourages **recidivism**—that is, *repeated criminal behavior after punishment.* All who are concerned with this issue agree that the juvenile courts are less than efficient, especially in the treatment of repeat offenders. One reason for this is that perhaps two-thirds of juvenile court time is devoted to processing children guilty of what are called status offenses, behavior that is criminal only because the person involved is a minor (examples are truancy and running away from home). Recognizing that status offenders clog the courts and add greatly to the terrible overcrowding of juvenile detention homes, states have sought ways to deinstitutionalize status offenders. One approach, known as diversion—steering youthful offenders away from the juvenile justice system to nonofficial social agencies—has been suggested by Edwin Lemert (1981).

### Violent Crime

The number of violent crimes continues to go down. The 2010 total was 13.2 percent below the 2006 total. Aggravated assaults accounted for the highest number of violent crimes reported to the police (62.5%). Robbery was second (29.5%), followed by forcible rape (6.8%) and murder (1.2%) (FBI, Uniform Crime Reports, 2011).

The violent crime rate in the United States also includes one of the highest homicide rates in the industrialized world. There are more homicides in any one of the cities of New York, Detroit, Los Angeles, or Chicago each year than in all of England and Wales combined. The majority of violent crime victims know their attackers. Nearly 70 percent of the rape and sexual assault victims know the offender as an acquaintance, friend, relative, or intimate.

### Property Crime

Seventy-five percent of all crime in the United States is what is referred to as crime against property as opposed

TABLE 7-2 Age Distribution of Arrests

| Age | Percentage of Arrests |
| --- | --- |
| 14 and younger | 3.8 |
| 15–19 | 20.1 |
| 20–24 | 19.6 |
| 25–29 | 14.5 |
| 30–34 | 10.3 |
| 35–39 | 8.6 |
| 40–44 | 7.8 |
| 45–49 | 7.0 |
| 50–54 | 4.4 |
| 55–59 | 2.2 |
| 60–64 | 1.0 |
| Over 65 | 0.8 |

*Source:* Federal Bureau of Investigation. Crime in the United States, 2009. Table 38 (www2.fbi.gov/ucr/cius2009/data/table_38.html), accessed June 22, 2012.

to crime against a person. In all instances of property crime, the victim is not present and is not confronted by the criminal. It is therefore considered a nonviolent crime.

The most significant nonviolent crimes are burglary, auto theft, and larceny-theft. In 2010, there were more than 9 million property crimes reported to the police. This is more than 12 percent lower than 2006. Larceny-theft accounted for more of the property crimes (68.1%), followed by burglary (23.8%) and auto theft (8.1%) (FBI, Uniform Crime Reports, 2011). (See "Thinking About Social Issues: How Bad Is the Crime Problem?" for more on this discussion.)

## White-Collar Crime

The label "white-collar crime" is often used to refer to either a certain type of offender (for instance, high socioeconomic status, occupation of trust, or both) or a certain type of offense (for instance, economic crime). The term **white-collar crime** was coined by Edwin H. Sutherland (1940) to refer to *the acts of individuals who, while occupying positions of social responsibility or high prestige, break the law in the course of their work for the purpose of illegal personal or organizational gain.* The FBI has decided to approach white-collar crime in terms of the offense, defining it as "those illegal acts

---

### THINKING ABOUT SOCIAL ISSUES

## How Bad Is the Crime Problem?

Most people think the crime problem is very bad. They think it is worse than it was last year, that it is worse on a national level, and that it is worse in their own neighborhoods. According to the Gallup poll between 2005 and 2011, an average of 70 percent of people believed there was more crime than the previous year (see Figure 7-9). Based on people's perceptions, you would think that the American public is under siege and that crime lurks around every corner.

How does this perception mesh with reality? Not very well. This situation is one of the strongest examples of how what people believe to be the case does not always match the sociological data. In this instance, people are woefully misinformed about the crime problem.

Imagine a politician or a criminologist 20 years ago saying, "In the next two decades, every type of crime will decrease by at least 50 percent." We would have said the statement was absurd and it would never happen. Yet, that is what has taken place. The rates of virtually every type of crime have dropped, and the crime rates continue to fall. Since 2001, rates of violent crimes have dropped 33.5 percent; rape, 24.1 percent; robbery, 23.8 percent; and car theft, 39.8 percent. These decreases were in addition to the even bigger

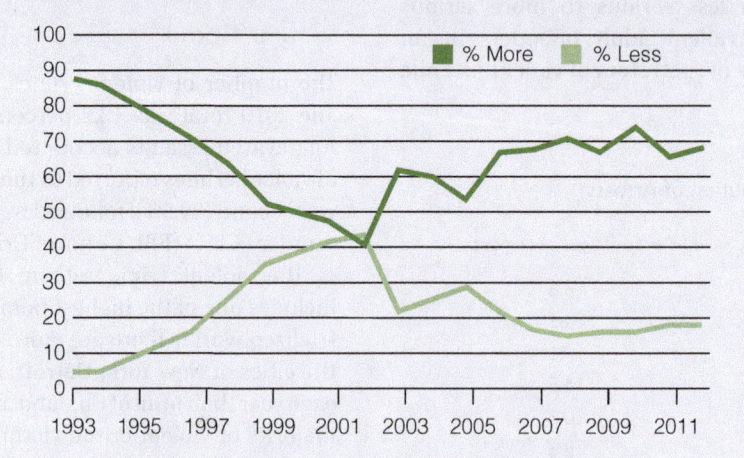

**FIGURE 7-9** Perceptions of Crime Problem, 1993–2011
Is there more crime in the United States than last year, or less?
*Source:* From Lydia Saad, "Most Americans Believe Crime in U.S. Is Worsening." Copyright © 2011 Gallup, Inc. All rights reserved. The content is used with permission; however, Gallup retains all rights of republication.

declines that occurred between 1993 and 2001. (See Figure 7-10.)

Why would people be so wrong in their perceptions of the crime problem? Earlier in Chapter 1, we mentioned that people use their personal experiences to come to conclusions about what is taking place in society. These experiences may be biased by what people see in the media. News programs often have stories about shootings, robberies, and general mayhem. Based on these reports, it may seem that there is a growing crime problem.

Sociologists look to data to find out what is really happening. As you can see, the data on crime do not support the perceptions of the public. We need to be very cautious whenever we use our personal experiences to draw conclusions about what is happening in society.

*Source:* Saad, Lydia. Gallup Poll. October 31, 2011. "Most Americans Believe Crime in U.S. Is Worsening" (www.gallup.com/poll/150464/Americans-Believe-Crime-Worsening.aspx), accessed June 21, 2012.

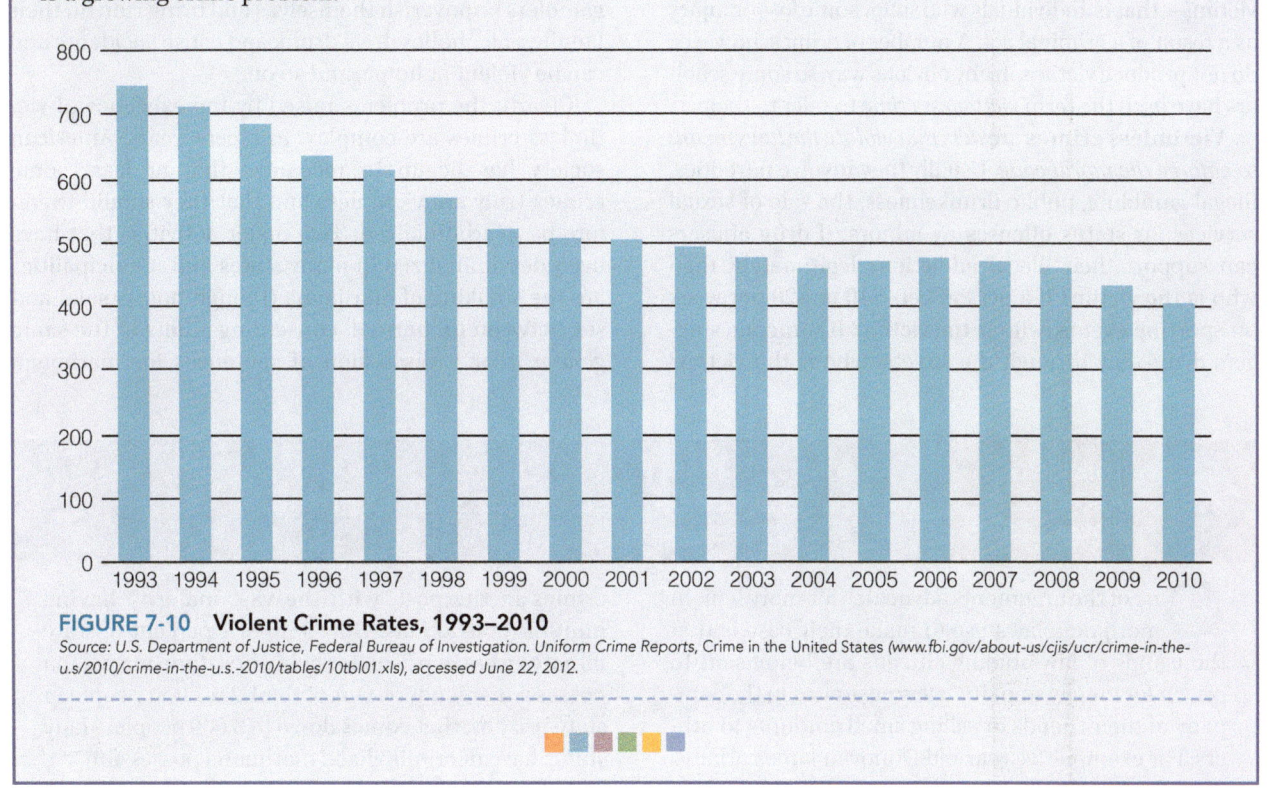

**FIGURE 7-10** Violent Crime Rates, 1993–2010
*Source: U.S. Department of Justice, Federal Bureau of Investigation. Uniform Crime Reports, Crime in the United States (www.fbi.gov/about-us/cjis/ucr/crime-in-the-u.s/2010/crime-in-the-u.s.-2010/tables/10tbl01.xls), accessed June 22, 2012.*

which are characterized by deceit, concealment, or violation of trust and which are not dependent upon the application or threat of physical force or violence" (Barnett, 2002).

White-collar crimes include such actions as embezzlement, bribery, fraud, theft of services, kickback schemes, and others in which the violator's position of trust, power, or influence has provided the opportunity for him or her to use lawful institutions for unlawful purposes. White-collar offenses frequently involve some sort of deception.

Although white-collar offenses are often less visible than crimes such as burglary and robbery, the overall economic impact of these crimes is considerably greater. Not only is white-collar crime very expensive, but it also is a threat to the fabric of society, causing some to argue that it causes more harm than street crime (Reiman, 1990). Sutherland (1961) has argued that because white-collar crime involves a violation of public trust, it contributes to a disintegration of social morale and threatens the social structure. This problem is compounded because, in the few cases in which white-collar criminals actually are prosecuted and convicted, punishment may be comparatively light (Barnett, 2002).

New forms of white-collar crime involving political and corporate institutions have emerged in the past decade. For example, the dramatic growth in high technology has brought with it sensational accounts of computerized heists by sophisticated criminals seated safely behind computer terminals. The possibility of electronic crime has spurred widespread interest in computer security by business and government alike. Crimes committed by corporate executives have also been highly publicized in recent years.

### Victimless Crime

Usually we think of crimes as involving culprits and victims—that is, individuals who suffer some loss or injury as a result of a criminal act. A number of crimes, however, do not produce victims in any obvious way, so some scholars have used the term *victimless crime* to refer to them.

**Victimless crimes** are *acts that violate the laws meant to enforce the moral code*. Usually they involve narcotics, illegal gambling, public drunkenness, the sale of sexual services, or status offenses by minors. If drug abusers can support their illegal addictions legitimately, then who is the victim? If a person bets $10 or $20 per week on sporting events, who is the victim? If someone staggers drunkenly through the streets, who is the victim? If a teenager runs away from home because conditions there are intolerable, who is the victim?

Some legal scholars argue that the perpetrators themselves are victims. Their behavior damages their own lives. This is, of course, a value judgment, but then the concept of deviance depends on the existence of values and norms (Schur and Bedau, 1974). Others note that such offenses against the public order do, in fact, contribute to the creation of victims, if only indirectly. Drug addicts rarely can hold jobs and eventually are forced to steal to support themselves, prostitutes are used to blackmail people and to rob them, chronic gamblers impoverish themselves and bring ruin on their families, alcoholics drive drunk and cause accidents and can be violent at home, and so on.

Clearly, the problems raised by the existence of victimless crimes are complex. In recent years, American society has begun to recognize that at least some crimes truly are victimless and that they should therefore be decriminalized. Two major activities that have been decriminalized in many states and municipalities are the smoking of marijuana (though not its sale) and sex between unmarried, consenting adults of the same gender. (For a discussion of sentences for marijuana

---

**THINKING ABOUT SOCIAL ISSUES**

### Are Peaceful Pot Smokers Being Sent to Prison?

One of the arguments advocates for more lenient marijuana laws use to make their case is that thousands of law-abiding citizens are being sent to prison for merely smoking a few cigarettes in the company of their friends or selling small amounts to others. For example, 24-year-old Donovan James Adams, described as a casual user, was tried in a Montana federal court and sentenced to 66 months in prison for selling three ounces of marijuana.

Critics of the nation's drug policies often point out that drug offenders typically serve longer sentences than people convicted of robbery, rape, or assault. Are overzealous police officers going overboard in enforcing marijuana laws, or is there more to the story?

The U.S. Bureau of Justice Statistics (BJS) investigated the matter and came up with a different story than the one put forth by people suggesting our marijuana laws are sending many first-time offenders to prison. Drug offenders can be tried in state or federal courts, depending on whether the offense took place locally or crossed state lines. The BJS found that of those serving time for any drug offense in state prisons, 83 percent were people who had committed crimes in the past, with the vast majority having multiple convictions. Only 3/10 of 1 percent of state inmates were in prison for being first-time marijuana possession offenders. Out of the 1.2 million people in state prisons, that comes down to 3,600 people. Many states have decriminalized marijuana possession.

In the federal prison system, the story is similar. Of all drug defendants convicted in federal court for marijuana violations, the overwhelming majority were guilty of drug dealing. According to the study, only 63 people were serving time in federal prison for simple possession. Yet, even these people were not necessarily people who had been smoking marijuana in their apartments or dorm rooms. The median amount of marijuana these 63 people were in possession of was 115 pounds. Using the number of 85 cigarettes to the ounce, that equals approximately 156,400 cigarettes. It would be a bit difficult for one person to smoke all those joints.

*Source:* "Untangling the Statistics: Numbers Don't Lie—But They Can Deceive" (www.whitehousedrugpolicy.gov/publications/whos_in_prison_for_marij/untangling_the_stats.pdf), accessed August 31, 2005.

possession, see "Thinking About Social Issues: Are Peaceful Pot Smokers Being Sent to Prison?")

### Victims of Crime

We have been discussing crime statistics, the types of crimes committed, and who commits them. But what about the victims of crime? Is there a pattern? Are some people more apt to become crime victims than others are? It seems that this is true; victims of crime are not spread evenly across society. Although, as we have seen, the available crime data are not always reliable, a pattern of victimization can be seen in the reported statistics. A person's race, gender, age, and socioeconomic status have a great deal to do with whether that individual will become a victim of a serious crime.

Statistics show that, overall, males are much more likely to be victims of serious crimes than females are. When we look at crimes of violence and theft separately, however, a more complex picture emerges. Younger people are much more likely than the elderly to be victims of crime. African Americans are more likely to be victims of violent crime than are whites or members of other racial groups. People with low incomes have the highest violent-crime victimization rates. Theft rates are the highest for people with low incomes and for those with higher incomes. Students and the unemployed are more likely than homemakers, retirees, or the employed to be victims of crime. Rural residents are less often crime victims than are people living in cities (Rennison, 2001).

Despite the growing, albeit unfounded, concern about crimes against the elderly, figures show that young people are most likely to be victims of serious crimes. For example, the violent victimization rates for people aged 16 to 19 are 15 to 20 times higher than for people 65 and older. Similarly, one in eight people murdered are younger than age 18 (Uniform Crime Reports, 2010).

When the elderly are victim of violent crime, it often takes place in their home and is perpetrated by a son or daughter or grandchild. This is the case for one-third of the violence against elderly males and nearly half the violence against elderly females (Violent Crime Against the Elderly, 2005–2009). The reason the elderly are less likely to be the victims of violent crime than the young is related in part to differences in lifestyle and income. Younger people more often might be in situations that place them at risk. They might frequent neighborhood hangouts, bars, and events that are likely places for an assault to occur. About 22 percent of the elderly reported that they never went out at night for entertainment, shopping, or other activities.

The only crime category that affected the elderly at about the same rate as most others (except those ages 12–24) was personal theft, which includes robbery, purse snatching, and pocket picking. Criminals might believe that the elderly are more likely to have large amounts of cash and are less likely to defend themselves, making them particularly vulnerable to these crimes (Rennison, 2001).

## CRIMINAL JUSTICE IN THE UNITED STATES

Every society that has established a legal code has also set up a **criminal justice system**—*personnel and procedures for arrest, trial, and punishment to deal with violations of the law*. The three main categories of our criminal justice system are the police, the courts, and the prisons.

### The Police

The police system developed in the United States is highly decentralized. It exists on three levels: federal, state, and local. On the federal level, the United States does not have a national police system. Congress, however, does enact federal laws. These laws govern the District of Columbia and all states when a federal offense has been committed, such as kidnapping, assassination of a president, mail fraud, bank robbery, and so on. The FBI enforces many of these laws and assists local and state law enforcement authorities in solving local crimes. If a nonfederal crime has been committed, the FBI must wait to be asked by local or state authorities before it can aid the investigation. If a particular crime is a violation of both state and federal law, state and local police often cooperate with the FBI to avoid unnecessary duplication of effort.

The state police patrol the highways, regulate traffic, and have primary responsibility for the enforcement of some state laws. They provide a variety of other services such as a criminal identification system, police training programs, and computer-based records systems to assist local police departments.

The jurisdiction of a police officer at the local level is limited to the state, town, or municipality in which the person is a sworn officer of the law. Some problems inevitably result from such a highly decentralized system. Jurisdictional boundaries sometimes result in communication problems and difficulty in obtaining assistance from another law enforcement agency.

Contrary to some expectations, the public has a great deal of confidence in the police. According to a Gallup poll, the majority of the public rated the police either "very high" or "high" for honesty and ethics. Only 11 percent had a negative view of the police. The situation changes, however, when we look at the numbers more closely. The perceptions of police held by African Americans and white Americans differed dramatically. Whereas 57 percent of whites had a high level of confidence in the police, only 32 percent of African Americans felt the same way. Thirty-one percent of Americans believe there is police brutality in their neighborhood (Jones, 2012).

People often wonder whether putting more police officers on the street reduces crime. It might seem that it should, but a number of studies have cast doubt on what would seem like a logical assumption. Cities with high crime rates have more police officers. Are they having an effect in deterring crime?

It turns out a good way to determine whether the police help stop crime is to look at police staffing during terror alerts. More police are on the street during terror alerts and fewer after them. Does the crime rate differ during these times? Two researchers (Klick and Tabarrok, 2005) looked at Washington, D.C., between March 2002 and July 2003. During that time, the terror alert level rose and fell four times. On high-alert days, total crimes decreased by an average of 6.6 percent. On high-alert days, the police officers spent an extra four hours on duty after their regular eight-hour shifts.

Were tourists and criminals avoiding the area during terror alert days? No, there were as many tourists as before, and crime was down throughout the city. A bigger police presence does not affect all crime levels the same. Murder levels, for example, were unchanged. Street crimes were down considerably. Theft from automobiles and car theft fell 40 percent during high-alert days. Burglaries dropped 15 percent.

Klick and Tabarrok estimated that every $1 spent to add police officers would reduce the costs of crime by $4. They suggested that a 10 percent increase in police nationally would mean about 700,000 fewer property crimes and 213,000 fewer violent crimes.

## The Courts

The United States has a dual court system consisting of state and federal courts, with state and federal crimes being prosecuted in the respective courts. Some crimes can violate both state and federal statutes. About 85 percent of all criminal cases are tried in the state courts.

The state court system varies from one state to another. Lower trial courts exist, mostly, to try misdemeanors and petty offenses. Higher trial courts can try felonies and serious misdemeanors. All states have appeal courts. Many have only one court of appeal, which is often known as the state supreme court. Some states have intermediate appeal courts.

The federal court system consists of three basic levels, excluding such special courts as the U.S. Court of Military Appeals. The U.S. district courts are the trial courts. Appeals may be brought from these courts to the appellate courts. There are 11 courts at this level, referred to as circuit courts. Finally, the highest court is the Supreme Court, which is an appeals court, although it has original jurisdiction in some cases. The lower federal courts and the state courts are separate systems. Cases are not appealed from a state court to a lower federal court. A state court is not bound by the decisions of the lower federal court in its district, but it is bound by decisions of the U.S. Supreme Court (Reid, 2008).

How a case progresses through the criminal justice system, or whether it is even addressed at all, depends on the decisions made by people along the way. Table 7-3 lists various ways a case can work its way through the criminal justice system.

## Prisons

Prisons are a fact of life in the United States. As much as we might wish to conceal them, and no matter how unsatisfactory we think they are, we cannot imagine doing without them. They represent such a fundamental defense against crime and criminals that we now keep a larger portion of our population in prisons than any other nation and for terms that are longer than in many counties. It is small wonder that Americans invented prisons as we know them. (For a look at incarceration in the United States compared to other countries, see "Our Diverse World: A Bad Country in Which to Be a Criminal.")

**TABLE 7-3** Who Decides?

These criminal justice officials must decide how to proceed with a case:

| | |
|---|---|
| **Police** | Enforce specific laws |
| | Investigate specific crimes |
| | Search people, vicinities, buildings |
| | Arrest or detain people |
| **Prosecutors** | File charges or petitions for adjudication |
| | Seek indictments |
| | Drop cases |
| | Reduce charges |
| **Judges or magistrates** | Set bail or conditions for release |
| | Accept pleas |
| | Determine delinquency |
| | Dismiss charges |
| | Impose sentences |
| | Revoke probation |
| **Correctional officials** | Assign to type of correctional facility |
| | Award privileges |
| | Punish for disciplinary infractions |
| **Paroling authorities** | Determine date and conditions of parole |
| | Revoke parole |

Before prisons, serious crimes were redressed by corporal or capital punishment. Jails existed mainly for pretrial detention. The closest thing to the modern prison was the workhouse, a place of hard labor designed almost exclusively for minor offenders, derelicts, and vagrants. The typical convicted felon was either physically punished or fined but not incarcerated. Today's system of imprisonment for a felony is a historical newcomer.

*Goals of Imprisonment* Prisons exist to accomplish at least four goals: (1) to separate criminals from society, (2) to punish criminal behavior, (3) to deter criminal behavior, and (4) to rehabilitate criminals.

1. *Separate criminals from society.* Prisons accomplish this purpose after felons reach the prison gates. Inasmuch as it is important to protect society from individuals who seem bent on repeating destructive behavior, prisons are one logical choice among several others, such as exile and capital punishment (execution). The American criminal justice system relies principally on prisons to segregate convicts from society, and in this regard they are quite efficient.
2. *Punish criminal behavior.* There can be no doubt that prisons are extremely unpleasant places in which to spend time. They are crowded, degrading, boring, and dangerous. Not infrequently, prisoners are victims of one another's violence. Inmates are constantly supervised, sometimes harassed by guards, and deprived of normal means of social, emotional, intellectual, and sexual expression. Prison undoubtedly is a severe form of punishment.
3. *Deter criminal behavior.* The general feeling among both the public and the police is that prisons have failed to achieve the goal of deterring criminal behavior. There are good reasons for this. First, by their very nature, prisons are closed to the public. Few people know much about prison life, nor do they often think about it. Inmates who return to society frequently brag to their peers about their prison experiences to recover their self-esteem. For the prison experience to be a deterrent, the very unpleasant aspects of prison life would have to be constantly brought to the attention of the population at large. To promote this approach, some prisons have allowed inmates to develop programs introducing high school students to the horrors of prison life. From the scanty evidence available to date, it is unclear whether such programs deter people from committing crimes.

    Another reason that prisons fail to deter crime is the funnel effect, discussed later. No punishment can deter undesirable behavior if the likelihood of being punished is minimal. Thus, the argument regarding the relative merits of different types of punishment is pointless until there is a high probability that whatever forms are used will be applied to all (or most) offenders. (For more on deterrence, see "Thinking About Social Issues: The Continuing Debate over Capital Punishment: Does It Deter Murderers?")

4. *Rehabilitate criminals.* Many Americans believe that rehabilitation—the resocialization of criminals to conform to society's values and norms and the teaching of usable work habits and skills—should be the most important goal of imprisonment. It is also the stated goal of almost all corrections officials, yet there can be no doubt that prisons do not come close to achieving this aim. According to the FBI, 43.3 percent of former inmates released from state prisons in 2004 committed at least one serious new crime within the following three years.

Sociological theory helps explain why rehabilitation is often ineffective. Sutherland's ideas on cultural transmission and differential association point to the fact that inside prisons, the society of inmates has a culture of its own in which obeying the law is not highly valued. New inmates are socialized quickly to this peer culture and adopt its negative attitudes toward the law. Further, labeling theory tells us that after someone has been designated as deviant, his or her subsequent behavior often conforms to that label. Prison inmates who are released find it difficult to be accepted in the society at large and to find legitimate work. Hence, former inmates quickly take up with their old acquaintances, many of whom are active criminals. It thus becomes only a matter of time before they are once more engaged in criminal activities.

This does not mean that prisons should be torn down and all prisoners set free. As we have indicated, prisons do accomplish important goals, although certain changes are needed. Certainly it is clear that the entire criminal justice system needs to be made more efficient and that prison terms as well as other forms of punishment must predictably follow the commission of a crime. Another idea, which gained some approval in the late 1960s but seems of late to have declined in popularity, is to create halfway houses and other institutions in which the inmate population is not so completely locked away from society. This way, they are less likely to be socialized to the prison's criminal subculture. Labeling theory suggests that if the process of de-labeling former prisoners were made open, formal, and explicit, released inmates might find it easier to win reentry into society. Finally, just as new prisoners are quickly socialized into a prison's inmate culture, released prisoners must be resocialized into society's culture. This can be accomplished only if means are found to bring ex-inmates into

## OUR DIVERSE WORLD: A Bad Country in Which to Be a Criminal

Do you think the United States is soft on crime? Think again; a criminal is more likely to get hard time in the United States than just about anywhere else. With only 5 percent of the world's population, the United States has one quarter of the world's inmates. Crimes such as writing a bad check or selling small quantities of drugs, which get a slap on the wrist in many countries, will produce a prison sentence in the United States.

The United States has more than 2.3 million people behind bars. People often assume that China has more inmates. China, however, with four times the U.S. population, has 1.64 million people in prison. Granted, they also have in prison, as part of "re-education" efforts, several hundred thousand political activists who have not committed any crimes. China also imposes and carries out death sentences in a rapid fashion, so they are not a model of judicial progress.

In the United States, 730 people are in prison or jail for every 100,000 in population. Russia, with 511 prisoners for every 100,000 people, is second. The other industrialized countries have much lower rates. For England, it is 154; France, 101; Germany, 83; and Japan, 55. The median rate for all nations is about a sixth of the American rate.

Why is the United States so convinced that locking criminals up and throwing away the key is the way to address crime? The United States has much higher levels of violent crime than most other countries, most likely an outgrowth of the ready availability of guns. Assault rates in the United States and Great Britain are similar, but the murder rate, particularly with firearms, is much higher in the United States. Even with the recent decline in the U.S. murder rate, it is still four times that of Western Europe.

The high U.S. incarceration rates are also due to the concerted effort to address drug trafficking with long sentences that are often mandatory and cannot be reduced by judges. We now have about 12 times as many people in prison for drug crime as we had in 1980. In addition, in many states, judges are elected and, to be reelected, must be seen as tough on crime.

Prison sentences in the United States are exceptionally long compared to those in the rest of the world. Nonviolent criminals often do not get prison time in many countries and, if they do, certainly not a very long sentence. Only in the United States, for example, can someone be imprisoned for writing bad checks. For crimes such as burglary, the average U.S. sentence is 16 months. In Great Britain, it is seven months and, in Canada, it is five months.

In 1831, Alexis de Tocqueville traveled from Europe to observe U.S. prisons and noted, "In no country is criminal justice administered with more mildness than in the United States." Nobody would make such a statement today.

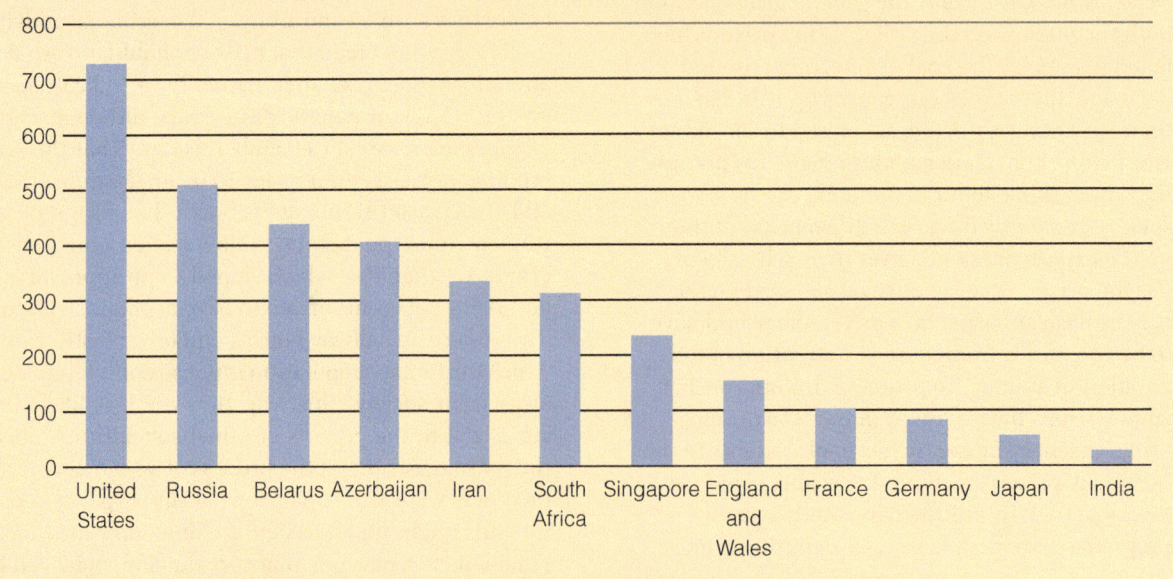

FIGURE 7-11  Prison Inmates by Country (Inmates per 100,000 People)
Source: International Centre for Prison Studies. World Prison Brief (www.prisonstudies.org/info/worldbrief/), accessed June 25, 2012.

### THINKING ABOUT SOCIAL ISSUES

## The Continuing Debate over Capital Punishment: Does It Deter Murderers?

Many countries throughout the world no longer use the death penalty. Among those that do, China stands out with the largest number of executions each year. China, Iran, Pakistan, Iraq, Sudan, and the United States account for 91 percent of the executions in the world.

The United States executed forty-three inmates in 2011, bringing the total number of U.S. executions to 1,234 since 1976, the year the Supreme Court reinstated the death penalty. Currently, 3,158 prisoners await execution (see Figure 7-12 and Figure 7-13). The average person executed spends nearly 15 years on death row. It seems obvious that the vast majority of inmates sentenced to death will not be executed.

The public supports the death penalty. The Gallup poll has been asking Americans about the death penalty for almost 50 years. As of 2011, 61 percent of Americans favored the death penalty in cases of murder, down from its high point of 80 percent in 1994 (Newport, Gallup Poll 2011).

But capital punishment has also been opposed for many years and for many reasons. In the United States, the Quakers were the first to oppose the death penalty and to provide prison sentences instead. Amnesty International USA calls capital punishment a "horrifying lottery" in which the penalty is death and the odds of escaping are determined more by politics, money, race, and geography than by the crime committed. The group bases its impression on the fact that black men are more likely to be executed than white men; and southern states, including Texas, Virginia, Missouri, Louisiana, and Florida, account for the majority of executions that have occurred since 1977.

It is also no surprise that nearly all death-row inmates are poor. They often had a public defender who might not have been qualified for the task. Even if the inmate's attorney made errors during the defense, the defendant's appellate attorney must demonstrate that the defense counsel's blunders directly affected

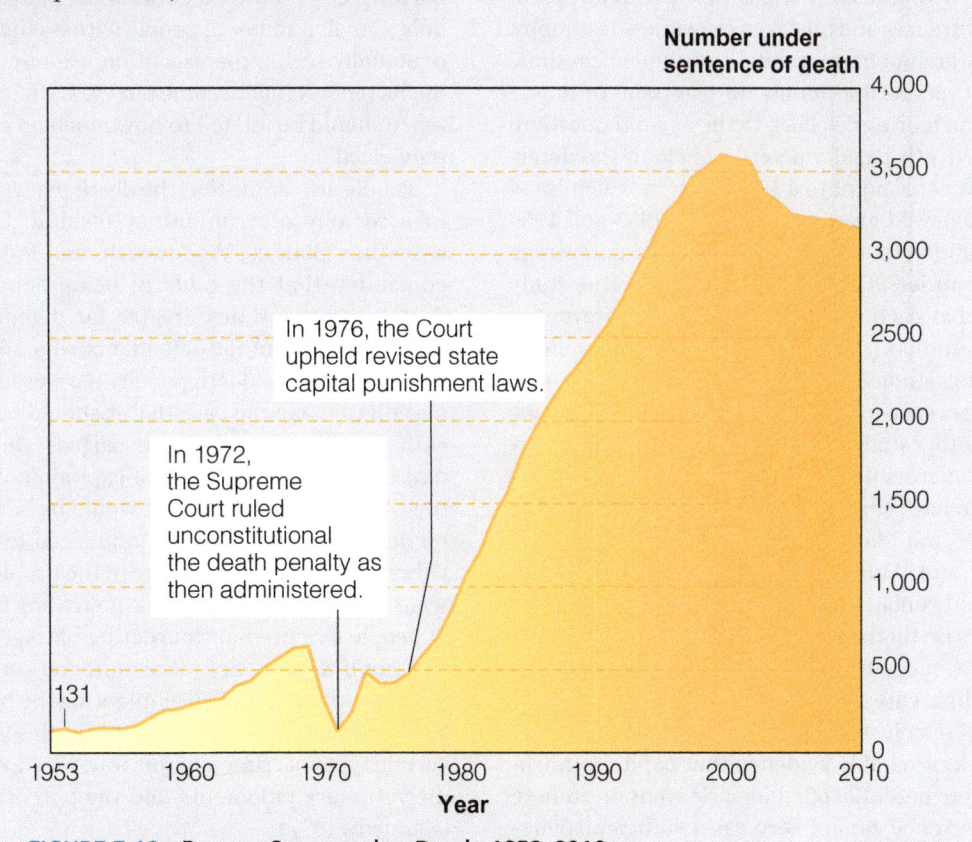

**FIGURE 7-12** Persons Sentenced to Death, 1953–2010
Source: U.S. Department of Justice, Bureau of Justice Statistics. December 2011. NCJ 236510. Capital Punishment, 2010—Statistical Tables (www.bjs.gov/content/pub/pdf/cp10st.pdf), accessed June 25, 2012.

(*Continued*)

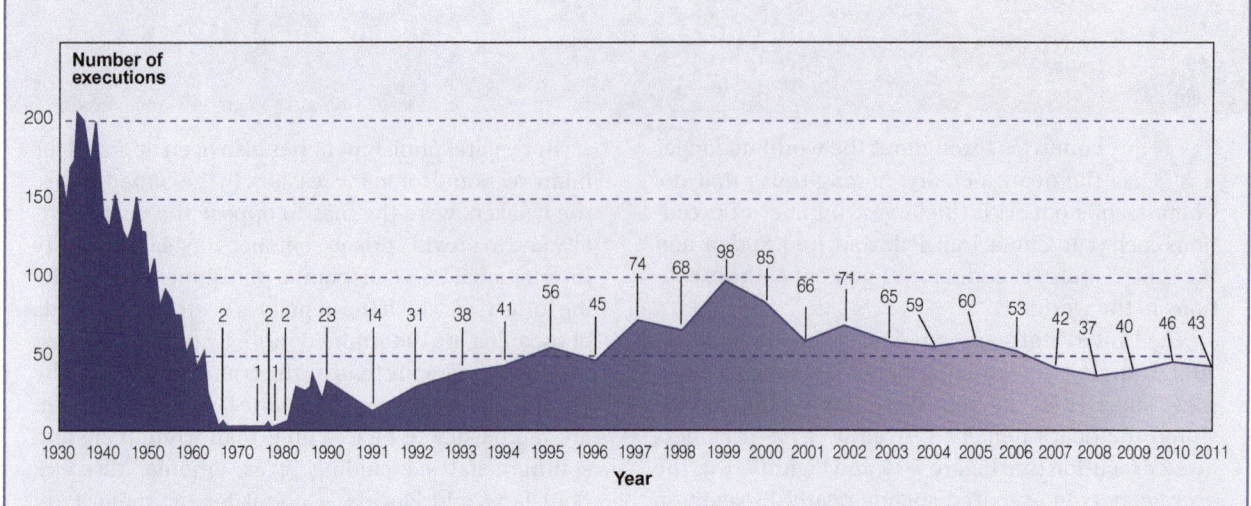

**FIGURE 7-13** Inmates Executed, 1930–2011
Source: U.S. Department of Justice, Bureau of Justice Statistics. December 2011. NCJ 236510. Capital Punishment, 2010—Statistical Tables (www.bjs.gov/content/pub/pdf/cp10st.pdf), accessed June 25, 2012.

the jury's verdict and that without those mistakes, the jury would have returned a different verdict (Prejean, 1993). The two most frequent causes of errors that produced wrongful convictions were perjury by prosecution witnesses and mistaken eyewitness testimony.

Yet, the arguments for capital punishment continue to mount, centering mainly on the issue of deterrence. Which brings us back to the age-old question: Does the death penalty deter homicide? The deterrence advocates point to a 1975 study in which Isaac Ehrlich reviewed executions between 1933 and 1969 and concluded that execution prevented an average of eight murders. Other studies followed. One study claimed that each execution, on average, prevented eighteen murders (Dezhbakhsh, Rubin, and Shepherd, 2002). Other studies have claimed that each execution prevents seventy-four murders (Adler and Summers, 2007). Another study reported that the unofficial moratorium on executions during most of 1996 in Texas appears to have contributed to additional homicides (Cloninger and Marchesini, 2001). Two economists, Donohue and Wolfers (2008), reviewed the various studies and concluded the "view that the death penalty deters is still the product of belief, not evidence."

In 2008, more than three decades after he cast the deciding vote to reinstitute the death penalty, Supreme Court Justice John Paul Stevens noted there was a lack of reliable evidence that capital punishment deters potential offenders. He went on to note that "deterrence cannot serve as a sufficient penological justification for this uniquely severe and irrevocable punishment" (*Baze v. Rees*, 2008).

More might be involved in deterrence than we think. Plato believed we are deterred from committing crimes by seeing others punished. He was referring to punishments administered in public, where everyone could see the gory details of torture and execution. Fortunately, today, executions are not public, and only a small number of people witness them. In place of actually seeing the execution, we now have mass media reports that become our eyes. Therefore, deterrence should be related to how much an execution is publicized.

People also argue that the death penalty is applied in a racially discriminatory fashion. One extensive study (Baldus, Woodworth, and Pulaski, 1990) concluded that the odds of being condemned to death were 4.3 times greater for defendants who killed whites than for defendants who killed blacks. Opponents of the death penalty have used this information to make the case that it should be abolished entirely on the grounds that racial bias is an inevitable part of the administration of capital punishment in the United States and that it would be better to have no death penalty than one influenced by prejudice. Others argue that the remedy to the problem is to do what is known as leveling up—increasing the number of people executed for murdering blacks. They point out that if we sentence more murderers of black people to death, we are then eliminating the bias. A third solution is to impose mandatory death sentences for certain types of crimes. In this way, we are eliminating discretionary judgments and the potential for bias (Kennedy, 1997).

In recent years, the number of executions in the United States has been declining, mainly because it is expensive. It costs $90,000 more per inmate per year to put an inmate on death row than in a maximum security prison. California alone spends more than $100 million a year on death penalty cases. Ultimately, the death penalty may disappear because the state will not be able to afford it.

*Sources:* U.S. Department of Justice, Bureau of Justice Statistics. December 2011. NCJ 236510. BJS Bulletin *Capital Punishment, 2010* (http://bjs.ojp.usdoj.gov/content/pub/pdf/cp10st.pdf), accessed July 5, 2012; Prejean, H. 1993. *Dead Man Walking*. New York: Random House, and an interview with the author, September 1993; Baldus, D. C., G. Woodworth, and C. A. Pulaski Jr. 1990. *Equal Justice and the Death Penalty: A Legal and Empirical Study*. Boston: Northeastern University Press; Kennedy, R. 1997. *Race, Crime and the Law*. New York: Pantheon Books, and an interview with the author, June 1997; Bedau, H. A. (Ed.). 1997. *The Death Penalty in America: Current Controversies*. New York: Oxford University Press, and an interview with the author, May 1997; Dezhbakhsh, H., P. H. Rubin, and J. M. Shepherd. January 2002. "Does Capital Punishment Have a Deterrent Effect? New Evidence from Post-Moratorium Panel Data." Emory University: Department of Economics; Cloninger, D. O. and R. Marchesini. 2001. "Execution and Deterrence: A Quasi-Controlled Group Experiment." *Applied Economics* 35(5):569–76; Mocan, Naci and Kaj Gittings. 2001. "Pardons, Executions and Homicide." Working Paper 8639, National Bureau of Economic Research; Fagan, Jeffrey, Franklin E. Zimring, and Amanda Geller. June 2006. "Capital Punishment and Capital Murder: Market Share and the Deterrent Effects of the Death Penalty." *Texas Law Review* 84:1803–1867; Adler, Roy D. and Michael Summers. November 2, 2007. "Capital Punishment Works." *Wall Street Journal*, p. A13.

---

frequent, supportive, and structured contact with stable members of the wider society (again, perhaps, through halfway houses). The simple separation of prisoners from society undermines this goal.

To date, no society has been able to come up with an ideal way of confronting, accommodating, or preventing deviant behavior. Although much attention has been focused on the causes of and remedies for deviant behavior, no theory, law, or social-control mechanism has yet provided a fully satisfying solution to the problem.

## A Shortage of Prisons

Today's criminal justice system is in a state of crisis over prison crowding. Even though our national prison capacity has expanded, it has not kept up with demands. The National Institute of Justice estimates that we must add 1,000 prison spaces a week just to keep up with the growth in the criminal population.

Many states have mandated prison terms for chronic criminals, drunken drivers, and those who commit gun crimes, compounding the problem of overcrowding. Yet, nearly every community will have an angry uprising if the legislature suggests building a new prison in its neighborhood. Given state financial pressures, community resistance, and soaring construction costs, people face a difficult choice. They must either build more prisons or let most convicted offenders go back to the community.

A key consideration in sending a person to prison is money. The custodial cost of incarceration in a medium-security prison is $15,000 a year. The cost is closer to $35,000 after adding to this the cost of actually building the prison and additional payments to dependent families. You can see why judges are quick to use probation as an alternative to imprisonment, particularly when the prisons are already overcrowded.

The other side of the question, however, is how much it costs us not to send this person to prison. Although it is easy to calculate the cost of an offender's year in prison, it is considerably more difficult to figure the cost to society of letting that individual roam the streets. Crime imposes an assortment of costs on the individual victim. It costs the offender his or her freedom and productivity. There are societal costs when people become fearful and society must spend time and money on crime prevention.

To come up with a number, researchers (DeLisi et.al., 2010) performed a survey and asked people how much they would pay to reduce the incidence of a crime in their community by 10 percent. For homicide and other forms of violent crime, Americans are willing to pay a great deal. The total cost of a crime includes the victim cost, criminal justice system costs, and lost productivity costs for both the victim and criminal. When all the costs were added up (DeLisi et.al., 2010), it was concluded that each murder costs $17,252,656, a rape costs $448,532, armed robbery costs $335,733, aggravated assault costs $145,379, and burglary costs $41,288.

Studies suggest that it is more expensive to release an offender than to incarcerate such a person when you weigh the value of crime prevented through imprisonment. How much does each crime cost the public? The National Institute of Justice has come up with a figure of $2,300 per crime. This number undoubtedly overestimates the value of petty larcenies and underestimates the cost of rapes, murders, and serious assaults. It is an average, however, and it does give us some way of comparing the costs of incarceration with the costs of freedom. Using the $2,300 per-crime cost, we can see that a typical inmate committing dozens of crimes a year can be responsible for a substantial amount of money lost per year. Sending 1,000 additional offenders to prison, instead of putting them on probation, could cost a substantial amount of money. The crimes averted, however, by taking these individuals out of the community could save society considerably more than that.

## Women in Prison

As prison reform began in this country, the practice was to segregate women into sections of the existing institutions. There were few women inmates, a fact that was used to justify not providing them with a matron. Vocational training and educational programs were not even considered. In 1873, the first separate prison for women, the Indiana Women's Prison, was opened, with its emphasis on rehabilitation, obedience, and religious education.

In contrast with institutions for adult males, institutions for adult women are generally more aesthetic and less secure. This is an outgrowth of the fact that in the past, women inmates were not considered high-security risks, nor have they proved to be as violent as male inmates. Women were more likely to commit property crimes such as larceny, forgery, and fraud. This trend in crimes has changed, however, and women now commit more violent crimes than property crimes. Still, three-quarters of the violent crimes committed by women are the less serious type known as simple assault. Drug offenses by women have also increased dramatically in recent years.

Two-thirds of women in prison are mothers, and the vast majority of their children are younger than age 18. Some prisons, such as the maximum security women's prison in Bedford Hills, New York, allow inmate mothers to keep their babies with them until the babies are 18 months of age.

With some exceptions, on the whole, women's institutions are built and maintained with the view that their occupants are not great risks to themselves or to others. Women inmates also usually have more privacy than men do while incarcerated, and women usually have individual rooms. With the relatively smaller number of women in prison, there is a greater opportunity for the inmates to have contact with the staff, and there is also a greater chance for innovation in programming (Reid, 2012).

The number of women in state and federal prisons was nearly 114,000 in 2009. Women still make up a relatively small segment (about 7%) of the total prison population. Relative to their numbers in the U.S. population, men are 14 times more likely than women to be incarcerated (West, Sabol, and Greenman, 2011) (see Figure 7-14).

Compared with male inmates, female inmates appear to have greater difficulty adjusting to the absence of their families, especially their children. Two-thirds of women in prison are mothers, and the majority (88%) of their children are younger than age 18. Only 25 percent of these children are cared for by the father while the mother is in prison. Most of the time, a grandparent cares for the children.

## The Funnel Effect

One complaint voiced by many of those concerned with our criminal justice system is the existence of the funnel effect, in which many crimes are committed, but few people ever seem to be punished. The funnel effect begins with the fact that fewer than 50 percent of all crimes committed are reported to the police (Bureau of Justice Statistics Sourcebook, 2002). Only about 26 percent lead to an arrest. Further, false arrests, lack of evidence, and plea bargaining (negotiations in which individuals arrested for a crime are allowed to plead guilty to a lesser charge of the crime, thereby saving the criminal justice system the time and money spent on a trial) considerably reduce the number of complaints that actually are brought to trial. To be fair, the situation is not quite as bad as it appears. The number of arrests for serious crimes is considerably higher than it is for crimes in general.

What about punishment? Those who criticize the system's funnel effect seem to regard only a term in prison as an effective punishment. Yet, the usual practice is to send to prison only those criminals whose terms of confinement are set at longer than one year. Many thousands of other criminals receive shorter sentences and serve them in municipal and county jails. Thus, if the numbers of people sent to local jails as well as to prison are counted, the funnel effect appears less severe than it often is portrayed. The question then becomes one of philosophy. Is a jail term of less than one year

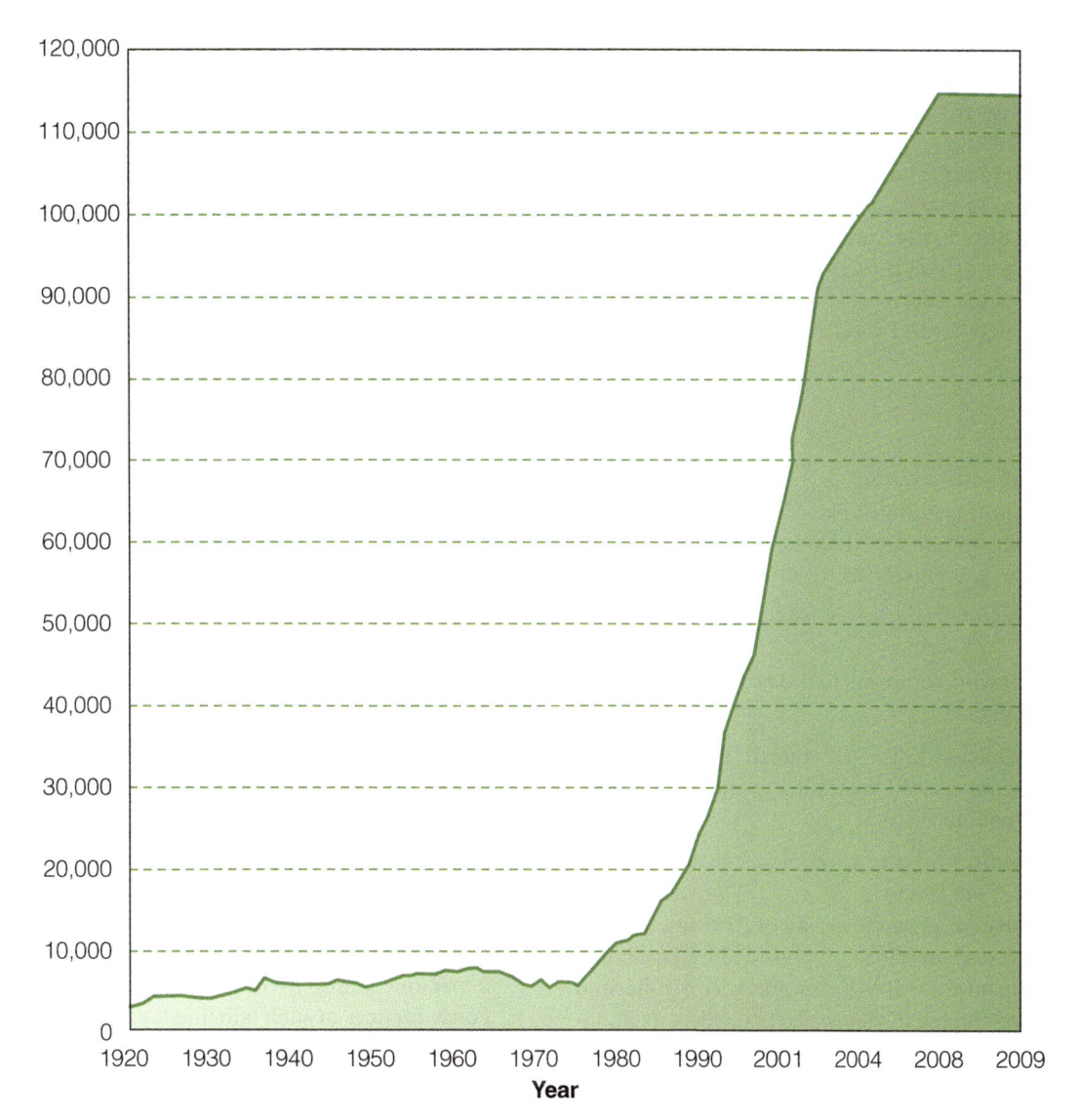

**FIGURE 7-14** Women Prisoners in State and Federal Institutions, 1925–2009
Source: West, Heather C., William J. Sabol, and Sarah J. Greenman. December 2010, revised October 27, 2011. NCJ 231675. "Prisoners in 2009." U.S. Department of Justice, Bureau of Justice Statistics (http://bjs.ojp.usdoj.gov/content/pub/pdf/p09.pdf).

an adequate measure for the deterrence of crime? Or should all convicted criminals have to serve longer sentences in federal or state prisons, with jails used primarily for pretrial detention?

### Truth in Sentencing

The amount of time offenders serve in prison is almost always shorter than the time they are sentenced to serve by the court. The public has been in favor of longer sentences and uniform punishments for prisoners. Prison crowding and reductions in prison time for good behavior have often resulted in the release of prisoners well before they have served their assigned sentences. In response to complaints that criminals were not paying for their crimes, many states enacted restrictions on the possibility of early release; these laws became known as "truth in sentencing."

The truth-in-sentencing laws require offenders to serve a substantial portion of the prison sentence imposed by the court before being eligible for release. The laws are based on the belief that victims and the public are entitled to know exactly what punishments offenders are receiving (Ditton and Wilson, 1999). The first truth-in-sentencing law was passed in 1984, and by 2008, thirty-five states had them. In 1994, a federal truth-in-sentencing law was passed requiring offenders to serve at least 85 percent of their sentence before becoming eligible for parole.

Truth-in-sentencing laws gained momentum with the help of the U.S. Congress, which authorized grants to expand or build correctional facilities if states would enact such laws. The laws have limited the powers of parole boards to "set release dates, or of prison managers to award good time, earned time, or both" (Mackenzie, 2000).

Sentencing reforms have also led to more blacks than whites going to prison after arrest. If current trends continue, a black male in the United States would have about a 1-in-3 chance of going to prison during his lifetime, whereas a Hispanic male would have a 1-in-6 chance, and a white male would have a 1-in-17 chance of going to prison.

## ■ SUMMARY

▸ **What is the moral code?**

A culture's norms and values make up its moral code, the symbolic system by which behavior is viewed as right or wrong, good or bad within that culture.

▸ **How is deviant behavior different from normal behavior?**

Normal behavior conforms to the norms of the group in which it occurs.

Deviant behavior fails to conform to the group's norms. Deviant, as well as criminal, behavior has been found throughout history.

▸ **How are biological and physiological theories used to explain crime?**

Scholars have proposed a variety of theories. Biological theories such as those propounded by Lombroso and Sheldon have stressed the importance of inherited factors in producing deviance. Psychological explanations have emphasized cognitive or emotional factors within the individual as the cause of deviance. Psychoanalytic theory has suggested that criminals act on the irrational impulses of the id because they have failed to develop a proper superego, or conscience, in the socialization process. Behaviorists have argued that crime is the product of conditioning.

▸ **How do sociological theories approach crime?**

Sociological theories of deviance rely on patterns of social interaction and the relationship of the individual to the group as explanations.

▸ **What is rational choice theory?**

Wilson and Herrnstein proposed that criminal activity, like all human behavior, is the product of a rational choice by the individual as a result of weighing the costs and benefits of alternative courses of action.

▸ **What is anomie theory?**

Émile Durkheim argued that in modern, highly differentiated and specialized societies, and particularly under conditions of rapid social change, individuals can become morally disoriented. This condition, which he called anomie, can produce deviance.

▸ **How do control theorists explain crime?**

Control theorists such as Hirschi have argued that everyone is a potential deviant. The issue, for such theorists, is not what causes deviance but what causes conformity. When individuals have strong bonds to society, their behavior will conform to conventional norms. When any of those bonds are weakened, however, deviance is likely.

▸ **What are techniques of neutralization?**

Sykes and Matza have argued that people become deviant as a result of developing techniques of neutralization or rationalizations to justify illegal or deviant behavior. Their view is that these techniques are learned as part of the socialization process.

▸ **Describe cultural transmission theory.**

Cultural transmission theory, pioneered by Shaw and McKay, emphasizes the cultural context in which deviant behavior patterns are learned.

▸ **How does differential association theory explain crime?**

Sutherland and Cressey suggested the theory of differential association, that individuals learn criminal techniques and attitudes through intimate contact with deviants.

▸ **What role does labeling theory play in crime?**

Labeling theory shifts the focus of attention from the deviant individual to the social process by which a person comes to be labeled as deviant and the consequences of such labeling for the individual.

▸ **How do we decide which theory of deviance is correct?**

In all likelihood, some combination of the various sociological theories is necessary for gaining a fuller understanding of the emergence and continuation of criminal and deviant behavior.

▸ **What is the difference between a crime and deviant behavior?**

Crime is behavior that violates a society's criminal laws.

▸ **How does violent crime differ from property crime?**

Violent crime can result in injury to a person; property crime is committed with the intent of obtaining property and does not involve the use or threat of force against an individual. Seventy-five percent of all crime in the United States is crime against property, not against a person. The U.S. violent crime rate includes one of the highest homicide rates in the industrialized world. Other violent crimes that affect American households include rape, aggravated assault, murder, and robbery.

▶ **How does the Uniform Crime Reports differ from the National Crime Victimization Survey?**

The FBI publishes statistics on the frequency of select reported crimes in the Uniform Crime Reports. These statistics are not always reliable, however. The National Crime Victimization Survey shows that only a small fraction of all crimes are reported to the authorities.

▶ **What comprises the criminal justice system?**

The criminal justice system consists of personnel and procedures to facilitate the arrest, trial, and punishment of those who violate the laws. The three main categories of this system are the police, the courts, and the prisons.

▶ **What are the goals of imprisonment?**

The goals of imprisonment include separating the criminal from society; punishing criminal behavior; deterring criminal behavior; and rehabilitating, or resocializing, criminals to conform to society's values and norms by teaching them usable work habits and skills.

## Media Resources

**CourseMate for *Introduction to Sociology*, Eleventh Edition**

Cengage Learning's Sociology CourseMate brings course concepts to life with interactive learning, study, and exam preparation tools that support the printed textbook. Access an integrated eBook, learning tools including glossaries, flashcards, quizzes, videos, and more in your Sociology CourseMate. Go to CengageBrain.com to register or purchase access.

# CHAPTER SEVEN STUDY GUIDE

## Key Concepts

*Match each of the following concepts with its definition, illustration, or explanation.*

a. Social control
b. Informal sanctions
c. Formal sanctions
d. Positive sanctions
e. Mesomorph
f. Strain theory
g. Innovators
h. Anomie
i. Techniques of neutralization
j. Differential association theory
k. Labeling theory
l. Secondary deviance
m. Violent crime
n. Property crime
o. White-collar crime
p. Felony
q. Recidivism
r. Status offense
s. Victimless crime
t. Diversion
u. Funnel effect
v. Rehabilitation
w. Deterrence
x. Broken-windows theory
y. Atavistic beings

____ 1. Acts of approval and disapproval applied in a public ritual, usually under the direct or indirect control of authorities
____ 2. Acts of approval or disapproval applied spontaneously by group members
____ 3. An approach to deviance that emphasizes the reaction to deviance and how agents of social control define some people and acts as deviant but not others
____ 4. Predatory crimes, such as theft, during which the criminal does not directly confront the victim
____ 5. The process by which a large number of crimes results in only a small number of offenders being sent to prison
____ 6. Ways of directing or influencing members to conform to the group's values and norms
____ 7. Crimes, such as drug use and gambling, that are not predatory but nevertheless violate the moral code
____ 8. The idea that if small instances of public disorder are ignored, more serious forms of deviance will follow
____ 9. The resocialization of criminals to conform to society's values and norms, and instruction in usable work habits and skills
____ 10. Deviant or criminal behavior that people develop as a result of having been labeled as deviant
____ 11. Crimes committed even after punishment has occurred
____ 12. A state of normlessness in which values and norms have little effect, and the culture no longer provides adequate guidelines for behavior
____ 13. A serious offense punishable by a year or more in prison
____ 14. In anomie theory, people who take illegal routes to socially approved goals
____ 15. Acts by individuals who, while occupying positions of social responsibility or high prestige, break the law in the course of their work
____ 16. The explanation of deviance emphasizing that people become deviant because they learn and adopt the behavior and the ideas of friends and other close contacts
____ 17. The explanation of crime and deviance that emphasizes that although most members of society share the same goals, some people have less access to legitimate routes to those goals
____ 18. The reduction in crime resulting from people's fear of punishment for that crime
____ 19. An offense that is punishable if committed by a juvenile but not by an adult
____ 20. Rewards given for good behavior
____ 21. A ruggedly muscular body type associated with being assertive and action-oriented
____ 22. According to Lombroso, evolutionary throwbacks whose behavior is more apelike than human
____ 23. Thought processes that justify illegal or deviant behavior
____ 24. Sending offenders, especially juveniles, to agencies outside of the justice system
____ 25. Crimes committed directly against a person in the perpetrator's presence, using force or threat of force

## Key Thinkers

*Match the thinkers with their main idea or contribution.*

a. Émile Durkheim
b. Cesare Lombroso
c. Sigmund Freud
d. James Q. Wilson and Richard Herrnstein
e. Edwin H. Sutherland
f. Travis Hirschi
g. Clifford Shaw and Henry McKay

____ 1. Argued that crime is produced by the unconscious impulses of the individual
____ 2. Argued that crime is the product of a rational choice by an individual as a result of weighing the costs and benefits of alternative courses of action
____ 3. Developed control theory, in which it is hypothesized that the strength of social bonds keeps most of us from becoming criminals
____ 4. Suggested that criminals are evolutionary throwbacks who can be identified by primitive physical features, particularly with regard to the head
____ 5. Argued that deviant behavior is an integral part of all healthy societies; developed the concept of anomie
____ 6. Used cultural transmission theory to explain why neighborhood crime rates persisted over decades even when the population of the neighborhood changed
____ 7. Developed the theory of differential association, emphasizing that people commit crime because they have learned "definitions" of behavior and law that are favorable to lawbreaking; coined the term *white-collar crime*

## Central Idea Completions

*Fill in the appropriate concepts and descriptions for each of the following questions.*

1. We usually think that deviance is dysfunctional for the society in which it occurs, but what functions can it serve for the society? The text mentions five. Cite them and give examples.

   a. _____
   Example _____

   b. _____
   Example _____

   c. _____
   Example _____

   d. _____
   Example _____

   e. _____
   Example _____

2. How have rates of crime and imprisonment in the United States changed over the course of the last three decades?

   _____

3. Describe four of the dysfunctions of deviance and give examples of each.

   a. _____

   Example _____

   b. _____

   Example _____

   c. _____

   Example _____

   d. _____

   Example _____

4. Apply Sykes and Matza's five techniques of neutralization to a situation involving cheating in a college or university community.

   a. _____

   b. _____

   c. _____

   d. _____

   e. _____

## Critical Thought Exercises

1. Some explanations of crime focus on the individual offender; others look more at the factors in the social environment. The same is true of proposals for policies designed to reduce crime. Compare the ways in which these two approaches might be applied in thinking about some form of deviance on your own campus (academic dishonesty, excessive drinking, and so on). Describe the possible sanctions, both formal and informal, that might be brought to bear on this form of deviance. How effective is each type of sanction?

2. Tischler points out that the United States imprisons more of its population and for longer terms than other advanced industrialized countries (Canada, Australia, Japan, European countries). Most of these other countries have also abolished the death penalty (or if they have it, almost never use it). Why is the United States so much more punitive?

## Internet Activity

1. Part A: Log on to the World Wide Web. Using any available search engine or browser, investigate one or more sites that demonstrate the following:

   a. An example of a behavior you personally would define as both deviant and harmful to society

   b. An example of a behavior you would define as deviant but not harmful to society

   c. A behavior you would define as deviant but not illegal

   d. A behavior you believe many people older than you might define as deviant but that you and members of your age cohort would not define as deviant

Part B: After you have completed Part A, download an example page for each of the sites you visited and discuss:

a. The common features of each form of deviance

b. Which aspects, if any, of these websites made you uncomfortable

c. Assuming the sites you defined as deviant are defined by others in a similar manner, the aspects of the behavior that lead to these definitions

d. The role new technologies, such as the Internet, might play in a society's shifting definitions of deviance

2. Learn about crime in your town. Go to the Uniform Crime Reports at www.fbi.gov/ucr/ucr.htm, and click the Crime in the United States link for the most recent year. Click City Agency and select your state from the list. Choose your town or city (or one close by), and look at the numbers of crimes in each of the categories (murder, rape, motor vehicle theft, and so on). Compute the crime rate per 100,000 population by dividing the number of crimes by the town population and multiplying by 100,000. Now go back and look at similar data from 1995 (http://www.fbi.gov/about-us/cjis/ucr/crime-in-the-u.s/1995/toc95.pdf, Table 8, "Number of Offenses Known to the Police, Cities and Towns 10,000 and over in Population, 1995"). How have crime rates changed? How does the change in your town compare with the change in the country as a whole?

3. The FBI's Sex Offender Registry (www.fbi.gov/scams-safety/registry) encompasses all 50 states, the District of Columbia, and all U.S. territories. Navigate the site. Go to the link for your state. What impact do you believe this database has had on the U.S. public since it began? How has it changed the lives of people who can now access it? Do you believe it has deterrent power? For those wrongly accused, getting their names purged from the list has proven to be nightmarish. Imagine you have been acquitted of a crime, yet your name still appears on the registry for months or years after the fact. What effect might this have on your life?

## Answers to Key Concepts

1. c;  2. b;  3. k;  4. n;  5. u;  6. a;  7. s;  8. x;  9. v;  10. l;  11. q;  12. h;  13. p;  14. g; 15. o;  16. j;  17. f;  18. w;  19. r;  20. d;  21. e;  22. y;  23. i;  24. t;  25. m

## Answers to Key Thinkers

1. c;  2. d;  3. f;  4. b;  5. a;  6. g;  7. e

Alex Segre/Alamy Limited

# 8 Social Class in the United States

## The American Class Structure
- The Upper Class
- The Upper-Middle Class
- The Middle-Middle Class
- The Lower-Middle Class
- The Lower Class
- Income Distribution

*Day-to-Day Sociology:* Would You Be Happier If You Were Richer?

## Poverty
- The Feminization of Poverty
- How Do We Count the Poor?
- Myths About the Poor

*Thinking About Social Issues:* Does the Income Gap Between the Rich and Poor Matter?
- Government Assistance Programs
- The Changing Face of Poverty

*Our Diverse World:* Rich Countries with Poor Children

## Consequences of Social Stratification
*Thinking About Social Issues:* Who Smokes?

## Why Does Social Inequality Exist?
- The Functionalist Theory
- Conflict Theory
- Modern Conflict Theory
- The Need for Synthesis

## Summary

## ■ LEARNING OBJECTIVES

After studying this chapter, you should be able to do the following:

- Explain the factors that affect a person's chances of upward social mobility.
- Describe the distribution of wealth and income in the United States.
- Summarize the functionalist and conflict theory views of social stratification.
- Describe the characteristics of each of the social classes in the United States.
- Describe differences in the poverty rate among various groups in American society.
- Compare poverty rates in the United States with those of other industrialized countries.
- Describe some of the personal and social consequences of a person's position in the class structure.

Americans like to think that social stratification and social class are minor issues. After all, we do not have inherited ranks, titles, or honors. We do not have coats of arms or rigid caste rankings. Besides, equality among men—and women—is an ideal guaranteed by our Constitution and summoned forth regularly in speeches from podiums and lecterns across the land. Yet, in the 1970s a long rapid increase in inequality began that persists today. Those in the lower income groups saw a decrease in their share of total national income. Those in the middle barely kept pace, while those at the top experienced vast increases in wealth.

Why did this take place? Opinions differ on the causes. Some point to technological changes and the increased value of higher education. Others mention globalization, which has exposed U.S. workers to low-wage competition from abroad and the diminished power of unions. Still others point to tax breaks for the well-off (Putnam and Campbell, 2011).

The lavish displays of wealth and the attempts of many people to obtain power and privilege make it difficult to ignore social inequality and the uneven distribution of material rewards. Once rewards are distributed unequally within a society, economic, political, and social stratification begin. In this chapter, we begin to see that social stratification is quite complex and open to many subtle variations. It does not always fit neatly into our stereotypes.

## ● THE AMERICAN CLASS STRUCTURE

A **social class** consists of *a category of people who share similar opportunities, similar economic and vocational positions, similar lifestyles, and similar attitudes and behaviors. A society that has several social classes and permits social mobility is based on a* **class system of stratification.** Class boundaries are maintained by limiting social interaction, intermarriage, and mobility into another class.

Some form of class system is usually present in all industrial societies, whether they are capitalist or communist. Social mobility in a class system is often the result of an occupational structure that opens up higher-level jobs to anyone with the education and experience required. A class society encourages striving and achievement. Here in the United States, we should find this concept familiar because ours is basically a class society.

There is little agreement among sociologists about how many social classes exist in the United States and what their characteristics might be. For our purposes here, however, we will follow a relatively common approach of assuming that there are five social classes in the United States: upper class, upper-middle class, middle-middle class; lower-middle class, and lower class (Rossides, 1990). Table 8-1 presents descriptions of each of these social classes.

### The Upper Class

Members of the upper class have great wealth, often going back for many generations. They recognize one another, and are recognized by others, by reputation and lifestyle. They usually have high prestige and a lifestyle that excludes those of other classes. Members of this class often influence society's basic economic and political structures. The upper class usually isolates itself from the rest of society by residential segregation, private clubs, and private schools. Historically, they have been Protestant, especially Episcopalian or Presbyterian. This is less true today. It is estimated that in the United States, the upper class consists of from 1 to 3 percent of the population.

Since the 1970s, the upper class also has come to include society's new entrepreneurs—people who have often made many millions, and sometimes billions, of dollars in business. In many respects, these people do not resemble the upper class of the past. Included in *Forbes* magazine's 2011 list are people such as William Gates, the founder of Microsoft Corporation, who has a net worth of $59 billion; Warren Buffett, the founder of Berkshire Hathaway, who is worth $39 billion; and Larry Ellison, the founder of Oracle Corporation, who has a net worth exceeding $33 billion. Mark Zuckerberg, the founder of Facebook, is worth $17.5 billion at age 28.

TABLE 8-1  Social Classes in the United States

| Class | Occupation | Education | Children's Education |
| --- | --- | --- | --- |
| Upper class | Corporate ownership; upper-echelon politics; honorific positions in government and the arts | Liberal arts education at elite schools | College and postcollege |
| Upper-middle class | Professional and technical fields; managers; officials; proprietors | College and graduate training | College and graduate training |
| Middle-middle class | Clerical and sales positions; small business semiprofessionals; farmers | High school; some college | Option of college |
| Lower-middle class | Skilled and semiskilled manual labor; craftspeople; foremen; nonfarm workers | Grade school; some or all of high school | High school; vocational school |
| Lower class | Unskilled labor and service work; private household work and farm labor | Grade school | Little interest in education; high school dropouts |

*Source:* Adapted from U.S. Bureau of the Census. 1981. *Statistical Abstract of the United States: 1981.* Washington, DC: U.S. Government Printing Office.

Not all billionaires lead opulent lifestyles, and many in the upper class do not approve of displaying wealth. For many, the money is merely a way of keeping score of how well they are doing at their chosen endeavors.

## The Upper-Middle Class

The upper-middle class comprises successful business and professional people and their families. They are usually just below the top in an organizational hierarchy but still command a reasonably high income. Many aspects of their lives are dominated by their careers, and continued success in this area is a long-term consideration. These people often have a college education, own property, and have a savings reserve. They usually live in comfortable homes in the more exclusive areas of a community, are active in civic groups, and carefully plan for the future. They very likely belong to a church. The most common denominations represented are Presbyterians, Episcopalians, Congregationalists, Jews, and Unitarian Universalists. In the United States, 10 to 15 percent of the population falls into this category.

A large percentage of the new upper-middle class are two-income couples, both of whom are college-educated and employed as corporate executives, high government officials, business owners, or professionals. These relatively affluent individuals are changing the face of many communities. They are gentrifying rundown city neighborhoods with their presence and their money.

## The Middle-Middle Class

The middle-middle class shares many characteristics with the upper-middle class, but its members have not been able to achieve the same kind of lifestyle because of economic or educational shortcomings. Usually high school graduates with modest incomes, they are semiprofessionals, clerical and sales workers, and upper-level manual laborers.

The people in this class emphasize respectability and security, have some savings, and are politically and economically conservative. They often would like to improve their standard of living, jobs, and family incomes. They are likely to be represented among the Protestant denominations such as Baptists, Methodists, and Lutherans, or they might be Catholic or Greek Orthodox. They make up 25 to 30 percent of the U.S. population.

## The Lower-Middle Class

The lower-middle class comprises skilled and semi-skilled laborers, factory employees, and other blue-collar workers. These are the people who keep the country's machinery going. They are assembly-line workers, auto mechanics, and repair personnel. They are the most likely to be affected by economic downturns. More than half belong to unions.

Social inequality involves the uneven distribution of privileges, material rewards, and power.

Members of the upper-middle class have jobs in professional and technical fields.

Lower-middle class people live adequately but have little for luxuries. They are less likely to vote than the higher classes, and they feel politically powerless. Although they have little time to be involved in civic organizations, they are very much involved with their extended families. The families are likely to be patriarchal with sharply segregated sex roles. They stress obedience and respect for elders. Many of them have not finished high school. The religious makeup is similar to that of the middle-middle class. They represent 25 to 30 percent of the U.S. population.

## The Lower Class

These are the people at the bottom of the economic ladder. They have little in the way of education or occupational skills and consequently are unemployed or underemployed. Lower-class families often have many problems, including broken homes, illegitimacy, criminal involvement, and alcoholism.

Members of the lower class have little knowledge of world events, are not involved with their communities, and usually do not identify with other poor people. They have low voting rates. Because of a variety of personal and economic problems, they often have no way of improving their lot in life. For them, life is a matter of surviving from one day to the next. Their dropout rate from school is high, and they have the highest rates of illiteracy of any of the groups. The lower class is disproportionately African American and Hispanic, but race and poverty do not define the class exclusively. Rather, it is defined by a set of characteristics and conditions that are part of a broader lifestyle. Lower-class people often belong to fundamentalist or revivalist religious sects. About 15 to 20 percent of the population falls into this class.

In a class society, the desire for wealth produces a variety of business ventures.

Money, power, and prestige are distributed unequally among these classes. However, members of all five classes share a desire to advance and achieve success, which makes them believe that the system is just and that upward mobility is open to all. Therefore, they tend to blame themselves for lack of success and for material need (Vanfossen, 1979).

## Income Distribution

The U.S. Bureau of the Census has published annual estimates of the distribution of family income since 1947. Those figures show a highly unequal distribution of wealth. In 2010, for example, the richest one-fifth of families earned 50.2 percent of the total income for the year, whereas the poorest one-fifth earned only 3.3 percent (U.S. Census Bureau, 2011). (See Table 8-2.)

Without further elaboration, this information allows us to imagine that the richest one-fifth of families

TABLE 8-2  Share Total Income of Various Earners, 2010

| | |
|---|---|
| Lowest 20% | 3.3 |
| Second 20% | 8.5 |
| Middle 20% | 14.6 |
| Fourth 20% | 23.4 |
| Highest 20% | 50.2 |
| Top 5% | 21.7 |

*Source:* U.S. Bureau of the Census. 2011. Current Population Reports, P60-239. Income, Poverty, and Health Insurance Coverage in the United States: 2010. Current Population Survey. Washington, D.C.: U.S. Government Printing Office.

consists of millionaire real estate moguls, Wall Street professionals, and CEOs of major companies. The image is somewhat misleading. In 2010, the richest one-fifth included all families with incomes of $113,774 or more (see Figure 8-1). Keep in mind that this is a family income derived from jobs held by husbands, wives, and all other family members. Family incomes for the richest 5 percent of the population began at $200,354 (U.S. Census Bureau, 2011).

This is not to imply, though, that there is not a significant difference in the distribution of wealth in the United States. Income is only part of the picture. Total wealth—in the form of stocks, bonds, real estate, and other holdings—is even more unequally distributed. The richest 20 percent of American families owns more than three-fourths of all the country's wealth. In fact, the richest 5 percent of all families owns more than half of America's wealth. There is also evidence to support the old adage that "The rich get richer, and the poor get poorer." The number of people in poverty grew from 24.5 million in 1978 to 46.2 million in 2010 (DeNavas et al., 2011).

Some of the growth in income inequality has resulted from demographic trends in society. As a population ages, more income inequality takes place as some people accumulate wealth and others stay at a modest level.

There is also more income inequality among more-educated groups than among less-educated groups. The less educated are more likely to be clustered with others of relatively low incomes. The educated have more diverse incomes with some highly motivated income seekers as well as those who pursue academic or artistic pursuits. The United States has been growing older and more educated, producing some of the income inequality. (See "Day-to-Day Sociology: Would You Be Happier If You Were Richer?")

**FIGURE 8-1  Family Income by Quintile, 2010**

| Quintile | Income |
|---|---|
| Top 5% | Over $200,354 |
| Top 20% | Over $113,774 |
| 4th 20% | $74,144 to $113,773 |
| 3rd 20% | $48,000 to $74,143 |
| 2nd 20% | $26,685 to $47,999 |
| Bottom 20% | Below $26,685 |

*Richest 5% of all families (included in the fifth quintile)

Source: U.S. Bureau of the Census, Current Population Reports. Annual Social and Economic (ASEC) Supplement (www.census.gov/hhes/www/cpstables/032011/faminc/new06_000.htm), accessed August 24, 2012.

---

### DAY-TO-DAY SOCIOLOGY

## Would You Be Happier If You Were Richer?

Most people believe that they would be happier if they were richer, but evidence does not support this view. Surveys in many countries conducted over decades indicate that, on average, reported happiness has not changed much over the last four decades, in spite of large increases in real income per capita.

People do not continuously think about their circumstances, whether positive or negative. Individuals who have recently experienced a significant life change—for example, becoming disabled, winning a lottery, or getting married—surely think of their new circumstances many times each day, but the amount of time they spend thinking about them decreases over time, so eventually they spend most of their time thinking about everyday events instead.

In addition, as society grows richer and you grow richer, your average rank does not change, so this may explain why the average happiness stays the same even as national income goes up.

Also, as you get richer, your perceived happiness may be offset by changes in your reference group: After a promotion and pay raise, your new peers, instead of previous peers, serve as a comparison point, making the improvement seem less important. People tend to adapt to the income they say they need to get along as it rises. In other words, the more you make, the more you start to believe you need more to get along.

Material goods yield only temporary joy for most individuals. Happiness is connected to how people spend their time, not how they spend their money.

Source: Kahneman, Daniel, Alan B. Krueger, David Schkade, Norbert Schwarz, and Arthur A. Stone. May 2006. "Would You Be Happier If You Were Richer? A Focusing Illusion." CEPS Working Paper No. 125.

## POVERTY

On a very basic level, poverty refers to a condition in which people do not have enough money to maintain a standard of living that includes the basic necessities of life. Depending on which official or quasi-official approach we use, it is possible to document that anywhere from 14 million to 45 million Americans are living in poverty. The fact is, we really do not have an unequivocal way of determining how many poor people there are in the United States.

Poverty seems to be present among certain groups much more than among others. In 2010, 15.1 percent of all Americans lived below the poverty level. That was the highest level since 1993. Although 13 percent of all whites were living in poverty, 27.4 percent of all blacks and 26.6 percent of Hispanic origin fell into this group. (See Figure 8-2 for the poverty rates by race and Hispanic origin.)

People living in certain regions of the United States are much more likely to live in poverty than those living in other U.S. regions. For example, the Southern and West Coast states have higher poverty rates than Northeast states.

Although poverty rates are higher in urban areas than in rural areas, a substantial proportion of the poor live in rural America—a reality that is often overlooked by those who focus only on the problems of the urban poor. Even worse, the economic conditions of the rural poor are expected to deteriorate along with the decline of unskilled manufacturing jobs and changes in the mining, agricultural, and oil industries. The problem is especially acute for those with little education or few marketable job skills.

### The Feminization of Poverty

Different types of families also have different earning potentials. In 2010, a married couple had a median income of $72,751. For an unmarried male householder, the figure was $35,627, and for an unmarried female householder, it was $25,456. This has caused some sociologists to note the "feminization of poverty," a phrase referring to the disproportionate concentration of poverty among female-headed families.

The real impact of these differences becomes even more striking when we look at single women with children. Whereas 6.2 percent of all married couple families lived below the poverty line in 2010, 31.6 percent of all single women with children were living in poverty. If present trends continue, 60 percent of all children born today will spend part of their childhood in a family headed by a mother who is divorced, separated, unwed, or widowed (Current Population Reports, 2011).

Not all female-headed families are the same, however. The feminization of poverty is both not as bad as and much worse than the previous statement suggests. Families headed by divorced mothers are doing better than the numbers suggest, whereas families headed by never-married mothers are doing much worse.

What accounts for the fact that never-married mothers are so much poorer than their divorced counterparts? Seventy percent of all out-of-wedlock births occur to young women between the ages of 15 and 24. They are, on average, ten years younger than divorced mothers. Never-married mothers are also, on average, much less educated. The gender gap in poverty rates is greatest at low educational levels. It is much narrower among those with a high school diploma and practically nonexistent among those with a college education. Single mothers without a high school diploma often have difficulty finding a job that pays enough to cover child-care costs, leading to a dependence on welfare programs (O'Hare, 1996).

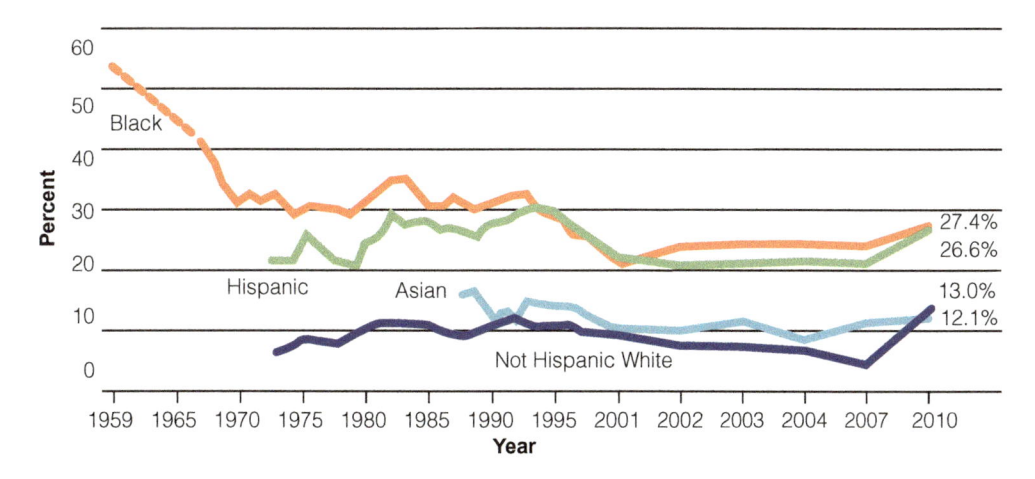

**FIGURE 8-2  Poverty Rates by Race and Hispanic Origin, 1959–2010**
Source: DeNavas-Walt, Carmen, Bernadette D. Proctor, and Jessica C. Smith. 2011. U.S. Bureau of the Census, Current Population Reports, P60–239. Income, Poverty, and Health Insurance Coverage in the United States: 2010. Washington, D.C.: U.S. Government Printing Office.

Out-of-wedlock births to young women have contributed to the feminization of poverty.

## How Do We Count the Poor?

To put a dollar amount on what constitutes poverty, the federal government has devised a poverty index of specific income levels, below which people are considered to be living in poverty. Many people use this index to determine how many poor people live in the United States. According to the index, the poverty level for a family of four in 2012 was $23,050 (see Table 8-3).

TABLE 8-3 Average Income Levels Below Which Families Are Considered to Be Living in Poverty (2012)

| Size of Unit | Income |
| --- | --- |
| One person | 11,1701 |
| Two people | 15,130 |
| Three people | 19,090 |
| Four people | 23,050 |
| Five people | 27,010 |
| Six people | 30,970 |
| Seven people | 34,930 |
| Eight people | 38,890 |

Source: *Federal Register*, Vol. 77, No. 17. January 26, 2012. Washington, DC: U.S. Government Printing Office, pp. 4034–35.

The poverty index is based solely on money income and does not reflect the fact that many low-income people receive noncash benefits such as food stamps, Medicaid, and public housing.

The way we measure poverty today is based on a 1965 study by an economist at the Social Security Administration, Mollie Orshansky. In an attempt to define poverty, Orshansky took the cost of a basic, low-cost, nutritionally adequate diet. She then multiplied it by three because, at the time, food accounted for a third of a family's expenses. Using this formula, there has been little change in the poverty rate since the 1970s.

Many people believe we need to overhaul how we calculate poverty because there have been many changes in society since that simple formula was developed. For example, some suggest we should expect to see less poverty today than in the past because in 1973, 40 percent of adults over 25 lacked a high school degree compared to today's less than 15 percent. In addition, spending on programs for the poor such as food stamps, housing subsidies, Medicaid, and earned income tax credits has tripled since the 1970s (Eberstadt, 2008).

The poverty index was not originally intended to certify that any individual or family was in need. In fact, the government specifically has warned against using the index for administrative use in any specific program. Despite this warning, people continue to use, or misuse, the poverty index and variations of it for a variety of purposes for which it was not intended. For example, those wanting to show that current government programs are inadequate for the poor will try to inflate the numbers of those living in poverty. Those trying to show that government policies are adequate for meeting the needs of the poor will try to show that the number of poor people is decreasing.

Those who think the poverty index overestimates the poor offer three major criticisms. First, when the federal government developed the poverty index in 1965, about one-quarter of federal welfare benefits were in the form of goods and services. Today, noncash benefits account for about two-thirds of welfare assistance. For example, looking at income alone ignores the effects of things like the Earned Income Tax Credit, Medicaid, food stamps, and housing subsidies, which are not considered income under existing poverty-index rules. Complicating the issue further, the market value of in-kind benefits—such as school lunch programs and health care services, among others—has been multiplied by a factor of 40. Some suggest that if the noncash benefits were counted as income, the poverty rate would be 3 percentage points lower.

Second, the poverty measure looks only at income, not at assets. If the value of a home or other assets were included, the poverty rate would also be lower.

Third, food typically accounts for a considerably smaller proportion of family expenses today than it did

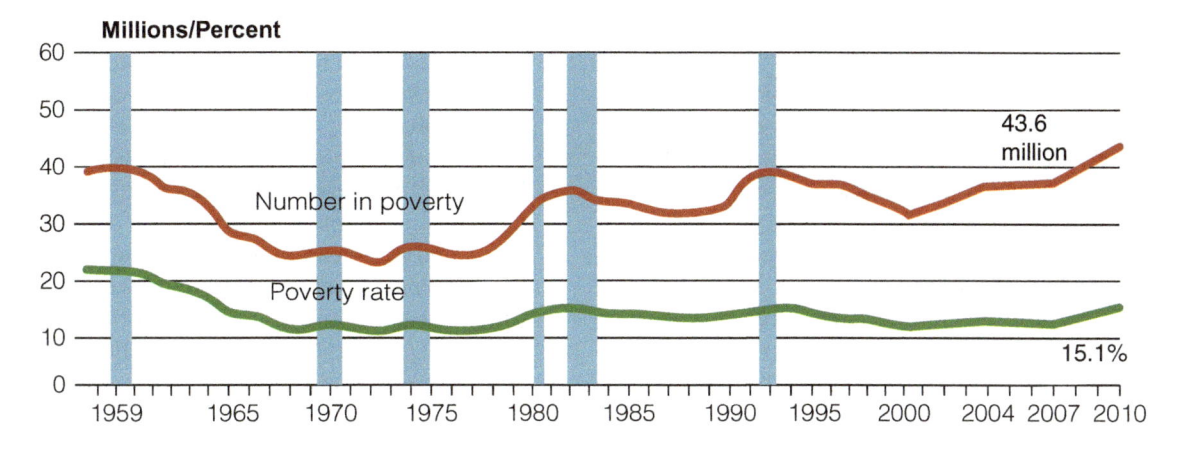

**FIGURE 8-3** Number in Poverty and Poverty Rates, 1959–2010
Source: U.S. Bureau of the Census. *Current Population Survey. 1960–2011 Annual Social and Economic Supplements*. Washington, D.C.: U.S. Government Printing Office.

previously. If we were to try to develop a poverty index today, we would probably have to multiply minimal food costs by a factor of 5 instead of 3.

Those who think the poverty figures underestimate the poor have their criticisms also. First, they point out that money used to pay taxes, alimony, child support, health care, or work-related expenses should be excluded when considering assets because these sums cannot be used to buy food or other necessities.

Second, there is no geographic cost-of-living adjustment. The federal government uses the same poverty-level figures for every part of the country. That means the poverty threshold is the same in rural Mississippi as it is in New York City.

Third, many believe that the poverty threshold is unrealistically low. Rather than use an absolute number, poverty status should be determined by comparing a person's financial situation with that of the rest of society.

The poverty index has become less and less meaningful. However, its continued existence over all these years has given it somewhat of a sacred character. Few people who cite it know how it is calculated, and they choose to assume it is a fair measure for determining the number of poor in the country. The poverty index has never been a sufficiently precise indicator of need to make it the perfect test for deciding which individuals and families are poor and which are not.

The number of people living in poverty also is distorted by the fact that the U.S. Census Bureau's *Current Population Survey* is derived from households. It excludes all the people who do not live in traditional housing, specifically, the growing numbers of the homeless, estimated at anywhere from 350,000 to 1 million. People in nursing homes and other types of institutions also are not included in the poverty figures because of surveying techniques.

This is not to downplay the number of poor people in the United States. The basic fact is that trying to determine how many poor people there are depends on whom you ask and what type of statistical maneuvering is involved (see Figure 8-3). The Census Bureau is taking steps to develop a new poverty measure that addresses some of the concerns people have raised.

## Myths About the Poor

We are presented with differing views on poverty and what should be done about it. One side argues that more government aid and the creation of jobs are needed to combat changes in the employment needs of the national economy. The other side contends that government assistance programs launched with the War on Poverty in the mid-1960s have encouraged many of the poor to remain poor and should be eliminated for the able-bodied poor of working age (Murray, 1994).

Our perceptions of the poor shape our views of the various government programs available to help them. It is important to have a clear understanding of who the poor are to direct public policy intelligently. Many Americans believe a number of common myths about the poor. Let us try to clear some of them up.

**Myth 1: People are poor because they are too lazy to work.** Half of the poor are not of working age. About 40 percent are younger than 18; another 10 percent are older than 65. Most of the able-bodied poor of working age are working or looking for work. Many of the poor adults who do not work have good reasons for not working; they may be ill or disabled, and many others are going to school (mostly those in their late teens from poor families). Many of the poor work, and many work year-round. However, a person working 40 hours a week, every week of the year, at minimum wage will not earn enough to lift a family of three out of poverty.

The numbers of the working poor are increasing. There are several reasons for this growth. First, although there are more jobs in the economy than ever before, many of these jobs are in low-paying service industries. A janitor or a cook at a fast-food restaurant earns no more than minimum wage. Second, the better jobs the poor used to hold are no longer part of the U.S. economy. Many companies, seeking sources of cheap labor, have set up manufacturing operations overseas to increase their ability to compete in the world market. Finally, many of the working poor are women or young people with few marketable skills. Often, they are forced to settle for poorly paid, part-time work.

In many ways, the working poor are in worse straits than those on the welfare rolls. For example, a mother on welfare might be eligible for public housing and a variety of services for which a working-poor, two-parent family might not be eligible.

It is easy for the government to ignore the plight of the working poor. Scattered throughout the country and with no collective voice to express protest, they are relatively invisible and, therefore, easily forgotten.

**Myth 2: Most poor people are minorities, and most minorities are poor.** Neither of these statements is true. Most poor people are white merely because many more whites than minorities live in the United States. The poverty rate, however, remains considerably higher for African Americans and Hispanics than for whites, 27.4 percent and 26.6 percent, respectively (*Current Population Survey*, 2011).

One of the reasons African Americans are associated with the image of poverty is that they make up more than half of the long-term poor. Another reason is that the War on Poverty was motivated in part, and occurred simultaneously with, the civil rights movement of the 1960s.

**Myth 3: Most of the poor are single mothers with children.** It is true that a disproportionate share of poor households are headed by women and that the poverty rate for female-headed families is extremely high. For example, about 60 percent of mothers receiving assistance have never been married. The majority of people in poverty, however, live in other family arrangements. About one-third of the poor live in married-couple families; nearly one-quarter live alone or with nonrelatives. The remainder live in a male-headed or other family setting (*Current Population Survey*, 2011).

**Myth 4: Most people in poverty live in the inner cities.** Historically, poverty has been more prevalent in rural areas than in urban areas. Rural residents have higher unemployment rates and earn lower wages than urban residents. Rural residents also tend to have below-average educational levels and limited job skills.

Much of rural poverty is invisible because it occurs in isolated pockets. Poverty rates are exceptionally high in rural counties in Appalachia, the Mississippi Delta, and American Indian reservations. Except for rural Appalachia, which is predominantly white, most rural pockets of poverty are disproportionately composed of African Americans, Hispanic Americans, and Native Americans.

**Myth 5: Welfare programs for the poor are straining the federal budget.** Since the passage of welfare reform in 1996, the number of families receiving aid has decreased by about 50 percent (Lichter and Crowley, 2002). Social assistance programs for low-income people cost the federal government only about a third as much as other types of social assistance such as Social Security and Medicare, which mainly go to middle-class Americans, not to the poor (O'Hare, 1996).

(See Thinking About Social Issues: Does the Income Gap Between the Rich and the Poor Matter?)

## Government Assistance Programs

The public appears to be quite frustrated and upset about the costs of poverty. Much of this frustration, however, stems from a misperception of what programs are behind the escalating government expenditures, a misunderstanding about who is receiving government assistance, and an exaggerated notion of the amount of assistance going to the typical person in poverty. Most government benefits go to the middle class. Many of the people reading this book will be surprised to know that they or their families actually might receive more benefits than those people typically defined as poor. The value of benefits going to the poor actually has fallen in recent years, whereas that going to the middle class has risen.

Government programs that provide benefits to families or individuals can be divided into two categories: (1) social insurance and cash benefits going to people of all income levels, and (2) means-tested programs and cash assistance going only to the poor. Social insurance benefits are not means tested, meaning that you do not have to be poor to receive them. They go primarily to the middle class. Many people receiving payments from social insurance programs, such as Social Security retirement and unemployment insurance, feel they are simply getting back the money they put into these programs. They accuse those receiving benefits from means-tested programs of getting something for nothing. This is not exactly true when we recognize that many social insurance recipients receive far more than they put in and

## THINKING ABOUT SOCIAL ISSUES: Does the Income Gap Between the Rich and Poor Matter?

Income inequality is a worldwide phenomenon. Three people, Bill Gates, Warren Buffett, and Carlos Slim Helu, possess more assets than the 48 poorest countries combined. The United States has experienced the greatest increase in income inequality among rich nations. Throughout much of U.S. history, there has been a large gap in the amount of wealth held by America's richest and poorest citizens. During the nineteenth and twentieth centuries, the richest 1 percent of U.S. adults held between 20 and 30 percent of all private wealth in the country.

Around the start of the 1970s, the income gap began to grow—slowly at first and then more rapidly during the early 1980s. Today, the gap between rich and poor Americans is at the highest level since the late 1970s. The top wage earners have seen large increases in their real incomes, whereas those at the bottom have actually experienced losses in real income.

According to the U.S. Census Bureau, the top 20 percent of Americans earned 50.2 percent of the nation's income, compared with the 3.3 percent earned by Americans living below the poverty line (roughly 15 percent of the population). This means that the total earnings of the top 20 percent are 14.5 times that of those in poverty. In 1968, the top 20 percent earned 7.69 times that of those in poverty.

Average income for a household in the top 1 percent has quadrupled, from $350,000 in 1979 to $1.5 million in 2011. These figures are adjusted for inflation. By comparison, the average household income for the middle 60 percent was $44,639 in 2011. For the bottom 20 percent, the average household income was $9,187 (Congressional Budget Office).

Why has income inequality grown? Several reasons have been suggested:

1. Unskilled workers have been losing jobs that have been shifted to low-wage workers in Asia and other countries.
2. The current economy requires highly skilled and educated people, lowering the value of less educated workers.
3. The high levels of immigration of low-skilled workers have reduced wages for less-skilled, American-born workers.
4. The U.S. tax code makes it possible for the wealthy to continue getting richer and to keep more of their earnings than in the past.
5. A rising share of national income has been going to corporations (as profits), while the portion that goes to pay workers has declined. That shift benefits those who are able to own stocks.

Does the U.S. government have an obligation to shrink the income gap? Or is a substantial gap between the highest and the lowest-paid workers simply a natural part of a capitalist economy?

In the United States, there is a strong belief that what matters most is the opportunity to move up rather than that the government is making sure there is an equality of outcomes. Everyone should have a chance to advance because of their talents and hard work, the thinking goes. People are likely to proclaim that if you are well-off, you earned it. If you are poor, you did not take advantage of the opportunities available to you.

This view is not necessarily shared in other middle-income and wealthy countries. For example, in an international survey, 69 percent of Americans agreed with the statement that "people are rewarded for intelligence and skill," compared with an average of 39 percent for the other twenty-five countries in the sample. Sixty-one percent of Americans believed that "people are rewarded for their efforts," but only 36 percent of people from other countries in the sample agreed with that statement. Despite rising income inequalities, Americans are still reluctant to let the government take responsibility for reducing income disparities.

*Sources:* Congressional Budget Office; Trumbull, Mark. February 14, 2012. "America's Big Wealth Gap: Is It Good, Bad, or Irrelevant?" *The Christian Science Monitor* (www.csmonitor.com/USA/Politics/2012/0214/America-s-big-wealth-gap-Is-it-good-bad-or-irrelevant/(page)/2); Sawhill, Isabell and John E Morton. May 2007. "Economic Mobility: Is the American Dream Alive and Well?" Economic Mobility Project. Washington, D.C.: Pew Charitable Trust.

---

that the poor, the majority of whom work, pay taxes that contribute to their own means-tested benefits.

Social insurance programs account for the overwhelming majority of federal cash assistance expenditures, and their share has been rising rapidly. Female-headed families in poverty, often portrayed as a heavy drain on the government treasury, account for only 2 percent of the federal outlays for human resources.

In contrast, Social Security for the retired elderly, the vast majority of whom are middle class, accounts for 38 percent.

## The Changing Face of Poverty

It appears that economic rewards are distributed more unequally in the United States than elsewhere in the Western industrialized world. In addition, the United States experiences more poverty than other capitalist countries with similar standards of living.

In one international study, the poverty rates for children, working-age adults, and the elderly were tabulated for a variety of countries. The results showed that the United States has been successful in holding down poverty among the elderly. The American elderly experience far less poverty than the elderly in Great Britain, approximately the same as the elderly in Norway and Germany, and far more poverty than the elderly in Canada and Sweden.

The United States has been much less successful in keeping children and working-age adults out of poverty. The U.S. child poverty rate is higher than the rate in Great Britain and more than double the rate in Norway, Sweden, and Germany. (See "Our Diverse World: Rich Countries with Poor Children.")

How has it happened that the United States has made progress in combating poverty among the elderly but not among other groups? Since 1960, a variety of social policies have been enacted that have improved the standard of living of the elderly relative to that of the younger population. Social Security benefits were increased significantly and protected against the threat of future inflation; Medicare provided the elderly with national health insurance; supplemental security income provided a guaranteed minimum income; special tax benefits for the elderly protected their assets during the later years; and the Older Americans Act supported an array of services specifically for this age group. As a consequence of these measures, poverty among the elderly has declined substantially. Although 24.6 percent of those families 65 and older lived below the poverty level in 1970, only 9 percent did so by 2010 (*Current Population Survey*, 2011).

To achieve this dramatic improvement in the conditions of the elderly, it has been necessary to increase greatly the federal money spent on this age group. If these arrangements are maintained, projections show that about 60 percent of the federal budget will be going to the elderly by the year 2030. A group that has suffered particularly under this shift in expenditures to the elderly is the young. Whereas 14 percent of children lived in poverty in 1970, 22 percent did so in 2010. (See Figure 8-4.)

It would also surprise many people to learn that not only are the elderly as a group not poor, but they are actually better off than most Americans. They are more likely than any age group to possess money market accounts, certificates of deposit, U.S. government securities, and municipal and corporate bonds. The median household net worth of those aged 65 to 69 is the highest of any age group, followed by those 70 to 74 years old. They have the highest rate of home ownership of any age group. Seventy-seven percent of those 65 to 74 own homes, and most of these homes are paid for in full.

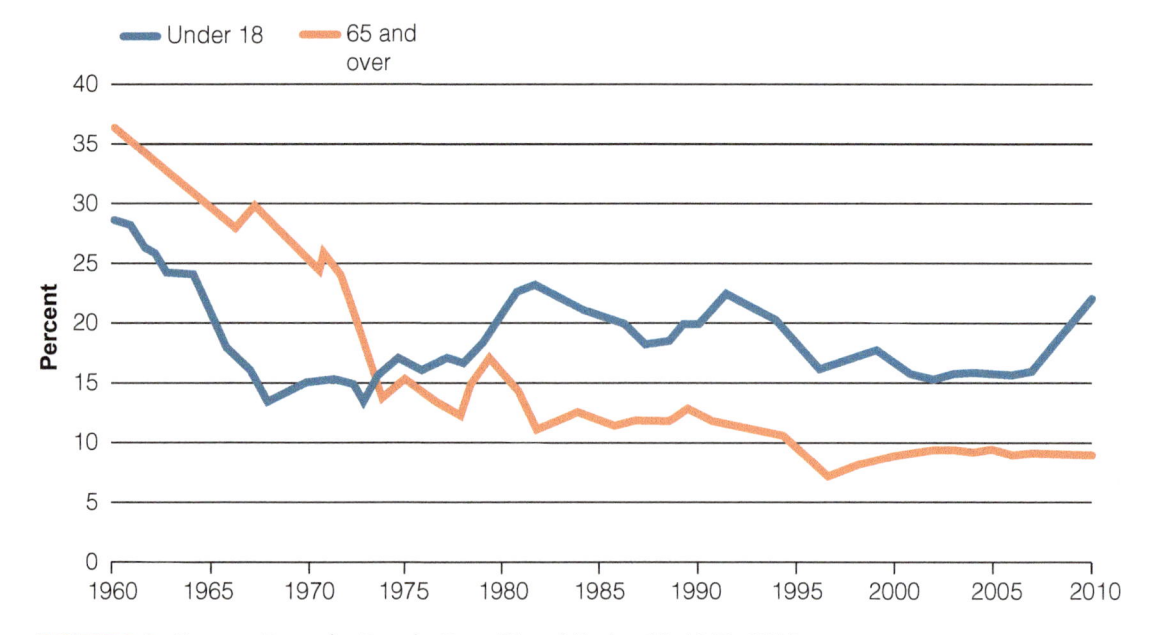

FIGURE 8-4  Poverty Rates for People Over 65 and Under 18, 1960–2010
*Source: U.S. Bureau of the Census. 2011. Current Population Reports, P60-239. Income, Poverty, and Health Insurance Coverage in the United States: 2010. Washington, D.C.: U.S. Government Printing Office.*

## OUR DIVERSE WORLD

### Rich Countries with Poor Children

We know that children have a miserable existence in third-world countries, but wealthy countries have trouble keeping children out of poverty also. The United States has been very successful at lifting the elderly out of poverty. It has not been as successful with children. In recent years, the U.S. child poverty rate has fluctuated between a high of 22.7 percent in 1993 to a low of 16.2 percent in 2000 (U.S. Bureau of the Census, 2011). Not only has child poverty remained stubbornly high, but when compared with other countries in the world, the United States appears to do less to improve the living conditions of its poor children than many other countries do.

Even though 27.7 percent of children in France start out in poverty, the government's assistance expenditures reduce that number to 7.5 percent with a variety of programs. In the United States, 26.6 percent of all children start out in poverty, but the government's programs reduce that number to only 16.9 percent (see Figure 8-5; UNICEF 2005).

The United States has such high rates of poverty among children for a number of reasons. First, many children are born to unmarried women. Nine percent of children in married-couple families live in poverty, but it rises to 42 percent of those in female-headed households. Second, American mothers with limited education or skills are less likely than European mothers to return to work quickly after childbirth because high-quality child care is comparatively expensive and the jobs will not cover the costs. Third, the United States has more poor immigrants than any other country. Many of the children in poverty are born to poor immigrants who have been in the country for only a few years. A fourth factor is an outgrowth of the realities of the American political system, which depends on advocates supporting programs for special constituencies. Children obviously do not vote, and the poor in general have low voting records. The elderly, on the other hand, have high voting records and therefore have substantial political clout. Politicians are usually not voted out of office for cutting benefits to children, whereas they are if they do so for the elderly.

Some (for example, Bradsher 1995) have suggested that other countries have avoided high levels of child poverty by limiting economic growth and lowering the living standard for everyone. Many European countries with generous social assistance programs also have high rates of unemployment and living standards that do not match those in the United States. These critics charge that instead of bringing everyone up, other countries have brought everyone down. They claim that the high rate of American child poverty appears greater because the gap between the rich and the poor is so much larger in the United States, and other countries just have gone further in redistributing income.

Despite the arguments over the data, the fact remains that substantial numbers of poor children live in the United States. Great harm is being done to these children when their poor living conditions are not addressed.

*Sources:* UNICEF. 2005. "Child Poverty in Rich Countries, 2005." Innocenti Report Card No. 6. Florence, Italy: UNICEF Innocenti Research Centre; U.S. Bureau of the Census. 2007. *Poverty Status of People, by Age, Race, and Hispanic Origin: 1959 to 2006* (www.census.gov/hhes/www/poverty/histpov/hstpov3.html), accessed May 13, 2009; UNICEF. 2010. "The Children Left Behind: A League Table of Inequality in Child Well-Being in the World's Rich Countries." Innocenti Report Card No. 9. Florence, Italy: UNICEF Innocenti Research Centre.

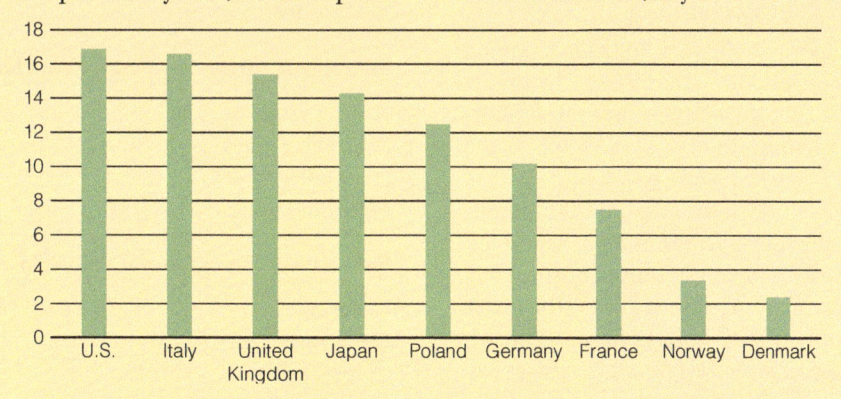

**FIGURE 8-5  Child Poverty Rates in Rich Countries**
Sources: UNICEF. 2005. "Child Poverty in Rich Countries, 2005." Innocenti Report Card No. 6. Florence, Italy: UNICEF Innocenti Research Centre; U.S. Bureau of the Census. 2007. Poverty Status of People, by Age, Race, and Hispanic Origin: 1959 to 2006 (www.census.gov/hhes/www/poverty/histpov/hstpov3.html), accessed May 13, 2009; UNICEF. 2010. "The Children Left Behind: A League Table of Inequality in Child Well-Being in the World's Rich Countries." Innocenti Report Card No. 9. Florence, Italy: UNICEF Innocenti Research Centre (www.unicef-irc.org/publications/619), accessed August 24, 2012.

## CONSEQUENCES OF SOCIAL STRATIFICATION

Studies of stratification in the United States have shown that social class affects many factors in a person's life. Striking differences in health and life expectancy are apparent among the social classes, especially between the lower-class poor and the other social groups. As might be expected, lower-class people are sick more frequently than are others.

Women living in poverty are more likely to have babies with low birth weights, putting them at higher risk for various cognitive and physical problems. Poor adults are four times as likely to regard themselves in fair or poor health compared with wealthier adults.

The poor of all races and ethnicities also experience lower life expectancy. The poor are more likely to develop illnesses that shorten the life span, such as heart disease, lung cancer, diabetes, and other degenerative diseases. Poor people are also more likely to die violent deaths, with homicide and suicide rates being substantially higher than among people with higher incomes (Lichter and Crowley, 2002). (See "Thinking About Social Issues: Who Smokes?")

Poverty particularly affects the health of the young. Babies born into poverty are significantly more likely to die before their first birthday than those born into families living above the poverty level. The infant mortality rate for poor children in the United States often is as high as that in third-world countries.

Babies born to African American girls between ages 15 and 19 are more than twice as likely to die as those born to white teenagers. In addition, an African American mother is more than three times as likely to die giving birth than a white mother. Diet and living conditions also give a distinct advantage to the upper classes because they have access to better and more sanitary housing and can afford more balanced and nutritious food. A direct consequence of this situation is seen in the life-expectancy pattern for each social class. Not surprisingly, lower-class people do not live as long as do those in the upper classes. White males have a life expectancy six years longer than for African American males, many of whom are concentrated in the lower income brackets (Bureau of the Census, 2010).

Family, childbearing, and childrearing patterns also vary according to social class. Women in the higher-income groups, who have more education, tend to have fewer children than do lower-class women with less schooling. Women more often head the family in the lower class, compared with women in the other groups. Middle-class women discipline their children differently than do working-class mothers. The former punish boys and girls alike for the same infraction, whereas the latter often have different standards for sons and daughters. Also, middle-class mothers judge the misbehaving child's intention, whereas working-class women are more concerned with the effects of the child's action.

Further, there is a direct relationship between a person's social class and the possibility of his or her arrest, conviction, and sentencing if accused of a crime. For the same criminal behavior, the poor are more likely to be arrested; if arrested, they are more likely to be charged;

---

### THINKING ABOUT SOCIAL ISSUES: Who Smokes?

Each day in the United States, nearly 4,000 teens between 12 and 17 years old smoke their first cigarette. The vast majority of teens who begin smoking during adolescence are addicted to nicotine by age 20. The prevalence of cigarette smoking among smokers has declined, but that decline has stalled during the past 5 years.

There are large social class, race, and ethnic differences in who smokes (see Table 8-4). American Indians/Alaska Natives have high rates of smoking. High smoking rates are also present among the lower social classes; people with histories of mental health illness and substance abuse; the lesbian, gay, bisexual, and transgender community; and people living in the South and Midwest regions of the United States.

Cigarette smoking remains the leading cause of preventable death and illness in the United States, resulting in an estimated 443,000 premature deaths and $193 billion in direct health care costs and productivity losses each year.

**TABLE 8-4** Education and Work Characteristics of Smokers

**Educational Attainment Percent Who Smoke**

| | |
|---|---|
| No high school diploma | 32 |
| High school graduate | 29.3 |
| Some college | 25.7 |
| College graduate | 13.3 |

**Employment Status Percent Who Smoke**

| | |
|---|---|
| Full-time | 27.8 |
| Part-time | 24.5 |
| Unemployed | 44.7 |

Source: Centers for Disease Control. January 14, 2011. *Health Disparities and Inequalities Report —United States, 2011.* Morbidity and Mortality Weekly Report, Supplement, Vol. 60.

if charged, they are more likely to be convicted; if convicted, they are more likely to be sentenced to prison; and if sentenced, they are more likely to be given longer prison terms than members of the middle and upper classes (Reiman, 1990).

The poor are singled out for harsher treatment at the very beginning of the criminal justice system. Although many surveys show that almost all people admit to having committed a crime for which they could be imprisoned, the police are prone to arrest a poor person and release, with no formal charges, a higher-class person for the same offense. A well-to-do teenager who has been accused of a criminal offense frequently is just held by the police at the station house until the youngster can be released to the custody of the parents; poorer teenagers who have committed the same kind of crime more often are automatically charged and referred to juvenile court.

The poor tend to commit violent crimes and crimes against property—they have little opportunity to commit such white-collar crimes as embezzlement, fraud, or large-scale tax evasion—and they are much more severely punished for their crimes than upper-class criminals are for theirs. Yet, white-collar crimes are far more damaging and costly to the public than are the crimes more often committed by poor people. The government has estimated that white-collar crimes cost more than $40 billion a year—more than ten times the total amount of all reported thefts and more than 250 times the amount taken in all bank robberies.

Even the language used to describe the same crime committed by an upper-class criminal and a poor one reflects the disparity in the treatment they receive. The poor thief who takes $2,000 is accused of stealing and usually receives a stiff prison sentence. The corporate executive who embezzles $200,000 merely has misappropriated the funds and is given a lighter sentence, or none at all, on the promise to make restitution. A corporation often can avoid criminal prosecution by signing a consent decree, which is in essence a statement that it has done nothing wrong and promises never to do it again. If this ploy were available to ordinary burglars, the police would have no need to arrest them; a burglar would merely need to sign a statement promising never to burgle again and file it with the court.

Once charged, the poor are usually dependent on court-appointed lawyers or public defenders to handle their cases. The better-off rely on private lawyers who have more time, resources, and personal interest in defending their cases.

If convicted of the same kind of crime as a well-to-do offender, the poor criminal is more likely to be sentenced and will generally receive a longer prison term. As for prison terms, the sentence for burglary, a crime of the poor, is generally more than twice as long as for fraud, and a robber will draw an average sentence more than six times longer than of an embezzler. The result is a prison system heavily populated by the poor.

Another serious consequence of social stratification is mental illness. Studies have shown that at least one-third of all homeless people suffer from schizophrenia, manic-depressive psychosis, or other mental disorders. Such people are the least likely to reach out for help and the most likely to remain on the streets in utter poverty and despair year after year (Jencks, 1994; Torrey, 1988).

Heavy drinking among the poor is often a way to escape the stress and lack of control in their lives.

The mentally ill among the homeless are the least likely to reach out for help and the most likely to remain on the streets.

Thus, social class has very real and immediate consequences for individuals. In fact, class membership affects the quality of people's lives more than any other single variable.

## WHY DOES SOCIAL INEQUALITY EXIST?

Sociologists and social philosophers before them have long tried to explain the presence of social inequality, when the very wealthy and powerful coexist with the poverty-stricken and powerless. Several theories have been proposed to explain this phenomenon.

### The Functionalist Theory

Functionalism is based on the assumption that the major social structures contribute to the maintenance of the social system (see Chapter 1, "The Sociological Perspective"). The existence of a specific pattern in society is explained in terms of the benefits that society receives because of that situation. In this sense, the function of the family is to socialize the young, and the function of marriage is to provide a stable family structure.

The functionalist theory of stratification as presented by Kingsley Davis and Wilbert Moore (1945) holds that social stratification is a social necessity. Every society must select individual members to fill a wide variety of social positions (or statuses) and then motivate those people to do what is expected of them in these positions—that is, to fulfill their role expectations. For example, our society needs teachers, engineers, janitors, police officers, managers, farmers, crop dusters, assembly-line workers, firefighters, textbook writers, construction workers, sanitation workers, chemists, inventors, artists, bank tellers, athletes, pilots, secretaries, and so on. To attract the most-talented individuals to each occupation, society must set up a system of differential rewards based on the skills needed for each position.

According to Davis and Moore, (1) different positions in society make different levels of contributions to the well-being and preservation of society, (2) filling the more complex and important positions in society often requires talent that is scarce and has a long period of training, and (3) providing unequal rewards ensures that the most-talented and best-trained individuals will fill the roles of greatest importance. In effect, Davis and Moore believe that people who are rich and powerful are at the top because they are the best qualified and are making the most significant contributions to the preservation of society (Zeitlin, 1981).

Many scholars, however, disagree with Davis and Moore, and their arguments generally take two forms. The first is philosophical and questions the morality of stratification. The second is scientific and questions its functional usefulness. Both criticisms share the belief that social stratification does more harm than good and that it is dysfunctional.

*The Immorality of Social Stratification* On what grounds, one might ask, is it morally justifiable to give widely different rewards to different occupations, when all occupations contribute to society's ongoing functioning? How can we decide which occupations contribute more? After all, without mail carriers, janitors, auto mechanics, nurse's aides, construction laborers, truck drivers, sanitation workers, and so on, our society would grind to a halt. How can the multimillion-dollar-a-year incomes of a select few be justified when the earnings of 13 percent of the American population fall below the poverty level, and many others have trouble making ends meet (U.S. Bureau of the Census, 2011)? Why are the enormous resources of our society not more evenly distributed?

In addition to the moral arguments against social stratification, there are other grounds on which stratification has been attacked—namely, that it is destructive for individuals and society as a whole.

*The Neglect of Talent and Merit* Regardless of whether social stratification is morally right or wrong, many critics contend that it undermines the very functions that its defenders claim it promotes. A society divided into social classes (with limited mobility among them) is deprived of the potential contributions of many talented individuals born into the lower classes. From this point of view, it is not necessary to do away with differences in rewards for different occupations. Rather, it is crucial to put aside all the obstacles to achievement that currently handicap the children of the poor.

*Barriers to Free Competition* It can also be claimed that access to important positions in society is not really open. That is, the members of society who occupy privileged positions allow only a small number of people to enter their circle, so shortages are created artificially. This, in turn, increases the perceived worth of those who are in the important positions. For example, the American Medical Association (AMA) is a wealthy and powerful group that exercises great control over the quality and quantity of physicians available to the American public. Historically, the AMA has directly influenced the number of medical schools in the United States and, thereby, the number of doctors produced each year, effectively creating a scarcity of physicians. A direct result of this influence is that medical-care costs have increased more rapidly than has the pace of inflation.

This situation is beginning to change, however. As more and more doctors fight for the same patient dollars, and as health maintenance organizations (HMOs) try to control costs, earnings might begin to suffer. Thus,

Neither functionalist theory nor conflict theory can fully explain why media people earn very large sums of money.

although barriers to free competition exist in our society, the marketplace often overrules them in the end.

*Functionally Important Jobs* When we examine the functional importance of various jobs, we become aware that the rewards attached to jobs do not necessarily reflect the essential nature of the functions. Why should a Hollywood movie star receive an enormous salary for starring in a film and a child-protection worker receive barely a living wage? It is difficult to prove empirically which positions are most important to society or what rewards are necessary to persuade people to want to fill certain positions.

## Conflict Theory

As we saw, the functionalist theory of stratification assumes society is a relatively stable system of interdependent parts in which conflict and change are abnormal. Functionalists maintain that stratification is necessary for the smooth functioning of society. Conflict theorists, in contrast, see stratification as the outcome of a struggle for dominance.

Current views of the conflict theory of stratification are based on the writings of Karl Marx. Later, Max Weber developed many of his ideas in response to Marx's writings.

*Karl Marx* Karl Marx believed that stratification emerges from the power struggles for scarce resources.

> The history of all hitherto existing society is the history of class struggles. [There always has been conflict between] freeman and slave, patrician and plebeian, lord and serf, guild-master and journeyman, in a word, oppressor and oppressed. (Marx and Engels, 1961)

The groups who own or control the means of production within a society obtain the power to shape or maintain aspects of society that favor their interests. They are determined to maintain their advantage. They do this by setting up political structures and value systems that support their position. In this way, the legal system, the schools, and the churches are shaped in ways that benefit the ruling class. As Marx and his collaborator Friedrich Engels put it, "The ruling ideas of each age have always been the ideas of its ruling class" (1961). Thus, the pharaohs of ancient Egypt ruled because they claimed to be gods.

In the first third of the twentieth century, America's capitalist class justified its position by misusing Charles Darwin's theory of evolution. The capitalists adhered to the view—called social Darwinism (see Chapter 1)—that those who rule do so because they are the most fit to rule, having won the evolutionary struggles that promote the survival of the fittest.

Marx believed that in a capitalist society, there are two great classes: the bourgeoisie, or the owners of the means of production or capital, and the proletariat, or the working class. Those in the working class have no resources other than their labor, which they sell to the capitalists. In all class societies, one class exploits another.

According to Marx, the capitalists will work to maintain and strengthen their position. The exploitative nature of capitalism is evident when the capitalists pay the workers a bare minimum wage, below the value of what the workers actually produce. The remainder is taken by the capitalists as profit and adds to their capital. Eventually, in the face of continuing exploitation, the working classes find it in their interest to overthrow the dominant class and establish a social order more favorable to their interests. Marx believed that with the proletariat in power, class conflict would finally end. The proletariat would have no class below it to exploit. The final stage of advanced communism would include an industrial society of plenty, where all could live in comfort.

In Marx's view, people's lives are influenced by how wealth is distributed among the people. Wealth can be distributed in at least four ways:

1. *To each according to need.* In this kind of system, the basic economic needs of all the people are satisfied. These needs include food, housing, medical care, and education. Extravagant material possessions are not basic needs and have no place in this system.
2. *To each according to want.* Here, wealth will be distributed according to what people desire and

request. Material possessions beyond the basic needs are included.
3. *To each according to what is earned.* People who live according to this system become the source of their own wealth. If they earn a great deal of money, they can lavish extravagant possessions upon themselves. If they earn little, they must do without.
4. *To each according to what can be obtained—by whatever means.* Under this system, everyone ruthlessly attempts to acquire as much wealth as possible without regard for the hardships that might be brought on others because of these actions. Those who are best at exploiting others become wealthy and powerful, and the others become the exploited and poor (Cuzzort and King, 1980).

In Marxist terms, the first of these four possibilities is what would happen in a socialist society. Although many readers will believe that the third possibility describes U.S. society (according to what is earned), Marxists would say that a capitalist society is characterized by the last choice—the capitalists obtain whatever they can get in any possible way.

*Max Weber* Weber agreed with Marx on many issues related to stratification, including the following:

1. Group conflict is a basic ingredient of society.
2. People are motivated by self-interest.
3. Those who do not have property can defend their interests less well than those who have property.
4. Economic institutions are of fundamental importance in shaping the rest of society.
5. Those in power promote ideas and values that help them maintain their dominance.
6. Only when exploitation becomes extremely obvious will the powerless object. (Vanfossen, 1979)

From those areas of agreement, Weber went on to add to and modify many of Marx's basic premises. Weber's view of stratification went beyond the material or economic perspective of Marx. He included status and power as important aspects of stratification as well as class.

Class, status, and power, although related, are not the same. One can exist without the others. To Weber, they are not always connected in some predictable fashion, nor are they always tied to the economic mode of production. An aristocratic Southern family might be in a condition that is often labeled genteel poverty, but the family name still elicits respect in the community. This kind of status sometimes is denied to the rich, powerful labor leader whose family connections and school ties are not acceptable to the social elite. In addition, status and power are often accorded to those who have no relationship to the mode of production. For example, Nobel Peace Prize winner Mother Teresa, known for her work with the poor in India, controlled no industry, nor did she have any great personal wealth, and yet her influence was felt by heads of state the world over.

Whereas Marx was somewhat of an optimist in that he believed that conflict, inequality, and exploitation eventually could be eliminated in future societies, Weber was much more pessimistic about the potential for a more just and humane society. (See Table 8-5 for a comparison of the functionalist and conflict theory views of social stratification.)

TABLE 8-5 Functionalist and Conflict Views of Social Stratification: A Comparison

| The Functionalist View | The Conflict View |
| --- | --- |
| 1. Stratification is universal, necessary, and inevitable. | 1. Stratification may be universal without being necessary or inevitable. |
| 2. Social organization (the social system) shapes the stratification system. | 2. The stratification system shapes social organizations (the social system). |
| 3. Stratification arises from the societal need for integration, coordination, and cohesion. | 3. Stratification arises from group conquest, competition, and conflict. |
| 4. Stratification facilitates the optimal functioning of society and the individual. | 4. Stratification impedes the optimal functioning of society and the individual. |
| 5. Stratification is an expression of commonly shared social values. | 5. Stratification is an expression of the values of powerful groups. |
| 6. Power usually is distributed legitimately in society. | 6. Power usually is distributed illegitimately in society. |
| 7. Tasks and rewards are allocated equitably. | 7. Tasks and rewards are allocated inequitably. |
| 8. The economic dimension is subordinate to other dimensions of society. | 8. The economic dimension is paramount in society. |
| 9. Stratification systems generally change through evolutionary processes. | 9. Stratification systems often change through revolutionary processes. |

*Source:* Adapted from "Some Empirical Consequences of the Davis-Moore Theory of Stratification," by A. L. Stinchcombe, 1969, in J. L. Roach, L. Gross, & O. R. Gursslin, eds., *Social Stratification in the United States,* Englewood Cliffs, NJ: Prentice-Hall, p. 55. Used by permission.

## Modern Conflict Theory

Conflict theorists assume that people act in their own self-interest in a material world in which exploitation and power struggles are prevalent. Modern conflict theory has five aspects:

1. Social inequality emerges through the domination of one or more groups by other groups. Stratification is the outgrowth of a struggle for dominance in which people compete for scarce goods and services. Those who control these items gain power and prestige.
2. Those who are dominated have the potential to express resistance and hostility toward those in power. Although the potential for resistance exists, it sometimes lies dormant. Opposition might not be organized because the oppressed groups might not be aware of their mutual interests. They might also be divided because of racial, religious, or ethnic differences.
3. Those in power will be extremely resistant to any attempts to share their advantages. Economic and political power are important advantages in maintaining a position of dominance.
4. What are thought to be the common values of society are really the values of the dominant groups. The dominant groups establish a value system that justifies their position. In this way, the subordinate groups come to accept a negative evaluation of themselves and to believe that those in power have a right to that position.
5. Because those in power are engaged in exploitative relationships, they must find mechanisms of social control to keep the masses in line. By holding out the possibility of a small amount of social mobility for those who are deprived, the power elite will try to induce them to accept the system's basic assumptions. Thus, the oppressed masses will come to believe that by behaving according to the rules, they will gain a better life (Vanfossen, 1979).

## The Need for Synthesis

Any empirical investigation will show that neither the functionalist theory nor the conflict theory of stratification is entirely accurate. This does not mean that both are useless in understanding how stratification operates in society. Ralf Dahrendorf (1959) suggested that the two really are complementary rather than opposed. We do not need to choose between the two but, instead, should see how each is qualified to explain specific situations. For example, functionalism can help explain why differential rewards are needed to serve as an incentive for a person to spend many years training to become a lawyer. Conflict theory would help explain why the offspring of members of the upper classes study at elite institutions and end up as members of prestigious law firms, whereas the sons and daughters of the middle and lower-middle classes study at public institutions and become overworked district attorneys.

## SUMMARY

▶ **Describe the social stratification situation in the United States.**

Despite the American political ideal of the basic equality of all citizens and the lack of inherited ranks and titles, the United States nonetheless has a class structure that is characterized by extremes of wealth and poverty. Class distinctions exist in the United States based on race, education, family name, career choice, and wealth.

▶ **How does social class affect people's lives?**

Social class affects many aspects of people's lives. For instance, lower-class people get sick more often and have higher infant mortality rates, shorter life expectancies, and larger families. The poor are more likely to be arrested, charged with a crime, convicted, and sentenced to prison and are likely to get longer prison terms than middle- and upper-class criminals.

▶ **How does functionalist theory view social stratification?**

The functionalist theory of stratification as presented by Davis and Moore holds that stratification is socially necessary. These authors argued that different positions in society make different levels of contributions to the well-being and preservation of society. Filling the more complex and important positions in society often requires talent that is scarce and has a long period of training. Providing unequal rewards ensures that the most-talented and best-trained individuals will fill the statuses of greatest importance and be motivated to carry out role expectations competently.

▶ **What are the criticisms of the functionalist theory of stratification?**

Critics of the functionalist view suggest that stratification is immoral because it creates extremes of wealth and poverty and denigrates the people at the bottom. In addition, it is dysfunctional in that it neglects the talents and merits of many people who are stuck in the lower classes. It also ignores the ability of the powerful to limit access to important positions, and overlooks the fact that the level of rewards attached to jobs does not necessarily reflect their functional importance.

▶ **How does the conflict theory view social stratification?**

Conflict theorists see stratification as the outcome of a struggle for dominance. Karl Marx believed that to understand human societies, one must look at the economic conditions surrounding production of the necessities of life. Marx believed the groups that own or control the means of production within a society also have the power to shape or maintain aspects of society to favor their interests.

## Media Resources

**CourseMate for *Introduction to Sociology*, Eleventh Edition**

Cengage Learning's Sociology CourseMate brings course concepts to life with interactive learning, study, and exam preparation tools that support the printed textbook. Access an integrated eBook, learning tools including glossaries, flashcards, quizzes, videos, and more in your Sociology CourseMate. Go to CengageBrain.com to register or purchase access.

# CHAPTER EIGHT STUDY GUIDE

## Key Concepts

*Match each of the following concepts with its definition, illustration, or explanation.*

a. Social class
b. Class system of stratification
c. Upper class
d. Upper-middle class
e. Middle-middle class
f. Lower-middle class
g. Lower class
h. Distribution of income
i. Distribution of wealth
j. Poverty
k. Poverty index
l. Feminization of poverty
m. Functionalist theory
n. Conflict theory

____ 1. A U.S. social class characterized by corporate ownership, elite schools, upper-echelon politics, and a higher education
____ 2. An explanation for the existence of social classes based on the idea that to attract talented individuals to each occupation, society must set up a system of differential rewards
____ 3. A category of people who share similar opportunities, similar economic and vocational positions, similar lifestyles, and similar attitudes and behaviors
____ 4. The phrase referring to the increasing concentration of poverty among female-headed households
____ 5. A U.S. social class that comprises skilled and semiskilled laborers, factory employees, and other blue-collar workers
____ 6. A U.S. social class characterized by unskilled labor, service work, farm labor, and little interest in education or high-school completion
____ 7. The U.S. government's specification of income levels below which people are considered to be living in poverty
____ 8. The degree to which all earnings in the nation are spread out among the population
____ 9. A system of stratification that includes several social classes and permits social mobility
____ 10. The condition in which people do not have enough money to maintain a standard of living that includes the basic necessities of life
____ 11. A U.S. social class characterized by professional and technical occupations and college and graduate-school training
____ 12. An explanation that says social class arises and persists because those with more wealth and power use their means to enhance their own position at the expense of others
____ 13. The degree of concentration or spreading out of property and other financial assets

## Key Thinkers

*Match the thinkers with their main idea or contribution.*

a. Max Weber
b. Karl Marx
c. Kingsley Davis and Wilbert Moore

____ 1. Developed the functionalist theory of social stratification
____ 2. Argued that class was based on ownership and that capitalism required a conflict between owners and workers
____ 3. Argued that social stratification was not just a matter of wealth but included prestige and political power as well

## Central Idea Completions

*Fill in the appropriate concepts and descriptions for each of the following questions.*

1. Assess each the following ideas about poverty in America.

    a. People are poor because they are too lazy to work. _____
    _____
    _____

    b. Most poor are minorities, and most minorities are poor. _____
    _____
    _____

    c. Most people in poverty live in inner-city areas. _____
    _____
    _____

    d. Welfare programs for the poor are straining the federal budget. _____
    _____
    _____

2. As your text notes, those arrested, accused, convicted, and imprisoned are disproportionately poor. Yet wealthier people also commit crimes. Compare and contrast the ways the criminal justice system operates for the poor and for the rich in each of the following areas.

    a. Type of crime: _____
    _____
    _____

    b. Legal representation: _____
    _____
    _____

    c. Conviction outcomes: _____
    _____
    _____

3. Summarize each of these theories of stratification in one or two sentences.

    a. Functionalist: _____
    _____

    b. Conflict: _____
    _____

    c. Modern conflict: _____
    _____

## Critical Thought Exercises

1. Assess the performance of the United States, compared with other Western industrialized countries, in reducing poverty among these three groups: children, the elderly, and working-age adults. Where does the United States fare best and where does it do the worst? Describe some of the major factors responsible for the United States' performance with regard to each of the three groups.

2. Try thinking of stratification in your school based on the classes of faculty, students, administrators, and perhaps other groups. How would a conflict theorist look at the actions of each of these groups? How would a functionalist look at them? To what extent do the class distinctions of the wider society come into play on campus and intertwine with the campus categories?

3. Some government policies benefit the wealthy and not the poor, such as certain tax policies, lower minimum wage, less spending for welfare, restrictions on food stamps, and so on. How do proponents of these policies justify them? Do these arguments provide examples of Marx and Engels's observation that "the ruling ideas of each age have always been the ideas of its ruling class"?

4. Develop a research paper within which you explore why economic rewards are distributed more unequally in the United States than elsewhere in the Western industrialized world. What are the positive and negative outcomes of this system?

## Internet Activities

1. Visit www.pbs.org/peoplelikeus/film/index.html. This site is connected with a video, "People Like Us," produced by PBS and often presented in sociology courses. For this exercise, you do not need to see the entire video. Rather, visit the section that has short video clips with commentaries from the persons interviewed in the complete film. The "People" section has an option to e-mail, through PBS, any or all of the persons appearing in the short video clips. After watching the clips, select any three, develop your question, and e-mail each. Report your findings in a short paper for the class.

2. *The New York Times* has published a series of articles on social class. The articles are on their website along with some interesting flash graphics on social class, income inequality, and social mobility. Visit www.nytimes.com/pages/national/class/.

## Answers to Key Concepts

1. c;  2. m;  3. a;  4. l;  5. f;  6. g;  7. k;  8. h;  9. b;  10. j;  11. d;  12. n;  13. i

## Answers to Key Thinkers

1. c;  2. b;  3. a

Mike Goldwater/Alamy Limited

# 9 Global Stratification

*Our Diverse World:* How Countries Differ—Japan and Nigeria

## Stratification Systems
The Caste System
The Estate System
The Class System

## Population Dynamics
Fertility
Mortality
Migration

## Theories of Population
Malthus's Theory of Population Growth
Marx's Theory of Population Growth
Demographic Transition Theory
A Second Demographic Transition

## Global Diversity
*Our Diverse World:* Life Chances of an Adolescent Girl in Liberia
World Health Trends
The Health of Infants and Children in Developing Countries
HIV/AIDS
Population Trends

*Thinking About Social Issues:* Where Are the Baby Girls?
*Thinking About Social Issues:* What If the Population Problem Is Not Enough People?

## Summary

## LEARNING OBJECTIVES

After studying this chapter, you should be able to do the following:

- Describe the caste, estate, and class systems of social stratification.
- Describe the phenomenon of exponential growth.
- Define the three major components of population change.
- Contrast the Malthusian and Marxist theories of population.
- Summarize the demographic transition model and explain why there might be a second demographic transition.
- Discuss the determinants of fertility and family size.
- Discuss the problems of overpopulation and possible solutions.
- Discuss world health trends.
- Understand the trends in global aging.

Should there be limits on inequality? Adam Smith certainly thought there had to be limits to deprivation. Smith noted, "No society can be flourishing and happy of which the far greater part of members are poor and miserable." Smith believed all members of society should have enough to enable them to walk down the street "without shame" (Smith, [1776] 1976). Nearly all religions note that it is a moral obligation to help the less fortunate. Yet, despite these expressions, global inequality is widespread.

For example, a girl born in Japan today might have a 50 percent chance of seeing the twenty-second century, but a newborn in Afghanistan has a 1 in 4 chance of dying before age five. The richest 5 percent of the world's people have incomes 114 times those of the poorest 5 percent. Every year, about 11 million children die of preventable causes, often because they lack simple and easily provided improvements in nutrition, sanitation, and maternal health and education. Every year more than 500,000 women die as a result of pregnancy and childbirth, with huge regional disparities. Fifteen million children have lost their mother or both parents to AIDS, and more than 40 million people have been infected with the human immunodeficiency virus (HIV), 90 percent of them in developing countries and 75 percent in sub-Saharan Africa.

In sub-Saharan Africa, human development has actually regressed in recent years, and the lives of its very poor people are getting worse. More than half of the world's people live on less than $2 a day, the internationally defined poverty line. (See Figure 9-1.) Because of population growth, the number of poor people in that region has increased.

Throughout the world, 1.2 billion people lack access to safe water, and 2.6 billion do not have access to any form of improved sanitation services. This situation causes almost 2 million children to die each year from infectious diseases. Unclean water is the second biggest killer of children (United Nations Development Report, 2006).

These examples demonstrate the extreme levels of social inequality that exist in the world today. And recent information indicates that the situation might be getting worse, not better. "In 73 countries representing 80 percent of the world's people, 48 have seen inequality increase since the 1950s, 16 have experienced no change, and only nine—with just 4 percent of the

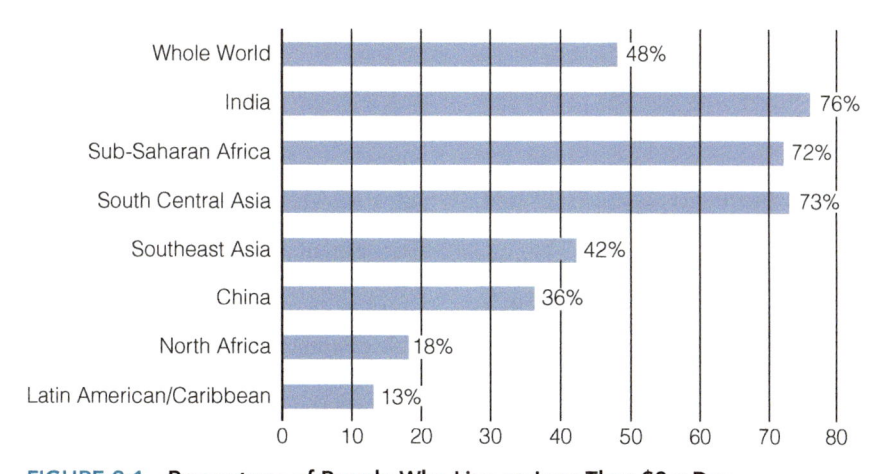

| Region | Percentage |
|---|---|
| Whole World | 48% |
| India | 76% |
| Sub-Saharan Africa | 72% |
| South Central Asia | 73% |
| Southeast Asia | 42% |
| China | 36% |
| North Africa | 18% |
| Latin American/Caribbean | 13% |

FIGURE 9-1 **Percentage of People Who Live on Less Than $2 a Day**

Source: Haub, Carl. "2011 World Population Data Sheet" (www.prb.org/pdf11/2011population-data-sheet_eng.pdf), accessed July 17, 2012.

## OUR DIVERSE WORLD

### How Countries Differ—Japan and Nigeria

There are enormous differences among the world's countries. On one side, we have mostly poor countries with high birthrates and low life expectancies. On the other side, we have wealthy countries with low birthrates and high life expectancies. These differences produce large economic, social, and political situations. People in these countries have sharply different living standards, health, and future prospects.

Two countries that provide an example of these differences are Japan and Nigeria, which currently have similarly sized populations. (See Table 9-1.)

#### Japan

Japan has the world's second-largest economy and enjoys a high per-capita income and standard of living. The Japanese are highly educated; most finish high school, and a third go on to college or university. The Japanese have the world's longest life expectancy—83 years—and one of the lowest rates of infant mortality. Japanese women have an average of 1.4 children, giving Japan one of the lowest fertility rates in the world. This situation causes the Japanese people to be one of the most rapidly aging populations in the world.

#### Nigeria

In Nigeria, the average birthrate is nearly six births per woman. The life expectancy is 51 years. With 162 million people, it is the most populous country in Africa. More than 84 percent of Nigerians live on less than US$2 per day. About one-half of Nigerian women are literate. Nigeria's population could double by 2050, and growth and poverty are expected to continue into the second half of the twenty-first century.

**TABLE 9-1** A Comparison of Two Countries: Japan and Nigeria

|  | Japan | Nigeria |
| --- | --- | --- |
| Population 2012 (millions) | 127.6 | 162.3 |
| Population 2050 (millions) | 95.2 | 402 |
| Lifetime births per woman | 1.4 | 5.6 |
| Under 15 (%) | 13 | 43 |
| Over 65 (%) | 24 | 3 |
| Life expectancy at birth (years) | 83 | 51 |
| Infant deaths per 1,000 births | 2.3 | 77 |
| Adults with HIV/AIDs, 2003 (%) | < 0.1 | 3.6 |
| Percentage of population living on less than US$2/day (%) | 0 | 84 |
| Probability of not living past 40 (%) | 1.3 | 39 |

Source: Haub, Carl and Toshiko Kaneda. July 2012. "2012 World Population Data Sheet" (www.prb.org/pdf12/2012-population-data-sheet_eng.pdf), accessed July 19, 2012.

---

world's people—have seen inequality decline" (United Nations, 2002).

Social inequality is the outgrowth of social stratification, which is the uneven distribution of privileges, material rewards, opportunities, power, prestige, and influence among individuals and groups. Social inequality exists in all societies. The inequality can occur because of wealth, prestige, or power. Societies differ in terms of what is unequal and how the inequality comes about. (See "Our Diverse World: How Countries Differ" for an example of how Japan and Nigeria differ.)

### STRATIFICATION SYSTEMS

Social inequality is the result of stratification systems that operate in two ways: (1) People can be assigned to societal roles based on an ascribed status—an easily identifiable characteristic, such as gender, age, family name, or skin color—over which they have no control. This will produce the caste and estate systems of stratification; or (2) people's positions in the social hierarchy can be based to some degree on their achieved statuses (see Chapter 5, "Social Interaction"), gained through their individual, direct efforts. This is known as the class system.

## The Caste System

The **caste system** is *a rigid form of stratification, based on ascribed characteristics such as skin color or family identity, that determines a person's prestige, occupation, residence, and social relationships.* Many people associate caste systems with India; however, numerous other countries have had caste-like systems where minorities have been discriminated against, denied civil rights, and ostracized. Examples include the Burakumin in Japan, the Al-Akhdam of Yemen, the Baekjeong of Korea, and West Indians in Great Britain.

Under a caste system, people are born into and spend their entire lives within a caste with little chance of leaving it. The caste is a closed group whose members are severely restricted in their choice of occupation and degree of social participation. Marriage outside the caste is prohibited. Social status is determined by the caste of one's birth, and it is very unusual for a person to overcome his or her origins.

Contact between castes is minimal and governed by a set of rules or laws. If interaction must take place, it is impersonal, and examples of the participants' superior or inferior status are abundant. Access to valued resources is extremely unequal. A set of religious beliefs often justifies a caste system.

Hindus have never placidly accepted the caste system. Since the nineteenth century, economic developments have made the caste system less stringent. In the 1930s, Mohandas Gandhi tried to change attitudes toward the untouchables, and untouchability was declared illegal in 1949 (Caste, 2002).

The scheduled castes are people who were known as untouchables. The contemporary term for this group is Dalit, and they represent 15 percent (160 million people) of the population. The Dalits have encountered extreme discrimination and social isolation. Even a Dalit's shadow was thought to pollute the higher castes. Dalits have been barred from entering the villages of the upper castes, drinking from public wells, or going to the temples of the upper castes. Dalit children have been discriminated against in schools (Haviland, 2005).

The government of India has an affirmative action system that it calls "reservations," which applies to certain government jobs and admission to all public and private educational institutions. Unlike affirmative action in the United States, the reservation system is based on specific quotas for groups defined as scheduled castes, scheduled tribes, or backward classes. For example, 27 percent of the seats in higher education are reserved for these groups.

Since 1950, India has put into place many laws to protect and improve the lives of its Dalit population. Of the highest-paying senior positions in government, over 10 percent were held by members of the Dalit community, a tenfold increase in 40 years. Inter-caste marriage is on the rise in urban India as female literacy and education, as well as influences from the media, have changed earlier caste views. In 2009, the Indian parliament unanimously elected Meira Kumar from the Dalit community as the first woman speaker.

## The Estate System

The **estate system** is *a closed system of stratification in which a person's social position is defined by law, and membership is determined primarily by inheritance.* An estate is a segment of a society that has legally established rights and duties. The estate system is similar to a caste system but not as extreme. Some mobility is possible but by no means as much as exists in a class system.

The estate system of medieval Europe is a good example of how this type of stratification system works. The three major estates in Europe during the Middle Ages were the nobility, the clergy, and, at the bottom of the hierarchy, the peasants. A royal landholding family at the top had authority over a group of priests and the secular nobility, who were quite powerful in their own right. The nobility were the warriors; they were expected to give military protection to the other two estates. The clergy not only ministered to the spiritual needs of all the people but also were often powerful landowners as well. The peasants were legally tied to the land, which they worked to provide the nobles with food and wealth. In return, the nobles were supposed to provide social order, not only with their military strength but also as the legal authorities who held court and acted as judges in disputes concerning the peasants who belonged to their land. The peasants had low social status, little freedom or economic standing, and almost no power.

Just above the peasants was a small but growing group, the merchants and craftspeople. They operated

*In some Middle Eastern societies, women are expected to cover themselves in public. Such a situation helps perpetuate inequality between men and women.*

somewhat outside the estate system in that, although they might achieve great wealth and political influence, they had little chance of moving into the estate of nobility or warriors. It was this marginal group, which was less constricted by norms governing the behavior of the estates, that had the flexibility to gain power when the Industrial Revolution, starting in the eighteenth century, undermined the estate system.

Individuals were born into one of the estates and remained there throughout their lives. Under unusual circumstances, people could change their estate, as, for example, when peasants—using produce or livestock saved from their own meager supply or a promise to turn over a bit of land that by some rare fortune belonged to them outright—could buy a position in the church for a son or daughter. For most, however, social mobility was difficult and extremely limited because wealth was permanently concentrated among the landowners. The only solace for the poor was the promise of a better life in the hereafter (Vanfossen, 1979).

### The Class System

In Chapter 8, "Social Class in the United States," we noted that a society that has several social classes and permits greater social mobility than a caste or estate system is based on a class system of stratification. Remember that a social class consists of a category of people who share similar opportunities, similar economic and vocational positions, similar lifestyles, and similar attitudes and behaviors. Some form of class system is usually present in all industrial societies, whether they are capitalist or communist. Mobility is greater in a class system than in either a caste or an estate system. This mobility is often the result of an occupational structure that allows higher-level jobs to be available to anyone with the education and experience required. A class society encourages striving and achievement.

### ● POPULATION DYNAMICS

Let's look at the countries experiencing population growth first. Africa is now the area of the world with the most rapid population expansion. The continent has a population of 1 billion and is growing at about 2.4 percent per year. At that rate, it takes 29 years for the population to double. In Uganda, for example, the average woman now has 6.4 children. When this fact is combined with the declining infant mortality rate, the country's population could balloon from 34.5 million in 2011 to 116.8 million in 2053. Enormous growth has also been projected for Mali and the Palestinian Territory (Population Reference Bureau, 2011). (See Table 9-2 for projected yearly population growth percentages for selected countries.)

**TABLE 9-2** Annual Percentage Increase and Years Needed to Double Current Population

| | Yearly Percentage Increase | Years Needed to Double the Population |
|---|---|---|
| World | 1.2 | 58 |
| More developed | 0.1 | 700 |
| Less developed | 1.4 | 50 |
| Least developed | 2.4 | 29 |
| Niger | 3.5 | 20 |
| Uganda | 3.3 | 21 |
| Mali | 3.2 | 22 |
| Jordan | 3.0 | 23 |
| Palestinian Territory | 2.9 | 24 |
| Chad | 2.8 | 25 |
| Afghanistan | 2.8 | 25 |
| Sudan | 2.4 | 29 |
| Israel | 1.6 | 44 |
| Mexico | 1.5 | 47 |
| Argentina | 1.1 | 64 |
| Ireland | 1.0 | 70 |
| France | 0.4 | 175 |
| United Kingdom | 0.4 | 175 |
| Japan | −0.2 | — |
| Bulgaria | −0.5 | — |

*Source:* Haub, Carl and Toshiko Kaneda. July 2012. "2012 World Population Data Sheet" (www.prb.org/pdf12/2012-population-data-sheet_eng.pdf), accessed July 19, 2012.

As you can probably guess, most of the world's growth in population in the next few decades will take place in developing countries. The population of the richer countries will increase by 200 million by the year 2050, whereas the developing areas will have added about 6 billion (see Figure 9-2).

Symbolizing the shift in population from the developed to the less-developed world is Sao Paulo, Brazil. In 1950, this city was smaller than Manchester, England, but by the year 2010, Sao Paulo's population had reached about 19 million, making it one of the ten largest cities in the world. London, which was the second-largest city in the world in 1950, is no longer in the top 20 largest cities in the world.

In certain parts of the world, population problems become increasingly pressing because of what is referred to as exponential growth. The yearly increase in population is determined by a continuously expanding base. Each successive addition of 1 million people to the population requires less time than the previous addition required, even if the birthrate does not increase. The best way to demonstrate the effects of exponential growth is to use a simple example.

Let us assume that you have a job that requires you to work 8 hours a day, every day, for 30 days. At the end of that time, your job is over. Your boss offers you a choice between two methods of payment. The first choice is

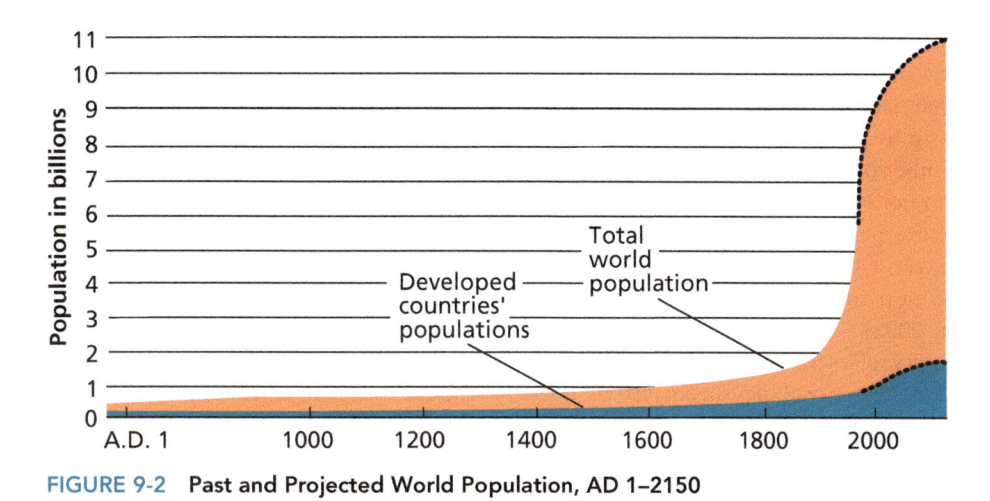

FIGURE 9-2  Past and Projected World Population, AD 1–2150

to be paid $100 a day for a total of $3,000. The second choice is somewhat different. The employer will pay you 1 cent for the first day, 2 cents for the second day, 4 cents for the third, 8 cents on the fourth, and so on. Each day, you will receive double what you received the day before. Which form of payment would you choose? The second form of payment would yield a significantly higher total payment—so high, in fact, that no employer could realistically pay the amount. Through this process of successive doubling, you would be paid $5.12 for your labor on the 10th day. Only on the 15th day would you receive more than the flat $100 a day you could have received from the first day under the alternative payment plan; the amount that day would be $163.84. However, from that day on, the daily pay increase is quite dramatic. On the 20th day, you would receive $5,242.88, and on the 25th day, your daily pay would be $167,772.16. Finally, on the 30th day, you would be paid $5,368,709.12, bringing your total pay for the month to more than $10 million.

This example demonstrates how the continual doubling in the world's population can produce enormous problems. The annual growth rate in the world's population has declined from a peak of 2.04 percent in the late 1960s to 1.2 percent in 2011. This difference between global birth and death rates means that the world's population now doubles every 58 years instead of every 35 years as it did just a few decades ago (see Table 9-2).

Our discussion of the population shifts in various countries falls into what is called demography. **Demography** is *the study of the size and composition of human populations as well as the causes and consequences of changes in these factors.* Demography is influenced by three major factors: fertility, mortality, and migration.

## Fertility

**Fertility** refers to *the actual number of births in a given population.* For most countries, population growth

In South Asia and sub-Saharan Africa, about half of all women between the ages of 15 and 19 are or have been married. Thus, fertility rates are extremely high in these regions.

depends on the natural increase resulting from more births than deaths. Most countries have little immigration or emigration, something that might seem surprising to people in the United States, where immigration is an important growth factor.

**Fecundity** is *the physiological ability to have children.* Most women between the ages of 15 and 45 are capable of bearing children. During this time, a woman potentially could have up to 30 children; however, the realistic maximum number of children a woman can have is about 15. This number is a far cry from real life, though, where health, culture, and other factors limit childbearing. Even in countries with high birthrates, the average woman rarely has more than seven children (Population Reference Bureau, 2008).

Whereas fecundity refers to the biological potential to bear children, a common way of measuring fertility is by using the **crude birthrate,** which is *the number of annual live births per 1,000 people in a given population.* The crude birthrate for the United States fell from 24.1 in 1950 to 14 in recent years. Only Europe has a lower crude birthrate than the United States, with many countries in the range of 10 to 11. In contrast to the United States and Europe, the crude birthrate is 40 in Nigeria, 27 in Pakistan, and 25 in the Philippines (The World Bank, 2012).

Many demographers think a better way to measure fertility is to use the **total fertility rate,** which is *the average number of children that would be born to a woman over her lifetime.* A total fertility rate of 2.1 is needed to keep the population unchanged. In the United States, the total fertility rate peaked at 3.8 in the late 1950s and is now at 2.1, indicating that the country would not be growing without immigration. In Africa, dozens of countries have total fertility rates between 5 and 7, producing large growth rates. In Western Europe, the rate is 1.7, and many countries have declining populations (Population Reference Bureau, 2011).

As you will see later in this chapter, the fertility rate is linked to industrialization. Fertility declines with modernization, but not immediately. This lag is a source of tremendous population pressure in developing nations that have benefited from the introduction of modern medical technology, which immediately lowers mortality rates.

## Mortality

People die eventually, but in some countries, they die much earlier than in others. **Mortality** is *the frequency of deaths in a population.* The most commonly used measure of this is the **crude death rate,** which is *the annual number of deaths per 1,000 people in a given population.* In 2011, the United States had a crude death rate of 8.07 per 1,000 (National Centers for Disease Control and Prevention, 2012).

Demographers also look at age-specific death rates. For example, one measure used is the **infant mortality rate,** which *measures the number of children who die within the first year of life per 1,000 live births.* Of 1,000 babies born this year in Eastern Africa, 69 will die within one year. In the world's more-developed countries, it will take 60 years for these 67 deaths to occur. The difference reflects a continuing gap in mortality levels between the world's more- and less-developed countries (Population Reference Bureau, 2011).

In the United States, the infant mortality rate dropped from 47 in 1940 to 6.1 in 2011 (Population Reference Bureau, 2011). This rate does not apply to all infants, however. In 1940, whites had an infant mortality rate of 43.2 (below the national average at that time), and African Americans had an infant mortality rate of 73.8. Today, the rates are 5.11 for whites and 11.42 for blacks (National Center for Health Statistics, 2012). The lowest infant mortality rates are for infants born to Chinese mothers (2.9) and Japanese mothers (3.4). Those figures suggest that differing cultural patterns of childrearing might affect infant mortality as well as the availability of medical care.

Infectious diseases caused most child deaths in the early part of the twentieth century. Millions of children died from respiratory diseases, gastrointestinal diseases, and tuberculosis. These deaths have been reduced through the introduction of penicillin and other antibiotics, and vaccines have helped prevent illnesses that killed or crippled large numbers of children in the past. Today, accidents, birth defects, and cancer are the greatest threats to children (AmeriStat, August 2010).

Although the infant mortality rate in the United States is considerably lower than the rates of developing countries, it is higher than those in some Asian or European countries such as China, Singapore, and Japan.

Mortality is also reflected in people's **life expectancy,** *the average number of years a person born in a particular year can expect to live.* The average life expectancy at birth in the United States is 77 years. The world's longest life expectancy is in Japan, at 83 years. The shortest life expectancies are in Sierra Leone, Liberia, and Djibouti (Population Reference Bureau, 2008). (See Table 9-3 for life expectancies in various countries.)

Maternal mortality is another factor that varies widely throughout the world. There are large differences

**TABLE 9-3** Countries with the Highest and Lowest Life Expectancies

| Highest | Years | Lowest | Years |
| --- | --- | --- | --- |
| Japan | 83 | Sierra Leone | 47 |
| Hong Kong | 83 | Swaziland | 48 |
| Singapore | 82 | Zambia | 48 |
| Italy | 82 | Lesotho | 48 |
| Canada | 81 | Zimbabwe | 48 |
| United States | 79 | Guinea-Bissau | 48 |
| France | 81 | Djibouti | 43.4 |
| Iceland | 80.7 | Liberia | 42 |
| Sweden | 80.9 | Sierra Leone | 41.2 |
| World Average 66.6 | | | |

*Source:* Haub, Carl and Toshiko Kaneda. July 2012. "2012 World Population Data Sheet" (www.prb.org/pdf12/2012-population-data-sheet_eng.pdf), accessed July 19, 2012.

in the risk of dying from pregnancy-related causes in developed versus developing countries. A woman in sub-Saharan Africa has a lifetime chance of maternal death of 1 in 22; in Asia, it is 1 in 90; and in Europe, 1 in 9,400 (Population Reference Bureau, 2008).

Life expectancies in the world vary greatly. They range from an average regional low of about 58 years in Africa to 77 years in Europe (Population Reference Bureau, 2012). Life expectancy is usually determined more by infant mortality than by adult mortality. Once an individual survives infancy, life expectancy improves dramatically. In the United States, for example, only when individuals reach their 60s do their chances of dying approximate those of their infancy (U.S. Bureau of the Census, 2008).

Disease took a dramatic toll on life expectancy in the United States in the not-too-distant past. For example, Abraham Lincoln's mother died when she was 35 and he was 9. Prior to her death, she had three children. Of them, Abraham Lincoln's brother died in infancy, and his sister died in her early 20s. Of the four sons born to Abraham and Mary Todd Lincoln, only one survived to maturity.

Where we see short life expectancies in countries throughout the world, we also see high infant mortality rates. If we compare Mali with Israel, for example, we will find that the two countries have very different health numbers. In Mali, the life expectancy is 51, and the infant mortality rate is 97 deaths per 1,000 live births. In Israel, the life expectancy is 82, and the infant mortality rate is 3.4. It would be easy to say that these numbers are because Israel is a wealthier country than Mali; however, some poor countries, such as China and Sri Lanka, have longer life expectancies than some richer countries, such as Brazil and South Africa. Other factors, such as education and cultural differences, also play a role (Population Reference Bureau, 2012).

A fact that is often overlooked is that the rapid increase in population growth in third-world countries is caused by sharp improvements in life expectancy, not by rising birthrates. A rapid decrease in infant mortality in a developing nation will result in a significant rise in the overall rate of population growth.

In developing countries, the proportion of infant and child deaths is quite high, resulting in a significantly lower life expectancy than that in developed countries. In Bangladesh, infant deaths account for more than one-third of all deaths. In the United States, that figure is about 1 percent. The high proportion of deaths in developing countries can be attributed to impure drinking water and unsanitary conditions. In addition, the diets of pregnant women and nursing mothers often lack proper nutrients, and babies and children are not fed a healthy diet. Flu, diarrhea, and pneumonia are common, as are typhoid, cholera, malaria, and tuberculosis. Many children are not immunized against common childhood diseases (such as polio, measles, diphtheria, and whooping cough), and the parents' income is often so low that when the children do fall ill, they cannot provide medical care.

When countries succeed in improving the health of the children, there may be a short-term spike in population growth.

## Migration

**Migration** is *the movement of populations from one geographical area to another.* We call it **emigration** *when a population leaves an area* and **immigration** *when a population enters an area.* All migrations, therefore, are both emigrations and immigrations.

Of the three components of population change—fertility, mortality, and migration—migration historically exerts the least impact on population growth or decline.

Most countries do not encourage immigration. When they do permit immigration, it is often viewed as a way to provide needed skilled labor or to provide unskilled labor for jobs the resident population no longer wishes to do. Exceptions to this trend are the traditional receiver countries, such as the United States, Australia, and Canada. These countries owe much of their growth to immigrant populations.

Most of the world's population growth in the next few decades will occur in developing countries.

Where migration is a significant factor, it is necessary to take into account the age and sex of the immigrants and emigrants as well as the number of migrants. Those characteristics tell us the number of potential workers among the migrants, the number of women of childbearing age, the number of school-age children, the number of elderly, and other factors that will affect society.

Sometimes it is important to distinguish the movements of populations that cross national boundary lines from those that are entirely within a country. To make this distinction, sociologists use the term **internal migration** for *movement within a nation's boundary lines,* in contrast to "immigration," by which boundary lines are crossed.

Since 1970, population growth in the United States has been greatest in the Sunbelt states, reflecting continued migration patterns toward the South and West. California, Texas, and Florida are growing significantly faster than the United States as a whole because they attract many Northeastern and Midwestern residents. There is some indication, however, that internal migration patterns might be starting to change and migration to the sunbelt slowing.

## ● THEORIES OF POPULATION

The study of population is a relatively new scholarly undertaking; it was not until the eighteenth century that populations as such were examined carefully. The first person to do so, and perhaps the most influential, was Thomas Malthus.

### Malthus's Theory of Population Growth

Malthus (1776–1834) was a British clergyman, philosopher, and economist who believed that population growth is linked to certain natural laws. The core of the population problem, according to Malthus, is that populations will always grow faster than the available food supply. With a fixed amount of land, farm animals, fish, and other food resources, agricultural production can be increased only by cultivating new acres, catching more fish, and so on—an additive process that Malthus believed would increase the food supply in an arithmetic progression (1, 2, 3, 4, 5, and so on). Population growth, by contrast, increases at a geometric rate (1, 2, 4, 8, 16, and so on) as couples have 3, 4, 5, and more children. (A stable population requires two individuals to produce no more than 2.1 children, two to reproduce themselves and 0.1 to make up for those people who remain childless.) Thus, if left unchecked, human populations are destined to outgrow their food supplies and suffer poverty and a never-ending "struggle for existence" (a phrase coined by Malthus that later became a cornerstone of Darwinian and evolutionary thought).

Malthus recognized the presence of certain forces that limit population growth, grouping these into two categories: preventive checks and positive checks. **Preventive checks** are *practices that would limit reproduction.* Preventive checks include celibacy, the delay of marriage, and such practices as contraception within marriage, extramarital sexual relations, and prostitution (if the latter two are linked with abortion and contraception). **Positive checks** are *events that limit reproduction either by causing the deaths of individuals before they reach reproductive age or by causing the deaths of large numbers of people, thereby lowering the overall population.* Positive checks include famines, wars, and epidemics. Malthus's thinking assuredly was influenced by the plague that wiped out so much of Europe's population during the fourteenth and fifteenth centuries.

Malthus refuted the theories of the utopian socialists, who advocated a reorganization of society to eliminate poverty and other social evils. Regardless of planning, Malthus assumed that population growth would lead to urban crowding, lower wages, infectious diseases, and other miseries. On the one hand is the constant threat that population will outstrip the available food supplies; on the other hand are the unpleasant and often devastating checks on this growth, which result in death, destruction, and suffering.

Malthus was wrong for two reasons. He adamantly rejected an obvious way of controlling population, namely, birth control. He was opposed to birth control on moral grounds and believed the only acceptable way to control population was later marriage. He was a clergyman for the Church of England, and that view was common at the time. He assumed that others would be as opposed to birth control as he was.

He was also wrong in that he had no understanding of how growth and the environment are related. Pollution, loss of space, and, obviously, global warming were never part of his ideas. Needless to say, what could be done to improve the situation never figured into his thinking either (Gilbert, 2005).

History proved Malthus wrong, at least for developed countries. Technological breakthroughs in the nineteenth century enabled Europe to avoid many of Malthus's predictions. The newly invented steam engine used energy more efficiently, and labor production was increased through the factory system. An expanded trade system provided raw materials for growing industries and food for workers. Fertility declined and emigration eased Europe's population pressures. By the end of the nineteenth century, Malthus and his concerns had been all but forgotten.

### Marx's Theory of Population Growth

Karl Marx and other socialists rejected Malthus's view that population pressures and the related miseries are inevitable. Marxists argue that the sheer number of people in a population is not the problem. Rather, they contend, it is industrialism (and, in particular, capitalism) that creates the social and economic problems associated with population growth. Industrialists need large populations to keep the labor force adequate, available, flexible, and inexpensive. In addition, the capitalistic system requires constantly expanding markets, which can be provided only by an ever-increasing population.

As the population grows, large numbers of unemployed and underemployed people compete for the few available jobs, which they are willing to take at lower and lower wages. Therefore, according to Marxists, the norms and values of a society that encourage population growth are rooted in its economic and political systems. Only by moving the political economy of industrial society in the direction of socialism, they contend, is there any hope of eliminating poverty and the miseries of overcrowding and scarce resources for the masses.

## Demographic Transition Theory

Sweden has been keeping records of births and deaths longer than any other country. Throughout the centuries, Swedish birth and death rates fluctuated widely through periods of rapid population growth followed by periods of slow growth and even population declines during famines. In the late 1800s, Sweden's death rate began a sustained decline, whereas the birthrate remained high. Eventually, Sweden's birthrate also declined, so that today its births and deaths are virtually in balance.

The shift in Sweden's population can be explained by a theory of population dynamics developed by Warren Thompson. According to the **demographic transition theory,** *societies pass through four stages of population change from high fertility and high mortality to relatively low fertility and low mortality.* During stage 1, high fertility rates are counterbalanced by a high death rate from disease, starvation, and natural disaster. The population tends to be very young, and there is little or no population growth. During stage 2, populations rapidly increase because of a continued high fertility that is linked to an increased food supply, development of modern medicine, and enhanced public health care. Slowly, however, the traditional institutions and religious beliefs that support a high birthrate are undermined and replaced by values stressing individualism and upward mobility. Family planning is introduced, and the birthrate begins to fall. This is stage 3, during which population growth begins to decline. Finally, in stage 4, both fertility and mortality are relatively low, and population growth once again stabilizes (see Figure 9-3).

As long as many developing nations remain in stage 2 of demographic transition (high fertility but falling mortality), they will continue to be burdened by overpopulation, which slows economic development and creates widespread severe hunger. Overpopulation undermines economic growth by disproportionately raising the **dependency ratio,** *the number of people of nonworking age in a society for every 100 people of working age.* Because populations at stage 2 have a high proportion of

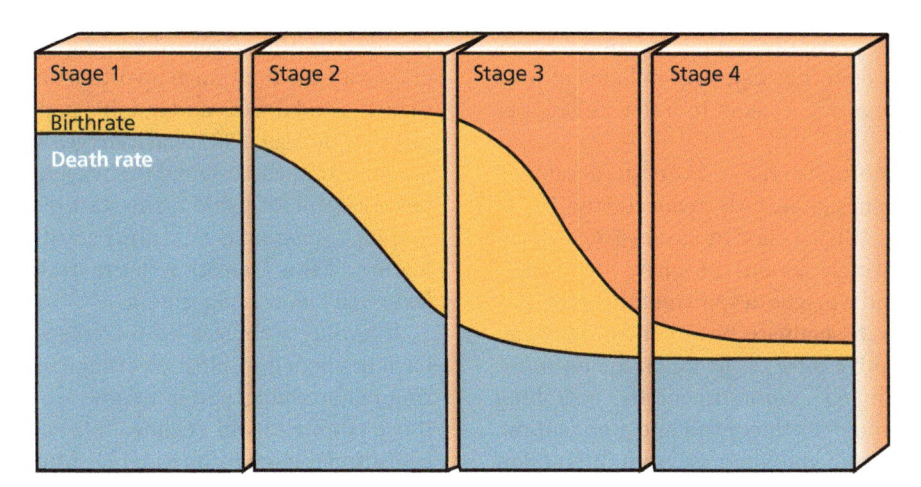

FIGURE 9-3 **The Demographic Transition Theory**

The demographic transition theory states that societies pass through four stages of population change. Stage 1 is marked by high birthrates and high death rates. In stage 2, populations rapidly increase as death rates fall, but birthrates stay high. In stage 3, birthrates begin to fall. Finally, in stage 4, both fertility and mortality rates are relatively low.

children as compared with adults, they have fewer able-bodied workers than they need. For example, 48 percent of Uganda's population is below the age of 15, compared to 20 percent in the United States (Population Reference Bureau, 2011). The economic development of countries with high dependency ratios is slowed further by the channeling of capital away from industrialization and technological growth and toward mechanisms for feeding the expanding populations.

*Applications to Industrial Society* The first wave of declines in the world's death rate came in countries experiencing real economic progress. Those declines gradually gained momentum as the Industrial Revolution proceeded. Advances in agriculture, transportation, and commerce made it possible for people to have a better diet, and advances in manufacturing made adequate clothing and housing more widely available. A rise in people's real income facilitated improved public sanitation, medical science, and public education.

Although the preceding explanation applies well to Western society, it does not explain the population trends in the less developed areas of today's world. Since 1920, those areas have experienced a much faster drop in death rates than Western societies without a comparable rate of increase in economic development. The rapid rate of decline in the death rate in these countries has been due primarily to the application of medical discoveries made in and financed by the industrial nations. For example, the most important death threat being eliminated is infectious disease. Those diseases have been controlled through the introduction of vaccines, antibiotics, and other medical advances developed in the industrial nations. Those who are accustomed to paying high prices for private medical care will find it hard to believe that preventive public health measures in underdeveloped countries can save millions of lives at costs ranging from a few cents to a few dollars per person.

Because the mortality rates in underdeveloped countries have been significantly reduced, the birthrate, which has not fallen as fast or as consistently, has become an increasingly serious problem. Often, this problem persists and worsens despite monumental government efforts to disseminate birth control information and contraceptive devices. In India, for example, despite the government's commitment to controlling population size through birth control and sterilization, the population is expected to increase from 717 million in 1982 to 1.3 billion in 2016. With the current rate of growth, India could surpass China as the most populated country sometime between 2030 and 2035 (Population Reference Bureau, 2008).

These failures have shown that the birthrate can be brought down only when attention is paid to the complex interrelationships of biological, social, economic,

Immigration has changed the face of many areas in New York City and Los Angeles.

political, and cultural factors. For example, a study in Pakistan revealed that even among women who already had six children, if all six were daughters, there was a 46 percent chance that the mother would want more children. If, however, the six children were boys, there was only a 4 percent chance that she would want additional children (Weeks, 1994).

## A Second Demographic Transition

At the beginning of the chapter, we mentioned that many areas of the world are experiencing a potential decline in population. The original demographic transition theory ends with the final stage involving an equal distribution of births and deaths and a stable population. Some people (van de Kaa, 1987) have suggested that Europe has gone beyond the original theory and entered a second demographic transition. The start of this second demographic transition is arbitrarily set at 1965; the principal feature of this transition is the decline in fertility to a level well below replacement.

If fertility stabilizes below replacement, as seems to have happened in most of Europe and Japan, and barring major changes in immigration, the populations of those countries will decline. This second demographic transition is already the case in Estonia, Latvia, Hungary, Bulgaria, the Ukraine, and Italy as well as in most of Eastern Europe (Population Reference Bureau, 2012).

The United States is not expected to experience a population decline in the near future. United Nations population projections to the year 2025 indicate that the United States will continue to grow modestly, even with continued low fertility, because of immigration levels.

## GLOBAL DIVERSITY

The world population, reached 7 billion in 2011 and has more than doubled since 1960. It is projected to be 9.3 billion by 2050. In the more-developed countries, the average fertility rate is about 1.7 births per woman—below the replacement level of 2.1 births. In the least-developed countries, the rate is about 4.5, with sub-Saharan Africa having a rate of 5.2. Worldwide, however, fertility rates have been gradually dropping since the middle of the last century. Reductions in fertility speed economic growth and reduce poverty.

Even with gradual global fertility declines, however, about 80 million people are added to the world each year, a number roughly equivalent to the population of Germany or Ethiopia. Population growth continues today because of the high numbers of births in the 1950s and 1960s, which produced millions of young people reaching their reproductive years over succeeding generations (2011 World Population Data Sheet).

The world's richest countries, with 20 percent of the global population, account for 86 percent of private consumption; the poorest 20 percent account for just 1.3 percent. A child born in an industrialized country will add more to consumption and pollution over his or her lifetime than 30 to 50 children born in developing countries (United Nations, 2012, *State of World Population 2011*). (For an example of what life is like in one of these countries, see "Our Diverse World: Life Chances of an Adolescent Girl in Liberia.")

Throughout the world, some 2 billion people are malnourished, one-quarter lack adequate housing, 20 percent do not have access to modern health services, and 20 percent of children do not attend school. In the next section, we will look at some of the vast differences in everyday life that exist throughout the world.

## World Health Trends

The World Health Organization defines health as "a state of complete mental, physical, and social well-being." This concept might appear straightforward, but it does not easily lend itself to measurement. Consequently, to describe the state of health in the world, we must look at trends. When we look at human health from this perspective, we find that the twentieth century has seen unprecedented gains in health and survival. On a worldwide basis, the average life expectancy for a newborn more than doubled, from 30 years in 1900 to 70 years in 2011 (Population Reference Bureau, 2012). For a country such as China, this has meant moving from conditions at the turn of the century, when scarcely 60 percent of newborns reached their fifth birthday, to the present, when more than 60 percent can expect to reach their seventieth birthday.

Health advances in some countries have reached the point at which it appears the population is approaching the upper limit of average life expectancy. In Japan, where life expectancy is 83 years, a newborn has only a 4 in 1,000 chance of dying before its first birthday and less than a 1 in 1,000 risk of dying by age 40.

Unfortunately, the same cannot be said for less-developed countries. More than 300 million people live in 24 countries where life expectancy is less than 50 years. In these countries, 1 of 10 newborns die by age 1, and 3 million a year do not survive for one week. In some African villages, deaths among infants and young children occur ten times more frequently than deaths among the aged.

Currently, 80 percent of the world's population does not have access to any health care. Malnutrition and parasitic and infectious diseases are the principal causes of death and disability in the poorer nations. The problems these diseases cause are largely preventable, and most of these conditions could be dramatically reduced at a relatively modest cost.

---

### OUR DIVERSE WORLD: Life Chances of an Adolescent Girl in Liberia

- Has probably not been to primary school
- Is at high risk of being illiterate
- Has a high risk of suffering rape
- Has a high probability of being married before the age of 18
- Has a 1 in 12 chance that she will die from pregnancy and childbirth
- Has a 1 in 10 chance that her child will die within the first year of life
- Will probably not have adequate support from her husband

*Source:* Gayflor, Vabah. 2009. "The Challenges Faced by Adolescent Girls in Liberia." In United Nations Children's Fund (UNICEF). February 2012. *The State of the World's Children 2012.* New York.

According to Thomas Malthus, population will always grow faster than the available food supply. Throughout the world many people depend on food banks to survive.

## The Health of Infants and Children in Developing Countries

When we look at infant and child health from a global perspective, we find that death among children is overwhelmingly a problem of the developing countries in Africa, Asia, and Latin America. Those countries account for 98 percent of the world's deaths among children younger than five. To make matters worse, UNICEF estimates that 95 percent of these deaths are preventable. Even though child mortality has fallen, more than 10.5 million children die each year before age five from preventable diseases.

*Child Killers* The preventable and treatable diseases children die from include diarrheal dehydration, acute respiratory infection, measles, and malaria. In half of the cases, illness is complicated by malnutrition. In addition, 30 million of the world's children are still not routinely vaccinated and continue to die as a result of major childhood killers such as diphtheria, tuberculosis, pertussis, measles, and tetanus.

Although infectious diseases account for only 1 percent of deaths in developed countries, they are the major killers in the developing world, accounting for a staggering 41.5 percent of all deaths. Malaria, which is extremely rare in developed countries, accounts for nearly 1 million deaths annually. (See Table 9-4 for a comparison of infant mortality throughout the world.)

Pneumonia is the leading cause of death in children worldwide. Pneumonia kills an estimated 1.4 million children under the age of five years every year—more than AIDS, malaria, and tuberculosis combined. Pneumonia accounts for 18 percent of all deaths of children under five years old worldwide. Pneumonia can be caused by viruses, bacteria, or fungi. The majority of these deaths could be prevented by a five-day treatment course of antibiotics costing 25 cents. Only 30 percent of children with pneumonia receive the antibiotics they need. Pneumonia also can be prevented by immunization, adequate nutrition, and by addressing environmental factors (World Health Organization Fact Sheet, October 2011).

Dirty drinking water and general unsanitary conditions bring on diarrhea, causing the body to lose large quantities of salt and water. Progress in addressing this condition has been made because of oral rehydration therapy.

In the countries of sub-Saharan Africa and South Asia, measles continues to be a major killer of children under five, although now 85 percent of the world's children are vaccinated against the disease. Measles is a highly contagious disease, and to control it, at least 90 percent of the children must be vaccinated. High doses of vitamin A also help with the disease. Children who get the disease and do not die are often left blind and deaf.

Malaria has reemerged as a major cause of child deaths. The disease causes severe anemia and is one of the main causes of low birth weight in infants. Malaria can be easily prevented by making sure that pregnant women and children sleep under insecticide-treated mosquito nets. This relatively simple action could prevent 400,000 child deaths in Africa each year. Yet, few children sleep under mosquito nets, and those nets that are used have not been treated.

TABLE 9-4  Best and Worst Infant Mortality Rates (Deaths per 1,000 Live Births)

| Best | Rate | Worst | Rate |
| --- | --- | --- | --- |
| Iceland | 0.9 | Afghanistan | 129 |
| Hong Kong | 1.3 | Chad | 128 |
| Singapore | 2.03 | Sierra Leone | 109 |
| Japan | 2.3 | Somalia | 107 |
| Finland | 2.4 | Guinea Bissau | 103 |

*Source:* Haub, Carl and Toshiko Kaneda. July 2012. "2012 World Population Data Sheet" (www.prb.org/pdf12/2012-population-data-sheet_eng.pdf), accessed July 19, 2012.

A child deficient in vitamin A is 25 times more likely to die of treatable diseases such as measles, malaria, and diarrhea. Vitamin A deficiency can also lead to irreversible blindness. Vitamin A is crucial in developing resistance to infection and in preventing anemia and night blindness, but poor families can rarely afford foods rich in this vitamin, such as meat, eggs, fruits, red palm oil, and green leafy vegetables. Many countries now fortify staple foods such as flour and sugar with vitamin A and other important micronutrients. Many children in poor countries are now receiving high-dose vitamin A capsules, which cost only a few cents per year.

Iodine deficiency, which results from insufficient iodized salt in the diet, is the world's single greatest cause of preventable mental retardation. Iodine deficiency can also produce stillbirth and miscarriage in women. During pregnancy, even mild iodine deficiency can hamper learning ability. More than 40 million newborns are still unprotected from learning disabilities, impaired speech, and hearing development problems because of iodine deficiency.

Yet, there are some success stories. Smallpox has been eradicated, and the World Health Organization believes that polio is on the verge of being eradicated, with only Afghanistan, Nigeria, and Pakistan having outbreaks. These success stories, however, are the exceptions.

*Maternal Health* The survival of infants and children depends critically on the nurturing the mother provides during pregnancy and childhood. When an expectant mother suffers from sexually transmitted illnesses, malaria, or malnutrition during pregnancy, she will be more likely to have a low-birth-weight baby. Low-birth-weight babies are defined as those weighing less than 5 pounds at birth. Newborns of low birth weight are more likely to die. Those who survive are at risk of having impaired immune systems and susceptibility to chronic illnesses. They are also likely to have lower intelligence and cognitive disabilities leading to learning difficulties. The majority of low-birth-weight babies are born in either South Asia or sub-Saharan Africa.

In addition to maternal malnutrition, other causes of low-birth-weight babies are the mother's excessively heavy workload, malaria infection, severe anemia related to an iron-deficient diet, and hookworm infestation.

*Maternal Age* An infant is at higher health risk if the mother is in her teens or older than 40, if she has given birth more than seven times, or if the interval between births is less than two years. The interval between births is the most important factor in infant and child mortality. Infants born to mothers at less than two-year intervals between births are 80 percent more likely to die than children born at birth intervals of two to three years.

This information suggests that family planning programs that discourage early childbearing can substantially reduce infant and child mortality by preventing births to high-risk mothers. When childbearing is spread out, the woman has more time and energy to devote to her own health and the care of her other children.

*Maternal Education* Children's chances of surviving improve as their mother's education increases. A mother's education affects her child's health in a number of ways. Better-educated mothers know more about good diet and hygiene. They are also more likely to use maternal and child health services—specifically, prenatal care, delivery care, childhood immunization, and other therapies

## HIV/AIDS

Approximately 34 million people worldwide are infected with the human immunodeficiency virus (HIV), the virus that causes AIDS, which has claimed 25 million lives since its identification in 1981. A growing gap is developing between the outcomes of infection with HIV in the developed world and in developing countries because, in North America, Western Europe, Australia, and New Zealand, antiretroviral drugs have reduced the speed at which HIV-infected people develop AIDS. In 2010, around 6.6 million people living with HIV were receiving antiretroviral therapy in low- and middle-income countries, but more than 7 million others were waiting for access to treatment.

Around the world, the effect of the epidemic varies. East Asia and the Pacific Rim countries are still keeping the epidemic at bay. Infection rates are lower, but the numbers are still large. More than half the world's population lives in the region, and even though the percentage of the population with HIV might be lower, the absolute number of people is very large. With some exceptions, such as sub-Saharan Africa, the bulk of new infections have been spread through sex workers, intravenous drug use, and men who have sex with men. One of the exceptions is China, where although "almost all cases of HIV/AIDS were previously transmitted through intravenous drug use and unsafe blood practices, the epidemic is now spreading through heterosexual contact" (UNAIDS, AIDS Epidemic Update 2011).

The epidemic is moving the fastest in sub-Saharan Africa. The region has about 10 percent of the world's population but fully 67 percent of people living with HIV in the world. Sub-Saharan Africa is faced with a tri-fold challenge, "providing care for the growing population of people infected with HIV, bringing down the number of new infections through more effective prevention, and coping with the effect that 20 million deaths has had on the continent" (UNAIDS, 2011).

The HIV epidemic is also growing in the Russian Federation and Eastern Europe. Previously, the disease was spread primarily through intravenous drug use, but now nearly 25 percent of new infections occur through sexual contact between heterosexuals. The disease is rapidly spreading from a select group to the wider population (UNAIDS, 2011).

In India, 2.5 to 3 million people are infected (UNAIDS, 2011; AIDS Epidemic Update, 2011). In Latin America and the Caribbean, Haiti shows the highest prevalence of the disease. Intravenous drug use, as well as unsafe heterosexual and homosexual sex, complicates the eradication of the disease. And although awareness of the epidemic is on the rise in the Caribbean, the region continues to accrue large numbers of new infections.

In the United States, African Americans comprise about 12 percent of the population but 50 percent of those with HIV/AIDS. Black women in the United States are 23 times more likely to be infected than white women. AIDS is the leading cause of death among black women aged 25 to 34 years old, and the second leading cause of death among black men aged 35 to 44 (Chase, 2008).

AIDS does not just affect the people with the disease; it also has a major impact on families. The disease of AIDS is shattering family structures. It has been estimated that since the beginning of the epidemic, more than 15 million children have lost one or both parents to AIDS. Increasingly, elderly grandparents and older children are the sole caretakers of scores of younger children (UN Population Newsletter, June 2005). In addition, as life expectancies fall, the social safety net provided by extended families breaks down, health and education systems falter, and businesses and governments lose their most productive staff.

Even though it appears that in the United States medical science is finding ways to control the spread of the disease, for the rest of the world, the picture is not as optimistic. The vast majority of people living with HIV live in the developing world where access to the new antiretroviral drugs is difficult or impossible.

### Population Trends

Global stratification is greatly influenced by population growth. During the first 2 million to 5 million years of human existence, the world population never exceeded 10 million people. The death rate was about as high as the birthrate, so there was no population growth. Population growth began around 8,000 BC, when humans began to farm and raise animals. In AD 1650, an estimated 510 million people lived in the entire world. One hundred years later, there were 710 million, an increase of some 39 percent. By 1900, there were 1.6 billion. Only 100 years later, the world population had spiraled to 6.08 billion, with 131.4 million people added each year (Population Reference Bureau, 2000).

Population growth has continued throughout the past three decades despite the decline in fertility rates that began in many developing countries in the late 1970s and, surprisingly enough, despite the toll taken by the HIV/AIDS epidemic. Although the rate of increase is slowing, in absolute terms, world population growth continues to be substantial. According to U.S. Census Bureau projections, world population will increase to a level of nearly 8 billion people by the end of the next quarter century and will reach 9.3 billion—a number almost half again as large as today's total—by 2050. Ninety-nine percent of global natural increase—the difference between numbers of births and deaths—occurs in the developing world.

Right now, the world population is doubling about every 58 years. If it continues to expand this way, it will quadruple within 116 years—a situation in which widespread poverty and famine are possible. In recent years, however, a small but significant slowing in the rate of world population growth has occurred. Between 1965 and 1970, the annual rate of growth was 2.1 percent, but by 2008, it had dropped to 1.2 percent (Population Reference Bureau, 2008). If this slowing trend continues, the world growth rate will be far smaller than previously predicted. This more hopeful pattern is contingent on the average family size being limited to two children. There are still significant differences in fertility in various parts of the world.

How many children a family has will affect that family's lifestyle in both developed and undeveloped countries. Many factors determine the typical family size in a country. In this section, we will examine some of those factors to gain an insight into the variety of economic, social, and psychological issues that have to be addressed when trying to limit population growth. (See Table 9-5 for a summary of some of the determinants of fertility.)

**TABLE 9-5** Factors Involved in Fertility Decisions

**Number of Children**

1. Women's average age at first marriage
2. Breast-feeding
3. Infant mortality

**Demand for Children**

1. Gender preferences
2. Value of children
    a. Children as insurance against divorce
    b. Children as securers of women's position in family
    c. Children's value for economic gain
    d. Children's value for old-age support
3. Cost of children

**Fertility Control**

1. Use of contraception
2. Factors influencing fertility decisions
    a. Income level
    b. Education of women
    c. Urban or rural residence

*Child Marriage and Early Marriage* Child marriage began as a way to protect unwelcome sexual advances and to gain economic security. That purpose is no longer being served, and child marriage often means the girl will have a life of sexual and economic servitude. The second-class status of women is both the cause and the result of child marriage.

A young bride is expected to have a child quickly after marriage. Her worth as a wife and daughter is determined by the early birth. In addition to the health risks of child marriage, it also means there will be little or no formal education for the girls.

Early marriage provides more years for conception to occur. It also decreases the parents' years of schooling and limits their employment opportunities. In South Asia and sub-Saharan Africa, nearly a quarter of all women between ages 15 and 19 are or have been married (see Figure 9-4). In areas of western and eastern Africa and South Asia, many marriages take place even before puberty.

To limit fertility, some countries have tried to establish a minimum age for marriage. In India, 65 percent of girls are married by the time they are 18. To address the problem, in 2006, India passed a law banning child marriage. In 1980, China raised the legal minimum marriage ages to 20 for women and 22 for men; it is one of the few countries that have been successful in raising the average age of marriage. When Chinese youth were asked what would be the ideal age for their own marriage, the responses averaged 24.3 years for women and 25.4 for men (Yu, 1995).

Death rates exceed birthrates in certain more-developed countries. As the growth rate in the world's more affluent nations comes to an end, *all* of the net annual gain in global population will, in effect, come from the world's developing countries. The issue is underscored by the increasing speed with which the world's population is multiplying.

*Breast-Feeding* Breast-feeding delays the resumption of menstruation and therefore offers limited protection against conception. In developing countries, it is much safer for infants to be breast-fed during the first six months of life than to be bottle-fed. Bottle-fed babies are more likely to experience respiratory infections or diarrhea than are breast-fed babies. These risks have produced an outcry against certain large companies that produce powdered milk and that have been trying to encourage bottle-feeding in less-developed countries. Bottle-feeding has been introduced into developing countries by hospitals that are trying to follow Western practices as well as by commercial concerns that aggressively promote breast-milk substitutes. The World Health Organization and UNICEF have tried to combat this trend with limited success.

In Bangladesh, Pakistan, Nepal, and most of sub-Saharan Africa, where very few women use contraception, breast-feeding is one of the few controls on fertility rates. In the United States, breast-feeding was relatively unpopular until the late 1960s. As the advantages of breast-feeding became known, however, better-educated women began to use it more. Today, college-educated women are the most likely to start breast-feeding and continue it for the longest periods.

*Infant and Child Mortality* High infant mortality promotes high fertility. Parents who expect some of their children to die might give birth to more babies than they really want as a way of ensuring a family of a certain size. This sets in motion a pattern of many children born close together, weakening both the mother and babies and producing more infant mortality.

In the short term, the prevention of ten infant deaths can produce only one to five fewer births. Initially, lower infant and child mortality will lead to somewhat larger families and faster rates of population growth. However, the long-term effects are most important.

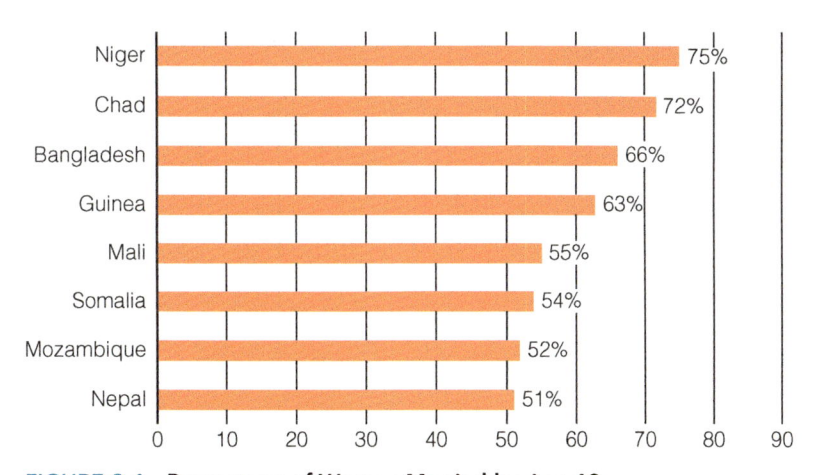

| Country | Percentage |
|---|---|
| Niger | 75% |
| Chad | 72% |
| Bangladesh | 66% |
| Guinea | 63% |
| Mali | 55% |
| Somalia | 54% |
| Mozambique | 52% |
| Nepal | 51% |

FIGURE 9-4 **Percentage of Women Married by Age 18**
*Source: United Nations Children's Fund (UNICEF). February 2012. The State of the World's Children 2012. New York.*

With improved chances of survival, parents devote greater attention to their children and are willing to spend more on their children's health and education. Eventually, lower mortality rates help parents achieve the desired family size with fewer births and lead them to want a smaller family.

Education and child health go hand-in-hand. Infant mortality rates drop as the mother's education level goes up. The mother's education also has an effect on a child's health and nutrition. Children whose mothers complete secondary education are much less likely to be underweight for their age than those whose mothers have less education. (See Table 9-6 for countries with the highest and lowest lifetime births per woman.)

*Gender Preferences* The effect of gender preference on fertility is actually more complicated than might appear at first glance. In most countries, there is a strong preference for male children. Logically, this does not make much sense. In most underdeveloped countries, daughters typically help their mothers with household chores. One would think this would increase their worth to the family. Clearly, there must be countervailing factors to reduce the desire for daughters.

Three sets of factors influence the desire for male children. They include

1. economic factors—the value assigned to women's work and the ability to contribute to family income
2. social factors—kinship, marriage patterns, and religion
3. psychological factors—influences on parents' decisions about size and composition of the family (Hudson and den Boer, 2004)

The preference for sons, however, might not be all that important in countries with high fertility rates. In countries with declining fertility rates, the preference for sons might cause the fertility rate to plateau above replacement level because couples might continue to have children until they have the desired son.

Girls have a better survival rate than boys, and women normally live longer than men. Even with the natural compensation of more male than female births, this still means that most nations have more women than men.

In a number of developing countries however, there are far fewer women than men. This is not what would normally be expected and is often due to willful acts or discriminatory customs. One explanation would be lower survival rates for girls—either because of female infanticide or because girls receive less health care than boys. It also is related to the estimated half a million women who die each year of pregnancy and birth complications. Unsafe abortions kill another 100,000 each year. Other factors include widow burnings, in which a woman is burned when her husband dies, and dowry deaths, in which family disputes over the dowry result in the woman being killed. Unfortunately, as appalling as these events sound, they occur regularly in some countries.

In some countries, fewer baby girls are born than would be expected. Figure 9-5 shows some of the countries with significantly fewer baby girls born than expected. In China, 118 boys are born for every 100 girls. Worldwide data show that on average about 105 to 107 boys are born for every 100 girls, implying that many Chinese baby girls (about 1.7 million) are missing annually (Hudson and den Boer, 2004). This does not mean that they are killed. Some are not reported to the authorities; others are adopted informally by friends or relatives (Population Data Sheet, 2012).

The major reason for the missing girls seems to be technology. As early as 1979, China began manufacturing ultrasound scanners. The scanners are meant to help doctors check whether fetuses are developing normally, but they can also be used to determine whether a fetus is male or female. Combination ultrasound/abortion clinics are common in China, South Korea, India, and Afghanistan. (See "Thinking About Social Issues: Where Are the Baby Girls?").

We would be wrong to assume that a strong preference for a son is limited just to less-developed

TABLE 9-6 Countries with the Highest and Lowest Fertility

| Lifetime Births per Woman ||||
|---|---|---|---|
| **Highest** | | **Lowest** | |
| Niger | 7.1 | Taiwan | 1.1 |
| Burundi | 6.4 | Latvia | 1.1 |
| Somalia | 6.4 | Singapore | 1.2 |
| Zambia | 6.3 | South Korea | 1.2 |
| Uganda | 6.2 | Japan | 1.4 |

Source: Haub, Carl and Toshiko Kaneda. July 2012. "2012 World Population Data Sheet" (www.prb.org/pdf12/2012-population-data-sheet_eng.pdf), accessed July 19, 2012.

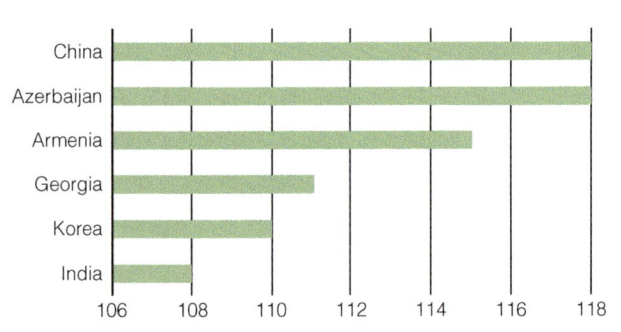

FIGURE 9-5 Countries with Fewer Female Babies than Expected (Male Babies per 100 Female Babies)
Source: United Nations, Population Division. April 2011. "World Population Prospects: The 2010 Revision" (http://esa.un.org/unpd/wpp/Excel-Data/fertility.htm), accessed July 17, 2012.

## THINKING ABOUT SOCIAL ISSUES

### Where Are the Baby Girls?

In many countries throughout the world, fewer baby girls are being born than expected. The issue seems to be very common throughout China and India. Ultrasound technology was introduced in those countries in 1986, and since then, it has become easy to find out the sex of the child before birth and have an abortion if it is a daughter.

In China, there are 32 million more boys than girls under the age of twenty. In the 1970s, the Chinese government instituted a strict one-child-per-family policy to prevent rapid population growth. At the same time, there is a long-standing Chinese preference for a male heir. In some parts of China, a second child is permitted if the first is a girl or if there is an unspecified hardship.

In a patriarchal society such as China, a daughter's responsibility to care for her parents lasts only until she is married. At that point, she becomes part of her husband's family and tends to her in-laws. A son, on the other hand, must care for his parents for life. Without a government-sponsored social security system, parents worry about who will support them in old age.

India is another patriarchal society that has seen a reduction in baby girls being born. A girl's parents must pay a large dowry when she is married, which has the potential to bankrupt many poor families. Consequently, the pressure is great to have a male child. Gender-based abortions have been illegal since 1994 but continue to occur, leading to 10 million fewer girls being born over the past two decades.

*Sources:* LaFraniere, Sharon. April 11, 2009. "Chinese Bias for Baby Boys Creates a Gap of 32 Million," *New York Times* (www.nytimes.com/2009/04/11/world/asia/11china.html?ref=global-home&pagewanted=print), accessed April 15, 2009; Associated Press. July 15, 2007. "India Tries to Stop Sex-Selective Abortions." *New York Times* (www.nytimes.com/2007/07/15/world/asia/15india.html), accessed April 4–15, 2009; CIA. *The 2008 World Factbook* (www.cia.gov/library/publications/the-world-factbook/fields/2018.html), accessed May 20, 2009.

---

countries. A recent Gallup poll showed that the American preference for male children is also quite strong. "Forty-two percent of Americans say they would prefer a boy if they could have only one child, while 27 percent said they would prefer a girl. One-quarter of Americans said it would not matter to them either way. Younger people between 18 and 29 are actually more likely to prefer a boy than those 65 and older" (Gallup, December 2000).

*Benefits and Costs of Children* In underdeveloped countries, the benefits for an individual family of having children have generally been greater than the costs. However, those costs and benefits change with the second, third, fourth, and fifth children.

For a rural sample in the Philippines, three-quarters of the costs involved in rearing a third child come from buying goods and services; the other quarter comes from costs in time (or lost wages). Receipts from child earnings, work at home, and old-age support offset about half of the total. By contrast, in the United States, almost half the costs of a third child are time costs. Receipts from the child's work offset only a tiny fraction of all costs.

Only economic costs and benefits are taken into account in these calculations. To investigate social and psychological costs, other researchers have examined how individuals perceive children. Economic contributions from children are clearly more important in the Philippines, where fertility is higher than in the United States. Concern with the restrictions children impose on parents is clearly greatest in the United States. The value of children changes with family growth and the economic development of the country. In the United States, the first child is important to cement the marriage and bring the spouses closer together as well as to have someone to carry on the family name if it is a boy. Thinking of the first child, couples also stress the desire to have someone to love and care for and the child's bringing play and fun into their lives.

In considering a second child, parents emphasize more the desire for a companion for the first child. They also place weight on the desire to have a child of the opposite sex from the first. Similar values are prominent in relation to third, fourth, and fifth children; emphasis is also given to the pleasure derived from watching children grow.

Throughout most of the world, economic considerations emerge again. After the fifth child, parents speak of the sixth child or later children in terms of their assistance around the house, contribution to the support of the household, and security during the parents' old age. For first to third children, the time taken away from work or other pursuits is the main drawback; for fourth and later children, the direct financial burden is more significant than the time costs.

*Contraception* Apart from the factors already mentioned, fertility rates eventually are tied to the increasing use of contraception. Contraception is partly a function of a couple's wish to avoid or delay having children and partly related to costs. People have regulated family size for centuries through abortion, abstinence, and even infanticide. In many countries, the costs of preventing a birth, whether economic, social, or psychological, might be greater than the risk of having another child.

Use of contraception varies widely from 19 percent or fewer for married women in almost all of sub-Saharan Africa to between 70 percent and 80 percent for women in Europe, Asia, and the United States (World Population Data Sheet, 2012).

Contraception is most effective when such programs are publicly subsidized. Not only do the programs address the economic costs of spreading contraception, but they also help communicate the idea that birth control is possible. In addition, these programs offer information about the private and social benefits of smaller families, which helps reduce the desired family size.

The support given to family planning programs differs dramatically from country to country. At one end of the family planning spectrum are the governments of India and China, which provide birth control information and devices and actively support abortion. At the other end are predominantly Muslim countries, such as Bangladesh, or predominantly Roman Catholic countries whose populations follow the anti-birth-control and antiabortion teachings of the religion.

Worldwide, abortion is the most widely used form of birth control, and it is common even when it is illegal. The demand for abortions is rising, and it has been estimated that, throughout the world, one in three pregnancies ends in abortion. Abortion is legal in the world's three most populous countries (China, India, and the United States) as well as in Japan and all of Europe except Belgium and Ireland, which have substantial restrictions. In Russia, where contraceptives are hard to find, there are more abortions than births. Recently Russia has made abortion more difficult in an attempt to combat the high abortion rate (Kishkovsky, 2011).

*Income Level* It is a well-established fact that people with higher incomes want fewer children. Alternative uses of time such as earning money, developing or using skills, and pursuing leisure activities become more attractive, particularly to women. The children's economic contributions become less important to the family welfare because the family no longer needs to think of children as a form of social security for old age. It is not the higher income itself but, rather, the life change it brings about that lowers fertility.

This relationship between income and fertility holds true only for those with an income above a certain minimum level. If people are extremely poor, increases in income will actually increase fertility. In the poorest countries in Africa and Asia, families are often below this threshold. Above the threshold, though, the greatest fertility reduction with rising income occurs among low-income groups.

*Education of Women* In nearly all societies, the amount of education a woman receives affects the number of children she has. Fertility levels are usually the lowest among the most highly educated women within a country. Despite this information, only about half of girls in the least-developed countries stay in school past fourth grade. In many African nations and Asian nations, school enrollment for girls is less than 80 percent that for boys. Two-thirds of the world's illiterate people are women (United Nations, March–May 2002).

Studies also show that level of education affects the fertility of women more than that of men. There are a number of reasons for this. In most instances, children have a greater effect on the lives of women, in terms of time and energy, than they do on the lives of men. The more educated a woman is, the more opportunities she might encounter that conflict with having children. Education also appears to delay marriage, which in itself lowers fertility.

Women with seven or more years of education marry at least 3.5 years later than do women with no education. Educated women are also more likely to know about and adopt birth control methods. Every additional year of schooling lowers desired family size by more than 0.1 children. (For another view on population growth, see "Thinking About Social Issues: What If the Population Problem Is Not Enough People?")

*Urban or Rural Residence* Rural lifestyles that involve farming and herding tend to produce high birthrates because children are regarded as contributing workers. Urban fertility tends to be lower in developing countries than rural fertility. Urban dwellers usually have access to better education and health services, a wider range of jobs, and more avenues for self-improvement and social mobility than do their rural counterparts. They are exposed to new consumer goods and are encouraged to delay or limit childbearing to increase their incomes. They also face higher costs in raising children. As a result, urban fertility rates are usually one to two children lower than rural fertility rates.

The urban woman marries, on the average, at least 1.5 years later than the rural woman does. She is more likely to accept the view that fertility should be controlled, and the means for doing so are more likely to be at her disposal.

Population growth is producing an increase in the percentage of people who live in urban environments.

*Global Aging* Population aging is having major consequences and implications in all areas of day-to-day human life, and it will continue to do so. Global aging will affect economic growth, savings, investment and consumption, labor markets, pensions, taxation, and the transfers of wealth, property, and care from one generation to another. The numerical and proportional growth of older populations around the world is indicative of major achievements—decreased fertility rates, reductions in infant and maternal mortality, reductions in infectious and parasitic diseases, and improvements in nutrition and education—that have occurred, although unevenly, on a global scale.

The rapidly expanding numbers of older people represent a social phenomenon without historical precedent. Worldwide, the number of persons aged 60 years or older will increase dramatically to nearly 1.9 billion by 2050. At that time, the number of people older than 65 in the world will exceed the number of young for the first time in the history of humankind.

In most countries, the elderly population is growing faster than the population as a whole. Almost half of the world's elderly live in China, India, the United States, and the countries of the former Soviet Union. China alone is home to more than 20 percent of the global total.

The oldest old (85 and older) are the fastest-growing segment of the population in many countries worldwide. Unlike the elderly as a whole, the oldest old today are more likely to live in developed than developing countries, although this trend, too, is changing.

For the time being, population aging has been a major issue mainly in the industrialized nations of Europe, Asia, and North America. For many of these countries, 15 percent or more of the entire population is 60 and older. Those nations have experienced intense public debate over elder-related issues such as how money will be found to pay for social security and health care costs.

---

## THINKING ABOUT SOCIAL ISSUES

### What If the Population Problem Is Not Enough People?

The birthrate is declining throughout the world. For a country to maintain its population, the average woman must have 2.1 children. This is known as the total fertility rate (TFR). The United States is the only developed country that is close to that number. Every other developed country is below 2.1 and consequently losing population. For example, Western European countries have low fertility rates, below the replacement rate of 2.1:

- Germany: 1.4 (total population: 81.9 million, of which 8.2% are foreigners)
- Holland: 1.8 (total pop.: 16.5 million, of which 4.4% are foreigners)
- Belgium: 1.8 (total pop.: 10.8 million, of which 9.8% are foreigners)
- Spain: 1.4 (total pop.: 46.1 million, of which 12.4% are foreigners)
- Italy: 1.4 (total pop.: 60.2 million, of which 7.1% are foreigners), birth control views notwithstanding
- Sweden: 1.9 (total pop.: 9.4 million, of which 6.4% are foreigners), which is a high TFR (the country provides deep support for parents), but still below the replacement rate
- Ireland and the U.K.: 2.07 and 1.94, respectively (total pop.: 4.7 million and 63.2 million, respectively), which are high TFRs, but these rates are derived from non-European immigrant parents

These low birthrates produce a number of changes in society. First, the median age of the population increases as older people outnumber the young. Second, the countries start to shrink. By 2050, Europe will have 100 million fewer people than in 2005. Japan will lose a quarter of its population during that time.

*(Continued)*

As the average age of the population increases, there are fewer workers and many more people who need to be supported through pension programs. The potential for economic problems also increases. Europe and Japan have experienced prolonged periods of economic stagnation with only limited prospects of recovery. An older population also increases the demands on the health care system and puts strains on the government and those who have to pay the increased costs.

The national identity and the political influence of the country that is losing population also suffer. Although Europe continues to think of itself as a major force in world politics, that idea can no longer be maintained with a substantial population decline.

There are a few things countries can do to deal with population decline:

1. Adopt policies that would encourage people to have children, perhaps through tax incentives or child-care benefits. France has a number of policies that encourage families to have children and consequently has the highest fertility rate in Europe.

2. Loosen up immigration policies so that more immigrants enter the country and help the population grow. To a certain extent, this is the case for the United States. European countries have been more wary of immigrants than the United States and worry about such a policy because it could threaten their national identities. Yet, some have been looking to immigrants to bolster their populations, as noted previously.

3. Raise the retirement age and limit pension and health benefits to the elderly. This would produce a significant outcry from those affected.

The continuing population decline in the developed world will raise a whole new set of population issues that are at odds with those of the developing world.

*Sources:* Population Reference Bureau. March 2004. *Transitions in World Population* 59(1); Longman, Phillip. 2004. *The Empty Cradle: How Falling Birthrates Threaten World Prosperity and What to Do About It.* New York: Basic Books; *Forbes* (www.forbes.com/forbes/2012/0507/current-events-population-global-declining-birth-rates-lee-kuan-yew_print.html).

## SUMMARY

### How does social stratification take place?

Social stratification can happen in two ways: (1) People can be assigned to societal roles, using as a basis for the assignment an ascribed status—an easily identifiable characteristic such as gender, age, family name, or skin color—over which they have no control. This will produce the caste and estate systems of stratification; or (2) people's positions in the social hierarchy can be based to some degree on achieved statuses gained through their individual, direct efforts. This is known as the class system.

### What is demography?

Demography is the study of the size and composition of human populations as well as of the causes. Demography is influenced by three major factors: fertility, mortality, and migration.

### How does life expectancy influence population growth?

Rapid population growth in the third world is largely due to an increase in life expectancy rather than to a rise in the birthrate. Life expectancy is usually determined more by infant mortality than by adult mortality.

### What is happening to the world's population?

The world population was 7 billion in 2011, and is projected to grow by almost half, to 9.3 billion, by 2050. Ninety-nine percent of global natural increase—the difference between numbers of births and deaths—occurs in the developing world. The death rates exceed birthrates for the world's more-developed countries. As the growth rate in the world's more affluent nations becomes negative, *all* of the net annual gain in global population will, in effect, come from the world's developing countries. Population growth has continued throughout the past three decades despite the decline in fertility rates that began in many developing countries in the late 1970s and, in some countries, despite the toll taken by the HIV/AIDS epidemic.

### What did Thomas Malthus say about population issues?

The core of the population problem, according to Thomas Malthus, is that populations will always grow faster than the available food supply because resources increase arithmetically, whereas population increases geometrically.

▸ **What is the demographic transition theory?**

According to demographic transition theory, societies pass through four stages of population change as they move from high fertility and mortality to relatively low fertility and mortality. This theory accurately describes the population changes that have occurred in Western society with the advance of industrialism; it does not, however, explain population trends in the underdeveloped world today.

▸ **How do the developed and less-developed countries differ?**

The world's richest countries, with 20 percent of the global population, account for 86 percent of private consumption; the poorest 20 percent account for just 1.3 percent.

▸ **How does health care differ throughout the world?**

Currently, 80 percent of the world's population does not have access to any health care. Malnutrition and parasitic and infectious diseases are the principal causes of death and disability in the poorer nations. More than 300 million people live in twenty-four countries where life expectancy is less than 50 years. In those countries, one of ten newborns dies by age 1, and 3 million a year do not survive for one week. Health advances in some countries have reached the point at which it appears that the population is approaching the upper limit of average life expectancy. That is not true for less-developed countries.

▸ **What are the health issues for infants throughout the world?**

When we look at infant and child health from a global perspective, we find that death among children is overwhelmingly a problem of the developing countries in Africa, Asia, and Latin America. Those countries account for 98 percent of the world's deaths among children younger than five years of age. Ninety-five percent of these deaths are preventable.

▸ **How is HIV infection influencing the world?**

A growing gap is developing between the outcomes of infection with the human immunodeficiency virus (HIV) in the developed world and in developing countries. One of the greatest causes for concern is that over the next few years, the epidemic is bound to get worse. Sub-Saharan Africa faces the triple challenge of providing care for the growing population of people infected with HIV, bringing down the number of new infections through more effective prevention, and coping with the effect that millions of deaths have had on the continent. AIDS is systematically reducing life expectancy in the countries where the disease is most common.

▸ **What factors do parents consider when thinking about their family's size?**

Factors that determine the typical family size in a country include: (1) average age of marriage, (2) breast-feeding, (3) infant and child mortality, (4) gender preferences, (5) benefits and costs of children, (6) availability and use of contraception, and (7) income level, education, and urban or rural residence of women.

▸ **What effect does population aging have on a country?**

Population aging is exerting major consequences and implications in all areas of day-to-day human life, and it will continue to do so. Global aging will affect economic growth, savings, investment, and consumption. By 2050, the number of people older than 65 in the world will exceed the number of young.

## Media Resources

**CourseMate for *Introduction to Sociology*, Eleventh Edition**

Cengage Learning's Sociology CourseMate brings course concepts to life with interactive learning, study, and exam preparation tools that support the printed textbook. Access an integrated eBook, learning tools including glossaries, flashcards, quizzes, videos, and more in your Sociology CourseMate. Go to CengageBrain.com to register or purchase access.

# CHAPTER NINE STUDY GUIDE

## Key Concepts

*Match each of the following concepts with its definition, illustration, or explanation.*

a. Caste system
b. Demography
c. Estate system
d. Social class
e. Class system
f. Global stratification
g. Crude birthrate
h. Dependency ratio
i. HIV
j. Acute respiratory illness
k. Fertility

___ 1. The virus that causes AIDS
___ 2. A closed system of social stratification in which a person's social position is defined by law, and membership is established mainly by inheritance
___ 3. A category of people who share similar opportunities and similar economic and vocational positions, lifestyles, attitudes, and behaviors
___ 4. The number of people of nonworking age in a society for every 100 people of working age
___ 5. The study of the make-up of human populations
___ 6. The number of annual live births per 1,000 people in a given population
___ 7. The *most* rigid form of social stratification based on ascribed characteristics
___ 8. One of the most prevalent child killers in developing nations
___ 9. The degree of difference in standard of living among the countries of the world
___ 10. Social hierarchy with greater mobility and based largely on occupation and achieved status
___ 11. A population's sum-total of births

## Central Idea Completions

*Fill in the appropriate concepts and descriptions for each of the following questions.*

1. Briefly describe the degree of inequality among the world's nations. Give three statistics that you think convey the extent of this inequality.

   a. _____

   b. _____

   c. _____

2. Describe the importance, distribution, and impact of HIV/AIDS.

   _____
   _____
   _____
   _____

3. Identify at least three diseases or types of diseases that are important in global stratification. Give estimates of their impact on children in poor countries and the cost to reduce the toll of each disease significantly.

   _____
   _____
   _____

4. List and explain two costs and two benefits of having children for people in underdeveloped countries.

   a. Costs: _____

   _____

   b. Benefits: _____

   _____

5. Briefly explain how maternal age, maternal health, and maternal education are related to infant and child mortality.

   Maternal age: _____

   Maternal health: _____

   Maternal education: _____

6. What factors help to account for developing countries having far fewer girls than boys?

   a. _____

   b. _____

   c. _____

7. What are two outcomes of educating women that affect a country's fertility rate?

   a. _____

   b. _____

8. Describe the four stages of population development according to the demographic transition theory.

   a. _____

   b. _____

   c. _____

   d. _____

9. Describe what is meant by the second demographic transition and present an example.

   _____

   _____

   _____

## ● Critical Thought Exercises

1. Consider the health factors that shorten the lives of children in developing countries and then contrast these factors with those factors that reduce the life expectancy among adults in the same countries.

2. Compare and contrast how gender preferences influence the ratio of boys to girls in Bangladesh and China. What factors affect the preference for girls or boys in each of these countries?

## Internet Activities

1. Gapminder (www.gapminder.org/) allows you to construct scatterplot-type charts with the countries of the world plotted on two variables (for example, average income and life expectancy). Press Play, and Gapminder shows the positions of the countries changing over the past 200 years.

2. The Atlas of Global Inequality (http://ucatlas.ucsc.edu/) at the University of California at Santa Cruz has a wealth of maps, charts, graphs, and information on several aspects of global inequality: income, development, trade, and foreign investment; health; gender issues; and so on.

3. For more on world population, go to www.geography.learnontheinternet.co.uk/topics/popn.html where you will find interesting graphics and useful links.

## Answers to Key Concepts

1. i;   2. c;   3. d;   4. h;   5. b;   6. g;   7. a;   8. j;   9. f;   10. e;   11. k

# 10 Racial and Ethnic Minorities

**The Concept of Race**
- Genetic Definitions
- Legal Definitions
- Social Definitions

*Thinking About Social Issues: How Have Public Attitudes About Racial Intermarriage Changed?*

**The Concept of Ethnic Group**

**The Concept of Minority**

**Problems in Race and Ethnic Relations**
- Prejudice

*Our Diverse Society: How Many Minorities Are There?*
- Discrimination
- Institutional Prejudice and Discrimination

**Patterns of Racial and Ethnic Relations**
- Assimilation
- Pluralism
- Subjugation
- Segregation
- Expulsion
- Annihilation

**Racial and Ethnic Immigration to the United States**
- Immigration Today Compared with the Past

*Our Diverse World: Whites: The New Minority*
- Illegal Immigration

**America's Ethnic Composition Today**
- White Anglo-Saxon Protestants
- African Americans
- Hispanics (Latinos)

*Thinking About Social Issues: Are You Hispanic, Latino, or Neither?*
- Asian Americans
- Native Americans
- A Diverse Society

**Summary**

## LEARNING OBJECTIVES

After studying this chapter, you should be able to do the following:

- Describe the genetic, legal, and social approaches to defining race.
- Explain the concept of ethnic group.
- Know how the sociological concept of minority is used.
- Understand the relationship between prejudice and discrimination.
- Recognize the effect of institutionalized prejudice and discrimination.
- Discuss the history of immigration to the United States.
- Describe the characteristics of the major racial and ethnic groups in the United States.

Paul and Philip Malone, fair-haired and light-skinned identical twin brothers, had always dreamed of becoming firefighters in the Boston Fire Department. The only problem was that the brothers received scores of 69 and 57, respectively, on the written portion of the entry test, when the passing grade was 82. At the time, Boston was under pressure to increase its minority population within the fire department and had separate, and lower, passing grades for nonwhites.

Paul and Philip were aware that they had a great-grandmother who was African American. The Malones had not given the issue of race much thought during their original application. They decided to apply again, only this time as African Americans. They retook the test, and this time their scores were considered passing because of the lower passing grades for minority applicants. The Malones were hired and were listed in the records as African American.

Once they were on the job, the issue of race appeared to be relatively unimportant. The Malones did their jobs well for ten years and then decided to take the civil service examinations for lieutenants. They scored exceptionally high, and their names were forwarded to the fire commissioner for promotion. As their applications were being reviewed, questions arose about their race because they did not look African American. They continued to insist that their claim to being African American was legitimate and produced photos of their great-grandmother.

Their protests were to no avail, however, and they were fired for misrepresenting their race on the original application. They appealed the decision in the courts, but eventually the Massachusetts Supreme Judicial Court ruled against them (Ford, 2003).

The Malones could have supported their claim to be black (1) by visual observation of their features; (2) by appropriate documentary evidence such as birth certificates, especially of black ancestry; or (3) by evidence that they or their families hold themselves to be black and are considered to be black in the community (Hernandez, 1988, 1989). The brothers went on to work for the Massachusetts Transportation Authority.

If you believe the Malones misrepresented their race, what would your view be of Walter White, who was head of the National Association for the Advancement of Colored People (NAACP) from 1931 to 1955? Although he thought of himself as a black man, it has been estimated that he had no more than one sixty-fourth African American ancestry, considerably less than that of the Malones. Both his parents would have been considered white by most observers. White would often go undercover and pass for white while investigating lynchings in the South for the NAACP. In 1923, he even deceived a Ku Klux Klan recruiter into inviting him to Atlanta to advise him on recruitment (Kilker, 1993).

Or what are we to make of Mark Linton Stebbins's claim of being black? In a close contest for a seat on the Stockton, California, city council, Ralph Lee White, an African American, lost his spot on the council to Stebbins, a pale-skinned, blue-eyed man with kinky reddish-brown hair. White claimed that Stebbins would not represent the minority district because he was white. Stebbins claimed that he was African American. Birth records showed that Stebbins's parents and grandparents were white. He had five sisters and one brother, all of whom were white. Yet, when asked to declare his race he noted: "First, I'm a human being, but I'm black."

Stebbins did not deny that he was raised as a white person. He began to consider himself black only after he moved to Stockton. "As far as a birth certificate goes, then I'm white; but I am black," he maintained. "There is no question about that."

Stebbins belonged to an African American Baptist church and to the NAACP. Most of his friends were African American. He had been married three times, first to a white woman, and then to two African American women. He had three children from the first two marriages; two from his white wife were being reared as whites, the third from his black wife as African American. He stated he considered himself black—"culturally, socially, genetically."

Ralph Lee White remained unconvinced, especially with his former council seat having gone to Stebbins. "Now, his mama's white and his daddy's white," White says, "so how can he be black? If the mama's an elephant and the daddy's an elephant, the baby can't be a lion. He's just a white boy with a permanent."

Stebbins believed the issue of race is tied to identifying with a community in terms of beliefs, aspirations, and concerns. He asserted that a person's racial identity depends on much more than birth records.

Linda Fay McCord would agree with Stebbins's view. McCord was adopted in the 1950s by a black couple. She grew up thinking she was black like her parents. When other children would make fun of her for being so white, she would insist she was black.

Then one day in 1998, a stranger called her on the phone claiming to be her aunt. Linda Fay was puzzled because the caller sounded white. After a few questions, the caller confirmed she was white but also noted, "So are you." Linda Fay continued to be skeptical so the aunt sent a copy of Linda Fay's birth certificate which stated that she was born in Toast, North Carolina, on November 18, 1946. The birth certificate clearly noted that her race was "white."

"I don't even know who I am," McCord then said. "I'm caught in the middle of something. My mind says I'm black. Then I look at my skin, and it says I'm white. I've come to the conclusion that color is just a state of mind" (Leland, 2002).

These four examples, which on one hand might seem lighthearted, are at the same time related to some very serious issues. Throughout history, people have gone to great lengths to determine what race a person belonged to. In many states in the United States, laws were devised to determine a person's race if, like the Malones, that person had mixed racial ancestry. Usually, these laws existed for the purpose of discriminating against certain minority groups. Our examples also raise the question of whether one can change one's race from white to black even if one has no black ancestors. Mark Stebbins seemed to think so.

In this chapter, we will explore these and other issues related to race and ethnicity as we try to understand how people come to be identified with certain groups and what that membership means.

## THE CONCEPT OF RACE

Although the origin of the word is not known, the term *race* has been a highly controversial concept for a long time. Many authorities suspect that it is of Semitic origin, coming from a word that some translations of the Bible render as "race," as in the "race of Abraham," but that is otherwise translated as "seed" or "generation." Other scholars trace the origin to the Czech word *raz*, meaning "artery" or "blood"; others to the Latin *generatio* or the Basque *arraca* or *arraze*, referring to a male stud animal. Some trace it to the Spanish *ras*, itself of Arabic derivation, meaning "head" or "origin." In all these possible sources, the word has a biological significance that implies descent, blood, or relationship.

We shall use the term **race** to refer to *a category of people who are defined as similar because of a number of physical characteristics.* Often the category is based on an arbitrary set of features chosen to suit the labeler's purposes and convenience. As long ago as 1781, German physiologist Johann Blumenbach realized that racial categories did not reflect the actual divisions among human groups. As he put it, "When the matter is thoroughly considered, you see that all [human groups] do so run into one another, and that one variety of mankind does so sensibly pass into the other, that you cannot mark out the limits between them" (Montagu, 1964a). Blumenbach believed that racial differences were superficial and changeable, and modern scientific evidence seems to support this view. Throughout history, races have been defined along genetic, legal, and social lines, each presenting its own set of problems.

### Genetic Definitions

Geneticists define race by noting differences in gene frequencies among selected groups. The number of distinct races that can be defined by this method depends on the particular genetic trait under investigation. Differences in traits, such as hair and nose type, have proved to be of no value in making biological classifications of human beings. In fact, the physiological and mental similarities among groups of people appear to be far greater than any superficial differences in skin color and physical characteristics.

Throughout history, races have been defined along genetic, legal, and social lines, each presenting its own set of problems. The U.S. Census Bureau relies on a self-definition of race.

Also, the various so-called racial criteria appear to be independent of one another. For example, any form of hair can occur with any skin color; a narrow nose gives no clue to an individual's skin pigmentation or hair texture. Thus, Australian aborigines have dark skins, broad noses, and an abundance of curly-to-wavy hair; Asiatic Indians also have dark skins but have narrow noses and straight hair. Likewise, if head form is selected as the major criterion for sorting, an equally diverse collection of physical types will appear in each category thus defined. If people are sorted on the basis of skin color, therefore, all kinds of noses, hair, and head forms will appear in each category.

### Legal Definitions

By and large, legal definitions of race have not been devised to determine who is black or of another race but, rather, who is not white. The laws were to be used in instances when separation and different treatments were to be applied to members of certain groups. Segregation laws are an excellent example. If railroad conductors had to assign someone to either the black or white cars, they needed fairly precise guidelines for knowing whom to seat where. Most legal definitions of race were devices to prevent blacks from attending white schools, serving on juries, holding certain jobs, or patronizing certain public places. The official guidelines could then be applied to individual cases. The common assumption that "anyone not white was colored," although imperfect, did minimize ambiguity.

There has been, however, very little consistency among the various legal definitions of race that have been devised. The state of Missouri, for example, made "one-eighth or more Negro blood" the criterion for nonwhite status. Georgia was even more rigid in its definition and noted:

> The term "white person" shall include only persons of the white or Caucasian race, who have no ascertainable trace of either Negro, African, West Indian, Asiatic Indian, Mongolian, Japanese, or Chinese blood in their veins. No person, any of whose ancestors [was]... a colored person or person of color, shall be deemed to be a white person.

Virginia had a similar law but made exceptions for individuals with one-fourth or more American Indian blood and less than one-sixteenth "Negro" blood. "Those Virginians were regarded as American Indians as long as they remained on an Indian reservation, but if they moved, they were regarded as blacks" (Berry and Tischler, 1978; Novit-Evans and Welch, 1983).

Most of these laws are artifacts of the segregation era. However, if people think that all vestiges of them have disappeared, they are wrong. As recently as September 1982, a dispute arose over Louisiana's law requiring anyone of more than one-thirty-second African descent to be classified as black.

Louisiana's one-thirty-second law is actually of recent vintage, having come into being in 1971. Before this law, racial classification in Louisiana depended on what was referred to as common repute. The 1971 law was intended to eliminate racial classifications by gossip and inference. In 1982, Susie Guillory Phipps obtained a copy of her birth certificate so that she could apply for a passport. She was surprised to see that her birth certificate classified her as black. Phipps, who at the time was 49, had lived her entire life as a white person. She requested that her race be noted as white. The state objected and produced an eleven-generation family tree with ancestors Phipps knew nothing about, including an early eighteenth-century black slave and a white plantation owner. Phipps responded, "My children are white. My grandchildren are white. Mother and Daddy were buried white." Louisiana was not convinced and calculated that she was three-thirty-seconds black, more than enough to make her black under state law (Cose, 1997; Nelson and Pang, 2006).

### Social Definitions

The social definition of race, which is the decisive one in most interactions, pays little attention to an individual's hereditary physical features or to whether his or her percentage of "Negro blood" is one-fourth, one-eighth, or one-sixteenth. According to social definitions of race, if a person presents himself or herself as a member of a certain race and others respond to that person as a member of that race, then it makes little sense to say that he or she is not a member of that race.

In Latin American countries, having African ancestry or African features does not automatically define an individual as black. For example, in Brazil, many individuals are listed in the census as white and are considered to be white by their friends and associates even if they had a grandparent who was of pure African descent. It is much the same in Puerto Rico, where anyone who is not obviously of African descent is classified as either mulatto or white.

The U.S. census relies on a self-definition system of racial classification and does not apply any biological, legal, or genetic rules. Since the 2000 census, the bureau has recognized that race may also include national or sociocultural groups. People also have been allowed to report more than one race. They can declare themselves as members of any one or more of five racial categories: American Indian/Alaskan Native, Asian, African American, Native Hawaiian/Pacific Islander, or white. Those listing themselves as white and a member of a minority have been counted as a minority. (See Table 10–1 for a listing of the various racial and ethnicity categories the Census Bureau has used over the years.)

*Multiracial Ancestry* The new census policy allowing people to choose more than one racial category has

TABLE 10-1  Race/Ethnicity Categories in the U.S. Census, 1790–2010

| Census | 1790 | 1860 | 1890* | 1900 | 1970 | 2010 |
|---|---|---|---|---|---|---|
| **Race** | **White** | White | White | White | White | White |
|  | Black | Black | Black | Black | Negro or | Black |
|  | Indian | Mulatto | Mulatto | (Negro Descent) | Black | African American or Negro |
|  |  |  | Quadroon |  |  |  |
|  |  |  | Octoroon |  |  |  |
|  |  |  | Chinese | Chinese | Chinese | Chinese |
|  |  |  | Japanese | Japanese | Japanese | Japanese |
|  |  |  | Indian | Indian | Indian (American) | American Indian or Alaska Native |
|  |  |  |  |  | Filipino | Filipino |
|  |  |  |  |  | Hawaiian | Native Hawaiian |
|  |  |  |  |  | Korean | Korean |
|  |  |  |  |  |  | Asian Indian |
|  |  |  |  |  |  | Vietnamese |
|  |  |  |  |  |  | Guamanian or Chamorro |
|  |  |  |  |  |  | Samoan |
|  |  |  |  |  |  | Other Asian (print race—example: Hmong, Laotian, Thai, Pakistani, Cambodian, etc.) |
|  |  |  |  |  |  | Other Pacific Islander (print race—example: Fijian, Tongan, etc.) |
|  |  |  |  |  |  | Some other race (print race) |
|  |  |  |  |  | Other | Some other race (print race) |
| **Hispanic Ethnicity** |  |  |  |  | Mexican | Mexican |
|  |  |  |  |  |  | Mexican American or Chicano |
|  |  |  |  |  | Puerto Rican | Puerto Rican |
|  |  |  |  |  | Central/South American |  |
|  |  |  |  |  | Cuban | Cuban |
|  |  |  |  |  | Other Spanish | Other Spanish/ |
|  |  |  |  |  |  | Hispanic/Latino origin (print origin—example: Argentinean, Columbian, Dominican, Nicaraguan, Salvadoran, Spaniard, etc.) |

*In 1890, mulatto was defined as a person who was three-eighths to five-eighths black. A quadroon was one-quarter black and an octoroon one-eighth black.

ignited debate among various interest groups. Parents of mixed-race children have liked the change because they no longer need to pick one race for their offspring. Others have been concerned that the new rules will decrease the official numbers of those considered African American or Native American and decrease their political clout (Holmes, 2000).

About 8 to 9 million people have identified themselves as belonging to two or more races. The most common combination has been "white" and "some other race" (32%). This has been followed by "white" and "American Indian/Alaska Native" (16%), "white" and "Asian" (13%), and "white" and "African American" (11%).

*Interracial Marriage* Attitudes toward interracial marriage are telling indicators of the social distance between racial groups. Between the 1660s and the 1960s, dozens of states enacted laws prohibiting interracial marriage. These laws did not just outlaw marriage between whites and blacks but also between whites and Native Americans or people of Chinese, Japanese, Korean, and Filipino ancestry (Kennedy, 2003).

In Dion Boucicault's play, *The Octoroon* (1859), a woman named Zoe who has one-eighth black ancestry is in love with George Payton and states:

> Of the blood that feeds my heart, one drop in eight is black—bright red as the rest may be, that one drop poisons all the blood; those seven bright drops give me love like yours—hope like yours—ambition like yours—life hung with passions like dew-drops on the morning flowers; but the one black drop gives me despair, for I'm an unclean thing—forbidden by the laws—I'm an Octoroon!

In June 1999, the Alabama Senate "voted to repeal the state's constitutional prohibition against interracial marriage." Even though the U.S. Supreme Court struck down prohibitions against interracial marriage 32 years earlier, in parts of rural Alabama, probate judges would still refuse to issue marriage licenses to interracial couples (Greenberg, 1999).

About 15 percent of all marriages in the United States in 2010 were between spouses of different races or ethnicities, more than double the share in 1980 (6.7%). Of those who married in 2010, 9 percent of whites, 17 percent of blacks, 26 percent of Hispanics, and 28 percent of Asians married someone of a different race (Pew Research Center, 2012). (See "Thinking About Social Issues: How Have Public Attitudes About Racial Intermarriage Changed?")

## THINKING ABOUT SOCIAL ISSUES: How Have Public Attitudes About Racial Intermarriage Changed?

Attitudes about marriage between whites and blacks are significantly more positive today than in the past (see Figure 10-1). Today, 86 percent of the American public approves of these interracial marriages. Among blacks, 96 percent approve. Back in 1958, only 4 percent of the American public approved of marriage between whites and blacks, and even as late as 1994 half the people still disapproved.

People tend to think that marriages between people of different races are a change for the better in society. The approval may grow out of the fact that more than a third of adults say they have an immediate family member or close relative who is married to someone of a different race. (See Figure 10-2.)

People who are the most approving of racial intermarriage tend to be younger, politically liberal, college educated, and living on the East or West Coasts of the country. However, even among the age group that is the least approving of racial intermarriage, those over 65, the majority still approve of black-white marriages.

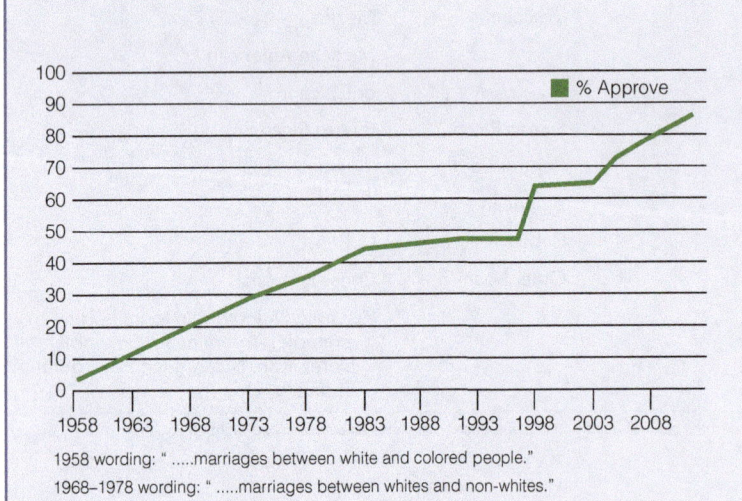

**FIGURE 10-1** Do You Approve or Disapprove of Marriage Between Blacks and Whites?

1958 wording: " .....marriages between white and colored people."
1968–1978 wording: " .....marriages between whites and non-whites."

Sources: Pew Research Center Race Survey. October 28–November 30, 2009; Jones, Jeffrey M. Gallup Poll. September 12, 2011. "Record-High 86% Approve of Black-White Marriages" (www.gallup.com/poll/149390/Record-High-Approve-Black-White-Marriages.aspx), accessed July 2, 2012; Wang, Wendy. Pew Research Center. February 16, 2012. "The Rise of Intermarriage: Rates, Characteristics Vary by Race and Gender" (www.pewsocialtrends.org/2012/02/16/the-rise-of-intermarriage/?src=prc-headline), accessed June 30, 2012.

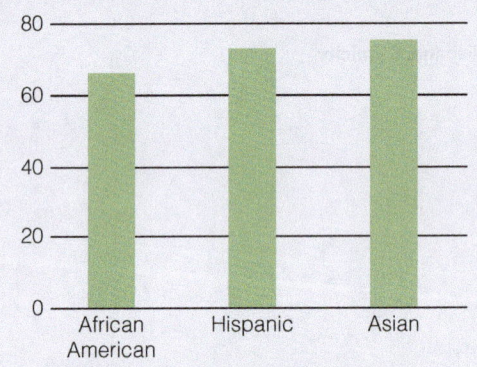

Percentage Approving of a Family Member Marrying Someone of a Different Race or Ethnicity

**FIGURE 10-2** How Would You React If a Member of Your Family Was Going to Marry Someone of a Different Race or Ethnicity?

Sources: Pew Research Center Race Survey. October 28–November 30, 2009; Jones, Jeffrey M. Gallup Poll. September 12, 2011. "Record-High 86% Approve of Black-White Marriages" (www.gallup.com/poll/149390/Record-High-Approve-Black-White-Marriages.aspx), accessed July 2, 2012; Wang, Wendy. Pew Research Center. February 16, 2012. "The Rise of Intermarriage: Rates, Characteristics Vary by Race and Gender" (www.pewsocialtrends.org/2012/02/16/the-rise-of-intermarriage/?src=prc-headline), accessed June 30, 2012.

Racial intermarriage is particularly high among native-born Asians, where the intermarried percentage reaches 40 percent. As more Asian Americans marry interracially, future generations of Asian Americans might increasingly blend with other racial and ethnic groups, mirroring the experience of European ethnic groups in the United States over the past century. Interestingly, the future size of the U.S. Asian population depends in part on whether the children of these marriages identify themselves as Asians. If they do, the Asian American population will increase faster than projected.

Most people with one white and one black parent, when given the opportunity to label themselves, have historically chosen (or been forced to choose, as noted previously in this chapter) one parent's racial identity, and that most often has been and continues to be black. Historically, one could not be white and have a black ancestor. In the United States, it is unusual for a person to say that he or she is half black. President Barack Obama is always referred to as the first black president even though he has one white parent.

The most multiracial U.S. states have significantly different racial combinations. In Hawaii, the most multiracial state, the most common combinations are Asian and Native Hawaiian or other Pacific Islander and white and Asian. In Oklahoma, two-thirds of the multiracial population is white and American Indian. In California, the most frequent combination is white and "some other race" (AmeriStat, November 2001).

Many ethnic groups form subcultures with a high degree of internal loyalty and adherence to basic customs.

set them apart as subcultures, many ethnic groups—for example, Arabs, French Canadians, Flemish, Scots, Jews, and Pennsylvania Dutch—are part of larger political parties.

## THE CONCEPT OF ETHNIC GROUP

An **ethnic group** *has a distinct cultural tradition that its own members identify with and that might or might not be recognized by others* (Glazer and Moynihan, 1975). An ethnic group need not necessarily be a numerical minority within a nation (although the term sometimes is used that way).

Many ethnic groups form subcultures (see Chapter 3, "Culture"). They usually possess a high degree of internal loyalty and adherence to basic customs, maintaining a similarity in family patterns, religion, and cultural values. They often possess distinctive folkways and mores; customs of dress, art, and ornamentation; moral codes and value systems; and patterns of recreation. The whole group is usually devoted to something such as a monarch, religion, language, or territory. Above all, members of the group have a strong feeling of association. The members are aware of a relationship because of a shared loyalty to a cultural tradition. The folkways might change, the institutions might become radically altered, and the object of allegiance might shift from one trait to another, but loyalty to the group and the consciousness of belonging remain as long as the group exists. Interestingly, despite the unique cultural features that

## THE CONCEPT OF MINORITY

Whenever race and ethnicity are discussed, it is usually assumed that the object of the discussion is a minority group. Technically, this is not always true, as we will see shortly. A minority is often thought of as being few in number. The concept of minority, rather than implying a small number, should be thought of as implying differential treatment and exclusion from full social participation by the dominant group in a society. In this sense, we will use Louis Wirth's definition of a **minority** as *a group of people who, because of physical or cultural characteristics, are singled out from others in society for differential and unequal treatment, and who therefore regard themselves as objects of collective discrimination* (Linton, 1936).

In his definition, Wirth speaks of physical and cultural characteristics and not of gender, age, disability, or undesirable behavioral patterns. Clearly, he is referring to racial and ethnic groups in his definition of minorities. Some people have suggested, however, that many other groups are in the same position as those more commonly thought of as minorities and endure the same sociological and psychological problems. In this light, women, gays and lesbians, adolescents, the aged, the handicapped, the radical right or left, and intellectuals can be thought of as minority groups. (See "Our Diverse World: How Many Minorities Are There?")

## PROBLEMS IN RACE AND ETHNIC RELATIONS

All too often, when people with different racial and ethnic identities come together, frictions develop among the groups. People's suspicions and fears are often aroused by those whom they feel to be different.

### Prejudice

There are many definitions of prejudice. People, particularly those with a strong sense of identity, often have feelings of prejudice toward others who are not like themselves. For example, in 1945, 56 percent of the American public said they opposed a law that would require employees to work alongside people of any race or color (Gallup Poll, 1972). In 1966, three in ten Americans (31%) thought houses for sale in their neighborhood should not be open to people regardless of race or nationality (Gallup Poll, 1972).

Literally, *prejudice* means a prejudgment or an attitude with an emotional bias (Wirth, 1944). However, this definition has a problem. All of us, through the process of socialization, acquire attitudes that might be in response not only to racial and ethnic groups but also to many things in our environment. We come to have attitudes toward cats, roses, blue eyes, chocolate cheesecake, television programs, and even ourselves. These attitudes run the gamut from love to hate, from esteem to contempt, from loyalty to indifference. How have we developed these attitudes? Has it been through the scientific evaluation of information, or by other, less logical means? For our purposes, we will define **prejudice** as *an irrationally based negative, or occasionally positive, attitude toward certain groups and their members.*

What is the cause of prejudice? Although pursuing this question is beyond the scope of this book, we can list some of the uses to which prejudice is put and the social functions it serves. First, a prejudice, simply because it is shared, helps draw together those who hold it. It promotes a feeling of we-ness, of being part of an in-group, and it helps define such group boundaries. Especially in a complex world, belonging to an in-group and consequently feeling special or superior can be an important social identity for many people.

Second, when two or more groups are competing for access to scarce resources (jobs, for example), it is easier on the conscience if one can write off his or her competitors as somehow less than human or inherently unworthy. Nations at war consistently characterize each other negatively, using terms that seem to deprive the enemy of any humanity whatsoever.

---

### OUR DIVERSE WORLD: How Many Minorities Are There?

Louis Wirth, when he defined the term *minority group,* used it to refer to physical and cultural characteristics that cause a group to receive different and unequal treatment. Wirth's definition never referred to the minority group's size, so that a minority group can be large or small in number. Since 1945, when Wirth first used it, many groups have expanded the definition to claim that they too are minority groups.

Why would these groups want to do that? As sociologists, we realize that labels matter. They cause people to see themselves in a certain way and they cause others to define the group in a certain way. A label can help develop a sense of pride and self-acceptance. It can also open the way for legal benefits. For example, the federal government might create policies or develop programs for minorities.

People claiming to be part of a minority group have usually noted that their group also suffers from oppression and discrimination. For example, many believe that women have been systematically victimized and disadvantaged as a group and deserve to be called a minority group. Others might point out that the group has a unique culture and lifestyle that has ethnic group-like qualities that makes it worthy of being treated as a minority group.

Being defined as a minority group implies that the members have been victims of discrimination and deserve certain rights. This realization can then be used for collective action and produce pride in the group's lifestyle and culture. During the past few decades, the term *minority group,* rather than being assigned to a group against its members' wishes, has been openly endorsed and embraced by such diverse groups as the disabled, gays and lesbians, and women, as well as by many other groups.

*Source:* Berbrier, Mitch. Winter 2004. "Why Are There So Many 'Minorities?'" *Contexts* 3(1):38–44.

Third, psychologists suggest that prejudice allows us to project onto others those parts of ourselves that we do not like and therefore try to avoid facing. For example, most of us feel stupid at one time or another. How comforting it is to know that we belong to a group that is inherently more intelligent than another group. Who does not feel lazy sometimes? But how good it is that we do not belong to that group—the one everybody knows is lazy.

Of course, prejudice also has many negative consequences, or dysfunctions, to use the sociological term. For one thing, it limits our vision of the world around us, reducing social complexities and richness to a sterile and empty caricature. Aside from this effect on us as individuals, prejudice also has negative consequences for the whole society. Most notably, it is the necessary ingredient of discrimination, a problem found in many societies, including our own.

## Discrimination

Prejudice is a subjective feeling, whereas discrimination is an overt action. **Discrimination** refers to *differential treatment, usually unequal and injurious, accorded to individuals who are assumed to belong to a particular category or group.* Discrimination against African Americans and other minorities has occurred throughout U.S. history. At the start of World War II, for example, blacks were not allowed in the Marine Corps; blacks could be admitted into the Navy only as mess stewards; and the Army had a 10 percent quota on black enlistments.

Prejudice does not always result in discrimination. Although our attitudes and our overt behavior are closely related, they are neither identical nor dependent on each other. We might have feelings of antipathy without expressing them overtly or even giving the slightest indication of their presence. This simple fact—namely, that attitudes and overt behavior vary independently—has been applied by Robert K. Merton (1949/1969) to the classification of racial prejudice and discrimination. There are, he believes, the following four types of people.

*Unprejudiced Nondiscriminators* These people are neither prejudiced against the members of other racial and ethnic groups nor do they practice discrimination. They believe implicitly in the American ideals of justice, freedom, equality of opportunity, and dignity of the individual. Merton recognizes that people of this type are properly motivated to spread the ideals and values of the creed and to fight against those forms of discrimination that make a mockery of them. At the same time, unprejudiced nondiscriminators have their shortcomings. They often believe their own spiritual house is in order, and thus they do not feel pangs of guilt and see little need for any collective effort to change things.

*Unprejudiced Discriminators* This type includes those who constantly think of expediency. Although they themselves are free from racial prejudice, they will keep silent when bigots speak out. They will not condemn acts of discrimination but will make concessions to the intolerant and will accept discriminatory practices for fear that to do otherwise would hurt their own position.

*Prejudiced Nondiscriminators* This category is for the timid bigots who do not accept the ideal of equality for all but conform to it and give it lip service when the slightest pressure is applied. Those who hesitate to express their prejudices when in the presence of those who are more tolerant belong in this category. Among them are employers who hate certain minorities but hire them rather than run afoul of affirmative action laws and labor leaders who suppress their personal racial bias when the majority of their followers demand an end to discrimination.

*Prejudiced Discriminators* These are the bigots, pure and unashamed. They do not believe in equality, nor do they hesitate to give free expression to their intolerance, both in their speech and in their actions. For them, attitudes and behavior do not conflict. They practice discrimination, believing that it is not only proper but in fact their duty to do so (Berry and Tischler, 1978).

Knowing a person's attitudes does not mean that that person's behavior always can be predicted. Attitudes and behavior are frequently inconsistent because of such factors as the nature and magnitude of the social pressures in a particular situation. Figure 10–3 shows the influence of situational factors on behavior.

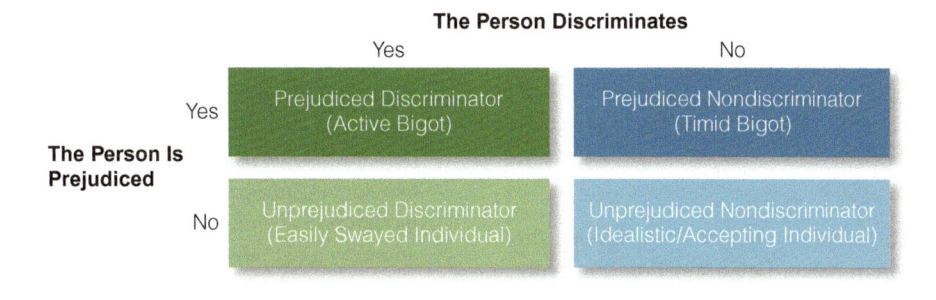

|  | **The Person Discriminates** | |
| --- | --- | --- |
|  | Yes | No |
| **The Person Is Prejudiced** — Yes | Prejudiced Discriminator (Active Bigot) | Prejudiced Nondiscriminator (Timid Bigot) |
| **The Person Is Prejudiced** — No | Unprejudiced Discriminator (Easily Swayed Individual) | Unprejudiced Nondiscriminator (Idealistic/Accepting Individual) |

**FIGURE 10-3** **The Interaction of Prejudice and Discrimination**
*As this diagram shows, the degree of social pressure being exerted can cause individuals of inherently dissimilar attitudes to exhibit relatively similar behaviors in a given situation.*

## Institutional Prejudice and Discrimination

Sociologists tend to distinguish between individual and institutional prejudice and discrimination. When individuals display prejudicial attitudes and discriminatory behavior, it is often based on the assumption of the outgroup's genetic inferiority. By contrast, **institutionalized prejudice and discrimination** refer to *complex societal arrangements that restrict the life chances and choices of a specifically defined group in comparison with those of the dominant group*. In this way, benefits are given to one group and withheld from another. Society is structured in such a way that people's values and experiences are shaped by a prejudiced social order. Discrimination is seen as a by-product of a purposive attempt to maintain social, political, and economic advantage (Davis, 1979).

Some people argue that institutionalized prejudice and discrimination are responsible for the substandard education that many African Americans receive in the United States. Schools that are predominantly black tend to be inferior at every level to schools that are predominantly white. The facilities for blacks are usually of poorer quality than are those for whites. Many blacks also attend unaccredited black colleges, where the teachers are less likely to hold advanced degrees and are poorly paid. The poorer education that blacks receive is one of the reasons they generally are in lower occupational categories than whites are. In this way, institutionalized prejudice and discrimination combine to maintain blacks in a disadvantaged social and economic position (Kozol, 1992).

## PATTERNS OF RACIAL AND ETHNIC RELATIONS

Relations among racial and ethnic groups seem to include an infinite variety of human experiences. They run the gamut of emotions and appear to be unpredictable, capricious, and irrational. They range from curiosity and hospitality, at one extreme, to bitter hostility at the other. In this section, we will show that a limited number of outcomes exist when racial and ethnic groups come into contact. These include assimilation, pluralism, subjugation, segregation, expulsion, and annihilation. In some cases, these categories overlap—for instance, segregation can be considered a form of subjugation—but each has distinct traits that make it worth examining separately.

### Assimilation

In 1753, twenty-three years before he signed the Declaration of Independence, Benjamin Franklin wondered about the costs and benefits of German immigration. On one hand, he commented that Germans are "the most stupid" and "great disorders may one day arise among us" because of these immigrants. Yet, on the other hand, he pointed out that they "contribute greatly to the improvement of a country." He finally decided that the benefits of German immigration could outweigh the costs if we "distribute them more equally, mix them with the English, establish English schools where they are now too thick settled" (Borjas, 1999).

Franklin was concerned with the assimilation of the German immigrants. **Assimilation** is *the process whereby groups with different cultures come to have a common culture*. It refers to more than just dress or language, including less tangible items such as values, sentiments, and attitudes. Assimilation is really referring to the fusion of cultural heritages.

Assimilation is the integration of new elements with old ones. Transferring culture from one group to another is a highly complex process, often involving the rejection of ancient ideologies, habits, customs, attitudes, and language. It also includes the elusive problem of selection. Of the many possibilities presented by a culture, which ones will another culture adopt? Why did some Native Americans, for example, when they were confronted with the white civilization, take avidly to guns, horses, rum, knives, and glass beads while showing no interest in certain other features to which whites themselves attached the highest value?

In the process of assimilation, one society sets the pattern because the give and take of culture seems never to operate on a 50–50 basis. Invariably, one group has a much larger role in the process than the other, and various factors interact to make it so. Usually, one of the societies enjoys greater power or prestige than the other, giving it an advantage in the assimilation process; one is better suited to the environment than the other; or one has greater numerical strength than the other. Thus, the pattern for the United States was set by the British colonists, and to that pattern the other groups were expected to adapt. This process has often been referred to as **Anglo conformity**—*the renunciation of the ancestral cultures in favor of Anglo-American behavior and values* (Berry and Tischler, 1978; Gordon, 1964).

The Anglo-conformity viewpoint was at its strongest around World War I, as demonstrated by this excerpt from a speech by President Woodrow Wilson:

> You cannot dedicate yourself to America unless you become in every respect and with every purpose of your will thorough Americans. You cannot become thorough Americans if you think of yourselves in groups. America does not consist of groups. A man who thinks of himself as belonging to a particular national group in America has not yet become an American, and the man who goes among you to trade upon your nationality is no worthy son to live under the Stars and Stripes. (Wilson, 1915)

Although assimilation frequently has been a professed political goal in the United States, it has seldom been fully achieved. Consider the case of the Native Americans. In 1924, they were granted full U.S. citizenship. Nevertheless, the federal government's policies regarding the integration of Native Americans into American society wavered back and forth until the Hoover Commission Report of 1946 became the guideline for all subsequent administrations. The report stated that:

> A program for the Indian peoples must include progressive measures for their complete integration into the mass of the population as full, tax-paying [members of the larger society].... Young employable Indians and the better cultured families should be encouraged and assisted to leave the reservations and set themselves up on the land or in business. (Shepardson, 1963)

However, to this day, Native American groups have yet to be fully integrated into the mainstream of American life. About 54 percent live on or near reservations, and most of the rest live in impoverished urban areas. In addition, many Native Americans who left the reservation for greater opportunity in the cities are returning to the reservation. Despite the economic and lifestyle hardships they face on the reservation, their ethnic pride overrides any desire to assimilate.

Other groups, whether or not by choice, also have not assimilated. The Amish, for instance, have steadfastly maintained their subculture in the face of the pressures of Anglo conformity from the larger American society.

China provides an interesting example of what might be called reverse assimilation. Usually, defeated minority groups are assimilated into the culture of a politically dominant group. In the seventeenth century, however, Mongol invaders conquered China and installed themselves as rulers. The Mongols were nomadic pastoralists. They were so impressed with the advanced achievements of the Chinese civilization that they gave up their own ways and took on the trappings of Chinese culture: language, manners, dress, and philosophy. During their rule, the Mongols fully assimilated the Chinese culture.

## Pluralism

**Pluralism,** or *the development and coexistence of separate racial and ethnic group identities within a society,* is a philosophical viewpoint that attempts to produce what is considered to be a desirable social situation. When people use the term *pluralism* today, they believe they are describing a condition that seems to be developing in contemporary American society. They often ignore the ideological foundation of pluralism.

The person principally responsible for the development of the theory of cultural pluralism was Horace Kallen, born in Germany. He came to Boston at the age of five and was raised in an Orthodox Jewish home. As he progressed through the Boston public schools, he underwent a common second-generation phenomenon—he started to reject his home environment and religion and developed an uncritical enthusiasm for the United States. As he put it, "It seemed to me that the identity of every human being with every other was the important thing, and that the term 'American' should nullify the meaning of every other term in one's personal makeup."

While Kallen was a student at Harvard, he experienced a number of shocks. Working in a nearby settlement house, he came in contact with liberal and socialist ideas and observed people expressing numerous ethnic goals and aspirations. This exposure caused him to question his definition of what it meant to be an American. This quandary was compounded by his experiences in the American literature class of Barrett Wendell, who believed that Puritan traits and ideals were at the core of the American value structure. The Puritans, in turn, had modeled themselves after the Old Testament prophets. Wendell even suggested that the early Puritans were largely of Jewish descent. These ideas led Kallen to believe that he could be an unassimilated Jew and still belong to the core of the American value system.

After discovering that he could be totally Jewish and still be American, he came to realize that the application could be made to other ethnic groups as well. All ethnic groups, he felt, should preserve their own separate cultures without shame or guilt. As he put it, "Democracy involves not the elimination of differences, but the perfection and conservation of differences."

Pluralism is a reaction against the assumption of assimilation and the idea of America as a melting pot in which immigrants from around the world combine into the new metal of the American. It is a philosophy that not only assumes that minorities have rights but also considers the lifestyle of the minority group to be a legitimate, and even desirable, way of participating in society. The theory of pluralism celebrates the differences among groups of people. The theory also implies a hostility to existing inequalities in the status and treatment of minority groups. Pluralism has provided a means for minorities to resist the pull of assimilation by allowing them to claim that they constitute the very structure of the social order. From the assimilationist point of view, the minority is seen as a subordinate group that should give up its identity as quickly as possible. Pluralism, on the other hand, assumes that the minority is a primary unit of society and that the unity of the whole depends on the harmony of the various parts.

Switzerland provides an example of balanced pluralism that, so far, has worked exceptionally well (Kohn, 1956). After a short civil war between Catholics and Protestants in 1847, a new constitution—drafted in 1848—established a confederation of cantons (states), and church-state relations were left up to the individual cantons. The three major languages—German, French, and Italian—were declared official languages for the whole nation, and their respective speakers were acknowledged as political equals (Petersen, 1975). Switzerland's linguistic regions are culturally quite distinctive. Italian-speaking Switzerland has a Mediterranean flavor; in French-speaking Switzerland, one senses the culture of France; and German-speaking Switzerland is distinctly Germanic. However, all three linguistic groups are fiercely pro-Swiss.

## Subjugation

In theory, we could assume that two groups may come together and develop an egalitarian relationship. However, racial and ethnic groups rarely have established such a relationship. One of the consequences of the interaction of racial and ethnic groups has been **subjugation**—*the subordination of one group and the assumption of a position of authority, power, and domination by the other.* The members of the subordinate group may, for a time, accept their lower status and even devise ingenious rationalizations for it. For the most part, this is so because there are few instances in which group contact has been based on the complete equality of power. Differences in power will invariably lead to a situation of superior and inferior positions. The greater the discrepancy in the power of the groups involved, the greater the extent and scope of the subjugation will be.

By the mid-1870s, the various Native American tribes were at the mercy of white Europeans. Native American traditions, beliefs, and ways of life were condemned as backward and immoral. It was thought that the best way to help Native Americans was to ensure that their tribal cultures were destroyed. The Indian peoples would have to be forced to assimilate into the mainstream culture. President Benjamin Harrison's commissioner of Indian affairs, Thomas Jefferson Morgan, expressed this view in 1889 when he noted:

> The logic of events demands the absorption of the Indians into our national life, not as Indians, but as American citizens.... The Indians must conform to "the white man's ways," peaceably if they will, forcibly if they must.... This civilization may not be the best possible, but it is the best the Indians can get. They cannot escape it, and must either conform to it or be crushed by it. (Josephy, 1994)

To subjugate and Americanize the Native Americans, the government banned their religions, rituals, and sacred ceremonies. Medicine men and shamans were either jailed or exiled. Attempts were even made to stop them from speaking their tribal languages. The final step was to send Native American children to boarding schools, where they were taught to become part of the white culture. As a Taos Pueblo youth noted:

> We all wore white man's clothes and ate white man's food and went to white man's churches and spoke white man's talk. And so after a while we also began to say Indians were bad. We laughed at our own people and their blankets and cooking pots and sacred societies and dances. (Embree, 1939/1967)

Why should different levels of power between two groups lead to the domination of one by the other? Gerhard Lenski (1966) proposed that it is because people have a desire to control goods and services. No matter how much they have, they are never satisfied. In addition, high status is often associated with the consumption of goods and services. When a racial or ethnic group is placed in an inferior position, its people often are eliminated as competitors. In addition, their subordinate position may increase the supply of goods and services available to the dominant group.

Quaker and missionary reformers wanted to use new methods to "civilize" American Indians. This resulted in schools like the Hampton Institute in Virginia (top) and the Carlisle School in Pennsylvania (bottom) where Native American children were to be taught to be like their "white brothers."

## Segregation

**Segregation,** *a form of subjugation, refers to the act, process, or state of being set apart.* It places limits and restrictions on the contact, communication, and social relations among groups. Many people think of segregation as a negative phenomenon—a form of ostracism imposed on a minority by a dominant group—and this is most often the case. However, for some groups, such as the Amish or Chinese in America, who wish to retain their ethnicity, segregation is voluntary.

The practice of segregating people is as old as the human race itself. Examples of it exist in the Bible and in preliterate cultures. American blacks originally were segregated by the institution of slavery and later by both formal sanction and informal discrimination. Although some African Americans formed groups that preached total segregation from whites as an aid to black cultural development, for most it was an involuntary and degrading experience. The word *ghetto* originally referred to the segregated quarter of a city where the Jews in Europe were often forced to live. Native American tribes were often forced to choose segregation on a reservation in preference to annihilation or assimilation. Segregation has operated in a wide range of circumstances.

## Expulsion

**Expulsion** is *the process of forcing a group to leave the territory in which it lives.* This can be accomplished indirectly by making life increasingly unpleasant for a group, as the Germans did for Jews after Adolf Hitler was appointed chancellor in 1933. Over the following six years, Jews were stripped of their citizenship, made ineligible to hold public office, removed from the professions, and forced out of the artistic and intellectual circles to which they belonged. In 1938, Jewish children were barred from public schools. At the same time, the government encouraged acts of violence and vandalism against Jewish communities. These actions culminated in Kristallnacht, November 9, 1938, when the windows in synagogues and Jewish homes and businesses across Germany were shattered and individuals were beaten and arrested. Under these conditions, Jews left Germany by the thousands. In 1933, some 500,000 Jews lived in Germany; by 1940, before Hitler began his final solution—that is, the murder of all remaining Jews—only 220,000 remained (Robinson, 1976).

Expulsion also can be accomplished through **forced migration,** or *the relocation of a group through direct action.* For example, forced migration was a major aspect of the U.S. government's policies toward Native American groups in the nineteenth century. When the army needed to protect its lines of communication to the West Coast, Colonel Christopher "Kit" Carson was ordered to move the Navajos of Arizona and New Mexico out of the way. He was instructed to kill all the men who resisted and to take everybody else captive. He accomplished this in 1864 by destroying their cornfields and slaughtering their herds of sheep, thereby confronting the Navajos with starvation. After a last showdown in Canyon de Chelly, some 8,000 Navajos were rounded up at Fort Defiance. They then were marched on foot 300 miles to Fort Sumner, where they were to be taught the ways of "civilization" (Spicer, 1962).

Although expulsion is an extreme attempt to eliminate a certain minority from an area, annihilation is the most extreme action one group can take against another.

## Annihilation

**Annihilation** refers to *the deliberate extermination of a racial or ethnic group.* In recent years, it has also been referred to as *genocide,* a word coined to describe the crimes committed by the Nazis during World War II—crimes that induced the United Nations to draw up a convention on genocide. Annihilation deprives an entire group of people of their right to life in the same way that homicide deprives one person of the right to life.

Sometimes annihilation occurs as an unintended result of new contact between two groups. For example, when the Europeans arrived in the Americas, they brought with them smallpox, a disease that was new to the people they encountered. Native American groups, including the Blackfeet, the Aztecs, and the Incas, who had no immunity against this disease, were nearly wiped out (McNeill, 1976). In most cases, however, the extermination of one group by another has been the result of deliberate action. Thus, the native population of Tasmania, a large island off the coast of Australia, was exterminated by Europeans during the 250 years after the island was discovered in 1642.

The largest, most systematic program of ethnic extermination was the killing of 11 million people, close to 6 million of whom were Jews, by the Nazis before and during World War II. In nearly every country the Germans occupied, the majority of the Jewish population was killed. Relatively few Jews anywhere in Europe escaped (Dwork and van Pelt, 2002). Thus, in the mid-1930s, before the war, about 3.3 million Jews lived in Poland, but at the end of the war in 1945, only 73,955 remained (Baron, 1976). Among them, not a single known family remained intact.

Although attempts have been made to portray this mass murder of Jews as a secret undertaking of the Nazi elite that was not widely supported by the German people, the historical evidence suggests otherwise. For example, during a wave of anti-Semitism (anti-Jewish prejudice accompanied by violence and repression) in Germany in the 1880s—long before the Nazi regime—only 75 German scholars and other distinguished citizens protested publicly. During the 1930s, the majority of German Protestant churches endorsed the so-called "racial" principles that the Nazis used to justify first the disenfranchisement of Jews, then their forced deportation, and finally their extermination. Jews were blamed for a bewildering combination of "crimes," including "polluting the purity of the Aryan race" and causing the rise of communism while at the same time manipulating capitalist economies through their "secret control" of banks.

It would seem, then, that the majority of Germans supported the Nazi racial policies or at best were apathetic (Goldhagen, 1996). Although, in 1943, both the Catholic Church and the anti-Nazi Confessing Church finally condemned the killing of innocent people and pointedly stated that race was no justification for murder, it is fair to say that even this opposition was "mild, vague, and belated" (Goldhagen, 2002). The fact that such objections were raised at all suggests that the Nazis' plan to exterminate all Jews was not a well-kept military secret. The measure of its success is that some 60 percent of all Jews in Europe—36 percent of all Jews in the world—were slaughtered (computed from figures in Baron, 1976).

Another "race" also slated for extermination by the Nazis were the Gypsies, small wandering groups who appear to be the descendants of the Aryan invaders of India, the central Eurasian nomads. For the past thousand years or so, Gypsy bands have spread throughout the continents, largely unassimilated (Ulc, 1969/1975). In Europe, they were widely disliked and constantly accused of small thefts and other criminal behavior.

The sheer magnitude and horror of the Nazi attempt to exterminate the Jews provoked outrage and efforts by the nations of the world to prevent such circumstances from arising again. On December 11, 1946, the General Assembly of the United Nations passed by unanimous vote a resolution affirming that genocide was a crime under international law and that both principals and accomplices alike would be held accountable and punished. The assembly called for a convention on genocide that would define the offense more precisely and provide enforcement procedures for its repression and punishment. After two years of study and debate, the draft of the convention on genocide was presented to the General Assembly and adopted (United Nations, 1948). Article II of the convention defines genocide as any of the following acts committed with intent to destroy, in whole or in part, a national, ethnic, racial, or religious group as such:

(a) Killing members of the group
(b) Causing serious bodily or mental harm to members of the group
(c) Deliberately inflicting on the group conditions of life calculated to bring about its physical destruction in whole or part
(d) Imposing measures intended to prevent births within the group
(e) Forcibly transferring children of the group to another group

The convention further provided that any of the contracting parties could call on the United Nations to take action under its charter for the "prevention and suppression" of acts of genocide. In addition, any of the contracting parties could bring charges before the International Court of Justice.

In the United States, President Harry Truman submitted the resolution to the Senate on June 16, 1949, for ratification. However, the Senate did not act on the measure, and the United States did not sign the document. In 1984, President Ronald Reagan again requested the Senate to hold hearings on the convention so that it could be signed. The United States finally signed the document in 1988.

In the more than 60 years of its existence, the Genocide Convention has never been used to bring charges of genocide against a country. Numerous examples of genocide have occurred during that period, however. It appears to serve more of a symbolic purpose, by asking nations to go on record as being opposed to genocide, than an effective means of dealing with actual instances of genocide (Berry and Tischler, 1978).

## ● RACIAL AND ETHNIC IMMIGRATION TO THE UNITED STATES

Since the settlement of Jamestown in 1607, well over 45 million people have immigrated to the United States. Until 1882, the policy of the United States was almost

These Jews are arriving at the Nazi concentration camp known as Auschwitz. Most of them were murdered and cremated a short time later as part of the German policy of genocide against the Jews.

one of free and unrestricted admittance. The country was regarded as the land of the free, a haven for those oppressed by tyrants, and a place of opportunity. The words of Emma Lazarus, inscribed on the Statue of Liberty, were indeed appropriate:

> Give me your tired, your poor,
> Your huddled masses yearning to breathe free;
> The wretched refuse of your teeming shore.
> Send these, the homeless, tempest-tost to me.
> I lift my lamp beside the golden door!

To be sure, some had misgivings about the immigrants. George Washington wrote to John Adams in 1794, "My opinion with respect to immigration is that except for useful mechanics and some particular descriptions of men or professions, there is no need for encouragement." Thomas Jefferson was even more emphatic in expressing the wish that there might be "an ocean of fire between this country and Europe, so that it would be impossible for any more immigrants to come hither." Such fears, however, were not widely felt. There was the West to be opened, railroads to be built, and canals to be dug; there was land for the asking. People poured across the mountains, and the young nation was eager for population.

Immigration of white ethnics to the United States can be viewed from the perspective of old migration and new migration. The old migration consisted of people from northern Europe who came before the 1880s. The new migration was much larger in numbers and consisted of people from southern and eastern Europe who came between 1880 and 1920. The ethnic groups that made up the old migration included the English, Dutch, French, Germans, Irish, Scandinavians, Scots, and Welsh. The new migration included Poles, Hungarians, Ukrainians, Russians, Italians, Greeks, Portuguese, and Armenians.

Figure 10–4 shows the number of immigrants who came to the United States in each decade from 1820 to 2011. The new migration sent far more immigrants to the United States than the old migration. The earlier immigrants felt threatened by the waves of unskilled and uneducated newcomers, whose appearance and culture were so different from their own. Public pressure for immigration restriction increased. After 1921, quotas were established limiting the number of people who could arrive from any particular country. The quotas were designed specifically to discriminate against potential immigrants from the southern and eastern European countries. The discriminatory immigration policy remained in effect until 1965, when a new policy was established.

Table 10–2 lists the people who were excluded from immigrating to the United States during each of the periods in its history. As you can see, the United States was much more lenient during the early days

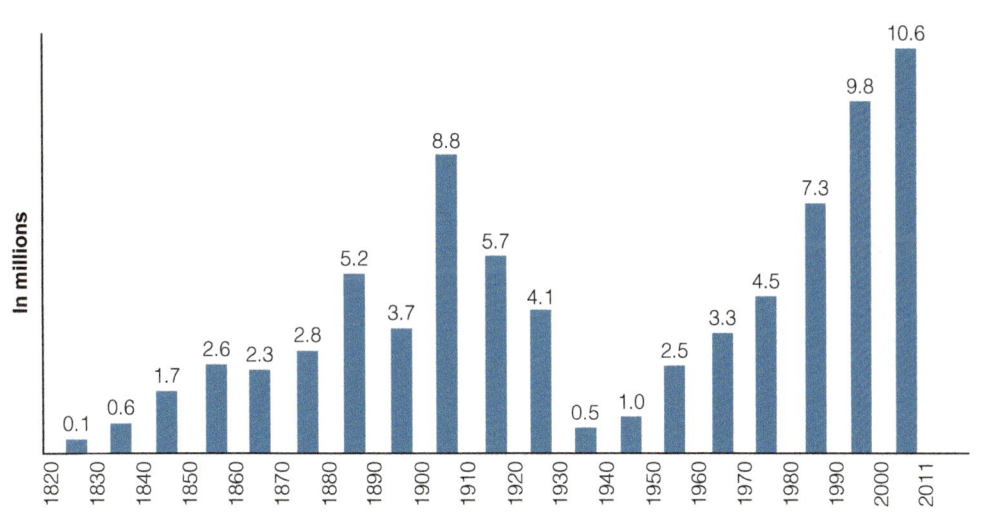

1882: Chinese immigration prohibited, more restrictions placed on immigration.

Immigration reached a peak between 1901 and 1910, when 8.8 million newcomers (8 million of them white Europeans) arrived in the United States.

1921: First immigration quotas (by country) announced.

1965: Quotas by hemisphere replaced limits by country.

1986: Immigration Reform Control Act created; means for illegal immigrants in U.S. since 1982 to be granted amnesty; led to legalization of many agricultural workers. Major beneficiaries: Mexicans (3 million have become legal immigrants since 1987), Cubans, Haitians.

**FIGURE 10-4** **Immigration to the United States, 1820–2011**
*Source: U.S. Department of Homeland Security. 2012. Annual Flow Reports.*

of its history. However, even with periods of restrictive immigration, the United States has had one of the most open immigration policies in the world, and it continues to take in more legal immigrants each year than the rest of the world combined (Kotkin and Kishimoto, 1988).

## Immigration Today Compared with the Past

The past 45 years have seen a marked shift in U.S. immigration patterns. From the beginning of the country's birth until the 1960s, most immigrants came primarily from northwestern European countries—Great Britain, Ireland, Germany, Scandinavia, France—and from Canada. In 1965, however, a major change in U.S. immigration policy occurred. The national origins quota system, which granted visas mainly to people coming from Western European countries, particularly Great Britain and Germany, was repealed. Under the new immigration system, family ties to people already living in the United States became the key factor that determined whether a person was admitted into the country. The number of people allowed to come into the country also was increased.

These shifts in immigration patterns have resulted in a much more racially and ethnically diverse foreign-born population. In 1890, only 1.4 percent of the foreign-born population was nonwhite. By 1970, 27 percent of this population was nonwhite, and by 2007, 54 percent was nonwhite. (For more on this topic, see "Our Diverse World: Whites: The New Minority.")

With the large number of people immigrating to the United States, 13 percent of U.S. residents are foreign-born. This is the highest percentage of foreign-born residents since World War II and more than double the 1970 level. Many of these people are recent arrivals. Of the 40 million foreign-born people living in the United States in 2006, half arrived after 1995.

TABLE 10-2  United States Immigration Restrictions

| | |
|---|---|
| 1769–1875 | No restrictions; open-door policy |
| 1875 | No convicts; no prostitutes |
| 1882 | No idiots; no lunatics; no people likely to need public care; Start of head tax |
| 1882–1943 | Chinese Exclusion Act, which barred Chinese immigrants for 10 years (later extended), and prohibited Chinese immigrants from naturalizing; provisions repealed in 1943 |
| 1885 | No gangs of cheap contract laborers |
| 1891 | No immigrants with dangerous contagious diseases; no paupers; no polygamists |
| | Start of medical inspections |
| 1903 | No epileptics; no insane people; no beggars; no anarchists |
| 1907 | No feeble-minded; no children under 16 unaccompanied by parents; no immigrants unable to support themselves because of physical or mental defects |
| | Gentlemen's Agreement, in which Japan and the U.S. agreed to stop issuance of passports for new Japanese laborers to come to the U.S. |
| | No adults unable to read or write; Start of literacy tests |
| 1917 | Immigration Act, which barred immigration from most countries in Asia |
| 1921 | No more than 3% of foreign-born of each nationality in U.S. in 1910; total about 350,000 annually |
| 1924–1927 | National Origins Quota Law; no more than 2% of foreign-born of each nationality already in U.S. in 1890; total about 150,000 annually |
| 1940 | Alien Registration Act; all aliens must register and be fingerprinted |
| 1950 | Exclusion and deportation of aliens dangerous to national security |
| 1965 | National Origins Quota system abolished; no more than 20,000 from any one country outside the Western Hemisphere; total about 170,000 annually |
| | Start of restrictions on immigrants from other Western Hemisphere countries; no more than 120,000 annually |
| | Preference to refugees, aliens with relatives here, and workers with skills needed in the United States |
| 1980 | Refugee Act of 1980, which repealed ideological and geographical preferences that had favored refugees fleeing communism and Middle Eastern countries |
| 1986 | The Immigration Reform and Control Act (IRCA) took effect, granting amnesty to illegal immigrants living in the U.S. since 1982 |
| | Increased sanctions against employers for hiring illegal immigrants |
| 1990 | Immigration Act, which increased ceiling on new immigrant visas, especially for family members of U.S. citizens and for skilled foreigners requested by U.S. |
| 1996 | Laws enacted to make it easier to deport immigrants who commit crimes; Tougher restrictions on ability of immigrants to collect welfare |

*Source:* Smithsonian Institution exhibit, Washington, D.C.

Today's immigrants are unique in their ethnic origins, education, and skills. Of the foreign-born population, 53 percent were born in Latin America, 28 percent in Asia, 12 percent in Europe, and the remaining 7 percent in the rest of the world. The waves of immigration during earlier periods in U.S. history were mostly from European countries. The education levels of today's foreign-born are at two extremes. Whereas most native-born people have completed high school, immigrants are less likely to have done so. At the same time, many other immigrants are highly educated, and immigrants are more likely than native-born people to have advanced college degrees (U.S. Bureau of the Census, 2012).

### Illegal Immigration

The U.S. census estimated that in 2011, about 11.5 million undocumented immigrants lived in the United States. Of that total, 6.8 million were from Mexico, representing 59 percent of the unauthorized population. From 2000 to 2011, the Mexican-born unauthorized population increased by 2.1 million, or an annual average of 190,000. The other main countries where illegal

---

**OUR DIVERSE WORLD**

## Whites: The New Minority

In 1965, Congress passed the U.S. Immigration and Nationality Act, which dramatically changed the U.S. immigration policy. The act removed specific country quotas and allowed people who had relatives living in the United States to be given immigration priority. This opened the door to new waves of immigrants from Latin America and Asia. These immigrants had higher birth rates than the U.S.-born population. Now, nearly 50 years since the change, there is an unmistakable trend toward fewer non-Hispanic white births and a rapid increase in births to Latino, Asian American, and interracial couples.

Minority populations are younger than whites, so are more likely to be having and raising children. Non-Hispanic whites have the oldest median age (42.3, in 2011), and Hispanics have the youngest (27.6). Between 2000 and 2010, racial and ethnic minorities accounted for nearly 92 percent of the nation's growth. Most of that increase was due to Hispanics, who have higher numbers of children than do non-Hispanic whites. Non-Hispanic whites accounted for only 8.3 percent of the growth over the decade. (See Figure 10–5.)

This trend is making the United States more racially and ethnically diverse. In the not too distant future, the United States will become a country where the majority of the people are part of a minority group. Minorities currently account for nearly 37 percent of the U.S. population but will become 50 percent of the population by 2042.

During the past few decades, the baby boom generation, which is predominantly white, has been a major force in the U.S. population and the labor force. When the members of that generation entered the workforce, their racial and ethnic characteristics were

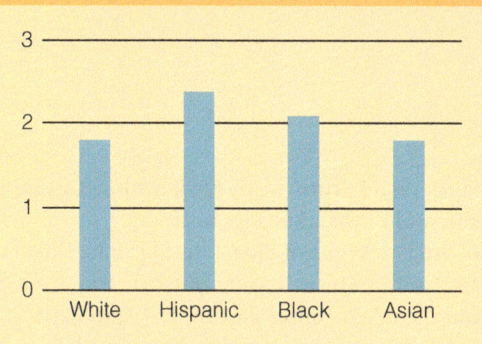

**FIGURE 10-5** Lifetime Number of Children per Woman
*Sources: Pew Research Center. "Analysis of the 2010 American Community Service IPUMS"; Passel, Jeffrey, Gretchen Livingston, and D'Vera Cohn. May 17, 2012. "Explaining Why Minority Births Now Outnumber White Births." Pew Research Center (www.pewsocialtrends.org/2012/05/17/explaining-why-minority-births-now-outnumber-white-births/), accessed July 25, 2012; U.S. Bureau of the Census Newsroom. May 17, 2012. "Most Children Younger Than Age 1 Are Minorities." U.S. Bureau of the Census Reports (www.census.gov/newsroom/releases/archives/population/cb12-90.html), accessed July 25, 2012.*

similar to those of previous generations; they were replacing people who were U.S.-born whites. As they have aged, they have been replaced by people who are much more likely to be Hispanic, multiracial, or Asian.

*Sources:* Passel, Jeffrey, Gretchen Livingston, and D'Vera Cohn. May 17, 2012. "Explaining Why Minority Births Now Outnumber White Births." Pew Research Center (www.pewsocialtrends.org/2012/05/17/explaining-why-minority-births-now-outnumber-white-births/), accessed July 25, 2012; U.S. Bureau of the Census Newsroom. May 17, 2012. "Most Children Younger Than Age 1 Are Minorities." U.S. Bureau of the Census Reports (www.census.gov/newsroom/releases/archives/population/cb12-90.html), accessed July 25, 2012.

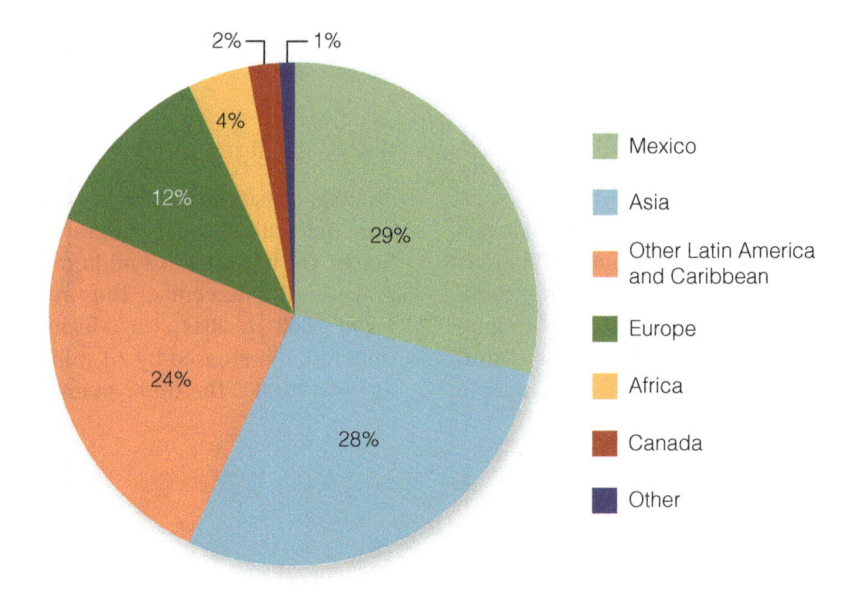

**FIGURE 10-6** Where Do Immigrants Come From?
Source: Jones-Correa, Michael. July 2012. Contested Ground: Immigration in the United States. Washington, D.C.: Migration Policy Institute.

- Mexico 29%
- Asia 28%
- Other Latin America and Caribbean 24%
- Europe 12%
- Africa 4%
- Canada 2%
- Other 1%

immigrants came from were El Salvador, Guatemala, Honduras, and China. (See Figure 10–6.)

There are two types of illegal immigrants: those who enter the United States legally but overstay their visa limits and those who enter illegally to begin with. Those who migrate over the border each day to work also must be figured into the total. Since 1970, illegal immigration has figured prominently in the ethnic make-up of several regions of the United States because illegal immigrants tend to settle in certain states. California is the leading state of residence for the illegal population with 25 percent, followed by Texas (14%), Florida (7%), and New York (6%) (Homeland Security, 2009).

In an attempt to control illegal immigration in 1986, the Immigration Reform and Control Act (IRCA) was passed, a law designed to control the flow of illegal immigrants into the United States. The law makes it a crime for employers, even individuals hiring household help, to knowingly employ an illegal immigrant. Stiff fines and criminal penalties can be imposed if they do so. The law also provided legal status to illegal immigrants who entered the United States before 1982 and who had lived here continuously since then. Between 1989 and 1993, 2.7 million people were granted legal resident status under special provisions of the IRCA. These people had lived and worked in the country illegally during the 1980s. Interestingly, only 4 percent of those who applied for amnesty under the 1986 act worked in farming, fishing, or forestry, running counter to the perception that most illegal immigrants are migrant farm workers. Needless to say, IRCA had little effect in stopping the flow of illegal immigrants.

Between 1892 and 1924, some 16 million immigrants came through Ellis Island in New York Harbor on the way to their new life in the United States.

## AMERICA'S ETHNIC COMPOSITION TODAY

What is America's racial and ethnic composition today? (See Figure 10–7.) The United States is perhaps the most racially and ethnically diverse country in the world. Unlike many other countries, it has no ethnic group that constitutes a numerical majority of the population. In the following discussion, we will examine the major groups in American society.

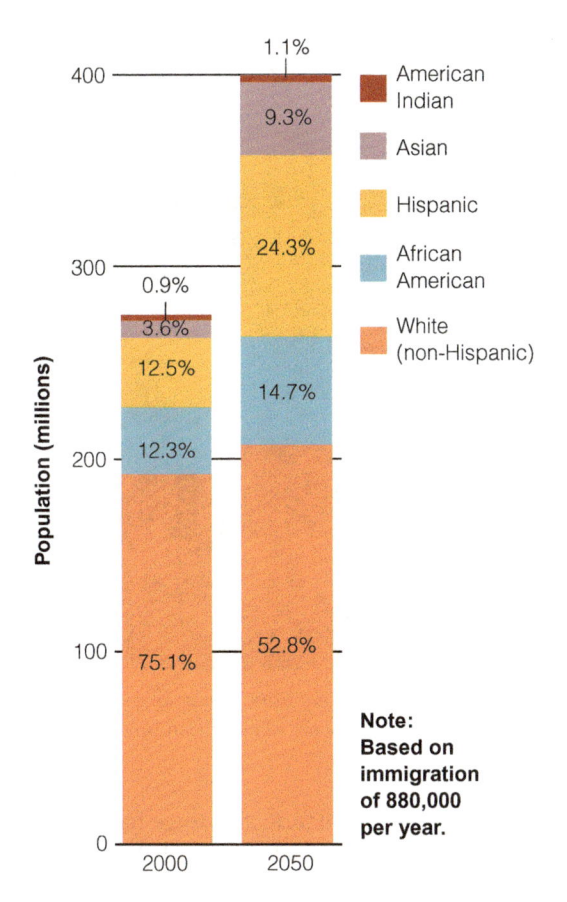

**FIGURE 10-7** Racial and Ethnic Make-up of U.S. Population, 2000 and 2050
*Sources: U.S. Bureau of the Census. March 2001. "Overview of Race and Hispanic Origin"; U.S. Bureau of the Census. 2000 Brief. Population Projections Program.*

## White Anglo-Saxon Protestants

About 41 million people claim some English, Scottish, or Welsh ancestors. These Americans of British origin are often grouped together and called white Anglo-Saxon Protestants (WASPs). Although in numbers they are a minority within the total American population, they have been in America the longest (aside from the Native Americans) and, as a group, have traditionally had the greatest economic and political power in the country. As a result, WASPs often have acted as if they were the ethnic majority in America, influencing other ethnic groups to assimilate or acculturate to their way of life, the ideal of Anglo conformity.

The Americanization of immigrant groups has been the desired goal of the dominant WASPs during many periods in American history. Contrary to the romantic sentiments expressed on the base of the Statue of Liberty, immigrant groups who came to America after the British Protestants had become established met with considerable hostility and suspicion.

The 1830s and 1840s saw the rise of the "native" American movement, directed against recent immigrant groups (especially Catholics). In 1841, the American Protestant Union was founded in New York City to oppose the "subjugation of our country to control of the Pope of Rome, and his adherents" (Leonard and Parmet, 1972). On the East Coast, Irish Catholics were feared and, in the Midwest, so were the German freethinkers. Protestant religious organizations across America joined forces and urged "native" Americans to organize to offset foreign voting blocs. They also conducted intimidation campaigns against foreigners and tried to persuade Catholics to renounce their religion for Protestantism (Leonard and Parmet, 1972).

As the twentieth century dawned, American sentiments against immigrants from southern and eastern Europe were running high. In Boston, the Immigration Restriction League was formed, which directed its efforts toward keeping out racially "inferior" groups, who were depicted as inherently criminal, mentally defective, and marginally educable. The league achieved its goal in 1924, when the government adopted a new immigration policy that set quotas on the numbers of immigrants to be admitted from various nations. Because the quotas were designed to reflect (and reestablish) the ethnic composition of America in the 1890s, they heavily favored the admission of immigrants from Britain, Ireland, Germany, Holland, and Scandinavia. This new policy was celebrated as a victory for the "Nordic" race (Krause, 1966).

Another expression of Anglo-conformity pressure was the Americanization Movement, which gained strength from the nationalistic passions brought on by World War I. Its stated purpose was to promote the very rapid acculturation of new immigrants. Thus, "federal agencies, state government, municipalities, and a host of private organizations joined in the effort to persuade the immigrant to learn English, take out naturalization papers, buy war bonds, forget his former origins and culture, and give himself over to patriotic hysteria" (Gordon, 1961/1975). From World War II until the early 1960s, Anglo conformity was essentially an established ideal of the American way of life. Since the upheavals of the 1960s, however, many ethnic groups have organized and reacted strongly against Anglo conformity. America once again is focusing on its ethnic diversity, and the assumptions of Anglo conformity are no longer dominant.

## African Americans

African Americans represent the second-largest race or ethnic group in the United States. According to the Census Bureau, 43.9 million African Americans live in the United States. Blacks now make up 13.5 percent of the U.S. population, the largest percentage since 1880 (U.S. Bureau of the Census, 2012).

The vast majority of all African Americans live in urban areas, and more than half live in the southern states. This is a significant shift from the 1940s, when roughly 80 percent of American blacks lived in the South and worked in agriculture. The states with the largest percentages of African Americans are Mississippi with 38 percent, followed by Louisiana (33%), Georgia (32%), Maryland (31%), South Carolina (29%), and Alabama (27%) (U.S. Bureau of the Census, 2012).

Immigrants have been accounting for a greater share of all blacks in the United States as their numbers have increased. These black immigrants have come to the United States from the West Indian countries of Jamaica, Haiti, the Dominican Republic, Barbados, and Trinidad. In addition, a significant number of African blacks are also entering the United States each year. About 2.8 million African Americans are foreign-born. More than one-fourth of the black population in New York, Boston, and Miami is foreign-born.

Even though the number of African immigrants from Ethiopia, Ghana, Kenya, and Nigeria has increased dramatically during the past decade, Latin America in general, and the Caribbean basin specifically, is the source of most U.S. immigrants of African descent. Of the African American foreign-born population, about 63 percent come from Caribbean nations (Kent, 2007).

The African American economic situation has been improving. Although 27.4 percent of the African American population fell below the poverty rate in 2010, that was lower than most years since the U.S. Census Bureau began collecting poverty data in 1959. The median African American household income was $32,068 in 2010. The income of a white family has been increasing more rapidly, however, and the ratio of African American to white earnings has actually fallen (U.S. Bureau of the Census, February 2012).

Why have African American families' earnings lost ground compared to white families' earnings over the past two decades? A major reason is the growth in female-headed families, who have only one-third the annual income of African American married-couple families. Only 45 percent of African American families are married-couple families. Another factor is that the average African American family has fewer members in the labor force than white families have. Even if African Americans and whites held comparable jobs and earned equal pay, the higher number of wage earners per family for whites would still keep the average African American family income below that of whites (U.S. Bureau of the Census, February 2009).

## Hispanics (Latinos)

"Hispanic," "Hispano," "Latino," "Latin," "Mejicano," "Spanish," "Spanish-speaking"—which is the right term to use when referring to the Spanish-speaking population in the United States? Political, racial, linguistic, and historical arguments have been advanced for using each of these terms. Historically, usage has gone from "Spanish" or "Spanish-speaking" to "Latin American," "Latino," or "Hispanic." Geographically, *Hispanic* is preferred in the Southeast and much of Texas. New Yorkers use both *Hispanic* and *Latino*. Some people in New Mexico prefer *Hispano*.

Politically, the term *Hispanic* has been identified with more conservative viewpoints, and the term *Latino* has been associated with more liberal politics. This is partially because Hispanic is an English word meaning "pertaining to ancient Spain." The U.S. Census Bureau has decided to settle on one term, *Hispanic*.

The Hispanic population represents the largest minority group in the United States. The 50.5 million Hispanics in the United States in 2010 comprised 16.3 percent of the total population. California (14.0 million) has the largest Hispanic population of any state, followed by Texas (9.5 million) and Florida (4.2 million). (See Table 10-3 for states with the largest concentrations of Hispanics.)

Immigrants of African descent from the Caribbean basin account for an increasing share of all African Americans in the United States.

TABLE 10–3   States with Large Hispanic Populations

| State | Hispanic Population (millions) | Percent of Total Population |
|---|---|---|
| California | 14.0 | 37.6 |
| Texas | 9.5 | 37.6 |
| Florida | 4.2 | 22.5 |
| New York | 3.4 | 17.6 |
| Illinois | 2.0 | 15.8 |

*Source*: U.S. Bureau of the Census. May 2011. "2010 Summary File 1: The Hispanic Population 2010." Census Brief (www.hacu.net/images/hacu/OPAI/2012_Virtual_Binder/2010%20census%20brief%20-%20hispanic%20population.pdf).

Nearly two-thirds (63%) of the nation's Hispanics are of Mexican origin. Another 9.2 percent are of Puerto Rican background, with 3.5 percent Cuban, 3.3 percent Salvadoran, and 2.8 percent Dominican. The rest are of Central American and South American origin. More than three-fourths of Hispanics live in the West or South (U.S. Bureau of the Census, 2011).

The Hispanic population in the United States is more diverse than it was in the past. The group of Hispanics of Mexican origin—the largest Hispanic group—has been growing substantially. Another fast-growing group, in percentage terms, is the group of "other" Hispanics. Many of those who choose "Other Spanish/Hispanic/Latino" on the U.S. census form are immigrants from Central or South America. "Other" Hispanics also include a growing number of people with multiethnic backgrounds who do not identify with a single country or region of origin. (For more on Hispanic diversity, see "Thinking About Social Issues: Are You Hispanic, Latino, or Neither?")

## THINKING ABOUT SOCIAL ISSUES: Are You Hispanic, Latino, or Neither?

Over the years, the U.S. Census Bureau has used a variety of methods to classify Hispanics and Latinos. In the 1930s, during the Mexican Revolution and prior to the Great Depression, as large numbers of Mexicans moved to the United States, the bureau designated "Mexican" as a racial category. Of the approximately 1.4 million people classified as Mexican in the 1930 census, most lived in Texas or California. As the Depression got worse, between 400,000 and 500,000 Mexicans moved back to Mexico.

The Hispanic/Latino population increased again during the Civil Rights era, causing the Census Bureau to renew its efforts to provide a common label. Using the labels "persons of Spanish surname" or "persons of Spanish language" to count the population seemed inadequate. In 1970, the Census Bureau decided that *Hispanic* and *non-Hispanic* were the most appropriate terms. Hispanics were people from Mexico, Puerto Rico, Cuba, Central or South America, and other Spanish-speaking countries.

Some people objected to the *Hispanic* term, claiming it emphasizes the Spanish element while neglecting the Latin American roots. They advocated for use of the "Latino" label. Opponents of the *Latino* term claimed it ignores the Spanish roots and excludes people originating from Spain. Today, both terms are accepted even though there is no agreement on a common term. A 2011 study by the Pew Hispanic Center found that 51 percent of respondents did not have a preference concerning the two terms, 33 percent preferred the *Hispanic* identity, and 14 percent opted for the *Latino* label. The use of the terms "Hispanic" and "Latino" to describe Americans of Spanish origin or descent is unique to the United States. Outside of the United States, these terms are not widely used.

The Census Bureau considers Hispanics or Latinos an ethnic and not a racial group, and asks them to select a racial category once they choose one or the other. In 2000, 42 percent of the Hispanic population that identified a race indicated "Some other race." Forty-six percent indicated they were white, but many Hispanics did not answer the race question at all. The Census Bureau has attempted to move people away from the "other" racial designation. The questionnaire emphasizes that "Hispanic origins are not races."

This effort changed the way Hispanics/Latinos identify themselves racially. For example, the percentage of Hispanics/Latinos selecting the white racial category increased from 48 percent in the 2000 census to 63 percent in 2010. Cubans have the strongest preference for the white racial category (88 percent). Dominicans have the weakest preference for this racial category (35 percent), opting instead for the "other" racial category (49 percent).

United States race relations have long been understood in terms of black and white. Until recently, many books on the subject did not even mention other races or did so only as a brief afterthought. Now Hispanics have replaced blacks as the primary minority. The Census Bureau appears to be suggesting that there is a shared common culture among U.S. Hispanics. Most Hispanics, however, do not agree; nearly seven in ten (69%) say Hispanics in the United States have many different cultures. Some people have expressed concern that efforts to make Hispanics a single minority group—for purposes ranging from elections to education to the allocation of public funds—are unnecessarily dividing American society along racial and ethnic lines.

*Sources:* Taylor, Paul, Mark Hugo Lopez, Jessica Hamar Martínez, and Gabriel Velasco. "When Labels Don't Fit: Hispanics and Their Views of Identity," April 4, 2012. Pew Hispanic Research Center. http://www.pewhispanic.org/2012/04/04/when-labels-dont-fit-hispanics-and-their-views-of-identity/ accessed November 23, 2012.

Hispanics are not a very well understood segment of the population. First, no one knows exactly how many Hispanics have crossed the border from Mexico as illegal immigrants. Second, many Americans of Hispanic descent do not identify themselves as Hispanic on census forms and are not counted as such.

People born in Latin America can be found all across the United States, but most live in only a few areas. The difference is the place of birth. For example, 75 percent of U.S. residents born in the Caribbean live either in the New York or the Miami metro areas. More than half of those born in Mexico live in the Los Angeles area or in Florida or Texas (U.S. Bureau of the Census, May 2008).

Interestingly, although the vast majority of Hispanics come from rural areas, 90 percent settle in America's industrial cities and surrounding suburbs. Living together in tightly knit communities, they share a common language and customs.

With the Latin migration north, the United States has experienced the largest migration in its history. Around one-half of Hispanic residents in the United States were born abroad. Newcomers to the United States begin their stay with many economic and educational disadvantages compared to the average American; more than half have not graduated from high school, and most are unskilled.

*Mexican Americans* In 2010, nearly 32 million Mexican immigrants lived in the United States. This is 17 times the number in 1970. Mexicans now account for 29.5 percent of all U.S. immigrants living in this country. The second-largest nationality group of immigrants, people from India, account for just 4.5 percent of all immigrants. More than half (55%) of the Mexican immigrants in this country are illegal immigrants.

As recently as 1960, Mexico ranked seventh as a source of immigrants to the United States, behind Italy, Germany, Canada, the Soviet Union, the United Kingdom, and Poland. By 1980, Mexican immigrants were in first place.

Such a high concentration of immigrants from one country is not unprecedented. Irish immigrants represented a third or more of the immigrant population from 1850 to 1870. Germans were 26 percent to 30 percent of the foreign-born population from 1850 to 1900.

Mexicans have lower levels of education, lower incomes, and higher poverty rates and unemployment rates than other groups. They are likely to work in lower-skilled occupations (Pew Hispanic Center, 2009).

Some people of Mexican origin call themselves *Chicanos,* although the term is somewhat controversial. It has long been used as a slang word in Mexico to refer to people of low social class. In Texas, it came to be used for Mexicans who illegally crossed the border in search of work. Many Mexican Americans have taken to using the term themselves to suggest a tough breed of individuals of Mexican ancestry who are committed to achieving success in this country and are willing to fight for it (Castillo, 2005).

*Puerto Ricans* Also included under the category of Hispanics are Puerto Ricans. In 1898, the United States fought a brief war with Spain and as a result took over the former Spanish colonies in the Pacific (the Philippines and Guam) and the Caribbean (Cuba and Puerto Rico). Puerto Ricans were made full citizens of the United States in 1917. Although government programs improved their education and dramatically lowered the death rate, rapid population growth helped keep the Puerto Rican people poor. American business took advantage of the large supply of cheap, nonunion Puerto Rican labor and built plants there under very favorable tax laws.

More than 4.6 million Puerto Ricans live in the United States. Approximately 2.5 million Puerto Ricans live in the Northeast part of the United States, comprising 54 percent of the population nationwide. New York City has the largest population of Puerto Ricans in the country, followed by Philadelphia. Many return frequently to the island to visit relatives and friends.

Puerto Ricans living in the United States have the second-lowest median family incomes of the various Hispanic groups (Dominicans have the lowest). Ironically, the poverty in many Puerto Rican families is due in part to the ease of going back and forth between their homeland and the United States. The desire to return permanently one day to Puerto Rico interferes with a total commitment to assimilate into American culture (U.S. Bureau of the Census, 2010 American Community Survey).

*Cuban Americans* Currently, there are nearly 1.9 million people of Cuban ancestry in the United States. The first immigrants from Cuba arrived in the area now known as Florida as early as the 1500s, but most have come to the United States since the late 1950s. Only in the 1970s did they begin to have a visible cultural and economic effect on the cities where they have settled in sizable numbers.

Many Cubans came to the United States as a result of the 1959 revolution that catapulted Fidel Castro into power. At that time, Castro's rebel forces overthrew the government of Fulgencio Batista. Castro, a Marxist closely aligned with the former Soviet Union, began to restructure the social order, including appropriating for the state privately owned land and property. Professionals and businesspeople who were part of the established Cuban society felt threatened by these changes and fled to the United States. More than 155,000 Cubans immigrated to the United States between 1959 and 1962. As a whole, these immigrants have done extremely well in American society; they had the distinct advantage of coming to the United States with marketable skills and money.

The second wave of Cubans immigrated between 1965 and 1974. This movement involved an orderly departure on "freedom flights" of middle- and working-class people.

Another large wave of Cuban immigration occurred in 1980, when Castro allowed people to leave Cuba from Mariel Harbor. The result was a flotilla of boats bringing 125,000 refugees to the United States. This wave of immigrants was poorer and less well educated than the first. It also included several thousand prisoners and mental patients, many of whom were imprisoned in the United States as soon as they arrived. Others fled to Miami and other cities to lead a life of crime. Serious friction existed between the *Marielitos* and other Cuban immigrants because of differences in background and social class.

The fourth wave, which continues to the present, includes people who arrive on improvised boats often by way of Mexico, as well as those who obtain special visas through a lottery system the governments of the United States and Mexico agreed on in 1994.

Cubans are relatively recent immigrants, and the first-generation foreign-born predominate. They are exiles who came to the United States not so much because they preferred the U.S. way of life but because they felt compelled to leave their country. For that reason, most Cuban immigrants fiercely attempt to retain the culture and way of life they knew in Cuba. Of all the Hispanic groups, they are the most likely to speak Spanish in the home; eight of ten families do so.

Cubans are largely found in a few major cities. Metropolitan Miami (Dade County, Florida) is the undisputed center, with nearly 65 percent of all Cubans in the United States living there. In Miami, where Latinos make up a larger percentage of the population than Anglos, the city has become distinctively Cuban. Most other Cubans live in New York City, Jersey City and Newark in New Jersey, Los Angeles, and Chicago, all of which are large centers for Hispanics in general.

Acculturation and assimilation have been slow in Miami because the community is so self-sufficient and has such a large immigrant base. There also appears to be a lack of social and cultural integration between Cubans and other Hispanic groups in U.S. cities with sizable and differentiated Spanish-speaking populations. Of the major Hispanic groups, only the Cubans have come as political exiles, and this has resulted in social, economic, and class differences. In the New York City area, Cubans and Puerto Ricans maintain a distinct social distance. Many Cubans feel or perceive that they have little in common with Puerto Ricans, Mexican Americans, or Dominicans.

Cubans who live in the United States have fared better than some other Hispanic immigrant groups. They are well educated, with 24 percent of Cubans having a bachelor's degree, more than double the rate for Mexicans, the largest Hispanic group. The median family income of Cubans is similar the earnings of the average Hispanic family (U.S. Bureau of the Census, 2010 American Community Survey).

## Asian Americans

Most Asian Americans are concentrated in the major metropolitan areas. California has the largest Asian population (5.8 million); New York (1.4 million) and Texas (915,000) follow in population. In Hawaii, Asians comprise the highest proportion of the total population (55%), with California (14%) and New Jersey and Washington (8% each) next. Asians are the largest minority group in Hawaii and Vermont.

The first Asians to settle in America in significant numbers were the Chinese. Some 300,000 Chinese migrated here between 1850 and 1880 to escape the famine and warfare that plagued their homeland. Initially, they settled on the West Coast, where they took back-breaking jobs mining and building railroads. However, they were far from welcome and were subjected to a great deal of harassment. In 1882, the government limited further Chinese immigration for ten years. This limitation was extended in 1892 and again in 1904, finally being repealed in 1943. The state of California imposed special taxes on Chinese miners, and most labor unions fought to keep them out of the mines because they took jobs from white workers. In the late 1800s and early 1900s, numerous riots and strikes were directed against the Chinese, who drew back into their Chinatowns for protection.

The harassment proved successful. In 1880, 105,465 Chinese lived in the United States; by 1900, the figure

These Japanese Americans are standing in line for a meal at the internment camp in Puyallup, Washington.

had dropped to 89,863 and, by 1920, to 61,729. The Chinese population in the United States began to rise again only after the 1950s (U.S. Bureau of the Census, 1976). As of 2010, 3.8 million people of Chinese ancestry lived in the United States (U.S. Bureau of the Census, 2011).

Japanese immigrants began arriving in the United States shortly after the Chinese and quickly joined them as victims of prejudice and discrimination. Feelings against the Japanese ran especially high in California, where one political movement attempted to have them expelled from the United States. In 1906, the San Francisco Board of Education decreed that all Asian children in that city had to attend a single, segregated school. The Japanese government protested, and after negotiations, the United States and Japan reached what became known as a gentlemen's agreement. The Japanese agreed to discourage emigration, and President Theodore Roosevelt agreed to prevent the passage of laws discriminating against Japanese in the United States.

Initially, Japanese immigrants were minuscule in number. In 1870, only 55 Japanese lived in America and, in 1880, a mere 148. By 1900, 24,326 lived in this country, and subsequently their numbers grew steadily. By 1970, they had surpassed the Chinese in number (U.S. Bureau of the Census, 1976), but later figures show that, despite a sharp increase after 1970, the number of ethnic Japanese has been far fewer than the number of Chinese. In 2010, approximately 1.3 million people of Japanese ancestry lived in the United States (U.S. Bureau of the Census, 2011).

Japanese Americans were subjected to especially vicious mistreatment during World War II. Fearing espionage and sabotage from the ethnic minorities whose home countries were at war with the United States, President Franklin D. Roosevelt signed Executive Order 9066, empowering the military to "remove any and all persons" from certain regions of the country. Although many German Americans actively demonstrated on behalf of Germany before the United States entered World War II, no general action was taken against them as a group. Several hundred noncitizen Italian Americans were interned for up to two years during the war. However, General John L. DeWitt ordered that all individuals of Japanese descent be evacuated from three West Coast states and moved inland to relocation camps for the duration of the war. In 1942, 120,000 Japanese, including some 77,000 who were American citizens, were moved and imprisoned solely because of their ethnicity—even though not a single act of espionage or sabotage against the United States ever was attributed to one of their number (Simpson and Yinger, 1972). Many lost their homes and possessions in the process. Even with such rampant racism, some Japanese Americans volunteered for the 442nd Regimental Combat Team, a fighting group composed solely of Japanese Americans. The 442nd became the most decorated unit in U.S. history.

In 1988, President Reagan signed legislation apologizing for the wartime internment of Japanese in the United States. The legislation attempted to "right a great wrong" by establishing a $1.25 billion trust fund as reparation for the imprisonment. Each eligible person was to receive a $20,000 tax-free award from the government. The president noted, as he signed the legislation, "Yes, the nation was then at war, struggling for its survival. And it's not for us today to pass judgment upon those who may have made mistakes while engaging in the great struggle. Yet we must recognize that the internment of Japanese Americans was just that, a mistake" (Johnson, 1988).

Compared with the earlier group of Asian immigrants, who came primarily from China and Japan, since the 1960s, many Asians in the United States have come from Vietnam, the Philippines, Korea, India, Laos, and Cambodia. Many of the Asian immigrants from Vietnam, Cambodia, and Laos were involuntary migrants who were forced to leave their homes because they feared persecution after the United States left Southeast Asia. Two distinct waves of refugees came from these countries. The first began in the 1960s and continued until the end of the Vietnam War in 1975. The first group of Southeast Asian immigrants were mainly middle- and upper-class Vietnamese who found ways to get their families and financial assets out of the country when it became clear that military victory for the United States was not going to be swift. This group numbered about 25,000.

The second wave of refugees was very different. Harsh economic conditions, political persecution within Vietnam, and widespread genocide by the Khmer Rouge government in Cambodia created a flood of refugees desperate to leave the area. Many crowded into unsafe boats and hoped to reach Hong Kong, Malaysia, and other neighboring countries. Some of these boat people eventually settled in the United States. Between 1975 and 1994, more than 700,000 Vietnamese refugees and 500,000 Cambodians and Laotians resettled in the United States (U.S. Office of Refugee Resettlement, 1995). Close to half (about 45%) of the nation's Asian-born population live in three metropolitan areas: Los Angeles, New York, and San Francisco.

"Despite their diversity, Asian Americans are often perceived as a single group, not only by other Americans but in their own minds as well" (Jacoby, 2000). What appears to unite them is their newness to this country. Approximately 60 to 70 percent of Asian Americans are foreign-born.

Most Asian immigrants are middle class and highly educated. Fifty percent of Asians 25 and older have a bachelor's degree or higher level of education. For all Americans 25 and older, the percentage is 28 percent.

(Immigrants from Vietnam, Cambodia, and other Indochinese countries are the exception.) Although Asian Americans comprised 5.6 percent of the total U.S. population, they accounted for nearly 16 percent of the 2010/2011 freshman class at Harvard and other elite universities. In 2010, the median household income of Asian Americans was over $67,000, the highest among all race groups (U.S. Bureau of the Census, 2012).

## Native Americans

Beginning with the 2000 census, it is now possible to know how many people in the United States think of themselves as only Native American and how many define themselves as Native American in combination with another race. The 1990 census showed nearly 2 million Native Americans. By 2011, those noting they were Native American not in combination with another race numbered 5.2 million (U.S. Bureau of the Census, 2012).

This increase cannot be attributed to just natural population growth. Other factors that might have contributed to the higher number include improvements in the way the U.S. Census Bureau counts people on reservations and a greater likelihood for people (especially those of mixed Native American and white parentage) to report themselves as American Indian.

As early Europeans first stepped ashore on what they considered the New World, they usually were welcomed by the Native Americans. The Indians seemed to regard their lighter-complexioned visitors as something of a marvel, not only for their dress, beards, and "winged" ships, but even more for the items they brought that were unusual to their way of life, such as steel knives and swords.

Early European colonists encountered Native American societies that in many ways were as advanced as their own. Especially impressive were their political institutions. For example, the League of the Iroquois, a confederacy that ensured peace among its five member nations and was remarkably successful in warfare against hostile neighbors, was the model on which Benjamin Franklin drew when he was planning the Federation of States (Kennedy, 1961).

Native Americans were surprised at European intolerance for native religious beliefs, sexual and marital arrangements, eating habits, and other customs. At the same time, they became perplexed when Europeans built permanent structures of wood and stone, thus precluding movement. Even village and town-dwelling Native Americans were used to relocating when local game, fish, and especially firewood gave out.

The colonists and their descendants never questioned the view that the land of the New World was theirs. They took land as they needed it—for agriculture, for mining, and later for industry—and drove off the native groups. Some land was purchased, some acquired through political agreements, some through trickery and deceit, and some through violence. In the end, hundreds of thousands of Native Americans were exterminated by disease, starvation, and deliberate massacre. By 1900, only about 250,000 Indians remained (perhaps one-eighth of their numbers in precolonial times) (McNeill, 1976).

On the Pine Ridge Indian reservation, many families lack electricity, telephone, running water, or sewers. Life expectancy of 48 for males is like that in sub-Saharan Africa.

To make up for past injustices perpetrated against Native Americans, the federal government pays for a number of programs to assist them with education, health care, and housing. The government has also granted Native Americans special rights to govern themselves, so that they are subject only to federal rather than state and local laws. These rights, based on hundreds of treaties made in the eighteenth and nineteenth centuries, provide for Native American sovereignty and give tribes independence from outside governments in return for land. These rights have enabled some tribes to open casinos in states that do not allow gambling establishments.

Interestingly, many people informally claim American Indian ancestry, yet fewer than one-third of them identify themselves as Native American. Most people who claim Native American ancestry do so in combination with another ancestry group such as the English or Irish.

Native Americans are often not well off economically. The median income of American Indian and Alaska Native households is considerably below that of the rest of the country, $35,020 compared to $50,046 for the nation as a whole. The percentage of American Indians and Alaska Natives that were in poverty in 2010 was 28.4 percent. For the nation as a whole, the corresponding rate was 15.3 percent (U.S. Bureau of the Census, 2011).

The five largest Native American tribes are the Cherokee, Navajo, Choctaw, Chippewa, and Sioux. Considerable differences exist among the various tribes.

For instance, the Iroquois and the Creek are much better off economically than the Navajo (U.S. Bureau of the Census, 2011).

More than half of all Native Americans live on or near reservations administered fully or partly by the federal Bureau of Indian Affairs. Conditions on many reservations are very bad. The unemployment rate is often very high, with more than 60 percent of residents living in poverty. For example, the Oglala Sioux live on the Pine Ridge Indian Reservation, which is located in Shannon County and Jackson County, South Dakota, two of the poorest counties in the United States. Adolescent suicides are four times as high as the national average. The infant mortality rate is five times higher than the national average. Life expectancies of 48 years for males are like those in sub-Saharan Africa (U.S. Bureau of the Census, 2011).

## A Diverse Society

As is evident by now, the many racial and ethnic groups in the United States present a complex and constantly changing picture. Some trends in intergroup relations can be discerned and are likely to continue; new ones might emerge as new groups gain prominence. The resurgence of ethnic-identity movements probably will spread and might be coupled with more collective protest movements among disaffected ethnic and racial minorities who might demand equal access to the opportunities and benefits of American society.

It is important to realize that the old concept of the United States as a melting pot is both simplistic and idealistic. Many groups have entered the United States. Most encountered prejudice, some severe discrimination, and others the pressures of Anglo conformity.

Contemporary American society is the outcome of all these diverse groups coming together and trying to adjust. Indeed, if these groups are able to interact on the basis of mutual respect, this diversity can offer America strengths and flexibility not available in a homogeneous society.

## ■ SUMMARY

▶ **How do people think about the concept of race?**
Race refers to a category of people who are defined as similar because of a number of physical characteristics. Racial characteristics are arbitrarily chosen to suit the labeler's purposes. Races have historically been defined according to genetic, legal, and social criteria, each with its own problems. Although genetic definitions center on inherited traits such as hair and nose type, in fact, these traits have been found to vary independently of one another. Moreover, all humans are far more genetically alike than they are different.

▶ **What do we mean when we speak of an ethnic group?**
An ethnic group is a group with a distinct cultural tradition with which its own members identify and that might or might not be recognized by others. Many ethnic groups form subcultures with a high degree of internal loyalty and distinctive folkways, mores, values, customs of dress, and patterns of recreation. Above all, members of the group have a strong feeling of association.

▶ **What constitutes a minority group?**
A minority is a group of people who, because of physical or cultural characteristics, are singled out for differential and unequal treatment and who therefore regard themselves as objects of collective discrimination.

▶ **What is the difference between prejudice and discrimination?**
Prejudice is an irrationally based negative, or occasionally positive, attitude toward certain groups and their members. Prejudice is a subjective feeling; discrimination is an overt action. Discrimination can be defined as differential treatment, usually unequal and injurious, accorded to individuals who are assumed to belong to a particular category or group.

▶ **What is involved in assimilation?**
Assimilation is the process whereby groups with different cultures come to share a common culture. Invariably, one group has a much larger role in the process than the other or others. The particular form of assimilation found in the United States is called Anglo conformity—the renunciation of ancestral cultures in favor of Anglo-American behavior and values.

▶ **How does pluralism differ from assimilation?**
Pluralism is the development and coexistence of separate racial and ethnic group identities within a society. Pluralism celebrates the differences among groups.

▶ **How has immigration to the United States changed over the years?**
Historically, the majority of immigrants to the United States have been from Europe. The old migration consisted of people from northern Europe who came before 1880. The new migration was far larger and consisted of people who came from southern and eastern Europe. Discriminatory quotas were set up at the beginning of the twentieth century to restrict the immigration of the latter groups. Today, the overwhelming majority of immigrants to the United States come from Latin America, Asia, and the Caribbean. Legal immigration to the United States has increased in recent decades.

▶ **How many illegal immigrants are in the United States?**

The U.S. census estimated that in 2011, about 11.5 million illegal immigrants lived in the United States. Of that total, 6.8 million were from Mexico, representing 59 percent of the unauthorized population. The other main countries from which illegal immigrants came were El Salvador, Guatemala, Honduras, and China. Illegal immigrants tend to be young and to settle in California, New York, Florida, Texas, and Illinois.

▶ **How does the racial and ethnic mix in the United States differ from that in other countries?**

With the large number of people immigrating to the United States, 13 percent of U.S. residents are foreign-born. The United States is perhaps the most racially and ethnically diverse country in the world; no single ethnic group comprises a majority of the population.

## Media Resources

**CourseMate for *Introduction to Sociology*, Eleventh Edition**

Cengage Learning's Sociology CourseMate brings course concepts to life with interactive learning, study, and exam preparation tools that support the printed textbook. Access an integrated eBook, learning tools including glossaries, flashcards, quizzes, videos, and more in your Sociology CourseMate. Go to CengageBrain.com to register or purchase access.

# CHAPTER TEN STUDY GUIDE

## Key Concepts

Match each of the following concepts with its definition, illustration, or explanation.

a. Race
b. Ethnic group
c. Minority
d. Prejudice
e. Discrimination
f. Institutional prejudice
g. Assimilation
h. Anglo conformity
i. Pluralism
j. Subjugation
k. Segregation
l. Expulsion
m. Annihilation (genocide)
n. Americanization movement

____ 1. The act, process, or state of being set apart
____ 2. A group of people who, because of physical or cultural characteristics, are singled out for differential and unequal treatment and who therefore regard themselves as objects of collective discrimination
____ 3. An irrationally based negative, or occasionally positive, attitude toward certain groups and their members
____ 4. A category of people who are defined as similar because of a number of physical characteristics
____ 5. The development and coexistence of separate racial and ethnic group identities within a society
____ 6. The deliberate extermination of a racial or ethnic group
____ 7. The process of forcing a group to leave the territory in which it lives
____ 8. The process whereby groups with different cultures come to share a common culture
____ 9. A group with a distinct cultural tradition with which its own members identify and that might or might not be recognized by others
____ 10. Complex societal arrangements that restrict the life chances and choices of a minority
____ 11. The domination of one group by another
____ 12. A social movement advocating the complete assimilation of immigrant groups to the Anglo culture
____ 13. The renunciation of other ancestral cultures in favor of Anglo-American behavior and values
____ 14. Differential treatment, usually unequal and injurious, accorded to individuals who are assumed to belong to a particular category or group

## Key Thinkers

Match the thinkers with their main idea or contribution.

a. Johann Blumenbach
b. Louis Wirth
c. Robert K. Merton
d. Horace Kallen
e. Gerhard Lenski
f. Benjamin Franklin
g. Hoover Commission

____ 1. Wondered in 1753 about the costs and benefits to the United States of German immigration and concluded Germans would "contribute greatly to the improvement of a country"
____ 2. Showed that there are various ways in which prejudice and discrimination can interact with each other
____ 3. Developed a definition of minority group that considers only race and ethnic status
____ 4. Stated that a program for Native Americans must include progressive measures for their complete integration as tax-paying members of the larger society
____ 5. Eighteenth-century German physiologist who realized that racial categories do not reflect the actual divisions among human groups
____ 6. Principally responsible for the development of the theory of cultural pluralism
____ 7. Proposed that dominance of one group over another arises because people have a desire to control goods and services

CHAPTER 10  RACIAL AND ETHNIC MINORITIES   255

● **Central Idea Completions**

*Fill in the appropriate concepts and descriptions for each of the following questions.*

1. Compare and contrast the basic features of legal and social definitions of race.

    a. Legal: _____

    _____

    b. Social: _____

    _____

2. Compare and contrast the concepts of minority group, ethnic group, and racial group.

    a. Minority group: _____

    _____

    b. Ethnic group: _____

    _____

    c. Racial group: _____

    _____

3. Compare and contrast the basic characteristics of the following minority groups.

    a. Asian Americans: _____

    _____

    b. Cuban Americans: _____

    _____

    c. Mexican Americans: _____

    _____

4. Which ethnic or racial group in the United States is *least* likely to marry interracially and which is *most* likely to do so? Why?

    Least: _____

    Most: _____

    Reasons: _____

5. Drawing on people you know personally or from news and the media, provide an example for each of Merton's categories of discriminators and nondiscriminators.

    a. Unprejudiced nondiscriminator: _____

    b. Unprejudiced discriminator: _____

    c. Prejudiced nondiscriminator: _____

    d. Prejudiced discriminator: _____

6. Briefly describe the unique features of the minority group experience of Native Americans relative to other minority groups' experiences in the United States.

7. What are the major periods in U.S. policy on immigration?

## ● Critical Thought Exercises

1. What are the essential elements of race and racial identity? Review the example of Mark Linton Stebbins's claim of being black. Do you feel he made an adequate case for his claim? What criteria would you use if you were faced with the task of determining a person's race? What is the relative importance of biology versus cultural experience in the self-definition people have or do not have regarding the racial category to which they claim membership?

2. Consider the different ways of defining race: genetic markers, ancestry (blood), social reaction, self-definition. Is one of these best in all situations? Describe how each definition might be useful in one circumstance but less so in others.

3. Using what you know about Hispanics (from this chapter and any other sources), discuss the difficulties in applying concepts such as race and ethnic group. How do these same difficulties affect thinking about other racial or ethnic groups?

4. At your campus, what organizations are based on ethnic or racial identities? What kind of definition of race or ethnicity do they seem to be using? Look at their literature (the group charter, for example) and activities. What do these reveal about the group's relation to the dominant culture?

5. Develop a critical thought paper in which you examine the changes in U.S. immigration policy since 1965.

6. What changes in residence, popular culture, and education do you hypothesize are likely to develop during the next 25 years given the changing face of U.S. immigration that you have just described?

7. How has the rise of the World Wide Web increased the potential for spreading messages of hate and bigotry? Visit at least six sites that have as their goal the promotion of prejudiced and resentful views about a particular group of people. What are the common features of your six sites? Some people see these sites as a threat to American society, whereas others view this issue as overblown. After your examination, what is your assessment of the seriousness of the threat these sites pose? What action, if any, would you propose to restrict the dissemination of information across the Internet?

## Internet Activities

1. Visit www.pbs.org/race/000_General/000_00-Home.htm, the companion site to a PBS series, which includes slide shows, quizzes, a sorting game in which you try to classify photos of people by race, and other fun stuff.
2. A psychology project at Harvard offers an interactive exercise to test your "Implicit Associations" to categories of race, age, gender, and so on at https://implicit.harvard.edu/implicit/. Take some of the "tests" and see what they reveal about attitudes you might not have realized you had.
3. *The New York Times* has an interactive map that displays where immigrants came from and the counties they settled in. The map is historical, so you can trace the years from 1880 to 2000. You can select a single country of origin, and when you run the pointer over a county, it gives the data for that county. Visit the website at www.nytimes.com/interactive/2009/03/10/us/20090310-immigration-explorer.html.

## Answers to Key Concepts

1. k; 2. c; 3. d; 4. a; 5. i; 6. m; 7. l; 8. g; 9. b; 10. f; 11. j; 12. n; 13. h; 14. e

## Answers to Key Thinkers

1. f; 2. c; 3. b; 4. g; 5. a; 6. d; 7. e

# 11 Gender Stratification

## Are the Sexes Separate and Unequal?
- Historical Views
- Religious Views

*Thinking About Social Issues:* Let Women Vote and You Will Get Masculine Women and Effeminate Men
- Biological Views

*Our Diverse World:* Why Do Women Live Longer Than Men?
- Gender and Sex
- Sociological View: Cross-Cultural Evidence

## What Produces Gender Inequality?
- The Functionalist Viewpoint
- The Conflict Theory Viewpoint

## Gender-Role Socialization
- Childhood Socialization
- Adolescent Socialization

## Gender Inequality and Work
*Day-to-Day Sociology:* Speaking, Writing, or Blogging—Nowhere to Hide Gender
- Job Discrimination

*How Sociologists Do It:* What Happened to the Men?
*Our Diverse World:* Who Is a Better Boss?

## Summary

## LEARNING OBJECTIVES

After studying this chapter, you should be able to do the following:

- Contrast biological and sociological views of sex and gender.
- Describe the concept of patriarchal ideology.
- Understand the functionalist and conflict theory viewpoints on gender stratification.
- Explain the process of gender-role socialization.
- Describe gender differences in the world of work.
- Be aware of the effect of changes in gender roles in U.S. society.

---

In a study of the top 100 most popular films in 2011, males accounted for 67 percent of all characters and 84 percent of the protagonists. Female characters were younger than their male counterparts. The majority of female characters were in their 20s and 30s. The majority of male characters were in their 30s and 40s.

Male characters were much more likely than females to be portrayed as leaders. Males were 93 percent of political and government leaders, 83 percent of business leaders, and 70 percent of scientific and intellectual leaders. This can be partially explained by the fact that most female characters were under the age of 40 (Lauzen, 2012).

Other research has shown that there are two common roles among the female characters: the traditional noncontroversial woman and the sexually suggestive woman. Females are five times as likely as males to be wearing revealing clothing.

When 1,034 children's television shows during a typical week were reviewed, it turned out again that male characters were much more common than female characters. Females were almost four times as likely as males to be shown in sexy attire. Females commonly had small waists and unrealistic figures. In cartoons, males often had a large chest and an unrealistically muscular physique (Smith and Cook, 2006).

From a very early age on, gender-role socialization is occurring, whether it is in the media or in real life. What does it mean to be socialized as a man or a woman? How much of gender is ascribed and how much is achieved?

In this chapter, we will look at some of the differences between the sexes, examine cross-cultural variations in gender roles, and try to understand how a gender identity is acquired. In the process, we will also focus on the changes that are occurring in gender roles in American society.

## ARE THE SEXES SEPARATE AND UNEQUAL?

Sociology makes an important distinction between sex and gender. **Sex** refers to *the physical and biological differences between men and women.* In general, sex differences are made evident by physical distinctions in anatomical, chromosomal, hormonal, and physiological characteristics. At birth, the differences are most evident in the male and female genitalia.

We also need to be taught how to be a man or a woman. **Gender** refers to *the social, psychological, and cultural attributes of masculinity and femininity that are based on the previous biological distinctions.* Gender pertains to the socially learned patterns of behavior and the psychological or emotional expressions of attitudes that distinguish males from females. Ideas about masculinity and femininity are culturally derived and determine the ways in which males and females are treated from birth onward. Gender is an important factor in shaping people's self-images and social identities. Sex is thought of as an *ascribed* status; a person is born either male or female (although transsexuals show that sex can be changed). Gender is learned through the socialization process and thus is an *achieved* status.

The dominant view in many societies is that gender identities are expressions of what is natural. People tend to assume that acting masculine or feminine is the result of an innate, biologically determined process rather than the result of socialization and social-learning experiences.

To support the view that gender-role differences are innate, people have sought evidence from religion and the biological and social sciences. Whereas most religions tend to support the biological view, biology and the social sciences provide evidence that suggests that what is natural about sex roles expresses both innate and learned characteristics.

### Historical Views

We must be careful when we use today's standards to evaluate the statements of people who lived during another time and in another place. Lifting these statements out of context and bringing them into the present has been compared to "trying to plant cut flowers" (Ellis, 1997). Yet, it is important to look at these views and see what has changed since they represented the common thinking. The ancient Greek philosopher Aristotle, for example, pointed out that:

> It is fitting that a woman of a well-ordered life should consider that her husband's wishes are as laws appointed for her by divine will, along with the marriage state and the fortune she shares. (Aristotle in Walford and Gillies, 1908)

The third-century Chinese scholar, Fu Hsuan, penned these lines about the status of women during his era:

> Bitter indeed it is to be born a woman,
> It is difficult to imagine anything so low!
> Boys can stand openly at the front gate,
> They are treated like gods as soon as they are born . . .
> But a girl is reared without joy or love,
> And no one in her family really cares for her,
> Grown up, she has to hide in the inner rooms,
> Cover her head, be afraid to look others in the face,
> And no one sheds a tear when she is married off.
> (Quoted in Bullough, 1973)

In traditional Chinese society, women were subordinate to men. Chinese women were often called *Neiren*, or "inside person." To keep women in shackles, Confucian doctrine created what were known as the three obediences and the four virtues. The three obediences were "obedience to the father when yet unmarried, obedience to the husband when married, and obedience to the sons when widowed." Thus, traditionally, Chinese women were placed under the control of men from cradle to grave.

The four virtues were "women's ethics," meaning a woman must know her place and in every way comply with the old ethical code; "women's speech," meaning a woman must not talk too much, taking care not to bore people; "women's appearance," meaning a woman must pay attention to adorning herself with a view to pleasing men; and "women's chores," meaning a woman must willingly do all the chores in the home.

A book of Chinese poetry, *The Book of Songs*, believed to date from 1000 to 700 BC, offers this advice to new parents:

> When a son is born
> Let him sleep on the bed,
> Clothe him with fine clothes.
> And give him jade to play with.
> When a daughter is born,
> Let her sleep on the ground,
> Wrap her in common wrappings,
> And give her broken tiles for playthings.

Thomas Jefferson believed in supreme personal liberty and the equal creation of all men. Yet, when it came to women, he thought very differently. He did not think women should be involved in politics or that they should have the right to vote. He felt strongly that women had a single purpose in life—marriage and the subordination to a husband. When his oldest daughter married, he wrote her, "The happiness of your life now depends on continuing to please a single person. To this all other objects must be secondary" (Nock, 1996).

In nineteenth-century Europe, attitudes toward women had not improved appreciably. The father of modern sociology, Auguste Comte (1968/1851), in constructing his views of the perfect society, also dealt with questions about women's proper role in society. Comte saw women as the mental and physical inferiors of men. "In all kinds of force," he wrote, "whether physical, intellectual or practical, it is certain that man surpasses women in accordance with the general law prevailing throughout the animal kingdom." Comte did grant women a slight superiority in the realms of emotion, love, and morality.

Comte believed that women should not be allowed to work outside the home, to own property, or to exercise political power. Their gentle nature, he said, required them to remain in the home as mothers tending to their children and as wives tending to their husbands' emotional, domestic, and sexual needs.

Comte viewed equality as a social and moral danger to women. He felt that progress would result only from making the female's life "more and more domestic; to diminish as far as possible the burden of out-door labour." Women, in short, were to be "the pampered slaves of men." (See "Thinking About Social Issues: Let Women Vote and You Will Get Masculine Women and Effeminate Men" for some of the barriers women encountered as they fought to gain the vote.)

### Religious Views

Many religions have overtly declared that men are superior to women. Men are thought within such religious outlooks to be spiritually superior to women, who are deemed dangerous and untrustworthy. For centuries of Judeo-Christian history, two short passages in scripture helped shape the common view of women's character and proper place: the creation of Eve from Adam's rib, and Eve's transgression and God's

*Society causes us to expect certain gender-role behaviors from males and females. A changing society, however, produces changes in how these roles are carried out.*

subsequent curse on her. The first of these passages presents Eve and women as an afterthought:

> And the Lord God said, It is not good that the man should be alone; I will make a help mate for him.... And the rib, which the Lord God had taken from man, made he a woman, and brought her unto the man. (Genesis, 2:18, 22)

The second passage relates how, after God made the world a paradise, Eve disobeyed God and helped to make the world the imperfect place we know:

> Unto the woman he said, I will greatly multiply thy sorrow and thy conception; in sorrow thou shalt bring forth children; and thy desire shall be to thy husband, and he shall rule over thee. (Genesis 3:16)

War, pestilence, famine, and every sin imaginable were the prices all humanity had to pay for Eve's disobedience. The biblical story of creation presents a God-ordained gender-role hierarchy, with man created in the image of God and woman as a subsequent and secondary act of creation. This account has been used as the theological justification for a **patriarchal ideology,** or *the belief that men are superior to women and should control all important aspects of society.* This kind of legitimization of male superiority is displayed in the following passage from the New Testament:

> But I would have you know, that the head of every man is Christ; and the head of the woman is the man;... For the man is not of the woman; but the woman of the man. Neither was the man created for the woman; but the woman for the man. (1 Cor. 11:3, 8–9)

## THINKING ABOUT SOCIAL ISSUES: Let Women Vote and You Will Get Masculine Women and Effeminate Men

"It seems to me," Jeannette Gilder wrote in 1894, "that it's a bigger feather in a woman's cap—a brighter jewel in her crown—to be the mother of George Washington than to be a member of Congress from the 32nd District."

Gilder was arguing that women shouldn't be allowed to vote. In her essay, "Why I Am Opposed to Women Suffrage," Gilder insisted that women belonged in the home, where they could exert more political influence by nurturing sons, fathers, and brothers than they could ever command with a single ballot. "Politics is too public, too wearing and too unfitted to the nature of women," Gilder concluded. "It is my opinion that letting women vote would loose the wheels of purgatory."

Until 1920, women—along with paupers, felons, and so-called idiots—couldn't vote in federal elections. At the time, it was believed that women simply couldn't be trusted to take the long, objective view. "The female vote . . . is always more impulsive and less subject to reason, and almost devoid of the sense of responsibility," wrote Francis Parkman, a historian and anti-suffragist.

Women, who were believed to be "too frail for rough usage," were also beleaguered by their household responsibilities to the point that many seemed to hover on the verge of constant breakdowns. "The instability of the female mind is beyond the comprehension of the majority of men," declared Edith Melvin, a Concord, Massachusetts, anti-suffragist.

Not surprisingly, many men agreed that females should not vote. One of their biggest fears was that women would outlaw drinking, and various breweries supported anti-suffrage political candidates. The men's anti-suffrage movement even went so far as to produce bogus statistics: "If women achieve the feministic idea and live as men do," wrote a male doctor who opposed female suffrage, "they would incur the risk of 25 percent more insanity than they have now."

But tens of thousands of women also enlisted in the war against women voting, claiming that it was a slippery slope from the ballot box to depravation. If women got the vote, they would have to serve in the army and on juries. There would be fewer children but more divorce. Men would become less chivalrous and reverent of womanhood. Women would take up men's occupations, and men would take up women's occupations; the result, according to an anti-suffrage booster, would be a "race of masculine women and effeminate men and the mating of these would result in the procreation of a race of degenerates."

And if women did run for office, wouldn't they invariably win? When all women can vote, wrote Goldwin Smith, "as the women slightly outnumber the men, and many men, sailors or men employed on railways or itinerants, could not go to the poll, the woman's vote would preponderate, and government would be more female than male."

Here the anti-suffragists couldn't have been more wrong. Of the 535 members of the 112th Congress, only 91 or just 17 percent, are women. Seventeen of the 100 senators are women.

**Source:** From Cynthia Crossen, "Even Women Didn't Want to Give Women the Vote," *Wall Street Journal*, March 5, 2003 (p. B1). Reprinted by permission.

In traditional India, the Hindu religion conceived of women as strongly erotic and thus a threat to male asceticism and spirituality. Women were cut off physically from the outside world. They wore veils and voluminous garments and were never seen by men who were not members of the family. Only men were allowed access to and involvement with the outside world.

Women's precarious and inferior position in traditional India is illustrated further by the ancient Manu code, which was drawn up between 200 BC and AD 200. The code states that if a wife had no children after eight years of marriage, she would be banished; if all her children were dead after ten years, she could be dismissed; and if she had produced only girls after eleven years, she could be repudiated.

Stemming from the Hindu patriarchal ideology was the practice of prohibiting women from owning and disposing of property. The prevalent custom in traditional Hindu India was that property acquired by the wife belonged to the husband. Similar restrictions on the ownership of property by women also prevailed in ancient Greece, Rome, Israel, China, and Japan. Such restrictions are still followed by fundamentalist Muslim states such as Saudi Arabia and Iran.

According to traditional Islamic law and tradition, three groups of people are not eligible for legal and religious equality: unbelievers, slaves, and women. Women were in the worst position of these three groups. The slave could become free, the unbeliever could become a believer, but the woman could do nothing to change her status. She was permanently doomed to her second-class status (Lewis, 2002).

## Biological Views

The view that male and female behavior is governed by biological differences is supported by examples that cover a wide spectrum. Males are more likely than females to use aggressive means against their rivals and to establish their position in a hierarchy. Establishing dominance through theft, violence, and warfare has been the domain of men throughout history. Homicide records show that a man killing a man is fairly common, while a woman killing a woman is a rare crime. Studies of shootings in a school or workplace find that nearly 95 percent of the perpetrators are men. Ninety-four percent of burglaries are committed by men.

Male aggressiveness seems to begin very early. Male toddlers tell aggressive stories 87 percent of the time, while females tell such stories 17 percent of the time. Boys are much more likely to be punished for aggressive behavior than girls. The peak of male aggressiveness is during the preschool years and over time boys become less aggressive. Boys are more likely to move around in class, be inattentive, call out impulsively, and prod others—especially boys for a reaction. Boys are four times as likely as girls to have language and reading problems. Girls speak earlier than boys and use more complex and varied language than boys. They have better reading comprehension, spelling, punctuation, and writing skills. The female advantage in writing is so large that a writing subtest was added to the Preliminary Scholastic Aptitude Test (PSAT) to offset the male advantage in math.

Supporters of the belief that the basic differences between males and females are biologically determined have sought evidence from two sources: studies of other animal species, including nonhuman primates (monkeys and apes), and studies of the physiological differences between men and women. We shall examine each.

*Animal Studies and Sociobiology* **Ethology** is *the scientific study of animal behavior*. Ethologists have observed that sexual differences in behavior exist throughout

From birth, parents all over the world treat boys and girls differently. This toy Escalade is part of this boy's gender-role socialization. Beginning at birth boys and girls worldwide are treated differently by their parents.

much of the nonhuman animal world. Evidence indicates that these differences are biologically determined—that in a given species, members of the same sex behave in much the same way and perform the same tasks and activities. Popularized versions of these ideas, such as those of Desmond Morris in *The Human Zoo* (1970) or Lionel Tiger and Robin Fox in *The Imperial Animal* (1971), generalize from the behavior of nonhuman primates to that of humans. They maintain that in all primate species, including *Homo sapiens,* fundamental differences exist between males and females. They try to explain human male dominance and the traditional sexual division of labor in all human societies on the basis of inherent male or female capacities. They even have extended their analysis to explain other human phenomena, such as war and territoriality, through evolutionary comparisons with other species. A more sophisticated treatment of this same theme is found in the field of sociobiology (see Chapter 1, "The Sociological Perspective") through the study of the genetic basis for social behavior (Wilson, 1975, 1978, 1994).

Sociobiologists believe that much of human social behavior has a genetic basis. Patterns of social organization such as family systems, organized aggression, male dominance, defense of territory, fear of strangers, the incest taboo, and even religion are seen to be rooted in the genetic structure of our species. The emphasis in sociobiology is on the inborn structure of social traits. Sociobiology is often thought to be similar to evolutionary psychology. Evolutionary psychology studies the animal world including neural mechanisms that cause behavior from an evolutionary perspective, while sociobiology focuses mainly on social behavior.

Critics contend that sociobiologists overlook the important role that learning plays among nonhuman primates in their acquisition of social and sexual behavior patterns (Montagu, 1973). They also assert that by generalizing from animal to human behavior, sociobiologists do not take into account fundamental differences between human and nonhuman primates such as the human use of a complex language system. While freely acknowledging the biological basis for sex differences, these critics assert that among humans, social and cultural factors overwhelmingly account for the variety in the roles and attitudes of the two sexes. Human expressions of maleness and femaleness, they argue, although influenced by biology, are not determined by it; rather, gender identities acquired through social learning provide the guidelines for appropriate gender-role behavior and expression.

*Gender and Physiological Differences* Even ardent critics cannot deny that certain genetic and physiological differences exist between the sexes—differences that influence health and physical capacity. Accordingly, the study of gender roles must take those differences into account in such areas as size and muscle development (both usually greater in males), longevity (females, with few exceptions, live longer in nearly every part of the world, sometimes as much as nine years longer on average), and susceptibility to disease and physical disorders (considerable variation exists between men and women). As you can see from Table 11-1, men and women are afflicted by very different chronic conditions.

There has been an explosion of interest in women's health after a highly publicized 1990 government study that raised concerns about the small numbers of women in clinical trials sponsored by the National Institutes of Health (NIH). In 1992, Congress made it illegal to exclude women as subjects in medical research. The National Institutes of Health now has an Office of Research on Women's Health to oversee the representation of women in the institutes' studies. As a result, women's health research has increased substantially.

The medical treatment of women also is getting more attention in the medical school curriculum after a House Appropriations Committee report noted that U.S. physicians are not trained adequately to address the needs of women. The NIH women's health research office and the Health Resources and Services Administration helped address gaps in medical education and recommended a model curriculum.

A number of medical schools and teaching hospitals have made a greater effort to rethink their curricula and to consider examples of women and illness in all their subjects, not just in obstetrics and gynecology.

The differences between men and women go far beyond the obvious, and findings from research in the emerging field of gender-based biology could one day lead to treatments that vary depending on the sex of the patient. Gender-based biology identifies the biological and physiological differences between men and women as well as differences in responses to drugs. According to the Society for the Advancement of Women's Health Research, there are at least ten diseases, including multiple sclerosis, diabetes, lupus, and rheumatoid arthritis, that disproportionately affect women.

At menopause, women's brains experience a major reduction in estrogen, whereas men, paradoxically, continue producing androgen, which metabolizes into an estrogen-like hormone in the brain. As a result, men maintain high levels of estrogen in their brains as they age. Such findings might explain why aging women experience a higher incidence of cognitive decline such as Alzheimer's disease than men.

Hormone differences between men and women affect how the brain functions. For example, researchers have found that estrogen affects the synthesis and metabolism of serotonin, a neurotransmitter that helps

## TABLE 11-1  Gender and Disease

**Heart Attack**—Men are more likely to suffer heart attacks. Heart disease is the leading cause of death for women. For reasons that are poorly understood, 42 percent of women will die within a year of having a heart attack compared to 24 percent of men.

**Cancer**—Cancers are the second leading cause of death for men and women. Lung and bronchus cancer is the leading cause of cancer death among females. Women smokers are 20 percent to 70 percent more likely to develop lung cancer than are male smokers.

**HIV/AIDS**— While HIV and AIDS disproportionately affect men who have sex with men, an increasing proportion of HIV/AIDS diagnoses occur among women, and particularly minority women. In 2009, adolescent and adult females accounted for about one-fourth of new HIV and AIDS diagnoses, up from 7 percent in 1985.

**Cardiovascular Disease**—Men have a greater prevalence of heart disease.

**Arthritis**— Arthritis is more common among women than men (25.9 versus 18.3 percent, respectively). The proportion of adults with arthritis increases dramatically with age for both sexes.

**Diabetes and Other Chronic Illnesses**— Diabetes prevalence does not vary by sex and generally increases with age for both men and women. There is a higher diabetes prevalence in certain minority groups, particularly Hispanic, non-Hispanic Black, and American Indian/Alaska Native populations.

**Hypertension**—High blood pressure is a risk factor for a number of conditions, including heart disease and stroke. Men have higher rates of hypertension than do women. The prevalence of hypertension varies by race and ethnicity. For example, 39.4 percent of non-Hispanic Black women had high blood pressure compared to 16.3 percent of Mexican American women

**Osteoporosis**—Osteoporosis is the most common underlying causes of fractures in the elderly. More than one in four women aged 65 and older had been diagnosed with osteoporosis, compared with 4.2 percent of men.

**Immunological Diseases**— Women are 2.7 times more likely than men to acquire an autoimmune disease, rheumatoid arthritis (RA), multiple sclerosis (MS), and myasthenia gravis. The sex distribution is 60-75 percent women.

**Mental Illness**—Mental illness affects both sexes, although many types of mental disorders are more prevalent among women. Women are more likely than men to experience an anxiety or mood disorder, such as depression, while men are more likely than women to experience an impulse-control or substance use disorder.

**Alzheimer's Disease**— Alzheimer's disease is the fifth leading cause of death among men and women aged 65 and older. The greater rates of Alzheimer's prevalence and mortality among women are related to their longer life expectancy rather than an increased age-specific risk of disease.

**Overweight and Obesity**—Women are less likely than men to be overweight (27.3 versus 39.6 percent, respectively) but more likely than men to be obese (34.9 versus 31.8 percent, respectively). Obesity is highest among women living in poverty. Among men, however, both overweight and obesity tend to increase with household income.

**Unintentional and Intentional Injuries**—Accidents and suicides are more common among men than among women.

**Chronic Fatigue Syndrome**—Although research indicates that any person can develop chronic fatigue syndrome, women are four times more likely to experience the disorder than are men.

**Asthma**—Asthma is a chronic inflammatory disorder of the airway. Women have higher rates of asthma than do men. This is true in every racial and ethnic group.

**Visual and Hearing Impairments**—Glaucoma can damage the optic nerve and result in vision loss or blindness. Men have nearly a 50 percent greater likelihood of experiencing these problems.

*Source:* From National Institutes of Health. "Women's Health USA 2011." Health Resources and Services Administration (http://mchb.hrsa.gov/whusa11/more/sitemap.html), accessed August 1, 2012.

---

control moods. Major estrogen receptors are scattered throughout the body but are concentrated in a region of the brain that controls thinking, emotion, sex, and eating (Shelton, 1999). Researchers will need to determine the significance of some of the gender differences they are starting to discover. (For more on this topic, see "Our Diverse World: Why Do Women Live Longer Than Men?")

*Responses to Stress* Gender differences also influence the way men and women react to stress—the fight-or-flight reaction is thought to play a part in heart disease, stroke, and coronary-artery disease, among other ailments. In earlier days, when a primitive man was threatened by wild animals while hunting, testosterone combined with adrenaline enabled him to react quickly to danger. This intense type of reaction is no longer important and may be part of the reason men suffer more heart attacks than women. Women, it appears, react more slowly to stress, putting less pressure on the blood vessels and the heart. Although learned behavior might play a role in women's response to stress, biology is no less important.

When women are confronted by stress, they tend to respond by seeking out contact with and support from others. The support they seek is usually from other women. This befriending behavior has been linked with the hormone oxytocin, which is released by the body during stress. It has been shown to make both rats and humans calmer, less fearful, and more social. Although men also secrete oxytocin, male hormones reduce its effects. Female hormones, on the other hand, amplify its effects (Jablon, 2000).

Although many differences between males and females have a biological basis, other physical conditions may be tied to cultural influences and variations in

## OUR DIVERSE WORLD

### Why Do Women Live Longer Than Men?

In nearly every country throughout the world, women live longer than men. In Russia, for example, the difference between male and female life expectancy is 12 years (64 vs. 76). In Japan, the gap is seven years. In the United States, it is 5.2 years. In some developing countries, such as Afghanistan, there is little or no difference between male and female life expectancies.

What's behind the male/female gap in life expectancy? Most likely, it is a combination of biological and sociological factors.

In developed countries, men engage in more risky and unhealthy behaviors than do women. They are more likely to smoke, drink heavily, own guns, work in hazardous occupations, and participate in risky leisure activities. Consequently, men have higher death rates than women due to lung cancer, accidents, suicide, and homicide.

The greater likelihood of risky behaviors also produces higher male mortality in developing countries, where the gender gap is often smaller. Unsafe water and poor nutrition increase death rates for both sexes. Women also face the additional risks of childbirth.

Socialization comes into play in several ways. In many cultures, drinking is condoned and even encouraged for men but discouraged for women. Women might be discouraged from working outside the home whereas men are expected to have a job.

Smoking, which is more common in men, is a major contributor to shorter life spans. In the United States in recent decades, women have smoked more as men began to smoke less. The result is that the gap in male and female life expectancy has narrowed in recent years, from 7.7 years in 1972 to 5.2 years in 2010. Now that smoking is decreasing for both men and women, mortality rates will probably decline further.

An interesting fact is that women rate their health worse than men and visit the hospital more often than men, yet they are less likely to die at each age. This situation is partially explained by the fact that women experience more chronic conditions such as pain, depression, or respiratory conditions than do men. Men, on the other hand, are more likely to suffer from hearing loss; smoking-related ailments such as emphysema and respiratory cancer; and circulatory problems, including cardiovascular disease and diabetes.

**Source:** Yin, Sandra. November 2007. "Gender Disparities in Health and Mortality." Population Reference Bureau (http://prb.org/Articles/2007/genderdisparities.aspx), accessed June 7, 2009; Central Intelligence Agency. World Factbook (www.cia.gov/library/publications/the-world-factbook/fields/2102.html), accessed August 10, 2012.

---

environment and activity. Changing cultural standards and patterns of social behavior have had a pronounced effect on other traits that formerly were thought to be linked to sex. For example, the rising incidence of lung cancer among women—a disease historically associated primarily with men—can be traced directly to changes in social behavior and custom, not to biology; women now smoke as freely as men.

In sum, differing learned behaviors contribute to the relative prevalence of certain diseases and disorders in each sex. However, not all male-female differences in disease rate and susceptibility can be attributed to these factors. In addition to genetically linked defects, differences in some basic physiological processes such as metabolic rates and adult secretions of gonadal hormones may make males more vulnerable than females to certain physical problems.

Even though physiological factors tend to play an influential part in gender-role differences, biology does not determine those differences. Rather, people acquire much of their ability to fulfill their gender roles through socialization.

### Gender and Sex

We know that men and women think differently about sex, but how differently? Is it just a stereotype that men seek sex more than women? Some researchers have tried to answer this question. In one study (Schmitt et al., 2003), people from ten areas of the world were asked, "Ideally, how many different sexual partners would you like to have in the next month?" In North America, 23% of the men and 3% of the women wanted more than one partner in the next month.

There were similar findings throughout the world. In South America, it was 35% for the men and 6.1% for the women. In Western Europe, 22.65% and 5.5%; in the Middle East, 33.1% and 5.9%; in Southeast Asia, 32.4% and 6.4%; and in East Asia, 17.9% and 2.6%. These findings held true for age and socioeconomic status.

Men and women also differ in how well they need to know a person before having sex. Participants were asked whether they would consider having sex with a desirable partner they had known for (a) one year, (b) one month, or (c) one week. Women preferred to know a person for a year, whereas men had no problem having sex with a woman they had known for a week (Buss, 1994).

The differences were even more striking in another study (Clark and Hatfield, 1989). Attractive men and women were hired to approach strangers of the opposite sex on college campuses and say to them, "I have been noticing you around campus; I find you very attractive," and then ask one of three questions: (a) "Would you go out with me tonight?" (b) "Would you come over to my apartment tonight?" or (c) "Would you go to bed with me tonight?" Half the men and half the women consented to a date, but that was where the similarity ended. Only 6% of the women consented to go to the confederate's apartment, compared with 69% of the men. None of the women consented to sex; however, 75% of the men consented to sex. Of the remaining 25%, many were apologetic, asking for a rain check or explaining that they could not do it now because they had to meet their girlfriend.

Another study was done to see if the results could be replicated with varying attractiveness of the male and female subjects. The researchers found that regardless of requestor attractiveness, men were far more interested in casual sex than women.

Of course, another way of interpreting this study would be that rather than showing that women are less interested in sex than men, it instead shows that young American males tend to be far more reckless than young women—as can be confirmed by anyone who has ever attended a frat party.

### Sociological View: Cross-Cultural Evidence

Most sociologists believe that the way people are socialized has a greater effect on their gender identities than do biological factors. Cross-cultural and historical research offers support for this view, revealing that different societies allocate different tasks and duties to men and women and that males and females have culturally defined views of themselves and of one another.

The World Economic Forum ranks 135 countries according to the global gender gap. The Global Gender Gap Index examines the gap between men and women

The medical treatment of women is now getting more attention in medical schools in response to studies that showed that U.S. physicians were not adequately trained in women's health issues.

in four fundamental categories: economic participation and opportunity, access to educational opportunity, health and survival, and political empowerment.

Iceland was ranked as the top-ranked nation in women's educational attainment and political representation. It was one of the first countries to give women the right to vote in 1915. It has had a female head of state for 18 of the past 50 years. With 81 percent of women in the workforce, it has one of the narrowest labor force participation gaps.

Other Nordic countries were at the top of the list: Norway (No. 2), Finland (No. 3), Sweden (No. 4), and Denmark (No. 7). All Nordic countries have generous paid maternity and paternity leave policies. In Sweden, women are offered 480 days of maternity time.

The worst countries for gender equality are Saudi Arabia (No. 131), Mali (No. 132), Pakistan (No. 133), Chad (No. 134), and Yemen (No. 135). The low-scorers have major barriers for women who wish to enter leadership or political positions. In Saudi Arabia, for example, women continue to fight for the right to drive, and only 22 percent of women are in the workforce.

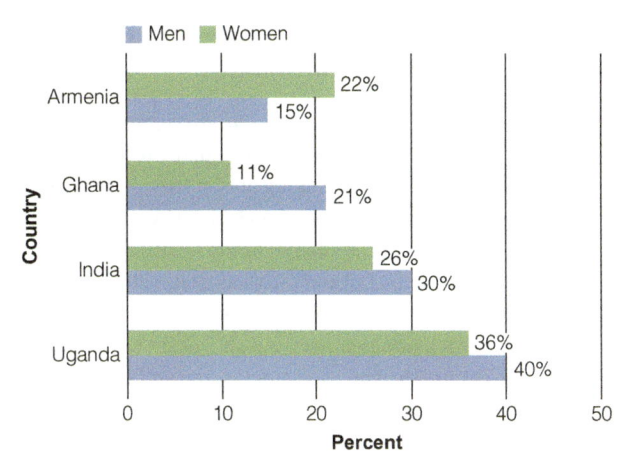

**FIGURE 11-1** Percentage of Men and Women Who Agree That Wife Beating Is Acceptable If a Wife Argues with Her Husband

| Country | Women | Men |
|---|---|---|
| Armenia | 22% | 15% |
| Ghana | 11% | 21% |
| India | 26% | 30% |
| Uganda | 36% | 40% |

Source: Population Reference Bureau. "The World's Women and Girls, 2011 Data Sheet" (www.prb.org/pdf11/world-women-girls-2011-data-sheet.pdf), accessed June 25, 2012.

Confining women to a secondary status in many parts of the world has had a profound effect on their lives. For example, gender-based violence against women—which includes female infanticide, sexual trafficking, and domestic violence—causes more death and disability worldwide among women aged 15 to 44 than cancer, malaria, and war combined (Center for Women Policy Studies, 2003). Two-thirds of the world's illiterate adults are women (UNDP Human Development Report, 2003). Women are not allowed to vote in Brunei, Vatican State, Saudi Arabia, or the United Arab Emirates (Women in Politics, 2012). Saudi Arabia will allow women to vote in 2015. (See Figure 11-1 for an example of countries where wife beating is acceptable.)

## ● WHAT PRODUCES GENDER INEQUALITY?

Sociologists have devoted much thought and research to answering this question. They also have tried to explain why males dominate in most societies. Two theoretical approaches have been used to explain dominance and gender inequality: functionalism and conflict theory.

### The Functionalist Viewpoint

From Chapter 1, you will recall that functionalists (or structural functionalists, to be more precise) believe that society consists of a system of interrelated parts that work together to maintain the smooth operation of society. Functionalists argue that it was quite useful to have men and women fulfill different roles in preindustrial societies. The society was more efficient when tasks and responsibilities were allocated to particular individuals who were socialized to fulfill specific roles.

The fact that the human infant is helpless for such a long time made it necessary that someone look after the child. It was also logical that the mother who gave birth to the child and nursed it was also the one to take care of it. Because women spent their time near the home, they also took on the duties of preparing the food, cleaning clothes, and attending to the other necessities of daily living. To the male fell the duties of hunting, defending the family, and herding. He also became the one to make economic and other decisions important to the family's survival.

This division of labor created a situation in which the female was largely dependent on the male for protection and food, so he became the dominant partner. This dominance, in turn, caused his activities to be more highly regarded and rewarded. Over time, this pattern came to be seen as natural and was thought to be tied to biological sex differences.

Talcott Parsons and Robert Bales (1955) applied functionalist theory to the modern family. They have argued that the division of labor and role differentiation by sex are universal principles of family organization and are functional to the modern family. They believe that the family functions best when the father assumes the *instrumental role,* which focuses on relationships between the family and the outside world. It mainly involves supporting and protecting the family. The mother concentrates her energies on the *expressive role,* which focuses on relationships within the family and requires her to provide the love and emotional support needed to sustain the family. The male is required to be dominant and competent, and the female to be passive and nurturing.

As you can imagine, the functionalist position has been much criticized. The view that gender roles and gender stratification are inevitable does not fit with the changing situation in American society (Crano and Aronoff, 1978). Critics contend that industrial society can be quite flexible in assigning tasks to males and females. Furthermore, they assert that the functionalist model was developed during the 1950s, an era of very traditional family patterns, and that rather than being predictive of family arrangements, it merely is representative of the era during which it became popular.

### The Conflict Theory Viewpoint

Although functionalist theory might explain why gender-role differences emerged, it does not explain why they persisted. According to conflict theory, males dominate females because of their superior power and control over key resources. A major consequence of this domination is the exploitation of women by men. By subordinating women, men gain greater economic, political, and social power. According to conflict theory, as long as the dominant group benefits from the existing relationship, it has little incentive to change it. The resulting inequalities are therefore perpetuated long after they might have served a functional purpose. In this way, gender inequalities resemble race and class inequalities.

Conflict theorists believe that the main source of gender inequality is the economic inequality between men and women. Economic advantage leads to power and prestige. If men have an economic advantage in society, that advantage will produce a superior social position in both society and the family.

Friedrich Engels (1942/1884) linked gender inequalities to capitalism, contending that primitive, non-capitalistic, hunting-and-gathering societies without private property were egalitarian. As these societies developed capitalistic institutions of private property, power came to be concentrated in the hands of a minority of men, who used their power to subordinate women and to create political institutions that maintained their power. Engels also believed that to free women from subordination and exploitation, society must abolish private property and other capitalistic institutions. He believed that socialism was the only solution to gender inequality.

Today, many conflict theorists accept the view that gender inequalities might have evolved because they initially were functional. Many functionalists also agree that gender inequalities are becoming more and more dysfunctional. They agree that the origins for gender inequalities are more social than they are biological.

## ● GENDER-ROLE SOCIALIZATION

**Gender-role socialization** is *a lifelong process whereby people learn the values, attitudes, motivations, and behavior considered appropriate to each sex according to their culture.* In our society, as in all others, males and females are socialized differently. In addition, each culture defines gender roles differently. This process is not limited to childhood but continues through adolescence, adulthood, and old age.

### Childhood Socialization

Even before a baby is born, its sex is a subject of speculation, and the different gender-role relationships it will form from birth already are being decided. Maccoby and Jacklin (1975) described this phenomenon by using a familiar cultural medium, the musical.

> A scene from the [1940s] musical *Carousel* epitomizes (in somewhat caricatured form) some of the feelings that parents have about bringing up sons as opposed to daughters. A young man discovers he is to be a father. He rhapsodizes about what kind of son he expects to have. The boy will be tall and tough as a tree, and no one will dare to boss him around; it will be all right for his mother to teach him manners, but she mustn't make a sissy out of him. He'll be good at wrestling and will be able to herd cattle, run a riverboat, drive spikes, etc. Then the prospective father realizes, with a start, that the child may be a girl. The music changes to a gentle theme. She will have ribbons in her hair; she will be sweet and petite (just like her mother); and suitors will flock around her. There's a slightly discordant note, introduced for comic relief from sentimentality, when the expectant father brags that she'll be half again as bright as girls are meant to be; but then he returns to the main theme: she must be protected, and he must find enough money to raise her in a setting in which she will meet the right kind of man to marry.

Parents carry in their minds images of what girls and boys are like, how they should behave, and what they should be in later life. Parents respond differently to girls and boys right from the beginning. Girls are caressed more than boys, whereas boys are jostled and rough-housed more. Mothers talk more to their daughters, and fathers interact more with their sons.

These differences are reinforced by other socializing agents—siblings, peers, educational systems, and the mass media. By the first two or three years of life, core gender identity—the sense of maleness or femaleness—is established as a result of the parents' conviction that their infant's assignment at birth to either the male or female sex is correct.

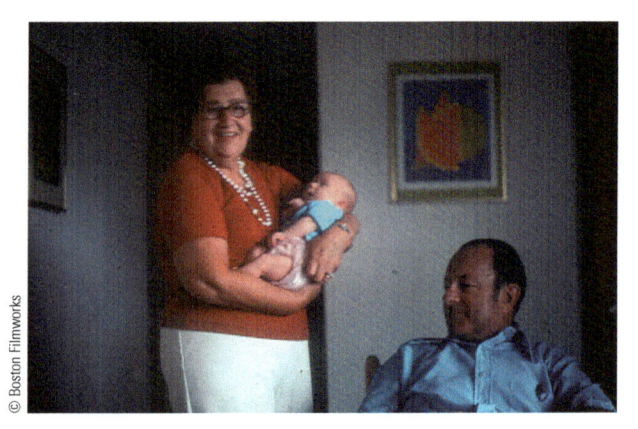

Parents and grandparents respond differently to boys and girls right from the beginning, and they carry in their minds images of what the child should be like, how he or she should behave, and what he or she should be in later life.

## Adolescent Socialization

Most societies have different expectations for adolescent girls and boys. Erik Erikson (1968) believed that the most important task in adolescence is the establishment of a sense of identity. During the adolescent stage, Erikson felt, both boys and girls undergo severe emotional crises centered on questions of who they are and what they will be. If the adolescent crisis is resolved satisfactorily, a sense of identity will be developed; if not, role confusion will persist.

According to Erikson, adolescent boys in our society generally are encouraged to pursue role paths that will prepare them for an occupational commitment, whereas girls generally are encouraged to develop behavior patterns designed to attract a suitable mate. Erikson observed that it is more difficult for girls than for boys to achieve a positive identity in Western society. This is because women are encouraged to be more passive and less achievement oriented than men and to stress the development of interpersonal skills—traits that are less highly valued in our society. Men, by contrast, are encouraged to be competitive, to strive for achievement, and to assert autonomy and independence—characteristics that are held in high esteem in our competitive society.

In the United States, boys are three times as likely to be placed in special education classes, twice as likely to repeat a grade, and a third more likely to drop out of high school. Tests of 15-year-olds in thirty European countries show girls far outstripping boys in reading and writing and holding their own in math. Boys are overrepresented in the top 1 percent of math achievers, but there are also more of them at the bottom.

Once they move from school to work, men on average earn more money and run more shows. They particularly dominate in national government, the corporate boardroom, and the science laboratory. Meanwhile, women are more likely to leave the labor force and to end up with lower pay and less authority if they come back.

Pinker (2008), a psychologist, does not discount sex discrimination or culture in shaping women's choices. She suggests that women care more about relationships because they are wired for empathy, and are better at gauging the effect they have on others. This is the result of genes and hormones. Beginning in utero, men are exposed to high levels of testosterone, driving them to be more competitive, assertive, vengeful, and daring. Women, meanwhile, are exposed to oxytocin and prolactin, which enable bonding. Pinker suggests that the reason women are not racing to the upper echelons of science, government, and the corporate world is because they are biologically predisposed to resist the demands at the top of those fields.

Nonetheless, female gender roles are changing rapidly, although traditional attitudes toward careers and marriage undoubtedly still remain part of the thinking of many people in our society. Girls are being encouraged not to limit themselves to these stereotyped roles and attitudes. More and more young women expect to pursue careers before and during marriage and childrearing. Work fields that in the past have been traditionally male, such as medicine, law, or pharmacy, are now seeing either equal enrollments of men and women or a majority of women in their training schools.

In every society, boys and girls are socialized differently, and as they mature, a variety of roles are tried out. In this picture, the boy is learning one aspect of the male role.

## GENDER INEQUALITY AND WORK

Women's numerical superiority over men has not enabled them as yet to avoid discrimination in many spheres of American society. In this discussion, we will focus on economic and job-related discrimination because these data are easily quantified and serve well to highlight the problem. Remember, though, that discrimination against women in America actually is expressed in a far wider range of social contexts and institutions. (For a discussion about how we display our gender wherever we go, see "Day-to-Day Sociology: Speaking, Writing, or Blogging—Nowhere to Hide Gender.")

### Job Discrimination

In 1970, about 43 percent of women aged 16 and older had paying jobs. By 2010, this number had increased to 58.6 percent of all American women. As women's labor

---

**DAY-TO-DAY SOCIOLOGY**

### Speaking, Writing, or Blogging—Nowhere to Hide Gender

Twenty years ago, Deborah Tannen, a sociolinguist and the author of the best seller, *You Just Don't Understand,* popularized the view that gender differences are widespread in everyday speech and that, in many ways, men and women are living in different worlds when they try to communicate with each other.

Tannen noted that men and women use language differently. Women use language primarily to create intimacy and connections to other people. For women, language is the glue that holds relationships together. Men, however, use language mainly to convey information. For men, activities are the things that hold people together, and in the absence of doing something with others, talking about an activity is the next best thing and involves talking about concrete events or facts. This explains the tendency among men to engage in endless discussions about sporting events, batting averages, and so on. Men are acutely aware of the status differences implied by knowledge or, in one way or another, of speaking. For men, language and information are used to attain status, not intimacy.

Tannen thought men were more comfortable with the language used in a public setting, whereas women find the language used in private settings more natural. Essentially, the public language is about information and one person attempting to attain status with knowledge, whereas private language is about personal feelings and sharing them with others.

What about writing? Do you give away your gender when you write? A group of Israeli researchers think so. They have created an algorithm to predict the gender of an author from a writing sample. It is called Gender Genie (www.bookblog.net/gender/genie.html) and allows anyone to paste in a few paragraphs and get a review of how masculine or feminine their writing style is.

The writing is analyzed based on the belief that women are far more likely than men to use personal pronouns ("I," "you," "she," and so on). Men use more words that identify or determine nouns ("a," "the," "that") or that quantify them ("one," "two," "more"). The authors assume, similarly to Tannen, that women are more comfortable writing about people and relationships, whereas men prefer to write about things.

The Web is also involved in writing, so it should come as no surprise that we display our gender there also. In one study (Herring and Paolillo, 2006), 68 percent of the messages posted by men used an adversarial style in which the poster "distanced himself from, criticized, or ridiculed other participants, often while promoting his own importance." The male Web style is often characterized by adversarial comments such as put-downs; strong, often contentious assertions; self-promotion; and sarcasm. Men are likely to take an authoritative, self-confident stance in which they represent themselves as experts.

The women, in contrast, displayed such features as "hedging, apologizing, asking questions rather than making assertions, and . . . revealing thoughts and feelings and interacting with and supporting others." The female style includes expressions of appreciation, thanking, and community-building activities that make the participants feel accepted and welcomed.

For most of us, it is difficult to be gender neutral whether we are speaking, writing, or blogging.

*Sources:* Herring, S. C., and J. C. Paolillo. 2006. "Gender and Genre Variation in Weblogs." *Journal of Sociolinguistics* 10(4):439–59; Argamon, Shlomo, Moshe Koppel, Jonathan Fine, and Anat Rachel Shimoni. 2003. "Gender, Genre, and Writing Style in Formal Written Texts." *Text* 23(3):321–46; Tannen, D. 1990. *You Just Don't Understand: Women and Men in Conversation.* New York: William Morrow.

force participation has increased, so has their employment in higher-paying occupations.

The good news is that women account for nearly 52 percent of all workers in the high-paying management, professional, and related occupations. They outnumber men in such occupations as registered nurses (91.7%); elementary and middle school teachers (81.2%); insurance underwriters (80.3%); medical and health services managers (69.4%); social and community service managers (68.1%); financial managers (54.8%); and medical scientists (52.3%).

The bad news is that men earn more than women in almost all occupations. In about 500 of the most common occupations, in only five occupations do women earn the same as or more than men. Male and female counselors bring home the same weekly pay check, and women earn more than men as special education teachers, an occupation in which most of the workers are women. (See Table 11-2 for the highest-paying jobs women hold.) The wage gap between men and women even holds up in the jobs that women are most likely to hold, such as secretaries, teachers, and nurses.

There is a large amount of occupational segregation in the workforce, with women working primarily with other women and men having mainly male coworkers. Two-fifths of women work in occupations in which at least three out of four workers are female. This is true for two of every four male workers also (Institute for Women's Policy Research, 2009). (See Table 11-3 for the most common male and female jobs.)

Job discrimination against women is a complicated phenomenon. Generally, women experience discrimination in the business world in three ways: (1) during the hiring process, when women are given jobs with lower occupational prestige than men who have equivalent qualifications; (2) through unequal wage policies, by which women receive less pay than men for equivalent work; and (3) in the awarding of promotions because women find it more difficult than men to move up the career ladder.

Discrimination against women in the economic sector is often quite subtle. Women are more or less channeled away from participation in occupations that are socially defined as appropriate to men. For example, it cannot really be argued that female bank presidents are paid less than men; instead, there are few woman bank presidents. Women and men often do not perform equal work; therefore, the phrase "equal pay for equal work" has little relevance. In some instances, men and women perform similar work, but there may be two different job titles and two pay scales.

Having painted a somewhat pessimistic picture here, we should note that recent years have shown some improvement. Women's share of jobs with high earnings has grown. Women also constitute a rising share of people being awarded college and postgraduate degrees. In 2010, they represented 57.5% of people

TABLE 11-2  Occupations in which Women Earn the Most

| Rank | Occupation | 2010 Median Weekly Earnings (US$) |
|---|---|---|
| 1. | Physicians and surgeons | 1,618 |
| 2. | Pharmacists | 1,605 |
| 3. | Chief executives | 1,598 |
| 4. | Lawyers | 1,461 |
| 5. | Computer software engineers | 1,445 |
| 6. | Computer and information systems managers | 1,415 |
| 7. | Physical therapists | 1,208 |
| 8. | Speech language pathologists | 1,184 |
| 9. | Computer programmers | 1,177 |
| 10. | Human resource managers | 1,170 |

Source: U.S. Department of Labor, Women's Bureau. "Women in the Labor Force in 2010" (www.dol.gov/wb/factsheets/Qf-laborforce-10.htm), accessed August 8, 2012.

TABLE 11-3  Most Common Male and Female Occupations, 2010

| Most Common Male Occupations | Percentage Male | Most Common Female Occupations | Percentage Female |
|---|---|---|---|
| Carpenters | 98.6 | Secretaries and administrative assistants | 96.1 |
| Construction workers | 97.3 | Dental hygienists | 95.1 |
| Drivers | 95.4 | Receptionists | 92.7 |
| Police officers | 87.0 | Registered nurses | 91.1 |
| Chefs and head cooks | 81.0 | Home health aids | 88.2 |
| Shipping and receiving clerks | 72.5 | Elementary and middle school teachers | 75.1 |
| Real estate brokers | 54.0 | Cashiers | 73.7 |

Source: U.S. Department of Labor. December 2011. "Women in the Labor Force: A Databook" (www.bls.gov/cps/wlf-databook-2011.pdf), accessed August 8, 2012.

awarded bachelor's degrees, 60.3% of those awarded master's degrees, and 51.7% of the doctorate recipients (U.S. Department of Education, 2008) (see Figure 11-2). In addition, young women and men (those younger than age 25) have fairly similar earnings; however, women's earnings are much lower than men's in older age groups. (For more on this topic, see "How Sociologists Do It: What Happened to the Men?")

Despite this progress, women still dominate low-paying fields. Five of the top ten occupations employing women are secretaries, receptionists, home health aids, cashiers, and customer service representatives. The two professional positions that women dominate are relatively low paying, namely, nursing and elementary school teaching (Bureau of Labor Statistics, 2007). (For more on gender and work, see "Our Diverse World: Who Is a Better Boss?")

## HOW SOCIOLOGISTS DO IT

### What Happened to the Men?

When American soldiers returned from World War II, many enrolled in college under what was known as the G.I. Bill, which made obtaining an education affordable. During this period and several decades before, college was where men went to prepare for entry into a good-paying job. Women to a lesser extent attended college also, often to meet a husband, as well as to obtain a teaching degree, which gave them the opportunity of a job until children came along. In 1947, more than twice as many men were in college as women.

Slowly things changed. By 1960, the number was down to 1.55 male college students for every female. In the late 1960s and early 1970s, women's expectations about their futures in the world of work changed and their college-enrollment and graduation numbers began to soar. No longer did young women limit their choice of majors to Education or Home Economics; instead, they joined their male classmates in the whole range of college possibilities.

Women also started to marry later, and their career aspirations quickly expanded. By 1980, there was no longer a college gender gap in enrollments. The male classmates continued to enroll in college at similar percentages as in the past as the wave of women heading to college continued to pick up speed. Women college enrollment eventually surged ahead of that for men; by 2003, there were 1.30 females for every male undergraduate.

Starting in elementary school, girls put more effort into their work than boys do. They continue to do better than boys in high school and college. Women are less likely to drop out of college and are continuing to increase their attendance and graduation rates in a wide range of fields. This trend will continue to have an effect on a wide range of work-, gender-, and family-related issues.

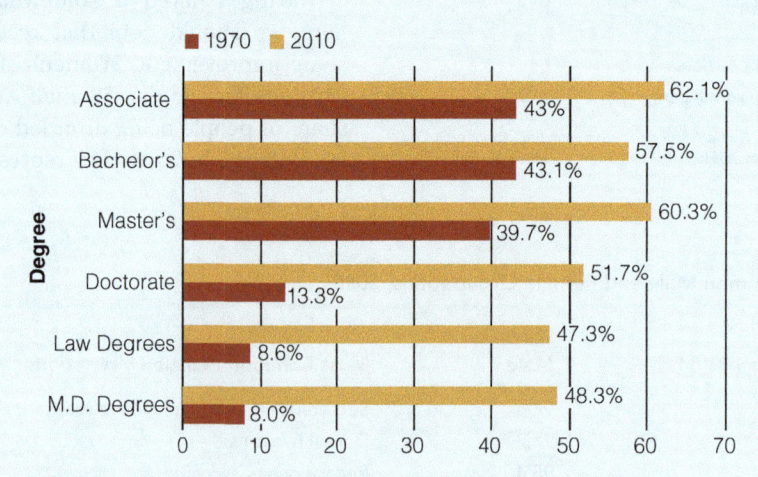

■ 1970   ■ 2010

| Degree | 1970 | 2010 |
|---|---|---|
| Associate | 43% | 62.1% |
| Bachelor's | 43.1% | 57.5% |
| Master's | 39.7% | 60.3% |
| Doctorate | 13.3% | 51.7% |
| Law Degrees | 8.6% | 47.3% |
| M.D. Degrees | 8.0% | 48.3% |

**FIGURE 11-2** Percentage of Advanced Degrees Awarded to Women
*Sources: U.S. Department of Education. "Digest of Education Statistics 2012"; Bell, Nathan E. September 2011. CGS/GRE Survey of Graduate Enrollment and Degrees, Graduate Enrollment and Degrees: 2000 to 2010.*

## OUR DIVERSE WORLD

### Who Is a Better Boss?

In 1953, when Americans were asked whether they preferred a male or female boss, 66% said they preferred a man and 5% favored a woman. Today, people continue to prefer male bosses, but the margin in men's favor is shrinking. In 2011, "32% of Americans expressed a preference for a male boss and 22% opted for a female one" (Gallup, September 2011).

The Bureau of Labor Statistics finds that nearly one of four chief executives is now a woman. Less obvious is the fact that most of these women run their own small business. Women still hold only about one out of 20 top-management positions in Fortune 500 corporations, representing only a minor change from twenty years ago.

The ability for men and women to lead successfully depends on being in the appropriate setting. Women are more likely to adopt a mentoring or coaching style when leading, which is more favorably received in female-dominated professions. Women also appear to reward good performance more than men. Men commonly use an assertive, take-charge style, which is well received in male-dominated professions. Men are also more likely to criticize subordinates and be less hands-on.

Some women using a participatory leadership style hope to serve as role models. They try to help the employees develop their skills and be productive and creative. This works if the organization is minimally hierarchical. The participatory style may not work in traditional male settings such as the military or organized sports. Conversely, the assertive style more typical of men may produce resentment in a social-service agency or retail setting.

On the one hand, this leads to an obvious conclusion. Women seem to be more effective leaders in female-oriented settings, and men are more effective in male-oriented settings. Contemporary work settings, however, are becoming more participatory and should be more conducive to female leaders. The most important consideration for someone stepping into a leadership position is to evaluate the setting and decide which style is appropriate and whether their personality lends itself to such a role.

**Source:** American Psychological Association. March 22, 2006. "When the Boss Is a Woman" (www.psychologymatters.org/womanboss.html), accessed June 1, 2009; Newport, Frank. "Americans Still Prefer Male Bosses; Many Have No Preference" (www.gallup.com/poll/149360/americans-prefer-male-bosses-no-preference.aspx), accessed August, 8, 2012.

## ■ SUMMARY

▸ **What is the difference between sex and gender?**
Sex refers to the physical and biological differences between men and women. Gender refers to the social, psychological, and cultural attributes of masculinity and femininity that are based on biological distinctions.

▸ **How do biological and socialization factors figure into gender and sex?**
Ideas about masculinity and femininity are culturally derived and are an important factor in shaping people's self-images and social identities. Whereas sex is generally an ascribed status, gender is learned through the socialization process and is thus an achieved status. Many people, on the other hand, believe that gender identities and masculine or feminine behavior result from an innate, biologically determined process.

▸ **What role does socialization play in gender?**
Gender-role socialization is a lifelong process whereby people learn the values, attitudes, motivations, and behavior deemed appropriate for males or females by their society.

▸ **How have societies viewed gender issues?**
Historically, societies in both the East and the West have viewed women as inherently inferior to men. Not only did intellectuals argue from the point of view of a patriarchal ideology, or a belief that men are superior to women and should control all important aspects of society, but religions in both the East and West supported this view.

▸ **How do functionalists view gender issues?**
Functionalists argue that the sexual division of labor was necessary for the efficient operation of preindustrial societies. Women, who birthed and nursed infants, remained involved in child care and the necessities of daily living. Men hunted, herded,

and defended the family. Functionalists argue that because the female was largely dependent on the male for protection and food, he became the dominant partner in the relationship. Functionalists maintain that this role differentiation is functionally necessary and efficient in modern society as well.

Critics contend that this view is outdated and that industrial society provides for more flexibility in gender-role assignment.

### How do conflict theorists view gender issues?

According to conflict theory, males dominate females because of their superior power and control over key resources. A major consequence of this domination is the exploitation of women by men. Conflict theorists contend that as long as men benefit from this arrangement, they have little incentive to change it.

## Media Resources

**CourseMate for *Introduction to Sociology*, Eleventh Edition**
Cengage Learning's Sociology CourseMate brings course concepts to life with interactive learning, study, and exam preparation tools that support the printed textbook. Access an integrated eBook, learning tools including glossaries, flashcards, quizzes, videos, and more in your Sociology CourseMate. Go to CengageBrain.com to register or purchase access.

# CHAPTER ELEVEN STUDY GUIDE

## Key Concepts

Match each of the following concepts with its definition, illustration, or explanation.

a. Sex
b. Gender
c. Patriarchal ideology
d. Sociobiology
e. Instrumental role
f. Expressive role
g. Gender-role socialization
h. Gender identity
i. Functionalism
j. Conflict theory viewpoint

____ 1. Provides a hypothesis of why males are dominant and females submissive
____ 2. Cultural attributes of masculinity and femininity
____ 3. Behaviors that focus on feelings and relationships among members of a group
____ 4. The process of acquiring and internalizing the ideas and behaviors deemed appropriate for males and females by their society
____ 5. The view that society is governed by interrelated parts that make up the smoothly functioning whole
____ 6. The individual's sense of maleness or femaleness
____ 7. The belief that men are superior to women and should control all important aspects of society
____ 8. A social science theory focusing on the inborn nature of social traits and the role of evolution in shaping human behavioral patterns
____ 9. Behaviors that focus on the accomplishment of specific tasks for a group rather than on personal feelings
____ 10. The physical and biological aspects of male and female

## Key Thinkers

Match the thinkers with their main idea or contribution.

a. Auguste Comte
b. Desmond Morris, Robin Fox, and Lionel Tiger
c. Susan Pinker
d. Talcott Parsons and Robert Bales
e. Friedrich Engels
f. Erik Erikson
g. Deborah Tannen

____ 1. States her belief that women, more so than men, are "wired" for empathy
____ 2. Conducted research on the conversational style and content differences among females and males
____ 3. Applied the functionalist idea about division of labor to sex roles in the modern family
____ 4. Colleague of Karl Marx; theorized that capitalism and private property were the sources of gender inequalities and the subordination of women
____ 5. The founder of sociology, he believed that women should not be allowed to work outside the home, to own property, or to exercise political power
____ 6. Writers who popularized the work of ethology and generalized from the behavior of nonhuman primates to that of humans
____ 7. Psychologist who pointed out that U.S. society pressures adolescent boys to base their identity on achievement but pressures girls to base their identity on finding a husband

## Central Idea Completions

*Fill in the appropriate concepts and descriptions for each of the following questions.*

1. Compare and contrast the biological and sociological perspectives on sex and gender.

   a. Biological: _____

   b. Sociological: _____

2. Keeping in mind the research discussed in this chapter, what can you conclude about the differences in how women and men think about sex?

   a. Women: _____

   b. Men: _____

3. Using examples from the chapter, such as the information about oxytocin, discuss the ways in which male-female differences in behavior and thought might be caused by biological differences.

   _____

4. Women earn less money than do men. Give both a functionalist and a conflict theory explanation for the earnings differential.

   a. Functionalist: _____

   b. Conflict: _____

## Critical Thought Exercises

1. How have ideas about sex, gender, and homosexuality changed in the United States over the past few decades? What factors have been responsible for these changes? What will affect whether similar changes happen with ideas about transgendered people?

2. How are male friendships different from female friendships? Design a mini-study of the gender friendships of your close associates. What variables and patterns would you look for in your research? How would you gather your data? Develop two hypotheses about what you expect to find.

3. With the agreement of the participants, record a discussion. It might be a class discussion, an informal chat among friends, or a formal public discussion. If you can't tape it, listen closely and observe. Describe the conversational styles of men and women in both single-sex groups and in groups with both men and women. Keep in mind the work of Deborah Tannen (read her book, *You Just Don't Understand*). Try to code each statement for its instrumental or expressive content. What male-female differences did you find?

## Internet Activities

1. Deborah Tannen says that men and women have different styles of speech. Do they also write differently? Two scientists have developed a computer program that supposedly can tell. Find a passage, preferably longer than 500 words, and paste it into the box at www.bookblog.net/gender/genie.html. Try it with a few passages and see how well the program does.
2. The blog Sociological Images (http://contexts.org/socimages/) often has posts related to the depiction of gender in the media, especially in advertisements. Check "gender" in posts by topic.
3. For information on gender inequality in pay and education, look at www.aauw.org/research/statedata/index.cfm, run by the American Association of University Women. Review the various research reports on the first page, choose one that you think is most interesting, and discuss why you chose it.

## Answers to Key Concepts

1. j;  2. b;  3. f;  4. g;  5. i;  6. h;  7. c;  8. d;  9. e;  10. a

## Answers to Key Thinkers

1. c;  2. g;  3. d;  4. e;  5. a;  6. b;  7. f

# 12 Marriage and Changing Family Arrangements

**The Nature of Family Life**
   Functions of the Family
   Family Structures

**Defining Marriage**
   Romantic Love
   Marriage Rules
   Marital Residence
   Mate Selection
*Day-to-Day Sociology:* Marriage and Divorce Quiz

**The Transformation of the Family**
   The Decline of the Traditional Family
   Changes in the Marriage Rate
*How Sociologists Do It:* Study, Graduate, and Be Married
   Childless Couples
   Changes in Household Size
   Women in the Labor Force
   Family Violence
   Divorce
   Divorce Laws

*How Sociologists Do It:* Do 50 Percent of All Marriages Really End in Divorce?
   Child Custody Laws
   Remarriage and Stepfamilies

**Family Diversity**
   The Growing Single Population
*Thinking About Social Issues:* Reluctant to Marry—The Men Who Want to Stay Single
   Single-Parent Families
   Gay and Lesbian Couples

**What Does the Future Hold?**

**Summary**

## ■ LEARNING OBJECTIVES

After studying this chapter, you should be able to do the following:

- Explain the functions of the family.
- Describe the major variations in family structure.
- Define marriage and describe its relationship to the phenomenon of romantic love.
- Describe the various rules governing marriage.
- Explain the ways in which mate selection is not random.
- Summarize recent changes in the family as an institution.
- Explain the impact of changes in divorce and child custody laws.
- Describe the various alternative family arrangements in contemporary American society.

In a national survey, Americans were asked whether they agreed with the statement, "Marriage is a lifetime relationship that should never be ended except under extreme circumstances." Seventy-six percent of the people agreed and only 6 percent disagreed. This belief is not limited to one class. When low-income unwed mothers were interviewed by two sociologists, they said they would only marry if they were sure the relationship would last for the rest of their lives. A little later in the survey, there was another statement, "When a marriage is troubled and unhappy, it is generally better for the children if the couple stays together." Now of the same people who just said marriage should be for life, only 25 percent agreed with this statement. Why did people express what appear to be such contradictory statements?

Andrew Cherlin, a sociologist who has spent his life studying the American family, believes the answer lies with two divergent beliefs in American culture. On the one hand, we think marriage is the best way to lead one's life. It should be a permanent, loving relationship with divorce as a last resort. On the other hand is the strong cultural belief in individualism that urges people to examine their personal situations and make changes if they are not satisfied. The individual pursuit of happiness has been elevated in American culture and is at the root of many of the changes in the American family (Cherlin, 2009).

These conflicting views may help to explain the major transformation that has taken place in marriage and family life during the last fifty years. Nearly everyone is marrying later today than in the past, but certain groups are not marrying at all. In 1960, 72 percent of all adults were married. By 2010, the percentage had dropped to 51 percent. Among people with a high-school degree or less, the number is 47 percent. This decline is especially dramatic for young adults: Twenty percent of 18- to 29-year-olds were married in 2010, compared with 59 percent in 1960 (Pew Research Center, 2011.)

Another big change that has taken place is that more children are being born to unwed mothers than in the past. In 1960, only 5 percent of all children were born to unmarried women. By 2010, the number had climbed to nearly 41 percent. Among Hispanic women the number is 53 percent, and among black women the number is 72 percent.

Instead of marrying, more people are cohabitating (living together as unmarried partners). The numbers have doubled since 1990, and 44 percent of people between 30 and 49 have cohabitated at some point (Pew 2010).

These numbers seem to be in line with the view that nearly four in ten Americans say marriage is becoming obsolete. Yet, if marriage is becoming obsolete, why would 62 percent of people who have never married say they would like to do so someday and only 12 percent say they would not like to marry?

It seems clear that many family forms are common today: single-parent families (resulting from either unmarried parenthood or divorce), remarried couples, unmarried couples, stepfamilies, and extended or multigenerational families. Mothers are more likely to be full- or part-time workers than they are to be full-time homemakers. This situation is an outgrowth of the greater move toward individualism and personal happiness. (See Figure 12-1 for how people define a family.)

The United States is not alone in experiencing such changes. Family patterns in the United States reflect broad social and demographic trends that are occurring in most industrialized countries around the world.

Has the American family always been in such flux? Although information about family life during the earliest days of our country is not very precise, it does appear quite clear, beginning with the 1790 census, that the American family was quite stable. Divorce and family breakup were not common. If a marriage ended because of desertion, death, or divorce, it was seen as a personal and community tragedy.

The American family has always been quite small. There has never been a strong tradition here of the extended family, in which relatives of several generations live within the same dwelling. Even in the 1700s, most American families consisted of a husband, wife, and approximately three children. This private, inviolate enclave made it possible for the family to endure severe circumstances and to help settle in the American frontier.

By the 1970s, radical changes were becoming evident. The marriage rate began to fall; the divorce rate, which had been fairly level, began accelerating upward; and fertility began to decline. The situation today shows that this trend has changed somewhat with the marriage rate

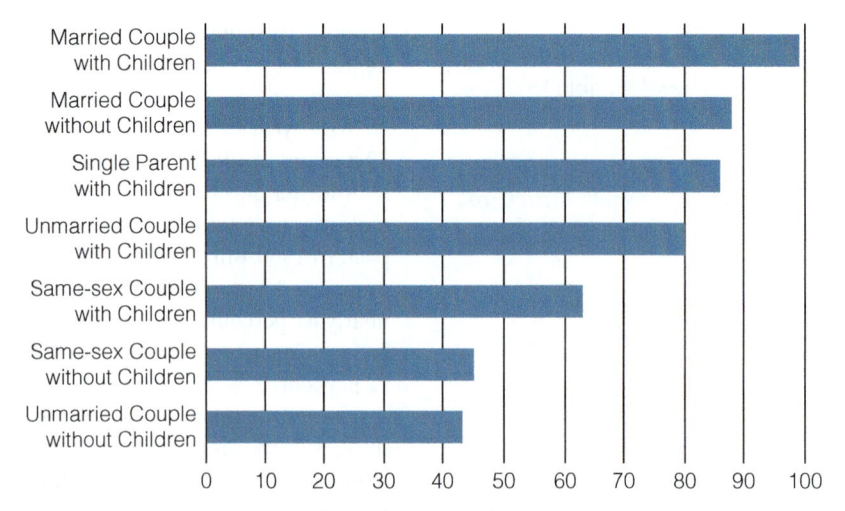

**FIGURE 12-1  Percentage of People Who Believe These are Families**
Source: Pew Research Center. November 18, 2010. Social & Demographic Trends Project, "The Decline of Marriage and Rise of New Families," p. 62 (www.pewsocialtrends.org/files/2010/11/pew-social-trends-2010-families.pdf), accessed August 23, 2012.

continuing to decline, but the divorce rate has started to level off and decline slightly.

The inescapable conclusion that can be drawn from all this is that the American family is in a state of transition. But transition to what? In this chapter, we shall study the institution of the family and look more closely at the current trends and what they forecast for the future.

The American family will continue to evolve and change even further during the next decade. Traditional families will account for only two out of three households, and married couples will make up slightly more than half of all households.

## THE NATURE OF FAMILY LIFE

The U.S. Bureau of the Census distinguishes between a *household* and a *family*. Households consist of people who share the same living space. A household can include one person who lives alone or several people who share a dwelling. A family, on the other hand, has a more precise definition. For example, in his classic study of social organization and the family, anthropologist George P. Murdock (1949) defined the family as:

> [A] social group characterized by common residence, economic cooperation, and reproduction. It includes adults of both sexes, at least two of whom maintain a socially approved sexual relationship, and one or more children, own or adopted, of the sexually cohabiting adults.

This definition has proved to be too limited. For one thing, it excludes many kinds of social groups that seem, on the basis of the functions they serve, to deserve the label of family. For example, in America, single-parent families are widely recognized. Despite Murdock's restricted definition of the family, it was his 1949 study of kinship and family in 250 societies, which showed the institution of the family to be present in every one of them, that led social scientists to believe that some form of the family is found in every known human society.

Perhaps a better understanding of the concept of the family can be gained by examining the various functions it performs in society.

Anthropologist George P. Murdock defined the family as a social group characterized by common residence, economic cooperation, and reproduction.

## Functions of the Family

Social scientists often assign to the fundamental family unit—married parents and their offspring—a number of the basic functions that it serves in most, if not all, societies. Although in many societies the basic family unit serves some of these functions, in no society does it serve all of them completely or exclusively. For example, among the Nayar of India, a child's biological father is socially irrelevant. Generally, another man takes on the social responsibility of parenthood (Gough, 1952). Or consider the Trobrianders, a people living on a small string of islands north of New Guinea. There, as reported by anthropologist Bronislaw Malinowski (1922), the father's role in parenthood was not recognized, and the responsibility for raising children fell to the family of their mother's brother.

The basic family is not always the fundamental unit of economic cooperation. Among artisans in preindustrial Europe, the essential economic unit was not the family but rather the household, typically consisting of the artisan's family plus assorted apprentices and even servants (Laslett, 1965). In some societies, members of the basic family group need not necessarily live in the same household. Among the Ashanti of western Africa, for example, husbands and wives live with their own mothers' relatives (Fortes et al., 1947). In the United States, a small but growing number of two-career families are finding it necessary to set up separate households in different communities, sometimes hundreds of miles apart. Husband and wife travel back and forth between these households to be with each other and the children on weekends and during vacations.

In all societies, however, the family does serve the basic social functions discussed next.

*Regulating Sexual Behavior* No society permits random sexual behavior. All societies have an **incest taboo,** which *forbids intercourse among closely related individuals,* although who is considered to be closely related varies widely. Almost universally, incest rules prohibit sex between parents and their children and between brothers and sisters. However, there are exceptions. The royal families of ancient Egypt, the Inca nation, and early Hawaii did allow sex and marriage between brothers and sisters.

In the United States, marriage between parents and children, brothers and sisters, grandparents and grandchildren, aunts and nephews, and uncles and nieces is defined as incest and is forbidden. In addition, approximately thirty states prohibit marriage between first cousins. The incest taboo usually applies to members of one's family (however the family is defined culturally), and thus it promotes marriage—and, consequently, social ties—among members of different families.

*Patterning Reproduction* Every society must replace its members. By regulating where and with whom individuals may enter into sexual relationships, society also patterns sexual reproduction. By permitting or forbidding certain forms of marriage (multiple wives or multiple husbands, for example), a society can encourage or discourage reproduction.

*Organizing Production and Consumption* In preindustrial societies, the economic system often depended on each family producing much of what it consumed. In almost all societies, the family consumes food and other necessities as a social unit. Therefore, a society's economic system and family structures often are closely correlated.

*Socializing Children* Not only must a society reproduce itself biologically by producing children, but it also must ensure that its children are encouraged to accept the lifestyle it favors, to master the skills it values, and to perform the work it requires. In other words, a society must provide predictable social situations within which its children are to be socialized (see Chapter 4, "Socialization and Development"). The family provides such a context almost universally, at least during the period when the child is dependent on the constant attention of others. The family is ideally suited to this task because its members know the child from birth and are aware of its special abilities and needs.

*Providing Care and Protection* Every human being needs food and shelter. In addition, we all need to be among people who care for us emotionally, who help us with the problems that arise in daily life, and who back us up when we come into conflict with others. Although many kinds of social groups are capable of meeting one or more of these needs, the family often is the one group in a society that meets them all.

*Providing Social Status* Simply by being born into a family, each individual receives both material goods and a socially recognized position defined by ascribed statuses (see Chapter 5, "Social Interaction"). These statuses include social class or caste membership and ethnic identity. Our inherited social position, or family background, probably is the single most important social factor affecting the predictable course of our lives.

Thus, we see that Murdock's definition of the family, based only on specific roles, is indeed too restrictive. A much more productive way of defining the family would be to view it as a universal institution that generally serves the functions previously discussed, although the way it does so may vary greatly from one society to another. The different ways various social functions are fulfilled by the institution of the family depend, in some instances, on the form the family takes.

## Family Structures

The **nuclear family** is *the most basic family form and is made up of a married couple and their biological or adopted children.* The nuclear family is found in all societies, and it is from this form that all other (composite) family forms are derived.

Two major composite family forms have been defined: polygamous and extended families. **Polygamous families** are *nuclear families linked together by multiple marriage bonds, with one central individual married to several spouses.* The family is **polygynous** *when the central person is male and the multiple spouses are female.* The family is **polyandrous** *when the central person is female and the multiple spouses are male.* Polyandry is rare; only one-half of 1 percent of all societies permit a woman to take several husbands simultaneously, and the women have to be wealthy members of those societies to do so. The Tlingit of southern Alaska are one example. Polyandry also occurs among well-to-do Tibetan families in the highlands of Limi, Nepal (Fisher, 1992).

**Extended families** *include other relations and generations in addition to the nuclear family, so that along with married parents and their offspring, there might be the parents' parents, siblings of the parents, the siblings' spouses and children, and in-laws.* All the members of the extended family live in one house or in homes close to one another, forming one cooperative unit.

Families, whether nuclear or extended, trace their relationships through the generations in several ways. Under the **patrilineal system,** *the generations are tied together through the males of a family; all members trace their kinship through the father's line.* Under the **matrilineal system,** exactly the opposite is the case: *The generations are tied together through the females of a family.* Under the **bilateral system,** *descent passes through both females and males of a family.* Although in American society descent is bilateral, the majority of the world's societies are either patrilineal or matrilineal (Murdock, 1949).

In patrilineal societies, social, economic, and political affairs usually are organized around the kinship relationships among men, and men tend to dominate public affairs. Polygyny often is permitted, and men also tend to dominate family affairs. Sociologists use the term **patriarchal family** to describe situations in which *most family affairs are dominated by men.* The **matriarchal family,** in which *most family affairs are dominated by women,* is relatively uncommon but does exist. Typically, it emerges in matrilineal societies. The matriarchal family is becoming increasingly common in American society, however, with the rise of single-parent families (most often headed by mothers).

Whatever form the family takes and whatever functions it serves, it generally requires a marriage to exist. Like the family, marriage varies from society to society in its forms.

## DEFINING MARRIAGE

**Marriage,** an institution found in all societies, is the *socially recognized, legitimized, and supported union of individuals of opposite sexes.* It differs from other unions (such as friendships) in that (1) it is initiated in a public (and usually formal) manner; (2) it includes sexual intercourse as an explicit element of the relationship; (3) it provides the essential condition for legitimizing offspring (i.e., it provides newborns with socially accepted statuses); and (4) it is intended to be a stable and enduring relationship. Thus, although almost all societies allow for divorce—that is, the breakup of marriage—no society endorses it as an ideal norm. Sociologists will need to revise the definition of marriage now that a number of European countries and several states in America have legalized same-sex marriages.

### Romantic Love

What would you think of someone who used the following proposal of marriage?

> It does not appear to me that my hand is unworthy of your acceptance, or that the establishment I can offer would be any other than highly desirable. My situation in life, my connections with the family, and my relationship to your own, are circumstances highly in my favor; and you should take it into further consideration, that in spite of manifold attractions, it is by no means certain that another offer of marriage may even be made to you.

This proposal, which would very likely produce outrage and be rejected by many women today, was made by Mr. Collins to Elizabeth in Jane Austen's novel, *Pride and Prejudice* (Austen, 1813). In addition to seeming exceedingly arrogant, it also appears to lack any sign that love exists.

Americans are strong believers in the power of love. More than half of American adults (52%) said they believed in love at first sight. Four in ten Americans said they have actually fallen in love at first sight, and almost three-quarters believed that one true love existed (Gallup Poll, February 2001).

We also expect to have an exceptionally close relationship with the person we marry. Emotional and spiritual connection rank far above the need for financial security in forming a romantic partnership. Eighty-one percent of women ages 20 to 29 reported that "it is more important to have a husband who can communicate about his deepest feelings than it is to have a husband who makes a good living" (Maybury, 2002).

Americans are also strongly wedded to the view that there is a compatibility between romantic love and the institution of marriage. Not only do we believe they are compatible, but we expect them to coexist. Our culture implies that without the prospect of marriage, romance is immoral, and that without romance, marriage is empty. This view underlies most romantic fiction and other media presentations.

When college students from ten countries were asked whether they would marry without love, 86 percent of American college students said they would not. Only 24 percent of college students from India said no to marriage without love. Cultures are likely to indulge in romantic love if they are wealthy and value individualism over the community (Levine, 1993).

Romantic love can be defined in terms of five dimensions: (1) idealization of the loved one, (2) the notion of a one and only, (3) love at first sight, (4) love winning out over all, and (5) an indulgence of personal emotions (Lantz, 1982).

In some cultures, people expect love to develop after marriage. Hindu children are taught that marital love is the essence of life. Men and women often enter married life enthusiastically expecting a romance to develop. As the Hindu saying goes, "First we marry, then we fall in love" (Fisher, 1992).

In many of the world's other societies, romantic love is unknown or seen as a strange maladjustment. It might exist, but it has nothing to do with marriage. Marriage in these societies is seen as an institution that organizes or patterns the establishment of economic, social, and even political relationships among families. Three families ultimately are involved: **families of origin** or **families of orientation,** *the two families that produced the two spouses,* and their **family of procreation,** *the family created by the spouses' union.*

In China, if the son was too much in love with his wife it was seen as a threat to the solidarity of the family. The parents might force the son to divorce his wife if her behavior was not what they expected and his love for her was an irrelevant consideration. The view was "you have only one family, but you can always get another wife" (Coontz, 2005).

A belief in romantic love helps make men and women more independent of their relatives. Romantic love weakens the emotional ties that bind people to their families. It makes it natural for partners to be committed to each other and provide mutual support. Romantic love provides a legitimate way to move out of the family without creating tensions and jealousies.

## Marriage Rules

In every society, marriage is the binding link that makes possible the existence of the family. All societies have norms or rules governing who may marry whom and where the newlywed couples should live. These rules vary, but certain typical arrangements occur in many societies around the world.

Almost all societies have two kinds of marriage norms or rules. *Rules of* **endogamy** *limit the social categories from within which one can choose a marriage partner.* For example, many Americans still attempt to instill in their children the idea that one should marry someone within the ethnic, religious, or economic group of one's family of origin. *Rules of* **exogamy,** by contrast, *require an individual to marry someone outside his or her culturally defined group.* For example, in many tribal groups, members must marry outside their lineage. In the United States, there are laws forbidding the marriage of close relatives, although the rules are variable.

These norms vary widely across cultures, but everywhere they serve basic social functions. Rules of exogamy determine the ties and boundaries between social groups, linking them through the institution of marriage and whatever social, economic, and political obligations go along with it. Rules of endogamy, by requiring people to marry within specific groups, reinforce group boundaries and perpetuate them from one generation to the next.

Marriage rules also determine how many spouses a person may have at one time. Among many groups, Europeans and Americans, for example, marriage is **monogamous**—that is, *each person is allowed only one spouse at a time.* However, many societies allow **multiple marriages,** in which *an individual may have more than one spouse* (polygamy). Polygyny, the most common form of polygamy, is found among such diverse peoples as the Swazi of Africa, the Tiwi of Australia, and, formerly, the Blackfeet Indians of the United States. Polyandry, as noted earlier, is extremely rare (Murdock, 1949).

As Marvin Harris (1975) has noted, "Some form of polygamy occurs in 90 percent of the world's cultures." However, within each such society, only a minority of people actually can afford it. In addition, the Industrial Revolution favored monogamy for reasons we shall discuss shortly. Therefore, monogamy is the most common and widespread form of marriage in the world today.

## Marital Residence

After two people are married, they must set up housekeeping. Most societies have strongly held norms that influence where a couple lives. **Marital residence rules** *govern where a couple settles down.* **Patrilocal residence** *calls for the new couple to settle down near or within the husband's father's household*—as among Greek villagers and the Swazi of Africa. **Matrilocal residence** *calls for the new couple to settle down near or within the wife's mother's household*—as among the Hopi Indians of the American Southwest.

Some societies, such as the Blackfeet Indians, allow **bilocal residence,** in which *new couples choose whether to live with the husband's or wife's family of origin.* In modern industrial society, newlyweds typically have even more freedom. With **neolocal residence,** *the couple may choose to live virtually anywhere,* even thousands of miles from their families of origin. In practice, however, it is not unusual for American newlyweds to set up housekeeping near one of their respective families.

Marital residence rules play a major role in determining the compositions of households. With patrilocal residence, groups of men remain in the familiar context of their father's home, and their sisters leave to join their husbands. In other words, after marriage, the women leave home to live as strangers among their husband's kinfolk. With matrilocal residence, just the opposite is true; women and their children remain at home, and husbands are the outsiders. In many matrilocal societies, this situation often leads to considerable marital stress, with husbands going home to their own mothers' families when domestic conflict becomes intolerable.

Bilocal residence and neolocal residence allow greater flexibility and a wider range of household forms because young couples can move to places in which the social, economic, and political advantages might be greatest. One disadvantage of neolocal residence is that a young couple cannot count on the immediate presence of kinfolk to help out in times of need or with demanding household chores (including the rearing of children). In the United States today, the surrogate or non-kin "family," made up of neighbors, friends, and colleagues at work, might help fill this void. In other societies, polygynous neolocal families help overcome such difficulties, with a number of wives cooperating in household work.

## Mate Selection

Like our patterns of family life, America's rules for marriage and mate selection spring from those of our society's European ancestors. Because we have been nourished, through songs and cinema, by the notions of "love at first sight," "love is blind," and "love conquers all," most of us probably believe that in the United States there are no rules for mate selection. Research shows, however, that this is not necessarily true.

If we think statistically about mate selection, we must admit that in no way is it random. Consider for a moment what would happen if it were. Given the population distribution of the United States, blacks would be more likely to marry whites than members of their race; upper-class individuals would have a greater chance of marrying a lower-class person; and various culturally improbable but statistically probable combinations of age, education, and religion would take place. In actual fact, **homogamy**—*the tendency of like to marry like*—is much more the rule.

There are numerous ways in which homogamy can be achieved. One way is to let someone older and wiser, such as a parent or matchmaker, pair up appropriately suited individuals. Throughout history, this has been one of the most common ways by which marriages

Despite high divorce rates, most Americans believe that marriage is a lifetime relationship that should only be ended under extreme circumstances.

have taken place. The role of the couple in question can range from having no say about the matter whatsoever to having some sort of veto power. This tradition is quite strong in Islamic countries such as Pakistan. Benazir Bhutto, the two-time Pakistani prime minister who was assassinated in 2007, agreed to an arranged marriage despite her Western education and feminist leanings. She submitted to Islamic cultural traditions, under which dating is not considered acceptable behavior for a woman.

In the United States, most people who marry do not use the services of a matchmaker, though the result in terms of similarity of background is so highly patterned that it could seem as if a very conscious homogamous matchmaking effort were involved.

*Age* In American society, people generally marry within their own age range. Comparatively few marriages have a wide gap between the ages of the two partners. In addition, only 23.2 percent of American women marry men who are younger than themselves. On the average, in a first marriage for both the man and the woman, the man tends to be about two years older than the woman. This is, however, related to age at the time of marriage. For example, 20-year-old men marry women with a median

As late as 1966, nineteen states sought to stop interracial marriage through legislation.

age only one month younger. Twenty-five-year-old men marry women with a median age of 11 months younger. For 30- to 34-year-old men, the median age for wives is 2.8 years younger; whereas for men over 65, the median age for wives is nearly 10 years younger (U.S. Bureau of the Census, 2007).

In 1890, the estimated median age at first marriage was 26.1 years for men and 22.0 years for women. At that time, a decline in the median age at first marriage began; it did not end until 1956, when the median age reached a low of 20.1 years for women and 22.5 years for men. This 66-year decline has been reversed, and in 2010, the median age at first marriage exceeded the 1890 level, standing at 28.7 years for men and 26.5 for women (Pew Research Center, 2011).

Age homogamy appears to hold for all groups within the population. Studies show that it is true for blacks as well as whites and for professionals as well as laborers. Clearly, the norms of our society are very effective in causing people of similar age to marry each other.

*Race* Homogamy is most obvious in the area of race. It was only about four decades ago that Thurgood Marshall, who was only months away from appointment to the Supreme Court, suffered an indignity that seems hard to believe today. Marshall and his wife had found a house in a Virginia suburb of Washington, D.C. There was one problem, however. Virginia did not allow interracial marriages, and Marshall was black and his wife was Asian. Just in the nick of time, the Supreme Court ruled that these laws were unconstitutional in January 1967. A Gallup poll from that time showed that 72 percent of Southern whites and 42 percent of Northern whites thought interracial marriage should be banned (Sailer, 1997).

The laws in nineteen states that sought to stop interracial marriage through legislation in the 1960s varied widely, and there was great confusion because of various court interpretations. In Arizona before 1967, it was illegal for a white person to marry a black, Hindu, Malay, or Asian person. The same thing was true in Wyoming, and residents of that state were also prohibited from marrying mulattos.

In 1966, Virginia's Supreme Court of Appeals had to decide on the legality of a marriage that had taken place in Washington, D.C., in 1958 between Richard P. Loving, a white man, and his part-Indian and part-black wife, Mildred Loving. The court unanimously upheld the state's ban on interracial marriages. The couple appealed the case to the U.S. Supreme Court, which agreed to decide whether state laws prohibiting racial intermarriage were constitutional. Previously, all courts had ruled that the laws were not discriminatory because they applied to both whites and nonwhites. However, on June 12, 1967, the Supreme Court ruled that states could not outlaw racial intermarriage.

Today, about 7 percent of U.S. married couples include spouses of a different race. This small percentage masks a 28 percent growth rate between 2000 and 2010 (U.S. Bureau of the Census, 2012). The most common marriage is between a spouse who is one race and the other spouse who is more than one race (16%). The next most common combination is a couple with one white spouse and one Asian spouse (14%), followed by couples with one white non-Hispanic spouse and one black non-Hispanic spouse (8%) (U.S. Bureau of the Census, 2012).

Interracial marriages are more likely to involve at least one previously married partner than are marriages between spouses of the same race. In addition, brides and grooms who marry interracially tend to be older than the national average.

The degree of education also differs in interracial marriages. White grooms in interracial marriages are more likely to have completed college than white grooms who marry within their race. In all white couples, 18 percent of the grooms hold college degrees. In a marriage involving a white groom and a black bride, 24 percent of the men finished college. Among black men married to women of other races, only 5 percent had completed college, whereas 13 percent of black men married to white women had done so. This compares with 9 percent of black men whose spouses were also black who held college degrees.

Although the rate of interracial marriage is growing, the proportions vary widely by state. Hawaii has the distinction of having both the highest number and the greatest proportion of interracial marriages. Almost one-fourth of all marriages there are interracial, the majority of which are Asian-Caucasian unions. Alaska is second to Hawaii in the proportion of interracial marriages; 13

percent of the state's marriages are interracial. Illinois has the largest number of white brides marrying black grooms of any state. The record for the reverse combination—black brides marrying white grooms—is held by New Jersey.

Even though their total numbers are still small, racially mixed marriages have become increasingly common in America's melting pot.

*Religion* Unlike many European nations, none of the American states has ever had legislation restricting interreligious marriage. Religious homogamy is not nearly as widespread as is racial homogamy, although most marriages still involve people of the same religion.

Attitudes toward religious intermarriage vary somewhat from one religious group to another. Almost all religious bodies try to discourage or control marriage outside the religion, but they vary in the extent of their opposition. Before 1970, the Roman Catholic Church would not allow a priest to perform an interreligious marriage ceremony unless the non-Catholic partner promised to rear the children of the union as Catholics, and the Catholic partner promised to encourage the non-Catholic to convert. However, in the late 1960s, the Catholic Church softened its policy, and since 1970, the pope has allowed local bishops to permit mixed-religion marriages to be performed without a priest and has eliminated the requirement that the non-Catholic partner promise to rear the children as Catholics. Nevertheless, the Catholic partner must still "ensure the baptism and education of the children in the Catholic Church" (Catechism of the Catholic Church, 1635).

Protestant denominations and sects also differ with regard to the barriers they place on the intermarriage of their members. At one extreme are the Mennonites, who excommunicate any member who marries outside the faith. At the other are numerous denominations that may encourage their members to marry within the faith but provide no formal penalties for those who do not (Heer, 1980).

Jewish religious bodies also differ in their degree of opposition to religious intermarriage. Orthodox Jews are the most adamantly opposed to intermarriage, whereas Conservative and Reform Jews, although by no means endorsing it, are more tolerant when it does take place. Jewish intermarriage rates, which were at 13 percent prior to 1970, have been increasing dramatically in recent years, with as many as 50 percent of all Jews marrying someone of another religion.

Religious leaders often are concerned about religious intermarriage for a variety of reasons. Some claim that one or both intermarrying parties are lost to the religion, and others believe that the potential for marital success is decreased greatly in intermarriage. Complex factors are involved in intermarriage, and simplistic and unequivocal predictions are not warranted.

Most marriages still involve those of the same religion. There is, however, a clear trend of more religious intermarriages in the United States today.

*Social Status* Level of education and type of occupation are two measures of social status. In these areas, there is usually a great deal of similarity between people who marry each other. Men tend to marry women who are slightly below them in education and social status, although these differences are within a narrow range. Wide-ranging differences between the two people often contain an element of exploitation. One partner might either be trying to make a major leap on the social class ladder or be looking for an easy way of taking advantage of the other partner because of unequal power.

The typical high-school environment often plays a major role in maintaining social status homogamy because it is in high school that students have to start making plans for their future careers. Some might go to college, and others might plan on going to work directly after graduation. This process causes the students to be divided into two groups: the college bound and the workforce bound. Although the lines separating these two groups are by no means impenetrable, in many high schools these two groups maintain separate social activities. In this way, barriers against dating and future marriage between those of unlike social status are set up. After graduation, those who attend college are more likely to associate with other college students and to choose their mates from that pool. Those who have joined the workforce are more likely to choose their mates from that environment. In this way, similarities in education between marriage partners really are not accidental.

As with several of the factors we have discussed already, education, social class, and occupation produce a similarity of experience and values among people. Just as growing up in an Italian American family can make one feel comfortable with Italian customs and traits, going to college can make one feel comfortable with those who have experienced that environment. Similarities in social status, then, are as much a result of socialization and culture as of conscious choice. We most likely will marry a person we feel comfortable with—a person who has had experiences similar to our own.

Coming to terms with the constraints on our marriage choices can be a sobering experience. What we thought to be freedom of choice in selecting a mate is revealed instead by various studies to be governed by rules and patterns. (For more information on marriage and divorce in the United States, see "Day-to-Day Sociology: Marriage and Divorce Quiz.")

## DAY-TO-DAY SOCIOLOGY

### Marriage and Divorce Quiz

1. **The so-called red states that encourage traditional families have lower divorce rates than the more liberal blue states.**

   *False.* Traditional family values do not seem to provide much protection against divorce. The highest divorce rates are found in the more conservative red states such as Arkansas, Oklahoma, and West Virginia in contrast to more liberal blue states such as Pennsylvania and Massachusetts.

2. **One of the reasons there is more divorce today than in the past is because people live longer and there is more time to get divorced.**

   *False.* Even though people live longer, they also marry later than in the past. About half of all divorces take place before the seventh year of marriage and are not influenced by life span. The length of the typical divorce-free marriage today is similar to what it was fifty years ago.

3. **If your parents divorced, your chances of divorcing are increased.**

   *True.* It appears that living in a family that has experienced divorce undermines the view that marriage is a lifelong commitment. The risk nearly triples if both partners come from homes that experienced a divorce.

4. **Because the high divorce rate weeds out the unhappy marriages, people who stay married have happier marriages than people did in the past when everyone stuck it out, no matter how bad the marriage.**

   *False.* The general level of happiness in marriages may have declined slightly. Marriages today have more work-related stress, more marital conflict, and less marital interaction than twenty to thirty years ago.

5. **An unmarried woman is more likely to experience domestic violence than a married woman.**

   *True.* Unmarried women, particularly those who cohabit with a man, are more likely to be the victims of domestic abuse.

6. **Second marriages are more successful than first marriages because people learn from their mistakes.**

   *False.* The divorce rate for second marriages is higher than for first marriages.

7. **Women are more likely than men to be the ones who initiate a divorce.**

   *True.* Women initiate two-thirds of all divorces. One reason may be that in most states, men stand a small chance of gaining custody of their children. A divorce usually means that they will have to pay child support but have limited contact with the children.

8. **Teenage marriages are fairly successful if they can get through the first year.**

   *False.* Marrying in your teens increases the likelihood of divorce two to three times over that of couples in their 20s and older.

*Source:* Popenoe, David. "The Top Ten Myths of Divorce." National Marriage Project (http://marriage.rutgers.edu/Publications/Print/Print%20Myths%20of%20Divorce.htm), accessed June 16, 2009.

## THE TRANSFORMATION OF THE FAMILY

The American family has gone through a variety of transformations. During preindustrial times, all members of the household, whether they were parents, children, or boarders, had to contribute in some way to help the family survive economically. Work was not defined by strict gender and age restrictions (Ruane and Cerulo, 2004).

The changes of the Industrial Revolution helped to produce the relatively isolated nuclear family with weak ties to an extensive kinship network. This type of family was well adapted to the pressures of industrialism.

First, industrialism demands that workers be geographically mobile so that a workforce is available wherever new industries emerge. The modern nuclear family, by having cut many of its ties to extended family networks, is freer to move than its predecessors. It was among laborers' families that extended kinship ties first were weakened. Eventually, middle- and upper-class families followed in the same direction.

Second, industrialism requires a certain degree of movement between the social classes (see Chapter 8, "Social Class in the United States") so talented workers can be recruited to positions of greater responsibility (with greater material rewards and increased prestige). A family that is too closely tied to other families in its kinship network will find it difficult to break free and climb into a higher social class. However, if families in the higher social classes are too tightly linked by kinship ties, newly arriving families will find it very difficult to fit into their new social environment. Hence, the isolated nuclear family is well suited to the social mobility required in an industrial society.

In the early 1950s, this steelworker in Youngstown, Ohio, earned $320 per month for a 40-hour workweek. His salary was enough to support a comfortable lifestyle for his wife and their two children.

Third, the modern nuclear family is more open to inheritance and descent through both sides of the family. Valuable material resources and social opportunities are not inherited mainly by the oldest males. This means that all children in a family will have a chance to develop their skills, which in turn means that industry will have a larger, more talented, and flexible labor force from which to hire workers.

By the early twentieth century, then, the nuclear family had evolved fully among the working classes of industrial society. It rested on (1) the child-centered family, (2) **companionate marriage** (i.e., *marriage based on romantic love*), (3) increased equality for women, (4) decreased links with extended families or kinship networks, (5) increased geographical mobility, (6) increased social mobility, and (7) the clear separation between work and leisure. Also, because much of the work was hard and boring, the nuclear family was expected to provide emotional support for its members.

The World War II years also had a profound effect on the American family. The war necessitated hundreds of thousands of women to work outside the home to support their families. They often had to take jobs that were vital to the American economy but had been vacated when their husbands went to fight overseas. After the war, an effort was made to "defeminize" the workforce. Nevertheless, many women remained on the job, and those who left knew what it was like to work for pay outside the home. Things were never the same again for the American family, and family life began to change.

The initial changes were not all that apparent. Indeed, by the 1950s, the United States had entered the most family-oriented period in its history. This was the era of the baby boom, and couples were marrying at the youngest ages in recorded American history. During the 1950s, 96 percent of women in the childbearing years married (Blumstein and Schwartz, 1983). The war years' experiences also paved the way for secondary groups and formal organization (see Chapter 6, "Social Groups and Organizations") to gradually take over many of the family's traditional activities and functions. The prime task of socializing the young was shifted from the family to the schools. Social workers intruded into the home, offering constantly expanding welfare services from outside agencies. The juvenile court system expanded, in the belief that deficiencies in the families of youthful offenders could be addressed by outside agencies.

Thus, the modern period has seen what sociologists refer to as the transfer of functions from the family to outside institutions. This transfer has had a great effect on the family, and it underlies the trends that currently are troubling many people.

There are several problems in trying to assess the prevailing state of the family. Some feel that the family is deteriorating, and they cite the examples of divorce rates and single-parent families to support their view. Others think of the family as an institution that is in transition but just as stable as ever. Was the family of the past with everyone working for the betterment of the whole better? Or has the family structure changed throughout history in response to the economic and political changes within society? In this next section, we shall explore these views and attempt to clarify the current direction of family life.

## The Decline of the Traditional Family

The number of "traditional" families, that is, married couples with children, has been decreasing steadily. With the aging of baby boomers and declining birth rates, the percentage of families with children at home has decreased.

The decline of the traditional family structure can also be attributed to the postponement of marriage and single-parent families resulting from divorce or out-of-wedlock births. People are also living longer than in the 1950s and 1960s. As people live longer, a larger share of the married-couple households are older and their children are living on their own. With the decline of the traditional family, alternative family structures have become more common.

## Changes in the Marriage Rate

The marriage rate has been going down. According to census data, barely half (51%) of adults ages 18 and older are married, compared with 72 percent in 1960. It is even more striking when we look at 18- to 29-year-old young adults, where only 20 percent were married in 2010 compared with 59 percent in 1960.

Much of this decline is due to the ages at which men and women marry for the first time. Since 1960, it has increased by nearly six years. It continues to rise and in 2011, the median age at first marriage was 28.7 for men and 26.5 for women. That means half of men don't marry until at least about age 29, and half of women don't marry until at least about age 27. In 1960, the median age at first marriage for both men and women was in the early 20s (Pew Research Center, 2011).

Are fewer people marrying now than in the past? The answer depends on how you evaluate the data. One way to look at it is to ask how many marriages there are. In 2010, there were 2.096 million marriages. A second way to look at it is to take the marriage rate per 1,000 of never-married women ages 15 and older (see Figure 12-2). This is the proportion of eligible women marrying. The marriage rate for unmarried women began to plummet around 1975, and today it is at an all-time low of 39.2 per 1,000.

The decline in marriage could be exaggerated. The age 15 cutoff point for comparing marriage rates can be deceptive because people now just do not wed at such an early age (Basharov, 1999). To measure marriage rates accurately today, we should look at the percentage of women who are "ever married." "Assuming that one should not count teenagers because so few marry, in 1960 the proportion of women 20 and older who had ever been married was 90 percent." Since then there has been little change in that number. Young people today marry later and often plan to make their marriage a partnership of equals, but they do marry (Besharov, 1999).

All of this may begin to explain why cohabitation has emerged as a major change in couple relationships. Although the marriage rate may be down, cohabitation is certainly up. (For more on factors affecting the marriage rate, see "How Sociologists Do It: Study, Graduate, and Be Married.")

**Cohabitation,** or *unmarried couples living together out of wedlock,* has increased dramatically in the past twenty years and is already having a significant impact on the American family. In 1988, fewer than one in five married Americans said they lived with their spouse before marriage (Jones, 2002).

Most adults in the United States eventually marry. In 2007, 91 percent of women ages 45 to 54 had been married at least once. The increasing social acceptance of cohabitation has resulted in definitions of being married or being single becoming less clear-cut. As the personal lives of unmarried couples start to resemble those of their married counterparts, the meaning and permanence of marriage changes also. The number of unmarried-couple households in 2010 was fifteen times as high as the number in 1970, when it was 523,000. Today, it is 7.7 million and continues to grow four times as fast as the overall household population. Maine has the highest percentage of opposite-sex unmarried partner households (U.S. Bureau of the Census, 2012).

An unmarried-couple household, as defined by the U.S. census bureau, contains only two adults with or without children younger than 15 years of age present. The adults must be of opposite sexes and not related to one another (U.S. Bureau of the Census, 2009).

It is possible that the increase in cohabitation figures represents better data collection as much as it represents an increase in couples living together. However, all signs point to an increase in cohabitation. More than four in ten women have been in an unmarried domestic partnership at some point in their lives.

For some, cohabitation is a prelude to marriage; for others, an alternative to marriage; and for still others, simply an alternative to living alone. Cohabitation is

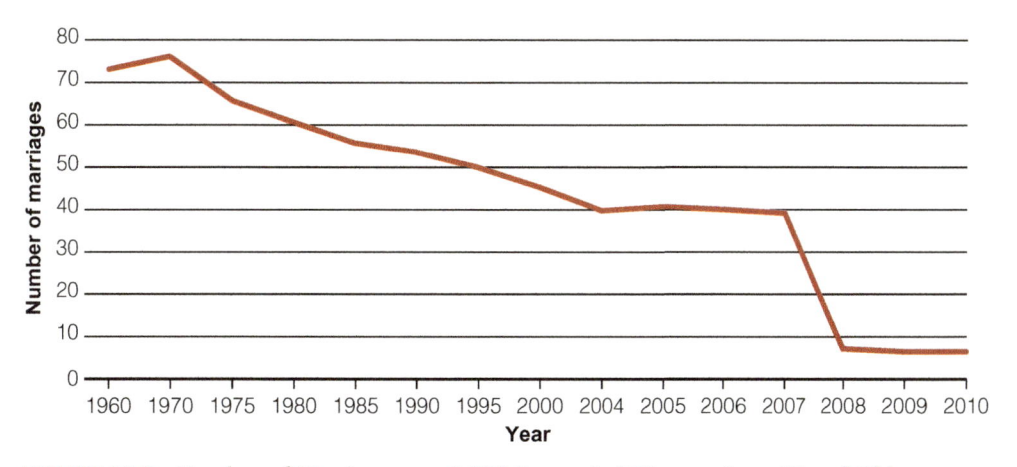

**FIGURE 12-2** Number of Marriages per 1,000 Unmarried Women, Ages 15 and Older
*Source: National Center for Health Statistics. National Marriage and Divorce Rate Trends, Provisional Number of Marriages and Marriage Rate: United States, 2000–2010 (www.cdc.gov/nchs/nvss/marriage_divorce_tables.htm), accessed August 15, 2012.*

## HOW SOCIOLOGISTS DO IT

### Study, Graduate, and Be Married

It used to be an unfortunate fact that the more education a woman had, the less likely she was to marry. Clichés like "It's not smart to be too smart" used to be thrown about, making some women wonder whether a high GPA would sink their chances of ever finding wedded bliss. Stereotypes like these were often given additional credence in the press. Maureen Dowd, a columnist for the *New York Times*, wrote, "The rule of thumb seems to be that the more successful the woman, the less likely it is that she will find a husband or bear a child. For men the reverse is true." Most of the respondents to her column agreed.

These stereotypes were true in the past. In 1950, only 67 percent of college-educated white women had married by the time they reached age 55. For those women without a college degree, 93 percent had married. This is a significant difference showing that there were significant factors preventing college-educated women from marrying in that time period. Even as late as 1980, a woman who did not have a high-school degree was more likely to be married than a woman with a college or graduate degree. By 2008, the gap had nearly disappeared, and some researchers are predicting that very shortly college-educated women will be more likely to marry than their less-educated peers. (See Figure 12-3.) College-educated women are already more likely to be married than high-school dropouts.

There have been significant changes for men also. Today, the college-educated male is more likely to marry than his less-educated peer, a reversal of what was the case in 1950.

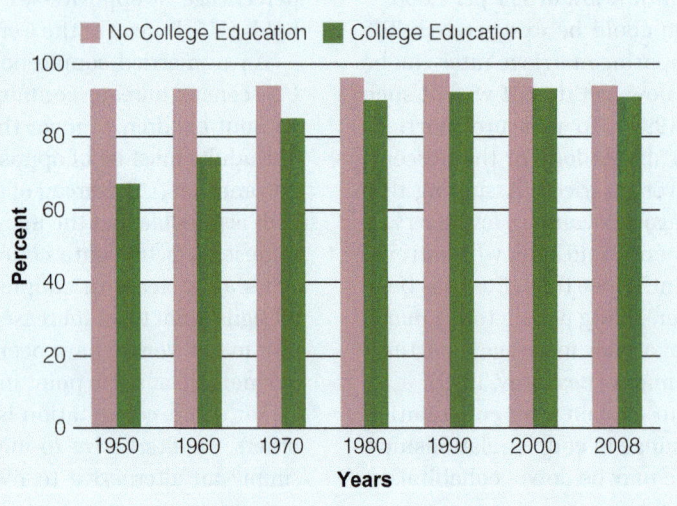

**FIGURE 12-3** Percentage of White Females Married by Age 50
Source: U.S. Bureau of the Census. 2008. American Community Survey.

---

more common among those of lower educational and income levels. Recent data show that among women in the 19 to 44 age range, 60 percent of high-school dropouts have cohabited compared to 37 percent of college graduates (Bumpass and Lu, 2000).

Cohabitation is also more common among those who are less religious than their peers, those who have been divorced, and those who have experienced parental divorce, fatherlessness, or high levels of marital discord during childhood (Smock, 2000).

Cohabitation is often seen as a prelude to marriage. Single people see it as a way to make sure that the couple is compatible before getting married. At least one of the partners expects the arrangement to result in marriage in 90 percent of cohabitations. The reality does not bear out these overly optimistic expectations, and at most 50 percent of women who cohabit end up marrying the man.

Many young people believe that living together before marriage is a good way to "find out whether you really get along," which would prevent a bad marriage and eventual divorce. "But the available data on the effects of cohabitation fail to confirm this belief. In fact, a substantial body of evidence indicates that those who

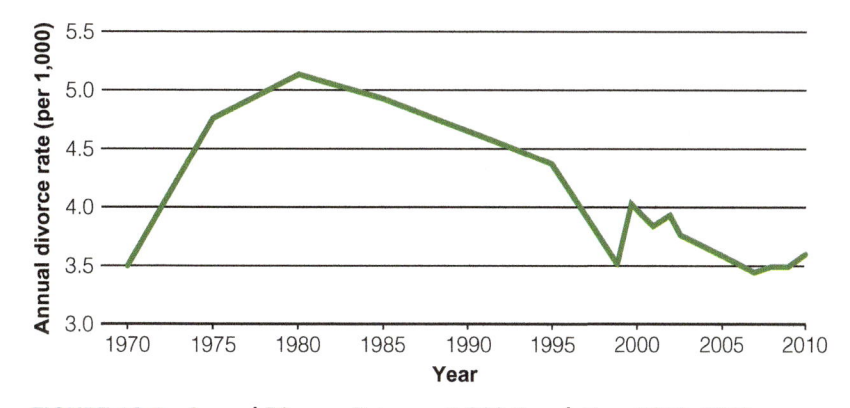

**FIGURE 12-4** Annual Divorce Rate per 1,000 Population, 1970–2010

Source: National Center for Health Statistics. National Marriage and Divorce Rate Trends, Provisional Number of Marriages and Marriage Rate: United States, 2000–2010 (www.cdc.gov/nchs/nvss/marriage_divorce_tables.htm), accessed August 15, 2012.

live together before marriage are more likely to break up after marriage." The probability of a first marriage ending in separation or divorce within five years is 20 percent, but the probability of a marriage involving cohabitation breaking up within five years is 49 percent. After ten years, the probability of a first marriage ending is 33 percent, compared with 62 percent for cohabitations (Bramlett and Mosher, 2002). (See Figure 12-4 for the annual divorce rate per 1,000 people in the population.)

This evidence is controversial because it is difficult to discount the fact that people who cohabit before marriage have different characteristics from those who do not. It can be these characteristics, and not the experience of cohabitation, that lead to marital instability. There is no clear evidence yet that those who cohabit before marriage have stronger marriages than those who do not (Whitehead and Popenoe, 2002).

## Childless Couples

Childlessness among married couples has been increasing in recent years. Many women in the childbearing years see postponement of marriage and childbearing as pathways to a good job and economic independence. For many women, this temporary postponement becomes permanent either by choice or chance. Women with the highest levels of education, those engaged in managerial and professional occupations, and those with the highest family incomes have the highest levels of childlessness.

Childlessness among women 40 to 44 years old (most women have completed childbearing by this age) increased from 10 percent in 1980 to 20 percent in 2008 (U.S. Bureau of the Census, 2010). This is the continuation of a trend that began in 1984. That was the year when there were more childless couples in the United States than couples with children younger than 18. This reversal of the ratio of couples with children to childless couples from what was the case throughout most of our history will continue into the foreseeable future (Fields, 2001).

Some of these couples plan to have children at some point. More than half of all women younger than 35 who are childless expect to have a child in the future. Others are nonparents either by choice or because of fertility problems.

## Changes in Household Size

Changes in household size need not be seen as either positive or negative. The American household of 1790 had an average of 5.8 members; by 2010, the average number had dropped to 2.6 (U.S. Bureau of the Census, 2011). The same trend also has been evident in other parts of the world. The average rural household in Japan in 1660 often had 20 or more members, but by the 1960s, the rural Japanese household averaged only 4.5 members. One reason for the reduction in size of the American household may be that today it is very unusual to house unrelated people. Until the 1940s, for a variety of reasons, it was common for people to have non-kin living with them, either as laborers for the fields or as boarders who helped with the rent payment.

The reduction in the number of non-relatives living with the family explains only part of the continuing reduction in the average household size. Another reason may be a rapid decrease in the number of aging parents living with grown children and their families. Some point to this as evidence of the fragmentation and loss of intimacy present in the contemporary family. At the turn of the century, more than 60 percent of those 65 or older lived with one or more of their children; today this figure is less than 10 percent.

How can we account for so many more older people living apart from their families? We might be tempted to say that the family has become so self-centered and so unable to fulfill the needs of its members that the elderly have become the first and most obvious castoffs. However, this trend of the elderly living away from their children can also be seen as a result of the increasing wealth of the population, including the elderly. In the past, many elderly lived with their children because they could not afford to do otherwise.

Another reason for the change in the size of households is the increasing divorce rate. As more families separate legally and move apart physically, the number of people living under one roof has fallen.

A further explanation for the smaller families of today is the tendency of young people to postpone marriage and the increase in the number of working women. As people marry later, they have fewer children. Many couples also are deciding to have no children as more and more women become involved in work and careers.

All these factors point to the most significant causes of the sharp decline in the average size of the American household: the decrease in the number of children per family and the increase in the number of people living alone. The number of Americans living alone has also increased substantially.

Nearly 30 million individuals live alone. These single people are not necessarily rejecting marriage. They are just a reflection of changing family and living arrangements. Many young people postpone marriage until they have completed college and found appropriate jobs. Older people are more likely to live in their own homes and apartments instead of nursing homes (AmeriStat, 2001). These changes ultimately will have important consequences for the entire structure of society.

## Women in the Labor Force

The period since World War II has seen a dramatic change in the labor force participation rates of American women. Nearly 65 million or 58.6 percent of all women had paying jobs in 2010, representing more than a 100-percent increase in forty years. The number of men in the labor force during this same period increased by less than 50 percent. The change in the labor force is probably the single most important recent change in American society (U.S. Department of Labor, 2011).

The level of education of women aged 25 to 64 in the labor force has risen substantially over the past forty years. In 2010, 36 percent of these women held college degrees, compared with 11 percent in 1970. Only 7 percent of women were high-school dropouts in 2010, down from 34 percent in 1970. In 2010, women accounted for 13 percent of architects and engineers, 32 percent of physicians and surgeons, 60 percent of accountants and auditors, and 82 percent of elementary and middle-school teachers.

Women have increased their numbers in the labor force because of a number of factors. After World War II, many women who took jobs in record numbers to ease labor shortages during the war remained on the job. Their numbers increased the social acceptability of the working woman. Widespread use of contraceptives is a second important factor. Effective contraception gave women the freedom to decide whether and when to have children. As a result, many women postponed childbearing and continued their educations. Baby boomers also had different economic expectations when they entered the workforce.

Having two incomes has become important to ensure the lifestyle and standard of living families have come to expect. Now husband and wife are earners in 55 percent of married-couple families, up from 44 percent in 1967 (U.S. Department of Labor, 2011).

## Family Violence

It appears that we are in greater danger of being the victim of violence when we are with our families than when we are in a public place. Recent years have seen a number of high-profile cases in which spouses have attacked each other, parents have attacked children, and children have attacked and killed parents.

Family violence is greater in poor and minority households, but it occurs in families at all socioeconomic levels. Research has shown the incidence of family violence to be highest among urban lower-class families. It is high among families with more than four children and in those in which the husband is unemployed. Families in which child abuse occurs tend to be socially isolated, living in crowded and otherwise inadequate housing. Research on family violence has tended to focus on lower socioeconomic groups, but scattered data from school counselors and mental health agencies suggest that family violence also is a serious problem among America's more affluent households.

Children who are victims of abuse are more likely to be abusive as adults than are children who did not experience family violence. Research shows that about 30 percent of adults who were abused as children are abusive to their own children (Gelles and Conte, 1990). In essence, they have been socialized to respond to frustrating situations with anger and violent outbursts.

Sociologists are trying to determine whether family violence is rising or decreasing in American society. Some sociologists believe there has been a decline due to the changing public attitudes toward family violence. Behavior that was formerly acceptable is now considered wrong. As a result, people may be less willing to accept violent behavior from a spouse or a parent, and outsiders are more willing to intervene when they suspect abuse.

Changes in society that have affected the family also may have contributed to a decline in family violence. People are getting married later and having fewer children. Many women are in the labor force. These facts lower the risk of family violence. Public policies toward family violence have changed also, and there are many more programs for treating victims and abusers. Shelters for battered women have increased dramatically. Police who used to talk to an abusive spouse and try to calm him or her down are now often required to make an arrest. Finally, the media have played a large role in defining family violence as a social problem deserving attention.

## Divorce

Until the past few decades, married couples could divorce only under certain special circumstances. Consequently, the divorce rate (the number of divorces per 1,000 people) was relatively low in the early twentieth century. In 1900, for example, the divorce rate was only 0.7, only about one-fifth the 2010 divorce rate of 3.6 (National Center for Health Statistics, 2011) (see Figure 12-4).

Rising divorce rates are not unique to the United States. Most industrialized nations have experienced similar patterns. Divorce rates in France and Great Britain have more than doubled in the past two decades. Divorce, however, is far more prevalent in the United States than elsewhere (Cherlin, 2009) (see Figure 12-5).

All segments of American society have been affected by the rising divorce rate. Among white and Hispanic women ages 15 to 44, more than one-third have experienced a divorce. Among African American women, the figure is one-half. Religious sanctions against divorce seem to have little influence. Catholics are no less likely to divorce than non-Catholics, despite strong opposition to divorce by the Catholic Church. Catholic clergy have adapted to the rising marital breakup rate among Catholics by approving many more annulments than in the past (Cherlin, 2009).

The likelihood of divorce varies considerably with several factors. For example, education levels seem to have a strong effect on divorce rates. The likelihood of a first marriage ending in divorce is nearly 60 percent for those people with some college education but no bachelor's degree. Those who have a college degree but no graduate-school training have nearly a 40 percent chance of divorce and are the least divorce-prone. We could argue that people with the personality traits and family background that lead them to achieve a college degree are also those most likely to achieve marital stability.

Although the divorce rate rose sharply during the 1970s, it has dropped off in recent years. In 2010, the divorce rate was the lowest it has been since 1969 (National Center for Health Statistics, 2011).

Even though the divorce rate has been dropping, the United States still has the highest divorce rate in the world. Current divorce rates imply that half of all marriages will end in divorce. Many argue that society cannot tolerate such a high rate of marital disruptions. Thirty years ago, few would have believed society could tolerate even one-third of all married couples divorcing, but that level has already been reached by some marriage cohorts.

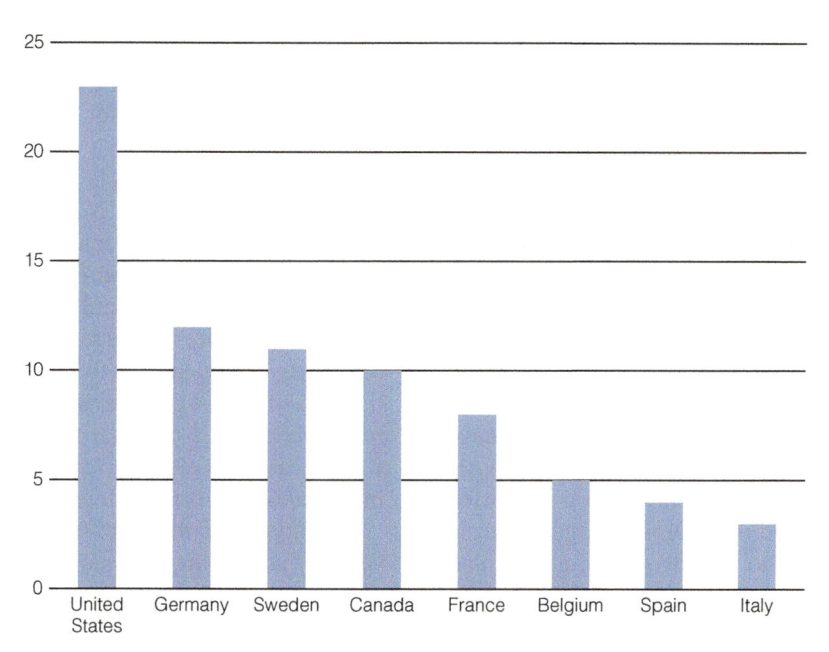

**FIGURE 12-5** Percentage of Marriages Ending in Divorce within First Five Years

*Source:* Cherlin, Andrew J. 2009. *The Marriage-Go-Round.* New York: Alfred A. Knopf, p. 206.

The large number of divorces is itself a force that helps keep the divorce rate high. Divorced people join the pool of available marriage partners, and a large majority remarry. Those remarriages then have a higher overall risk of divorce and, thus, an impact on the divorce rate.

Even though divorce rates were lower in 1910, 1930, and 1950 than they are today, can we assume that family life then was happier or more stable? Divorce during those periods was expensive, legally difficult, and socially stigmatized. Many who would have otherwise considered divorce remained married because of those factors. Is it thus accurate to say that it was better for the children and society for the partners to maintain these marriages?

As divorce becomes more common, it also becomes more visible, and such visibility actually can produce more divorces. Others become a model of how difficult marriages are handled. The model of people suffering in an unhappy marriage is being replaced by one in which people start new lives after dissolving a marriage.

Divorce also can be encouraged by the increasing tendency, mentioned earlier, for outside social institutions to assume traditional family functions that once helped hold the family together. Then again, divorce has become a viable option because people can look forward to living longer today, and they may be less willing to endure a bad marriage if they feel there is time to look for a better way of life.

Another reason for today's high divorce rate is that we have come to expect a great deal from marriage. It is no longer enough, as it might have been at the turn of the century, for the husband to be a good provider and the wife to be a good mother and family caretaker. We now look to marriage as a source of emotional support in which each spouse complements the other in a variety of social, occupational, and psychological endeavors.

Divorce rates are high also because the possibilities for women in the workforce have improved. During earlier eras, divorced women had great problems contending with the financial realities of survival, and many were discouraged from seeking a divorce because they could not envision a realistic way of supporting themselves. With their greater economic independence, many women can now consider divorce as an option.

Today's high divorce rates can also be traced to a number of legal changes that have taken place to make divorce a more realistic possibility for couples who are experiencing difficulties. Most states have instituted no-fault divorce laws, and many others have liberalized the grounds for divorce to include mental cruelty and incompatibility. These are rather vague terms and can be applied to many problem marriages. Even changes by the American Bar Association, which now allows lawyers to advertise, contribute to the increased divorce rate.

These legal changes are but a reflection of society's attitudinal changes toward divorce. We are a far cry from the 1960s when divorce was to be avoided at all costs and, when it did occur, it became a major source of embarrassment for the entire extended family. The fact that Ronald Reagan was divorced and remarried had little impact on his election to the presidency in 1980 and 1984. With so little public concern being shown, the role of peer-group pressure and public opinion in preventing divorce has been diminished greatly. (For more on factors that influence divorce, see "How Sociologists Do It: Do 50 Percent of All Marriages Really End in Divorce?")

## Divorce Laws

Before 1970, divorces in most American states could be obtained only on fault grounds. Fault grounds are those that assess blame against one of the spouses and typically consist of adultery, desertion, physical and mental cruelty, long imprisonment for a felony, and drunkenness (Scott, 1990). The fault laws led to wide variations among the states in how difficult or easy it was to obtain a divorce. New York State had a restrictive law that required one spouse to demonstrate that the other had engaged in adultery. No other grounds for divorce were accepted. Those restrictions led to migratory divorces (getting a divorce in another state) or couples inventing nonexistent fault grounds.

Those issues led to attempts to change divorce laws. California introduced the Family Law Act of 1969, which allowed divorce for irreconcilable differences between the spouses. Thus, California became the first state to enact a no-fault divorce law. Other states quickly followed suit, and by 1985, 49 states had some form of no-fault divorce law. In 2010, New York became the last state to enact no-fault divorce laws. Previously, New York law required both parties to agree to a separation agreement and live apart for a year, before a judge would grant a divorce.

No-fault divorce reflects changes in the traditional view of legal marriage. No-fault divorce laws allow couples to dissolve their marriage without either partner having to assume blame for the failure of the marriage. By eliminating the fault-based grounds for divorce and the adversary process, no-fault divorce laws recognize that frequently both parties are responsible for the breakdown of the marriage. Further, these laws are responding to the fact that the divorce procedure may worsen the situation by forcing potentially amicable individuals to become antagonists.

All fifty states make no-fault divorces possible with virtually no waiting periods, making it possible for couples to divorce much more quickly now than under the previous fault-based system (Walker and Elrod, 1993). The framers of the no-fault divorce laws did not intend to make divorce easy to obtain, however. Their intent was to permit divorce only after considerable evidence showed that the marriage could not be salvaged (Jacob, 1988).

## HOW SOCIOLOGISTS DO IT

### Do 50 Percent of All Marriages Really End in Divorce?

You have probably heard the statement that one out of every two American marriages ends in divorce. Is this really true? It is an informed estimate of how many marriages sociologists think are likely to end in divorce or separation before one of the spouses dies. Of course, every married couple is different and divorce rates are influenced by your age at marriage, income, education, religious observance, race, and whether there was a premarital pregnancy.

So let us consider some of these factors and see how they might influence the possibility of divorce. In Table 12-1, you can see the personal factors that produce percentage-point decreases in the probability of divorce or separation during the first ten years of marriage.

If a person is a college-educated, church-affiliated, Midwestern suburb-dweller from an intact family, he or she would have a relatively low probability of divorce. We would also be interested in how long people have been married. Half of all divorces occur within the first seven years of marriage.

The lifetime probability of divorce may actually be falling. This is due to the rising average age at marriage and increased education levels. The likelihood of divorce among women with college degrees has fallen sharply in recent decades. College-educated women have divorce rates 50 percent lower than women with no college education.

It is less clear that cohabitation is still a divorce risk factor. Cohabitation has become more common in the past twenty years, and may not be automatically linked to a higher chance of divorce as in the past.

TABLE 12-1  Factors That Reduce the Risk of Divorce

| Risk | Percent Decreased |
|---|---|
| Annual income over $50,000 vs. under $25,000 | = 30 |
| Having a baby seven months or more after marriage vs. before marriage | = 24 |
| Marrying after age 25 vs. before age 18 | = 24 |
| Family of origin intact vs. parents divorced | = 14 |
| Having a religious affiliation | = 14 |
| Some college vs. high-school dropout | = 13 |
| Live in the Northeast or Midwest vs. the South | = 10 |
| Suburbanite vs. central-city dweller | = 9 |

Source: National Center for Health Statistics.

*Source:* Kreider, Ross M., and Jason M. Fields. 2005. "Number, Timing and Duration of Marriages and Divorces, 2001." U.S. Bureau of the Census, *Current Population Reports*, pp 70–80.

---

Critics of no-fault divorce believe it has led to a skyrocketing divorce rate. Within ten years of the no-fault laws being passed, the divorce rate jumped from 3.2 to 5.3. Critics believe the laws make divorce too easy. They also lament the fact that most states allow a partner to receive a divorce without the spouse also wanting one.

In some states such as Arkansas, Arizona, and Louisiana, people opt for a covenant marriage. In such a marriage, the marrying couple agrees to more limited grounds for divorce. The movement has been adopted mostly by evangelical Christians. In a covenant marriage, the grounds for divorce are abuse, a felony with jail time, or adultery. Couples choosing a covenant marriage still represent only between 1 and 3 percent of all marriages in those states where it is exists (Brown, 2008).

At the same time that some lament the high divorce rate, others praise no-fault divorce for easing the pain of divorce for couples by preventing drawn-out and expensive legal proceedings. Defenders of no-fault divorce say it is easier for victims of mental and physical abuse to obtain a divorce without having to go through a long, agonizing process.

Despite the criticism, efforts to repeal no-fault divorce laws have been largely unsuccessful. Legislation that would modify or repeal no-fault divorce has failed in at least twenty-two states (Baker et al. 2008).

The financial aspect of a divorce under no-fault laws is based on equity, equality, and economic need rather than on fault- or gender-based role assignments. Alimony also is to be based on the respective spouses' economic circumstances and on the principle of social equality, not on the basis of guilt or innocence. No longer is alimony automatically awarded to the injured party, regardless of that person's financial needs; no-fault divorce does not recognize an injured party. Instead, these laws seek to reflect the changing circumstances of women and their increased participation in the labor force. Under no-fault divorce law, women are encouraged to become self-supporting and husbands are not automatically expected to continue to support their ex-wives throughout their lives.

Some see no-fault divorce legislation as a redefinition of the traditional marital responsibilities of men and women with a new norm of equality between women and men. Husbands are no longer automatically designated as the head of the household, solely responsible for support, nor are wives alone expected to assume the responsibility of household activities and childrearing. Gender-neutral obligations fall equally upon the husband and wife with no-fault laws. Alimony is no longer automatic and property is to be divided equally. Child-support expectations and child custody also reflect the new egalitarian criteria of no-fault divorce legislation. Today, both father and mother are expected to be equally responsible for the financial support of their children after divorce. In theory, mothers no longer receive custody of the child automatically; rather, a gender-neutral standard instructs judges to award custody in the best interests of the child.

No-fault has worked well for some divorcing couples, yet it has had devastating consequences for many others. Problems often emerge for older homemakers married thirty-five years or more who lack any labor-force experience or skills but who may be awarded short-term settlements, ordered to sell the family home, and instructed by the court to pursue job training. Similarly, mothers with toddlers are routinely left with virtually full responsibility for their care and former husbands must cover costs.

No-fault divorce laws are based on an idealized picture of women's social, occupational, and economic gains in achieving equality that, in fact, may not reflect their actual conditions and circumstances. This discrepancy between reality and the ideal can have extremely detrimental effects on women's ability to become self-sufficient after divorce.

## Child Custody Laws

The view that children would benefit most from living with their mother following divorce was widely accepted and seen in divorce and child custody laws until the 1970s. Fathers often were advised by legal counsel of the futility of contesting custody, and the burden of proof was on the father to document the unfitness of the mother or to prove his ability to be a better parent than the mother. However, there has been an increased recognition of fathers' rights regarding custody, reflecting the changing role of American fathers and the reevaluation of the judicial practice of automatically awarding custody to the mother.

Also, increasing numbers of states are allowing and encouraging joint custody, which involves both parents having decision-making authority for their children even if the child lives predominantly with one parent. Joint custody grants the parents equal rights in making decisions regarding the child's upbringing. Courts award joint custody for cases in which both parents can properly perform their duties as parents. If one parent sues for exclusive custody, the suing parent must prove that joint custody is not in the child's best interests. However, despite state laws emphasizing joint custody to varying degrees, an overwhelming majority of custody awards are still made solely to mothers.

In a legal sense, joint custody means that parental decision-making authority has been given equally to both parents after a divorce. It implies that neither parent's rights will be considered paramount. Both parents will have an equal voice in the children's education, upbringing, and general welfare. Although alternating living environments may accompany joint custody decisions, in most instances, they do not. In 90–95 percent of joint custody awards, the living arrangements are exactly the same as those under sole-custody orders; namely, the child physically resides with only one parent. However, both parents make decisions regarding the welfare of the child.

Those who believe joint legal custody is a good idea cite a variety of reasons. They assert that sole custody arrangements, which almost always involve the child's living with the mother, weaken father-child relationships. They create enormous burdens for the mothers and tend to exacerbate hostilities between the custodial parent and the visiting parent. They continue to perpetuate outmoded gender-role stereotyping. Studies also show that sole-custody arrangements are associated with poverty, antisocial behavior in boys, depression in children, lower academic performance, and juvenile delinquency. Advocates of joint custody assume that if fathers are given the opportunity to be available as nurturers, to be accessible, they will begin to participate more in the lives of their children. Furthermore, advocates say, such participation will have beneficial effects on children.

Before we too quickly assume that joint custody alleviates problems and produces benefits, we should note that it is far from being a panacea. If couples had trouble communicating and agreeing on things before the divorce, there is no reason to assume that they will have an easier time of it afterward. Most joint-custody orders are vague and do not decide at what point the joint-custodial parent's rights end and those of the parent with the day-to-day care of the child begin. What sorts of responsibilities can one parent require of the other parent? Issues such as these can easily erupt into disputes, particularly when a history of disagreement and distrust has preceded the joint-custody arrangement.

Joint custody does not give either parent the right to prevail over the other. To solve serious disputes, the parents must return to court, where they must engage in litigation to prove that one or the other is unfit—the very process that the original decision of joint custody was to have avoided.

Joint custody appears to work best for those parents who have the capacity, desire, and energy to make it work—and for the children whose characteristics and desires allow them to expend the effort necessary to make it work and to thrive under it.

## Remarriage and Stepfamilies

Throughout history, the reputation of stepfamilies has been surprisingly negative. The general thinking has been that the stepchild suffers in such families and that the suffering is caused by the stepmother. The stepmother has come down to us as a figure of cruelty and evil, constantly plotting to harm her stepchildren in a variety of ways. Just think of the Cinderella or Snow White fairy tales. The French word *maratre* means both "stepmother" and a "cruel and harsh mother." In English literature, "stepmother" is often preceded by "wicked." The stepchild was often a child who did not belong or whose status was similar to that of orphan. In fact, "stepchild" originally meant orphan.

Much of this has changed today, and stepfamilies have become a common sight on the American family landscape. The United States has the highest incidence of stepfamilies in the world. It is estimated that about 17 percent of married-couple households involve a stepparent. About one child in six is a stepchild.

The stepfamilies of today are different from those of the past in how they have come into being. The vast majority of stepfamilies now come about because of the marriage or cohabitation of mothers and fathers of children whose other parent is still living. Of these, the largest group by far is composed of families formed by the remarriage of divorced men and women. The high divorce rate is the key factor in the rise of the modern stepfamily.

In the past, stepfamilies resulted from quite different conditions and had different implications. Generally, stepfamilies were the product of remarriages after the death of a spouse. Mortality and the frequency of remarriage by the widowed spouse determined the number of stepfamilies. Moreover, unlike today, children who lived in stepfamilies rarely had more than one living parent. In the past, a stepparent replaced a deceased parent.

Today, a stepparent is often an additional parent figure that a child must incorporate. The new marriage partners must establish their own relationship in an existing family structure. They must create new rules for how the family is to run. Such changes can introduce disadvantages or tensions among the family members.

Remarriage makes parenthood and kinship an achieved status rather than an ascribed status. Traditionally, a person became a father or a mother at the birth of their child. They did not have to do anything else to be a parent, nor could they easily resign from the job, especially in a family system that strongly discouraged divorce.

Remarriage after divorce, though, adds a number of other, potential kinship positions. The new parent must now achieve the status of parent or some other role that can resemble it. Working out these arrangements presents many more problems than are usually encountered in a first-marriage family (Cherlin and Furstenberg, 1994).

When people get divorced, they are not rejecting the institution of marriage but rather a specific marriage partner. The median time between divorce and a second marriage has been about three and a half years. About 52 percent of divorced men and 44 percent of divorced women are currently married (U.S. Bureau of the Census, 2007).

Today, more than four in ten American adults have at least one step relative in their family—either a stepparent, a step or half sibling, or a stepchild. People with step relatives are just as likely as others to say that family is the most important element of their life. However, they typically feel a stronger sense of obligation to their biological family members (be it a parent, a child, or a sibling) than to their step relatives.

Young people, blacks, and those without a college degree are more likely to have step relatives. Among those under age 30, more than half (52%) report that they have at least one step relative. This compares with 40 percent of those who are age 30 and older. This is not surprising given that young adults are much more likely than their older peers to have grown up with parents who were not married.

A higher share of blacks than whites or Hispanics have step family members. Among black adults, 60 percent have at least one step relative. This compares with 46 percent of Hispanics and 39 percent of whites.

In addition to age and race, socioeconomic status is related to the prevalence of stepfamilies. Only a third of college graduates say they have at least one step relative. By contrast, 46 percent of those without a college degree have a step relative. Similarly, stepfamilies are less common among those with higher household incomes. Fully half of those with annual incomes of less than $30,000 say they have step family members (Pew Research Center, 2011).

With high divorce rates come high remarriage rates. The resulting blended families produce new relationships that need to develop over time.

Stepfamilies are changing such businesses as the greeting card industry. We now have birthday wishes to stepmothers and thank-you cards to stepfathers. Schools must now ask for information on stepparents as well as on biological parents.

Stepfamilies, also known as blended families, are transforming basic family relationships. Where there were once two sets of grandparents, there now can be four; an only child may acquire siblings when his mother remarries a man with children. Stepparents are taking on roles formerly held by biological parents. They are staying up nights with the spouse's sick children, attending class recitals, and having heated battles over such essential issues of childhood as curfews, TV, homework, and rights to the family car.

The most difficult stage occurs during the early years, when stepparents want everything to go right. Once stepparents realize that relationships with stepchildren build over time, and that their potential network of allies includes all the other adults in the stepfamily relationship, the adjustment for all is faster and healthier.

## FAMILY DIVERSITY

A number of options are increasingly available to people who, for various reasons, find the traditional form of marriage impractical or incompatible with their lifestyles. More young people are selecting cohabitation as a permanent alternative to marriage (although many consider it a prelude to marriage). In addition, some older men and women are opting to live together in a permanent relationship without getting married. These people choose cohabitation primarily for economic reasons—many would lose sources of income or control of their assets if they entered into a legal marriage. Several other trends are discussed next.

### The Growing Single Population

Slightly more than 31 million households or nearly 27 percent of all households consist of one person living alone. This is an increase of 4 million one-person households since 2000. For the first time since at least 1940, husband-wife households represent less than 50 percent of all households (U.S. Bureau of the Census, 2012).

There are a number of reasons for this increase. Some people are choosing not to marry. Others are marrying later than in the past. Working women do not need the financial security that a traditional marriage brings, and sex outside of marriage has become much more widespread. Moreover, many singles view marriage as merely a prelude to divorce and are unwilling to invest in a relationship that is likely to fail. As sociologist Frank Furstenburg notes, "Men who weren't married by their late 20s in the '60s were oddballs. Now they're just successful 29-year-olds." (See "Thinking About Social Issues: Reluctant to Marry—The Men Who Want to Stay Single.")

---

**THINKING ABOUT SOCIAL ISSUES**

### Reluctant to Marry—The Men Who Want to Stay Single

Most men marry by about age 30, but there is a significant group of about 20 percent of young men who are reluctant to commit to marriage. Who are they and how do they differ from men who marry earlier?

A number of key beliefs seem to be connected to a significantly more negative attitude toward marriage. Compared to men who marry earlier, these men are more likely to:

- Worry about the risks of divorce.
- Not want children.
- Believe women cannot be trusted to tell the truth about past relationships.
- Think single men have better sex lives than married men.
- Believe marriage will reduce their personal freedom.

These men are also:

- Less likely to come from traditional families.
- More likely to have had fathers who were not involved in their lives very much.
- More likely to question the value of marriage.

For many of these men, there are attractive alternatives to marriage. The urban single lifestyle provides the potential for sexual partners. Today, there is little social disapproval of being a single man in your 30s or 40s. The increase in cohabitation rates has also made it an attractive alternative to the long-term commitment of marriage. These men do not need to worry about the biological clock that influences women's lives. It is not clear that these men will never marry, but for the moment, they are happy to postpone it indefinitely.

**Source:** Popenoe, David, and Barbara Dafoe Whitehead. June 2004. "The State of Our Unions 2004." The National Marriage Project (http://marriage.rutgers.edu/Publications/SOOU/SOOU2004.pdf), accessed October 14, 2005.

Clearly, many singles would gladly change their marital status if the right person came along. The ratio of unmarried men to unmarried women varies by age group. There are more unmarried men than unmarried women at the younger ages, reflecting the fact that men have a later age at first marriage than women do. After age 40, there are fewer unmarried men than unmarried women because women tend to live longer and are less likely to remarry after divorce and widowhood.

### Single-Parent Families

What do Steve Jobs, Alexander Hamilton, Larry Ellison (CEO of Oracle), actor Sir Alec Guinness, Nobel prize winner Paul Nurse, and artist/inventor Leonardo da Vinci all have in common? They were all born to unmarried mothers.

In the early years of the twentieth century, it was not uncommon for children to live with only one parent because of the high death rate. Even by 1960, nearly one-third of all single mothers with children under 18 were widows. As divorce rates started to rise dramatically in the 1970s, the situation started to change. By 1980, only 11 percent of single mothers were widowed, and two-thirds were divorced or separated. The path to single motherhood changed again when many women bypassed marriage altogether.

The number of children born to unmarried women in the United States has increased dramatically. Figure 12-6 shows how large the increase in the percentage of unmarried births has been over the years. In 1950, only 4 percent of children were born to unmarried women. In 2010, nearly 41 percent of births in the United States were to unmarried women. For certain groups, the numbers are even higher. Among blacks, 72.5 percent of mothers are unmarried, and among Hispanics, 53.3 percent are unmarried.

Single motherhood has historically been more common among women with less education than among well-educated women, but this has become even more true in recent years. It also is not teenagers, but rather women in their early twenties who have the highest percentage of births outside of marriage. Today, 60 percent of all babies born out of wedlock are to women in their 20s, and only 23 percent are born to teenagers.

Unmarried births and single-parent households have both been increasing. The growth in single-parent households has been influenced by three trends. First, high divorce rates have increased the number of single-parent families. Most divorced parents now set up new households, whereas in earlier times many would have returned to their own parents' households. The incidence of divorce began to increase rapidly during the 1970s. Now about one million children under age 18 lose a resident father each year because of divorce.

The second reason for the shift in family structure is an increase in the percentage of babies born to unwed mothers, which suddenly and unexpectedly began to increase rapidly in the 1970s and has continued to do so.

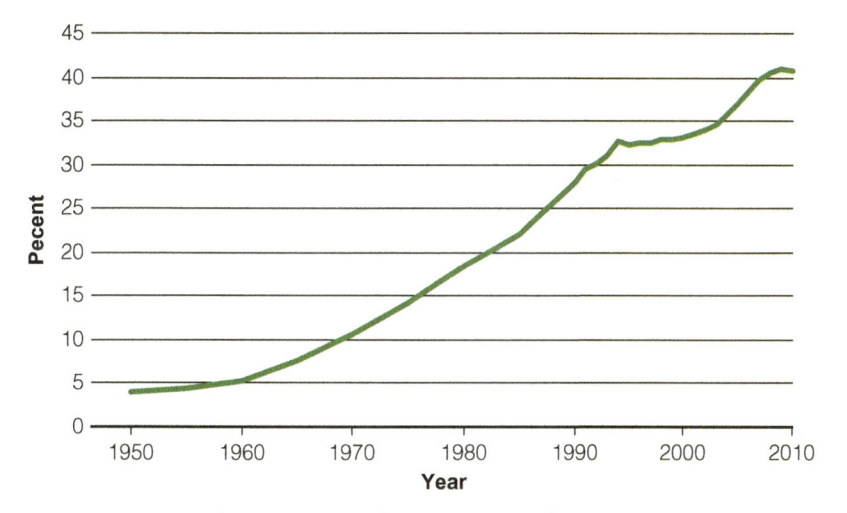

**FIGURE 12-6** Births to Unmarried Women: United States, 1950–2010

*Sources: Ventura, Stephanie J., and Christine A. Bachrach. October 18, 2000. "Nonmarital Childbearing in the United States, 1940–99." National Center for Health Statistics, National Vital Statistics Reports 48(16, Revised) (www.cdc.gov/nchs/data/nvsr/nvsr48/nvs48_16.pdf), accessed August 15, 2012; Martin, Joyce A., Brady E. Hamilton, Stephanie J. Ventura, Michelle J. K. Osterman, Sharon Kirmeyer, T. J. Mathews, and Elizabeth C. Wilson. November 3, 2011. "Births: Final Data for 2009." National Center for Health Statistics, National Vital Statistics Reports 60(1) (www.cdc.gov/nchs/data/nvsr/nvsr60/nvsr60_01.pdf), accessed August 15, 2012; Hamilton, Brady E., Joyce A. Martin, and Stephanie J. Ventura. November 17, 2011. "Births: Preliminary Data for 2010." National Center for Health Statistics, National Vital Statistics Reports 60(2) (www.cdc.gov/nchs/data/nvsr/nvsr60/nvsr60_02.pdf), accessed August 10, 2012; Solomon-Fears, Carmen. November 20, 2008. "Nonmarital Childbearing: Trends, Reasons, and Public Policy Interventions." Congressional Research Service (www.fas.org/sgp/crs/misc/RL34756.pdf), accessed August 15, 2012.*

The third reason for the growth in unmarried births reflects an increase in the number of couples who are in cohabiting relationships and having children. About 9 percent of white births, 15 percent of black births, and 22 percent of Hispanic births are to cohabitating women.

Even if the parents are cohabitating and not married, married-couple families are more stable and longer lasting than cohabitating-couple families. About one-fourth of cohabiting biological parents are no longer living together one year after the child is born.

Although most children who grow up in mother-only families become well-adjusted, productive adults, unfortunately a large number of children who grow up with only one biological parent at home are at greater risk for a variety of problems. These children are more likely to do poorly in school, have emotional and behavioral problems, become teenage parents, and are six times more likely to be poor than children in two-parent families.

The rise in the number of births to unmarried women has also been occurring in a number of other industrialized countries. Iceland, Sweden, Norway, France, Denmark, and the United Kingdom all have higher percentages of out-of-wedlock births than the United States. However, levels in the United States are much higher than in some industrialized countries such as Italy, Spain, and Japan (Ventura, 2009).

## Gay and Lesbian Couples

A phenomenon that is not new but one that has become increasingly visible is the household consisting of a gay or lesbian couple. Before 1970, almost all gay people wished to avoid the risks that would come with disclosure of their sexual orientation.

It was not until 1990 that the U.S. census attempted to calculate the number of gay and lesbian households. Through a self-identification process, nearly 1.2 million people say they are part of gay and lesbian couples in the United States, and although most live in four metropolitan areas—San Francisco, New York, Los Angeles, and Washington, D.C.—nearly one in six live in a rural community. There has also been an increase in the number of recorded lesbian couples. There appear to be more male couples than lesbian couples, but new data show that the numbers are beginning to even out.

It is important to note that census data include only that portion of the gay population who are part of a couple, which studies show is only about one-third of gay men and lesbians. Furthermore, the total number of gay and lesbian couples is underreported in the census figures because some are hesitant about sharing information about personal relationships with the government, even though the forms are confidential (Cohn, 2001).

Traditionally, researchers and the media have concentrated on the ways in which gays and lesbians are different from heterosexuals. Yet, gay and lesbian couples are similar to heterosexual couples in many ways. They form long-term relationships and have problems similar to those of heterosexual couples. Gay and lesbian couples are concerned with receiving the same family benefits as traditional couples, such as health insurance, life insurance, and family leave. Many are also involved in extending adoption rights and same-sex marriage recognition to gay and lesbian couples.

This gay couple is seen with their two-year-old twin children as they celebrate their wedding anniversary in California.

The federal government does not recognize same-sex marriage. On a state level, same-sex marriage has been legalized by six states: Massachusetts, Connecticut, Iowa, Vermont, Maine, and New Hampshire. New York and Pennsylvania may be the next states to legalize same sex marriage.

In the other forty-four states, gay and lesbian partners are usually not eligible for health care or other benefits that might be provided to married employees and their families. A number of states and municipalities, however, have passed domestic-partnership statutes to make such benefits available. In addition, some states, the District of Columbia, and hundreds of cities and counties have adopted policies that provide varying degrees of civil rights protection for gays and lesbians.

Fifty percent of Americans believe same-sex marriages should be recognized as legal marriages. Forty-eight percent say such marriages should not be legal. In 1996, when Gallup first asked the question, 27 percent supported it, while 68 percent were opposed. (See Figure 12-7.)

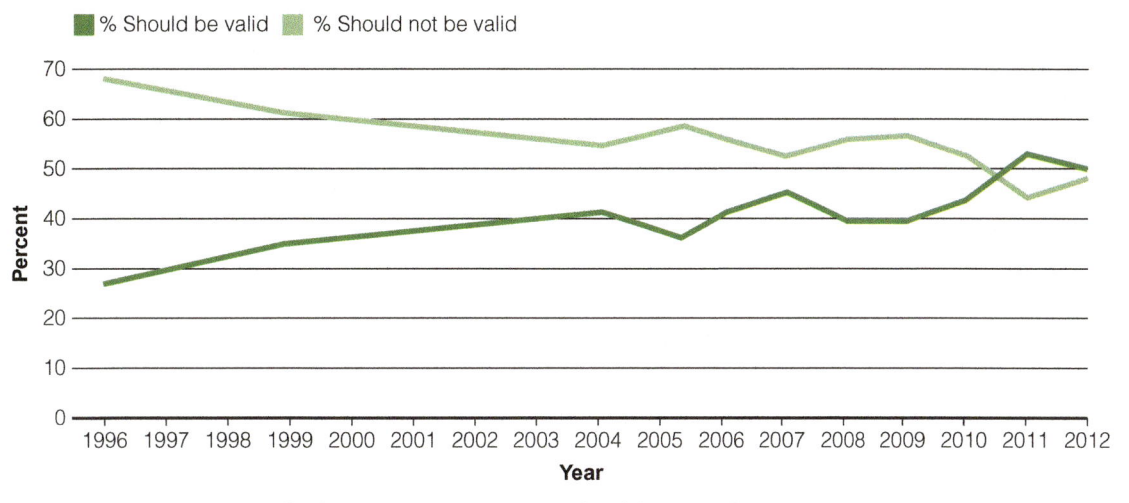

**FIGURE 12-7** Views on Whether Same-Sex Marriage Should Be Legal
Source: Newport, Frank. May 8, 2012. "Half of Americans Support Legal Gay Marriage." Gallup Poll (www.gallup.com/poll/154529/Half-Americans-Support-Legal-Gay-Marriage.aspx), accessed August 10, 2012.

Almost two-thirds of Democrats support legalizing same-sex marriage, compared with 57 percent of independents and 22 percent of Republicans. Catholics are right at the overall average in their views on same-sex marriage. Even though the Catholic Church officially opposes same-sex marriage, Catholic views are similar to those of the general population. Protestants are actually less likely to approve of same-sex marriage than Catholics. Those who identify with no religion overwhelmingly approve of same-sex marriage.

Even though a significant number of people oppose same-sex marriage, most Americans believe the law should not discriminate against gay individuals and gay couples (Gallup Poll, 2009).

Gays and lesbians in couple relationships have higher educational levels than men and women in heterosexual marriages. Cohabiting gay men 25 to 34 years old are twice as likely to have a postgraduate degree as married men. The differences are even greater for lesbian couples in this age group, who are three times as likely to have postgraduate education compared to married women (U.S. Bureau of the Census, 2009).

## ● WHAT DOES THE FUTURE HOLD?

Given all these changes in the American family, should we be concerned that marriage and family life as we know them will one day disappear? Probably not. The divorce rate continues to be high, but has come down in recent years. Divorce is just as much a social universal as is marriage. Anthropologist Margaret Mead noted that "no matter how free divorce" or "how frequently marriages break up," most societies have assumed that marriages would be permanent. Despite this belief, societies have also recognized that some marriages are incapable of lasting a lifetime and have provided mechanisms for dissolution (Riley, 1991).

Even though the divorce rate is high, the remarriage rate is also very high. The vast majority of people who divorce remarry, usually within a short time after they divorce. The high divorce rate does not necessarily mean that people are giving up on marriage. It just means there is a growing belief that marriage can be better. The high remarriage rate indicates that people are willing to continue trying until they reach their expectations. Obtaining a divorce does not mean the person believes that the idea of marriage is a mistake, only that a particular marriage was a mistake.

Despite claims to the contrary, there is little evidence that the family as an institution is in decline or any weaker today than a generation ago. Nor is there any indication that people place less value on their own family relationships, or on the role of the family within society, than they once did.

The traditional family is being replaced by family arrangements that may better suit today's lifestyles: There are fewer full-time homemakers because more women are in the workforce. Nonfamily households have increased substantially. The typical family with a working dad, homemaker mom, and two or more children is now a distinct minority.

The institutions of marriage and the family have proved to be extremely flexible and durable and have flourished in all human societies under almost every imaginable condition. As we have seen, these institutions take on different forms in differing social and economic contexts, and there is no reason to suspect that they will not continue to do so.

## SUMMARY

### What do we see happening with the American family today?
The American family has changed dramatically in the past forty years. The marriage rate is down, the divorce rate is high but leveling off, and more children are being born to single women.

### What functions does the family fulfill?
In all societies, the family serves certain basic social functions. These include regulating sexual behavior, patterning reproduction, organizing production and consumption, socializing children, providing care and protection, and providing social status.

### What is the basic family structure?
The nuclear family is the most basic family form and is made up of a married couple and their biological or adopted children. The nuclear family is found in all societies, and it is from this form that all other family forms are derived.

### What other family forms exist?
There are two major composite family forms, polygamous and extended families. Polygamous families are nuclear families linked together by multiple marriage bonds, with one central individual married to several spouses. The family is polygynous when the central person is male and the multiple spouses are female. The family is polyandrous when the central person is female and the multiple spouses are male. Extended families include other relations and generations in addition to the nuclear family, so that along with married parents and their offspring, there might be the parents' parents, siblings of the parents, the siblings' spouses and children, and in-laws.

### Why do families exist throughout all societies?
Sociologists generally agree that some form of family is found in every known human society and serves several important functions: regulating sexual behavior, patterning reproduction, organizing production and consumption, socializing children, providing care and protection, and providing social status.

### How does society regulate the family?
Every society must replace its members. By regulating where and with whom individuals may enter into sexual relationships, society patterns sexual reproduction. No society permits random mating; all societies have an incest taboo, which forbids sexual intercourse among closely related individuals. Sex between parents and their children is universally prohibited, but who else is considered to be closely related varies widely among societies.

### Do all cultures believe romantic love is necessary for marriage?
American culture is relatively unique in linking romantic love and marriage. Romantic love—which involves the idealization of the loved one, the notion of a one and only, love at first sight, love winning out over all, and an indulgence of personal emotions—has nothing to do with marriage throughout most of the world.

### What role does marriage play in society?
Marriage in many societies establishes social, economic, and even political relationships among families. Three families are ultimately involved in a marriage: the two spouses' families of origin or families of orientation—the families in which they were born or raised—and the family of procreation, which is created by the union of the spouses.

### How do societies view divorce?
Although almost all societies allow for divorce, or the breakup of marriage, none endorse it as an ideal norm.

## Media Resources

**CourseMate for *Introduction to Sociology*, Eleventh Edition**
Cengage Learning's Sociology CourseMate brings course concepts to life with interactive learning, study, and exam preparation tools that support the printed textbook. Access an integrated eBook, learning tools including glossaries, flashcards, quizzes, videos, and more in your Sociology CourseMate. Go to CengageBrain.com to register or purchase access.

# CHAPTER TWELVE STUDY GUIDE

## ● Key Concepts

Match each of the following concepts with its definition, illustration, or explanation.

a. Incest taboo
b. Nuclear family
c. Polygamous family
d. Polygyny
e. Polyandry
f. Extended family
g. Patrilineal system
h. Matrilineal system
i. Bilateral system of descent
j. Patriarchal family
k. Matriarchal family
l. Family of origin
m. Family of procreation
n. Rules of endogamy
o. Rules of exogamy
p. Patrilocal residence
q. Matrilocal residence
r. Bilocal residence
s. Neolocal residence
t. Homogamy
u. Companionate marriage
v. No-fault divorce

\_\_\_\_ 1. A requirement that a new couple settle down near or within the husband's father's household
\_\_\_\_ 2. A system in which a newly married couple may live virtually anywhere
\_\_\_\_ 3. A marriage based on romantic love
\_\_\_\_ 4. A set of nuclear families linked by multiple marriage bonds, with one central individual married to several spouses
\_\_\_\_ 5. A family in which a central female has several husbands
\_\_\_\_ 6. A rule that forbids sexual intercourse among closely related individuals
\_\_\_\_ 7. A family that consists of two spouses and their children
\_\_\_\_ 8. Divorce granted without one partner having to prove adultery, desertion, abuse, or other dereliction on the part of the other
\_\_\_\_ 9. A family in which a central male has several wives
\_\_\_\_ 10. The family in which a person was born and raised
\_\_\_\_ 11. A family in which power is seen as held mostly by the males
\_\_\_\_ 12. The family created by marriage
\_\_\_\_ 13. Family consisting of relatives in addition to the spouses and their children
\_\_\_\_ 14. Marriage outside one's own culturally defined group
\_\_\_\_ 15. Families where the female is the dominant figure in decision making
\_\_\_\_ 16. Kinship traced through the mother's family (i.e., your last name would be your mother's maiden name, and you would be considered primarily a part of your mother's side of the family)
\_\_\_\_ 17. Marrying someone with similar social characteristics
\_\_\_\_ 18. A newly married couple lives with or near the bride's mother's household
\_\_\_\_ 19. Kinship traced through the father's family (i.e., your last name would be the same as your father's)
\_\_\_\_ 20. A system in which a newly married couple may choose to live with either the husband's family or the wife's family
\_\_\_\_ 21. A kinship system in which people are considered to belong to the families of both parents
\_\_\_\_ 22. Requirement that people marry within a defined group

## ● Central Idea Completions

Fill in the appropriate concepts and descriptions for each of the following questions.

1. How have divorce rates changed over the last half century, and what factors have affected those rates?

_____
_____
_____

PART 4 INSTITUTIONS AND SOCIAL ISSUES

2. Identify at least three aspects of the Industrial Revolution that promoted the shift from extended families to nuclear families.

   a. _____

   b. _____

   c. _____

3. The single population in the United States is growing. Identify some of the causes and consequences of this change.

   _____

   _____

   _____

4. Identify three trends that have caused the dramatic shifts in the American family structure, and name which shifts they have fostered.

   a. _____

   b. _____

   c. _____

5. List five functions usually fulfilled by the family; in each case, describe how these functions have been fulfilled by your own family.

   a. _____

   b. _____

   c. _____

   d. _____

   e. _____

6. We often think that falling in love and selecting a marriage partner are matters of the heart, of individual preferences, or "chemistry." List some of the ways that these matters are in fact socially structured. What social forces constrain who we fall in love with and marry?

   a. _____

   b. _____

7. Family violence is a major concern. What socio-demographic variables are associated with family violence? What changes in society have affected rates of domestic violence in the past 20 years?

   **a.** Socio-demographic variables related to violence: _____

   _____

   **b.** Social changes affecting levels of violence: _____

   _____

## ● Critical Thought Exercises

1. How do Americans find someone to marry? Present an in-depth discussion of the process of mate selection in the United States. Be sure to devote attention to the effects of such variables as age, race, religion, and social status on how Americans generally choose their mates.

2. Americans, who overwhelmingly favor marriage based on romantic love, can easily see the problems that can arise in a system of arranged marriage. You could quickly make a list of several advantages of romantic-love marriages and difficulties inherent in arranged marriage.

    But imagine that you are from an arranged-marriage society. From your new perspective, make a list of the advantages of arranged marriage and the problems that can easily arise in a system of marriage based on love.

    Imagine the things that you personally would find unpleasant about an arranged marriage. Do you think people in arranged-marriage societies have similar feelings about these things and, if so, how might they deal with them? What institutions in American society are there to deal with the problems of our system of marriage? (For example, finding a mate, which is not problematic in arranged-marriage societies, has given rise in our own society to the institution of dating as well as to a variety of more formal and sometimes commercial solutions to this problem.)

3. Tischler describes "the decline of the traditional family." Yet despite this decline, Americans still claim to live in families and to value family. Obviously, there has been a change in ideas about what a family is and what family values are. Will we see further changes in the definition of family? If so, what will these changes be? What social factors will affect these changes and the speed at which they take place?

## ● Internet Activities

1. The world of the family is strongly affected by the world of work. Families need the income they get from work, but they suffer if work demands too much of their time and energy. The Work and Family Equity Index is a measure developed to assess the impact of work on family, especially in regard to government policies such as requiring employers to provide paid maternity leave. Find out more about it at http://www.hsph.harvard.edu/globalworkingfamilies/index.html and see how the United States compares with other countries (not too well).

2. YouTube (http://YouTube.com) has several clips of interviews with members of polygamous families in the United States. Search for "polygamy family." Describe three areas of family life that are violated by this practice.

## ● Answers to Key Concepts

1. p;   2. s;   3. u;   4. c;   5. e;   6. a;   7. b;   8. v;   9. d;   10. l;   11. j;   12. m;   13. f;   14. o;   15. k;   16. h;   17. t;   18. q;   19. g;   20. r;   21. i;   22. n

Steve McCurry/Magnum Photos

# 13 Religion

### The Nature of Religion
   The Elements of Religion
*Our Diverse World:* Who Is God?
### Magic
### Major Types of Religions
   Supernaturalism
   Animism
   Theism
   Monotheism
   Abstract Ideals
### A Sociological Approach to Religion
   The Functionalist Perspective
   The Conflict Theory Perspective
*Our Diverse World:* The Worst Offenders of Religious Freedom
### Organization of Religious Life
   The Universal Church
   The Ecclesia
   The Denomination
   The Sect
   Millenarian Movements
*Day-to-Day Sociology:* Today's Cult Might Be Tomorrow's Mainstream Religion
### Aspects of American Religion
   Religious Diversity
   Widespread Belief
   Secularism
*Thinking About Social Issues:* The Rise of No Religious Affiliation
   Ecumenism
### Major Religions in the United States
*How Sociologists Do It:* Is Your Professor an Atheist?
   Protestantism
*Day-to-Day Sociology:* Changing Religion Early and Often
   Catholicism
   Judaism
   Islam
   Social Aspects of Religious Affiliation
### Summary

## LEARNING OBJECTIVES

After studying this chapter, you should be able to do the following:

- Define the basic elements of religion.
- Differentiate among the major types of religion.
- Describe the functions of religion according to the functionalist perspective.
- Explain the conflict theory perspective on religion.
- Describe the basic types of religious organization.
- Describe important aspects of contemporary American religion.
- Describe the major religions in the United States.

The belief in one God is a basic tenet of Christianity as well as the cornerstone of Judaism. Islam also proclaims, "There is no God but Allah, and Muhammad is His Prophet."

Hinduism, however, expresses a very different concept: the idea of the "manyness" of God. Hinduism does not have one creed, one founder, one prophet, or one central moment of revelation. It is an expression of 5,000 years of religious and cultural development in Asia. During this development, Hinduism has assimilated many ideas and ideologies. In Hinduism, whether there is one god or many gods is unimportant. Rather than worshipping a single god, Hinduism involves worshipping one god at a time. Each Hindu is free to choose his or her own god or goddess.

In his book, *All Religions Are True*, Mohandas Karamchand Gandhi wrote:

> It has been a humble but persistent effort on my part to understand the truth of all the religions of the world, and adopt and assimilate in my own thought, word, and deed all that I have found to be best in those religions.

Gandhi's acceptance of diversity, however, has a downside. If all religions are true, then they all must be imperfect. After all, if any one religion were perfect, it would be better than all others.

The idea of the manyness of God is alien not only to Western religion but also to Western culture. Our monotheism leads us to believe in one Truth. In our minds, the ultimate or best can stand out in many areas, whether it is the best car, the best university, or the ultimate religion (Goldman, 1991).

The important thing to realize is that although religion assumes many forms, it is a universal human institution. To appreciate the many possible kinds of religious experiences, from the belief in one God to the belief in the manyness of God, requires an understanding of the nature and functions of religion in human life and society.

## THE NATURE OF RELIGION

**Religion** is *a system of beliefs, practices, and philosophical values shared by a group of people; it defines the sacred, helps explain life, and offers salvation from the problems of human existence.* It is recognized as one of society's important institutions.

In his classic study, *The Elementary Forms of Religious Life*, first published in 1915, Émile Durkheim observed that all religions divide the universe into two mutually exclusive categories, the profane and the sacred. The **profane** consists of *all empirically observable things—that is, things that are knowable through common, everyday experiences.* In contrast, the **sacred** consists of *things that are awe inspiring and knowable only through extraordinary experiences.*

The sacred can consist of almost anything: objects fashioned just for religious purposes (such as a cross), a geographical location (Mount Sinai), a place constructed for religious observance (a temple), a word or phrase ("Our Father, who art in heaven . . ."), or even an animal (the cow to Hindus, for example). To devout Muslims, the Sabbath, which falls on Friday, is a sacred day. To Hindus, the cow is holy, not to be killed or eaten. These are not ideas to be debated; they simply exist as unchallengeable truths. Similarly, to Christians, Jesus was the Messiah; to Muslims, Jesus was a prophet; to sociologists, the person of Jesus is a religious symbol. Religious symbols acquire their particular sacred meanings through the religious belief system of which they are a part.

Durkheim believed that every society must distinguish between the sacred and the profane. This distinction is essentially between the social and nonsocial. What is considered sacred has the capacity to represent shared values, sentiments, power, or beliefs. The profane is not supported in this manner; it may have utility for one or more individuals, but it has little public relevance.

We can look at the bat of the famous baseball player, Babe Ruth, as an example of transformation of the profane to the sacred. At first, it was merely a profane object that had little social value in itself. Today, however, one of Babe Ruth's bats is enshrined in baseball's Hall of Fame. It no longer is used in a profane way but instead is seen as an object that represents the values, sentiments, power, and beliefs of the baseball community. The bat has gained some of the qualities of a sacred object, thus changing from a private object to a public object.

In addition to sacred symbols and a system of beliefs, religion also includes specific rituals. **Rituals** are *patterns of behavior or practices that are related to the sacred.* For example, the Christian ritual of Holy Communion is

One of the functions of ritual and prayer is to produce an appropriate emotional state.

much more than the eating of wafers and the drinking of wine. To many participants, these substances are the body and blood of Jesus Christ. Similarly, the Sun Dance of the Plains Indians was not merely a group of men dancing around a pole to which they were attached by leather thongs that pierced their skin and chest muscles. It was a religious ritual in which the participants were seeking a personal communion.

## The Elements of Religion

All religions contain certain shared elements, including ritual and prayer, emotion, belief, and organization.

*Ritual and Prayer* All religions have formalized social rituals, but many also feature private rituals such as prayer. Of course, the particular events that make up rituals vary widely from culture to culture and from religion to religion.

All religions include a belief in the existence of beings or forces that human beings cannot experience. In other words, all religions include a belief in the supernatural. Hence, they also include **prayer,** or *a means for individuals to address or communicate with supernatural beings or forces*, typically by speaking aloud while holding the body in a prescribed posture or making stylized movements or gestures.

In the United States, three-quarters of all Americans pray at least weekly. Less than one in ten claim never to pray. Interestingly, 12 percent of atheists and 25 percent of agnostics also pray at least a few times a month. Forty-nine percent of the people who pray receive answers to their prayers (Pew Forum on Religion and Public Life, 2008).

*Emotion* One of the functions of ritual and prayer is to produce an appropriate emotional state. This can be done in many ways. In some religions, participants in rituals deliberately attempt to alter their state of consciousness through the use of drugs, fasting, sleep deprivation, and induction of physical pain. Thus, Scandinavian groups ate mushrooms that caused euphoria, as did many native Siberian tribes. Various Native American religions use peyote, a button-like mushroom that contains a hallucinogenic drug.

There are approximately 250,000 members of the Native American church who believe that the use of peyote brings them closer to God. Even though it is illegal to use peyote in most states, Congress made it possible for federally recognized tribes to practice traditional Indian ceremonies even if it violates local laws (Gehrke, 2001).

Although not every religion tries to induce altered states of consciousness in believers, all religions do recognize that such states may happen and believe that they can be the result of divine or sacred intervention in human affairs. Prophets, of course, are thought to receive divine inspiration. Religions differ in the degree of importance they attach to such happenings.

*Belief* All religions endorse a belief system, usually one that includes a supernatural order and, often, a set of values to be applied to daily life. Belief systems can vary widely. Some religions believe that a valuable quality can flow from a sacred object—animate or inanimate, part

Only three religions are known to be monotheistic: Judaism, Christianity, and Islam.

## OUR DIVERSE WORLD: Who Is God?

Why does God exist? How have the three dominant monotheistic religions—Judaism, Christianity, and Islam—shaped and altered the conception of God? Karen Armstrong, one of Britain's foremost commentators on religious affairs, traced the history of how men and women have perceived and experienced God from the time of Abraham to the present.

Armstrong points out that what the idea of God means depends on the time, place, and people who are making the interpretation. How God is viewed by one group might be meaningless to another group. Even the statement, "I believe in God," means something only when it is understood in the context of the time period and who is making the statement.

Armstrong notes that what people imply when they refer to "God" can have many meanings, which can often be contradictory. In fact, she believes that this flexibility is necessary to keep the notion of God alive. It allows discarding and replacing conceptions of God when they cease to have meaning or relevance. She points out that Christianity, Judaism, and Islam have each had to re-create the image of God as times have changed.

As strange as it might seem, the same is true for atheism. Stating "I do not believe in God" has also had a variety of meanings throughout history. Atheists are denying a particular conception of God that is rooted to a particular time and place. As Armstrong notes, "Is the 'God' whom atheists reject today the God of the patriarchs, the God of the prophets, the God of the philosophers, the God of the mystics, or the God of the eighteenth century?" These ideas of God are all very different from one another. In addition, "Jews, Christians, and Muslims were all called 'atheists' by their pagan contemporaries because they had adopted a revolutionary notion of divinity and transcendence." The atheist of today is denying the existence of a God adequate to address contemporary issues.

Armstrong believes religion has had to be highly practical. As soon as particular views about God ceased to be effective, they were adjusted to accommodate changing times. People had less trouble accepting this view in the past than they do today. She points out that our ancestors knew that ideas about God were provisional and open to reinterpretation.

**Sources:** Adapted from Armstrong, Karen. 1994. *A History of God*. New York: Ballantine Books, pp. xx–xxi; and interviews with the author, October 1994 and 1996.

---

or whole—to a lesser object. Numerous Christian sects, for instance, practice the laying on of hands, whereby a healer channels divine energy into afflicted people and thus heals them.

Some Christians also believe in the power of relics to work miracles simply because those objects once were associated physically with Jesus or one of the saints. Such beliefs are quite common among the world's religions. Native Australians have their sacred stones, and shamans among African, Asian, and North American societies try to heal through sympathetic touching. In some religions, the source of the valued quality is a personalized deity. In others, it is a reservoir of supernatural force that is tapped.

People's belief can be very strong. At Mercy Medical Center in Springfield, Massachusetts, thousands came to view a second-story window where people said they saw an image of the Virgin Mary. The image recently appeared in a window that had been put in place in the 1970s when the building was constructed. Glass specialists thought water seeped between the panes of the double-paned glass and that minerals caused an acidic reaction that appears to have etched the glass (Finucane, 2008).

In Chicago, a steady stream of people flocked to the Kennedy Expressway underpass to view a yellow and white stain on a concrete wall that many believed was also an image of the Virgin Mary. Police had to patrol the area under the Kennedy Expressway after hundreds of people gathered to see the image. Some women knelt with rosary beads while other people stood praying (CBS News website, April 20, 2005).

One of the best known sightings of the Virgin Mary appeared in office windows in Clearwater, Florida. Within weeks, a half million people traveled to the site. Many of those who came hoped for a miracle that would cure their ills. Others wanted to receive their own message from Mary, to see visions of Jesus, or to strengthen their faith.

It is not just Christians who report sightings. In the United Kingdom, people flocked to a pet store after it was noticed that the markings on the scales on one side of an albino Oscar fish looked like the Arabic script for Allah. On the other side of the fish people claimed to see the word Muhammad. Within hours, hundreds of people, photographers, and television crews gathered to view the fish (Lewis, 2006).

*Organization* Many religions have an organizational structure through which specialists can be recruited and trained, religious meetings conducted, and interaction facilitated between society and the members of the religion.

The organization also promotes interaction among the members of the religion to foster a sense of unity and group solidarity. Rituals may be performed in the presence of other members or limited to certain locations such as temples, or to processions from one place to another. Although some religious behavior may be carried out by individuals in private, all religions demand some public, shared participation. (See "Our Diverse World: Who Is God?")

## MAGIC

In some societies, magic serves some of the functions of religion, although there are essential differences between the two. **Magic** is *an active attempt to coerce spirits or to control supernatural forces. It differs from other types of religious beliefs in that one god or gods are not worshipped.* Magic is used to manipulate and control matters that seem to be beyond human control and that may involve danger and uncertainty. It is usually a means to an end, whereas religion is usually an end in itself, although prayer might be regarded as utilitarian when a believer asks for a personal benefit. In most instances, religion unifies a group of believers, whereas magic is designed to help the individual who uses it. Bronislaw Malinowski (1954) explains:

> We find magic wherever the elements of chance and accident, and the emotional play between hope and fear have a wide and extensive range. We do not find magic wherever the pursuit is certain, reliable, and well under the control of rational methods and technological processes. Further, we find magic where the element of danger is conspicuous.

During the Middle Ages, when most of the population was illiterate, the belief in magic was quite extensive. Almost everyone believed in sorcery, werewolves, witchcraft, and black magic. If a noblewoman died, her servants ran around the house emptying all containers of water so her soul would not drown. Bloodletting to cure illnesses was popular. Plagues were believed to be the result of an unfortunate conjuncture of the stars and planets. The air was believed to be infested with such soulless spirits as unbaptized infants, ghouls who pulled out cadavers in graveyards and gnawed on their bones, and vampires who sucked the blood of stray children. For the medieval mind, magic provided an understanding of how the world worked. It helped relieve anxiety and allowed people to blame events on bad luck or evil spirits, and it permitted them to cast blame on curses and witchcraft. Astrology was the most popular science of that time. Only religion could rival astrology as an all-embracing explanation for the unpredictability of life (Shermer, 2000).

Rodney Stark and William Bainbridge (1985) have noted that a belief in magic has always been a major part of Christian faith. A common theme throughout the centuries has been the effort of organized religion to prohibit unorthodox practices and practitioners and to monopolize magic. Non-church magic was identified as superstition. Serious efforts to root out magic once and for all emerged in the fifteenth century. Eventually, as many as 500,000 people may have been executed for witchcraft. "In order to monopolize religion," Stark and Bainbridge wrote:

> [A] church must monopolize all access to the supernatural.... But if the church is to deny others access to the supernatural, it must remain in the magic business. The demand for magic is too great to be ignored....
> Thus the Catholic Church remained deeply involved in dispensing magic. Immense numbers of magical rites and procedures were developed.... Saints and shrines that performed specialized miracles proliferated, and new procedures for seeking saintly intercession abounded. Many forms of illness, especially mental illness, were defined as cases of possession, and legions of official exorcists appeared to treat them.

Stark and Bainbridge noted that magic's respectability has decreased as more scientific attitudes have proliferated. Magic, especially magical healing, is now found mostly among sectarians and cultists. This fact makes the religious beliefs of sects and cults particularly vulnerable to criticism and refutation (Beckwith, 1986).

## MAJOR TYPES OF RELIGIONS

The earliest evidence of religious practice comes from the Middle East. In Shanidar Cave in Iraq, archaeologist Ralph Solecki (1971) found remains of burials of Neanderthals—early members of our own species, *Homo sapiens*—dating from between 60,000 and 45,000 years ago. Bodies were tied into a fetal position, buried on their sides, provided with morsels of food placed at their heads, and covered with red powder and sometimes with flower petals. Those practices—the food and the ritual care with which the dead were buried—point to a belief in some kind of existence after death.

Using studies of present-day cultures as well as historical records, sociologists have devised a number of ways of classifying religions. One of the simplest and most broadly inclusive schemes recognizes four types of religion: supernaturalism, animism, theism, and abstract ideals.

## Supernaturalism

**Supernaturalism** *postulates the existence of nonpersonalized supernatural forces that can, and often do, influence human events.* These forces are thought to inhabit animate and inanimate objects alike—people, trees, rocks, places, even spirits or ghosts—and to come and go at will. The Melanesian/Polynesian concept of *mana* is a good example of the belief in an impersonal supernatural power.

**Mana** is *a diffuse, nonpersonalized force that acts through anything that lives or moves,* although inanimate objects such as an unusually shaped rock also can possess *mana*. The proof that a person or thing possesses *mana* lies in its observable effects. A great chief, merely by virtue of his position of power, must possess *mana*, as does the oddly shaped stone placed in a garden plot that then unexpectedly yields huge crops. Although it is considered dangerous because of its power, *mana* is neither harmful nor beneficial in itself, but it may be used by its possessors for either good or evil purposes. An analogy in our culture might be nuclear power, a natural force that intrinsically is neither good nor evil but can be turned to either end by its possessors. We must not carry the analogy too far, however, because we can account for nuclear power according to natural, scientific principles and can predict its effects reliably without resorting to supernatural explanations. A narrower, less comprehensive, but more appropriate analogy in Western society is our idea of luck, which can be good or bad and over which we feel we have little control.

Although certain objects possess *mana*, taboos can exist in relation to other situations. A **religious taboo** is *a sacred prohibition against touching, mentioning, or looking at certain objects, acts, or people.* Violating a taboo results in some form of pollution. Taboos can exist in reference to foods not to be eaten, places not to be entered, objects and people not to be touched, and so on. Even a person who becomes a victim of some misfortune may be accused of having violated a taboo and can also become stigmatized.

Taboos exist in a wide variety of religions. Polynesian peoples believed that their chiefs and noble families were imbued with powerful *mana* that could be deadly to commoners. Hence, elaborate precautions were taken to prevent physical contact between commoners and nobles. The families of the nobility intermarried (a chief often would marry his own sister), and chiefs actually were carried everywhere to prevent them from touching the ground and thereby killing the crops. Many religions forbid the eating of selected foods. Jews and Muslims have taboos against eating pork at any time, and up until the early 1960s, Catholics were forbidden to eat meat on Fridays. Most cultures forbid sexual relations between parents and children and between siblings (the incest taboo) (see Chapter 3, "Culture").

Supernatural beings fall into two broad categories, those of nonhuman origin, such as gods and spirits, and those of human origin, such as ghosts and ancestral spirits. Chief among those of nonhuman origin are the gods who are believed to have created themselves and might have created or given birth to other gods. Although gods may create, not all religions attribute the creation of the world to them.

Many of the gods thought to have participated in creation have retired, so to speak. Having set the world in motion, they no longer take part in day-to-day activities. Other creator gods remain involved in ordinary human activities. Whether or not a society has creator gods, many other affairs are left to lesser gods. For example, the Maori of New Zealand have three important gods: a god of the sea, a god of the forest, and a god of agriculture. They call upon each god for help in the appropriate area.

Below the gods in prestige, but often closer to the people, are the unnamed spirits. Some of these can offer constructive assistance, and others take pleasure in deliberately working evil for people.

Ghosts and ancestor spirits represent the supernatural beings of human origin. Many cultures believe that everyone has a soul, or several souls, which survive after death. Some of those souls remain near the living and continue to be interested in the welfare of their kin (Ember and Ember, 1981).

## Animism

**Animism** is *the belief in inanimate, personalized spirits or ghosts of ancestors that take an interest in, and actively work to influence, human affairs.* The souls or spirits may inhabit the bodies of people and animals as well as inanimate objects such as plants or rocks. They may also be present in winds, rivers, or mountains. The souls or spirits are unique beings with feelings, motives, and a will of their own. Unlike *mana*, spirits may be intrinsically good or evil. Although they are powerful, they are not worshipped as gods, and because of their humanlike qualities, they can be manipulated—wheedled, frightened away, or appeased—by using the proper magic rituals. For example, many Native American and South American Indian societies (as well as many other cultures in the world) think sickness is caused by evil spirits. Shamans, or medicine men or women, supposedly can effect cures because of their special relationships with these spirits and their knowledge of magic rituals. If the shamans are good at their jobs, they supposedly can persuade or force the evil spirit to leave the sick person or to discontinue exerting its harmful influence.

In our own culture, some people consult mediums, spiritualists, and Ouija boards in an effort to contact the spirits and ghosts of departed loved ones.

## Theism

**Theism** is *the belief in divine beings—gods and goddesses—who shape human affairs.* Gods are seen as powerful beings worthy of worship. Most theistic societies practice **polytheism,** *the belief in a number of gods.* Each god or goddess usually has particular spheres of influence such as childbirth, rain, or war, and generally one is more powerful than the rest and oversees the others' activities. In the ancient religions of Mexico, Egypt, and Greece, for instance, we find a pantheon, or a host, of gods and goddesses.

## Monotheism

**Monotheism** is *the belief in the existence of a single god.* Only three religions are known to be monotheistic, Judaism and its two offshoots, Christianity and Islam. These three religions have the greatest number of believers worldwide (see Table 13-1). Even these faiths are not purely monotheistic, however. Christianity, for example, includes belief in such divine or semi-divine beings as angels, the devil, saints, and the Virgin Mary. Nevertheless, because all three religions contain such a strong belief in the supremacy of one all-powerful being, they are considered to be monotheistic.

## Abstract Ideals

Some religions are based on abstract ideals rather than a belief in supernatural forces, spirits, or divine beings. **Abstract ideals** *focus on the achievement of personal awareness and a higher state of consciousness through*

**TABLE 13-1** Major Religions of the World

|  |  | Percentage of World Population |
|---|---|---|
| Christians | 2,039,000,000 | 32.0 |
| Muslims | 1,600,000,000 | 22.0 |
| Protestants | 342,002,000 | 5.6 |
| Orthodox | 215,129,000 | 3.7 |
| Anglican | 79,650,000 | 1.3 |
| Unaffiliated Christians | 111,125,000 | 1.8 |
| Baha' | 7,106,000 | 0.1 |
| Buddhists | 353,794,000 | 6.0 |
| Chinese folk-religionists | 359,982,000 | 5.9 |
| Ethnic religionists | 228,367,000 | 3.8 |
| Hindus | 811,336,000 | 13.3 |
| Jews | 14,434,000 | 0.2 |
| Sikhs | 23,258,000 | 0.4 |
| Nonreligious | 768,159,000 | 12.7 |
| Atheists | 150,090,000 | 2.5 |

*Source: World Christian Trends: AD 30–AD 2200.* Pasadena, CA: William Carey Library, 2001; "Major Religions of the World Ranked by Number of Adherents." Adherents.com (www.adherents.com/Religions), accessed August 29, 2012.

Religious rituals fulfill a number of social functions. They bring people together physically, promote social cohesion, and reaffirm a group's beliefs and values.

*correct ways of thinking and behaving rather than by manipulating spirits or worshipping gods.* Such religions promote devotion to religious rituals and practices and adherence to moral codes of behavior. Buddhism is an example of a religion based on abstract ideals. The Buddhist's ideal is to become one with the universe, not through worship or magic, but by meditation and correct behavior.

## A SOCIOLOGICAL APPROACH TO RELIGION

When sociologists approach the study of religion, they focus on the relationship between religion and society. The functionalist sociologists have examined the functions religion plays in social life, whereas conflict theorists have viewed religion as a means for justifying the political status quo.

### The Functionalist Perspective

Since at least 60,000 years ago, as indicated by the Neanderthal burials at Shanidar Cave, religion has played a role in all known human societies. The question that interests us here is, what universal functions does religion have? Sociologists have identified four categories of religious function: satisfying individual needs, promoting social cohesion, providing a worldview, and helping adaptation to society.

**Satisfying Individual Need** Religion offers individuals ways to reduce anxiety and promote emotional integration. Although Sigmund Freud (1918, 1928) thought religion to be irrational, he saw it as helpful to the individual in coming to terms with impulses that induce guilt and anxiety. Freud argued that a belief in lawgiving, powerful deities could help people reduce their anxieties by providing strong, socially reinforced inducements for controlling dangerous or immoral impulses.

Further, in times of stress, individuals can calm themselves by appealing to deities for guidance or even for outright help, or they can calm their fears by trusting in God. In the face of so many things that are beyond human control and yet can drastically affect human fortunes (such as droughts, floods, or other natural disasters), life can be terrifying. It is comforting to "know" the supernatural causes of both good fortune and bad. Some people attempt to control supernatural forces through magic rituals.

Perhaps this is why the deceased best-selling author Michael Crichton (a medical doctor and author of *Jurassic Park*) ate the same meal for lunch every day while working on a new novel and why former New York Giants football coach Bill Parcells would stop and buy coffee at two different coffee shops on his way to the stadium before every game.

**Social Cohesion** Émile Durkheim, one of the earliest functional theorists, noted the ability of religion to bring about group unity and cohesion. According to Durkheim, all societies have a continuing need to reaffirm and uphold their basic sentiments and values. This is accomplished when people come together and communally proclaim their acceptance of the dominant belief system. In this way, people are bound to one another, and as a result, the stability of the society is strengthened.

Not only does religion in itself bring about social cohesion, but often, the hostility and prejudice directed at its members by outsiders also helps strengthen bonds between those members. For example, during the 1820s, Joseph Smith, a young farmer from Vermont, claimed that he had received visits from heavenly beings. He said the apparitions enabled him to produce a 600-page history, known as the *Book of Mormon,* of the ancient inhabitants of the Americas. Shortly after the establishment of the Mormon Church, Smith had a revelation that Zion, the place where the Mormons would prepare for the millennium, was to be established in Jackson County, Missouri. Within two years, 1,200 Mormons had bought land and settled in Jackson County. The other residents in the area became concerned about the influx, and in 1833, they published their grievances in a document that became known as the manifesto, or secret constitution. They charged the Mormons with a variety of transgressions and pledged to remove them from Jackson County.

Several episodes of conflict followed, which eventually forced the Mormons to move into an adjoining county. These encounters with a hostile environment produced a sense of collective identity at a time when it was desperately needed. The church was less than two years old and included individuals from diverse religious backgrounds. There was a great deal of internal discord, and without the unity that resulted from the conflict with the townspeople, the group might have disappeared altogether (MacMurray and Cunningham, 1973).

Durkheim's interest in the role of religion in society was aroused by his observation that religion, like the family, seemed to be a universal human institution. This universality meant that religion must serve a vital function in maintaining the social order. Durkheim felt that he could best understand the social role of religion by studying one of the simplest kinds—the totemism of the aboriginal Australian. A **totem** is *an ordinary object such as a plant or animal that has become a sacred symbol to and of a particular group or clan who not only revere the totem but also identify with it.* Thus, reasoned Durkheim, religious symbols such as totems, as well as religion itself, arose from society itself, not from outside it. When people recognize or worship supernatural entities, they are really worshipping their own society. Thus, society is

Despite their doubts about the existence of God, 12 percent of atheists and 25 percent of agnostics pray a few times a month.

the origin of the shared religious beliefs, which in turn helps solidify society.

Durkheim saw religious ritual as an important part of this social cement. Religion, through its rituals, fulfills a number of social functions. It brings people together physically, promoting social cohesion; it reaffirms the group's beliefs and values; it helps maintain norms, mores, and prohibitions so that violation of a secular law—murder or incest, for instance—is also a violation of the religious code and can warrant ritual punishment or purification; it transmits a group's cultural heritage from one generation to the next; and it offers emotional support to individuals during times of stress and at important stages in their lives such as puberty, marriage, and death.

In Durkheim's view, these functions are so important that even a society that lacks the idea of the sacred must substitute some system of shared beliefs and rituals. Indeed, some theorists see communism as such a system. Soviet communism had its texts and prophets (Karl Marx, Friedrich Engels), its shrines (V. I. Lenin's tomb), its rituals (May Day parade), and its unique moral code. Durkheim thought that much of the social upheaval of his day could be attributed to the fact that religion and ritual no longer played an important part in people's lives and that, without a shared belief system, the social order was breaking down.

Although many sociologists today take issue with Durkheim's explanation of the origins of religion based on totemism, they nevertheless recognize the value of his functional approach to understanding the vital role of religion in society.

Secular society depends on external rewards and pressures for results, whereas religion depends on the internal acceptance of a moral value structure. Durkheim believed that because religion is effective in bringing about adherence to social norms, society usually presents those norms as an expression of a divine order. For example, in ancient China, as in France until the late eighteenth century, political authority—the right to rule absolutely—rested securely on the notion that emperors and kings ruled because it was divine will that they do so—the divine right of kings. In Egypt, the political authority of pharaohs was unquestioned because they were more than just kings; they were believed to be gods in human form.

Religion legitimizes more than just political authority. Although many forms of institutionalized inequality do not operate to the advantage of the subgroups or individuals affected by them, they help perpetuate the larger social order and often are justified by an appeal to sacred authority. In such situations, although religion serves to legitimize social inequality, it does function to sustain societal stability. Thus, the Jews in Europe were kept from owning land and were otherwise persecuted because of the myth "they had killed Christ" (Trachtenberg, 1961); and even slavery has been defended on religious grounds. In 1700, Judge John Saffin of Boston wrote of the:

> Order that God hath set in the world, who hath Ordained different degrees and orders of men, some to be High and Honorable, some to be Low and Despicable . . . yea, some to be born slaves, and so to remain during their lives, as hath been proved. (Montagu, 1964a)

Religions do not always legitimize secular authority. In feudal Europe, the church had its own political structure, and there often was tension between church and state. Indeed, just as the church often legitimized monarchs, it also excommunicated those who failed to take its wishes into account. However, the fact remains that religious institutions usually do dovetail neatly with other social institutions, legitimizing and helping to sustain them.

*Establishing Worldviews* According to Max Weber, in his classic book *The Protestant Ethic and the Spirit of Capitalism,* religion responds to the basic human need to understand the purpose of life. In doing so, religion must give meaning to the social world within which life occurs. This means creating a worldview that can have social, political, and economic consequences. For instance, consider the issue of whether salvation can be achieved through active mastery (hard work, for example) or through passive contemplation (meditation). Calvinism focuses on the total depravity of human beings; God's salvation is required because humans are morally and spiritually incapable of following God or redeeming themselves. Divine intervention will change their unwilling hearts from rebellion to obedience. Some people will be saved while others will be condemned. People had to believe the gospel and repent to be saved. Many Eastern religions view passive contemplation and meditation as the road to salvation.

Using these ideas, Weber theorized that Calvinism fostered the Protestant ethic of hard work and asceticism and that Protestantism was an important influence on the development of capitalism. Calvinism is rooted in the concept of predestination, which holds that before they are born, certain people are selected for heaven and others for hell. Nothing anyone does in this world, Calvinists believe, can change this. The Calvinists consequently were eager to find out whether they were among those chosen for salvation. Worldly success—especially the financial success that grew out of strict discipline, hard work, and self-control—was regarded as proof that a person was among the select few. Money was accumulated not to be spent but to be displayed as proof of being among the chosen. Capitalist virtues became Calvinist virtues. It was Weber's view that even though capitalism existed before Calvinist influence, it blossomed only with the advent of Calvinism.

Weber's analysis has been criticized from many standpoints. Calvinist doctrines were not as uniform as Weber pictured them, nor was the work ethic confined to the Protestant value system. Rather, it seems to have been characteristic of the times, promoted by Catholics as well as by Protestants. Finally, one could just as well argue the reverse, that the social and economic changes leading to the rise of industrialism and capitalism stimulated the emergence of the new Protestantism—a position that Marxist analysts have taken. Today, it is generally agreed that although religious beliefs did indeed affect economic behavior, the tenets of Protestantism and capitalism tended to support each other. However, the lasting value of Weber's work is his demonstration of how religion creates and legitimizes worldviews and how important these views are to human social and political life.

*Adaptations to Society* Religion can also be seen as having adaptive consequences for the society in which it exists. For example, many would view the Hindu belief in the sacred cow, which may not be slaughtered, as a strange and not particularly adaptive belief. The cows are permitted to wander freely and defecate along public paths.

Marvin Harris (1966) has suggested that there might be beneficial economic consequences in India from not slaughtering cattle. The cows and their offspring provide a number of resources that could not be provided easily in other ways. For example, a team of oxen is essential to India's many small farms. Oxen could be produced with fewer cows, but food production would have to be devoted to feeding those cows. With the huge supply of sacred cows, although they are not well fed, the oxen are produced at no cost to the economy.

Cow dung is also necessary in India for cooking and as fertilizer. It is estimated that dung equivalent to 45 million tons of coal is burned annually. Alternative sources of fuel, such as wood or oil, are scarce or costly.

Although most Hindus do not eat beef, cattle that die naturally or are slaughtered by non-Hindus are eaten by certain groups. Without the Hindu taboo against eating beef, these other members of the Indian hierarchy would not have access to this food supply. Therefore, because the sacred cows do not compete with people for limited resources and because they give birth to the oxen that are a cheap source of labor, fuel, and fertilizer, the taboo against slaughtering cattle appears to be quite adaptive.

When societies are under great stress or attack, their members sometimes fall into a state of despair analogous, perhaps, to that of a person who becomes depressed. Institutions lose their meaning for people, and the society is threatened with what Durkheim called anomie, or "normlessness." If this continues, the social structure can break down, and the society may be absorbed by another society unless the culture can regenerate itself. Under these conditions, revitalization movements sometimes emerge. **Revitalization movements** are *powerful religious movements that stress a return to the traditional religious values of the past.* Many of them can be found in the pages of history and even exist today.

In the 1880s, for instance, the once free Plains Indians lived in misery, crowded onto barren reservations by soldiers of the U.S. government. Cheated out of the pitiful rations that had been promised them, they lived in hunger—and with memories of the past. Then a Paiute by the name of Wovoka had a vision, and he traveled from tribe to tribe to spread the word and demonstrate his Ghost Dance.

> Give up fighting, he told the people. Give up all things of the white man. Give up guns, give up European clothing, give up alcohol, and give up all trade goods. Return to the simple life of the ancestors. Live simply—and dance! Once the Indian people are pure again, the Great Spirit will come, all Indian ancestors will return, and all the game will return. A big flood will come, and after it is gone, only Indians will be left in this good time. (Brown, 1971)

Wovoka's Ghost Dance spread among the defeated tribes. From the Great Plains to California, Native American communities took up the slow, trancelike dance. Some believed that the return of the ancestors would lead to the slaughter of all whites. For others, the dance just rekindled pride in their heritage. For whatever reasons, the Ghost Dance could not be contained, despite the government's attempts to ban it.

On December 28, 1890, the people of a Sioux village camped under federal guard at Wounded Knee, South Dakota, and began to dance. They ignored orders to stop and continued to dance until someone suddenly fired a shot. The soldiers opened fire, and soon more than 200 of the original 350 men, women, and children were killed. The soldiers' losses were 29 dead and 33 wounded, mostly from their own bullets and shrapnel. This slaughter was the last battle between the Indians of the Great Plains and the soldiers of the dominant society (Brown, 1971). (For a discussion of religious tolerance in other countries, see "Our Diverse World: The Worst Offenders of Religious Freedom.")

## The Conflict Theory Perspective

Karl Marx asserted that the dominant ideas of each age have always been the ideas of the ruling class (Marx and Engels, 1848/1961), and from this it was a small step to his assertion that the dominant religion of a society is that of the ruling class, an observation that has been borne out by historical evidence. Marxist scholars emphasize religion's role in justifying the political status quo by cloaking political authority with sacred legitimacy and thereby making opposition to it seem immoral.

The concept of alienation is an important part of Marx's thinking, especially in his ideas of the origin and functions of religion. **Alienation** is *the process by which people lose control over the social institutions they themselves invented.* People begin to feel like strangers—aliens—in their own world. Marx further believed that religion is one of the most alienating influences in human society, affecting all other social institutions and contributing to a totally alienated world.

According to Marx, "Man makes religion, religion does not make man" (Marx, 1867a/1967). The function of God thus was invented to serve as the model of an ideal human being. People soon lost sight of this fact, however, and began to worship and fear the ideal they had created as if it were a separate, powerful supernatural entity.

Thus, religion, because of the fear people feel for the god they themselves have created, serves to alienate people from the real world.

Marx saw religion as the tool that the upper classes used to maintain control of society and to dominate the lower classes. In fact, he referred to it as the "opiate of the masses," believing that through religion, the masses were kept from actions that might change their relationship with those in power. The lower classes were distracted from taking steps toward social change by the promise of happiness through religion. If they followed the rules established by religion, they expected to receive their reward in heaven, so they had no reason to try to change or improve their condition in this world. These religious beliefs helped the ruling classes exploit the lower classes; religion legitimized upper-class power and authority. Although modern political and social thinkers do not accept all of Marx's ideas, they recognize his contribution to the understanding of the social functions of religion.

Although religion performs a number of vital functions in society—helping maintain social cohesion and control while satisfying the individual's need for emotional comfort, reassurance, and a worldview—it also has negative, or dysfunctional, aspects.

Marx would be quick to point out a major dysfunction of religion. Through its ability to make the existing social order seem the only conceivable and acceptable way of life, it obscures the fact that people construct society and therefore can change society. Religion, by imposing acceptance of supernatural causes of conditions and events, tends to conceal the natural and human causes of social problems. In fact, in this role of justifying or legitimating the status quo, religion may very well hinder much-needed changes in the social structure. By diverting attention from injustices in the existing social order, religion discourages the individual from taking steps to correct these conditions.

An even more basic and subtle dysfunction of religion is its insistence that only one body of knowledge and only one way of thinking are sacred and correct, thereby limiting independent thinking and the search for further knowledge.

## ● ORGANIZATION OF RELIGIOUS LIFE

Religious groups can be organized in a variety of ways. In the next section, we will examine some of these.

### The Universal Church

A **universal church** *includes all the members of a society within one united moral community* (Yinger, 1970). It is fully part of the social, political, and economic status quo and therefore accepts and supports (more or less) the secular culture. In a preliterate society, in which religion is not really a separate institution but rather part of the entire fabric of social life, a person belongs to the church simply by being a member of the society. In more complex societies, the church cuts across divisions and binds all believers into one moral community. A universal church, however, does not seek to change the conditions of social inequality created by the secular society and culture, and indeed, it may even legitimize them. (An example is the Hindu religion of India, which in the past perpetuated a rigid caste system.)

### The Ecclesia

An **ecclesia** is *a church that shares the same ethical system as the secular society and has come to represent and promote the interest of the society at large.* Like the universal church, an ecclesia extends itself to all members of a society, but because it has so completely adjusted its ethical system to the political structure of the secular society, it comes to represent and promote the interests of the ruling classes. In this process, the ecclesia loses some adherents among the lower social classes, who increasingly reject it for membership in sects, whether sacred or civil (Yinger, 1970). An ecclesia is usually the official or national religion. For most people, membership is by birth rather than by conscious decision. Ecclesias have been common throughout human history. Examples include the Catholic Church in Spain and the Roman Empire; the Anglican Church, which is now the official Church of England; Islam in Saudi Arabia; and Confucianism, which was the state religion in China until early in the twentieth century.

## OUR DIVERSE WORLD

## The Worst Offenders of Religious Freedom

Religious freedom is central to American identity and a core objective of U.S. foreign policy. Many countries severely restrict and deny people the ability to practice their religion. For example, in 2011, governments increasingly used blasphemy, apostasy, and defamation laws to restrict the rights of religious minorities, and limit freedom of expression. In 2007, the Organization of the Islamic Conference, comprising fifty-seven states with significant Muslim populations, declared that the group does not recognize the right of individuals to freely change their religion.

Even with widespread religious intolerance throughout the world, the United States has singled out seven of the worst offenders that it calls Countries of Particular Concern, or CPCs. These countries stand out in their refusal to allow religious freedom. They include:

**China,** an officially atheist country. Members of the Chinese Communist Party (CCP) are required to be atheists and are discouraged from participating in religious activities. China only allows groups belonging to one of the five state-sanctioned "patriotic religious associations" (Buddhist, Taoist, Muslim, Roman Catholic, and Protestant) to register with the government and legally hold worship services. Religious activities by groups that have not registered are illegal and may be punished. The government is particularly harsh on Tibetan Buddhists, Catholics faithful to the Vatican, underground Protestants, Muslims, and Falun Gong. Tibetan monks and nuns have been forced to sign statements personally denouncing the Dalai Lama. Many religious believers are in prison for their faith, and beatings, torture, and the destruction of places of worship have taken place.

**North Korea,** where religious freedom does not exist. Religious believers, particularly Christians, often face imprisonment, torture, or even execution. Christians have been imprisoned and tortured for reading the Bible and talking about God. Some reports indicate that Christians endured biological warfare experiments.

**Iran,** where discrimination against religious minorities continues to get worse. The government created a threatening environment for nearly all non-Shia religious groups, most notably for Baha'is, as well as for Sufi Muslims, evangelical Christians, Jews, Sunni, and Zoroastrians. These groups face imprisonment, harassment, intimidation, and discrimination based on their religious beliefs. There has been a rise in officially sanctioned anti-Semitic propaganda, creating a threatening atmosphere for the Jewish community.

**Sudan,** where the government continues to attempt to impose strict Muslim law on non-Muslims in some parts of the country, and non-Muslims face discrimination and restrictions on the practice of their faith. Students must study Islam even if they are enrolled in private Christian schools. Conversion from Islam, blasphemy, and some interfaith marriages are prohibited.

**Eritrea,** where, since 2002, only four religious groups have been officially recognized. Activities by other religious groups are banned. More than 200 Protestant Christians and Jehovah's Witnesses have been imprisoned for their faith. Severe torture has been used to pressure believers to renounce their faith. Thousands of others continue to be detained in harsh conditions.

**Saudi Arabia,** where freedom of religion does not exist. The government requires all citizens to be Muslim and prohibits all public manifestations of non-Muslim religions. Islamic practice generally is limited to that of a school of the Sunni branch of Islam as interpreted by Muhammad Ibn Abd Al-Wahhab, an eighteenth-century Arab religious reformer, and practices contrary to this interpretation are suppressed. Non-Muslim worshippers risk arrest, imprisonment, or deportation for engaging in religious activities. Arabic and religion textbooks contain overtly hostile statements against Jews and Christians as well as Shia and Sufi Muslims.

**Uzbekistan,** which requires religious groups to register and prohibits such activities as proselytizing, as well as publishing, importing, and distributing religious materials without a license. It is difficult to meet the registration requirements. In some cases, members faced heavy fines and even jail terms for violations of the state's religion laws. The government restricts religious activities that conflict with national security and prohibits religious groups from forming political parties or teaching religious principles in private.

**Source:** Office of International Religious Freedom. 2012. *The International Religious Freedom Report for 2011*. Washington, D.C.: Department of State (www.state.gov/j/drl/rls/irf/religiousfreedom/index.htm#wrapper), accessed August 29, 2012.

## The Denomination

A **denomination** *tends to limit its membership to a particular class, ethnic group, or religious group or, at least, to have its leadership positions dominated by members of such a group.* It has no official or unofficial connection with the state, and any political involvement is purely a matter of choice by the denomination's leaders, who may either support or oppose any or all of the state's actions and political positions. Denominations do not withdraw themselves from the secular society. Rather, they participate actively in secular affairs while tending also to cooperate with other religious groups. These two characteristics distinguish them from sects, which are separatist and unlikely to be tolerant of other religious persuasions (Yinger, 1970). In America, Lutheranism, Methodism, other Protestant groups, Catholicism, and Judaism embody the characteristics of a denomination.

## The Sect

A **sect** is *a small group that adheres strictly to religious doctrine and often claims that they are the authentic version of the faith from which they split.* Sects generally represent a withdrawal from secular society and an active rejection of secular culture (Stark and Bainbridge, 1979). For example, the Dead Sea Scrolls show clearly that the beliefs of both early Christian and Jewish sects, such as the Essenes, were rooted in a disgust with society's self-indulgent pursuit of worldly pleasures and in a rejection of the corruption perceived in the prevailing religious hierarchy (Abegg, Jr. et al., 2002).

Early in their development, sects often are so harsh in their rejection of society that they invite persecution. Some actually thrive on martyrdom, which causes members to intensify their fervent commitment to the faith. Consider, for example, the Christian martyrs in Rome before Emperor Constantine's conversion to Christianity. (For a discussion of the continuing evolution of religion, see "Day-to-Day Sociology: Today's Cult Might Be Tomorrow's Mainstream Religion.")

## Millenarian Movements

**Millenarian movements** *typically prophesy the end of the world, the destruction of all evil people and their works, and the saving of the just.* Millenarian (from the Latin word for "thousand") prophecies often are linked with the symbolic number 1,000 or multiples thereof. Throughout human history, religious leaders have emerged in times of stress, foretelling the end of the world and asking everyone to stop whatever they are doing to follow the bearers of the message.

With the advent of AD 1000, Christendom in medieval Europe was thrown into a panic by religious doomsday preachers who predicted that the end of the world was imminent. People abandoned their homes and crops, and mobs of the devout took refuge in churches or fled on pilgrimages to the Holy Land.

Similar predictions occurred with the advent of the year 2000. Many Christian groups believed the new millennium would see monumental events predicted in the New Testament book of Revelation, which predicts a thousand-year kingdom of peace and plenty and a new heaven and earth.

Why would anyone want to believe the end of the world is near? Part of the answer is that the apocalypse usually does not mean pure destruction but, rather, the destruction of evil and the victory of good. Believers feel they are the center of the universe and the cosmic drama is coming to its ultimate conclusion in their time. Life is filled with meaning.

Apocalyptic beliefs also appeal to people who feel that life has not been fair to them. One of the great appeals of apocalyptic views is that they help make sense of the world. All events start to have significance, but they also can lead to conspiracy theories because every detail is linked to a much larger drama in which good and evil are at odds. There is no such thing as chance.

What happens when the chosen date for the end of the world or other millennial event finally arrives and nothing happens? Some people might just accept it and move on. More often, they will try to find a way to prove that their prophecy did come true but not exactly in the way predicted. Others may reset their apocalyptic clocks to another date.

We should not be too quick to dismiss millennial movements. Christopher Columbus believed the world would end in 1650. He considered his discovery of the New World part of a divine plan to establish a millennial

In the United States, Lutheranism and Methodism (and other Protestant groups), Catholicism, and Judaism all embody the characteristics of a denomination.

> ### DAY-TO-DAY SOCIOLOGY
>
> ## Today's Cult Might Be Tomorrow's Mainstream Religion
>
> It's tempting to conceive of the religious world as being made up primarily of a few well-delineated and static religious blocs: Christians, Jews, Muslims, Buddhists, Hindus, and so on. But that's dangerously simplistic. It assumes a stability in the religious landscape that is completely at odds with reality. New religions are born all the time. Old ones transform themselves dramatically. Schism, evolution, death, and rebirth are the norm.
>
> And this does not apply only to religious groups that one often hears referred to as cults. Today, hundreds of widely divergent forms of Christianity are practiced around the world. Islam is usually talked about in monolithic terms (or, at most, in terms of the Shia-Sunni divide), but one almost never hears about the 50 million or so members of the Naqshabandiya order of Sufi Islam, which is strong in Central Asia and India, or about the more than 20 million members of various schismatic Muslim groups around the world. Think, too, about the strange rise and fall of the Taliban. Buddhism, far from being an all-encompassing glow radiating benignly out of the East, is a vast family of religions made up of more than 200 distinct bodies, many of which don't see eye-to-eye at all. Major strands of Hinduism were profoundly reshaped in the nineteenth century, revealing strong Western and Christian influences.
>
> There's no reason to think that the religious movements of today are any less subject to change than were the religious movements of hundreds or even thousands of years ago. History bears this out. Early Christianity was deemed pathetic by the religious establishment: Pliny the Younger wrote to the Roman Emperor Trajan that he could get nothing out of Christian captives but "depraved, excessive superstition." Islam, initially the faith of a band of little-known desert Arabs, astonished the whole world with its rapid spread.
>
> Protestantism started out as a note of protest nailed to a door. In 1871, Ralph Waldo Emerson dismissed Mormonism as nothing more than an "afterclap of Puritanism." Up until the 1940s, Pentecostalists were often dismissed as "holy rollers," but today the *World Christian Encyclopedia* suggests that by 2050 there may be more than a billion people affiliated with the movement. The implication is clear: what is now dismissed as a fundamentalist sect, a fanatical cult, or a mushy New Age fad could become the next big thing.
>
> *Source:* Excerpted from Lester, Toby. February 2002. "Oh, Gods!" *Atlantic Monthly,* pp. 37–45. Used by permission of the author.

paradise. "God made me the messenger of the new heaven and the new Earth of which he spoke in the Apocalypse of St. John," Columbus wrote in his journal, "and he showed me the spot where to find it" (Sheler, 1997). Many major religious movements received attention because of apocalyptic predictions. Ironically, although the apocalypse that millenarians prophesy may not come true, they often succeed; the world is a different place after them. It really was the end of the world as we had known it.

## ● ASPECTS OF AMERICAN RELIGION

The Pilgrims in 1620 sought to build a sanctuary where they would be free from religious persecution, and the Puritans who followed ten years later intended to build a community embodying all the virtues of pure Protestantism, a community that would serve as a moral guide to others.

Thus, religion pervaded the social and political goals of the early English-speaking settlers in America and played a major role in shaping colonial society. Today, the four main themes that characterize religion in America are religious diversity, widespread belief, secularism, and ecumenism.

### Religious Diversity

The United States has always been a land of many religions. The European settlers encountered a wide diversity of native religions. The early immigrants included British Anglicans, Spanish and French Catholics, and Quakers. The Chinese and Japanese who came to work on the West Coast brought with them Buddhist, Taoist, and Confucian traditions. European Jews and Irish and Italian Catholics arrived in large numbers with the nineteenth-century immigration. The past forty years have seen an even greater expansion in American religious diversity. Buddhists have come from Thailand, Vietnam,

No other country has the level of religious pluralism found in the United States.

and Cambodia; Hindus from India and East Africa; and Muslims from Indonesia, Bangladesh, and the Middle East. Immigrants from Haiti and Cuba have brought religious traditions that blend Catholic and African beliefs. Even though the United States continues to be a predominantly Christian country, it has become the world's most religiously diverse nation (Eck, 2001).

## Widespread Belief

Americans generally take religion for granted. Although the religious affiliation and degree of church attendance differ widely, almost all Americans (92%) claim to believe in God. About 74 percent believe in life after death, and 63 percent believe that Scripture is the word of God. Even 41 percent of those who are not affiliated with a particular religious tradition say religion is at least "somewhat important" in their lives. Americans also attend church or synagogue regularly, with 39 percent claiming to do so each week (Pew Forum on Religion and Public Life, 2008).

More than three-quarters of all Americans say that the Bible is the "word of God." Nearly half of Americans believe that the Bible "is to be taken literally, word for word" and "the Bible is totally accurate in all of its teachings." There is also overwhelming belief that the Bible "answers all or most of the basic questions of life" (Pew Forum on Religion and Public Life, 2010).

Evidence of whether America is experiencing a religious revival, as some have claimed, is contradictory. When asked whether they think religion's influence on American life is increasing or decreasing, 35 percent of respondents said it was increasing, and 58 percent said it was decreasing. This contrasts with 1957 when 69 percent said religion was increasing in influence and 14 percent said it was decreasing. Yet, nearly half of Americans are devout practitioners of their faith by, as already noted, attending church or synagogue at least once a week. Two-thirds believe the Bible answers all or most of the basic questions of life. More than three-fourths of the Americans who believe in heaven think their chances of getting there are good or excellent. By and large, for most Americans, religion is a very important part of their lives (Pew Forum on Religion and Public Life, 2010). (See Table 13-2 for a comparison of how people in various countries see America's religiosity.)

In recent years religiosity and conservatism have been seen as related. This has given many people the impression that the influence of religion was growing, when in actual fact certain groups have gained visibility. Another development that could give the impression that religion is on the rise is the growth of mega churches, with large numbers of worshipers. These trends have made religion among conservative groups more intense, but caused more liberal churches to lose members. Also, as the family structure has changed with more single-parent families, the transfer of religion to the next generation has been weakened. This change has touched liberal churches the most (Chaves, 2011.)

More than half of all religiously affiliated individuals belong to a Protestant denomination, clearly reflecting America's colonial history. However, other denominations are also well represented, especially Catholicism and Judaism. There are more than 200 formally chartered religious organizations in America today. Such pluralism is not typical of other societies and has resulted primarily from the waves of European immigrants who began to arrive in the postcolonial era. Americans' traditional tolerance of religious diversity reflects the constitutional separation of church and state, so that no one religion is recognized officially as better or more acceptable than any other. (For more on

**TABLE 13-2** World Views of American Religiosity

|  | Too Religious | Not Religious Enough |
|---|---|---|
| United States | 21 | 58 |
| France | 61 | 26 |
| Netherlands | 57 | 25 |
| Great Britain | 39 | 28 |
| Germany | 39 | 31 |
| Canada | 35 | 38 |
| India | 32 | 57 |
| Spain | 31 | 40 |
| Russia | 27 | 38 |
| Poland | 6 | 56 |
| Jordan | * | 95 |

*Source:* Pew Forum on Religion and Public Life. June 23, 2005. *Global Attitudes Survey.*

## HOW SOCIOLOGISTS DO IT

### Is Your Professor an Atheist?

During the early years of American higher education, colleges and universities had strong ties to religion. Harvard University, started in 1636, was founded to train ministers. Even by the early nineteenth century, most college and university professors believed their mission was to teach lessons in theology and moral philosophy that would prepare students for life.

By the end of the nineteenth century, things started to change. Professors began thinking of themselves as scientists and scholars who were seeking truth rather than teaching religious values. They began to specialize and established their own criteria for evaluating academic work. Many schools severed their ties to religious denominations, and the concept of academic freedom was advanced.

A major change took place in the mid-twentieth century when enrollment increased and students and faculty members from a variety of ethnic, religious, and class backgrounds entered the academic world. The secularization of American higher education continued in the 1960s when many professors suspicious of religion joined the academy.

A common stereotype of the college or university professor today is that she or he is an atheist who rejects religion in favor of science or critical inquiry. Is this true? Two sociologists used a nationally representative sample of professors in all fields and all types of higher education institutions to try to answer this question.

A quick review of the results confirms the view that religion does not seem to be a big part of the academic world. More than 31 percent of the professors in the survey, when asked what their religion was, said "none." The professors were also asked about their belief in God. Ten percent agreed with the statement, "I don't believe in God," whereas 13.4 percent chose the statement, "I don't know whether there is a God, and I don't believe there is any way to find out." Taken together, those two numbers mean that 23.4 percent of all professors are either atheists or agnostics. This percentage is much higher than for the U.S. population as a whole.

The General Social Survey of the U.S. found that only 2.8 percent of the people are atheists, whereas 4.1 percent are agnostic and do not know whether God exists.

When we look at professors at elite doctoral universities, we find that they are much less religious than professors teaching in other kinds of institutions. Nearly 37 percent of these professors are either atheists or agnostics, compared to about 15 percent of the professors teaching in community colleges.

Atheists and agnostics are more common in some disciplines than others. The fields of psychology and biology have the highest proportion of professors who are atheists and agnostics, with 61 percent. At the other end of the spectrum, 63 percent of accounting professors and nearly 59 percent of elementary education professors say they have no doubt that God exists.

As high as these numbers are and contrary to popular opinion, most professors are not atheists and agnostics even at elite schools. Professors are certainly less religious than other Americans, but there is substantial variation from discipline to discipline and across types of institutions.

**TABLE 13-3** Professors and Belief in God

| Belief Statement | Percent Who Agreed |
|---|---|
| I don't believe in God. | 10.0 |
| I don't know whether there is a God and I don't believe there is any way to find out. | 13.4 |
| I don't believe in a personal God, but I do believe in a Higher Power of some kind. | 19.6 |
| I find myself believing in God some of the time but not at others. | 4.4 |
| Although I have my doubts, I feel that I do believe in God. | 16.9 |
| I know God really exists and I have no doubts about it. | 35.7 |

*Source:* Gross, Neil, and Solon Simmons. 2009. "The Religiosity of American College and University Professors." *Sociology of Religion* 70(2):101–129.

---

widespread belief, see "How Sociologists Do It: Is Your Professor an Atheist?")

### Secularism

Many scholars have noted that modern society is becoming increasingly **secularized**, that is, *less influenced by religion*. Religious institutions are being confined to ever-narrowing spheres of social influence while people turn to secular sources for moral guidance in their everyday lives (Berger, 1967). This shift is reflected in Americans' lack of religious knowledge; they are, for the most part, notoriously indifferent to, and ignorant of, the basic doctrines of their faiths.

Of course, social and political leaders still rely on religious symbolism to influence secular behavior.

## THINKING ABOUT SOCIAL ISSUES

## The Rise of No Religious Affiliation

Throughout U.S. history, the vast majority of Americans have identified with a religion. When asked "what is your religious preference?", no more than 2 to 5 percent of the population responded with "none" or "nothing in particular." The percentage of Americans claiming no religious identity started to increase in the 1970s, and reached 11 percent by 1990. Today, 16 percent of Americans say they have no religious identity, and the numbers appear to be increasing. (See Figure 13-1.)

The growth seems to be driven by younger people. Whereas only 5 percent of older groups claim no religion, as many as 25 percent of younger groups claim no religion, causing the average for the population to increase rapidly.

What are some of the characteristics of the people who claim no religion?

1. Men, whites, and non-Southerners are more likely to be in this group than women, nonwhites, and Southerners.
2. They have not discarded all religious beliefs. Few claim to be atheists or agnostics. Many express a belief in God and the afterlife.
3. They were often raised in nonreligious homes. Many were raised with no religion.
4. They are more likely to be in the center or left of the political spectrum. Few are from the conservative political backgrounds.

Many of the people with no religion became unaffiliated because they think religious people are hypocritical, judgmental, and insincere. Many say they became unaffiliated because religious organizations focus too much on rules and not enough on spirituality.

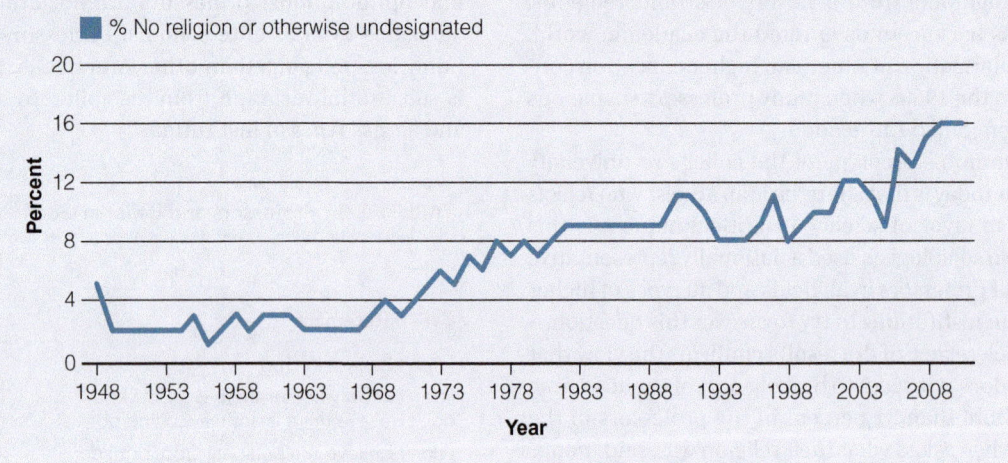

**FIGURE 13-1** Percentage of Americans Claiming No Religious Identity
Source: Newport, Frank. May 21, 2010. "In U.S., Increasing Number Have No Religious Identity." Gallup Poll (www.gallup.com/poll/128276/Increasing-Number-No-Religious-Identity.aspx), accessed August 27, 2012; Putnam, Robert D., and David E. Campbell. 2010. New York: Simon and Schuster, pp. 121–29.

---

The American Pledge of Allegiance tells us that we are "one nation, under God, indivisible," and our currency tells us that "In God We Trust." Since the turn of the century, however, modern society has turned increasingly to science, rather than to religion, to point the way. Secular political movements have emerged that attempt to provide most, if not all, of the functions that religion traditionally fulfilled. For example, communism prescribes a belief system and an organization that rival those of any religion. Like religions, communism offers a general concept of the nature of all things and provides symbols that, for its adherents, establish powerful feelings and attitudes and supply motivation toward action. Thus, some political movements lack only a sacred or supernatural component to qualify as religions. In this increasingly secular modern world, however, sacred legitimacy appears to be unnecessary for establishing meaning and value in life. (See "Thinking About Social Issues: The Rise of the No Religious Affiliation" for a discussion of the increasing numbers of Americans who report no religious affiliation.)

## Ecumenism

**Ecumenism** refers to *the trend among many religious communities to draw together and project a sense of unity and common direction*. It is partially a response to secularism and a tendency evident among many religions in the United States.

Unlike religious groups in Europe, where issues of doctrine have fostered sect-like hard-line separatism among denominations, most religious groups in America have focused on ethics—that is, how to live an ethical and moral life. There is less likelihood of disagreement over ethics than over doctrine. Hence, American Protestant denominations typically have had rather loose boundaries, with members of congregations switching denominations rather easily and churches featuring guest appearances by ministers of other denominations. In this context, ecumenism has flourished in the United States far more than in Europe.

## MAJOR RELIGIONS IN THE UNITED STATES

Nowhere is the diversity of the American people more evident than in their religious denominations. There are hundreds of different religious groups in the United States, and they vary widely in practices, moral views, class structure, family values, and attitudes. A recent survey found surprisingly large and persistent differences among even the major religious groups.

The U.S. census is prohibited from asking about religion, so the U.S. government generally has little to say on the matter. However, since 1972, the National Opinion Research Center has been conducting the General Social Surveys, which do give us a way of examining American religious attitudes and practices. This group has correlated information on a variety of issues with religious affiliation. Some of its findings are summarized here. (See Table 13-4 for the major religions in the United States.)

It is useful to think of American Protestant religious denominations as ranked on a scale measuring their degree of traditionalism. Conservative Protestant denominations include the fundamentalists (Pentecostals, Jehovah's Witnesses, and so on), Southern Baptists, and other Baptists. The moderates include Lutherans, Methodists, and interdenominationalists or nondenominationalists. Liberal Protestants are represented by Unitarian Universalists, Congregationalists, Presbyterians, and Episcopalians. This distinction among Protestants is important because the various branches often differ so markedly in their attitudes, especially toward social issues, that they resemble other religions more than the various denominations of their own.

For example, a 2007 survey showed that 44 percent of evangelicals believed that most men were

**TABLE 13-4** Major Religions in the United States

|  | Percentage of Adults |
|---|---|
| **Christian** | 78.4 |
| Protestant | 51.3 |
|   Evangelical churches | 26.3 |
|   Mainline churches | 18.1 |
|   Hist. black churches | 6.9 |
| Catholic | 23.9 |
| Mormon | 1.7 |
| Jehovah's Witness | 0.7 |
| Orthodox | 0.6 |
| Other Christian | 0.3 |
| **Other religions** | 4.7 |
| Jewish | 1.7 |
| Buddhist | 0.7 |
| Muslim | 0.6 |
| Hindu | 0.4 |
| Other faiths | 1.2 |
| **Unaffiliated** | 16.1 |
| **Don't know/Refused** | 0.8 |
| **Total** | 100 |

*Source:* Pew Forum on Religion and Public Life. 2008. *U.S. Religious Landscape Survey.*

better suited emotionally for politics than most women. Forty-eight percent of evangelicals believed "It's God's will that women care for children" (Baylor Religion Survey, 2008). Reflecting a similar viewpoint, in 2008, Lifeway Christian Bookstores, a chain run by the Southern Baptist Convention, pulled the magazine *Gospel Today* from company stores because the magazine featured five women who are church pastors on the cover. Southern Baptists do not believe women should be pastors (Pulliam, 2008). This is at a time when 71 percent of Americans are in favor of having women pastors, ministers, priests, or rabbis (Gallup Poll, August 2000).

With respect to the hereafter, 79 percent of Americans believe that a day will come when God judges whether one goes to heaven or hell (Gallup Poll, August 2000). Nearly 90 percent of fundamentalists and Baptists believe in an afterlife. This falls to 80 percent among the moderate and liberal denominations. Catholics are similar to liberal Protestants in that 75 percent believe in an afterlife. Among people with no religious affiliation, 46 percent believe in an afterlife.

A strong belief in sin is typical of fundamentalists and Baptists. This causes them to condemn extramarital and premarital sex and homosexuality and to favor outlawing pornography. There is greater sexual permissiveness among the moderate denominations and considerably more among the liberal denominations. Catholics tend to resemble the Protestant moderates, and Jews tend to be more liberal than the liberal Protestant denominations in this area. Attitudes toward drugs and alcohol follow the same pattern, with smoking, drinking, or the frequenting of bars least acceptable to fundamentalists and Baptists.

There are substantial class differences among the major denominations. Jews and Episcopalians have the highest median annual household incomes, and Baptists have the lowest. The pattern is the same for occupational prestige and education, with Jews and Episcopalians averaging three more years of education than fundamentalists and Baptists.

Given the wide differences in values and attitudes among religious groups, the relative proportion of the population that belongs to each group helps determine the shape of society. Protestants make up nearly 52 percent of the adult population. Among the major Protestant groups, the largest is composed of Baptists, who account for 18 percent of the adult population. Second are the Methodists with 9 percent (George W. Bush was the first Methodist to be elected U.S. president since William McKinley, who served from 1897 to 1901); next are the Lutherans with 7 percent, followed by Episcopalians at 3 percent, and the Church of Christ at 2 percent.

Roman Catholics, representing 23.9 percent of the population, make up the largest single religious denomination. Jews and Mormons (72 percent live in Utah) each represent 1.7 percent, followed by Buddhists at 0.7 percent, Eastern Orthodox at 0.6 percent, Muslim at 0.6 percent, Hindu at 0.4 percent, and a host of other religions (Pew Forum on Religious Life, 2008).

These percentages are in constant flux, however, because demographic factors such as birthrates and migration patterns can influence the number of people in any given religion. Religious conversion can also affect these numbers. Fundamentalism, for example, is gaining among the young and winning converts. (See "Day-to-Day Sociology: Changing Religion Early and Often" for a discussion of how changing one's religious affiliation is common in the United States.)

Despite trends toward ecumenism, it seems that the magnitude of religious differences, the persistence of established faiths, and the continual development of new faiths will ensure that this pattern of religious diversity will continue (Smith, 1984).

## Protestantism

The United States is a majority Protestant country. That is changing, however, and the percentage of the population that is Protestant has been falling. In 1993, 63 percent of Americans were Protestant. That percentage is now 51.3 percent. There are a number of reasons for why this is happening. Part of it is due to immigration, which has been largely from Asian and Latin American countries in recent decades. A second reason is that the number of people raised Protestant and remaining so in adulthood has declined to 83 percent. In addition, the percentage of people claiming no religion has jumped from 9 to 16 percent in the past decade (NORC, 2004).

Because American Protestantism is so fragmented, many sociologists simply have classified all non-Catholic Christian denominations in the general category of Protestant. However, differences exist among the various denominations. The number of members in historic mainline Christian churches has consistently decreased since the 1950s, and has really seen a significant drop since 2001. The Methodists and Episcopalians have seen large drops in membership. The decline is due to the growing public preference for the generic "Christian" or "non-denominational Christian" identification. Fewer than 200,000 people favored these terms in 1990, but today they account for more than 8 million Americans. The preference to identify as "born again" or "evangelical" rather than with any Christian tradition, church, or denomination is another factor (Kosmin and Keysar, 2009).

Since the late 1980s, fundamentalist and evangelical Christians have become a visible and vocal segment of the Protestant population, and their presence has been felt through the media and through their support of political candidates. Why are the fundamentalist and evangelical churches gaining in popularity? Some of their appeal might lie in the sense of belonging and the comfort they offer through their belief in a well-defined and self-assured religious doctrine—no ambiguities and, hence, few moral choices to be made.

The growth of these churches reflects religion's role as a social institution, changing over time and from place to place, partly in response to concurrent social and cultural changes and partly itself acting as an agent of social change.

## Catholicism

Approximately 24 percent of the American population is Roman Catholic, representing the largest single religious denomination in the United States. The Catholic population, which has traditionally been based in the Midwest and Northeast, has started to move to the South and Southwest. Rhode Island is the most Catholic state with almost 52 percent. The Catholic Church and evangelical churches have been competing for the allegiance of the growing Hispanic community (Religious Congregations and Membership in the United States, 2000).

One of the most striking things about Catholics in the United States is their youth. Twenty-nine percent are younger than 30, 36 percent are between 30 and 49, and 35 percent are older than 50. In contrast, 24 percent of Protestants are younger than 30, and 41 percent are older than 50. The higher birthrates among Hispanic Catholics account for a large part of this difference. Another part of the explanation is the difficulty that mainline Protestant denominations have had in retaining young people.

## DAY-TO-DAY SOCIOLOGY

### Changing Religion Early and Often

About half of all American adults have changed their religion at least once during their lives (see Table 13-5). Most people who change their religion do so before age 24, and very few do so after age 50. Many of those who change religion do so more than once.

The reasons people give for changing their religion—or leaving religion altogether—differ widely. Many of those who leave the religion of their childhood become unaffiliated. Two-thirds of former Catholics and half of former Protestants who became unaffiliated say they left their childhood faith because they stopped believing in the teachings of their religion. Others left because they do not believe in God. Many people who left also said they think of religious people as hypocritical or judgmental and that religious organizations focus too much on money or rules.

One in ten American adults is a former Catholic. Catholicism has suffered the greatest loss due to religious change. Former Catholics are about evenly divided between those who have become unaffiliated and those who have become Protestant. Many people who leave the Catholic Church do so for religious reasons. More than half of former Catholics left because they stopped believing in its teachings. A small number say the clergy sexual abuse scandal factored into their decision.

In contrast with other groups, those who switch from one Protestant denomination to another (for instance, were raised Lutheran and are now Methodist) do so in response to changes in their lives. Nearly 40 percent changed religious affiliation because they relocated to a new community, and about another 40 percent did so because they married someone from a different religious background.

**TABLE 13-5** Practicing the Religion of One's Youth

|  | Percent of U.S. Population |
|---|---|
| **Do not currently belong to childhood faith** | 44 |
| Raised Catholic, now unaffiliated | 4 |
| Raised Catholic, now Protestant | 5 |
| Raised Protestant, now unaffiliated | 7 |
| Raised Protestant, now different Protestant faith | 15 |
| Raised unaffiliated, now affiliated | 4 |
| Other change in religious affiliation | 9 |
| **Same faith as childhood** | 56 |
| Changed faith at some point but returned | 9 |
| Have not changed faith | 47 |

Source: Pew Forum on Religious Life. June 2008. *Faith in Flux: Changes in Religious Affiliation in the U.S.*

---

American Catholics have long been an immigrant people, and that tradition is continuing. One in five Catholics is a member of a minority group. Hispanics now make up 16 percent of American Catholics; another 3 percent are black, and an additional 3 percent describe themselves as nonwhite. Very few Hispanics identify themselves as nonwhite, so the data suggest that the influx of Catholic immigrants from Southeast Asia is starting to show in national surveys. Among Protestants, 14 percent are black and 2 percent are Hispanic. This means that although the percentage of blacks among Protestants is five times that among Catholics, a higher percentage of Catholics overall come from minority groups.

Since the mid-1960s, Catholics have equaled Protestants in levels of education and income, but the overall figures for Protestants mask significant differences between denominations. When we compare Catholics with individual Protestant denominations, we find them still ranking behind Presbyterians and Episcopalians, about on a par with Lutherans and Methodists, and well ahead of Baptists on scales that measure education and income. This comparison is striking in light of the large numbers of lower-income minorities included in the overall Catholic figures.

Catholics remain an urban people, with only one in four living in rural areas. A higher percentage of Catholics (39%) than of any other major denomination live in central cities, and 35 percent live in suburbs. The vast majority of Catholics are concentrated in the Northeast and Midwest.

Catholics have historically favored larger families than have other Americans, but by 1985, the difference in ideal family size between Catholics and Protestants had disappeared, with both groups considering two children the ideal. Despite the Catholic Church's condemnation of artificial means of birth control, American Catholics have favored information about and access to contraceptives in the same proportion as the rest of the population since the 1950s (Gallup and Castelli, 1987).

One of the most important developments in the recent history of Catholicism was the ecumenical council called by Pope John XXIII (Vatican II), which met from 1962 to 1965 and thoroughly reexamined Catholic doctrine. This ecumenical council led to many changes, often referred to as liberalization and including the substitution of common language for Latin in the Mass. One unintended consequence (or latent function) of Vatican II was that the centralized authority structure of the Catholic Church was questioned. Laypeople and priests felt free to dispute the doctrinal pronouncements of bishops and even of the pope. What began in the 1960s with a seemingly modest effort of reform has ended with every aspect of Catholic tradition under question. America's biggest single denomination now consists of traditional Catholics and "cafeteria Catholics," who pick and choose what to practice.

Under the leadership of Pope John Paul II, the Catholic Church took a more conservative turn. It continued to condemn all forms of birth control except the rhythm method and rejected high-technology aids to conception, such as artificial insemination, *in vitro* fertilization, and surrogate motherhood. Calls for a greater role for women in the church or their ordination to the priesthood were rejected. Women were told to seek meaning in their lives through motherhood and giving love to others.

The Catholic Church's opposition to birth control is opposed by two-thirds of American Catholics (Putnam and Campbell, 2010).

## Judaism

There is a strong identification among Jews on both cultural and religious levels. This sense of connectedness is an important factor in understanding current trends within the religion.

Jews can be divided into three groups on the basis of the manner in which they approach traditional religious precepts. Orthodox Jews observe traditional religious laws very closely. They maintain strict dietary laws and do not work, drive, or engage in other everyday practices on the Sabbath. Reform Jews, by contrast, allow for major reinterpretations of religious practices and customs, often in response to changes in society. Conservative Jews represent a compromise between the two extremes. They are less traditional than the Orthodox Jews are but not as willing to make major modifications in religious observance as the Reform Jews. In addition, a large secularized segment of the Jewish population still identifies itself as Jewish but refrains from formal synagogue affiliation.

As among Protestants, social class differences exist among the various Jewish groups. Reform Jews are the best educated and have the highest incomes. For Orthodox Jews, religious rather than secular education is the goal. They have the lowest incomes and the least amount of secular education. As might be expected, Conservative Jews are situated between these two poles.

The state of Israel has played a major role in shaping current Jewish thinking. For many Jews, identification with Israel has come to be a secular replacement for religiosity. Support for the country is tied to many deep psychological and emotional responses. To many, Israel and its continued existence represent a way of guaranteeing that never again will millions of Jews perish in a holocaust. The country is seen as a homeland that can help defend world Jewry from the unwarranted attacks that have occurred throughout history. For many Jews, identification with, and support for, Israel is important to the development of cultural or religious ties.

The Jewish community has had to deal with the issue of ordaining woman rabbis. Reform Jews have moved in this direction without a great deal of difficulty, and today women lead Reform congregations around the country. However, the issue of woman rabbis initially produced a bitter fight among Conservative Jews. Today, an increasing number of Conservative woman rabbis have become the heads of congregations. Orthodox Jews have not had to address the issue because, for them, the existence of a female rabbi would represent too radical a departure from tradition to be contemplated.

According to the Bible, God told the Jewish people to be fruitful and multiply. In the United States, it seems the group is not doing so. Recent surveys estimate that 5.2 million Jews live in the United States, a decline from 5.5 million in 1990 (National Jewish Population Survey, 2002). Demographers predict that the population will continue to decline. The American Jewish community is thus facing a crisis.

The reasons for the lack of growth in the Jewish community are varied. For one thing, Jews in the United States are not bearing enough children to replace themselves. The Jewish birthrate is 1.8 children per woman per

Orthodox Jews observe traditional religious laws very closely. They maintain strict dietary laws and refrain from work on the sabbath.

lifetime—well below the replacement rate of 2.1. The Jewish population is also quite old, with a median age of 41 and about 19 percent of American Jews older than 65. Jewish immigration from the former Soviet Union and Israel has increased the population somewhat but is unpredictable. Finally, substantial numbers of young Jews are choosing to marry outside the faith, although this does not necessarily lead to a loss of Jewish identity.

Whether population erosion is occurring, and what should be done about it, is an ongoing debate in the American Jewish community. Jewish religious groups have attempted to liberalize the definition of who is a Jew, and—in a radical break with tradition—have decided to seek converts.

A common fear is that a further decline in the number of American Jews would lessen their ability to defend their political interests. Others suggest that the United States would lose the contributions of Jewish scientists, artists, and performers. Even though Jews account for less than 2 percent of the U.S. population (Harris Poll, 2000), they make up 20 percent of America's Nobel laureates.

Although many of the reasons for the shrinking Jewish population are demographic, the issue of how to stem the decline has been the cause of major rifts among the branches of the religion. Recent statistics, for example, put the rate of Jewish intermarriage at 50 percent. Jewish opinion is not uniform on the subject but shows a sharp division between Orthodox Jews, the most strictly observant branch of Judaism, whose members are a small minority of the American Jewish population, and non-Orthodox. "For example, 64 percent of Orthodox Jews surveyed said they strongly disapproved of interfaith marriages, as opposed to 15 percent of Conservative Jews, 3 percent of Reform Jews, and 2 percent of those who identified themselves as 'just Jewish' " (Niebuhr, 2000). Interfaith couples have little difficulty today finding rabbis willing to officiate at the marriage ceremony.

A subject that has provoked even more controversy is the Reform movement's break with the tradition of matrilineal, or motherly, descent. In 1983, the Reform movement declared that people could be considered Jewish if either the father or mother was Jewish. Before that time, Judaism could be passed on to the child only from the mother.

American Jews also differ with respect to converts. The Reform movement's outreach goes against centuries of Jewish tradition in which proselytizing was disdained. The outreach program is quite low-key. There is no advertising or airwave sermonizing, which is more common with Christian evangelical movements. Instead, outreach sessions resemble comparative-religion discussion groups in which introductory classes in Judaism are available to interfaith couples as well as to converts.

Jews are dispersed across the United States, but the largest percentage (43%) of the Jewish population is concentrated in the Northeast. Only 13 percent of Jews live in the Midwest, and the Jewish population is relatively small in the South and West.

The Jewish population is very well educated. A quarter (24%) of Jewish adults 18 years of age and older have received a graduate degree, and 55 percent have earned at least a bachelor's degree. This compares with 5 and 28 percent for the general population, respectively. The additional years of education produce a median household income that is 16 percent higher than the median for all U.S. households (United Jewish Communities, 2002; Zoll, 2002).

## Islam

Islam, an Arabic word that means "surrender" or "submission," is the name given to the religion preached by the Prophet Muhammad, who taught from AD 612 until his death in AD 632. A Muslim is someone who has accepted the Islamic declaration of faith, or *shahadah*, that "There is no god but Allah, and Mohammed is His Prophet." Like Jews and Christians, Muslims are monotheists, and all three religions share the prophets of the Old Testament. (See Table 13-6 for the world's Muslim population.)

From its roots in the Arab world, Islam has spread in virtually every direction and is the world's second-largest religion behind Christianity, with 1.6 billion followers. Almost one-fourth of the world population might be Muslim in the near future if current high levels of fertility continue. Today, Muslims live in every country in the world. Although Islam began in Arabia, more than half of the world's Muslims live in South and Southeast Asia. The countries with the largest Muslim populations are Indonesia, India, Bangladesh, and Pakistan. Only about one-fifth of all Muslims live in the Middle East (see Table 13-6).

The world's 1.6 billion Muslims are united in their belief in God and the Prophet Muhammad and are bound together by such religious practices as fasting during the holy month of Ramadan and almsgiving to assist people in need. There are widely differing views

TABLE 13-6  The Muslim Population

|  |  | Percentage of Muslim Population |
|---|---|---|
| World | 1,619,314,000 |  |
| Asia-Pacific | 1,005,507,000 | 62.1 |
| Sub-Saharan | 321,869,000 | 19.9 |
| Africa | 242,544,000 | 15.0 |
| Europe | 44,138,000 | 2.7 |
| Americas | 5,256,000 | 0.3 |

*Source:* Pew Research Center's Religion and Public Life. January 2011. "The Future of the Muslim Population" (www.pewforum.org/The-Future-of-the-Global-Muslim-Population.aspx), accessed August 27, 2012.

about many other aspects of their faith, including how important religion is to their lives, who counts as a Muslim, and what practices are acceptable in Islam.

Islam is built around several articles of faith. These include: (1) there is only one God; (2) God has sent several messengers, but Muhammad was His final Prophet; (3) God revealed the Quran; (4) God's angels exist, even if people cannot see them; (5) there will be a Day of Judgment, when God will decide who will go to heaven or hell; and (6) God's will and knowledge are absolute.

The formal acts of worship called the Five Pillars of Islam provide the framework for all aspects of a Muslim's life. The pillars consist of (1) *shahadah*, or faith; (2) prayer; (3) almsgiving; (4) fasting; and (5) pilgrimage.

There are three historic divisions in Islam. The great majority of Muslims belong to the Sunni division. They follow a traditional interpretation of Islam. Most of the conservative Muslims that Westerners call fundamentalists are Sunnis. The next largest division is the Shiah i-Ali, whose members are called Shii Muslims or Shiites. Shiites honor Ali, the cousin and son-in-law of Muhammad, and Ali's descendants, whom they believe should be the leaders of the Muslim community. The Kharijites make up the smallest division of Islam. Their name is based on an Arabic word that means "secessionists." They were former followers of Ali who broke away in AD 657. Kharijites are strict Muslims whose beliefs are based on precise adherence to the teachings of the Quran, or Muslim scripture, and *sunnah* (customs) as their community interprets them (Cornell, 2001).

Among the five most populous nations, the United States has by far the fewest Muslims. It is, however, the third-largest faith in the United States, and there are already more Muslims than Presbyterians nationwide. Nonetheless, it is difficult to ascertain how many Muslims live in the United States. The census does not collect reliable data on religious preference. Some data do exist on church attendance, but Islam does not require attendance for formal membership in the way that many Christian groups do. Estimates of the number of Muslims in the United States note that the number could be as high as 4.7 million.

Yet, despite the estimates of millions of U.S. Muslims, there are only a thousand mosques and community centers to serve them. Islam faces the same challenge as Judaism faced with the influx from Eastern Europe toward the end of the nineteenth century. New-style congregations had to be invented, new buildings built, and new schools started to train a new type of rabbi. United States Islam is just beginning to create the communal organizations that have served Judaism so well. There is no coherent association of mosques to unite immigrants with native-born blacks, and national organizations of other types are young. Muslims are divided the way Protestants have been, by ethnicity, race, and language. Only in 1996 did U.S. Muslims establish a school for training clergy at the graduate level to parallel the Jewish and Christian seminaries (Ostling, 1999).

Islam is the third-largest faith in the United States. Despite estimates of millions of U.S. Muslims, there are only a thousand mosques and community centers to serve them.

About one in four Muslims in the United States is African American. African American Muslims are of two types—"those who follow mainstream Islamic doctrine and those who follow the teachings of the Nation of Islam. This is an important distinction because the Nation of Islam's teachings are very different from those in the Quran" (El-Badry, 1994).

The American public has a more favorable view of Muslim Americans than Islam as a religion. Fifty-one percent of the public has a favorable view of Muslim Americans, but only 39 percent has a favorable view of Islam.

Muslims tend to be socially conservative, favor a close-knit family, and support religious education. The conservatism is based partly on religious beliefs but also arises from cultural mores. Samia El-Badry (1994) explains that to some Muslims "sexual permissiveness is seen as a reflection on the family, rather than on the individual. The family is seen as the ultimate authority and is therefore responsible for the individual's behavior." Many Muslims also believe that the eldest man is the head of the family.

Muslims living in America often have difficulty reconciling the American way of life with the traditions and ideas about morality that exist within their belief. Immigrant Muslim parents are often at odds with their Americanized offspring about the use of alcohol, which is banned in Islam; dating, which is forbidden; and lack of respect for elders (El-Badry, 1994).

The world's Muslim population is expected to increase by about 35 percent in the next 20 years, rising from 1.6 billion in 2010 to 2.2 billion by 2030. The global Muslim population is expected to grow at about twice the rate of the non-Muslim population. By 2030, Muslims will make up 26.4 percent of the world's total population, up from 23.4 percent (Pew Forum on Religion and Public Life, 2012).

All religions and denominations are affected by the current mood of the country. A heightened social consciousness results in demands for reform, whereas stressful times often produce a movement toward the personalization of religion. In any event, although traditional forms and practices of religion might be changing in the United States, religion itself is likely to continue to function as a basic social institution.

## Social Aspects of Religious Affiliation

Religious affiliation seems to be correlated strongly with many other important aspects of people's lives. Direct relationships can be traced between membership in a particular religious group and a person's politics, professional and economic standing, educational level, family life, social mobility, and attitudes toward controversial social issues. For example, Jews are proportionally the best-educated group; they also have higher incomes than Christians in general; and a greater proportion are represented in business and the professions. Despite their high socioeconomic and educational levels, Jews, like Catholics, occupy relatively few of the highest positions of power in the corporate world and politics; these fields generally are dominated by white Anglo-Saxon Protestants.

Attitudes toward social policy also seem to be correlated, to some extent, with religious affiliation. The fundamentalist and evangelical Protestant sects generally are more conservative on key issues than are the major Protestant sects.

Although it is clear that religious associations show definite correlations with people's political, social, and economic lives, we must be careful not to ascribe a cause-and-effect relationship to such data, which at most can be considered an indicator of an individual's attitudes and social standing. However, the social and political correlates of religious affiliation have had a significant effect on the directions of the various religious denominations and sects in the United States.

## ■ SUMMARY

▸ **What is religion?**
Religion is a system of beliefs, practices, and philosophical values shared by a group of people that defines the sacred, helps explain life, and offers salvation from the problems of human existence. Religion is one of society's most important institutions. All religions endorse a belief system, which usually includes a supernatural order and a set of values to be applied to daily life.

▸ **How did Émile Durkheim view religion?**
Durkheim thought that all religions divide the universe into two mutually exclusive categories. The profane consists of empirically observable things, or things knowable through common, everyday experiences. In contrast, the sacred consists of things that are awe inspiring and knowable only through extraordinary experience. Almost anything may be designated as sacred. Sacred traits or objects symbolize important shared values.

▸ **What are rituals?**
Patterns of behavior or practices that are related to the sacred are known as rituals.

All religions have formalized social rituals, but many also feature private rituals such as prayer, which is a means for individuals to address or communicate with supernatural beings or forces. One of the functions of ritual and prayer is to produce an appropriate emotional state.

▸ **How do structural functionalists view religion?**
Structural functionalists examine the role of religion in social life. Religion, they say, offers individuals ways to reduce anxiety and promote emotional integration. They also believe religion provides for group unity and cohesion not only through its own practices but also, sometimes, through the hostility and prejudice directed at members of a religious group by outsiders.

▸ **How did being a structural functionalist influence Émile Durkheim's view of religion?**
Durkheim believed religion served a vital function in maintaining the social order.

He believed religion frequently legitimizes the structure of the society within which it exists. It also establishes worldviews that help people to understand the purpose of life. These worldviews can have social, political, and economic consequences. Religion can also help a society adapt to its natural environment or to changing social, economic, and political circumstances. Durkheim believed that when people recognize or worship supernatural entities, they are really worshipping their own society.

▸ **How do conflict theorists view religion?**
Conflict theorists emphasize religion's role in justifying the political status quo by cloaking political authority with sacred legitimacy and thereby making opposition to it seem immoral. Alienation, the process by which people lose control over the social institutions they themselves have invented, plays an important role in the origin of religion according to this view. Conflict theorists believe religion tends to conceal the natural and human causes of social problems in the world and discourages people from taking action to correct these problems.

▶ **What themes can be seen in religion in America?**
There are four main themes that characterize religion in America. They are religious diversity, widespread belief, secularism, and ecumenism.

▶ **What else does religious affiliation tell us about people?**
Religious affiliation seems to be correlated with other important aspects of people's lives. Direct relationships can be traced between membership in a particular religious group and a person's politics, professional and economic standing, educational level, family life, social mobility, and attitudes toward controversial social issues.

## Media Resources

**CourseMate for *Introduction to Sociology*, Eleventh Edition**

Cengage Learning's Sociology CourseMate brings course concepts to life with interactive learning, study, and exam preparation tools that support the printed textbook. Access an integrated eBook, learning tools including glossaries, flashcards, quizzes, videos, and more in your Sociology CourseMate. Go to CengageBrain.com to register or purchase access.

# CHAPTER THIRTEEN STUDY GUIDE

## Key Concepts

Match each of the following concepts with its definition, illustration, or explanation.

a. Religion
b. Profane
c. Sacred
d. Rituals
e. Magic
f. Supernaturalism
g. Mana
h. Taboo
i. Animism
j. Polytheism
k. Revitalization movement
l. Millenarian movement
m. Alienation
n. Universal church
o. Ecclesia
p. Denomination
q. Sect
r. Secularism
s. Ecumenism

____ 1. A church that includes all members of a society
____ 2. Belief in inanimate, personalized spirits or ghosts of ancestors that take an interest in and actively work to influence human affairs
____ 3. In Marxian theory, the process by which people lose control over the social institutions they themselves invented
____ 4. A system of beliefs, practices, and philosophical values shared by a group of people that defines the sacred, helps explain life, and offers salvation from the problems of human existence
____ 5. A diffuse, nonpersonalized force that acts through anything that lives or moves
____ 6. A small religious group that usually includes unconventional beliefs and practices
____ 7. The emphasis on nonreligious ideas and beliefs
____ 8. The trend among many religious communities to draw together and project a sense of unity and common direction despite doctrinal differences
____ 9. Pertaining to things that are awe inspiring and knowable only through extraordinary experience
____ 10. Regularized and prescribed patterns of behavior or practices that are related to the sacred
____ 11. A national or society-wide religion that is separate from civil society but closely parallel to it
____ 12. The use of spirits to control supernatural forces
____ 13. Belief in the existence of nonpersonalized, extraordinary, nonhuman forces that can and often do influence human events
____ 14. Pertaining to things that are part of common, ordinary, everyday experience
____ 15. A type of religion in which there are several gods
____ 16. One of several established churches in a society that tends to limit its membership to a particular class, ethnic, or religious group
____ 17. Religious movements that stress a return to traditional religious values and practices of the past
____ 18. A sacred prohibition against touching, mentioning, or looking at certain objects, acts, or people
____ 19. Movements that typically prophesy the end of the world, the destruction of all evil people and their works, and the saving of the just

## Key Thinkers

Match the thinkers with their main idea or contribution.

a. Max Weber
b. Bronislaw Malinowski
c. Marvin Harris
d. Karl Marx
e. Émile Durkheim

____ 1. The first sociologist to distinguish between the sacred and the profane and who discussed religion's role in promoting social cohesion
____ 2. Proposed, in *The Protestant Ethic and the Spirit of Capitalism*, the idea that the ideology of Calvinism promoted the development of capitalism
____ 3. Argued that the Hindu belief in the sacredness of cows is a practical strategy for adapting to the environment in India and therefore quite rational

____ 4. Saw religion as a tool that the upper classes use to maintain control of society and to dominate the lower classes

____ 5. An anthropologist who explained the functional differences between religion and magic, with the former uniting a group of believers and the latter helping the individual who used magic

## Central Idea Completions

*Fill in the appropriate concepts and descriptions for each of the following questions.*

1. Define and provide a brief example of the sacred and the profane.
   a. Sacred: _____
   b. Profane: _____

2. The functionalist view of religion holds that religion persists because it serves some important functions for individuals and for the society. What are some of these?
   a. _____
   b. _____
   c. _____
   d. _____

3. Define and provide an example of each of the following basic religious organizations:
   a. Sect: _____
   b. Church: _____
   c. Ecclesia: _____
   d. Denomination: _____
   e. Millenarian movement: _____

4. Define and provide an example of each of the following types of religions:
   a. Supernaturalism: _____
   b. Animism: _____
   c. Theism: _____

5. Do we believe in magic? Distinguish the similarities and differences between magic and religion. Does religion have elements of magic?
   _____
   _____

6. Millenarian movements stress the end of the world. What reasons does your text give for why one would want to believe the end of the world is near?
   _____
   _____

7. What are the characteristics of the three major religions in the United States?
   _____
   _____

8. Conflict theory focuses on conflicts between the interests of different socioeconomic groups. According to conflict theory, what part does religion play in this conflict?

___

___

## ● Critical Thought Exercises

1. **Leaving My Religion.** Given the information in the "Changing Religion Early and Often" box, it's likely that you know some people who have changed religion. Maybe you yourself do not belong to the religion of your parents. Why do people change religions? In the decision to change, how important are beliefs and religious doctrine? How important are nondoctrinal, social factors?

2. Mega churches and televangelists are fairly recent phenomena, and some people do not care much for them. What might some of their objections be, and how do these criticisms of these new forms of worship fit with functionalist and conflict theory ideas about religion?

3. Given that each individual has his or her own beliefs, how can a sociologist hope to investigate the phenomenon of religion empirically? In what ways can the idea of the sociological imagination discussed in Chapter 1, "The Sociological Perspective," be a useful tool for studying religion?

4. Durkheim argued that rituals and the sacred were the essential elements of religion, and Tischler points out that even nonreligious objects, such as Babe Ruth's bat, can take on sacred qualities. Keeping this in mind, compare a secular ritual (a pep rally, a birthday party, a parade, and so on) with religious rituals. How are they the same and how are they different?

## ● Internet Activities

1. Visit www.hartfordinstitute.org, the home page of the Hartford Institute for Religion Research. Explore the site. What topics of research and theory are presented? What can sociologists provide for nonsociologists, regarding the study of religion?

2. Visit www.pluralism.org, the site of the Pluralism Project. Compare the topics listed at this site with those from the site noted in activity 1. What are the similarities and differences between these two sites devoted to the sociological analysis of religion?

3. The website at www.adherents.com is not exactly sociological, but it provides information on religious affiliation. List three unusual facts you discovered on this website.

## ● Answers to Key Concepts

1. n;  2. i;  3. m;  4. a;  5. g;  6. q;  7. r;  8. s;  9. c;  10. d;  11. o;  12. e;  13. f;  14. b;  15. j; 16. p;  17. k;  18. h;  19. l

## ● Answers to Key Thinkers

1. e;  2. a;  3. c;  4. d;  5. b

Boston Filmworks

# 14 Education

### Education: A Functionalist View
- Socialization
- Cultural Transmission
- Academic Skills

*Thinking About Social Issues:* Is Education the Great Equalizer?
- Innovation
- Child Care
- Postponing Job Hunting

### The Conflict Theory View
- Social Control

*Our Diverse World:* Illiteracy Is Common Throughout the World
- Screening and Allocation: Tracking
- The Credentialized Society

*Day-to-Day Sociology:* Is a College Degree Worth the Trouble?

### Issues in American Education
- Unequal Access to Education
- Students Who Speak English as a Second Language
- High-School Dropouts
- Violence in the Schools
- Home Schooling
- Standardized Testing
- The Gifted

### Summary

## LEARNING OBJECTIVES

After studying this chapter, you should be able to do the following:

- Describe the manifest and latent functions of education.
- Explain the nature of education from the conflict theory view.
- Explain the causes and effects of racial segregation in the public schools.
- Identify issues related to students who speak English as a second language.
- Discuss the extent to which high-school dropouts are a social problem.
- Discuss the issue of standardized testing.

We all know certain proverbs and might, in fact, use them when we need to make a decision. Cervantes called proverbs "short sentences drawn from long experience." Consider the proverb, "A bird in the hand is worth two in the bush." It tells us not to give up a sure thing for something that might not happen. This proverb seems to apply to so many situations that it can be found in many languages. In Sweden, people say, "Rather one bird in the hand than ten in the woods." In Spain, we find people saying, "A bird in the hand is better than a hundred flying birds." Poles are likely to say, "A sparrow in your hand is better than a pigeon on the roof." People in Russia will say, "Better a titmouse in the hand than a crane in the sky." Other variations exist in Italy, Romania, Germany, and Iceland. This proverb might even go back 2,500 years because in one of Aesop's fables, we find a hawk who has captured a nightingale. The nightingale pleads that it should be released because it is too small to be much of a meal for a hawk. The hawk answers, "I would be foolish to release the bird I have in my hand for a bird that is not in sight." Proverbs help educate us about simple truths that we can apply to life (Heath and Heath, 2007).

Herbert Spencer would have agreed that proverbs are an important form of education. Spencer believed the purpose of education was "to prepare us for complete living." Every study must be judged by whether it had "practical value." The most important knowledge helped us live our lives or find a job or trade. Spencer believed all subjects had to be judged by how useful they would be in later life. Spencer's emphasis on practical education was applauded by a population that doubted the value of book learning.

Lester Frank Ward, the first president of the American Sociological Society, did not agree with Spencer. He believed the main purpose of education was to equalize society. Ward believed the source of inequality was the unequal distribution of knowledge. The main purpose of education was to equalize society by making knowledge available to all. The greatest advances in civilization had been made by people who had the opportunity for an education and the leisure to think. Ward maintained that the potential giants of the intellectual world could be the unskilled workers of today. He wrote, "[T]he number of individuals of exceptional usefulness will be proportionate to the number possessing the opportunity to develop their powers." Ward believed the entire society would benefit if there were more and better educational opportunity. Against both scholarly and popular opinion, he defended "intellectual egalitarianism." The differences between those at the top and the bottom of the social ladder were not due to any difference in intellect, he said, but to differences in knowledge and education. Ward believed the main job of education was to ensure that the heritage of the past was transmitted to all members of society (Ravitch, 2000).

In this chapter, as we examine the role of education in our society, we will contrast the functionalist and conflict theorist approaches to understanding the American educational system. Functionalists stress the importance of education in socializing the young, transmitting the culture, and developing skills. Conflict theorists, on the other hand, note that education preserves social class distinctions, maintains social control, and promotes inequality. We will also examine the impact of some of the contemporary issues facing education.

## EDUCATION: A FUNCTIONALIST VIEW

What social needs does our education system meet? What are its tasks and goals? Education has several *manifest functions* (Chapter 1, "The Sociological Perspective"), intended and predetermined goals such as socializing the young or teaching academic skills. It also has some latent functions, which are unintended consequences of the educational process. These can include child care, the transmission of ethnocentric values, and respect for the American class structure.

### Socialization

In the broadest sense, all societies must have an educational system. That is, they must have a way of teaching the young the tasks that are likely to be expected of them as they develop and mature into adulthood. If we accept this definition of an educational system, then we must believe that there really is no difference between education and socialization. As Margaret Mead (1943) observed, in many preliterate societies no such distinction is made. Children learn most things informally, almost incidentally, simply by being included in adult activities.

Traditionally, the family has been the main arena for socialization. As societies have become more complex, the family has been unable to fulfill all aspects of its socialization function. Thus, the formal educational system must extend the socialization process that starts in the family. Modern industrialized societies draw a distinction between education and socialization. In ordinary speech, we differentiate between socialization and education by talking of bringing up and educating children as separate tasks. In modern society, these two aspects of socialization are quite compartmentalized. Whereas rearing children is an informal activity, education or schooling is formal. The role prescriptions that determine interactions between students and teachers are clearly defined, and the curriculum to be taught is explicit. Obviously, the educational process goes far beyond just formalized instruction. In addition, children learn things in their families and among their peers. In school, children's master status (see Chapter 6, "Social Groups and Organizations") is that of student, and their primary task is to learn.

Schools, as formal institutions of education, emerged as part of the evolution of civilization. However, until about 200 years ago, education did not help people become more productive in practical ways; thus, it was a luxury that very few could afford. This changed dramatically with the industrialization of Western culture. Workers with specialized skills were required for production jobs, as were professional, well-trained managers.

When the Industrial Revolution moved workers out of their homes and into factories, the labor force consisted not only of adults but also of children. Subsequently, child labor laws were passed to prohibit children from working in factories. Public schools eventually emerged as agencies dedicated to socializing students, teaching them proper attitudes and behaviors, and encouraging conformity to the norms of social life and the workplace.

## Cultural Transmission

The most obvious goal of education is **cultural transmission,** *in which major portions of society's knowledge are passed from one generation to the next.* In relatively small, homogeneous societies, in which almost all members share the culture's norms, values, and perspectives, cultural transmission is a matter of consensus and needs few specialized institutions. In a complex, pluralistic society such as ours, with competition among ethnic and other minority groups for economic and political power, the decision about which aspects of the culture will be transmitted is the outgrowth of a complicated process.

As early as 1753, Benjamin Franklin was concerned about the different immigrant ethnic groups in the country and what could be done to educate them about American cultural values.

> Now they come in droves.... Few of their children in the country learn English; they import many books from Germany; and of the six printing houses in the Province, two are entirely German, two half German half English, and two entirely English. (Franklin, 1753)

It became clear that schools needed to be the means of cultural transmission. The view emerged that for a society to hold together, there must be certain core values and goals—some common traits of culture—that the different groups share to a greater or lesser degree (see Chapter 3, "Culture").

This core culture can also be open to change. A school's curriculum often reflects the ability of organized groups of concerned citizens to impose their views on an educational system, whether local, statewide, or nationwide. Thus, it was a political process that caused African American history to be introduced into elementary, high-school, and college curricula during the 1960s. Similarly, political activism caused the creation of women's studies programs in many colleges. Moreover, even though the concept of evolution is a cornerstone of modern scientific knowledge, political pressure (from Christian fundamentalists) causes some textbooks to refer to it as the "theory" of evolution and prevents it from being taught in certain counties.

In recent years, bilingual education has become an educational and political issue. Proponents believe that it is crucial for children whose primary language is not English to be given instruction in their native tongues. They believe that by acknowledging students' native languages, the school system is helping them make the transition into the all-English mainstream and is also helping preserve the diversity of American culture.

Others see a danger in these programs. They believe that many bilingual education programs never provide for the transition into English, leaving many youngsters without the basic skills needed to earn a living and participate in our society.

In the end, the debate centers on how closely our sense of who we are as a nation hinges on the language our children speak in school. For the time being, the only agreement between the two sides is that language is the cornerstone for cultural transmission.

## Academic Skills

Another crucial function of the schools is to equip children with the academic skills they need to function as adults to hold down a job, to balance a checkbook, to evaluate political candidates, to read a newspaper, to analyze the importance of a scientific advance, and so on. Have the schools been successful in this area? Most experts believe they have not.

In 1983, the National Commission on Excellence in Education issued a report titled "Nation at Risk" that bitterly attacked the effectiveness of American education. The message of the report was that the American education system was drowning in a tide of mediocrity that

An important function of education is to equip children with the academic skills that they need to function in society.

was threatening the future of the country. It also contained this memorable quote, "If an unfriendly foreign power had attempted to impose on America the mediocre educational performance that exists today, we might as well have viewed it as an act of war. As it stands, we have allowed this to happen to ourselves."

As a result of this report, reforms were instituted in all fifty states, which stressed the teaching of the "three Rs" and the elimination of frivolous electives that waste valuable student and teacher time. In addition, high-school graduation requirements were raised in forty states, and in nineteen states, students were required to pass minimum competency tests before they could receive their high-school diplomas. Forty-eight states also required new teachers to prove their competence by passing a standardized test.

Although the back-to-basics movement has grown, its success has been limited. Five years after "A Nation at Risk" appeared, a follow-up report was issued titled "American Education: Making It Work" (1988). According to the report, "the precipitous downward slide of previous decades has been arrested and we have begun the long climb back to reasonable standards." However, despite this progress, especially among minority groups, the report condemned the performance of American schools as unacceptably low. "Too many students do not graduate from our high schools, and too many of those who do graduate have been poorly educated.... Our students know little, and their command of essential skills is too slight," the report stated.

Particularly troublesome was student performance in math and science. According to a study done by the Nation's Report Card (1988), an assessment group that is part of the Educational Testing Service, the math performance of 17-year-olds is "dismal." At fault might be the very back-to-basics movement that was supposed to rescue our educational system from failure in the early 1980s. Although rote learning has helped improve the scores of the lowest-level students, it has left others totally unprepared to analyze complex problems. The Nation's Report Card also found American students' understanding of science "distressingly low."

By 2002, not much had changed. Between 1966 and 2002, the federal government had spent $321 billion (in today's dollars) to help educate disadvantaged children. Yet, despite increased spending:

- Fewer than one-third of U.S. fourth graders could read proficiently.
- Reading performance had not improved in more than fifteen years.
- Fewer than 20 percent of U.S. twelfth graders were proficient in math.
- And, among the industrialized nations of the world, U.S. twelfth graders ranked near the bottom in science and math.

In response to these facts in 2002, President George W. Bush signed into law the No Child Left Behind Act. The assumption behind this act is that setting high standards and establishing measurable goals will produce positive learning outcomes. States are required to develop ways of measuring the basic skills of all students in certain grades. The act was enforced by the threat of withholding federal funds for schools. By 2007, funding for the act exceeded $50 billion.

The act gave the federal government a role in kindergarten-through-grade-12 education and contains four basic education reform principles: stronger accountability for results, increased flexibility and local control, expanded options for parents, and an emphasis on teaching methods that have been proven to work. In an effort to hold education systems responsible for children's education, all states must implement statewide accountability systems, which will:

- Set academic standards for what students should know and be able to do in each content area.
- Gather specific, objective data through tests aligned with those standards.
- Use test data to identify strengths and weaknesses in the system.
- Report school academic achievement to parents and communities.
- Empower parents to take action based on school information.
- Recognize schools that make real progress.

The law requires real consequences for districts and schools that fail to make progress. It also enables parents with a child enrolled in a failing school to transfer their child to a better-performing public school or public charter school.

Some educators have criticized the act for emphasizing reading, writing, and math and paying less attention to the arts and foreign languages. The American Heart Association even blamed the act for increased childhood obesity by diminishing the importance of physical education classes (Trickey, 2006).

Education writer Jonathan Kozol was so upset by the No Child Left Behind Act that he spent more than two months on a partial hunger strike to protest the harm done by the act. He saw it as a racially punitive piece of legislation.

Every three years, the Programme for International Student Assessment (PISA) conducts a worldwide study of 15-year-olds in science and math literacy. In the most recent survey, 57 countries participated. United States students did not score in the top 20 countries on science or math literacy. They did do so in reading literacy. The countries with the top-performing students were Taiwan, Finland, South Korea, Hong Kong, and Canada (PISA 2006 Science Competencies for Tomorrow's World, 2007).

In 2010, President Obama changed the No Child law to provide funding so that students could learn a broad range of academic skills such as conducting research, using technology, solving problems, and communicating effectively. President Obama also imposed less stringent punishments for failing states and made allowances for students who are not English proficient, minorities, and those with special needs (Ravitch, 2010). By 2012, President Obama had granted waivers on meeting the requirement to twenty-six states.

The future of No Child Left Behind is going to depend on politicians as they continue to debate what is best for the nation's educational system. (For more on the role education can play in society, see "Thinking About Social Issues: Is Education the Great Equalizer?")

## THINKING ABOUT SOCIAL ISSUES: Is Education the Great Equalizer?

In the past people could be part of the middle class without a college education. In 1970, 74 percent of the people who had no more than a high school education were part of the middle class. Today, fewer than 40 percent of people without a college degree are part of the middle class. During that same period people with a college degree have continued to move into the middle class and beyond.

This information is not lost on people and college enrollment has surged. The vast majority of Americans believe that access to a college education is crucial for career success. People also believe no one should be denied access to college because of cost.

Where students go to college has changed in the last few decades, however, Lower-income students have been enrolling in the least selective four-year colleges and community colleges. At the same time, the affluent students have been going to the more selective four-year institutions. America's higher education system has become a two-tier system which is separate and unequal.

Half of annual college enrollments are concentrated in what Barron's Profile of American Colleges calls "competitive" four-year institutions, while the other half are concentrated in community colleges and other non four-year colleges at the bottom of the selectivity scale.

The more selective schools are not the common Ivy League college like Harvard, Yale, or Princeton. Barron's Profile lists 1,044 "competitive" four- year college. Over 75 percent of freshman going to these colleges are from families in the upper half of the income scale.

Barron's lists 1,000 community colleges and 2,999 less-prestigious four-year colleges that are part of the second half of the selectivity scale. Nearly 80 percent of the lowest-income students go to these schools.

Why do we see this difference? Sociologist Sean Reardon examined this question and concluded the gap is not due to the fact that students may come from families with highly and less educated parents. Reardon claims this factor has had the same effect for the last fifty years. Instead what has changed he believes is that higher income parents are more involved in their children's educational development.

Students going to the more competitive schools have a greater chance of completing their degree and are often looked on more favorably by employers. The view that education is the means for upward mobility and that it will be the great equalizer in American society is starting to be reexamined.

*Sources:* Sean F. Reardon, Whither Opportunity? Rising Inequality, Schools, and Children's Life Chances, New York: Russell Sage Foundation, 2011; Anthony P. Carnevale and Jeff Strohl, "Our Economically Polarized College System: Separate and Unequal," The Chronicle of Higher Education, September 25, 2011, http://chronicle.com/article/Our-Economically-Polarized/129094/, accessed 11-16-12.

## Innovation

A primary task of educational institutions is to transmit society's knowledge, and part of that knowledge consists of the means by which new knowledge is to be sought. Learning how to think independently and creatively is probably one of the most valuable tools the educational institution can transmit. This is especially true of the scientific fields in institutions of higher education.

Until well into this century, scientific research was undertaken more as a hobby than as a vocation. This was because science was not seen as a socially useful pursuit. Gregor Mendel (1822–1884), who discovered the principles of genetic inheritance by breeding peas, worked alone in the gardens of the Austrian monastery where he lived. Albert Einstein supported himself between 1905 and 1907 as a patent office employee while making several trailblazing discoveries in physics, the most widely known of which is the theory of relativity.

Today, science obviously is no longer the undertaking of part-timers. Modern scientific research typically is pursued by highly trained professionals, many of whom frequently work as teams, and the technology needed for exploration of this type has become so expensive that most research is possible only with extensive government or corporate funding. The United States continues to spend hundreds of billions of dollars on research and development funding. The areas of national defense, space exploration, and health research receive by far the greatest amount of support.

The achievements of government and industrial research and development notwithstanding, the importance of the contributions to science by higher academic institutions cannot be overestimated. First, there could be no scientific innovations—no breakthroughs—without the training provided by these schools. Second, the universities of the highest caliber continue to generate some of the most significant research in the biological and physical sciences.

In addition to their manifest or intended functions, the schools in America have come to fulfill a number of functions that they were not originally designed to serve.

## Child Care

One latent function of many public schools is to provide child care outside the nuclear family. This has become increasingly important since World War II, when women began to enter the labor force in large numbers. As of 2010, 58.6 percent of women were in the labor force. In addition, women with children worked more hours each week on average than they did in the past 1969 (U.S. Bureau of Labor Statistics, *Monthly Labor Review,* February 2011).

A related service of schools is to provide children with at least one nutritious meal per day. In 1975, the federal government spent $1.28 billion on federally funded school lunch programs. By 2011, the program served 31.8 million children with a cost of $11.1 billion (U.S. Department of Agriculture, 2012).

## Postponing Job Hunting

More and more young American adults are choosing to continue their education after graduating from high school. In 2011, 64.6 percent of male and 72.3 percent of female recent high-school graduates were enrolled in college (U.S. Bureau of Labor Statistics, 2012). Even though some of these individuals also work at part-time and even full-time jobs, an important latent function of the American educational system is to slow the entry of young adults into the labor market. This helps keep down unemployment as well as competition for low-paying unskilled jobs.

Originally, two factors pointed to the possibility that college enrollments would not continue to increase. Because of low birthrates, the number of high-school graduates peaked at 3.2 million in 1977 and began a 14-year decline, with only 2.28 million graduating in 1991. That trend has been reversed, and 3.1 million graduated in 2011 (U.S. Bureau of Labor Statistics, 2012).

Learning how to think independently and creatively is probably one of the most valuable tools that our educational system can transmit.

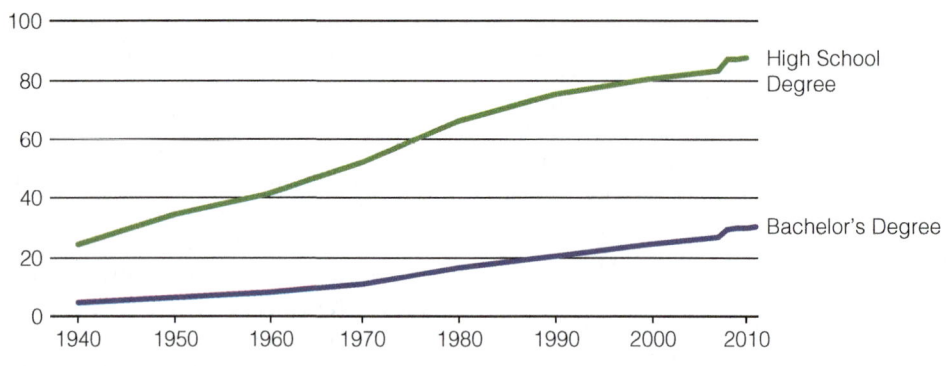

**FIGURE 14-1** Percentage of 25- to 29-Year-Olds Who Attained High-School and Bachelor's Degrees, 1940–2011

Source: Department of Commerce, U.S. Bureau of the Census. U.S. Census of Population: 1960, *vol. I, part 1*; Folger, J. K., and C. B. Nam. Education of the American Population *(1960 Census Monograph)*; Current Population Reports, *Series P-20, various years*; and Current Population Survey *(CPS), March 1970 through March 2011.*

Colleges and universities, anticipating enrollment problems, embarked on concerted efforts to ward off disaster. Through hard work and luck, they have succeeded. Total enrollment in two- and four-year colleges rose from 11.5 million in 1977 to more than 19.7 million in 2011 (U.S. Bureau of the Census, 2012). (See Figure 14-1 for the percentage of male and females with high-school diplomas and bachelor's degrees.)

Colleges have also benefited because the U.S. economic base has shifted from manufacturing jobs to service jobs. This has caused the demand for professionals and technicians to grow. The salaries for those types of positions are considerably higher than salaries for manufacturing jobs.

Two other trends have also benefited colleges; the first is the increase in the number of women going to college, and the second is the increase in older students. Since 1980, the majority of college students have been women, an outgrowth of changing attitudes about the status of women in our society and the breakdown in gender-role stereotypes (see Chapter 11, "Gender Stratification"). The second trend is because people 25 and older represent the most rapidly growing group of college students, accounting for 45 percent of all undergraduate and graduate students. Women are overrepresented in this group, as are part-time students. Many returning women students are also responding to changing gender-role expectations.

## THE CONFLICT THEORY VIEW

To the conflict theorist, society is an arena for conflict, not cooperation. In any society, certain groups come to dominate others, and social institutions become the instruments by which those in power control the less powerful. The conflict theorist thus sees the educational system as a means for maintaining the status quo, carrying out this task in a variety of ways. The educational system socializes students into values dictated by the powerful majority. Schools are seen as systems that stifle individualism and creativity in the name of maintaining order. To the conflict theorist, the function of school "is to produce the kind of people the system needs, to train people for the jobs the corporations require and to instill in them the proper attitudes and values necessary for the proper fulfillment of one's social role" (Szymanski and Goertzel, 1979). (For a discussion of how many of the people in the world are unable to control decisions about their lives, see "Our Diverse World: Illiteracy Is Common Throughout the World.")

### Social Control

In the United States, schools have been assigned the function of developing personal control and social skills in children. Although the explicit, formally defined school curriculum emphasizes basic skills such as reading and writing, much of what is taught is oriented away from practical concerns. Many critics point out that much of the curriculum (other than in special professional training programs) has little direct, practical application to everyday life. This has led conflict theorists and others to conclude that the most important lessons learned in school are not those listed in the formal curriculum but, rather, those that involve a hidden curriculum. The **hidden curriculum** refers to *the social attitudes and values taught in school that prepare children to accept the requirements of adult life and to fit into the social, political, and economic statuses the society provides.*

To succeed in school, a student must learn both the official (academic) curriculum and the hidden (social) curriculum. The hidden curriculum is often an outgrowth of the structure within which the student is asked to learn. Within the framework of mass education, it would be impossible to provide instruction on a one-to-one basis or even in very small groups. Consequently, students are usually grouped into relatively larger classes. Because this system obviously demands a great deal of social conformity by the children, those who divert attention and make it difficult for the teacher to proceed are punished.

## OUR DIVERSE WORLD

### Illiteracy Is Common Throughout the World

There are about 774 million men and women in the world who cannot read. About 75 million children do not attend school, and millions more young people leave school without adequate reading skills. These children then become adults who become dependent on others who can read to make important decisions about their lives. These issues become even more striking in light of today's information age and the Internet.

Eighty-five percent of the world's illiterate populations live in just thirty-five countries. (See Figure 14-2.) Each of those countries has a literacy rate of less than 50 percent of its population. Two-thirds of the illiterate in these countries are women and girls.

*Source:* UNESCO. 2004. *EFA Global Monitoring Report 2005: Education for All, the Quality Imperative.* Statistical Annex, Table 2. Paris: UNESCO; U.S. Central Intelligence Agency. *The World Factbook* (www.cia.gov/library/publications/the-world-factbook/fields/2103.html), accessed August 30, 2012.

Pie chart values:
- Afghanistan: 71.8%
- South Sudan: 73.0%
- Ethiopia: 57.3%
- Pakistan: 45.1%
- Bangladesh: 43.2%
- India: 39.0%
- Nigeria: 38.7%
- Egypt: 28.0%
- Brazil: 11.4%
- Indonesia: 9.6%
- China: 7.8%

**FIGURE 14-2** Percentage of People in Each Country Who Are Illiterate
*Source: U.S. Central Intelligence Agency. The World Factbook (www.cia.gov/library/publications/the-world-factbook/fields/2103.html), accessed August 30, 2012.*

---

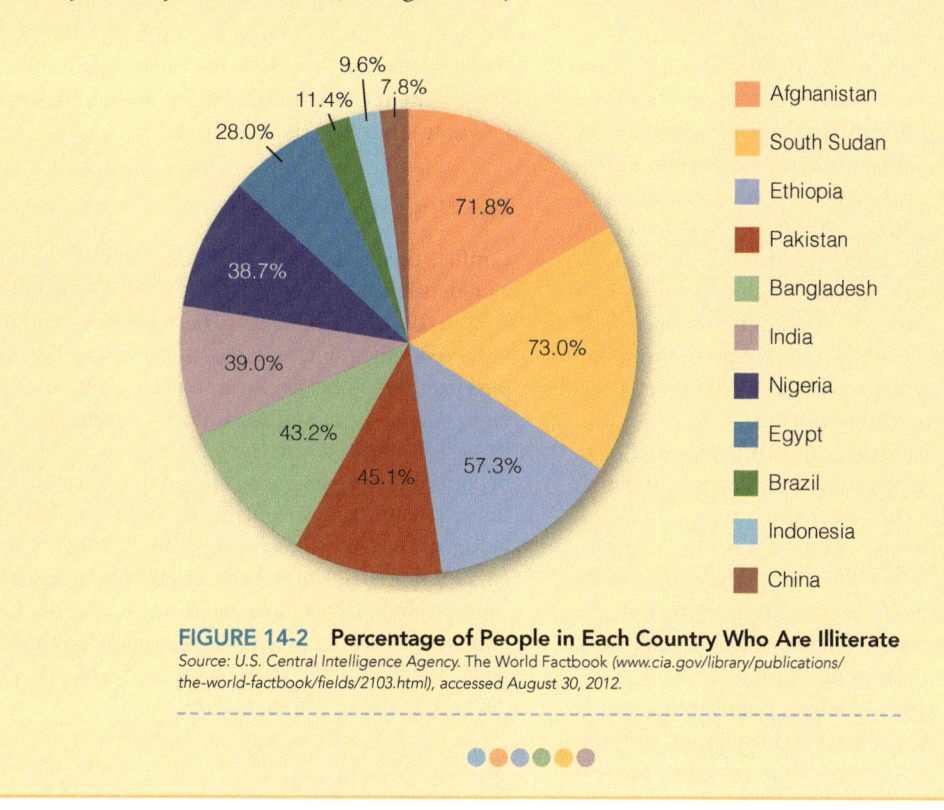

In many respects, the hidden curriculum is a lesson in being docile. For example, an article in *Today's Education,* the journal of the National Education Association, gives an experienced teacher's advice to new teachers:

> During the first week or two of teaching in an inner-city school, I concentrate on establishing simple routines, such as the procedure for walking downstairs. I line up the children, and . . . have them practice walking up and down the stairs. Each time the group is allowed to move only when quiet and orderly.

Social skills are highly valued in American society, and a mastery of them is widely accepted as an indication of a child's maturity. The school is a miniature society, and many individuals fail in school because they are either unable or unwilling to learn or use the values, attitudes, and skills contained in the hidden curriculum. We do a great disservice to these students when we make them feel that they have failed in education when they have, in fact, only failed to conform to the school's socialization standards.

### Screening and Allocation: Tracking

Tracking systems were common in England and Wales until the 1970s, and in Northern Ireland until 2009. Germany also uses a tracking system. In Germany, students' achievements in their last four years of primary school determine the type of secondary school they will be permitted to attend. Tracking has existed in American classrooms since the beginning of the twentieth century.

First introduced in Britain in the 1920s as "streaming," tracking became widespread in the United States after World War II ("Tracking," 2000).

From its beginning, the American school system, in principle, has been opposed to tracking or the stratification of students by ability, social class, and various other categories. Educators saw in compulsory education a way to diminish the grip of inherited social stratification by providing the means for individuals to rise as high as their achieved skills would allow. In the words of Horace Mann, an influential American educator of the late nineteenth century, public education was to be "the great equalizer of the conditions of men." Despite the principles on which the American educational system is based, at least two-thirds of U.S. high schools use tracking.

Although tracking is not as formally structured or as completely irreversible in America as in most other industrial societies, it is influenced by many factors, including socioeconomic status, ethnicity, and place of residence. It is also consistently expressed in the differences between public and private schools as well as in the differences among public schools. (In New York City, for example, there are highly competitive math- and science-oriented and arts-oriented high schools, neighborhood high schools, and vocational high schools.) Of course, tracking occurs in higher education in the selection of students by private colleges and universities, state colleges, and community colleges.

Tracking begins with stratifying students into "fast," "average," and "slow" groups, from first grade through high school. It can be difficult for a student to break out of an assigned category because teachers come to expect a certain level of performance from an individual. The student, sensing this expectation, will often give the level of performance that is expected. In this way, tracking becomes a self-fulfilling prophecy.

In one study of this phenomenon, R. Rosenthal and L. Jacobson (1966) gave IQ tests to 650 lower-class elementary school pupils. Their teachers were told that the test would predict which of the students were the "bloomers" or "spurters." In other words, the tests would identify the superior students. This approach was, in fact, not the one employed. Twenty percent of the students were randomly selected to be designated as superior, even though there was no measured difference between them and the other 80 percent of the school population. The point of the study was to determine whether the teacher's expectations would have any effect on the "superior" students.

At the end of the first year, all the students were tested again. There was a significant difference in the gain in IQ scores between the "superior" group and the control group. This gain was most pronounced among students in the first and second grades. Yet, the following year, when these students were promoted to another class and assigned to teachers who had not been told that they were "superior," they no longer made the sort of gains they had experienced during the previous year.

Nonetheless, the "superior" students in the upper grades continued to gain during their second year, showing that there had been long-term advantages from positive teacher expectations for them. Apparently, the younger students needed continuous input to benefit from the teachers' expectations, whereas the older students needed less.

The most common argument against tracking notes that tracking fosters race and class segregation. Studies have shown that black and Hispanic students are overrepresented in low tracks and underrepresented in high tracks. Critics of tracking argue that it reflects the effects of placement rather than the students' true abilities. Critics also claim that tracking can have a negative impact on self-esteem by separating students into winners and losers. Others counter that throwing all students into a mixed-ability classroom depresses the achievement of high-ability students and frustrates the less able students (Hyland, 2006.)

## The Credentialized Society

Conflict theorists would also argue that we have become a credentialized society (Collins, 1979). A degree or certificate has become necessary to perform a vast variety of jobs. This credential might not necessarily cause the recipient to perform the job better. Even in professions such as medicine, engineering, and law, most knowledge is acquired by performing tasks on the job. However, credentials have become a rite of passage and a sign that a certain process of indoctrination and socialization has taken place. The individual is recognized as having gone through a process of educational socialization that constitutes adequate preparation to hold the occupational status. Therefore, colleges and universities act as gatekeepers, allowing those who are willing to play by the rules to succeed while barring those who might disrupt the social order.

At the same time, advanced degrees are undergoing constant change and becoming less specialized. A law degree from Harvard, Yale, or Columbia is less a measure of the training of a particular candidate than a basis on which leading corporations, major public agencies, and important law firms can recruit those who will maintain the status quo. The degree signifies that the candidate has forged links with the established networks and achieved grades necessary to obtain a degree. See Figure 14-3 for bachelor's degree earnings by field of training.

Colleges and universities are miniature societies more than centers of technical and scientific education. In these environments, students learn to operate within the established order and to accept traditional social hierarchies. In this sense, they provide the power structure with a constantly replenished army of defenders of the status quo. According to this view, those who could disrupt the established order are not permitted to enter positions of power and responsibility (For more on this topic, see "Day-to-Day Sociology: Is a College Degree Worth the Trouble?")

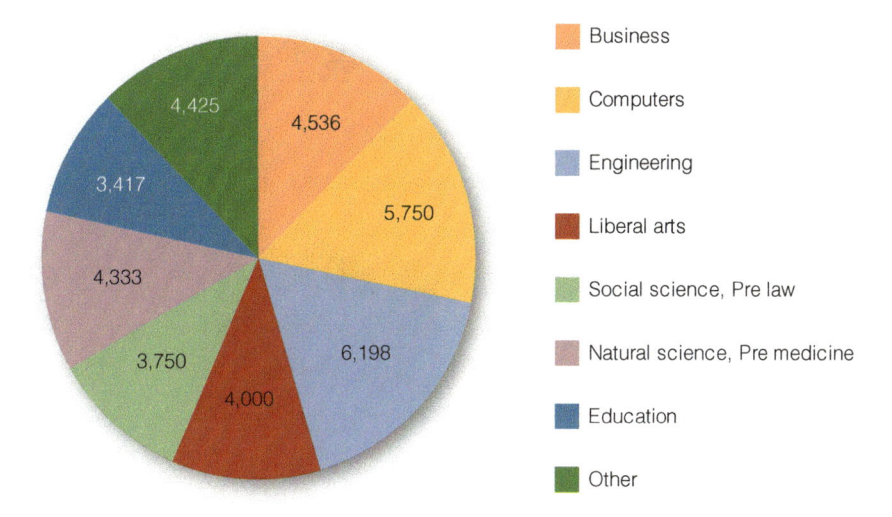

**FIGURE 14-3** Monthly Bachelor's Degree Earnings by Field of Training
*Source: U.S. Bureau of the Census. February 2012. Current Population Reports, "What It's Worth: Field of Training and Economic Status in 2009," accessed August 30, 2012.*

- Business: 4,536
- Computers: 5,750
- Engineering: 6,198
- Liberal arts: 4,000
- Social science, Pre law: 3,750
- Natural science, Pre medicine: 4,333
- Education: 3,417
- Other: 4,425

---

### DAY-TO-DAY SOCIOLOGY

## Is a College Degree Worth the Trouble?

Most high-school students have heard the message that they should go to college. The message certainly seems to be getting through. During the 2012–2013 academic year, 21.6 million students attended colleges and universities, and 1.8 million bachelor's degrees were expected to be awarded (National Center for Education Statistics, 2012).

Should we see this as a great accomplishment? Charles Murray, the author of *Real Education*, does not think so. First of all, he thinks most college students are not smart enough to be there. Only 10–20 percent of college students should really be occupying a class seat. Murray believes college work requires an intelligence quotient (IQ) of at least 115. Only 16 percent of the population scores that high, but 39 percent of all 18- to 24-year-old people are in college. In the 1950s, the average college graduate had an IQ of 115, but today the average would be far below that.

Is it possible for someone with average reading ability to sit through lectures and read textbooks and get something out of that experience? Murray says sure they can, but they do not gain much from it. "They take away a mishmash of half-understood information and outright misunderstandings that probably leave them under the illusion they know something they do not," he claims.

The second point Murray raises is why we think spending four years taking classes makes sense. Who really enjoys reading complex material for hours at a time? To be willing to spend days writing papers or studying for exams seems at least peculiar, if not masochistic.

In addition, four years of taking classes just seems too long. You will have taken at least 32 semester-long courses during those years. How many jobs really require knowledge that can be obtained only by taking all those courses?

What about employers? They want you to have a college degree. Yes, but not because of what you have learned, just the fact that you have one. It is a no-cost screening device for perseverance and fortitude. With many people having a college degree, it makes sense for employers to require one. Employers can limit their hiring pool to college graduates and not waste time and money identifying talented students who do not have a BA.

So we have a dilemma here. College is not going to teach you that much, but employers will not interview you if you do not have a degree. If everyone is supposed to have a college degree, we punish all the people who do not go to college. Bill Gates dropped out of college and never finished his degree. Unless you think you are as talented as Gates, you probably need to get the degree and enjoy the experience as much as you can.

*Source:* Murray, Charles. 2008. *Real Education*. New York: Random House; U.S. Department of Education. March 2009. "Digest of Education Statistics, 2008" (http://nces.ed.gov/Pubsearch/pubsinfo.asp?pubid=2009020), accessed June 14, 2009.

## ISSUES IN AMERICAN EDUCATION

How well have American schools educated the population? The answer depends on the standards one applies. Americans take it for granted that everyone has a basic right to an education and that the state should provide free elementary and high-school classes. The United States pioneered this concept long before similar systems were introduced in Europe.

As we have attempted to provide formal education to everyone, we also have had to contend with a wide variety of problems stemming from the diverse population. In this section, we examine some of the concerns in contemporary American education.

### Unequal Access to Education

American minorities have sought equal access to public schools for two centuries. Tracing those efforts over the generations reveals a pattern of dissatisfaction with integrated as well as segregated schools.

African American parents attributed the ineffective instruction at the schools attended by their children to one of two causes. If the schools were all black, failure was attributed to the racially segregated character of those schools. If whites were attending the schools, African American parents concluded, conditions would be better. This was the dominant theme in the nineteenth and twentieth centuries.

Discontent also has occurred when African American children have attended predominantly white schools. In those instances, the racially integrated character of the school was seen as a problem because white students were thought to be favored by the teachers.

In the 1954 case of *Brown v. Board of Education of Topeka, Kansas,* the Supreme Court ruled that school segregation was illegal. The court held that "in the field of public education, the doctrine of separate but equal has no place. Separate educational facilities are inherently unequal." Segregating African American schoolchildren from white schoolchildren was a violation of the equal-protection clause of the Constitution. However, although the court's verdict banned **de jure segregation,** or *laws prohibiting one racial group from attending school with another,* it had little effect on **de facto segregation,** or *segregation resulting from residential patterns.* For example, minority groups often live in areas of a city where there are few, if any, whites. Consequently, when children attend neighborhood schools, they usually are taught in an environment that is racially segregated.

Ten years after the 1954 ruling, the federal government attempted to document the degree to which equality of education had been achieved. It financed a cross-sectional study of 645,000 children in grades 1, 3, 6, 9, and 12 attending some 4,000 schools nationwide. The results, appearing in James S. Coleman's now-famous

The Coleman study provided evidence that the home environment, the quality of the neighborhood, and the types of friends one has are more influential in school achievement than is the quality of the school facilities or the skill of the teachers.

report, "Equality of Educational Opportunity" (1966), supported unequivocally the conclusion that American education remains largely unequal in most parts of the country, including those where "Negroes form any significant proportion of the population." Coleman noted further that on all tests measuring pupils' skills in areas crucial to job performance and career advancement, not only did Native Americans, Mexican Americans, Puerto Ricans, and African Americans score significantly below whites, but also that the gaps widened in the higher grades.

Now for a subtle but extremely important point: Although wide inequalities of educational opportunity are acknowledged throughout the United States, the discrepancies between the skills of minorities and those of their white counterparts could not be accounted for in terms of how much money was spent on education per pupil, quality of school buildings, number of laboratories or libraries, or even class sizes. In spite of good intentions, a school presumably cannot usually outweigh the influence of the family backgrounds of its individual students and of its student population as a whole. The Coleman study thus provided evidence that schools, per se, do not play as important a role in student achievement as was once thought. It appears that the home environment, the quality of the neighborhood, and the types of friends and associates one has are much more influential in school achievement than is the quality of the school facilities or the skills of teachers.

In effect, then, the areas that schools have least control over—social influence and development—are the most important in determining how well an individual will do in school.

The Coleman report pointed out that lower-class, nonwhite students showed better school achievement when they went to school with middle-class whites. Racial segregation, therefore, hindered the educational attainments of nonwhites.

A direct outgrowth of the Coleman report pointing to the harm of de facto segregation was the busing of children from one neighborhood to another to achieve racial integration in the schools. The fundamental assumption underlying school busing was that it would bring about improved academic achievement among minority groups. Nationwide, many parents, black and white, responded negatively to the idea that their school-age children must leave their neighborhoods. For all practical purposes, busing is no longer a major issue.

One factor that has increased the difficulty of integrating public schools is white flight, the continuing exodus of white Americans by the hundreds of thousands from the cities to the suburbs. White flight has been prompted partly by the migration of African Americans from the South to the inner cities of the North and Midwest since the 1960s, but some authorities strongly maintain that it also is related closely to school-desegregation efforts in the large cities.

For example, in a later view of desegregation attempts (1977), Coleman vastly revised the position in his 1966 report, stating that urban desegregation has in some instances had the self-defeating effect of emptying the cities of white pupils. Some authorities (Pettigrew and Green, 1975) took exception to the Coleman thesis, and others believe that what might appear to be flight is related more directly to the characteristic tendency of the American middle class to be upwardly mobile and constantly seeking a better lifestyle. Even though there is some evidence of a countertrend, in which middle-class whites are beginning to re-gentrify inner cities, this migration does not seem to be abating. Nor have most established communities relinquished the ideal of self-determination as embodied in the right to maintain neighborhood schools.

There are currently many more minority and far fewer white students in our public schools than in the past. In 1991, the Supreme Court allowed for segregated neighborhood schools. Since then there has been a continuing increase in school segregation. About 40 percent of black and Hispanic students attend segregated schools. The average black or Hispanic student attends a school where the majority of students are near or below the poverty line. Residential segregation continues to be the key factor in determining the racial composition of the schools (Orfield, 2009).

Colleges and universities have had mixed results in increasing minority enrollment. Since the mid-1980s, the number of black undergraduates has not increased appreciably. Many qualified students do not apply because their families cannot afford tuition, even though complete aid packages are available. To overcome this problem, many schools have taken an aggressive recruitment stance, believing that once they find qualified candidates, they can persuade them to attend. At Harvard University, 6.2 percent of the freshman class of 2015 was African American. Harvard has the luxury of not needing to turn any student away because of financial need.

Still, thousands of qualified minority students never attend college, and many who do attend fail to graduate. The reasons for low graduation rates include financial problems, poor preparation, and the feeling of being unwelcome. Many cannot afford the loss of income that comes with being a full-time student. For many, family survival depends on the money they contribute. Others are victims of inadequate schools. They simply do not have the skills needed to complete college. Still others drop out because they feel out of place in the predominantly white world of higher education.

### Students Who Speak English as a Second Language

The Department of Education reports that 11.2 million children aged 5 to 17 speak a language other than English at home. Millions more students are classified as having limited English proficiency. The majority of these students are of Hispanic origin. Many of these students are enrolled in remedial language programs (U.S. Department of Education, National Center for Education Statistics, 2012).

The goal of bilingual education is to give immigrant children the opportunity to learn academic subjects in their native languages at the same time as they are learning English. After they become fluent in English, they can join their English-speaking peers at the appropriate grade level. In this way, they do not fall behind other students as they learn English. With bilingual education, they are not made to feel that their native language and culture is holding them back.

Critics of bilingual education contend the United States has a long history of using the education system to help students assimilate into American society. Immersing students in the language and culture of their new society speeds up the assimilation process. Critics see supporters of bilingual education as people determined to preserve the culture and traditions of another society at the expense of the students. They point to high dropout rates and slow progress in transitioning to English-only classes as evidence that this approach is flawed.

In general, the younger foreign-born children are when they enter school, the better they will perform. Immigrant children's chances for succeeding in school also improve in direct relation to their parents' educational attainment and income. The obstacles are especially great for children whose parents are illiterate and cannot help with homework.

Bilingual educational research involves many complicated background factors that affect the outcomes of academic studies. Because of the complex nature of educational studies, no single body of research can accurately account for the impact of bilingual education on various aspects of society.

The passage of No Child Left Behind decreased the emphasis on bilingual education and required that all students be tested yearly in English. English-language proficiency is important today because the work world is more complex than during previous decades. Immigrants now come from a broad range of backgrounds and enter a more complex culture. In the past, less-educated people could find well-paid factory jobs. In today's more complex economy, the poorly educated and illiterate are often unemployed.

## High-School Dropouts

Dropping out of high school has long been viewed as a serious educational and social problem. By leaving high school before graduation, dropouts risk serious educational deficiencies that severely limit their economic and social well-being. Over the past century, the proportion of people in the adult population who have failed to finish high school has decreased substantially. In 1910, the proportion of the adult population (ages 25 and older) that had completed at least four years of high school was 13.5 percent (U.S. Bureau of the Census, *Current Population Survey*, January 2009).

The rate of students dropping out of high school has been declining in recent decades. In 1970, the percentage of 18- to 24-year-olds who dropped out of high school was 17.3 percent. In 2010, that number fell to 7.4 percent. Gender differences between high-school dropouts have also changed in recent years. In the early 1970s, the trend was for girls to drop out of high school at a higher rate than boys did. But in the late 1970s, this trend reversed so that now, the dropout rate for boys is much higher than for girls.

In 2011, 87.6 percent of American adults ages 25 and older had at least completed high school; 30.4 percent had a bachelor's degree or higher. In 1975, 63 percent of adults had a high-school diploma, and 14 percent had obtained a bachelor's degree. Much of the increase in educational attainment levels of the adult population is due to a more educated younger population replacing an older, less-educated population (U.S. Bureau of the Census, "Educational Attainment in the United States: 2011").

Despite these long-term declines in dropout rates, interest in the dropout issue among educators and policymakers has increased substantially in recent years. Legislators and education officials are devoting ever more time and resources to dealing with the issue.

If the long-term trend is that dropout rates are declining, why has the concern for this problem increased lately? First, although the long-term dropout trend has declined, the short-term trend has remained steady and even increased for some groups.

A second reason is that minority populations, who always have had higher dropout rates than whites, are increasing as a proportion of the public high-school population. For white students, the dropout rate is 5.1 percent, but for black students, it is 8.0 percent and for Hispanic students, it is 15.1 percent (U.S. Department of Education, 2012).

Racial and ethnic minorities represent the majority of students enrolled in most large U.S. cities and more than 90 percent of all students in such cities as Newark, New Jersey; Atlanta; and San Antonio (U.S. Bureau of the Census, 2000). Dropout rates are higher for members of racial, ethnic, and language minorities; higher for males than females; and higher for people from the lower socioeconomic classes.

Factors associated with dropping out include low educational and occupational attainment levels of parents, low family income, speaking a language other than English in the home, single-parent families, and poor academic achievement.

The influence of peers is also important, but it has not received much attention in previous research. Many dropouts have friends who are dropouts, but it is not clear to what extent and in what ways a student's friends and peers influence the decision to leave school.

High-school dropouts do more than just damage their employment and earnings potential (see Figure 14-4). Dropping out of high school affects not only those who leave school but also society in general for the following reasons:

1. Dropouts pay less in taxes because of their lower earnings.
2. Dropouts increase the demand for social services, including welfare, medical assistance, and unemployment compensation.
3. Dropouts are less likely to vote.
4. Dropouts have poorer health.
5. Half of all state prison inmates did not complete high school.

Given these facts, it is small wonder that the U.S. Department of Education has focused an increasing amount of attention on how to improve high-school completion rates.

## Violence in the Schools

Nothing undermines the effectiveness of our educational system more than unsafe schools. Throughout the country, students bring to school drugs, guns, knives,

## Median Income by Education Level

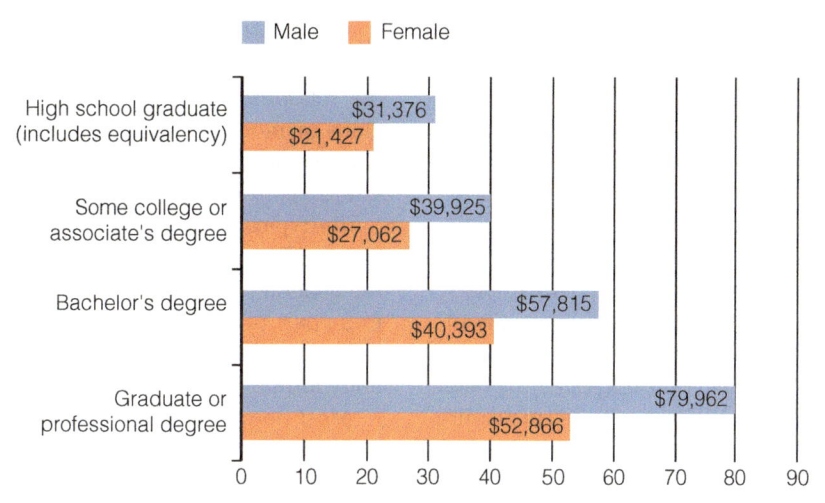

| Education Level | Male | Female |
|---|---|---|
| High school graduate (includes equivalency) | $31,376 | $21,427 |
| Some college or associate's degree | $39,925 | $27,062 |
| Bachelor's degree | $57,815 | $40,393 |
| Graduate or professional degree | $79,962 | $52,866 |

**FIGURE 14-4** **Median Income by Education Level**
*Source: U.S. Bureau of the Census. 2010. American Community Survey.*

---

and other paraphernalia of destruction. Many urban school systems screen students with metal detectors when entering school grounds.

In 2010, students aged 12 through 18 were victims of about 828,000 crimes, including twenty-five homicides. Ten percent of male students in grades nine through twelve reported being threatened or injured with a weapon on school property. Eighteen percent of students in those same grades reported they had carried a weapon anywhere, somewhere, and 6 percent reported they had carried a weapon on school property during the previous thirty days.

Gangs have also become a major problem in many schools. Twenty-three percent of students said their school had gangs. In some schools, where gangs freely sell drugs to students within school buildings, principals vainly chain doors in an effort to keep dealers out and students in. In other schools, students are afraid to use the filthy bathrooms because gang members hang out there. Twenty-two percent of all students in grades nine through twelve reported that someone had offered, sold, or given them an illegal drug on school property in the past twelve months.

The good news is that the percentage of students being victimized at school has declined over the past few years, and the percentage of students who reported being victims of crime at school also has decreased. However, the prevalence of other problem behavior at school has increased. For example, one-quarter of schools report student bullying is a problem. As the rates of criminal victimization in schools have declined or remained constant, however, students seem to feel more secure at school now than just a few years ago. Fewer students are reporting that they avoid places at school for their own safety (U.S. Department of Education, *Indicators of School Crime and Safety*, 2012).

Nothing undermines the effectiveness of our education system more than unsafe schools.

### Home Schooling

Home schooling is emerging as one of the most significant social trends in education. It is an alternative to traditional schooling in which parents assume the primary responsibility for the education of their children. This trend toward an old practice has occurred for a distinctly modern reason: a desire to wrest control from public education and reestablish the family as central to a child's learning.

Home schooling is almost always a matter of choice and not a necessity caused by the unavailability of schools. Public schools are there, but home-schooling families choose not to use them. In the past decade, there has been an explosive growth in this type of schooling and the numbers are still growing. The number of home schoolers is estimated at about 1.5 million students, representing about 2.9 percent of all students (U.S. Department of Education, 2008).

The contemporary home-schooling trend began as a liberal, not a conservative, alternative to the public school. Some families in the late 1950s and early 1960s found schools were too rigid and conservative. They instead wanted to pursue a more liberal philosophy of education as advocated by educators such as John Holt, the author of *Why Children Fail*. Holt suggested the best learning took place when children were allowed to pursue their own interests without an established curriculum.

Conservative and religious families joined the home-schooling trend in the 1980s when they decided public schools were undermining their values. Some believed religious duty required them to teach their own children; others sought to integrate religion, learning, and family life. Joining the liberal and conservative wings of the home-schooling movement are families who simply seek the highest-quality education for their child, which they believe public and even private schools can no longer provide.

For centuries, children have learned outside formal school settings. Compulsory schooling is relatively new.

> Not until the nineteenth century did state legislatures begin requiring local governments to build schools and parents to enroll their children in them. Even then, compulsory requirements extended to only a few months a year.... Only recently have we begun to treat schooling as a full-time affair entrusted to professional teachers. Yet in such a short span of time, most of the nation has come to accept classroom schooling as the norm, and so the recent upsurge in home schooling has come to many as a surprise. (Lines, 2000)

Approximately 1.5 million students are being educated at home, up from 15,000 in the early 1980s.

There are three main reasons parents choose to home school. Two-thirds of the parents are concerned about the school environment. Others want to provide religious or moral instruction. Still others are dissatisfied with the academic instruction available at the available schools (U.S. Department of Education, 2008).

The typical home-schooling family is religious, conservative, white, middle-income, and better educated than the general population. Home schoolers are more likely to be part of a two-parent family, and the mother typically assumes the largest share of the teaching responsibility, although fathers almost always are involved also.

Most home-schooling children spend time at libraries, museums, or classes offered at a local public school. Normally, parents plan and implement the learning program. The Internet has provided an important resource for these parents and enables them to share information on books and learning opportunities.

Perhaps the largest impact of home schoolers has been the entry of new educational organizations into the field. Many private organizations and enterprises have entered the K–12 distance education field with their sights set on home schoolers as a primary audience (Hill, 2000). The State of Florida has developed an extensive set of courses that can be taken by home schoolers over the Internet for high-school credit and others who choose to use this resource, and Illinois is developing a similar program (Carothers, 2000; Trotter, 2001). Meanwhile, several for-profit ventures have entered the field, offering courses and, in one case, accredited diplomas over the Internet.

How well do home schoolers do? Colleges have been fairly open to admitting home-schooled students. Home schoolers have been admitted to more than 900 colleges and universities, including highly selective universities and the U.S. military academies (Cholo, 2007).

The most frequent criticism of home schooling is that the children are not socialized to deal with diversity and the social aspects of life. But there is plenty of evidence to the contrary. Almost all home-schooled children participate in extracurricular activities. Many public schools allow home schoolers to participate in team sports, science labs, or social organizations (Hammons, 2001).

Other criticisms focus on the inadequate standards of academic quality among the parents. Home schooling also has the potential to produce religious and social extremism and can also reduce funding for public schools.

The success or failure of home schooling depends on the success or failure of the family's interpersonal relationships. Home schooling is a complex issue and represents a tremendous commitment on the part of the parents. In most cases, the father is the sole earner in the family and the mother spends her time instructing the children.

More research on home schooling is necessary. Up to now, research has been limited to case studies of families or self-reports from participants in home schooling. We need a more accurate and thorough assessment of this growing trend in education (U.S. Department of Education, *Issue Brief,* December 2008).

## Standardized Testing

In American schools, the standardized test is the most frequently used means of evaluating students' aptitudes and abilities. Every year, more than 100 million standardized tests are administered, ranking the mental talents of students from nursery school to graduate school.

Children encounter standardized tests almost from the first day they go to school. Usually, their first experience with testing is an intelligence test. These are given to more than 2 million youngsters each year. Students are also required to take a number of achievement tests, beginning in elementary school. High-school and college seniors take college admissions tests that decide whether they will be accepted at universities and graduate schools, respectively.

Much criticism has been leveled at standardized tests. The testing services say the tests merely try to chart, scientifically and objectively, different levels of mental achievement and aptitude. The critics assert that the tests are invalid academically and biased against minorities.

The Educational Testing Service (ETS) Scholastic Aptitude Test (SAT) is the best-known college admissions test and is required by about 1,200 U.S. colleges and universities. Another 2,800 American colleges require or recommend the American College Test (ACT). Students wishing to go to graduate school are required to take other exams, which measure the ability and skills used in the fields that they wish to enter.

The ETS professes to be meticulous in its test construction. It hires college students, teachers, and professors to assist its staff in writing questions. Each of the approximately 3,000 questions created each year are reviewed by about fifteen people for style, content, or racial bias.

The criticism of standardized tests, however, continues to grow. Many assert that all standardized tests are biased against minorities. The average African American or Hispanic youngster encounters references and vocabulary on a test that are likely to be more familiar to white, middle-class students. Many others oppose the secrecy surrounding the test companies. Groups have pushed for truth in testing, meaning that the test makers must divulge all exam questions and answers shortly after the tests are given. This would enable people to evaluate the tests more closely for cultural bias and possible scoring errors. The testing industry opposes such measures, which would force it to create totally new tests for each administration without the possibility of reusing valid and reliable questions.

No one would contend that standardized tests are perfect measuring instruments. At best, they can provide an objective measure to be used in conjunction with teachers' grades and opinions. At worst, they might discriminate against minorities or incorrectly measure potential ability. Nonetheless, college admissions officers insist that results from standardized college admissions tests give them a significant tool for evaluating students from a variety of backgrounds and many parts of the country (see Figure 14-5).

## The Gifted

The very term *gifted* is emotionally loaded. The word can evoke feelings that range from admiration to resentment and hostility. Throughout history, people have displayed

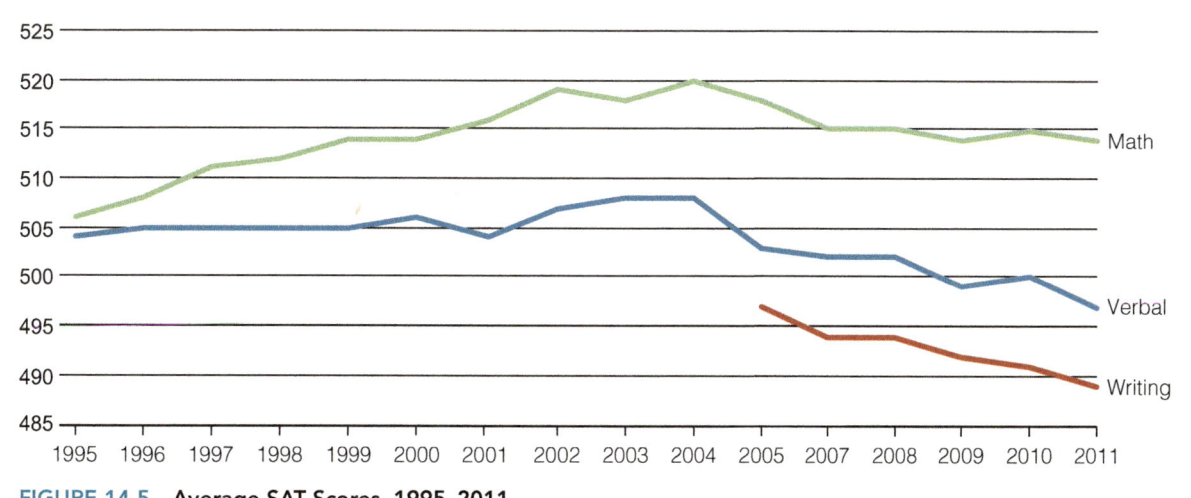

**FIGURE 14-5**  **Average SAT Scores, 1995–2011**
*Source: The College Board, 2012.*

a marked ambivalence toward the gifted. It was not unusual to view giftedness as either divinely or diabolically inspired. Genius was often seen as one aspect of insanity. Aristotle's observation, "There was never a great genius without a tincture of madness," continues to be believed as common folklore.

People also tend to believe that intellectualism and practicality are incompatible. That belief is expressed in such sayings as, "He (or she) is too smart for his (her) own good," or, "It's not smart to be too smart." High intelligence is often assumed to be incompatible with happiness.

There is little agreement on what constitutes giftedness. The most common measure is performance on a standardized test. All those who score above a certain level are defined as gifted, although there are serious problems when this criterion alone is used. Arbitrary approaches to measuring giftedness tend to ignore the likelihood that active intervention could increase the number of candidates among females, minorities, and the disabled, groups that are often underrepresented among the gifted.

Ellen Winner (1996) has proposed that gifted children have three atypical traits. These include:

1. *Precociousness.* Gifted children begin early to master some domain.
2. *Nonconformity.* Gifted children insist on doing things according to their own specific rules.
3. *A rage to master.* Gifted children exhibit a passion to know everything there is to know about a subject.

Females tend to be underrepresented among the gifted because popular culture holds that high intelligence is incompatible with femininity; thus, some girls quickly learn to deny, disguise, or repress their abilities. Minorities are hindered because commonly used assessment tools discriminate against ethnic groups whose members have had different cultural experiences or speak English as a second language. The intellectual ability of disabled youngsters is often overlooked. Their physical handicaps can mask or divert attention from their mental potential, particularly when communication is impaired, because this is a key factor in assessment procedures.

Teachers often confuse intelligence with unrelated school behaviors. Children who are neat, clean, and well mannered, have good handwriting, or manifest other desirable but irrelevant classroom traits can often be thought very bright. Teachers often associate giftedness with children who come from prominent families, have traveled widely, and have had extensive cultural advantages. They are likely to discount high intelligence when it is present in combination with poor grammar, truancy, aggressiveness, or learning disabilities.

The first attempt to deal with the gifted in public education took place in the St. Louis schools in 1868. The program involved a system of flexible promotions enabling high-achieving students to advance from grade to grade at their own pace. By the early 1900s, special schools for the gifted began to appear.

There has never been a consistent, cohesive national policy or consensus on how to educate the gifted. Special programs that have been instituted have reached only a small fraction of those who conceivably could benefit from them. A serious problem with the education of the gifted arises from philosophical considerations. Many teachers are reluctant to single out the gifted for special treatment because they feel that the children are already naturally privileged. Sometimes, attention given to gifted children is seen as antidemocratic.

No matter how inadequate it can seem, the effort to provide for the educational needs of learning-disabled children has far exceeded that expended for the gifted. Similarly, the time and money spent on research into educating the slower children far outstrip that set aside for research on materials, methodology for teaching, and so on for the gifted.

When schools do have enrichment programs, they are rarely monitored for effectiveness. Enrichment programs are often provided by teachers totally untrained in dealing with the gifted because it is assumed that anyone qualified to teach is capable of teaching the gifted. Yet, most basic teacher-certification programs do not require even one hour's exposure to information on the theory, identification, or methodology of teaching such children. Most administrators do not have the theoretical background or practical experience necessary to establish and promote successful programs for the gifted.

The implementation of the No Child Left Behind program did not require any programs for gifted, talented, and other high-performing students. Federal funding of gifted education decreased by a third over the law's first five years.

There is some evidence that the nation's population of gifted children—and, possibly, prodigies—is growing. Researchers who test large numbers of children have detected a startling proportion in the 170 to 180 IQ range.

Although psychologists agree that early exceptional ability should be nurtured to thrive, they do not necessarily think the current movement to produce super-babies by force-feeding a diet of mathematics and vocabulary to infants is a good idea. Pediatricians have begun seeing children with backlash symptoms—headaches, stomachaches, hair-tearing, anxiety, depression—as a result of this pressure to perform.

History has shown that being an authentic child prodigy creates problems enough of its own. The fine line between nurturing genius and trying to force a bright but not brilliant child to be something he or she is not is clearly one that must be walked with care.

Karen Stone McCown (1998), working with a group of Nobel prize winners, found many reported that their social-emotional development was shortchanged. They said that they were so self-motivated to pursue their intellectual passions that almost nothing would have stopped that work, but missing from their lives were the social skills that would help them interact with and connect to family, friends, and the larger world.

It appears that more than 2.5 million schoolchildren in the United States can be described as gifted, or about 3 percent of the school population. Giftedness is essentially potential. Whether these children will achieve their potential intellectual growth will depend on many factors, not the least of which is the level of educational instruction they receive. Several universities have developed programs for gifted children. These include the Johns Hopkins University Center for Talented Youth and Stanford University's Education Program for Gifted Youth. Both provide course materials for students, including both self-paced courses with tutor support and online classroom-based courses.

## SUMMARY

**How do structural functionalists view education?**

Functionalist sociologists believe education consists of activities that are functional for the society as a whole. One of the manifest functions of education is socialization. Another function of education is cultural transmission, in which major portions of society's knowledge are passed from one generation to the next. A third function of schools is to equip children with the academic skills needed to function as adults.

**Are there latent functions in education also?**

In addition to its manifest functions, schooling in America has developed a number of unintended consequences as well. One latent function of many public schools is to provide child care outside of the nuclear family. This function has become increasingly important in recent years with the growing number of women in the labor force and the dramatic increase in single-parent families. A related service is to provide students with at least one nutritious meal a day.

**What is the relationship between education and socialization?**

In nonindustrial societies, no real distinction is made between education and the socialization that occurs within the family.

**How do conflict theorists view education?**

Conflict theorists view society as an arena of conflict in which certain groups dominate others, and social institutions become the instruments by which those in power can control the less powerful. Conflict theorists see education as a means for maintaining the status quo by producing the kinds of people the system needs. This is accomplished through teaching the hidden curriculum—attitudes and values that prepare children to accept the requirements of adult life and to fit into the social, political, and economic statuses the society provides.

**What are some of the major issues in education today?**

Dropout rates—Although the overall high-school graduation rate has been increasing, the dropout rate for minorities has remained high. Factors associated with dropping out include low educational and occupational attainment levels of parents, and low family income.

English proficiency—The Department of Education reports that 11.2 million children aged 5 to 17 speak a language other than English at home. Millions more students are classified as having limited English proficiency. The majority of these students are of Hispanic origin.

Violence in schools—Violence causes many urban schools to operate under a siege mentality, with doors chained shut and students afraid to be in the wrong place at the wrong time, inside of school or out.

Standardized testing—In American schools, the standardized test is the most frequently used means of evaluating student aptitude, ability, and performance. Critics assert that these tests are academically invalid and culturally biased against minorities and the lower class.

### Media Resources

**CourseMate for *Introduction to Sociology*, Eleventh Edition**

Cengage Learning's Sociology CourseMate brings course concepts to life with interactive learning, study, and exam preparation tools that support the printed textbook. Access an integrated eBook, learning tools including glossaries, flashcards, quizzes, videos, and more in your Sociology CourseMate. Go to CengageBrain.com to register or purchase access.

# CHAPTER FOURTEEN STUDY GUIDE

## Key Concepts

Match each of the following concepts with its definition, illustration, or explanation.

a. cultural transmission
b. "Nation at Risk"
c. No Child Left Behind
d. hidden curriculum
e. tracking
f. the credentialized society
g. *Brown v. Board of Educ./Topeka*
h. de jure segregation
i. de facto segregation
j. white flight

\_\_\_\_ 1. Supreme Court decision that ended de jure segregation in public schools
\_\_\_\_ 2. Separation of races based on laws prohibiting interracial contact
\_\_\_\_ 3. Separation of races resulting from residential housing patterns
\_\_\_\_ 4. The social attitudes and values taught in school that prepare children to accept the requirements of adult life and to "fit into" the social, economic, and political statuses the society provides
\_\_\_\_ 5. The migration of large numbers of white Americans from the central cities to the suburbs
\_\_\_\_ 6. The process in which major portions of a society's knowledge are passed from one generation to the next
\_\_\_\_ 7. The increasing trend in the United States for more and more jobs to require a degree regardless of whether possession of a degree increases job performance
\_\_\_\_ 8. President George W. Bush's education plan requiring extensive standardized testing and allowing parents of children in inferior schools to transfer the children to another school
\_\_\_\_ 9. The stratification of students by ability, social class, and various other categories
\_\_\_\_ 10. 1983 report detailing a "rising tide of mediocrity" in U.S. education

## Key Thinkers

Match the thinkers with their main idea or contribution.

a. James Coleman
b. Charles Murray
c. R. Rosenthal and L. Jacobson
d. Herbert Spencer
e. Lester Frank Ward

\_\_\_\_ 1. Conducted a famous study on the effect teachers' expectations had on student performance
\_\_\_\_ 2. Questioned the value of encouraging large numbers of people to get a college or university degree
\_\_\_\_ 3. Author of several books criticizing public education for shortchanging the poor, especially black and Hispanic children
\_\_\_\_ 4. Social Darwinist who argued that little could be done to help those at the bottom of the social class ladder and that a school curriculum should be judged on its practical value
\_\_\_\_ 5. Sociologist who maintained that the purpose of education was to equalize society

## Central Idea Completions

Fill in the appropriate concepts and descriptions for each of the following questions.

1. Use your own high school as an example and describe the manifest and latent functions of education.

   a. Manifest: _____

   b. Latent: _____

CHAPTER 14   EDUCATION   353

2. Higher education can be seen as job postponement. What consequences does this function have for the individual and the society?

   a. Individual: _____

   b. Society: _____

3. What is standardized testing? What are the purposes that its proponents claim for it? What are some of the problems with standardized testing?

   a. Definition: _____

   b. Purposes: _____

   c. Problems: _____

4. Who home schools and why? Why has home schooling become more popular in recent years?

   a. Characteristics of home schoolers: _____

   b. Popularity factors: _____

5. What are the major factors contributing to the dropout rate for high-school students?

   _____
   _____
   _____

6. To what extent do American minorities continue to face unequal access to education?

   _____
   _____
   _____

7. How would a conflict theorist evaluate education in American public high schools?

   _____
   _____
   _____

## ● Critical Thought Exercises

1. As an agent of socialization, the school fosters the attitudes, behaviors, norms, and values of the dominant society. But students themselves are also agents of socialization. Does student culture encourage the same attitudes and behaviors as those promoted by the school? In your high school, how did student culture differ from official school culture, and what forms of social control did each group use to further its goals and ideas?

2. Do schools "shortchange" girls? In what ways is primary and secondary education slanted in favor of boys? But if the education system favors boys, why do more girls go on to college? In what ways does the system favor girls?

3. What implications does the Rosenthal-Jacobson study have for the policy of "tracking"?

## Internet Activities

1. The National Center for Education Statistics (http://nces.ed.gov/surveys/) is a good gateway to a wealth of data on education. One link ("Data and Tools") takes you to a page where you can create your own table including only the variables in which you are particularly interested. Take a look at the Postsecondary survey link. Choose one of the studies, and discuss what you found interesting about it.

2. *American Educator* is the magazine of the AFT (the nation's leading teacher's union). The website (www.aft.org/newspubs/periodicals/ae/) has articles about primary and secondary education. What are three of the current issues this group is concerned about?

3. For a view the AFT probably would disagree with, the Center for Education Reform (www.edreform.com/) provides support for charter schools, No Child Left Behind, and other conservative positions. It frequently criticizes the AFT. Their "Issue Accuracy" link takes you to articles supporting their own view on current controversies in education. Is their position biased? Why or why not?

## Answers to Key Concepts

1. g;  2. h;  3. i;  4. d;  5. j;  6. a;  7. f;  8. c.;  9. e;  10. b

## Answers to Key Thinkers

1. c;  2. a;  3. b;  4. d;  5. e

# 15 Political and Economic Systems

### Politics, Power, and Authority
- Power
- Political Authority

### Government and the State
- Functions of the State

### Types of States
- Autocracy
- Totalitarianism
- Democracy

### Functionalist and Conflict Theory Views of the State

### The Economy and the State
- Capitalism

*Day-to-Day Sociology:* Eat Your Fresh Fruit and Vegetables or Pay a Fine
- The Marxist Response to Capitalism
- Socialism
- The Capitalist View of Socialism
- Democratic Socialism

### Political Change
- Institutionalized Political Change
- Rebellions

*Our Diverse World:* Does Suicide Terrorism Make Sense?
- Revolutions

### The American Political System
- The Two-Party System

*Thinking About Social Issues:* I Know It's Not True, But I'm Not Voting for Him Anyway
- Voting Behavior
- African Americans as a Political Force
- Hispanics as a Political Force
- The Role of the Media
- Special-Interest Groups

### Summary

## LEARNING OBJECTIVES

After studying this chapter, you should be able to do the following:

- Distinguish between authority and coercion.
- Understand the basic functions of the state.
- Know the basic features of capitalism.
- Distinguish between capitalism, socialism, and democratic socialism.
- Describe the basic features of political democracy.
- Contrast the functionalist and conflict theory views of the state.
- Describe the major features of the American political system.

After every presidential election, we require the loser to engage in a public declaration of defeat. Why do we require this invasion of privacy? Why do we not allow the defeated candidate to suffer defeat away from the glare of the media? Could it be that this ritual serves an important social function?

Dwelling on defeat contradicts a basic American commitment to success. After a presidential election, our gaze is on the triumphant winner; but instead of drawing a veil of silence over the crushed hopes of the losing candidate, we require one final ordeal, the concession speech.

The words themselves are less important than the larger purpose they serve. They ritualize the passing of power and the legitimacy of the new authority. As Mitt Romney noted a few hours after being defeated for president:

> The nation, as you know, is at a critical point. At a time like this, we can't risk partisan bickering and political posturing. Our leaders have to reach across the aisle to do the people's work.
>
> And we citizens also have to rise to the occasion. We look to our teachers and professors, we count on you not just to teach, but to inspire our children with a passion for learning and discovery.
>
> We look to our pastors and priests and rabbis and counselors of all kinds to testify of the enduring principles upon which our society is built: honesty, charity, integrity and family.
>
> We look to our parents, for in the final analysis everything depends on the success of our homes.
>
> We look to job creators of all kinds. We're counting on you to invest, to hire, to step forward.
>
> And we look to Democrats and Republicans in government at all levels to put the people before the politics.
>
> I believe in America. I believe in the people of America.

Throughout the world, the yielding of power is often a matter of life or death. The concession speech is not merely a report of an election result; it is a reaffirmation of the democratic process (Corcoran, 1994). It affirms the view that the control of the political process must be in the hands of the people and that we are quite fearful of concentrated political power.

The founders of the United States distrusted a strong unified government. The U.S. Constitution, the oldest of its kind in the world, established a divided form of government. There was to be a president, two houses of Congress, and a federal high court. These actions represented a deliberate decision to create a weak political system. There were to be varying terms of office. The president was to be chosen every four years. Two senators from each state were to be chosen for six-year terms, with one-third of the seats open every two years. Members of the House of Representatives were to be elected every two years, with the number allotted to each state roughly proportional to its share of the national population (Lipset, 1996).

Thomas Jefferson thought that even this arrangement had to be reexamined frequently. He suggested that every two decades or so, a new generation should be required collectively to define its own values and redefine those of its forebears. Jefferson believed that a political system had to be taken out periodically, inspected, examined in the harsh daylight, and changed before it was bequeathed to a new generation. If it did not pass muster, then some other political system had to be found that could guarantee life, liberty, and the pursuit of happiness better. A new generation should not be burdened by the traditions of the past or the comfort of well-worn customs (Hart, 1993).

The founders of the United States realized that, in most societies, what laws are passed or not passed depend to a large extent on which categories of people have the power. The powerful in a society work hard to pass laws to their liking. The founders were trying to provide a prescription for how power was to be used and how it was to be passed from one generation to the next. It sounds like a fairly radical and idealistic view of how a country should be governed. However, the United States was born out of radical and idealistic conceptions of politics.

In this chapter, we begin by examining the political institution. Ultimately, the economy is intimately tied to the political system, and we will examine that connection also. **Politics** is *the process by which power is distributed and decisions are made.* This chapter clarifies what is unique about our two-party system and where the American political system fits into the whole spectrum of political institutions. We then move on to examine the relationship between the political system and the economic system.

## POLITICS, POWER, AND AUTHORITY

More than 800 candidates run for the U.S. House of Representatives every other year. Americans also select senators, governors, and a host of other officials on a regular basis. Candidates ring our doorbells, shake our hands, stuff our mailboxes, and exhort us through our television sets. They make promises they often cannot keep. This is politics, American style. Small wonder it has been said that politics, like baseball, is the great American pastime. Running for president of the United States is a political activity; so is enacting legislation; so is taxing property owners to subsidize digging sewers; so is going to war. The study of the political process, then, is the study of power.

### Power

Max Weber (1958a) referred to **power** as *the ability to carry out one person's or group's will, even in the presence of resistance or opposition from others.* In this sense, power is the ability to make others comply with one's decisions, often through the threat or actual use of sanctions, penalties, or force.

In some relationships, the division of power is spelled out clearly and defined formally. Employers have specific powers over employees, army officers over enlisted personnel, ship captains over their crews, professors over their students. In other relationships, the question of power is defined less clearly and can even shift back and forth, depending on individual personalities and the particular situation, between wife and husband, among sisters and brothers, or among friends in a social clique.

Power is an important part of many types of relationships that covers a broad spectrum of interactions. At one pole is **authority**, *power that is regarded as legitimate by those over whom it is exercised, who also accept the authority's legitimacy in imposing sanctions or even in using force if necessary.* For example, in the United States, few people are eager to pay income taxes, yet most do so regularly. Most taxpayers accept the authority of the government not only to demand payment but also to impose penalties for nonpayment.

At the other extreme is **coercion**, *power that is regarded as illegitimate by those over whom it is exerted.* The people comply because they fear reprisals that are not considered acceptable or legitimate. Power based on authority is quite stable, and obedience to it becomes a social norm. Power based on coercion, in contrast, is unstable. People will obey only out of fear, and any opportunity to test this power will be taken. Power based on coercion will fail in the long run. The American Revolution, for example, was preceded by less and less acceptance of the legitimacy of the existing system. The authority of the king of England was questioned, and his power, based increasingly on coercion rather than on acceptance as a social norm, inevitably crumbled.

### Political Authority

An individual's authority often will apply only to certain people in certain situations. For example, a professor has the authority to require students to write term papers but no authority to demand the students' votes should he or she run for public office.

In the same sense, Weber pointed out that the most powerful states do not impose their will by physical force alone but by ensuring that their authority is seen as legitimate. In such a state, people accept the idea that the allocation of power is as it should be and that those who hold power do so legitimately. Weber (1957) identified three kinds of authority: legal-rational authority, traditional authority, and charismatic authority.

*Legal-Rational Authority* **Legal-rational authority** *is derived from the understanding that specific individuals have clearly defined rights and duties to uphold and implement rules and procedures impersonally.* Indeed, that is the key. Power resides not in individuals but in particular positions or offices. There usually are rules and procedures designed to achieve a broad purpose. Rulers acquire political power by meeting requirements for office, and they hold power only as long as they obey the laws that legitimize their rule.

*Traditional Authority* **Traditional authority** *is rooted in the assumption that the customs of the past legitimate the present,* that things are as they always have been and basically should remain that way. Usually, both rulers and ruled recognize and support the tradition that legitimizes such political authority. Typically, traditional authority is hereditary, although this is not always the case. For example, throughout most of English history, the English crown was the property of various families. As long as tradition is followed, the authority is accepted.

*Charismatic Authority* **Charismatic authority** *derives from a ruler's ability to inspire passion and devotion among followers.* Weber noted that a charismatic leader—who is most likely to appear during a period of crisis—will emerge when followers (1) perceive a leader as somehow supernatural, (2) blindly believe the leader's statements, (3) unconditionally comply with the leader's directives, and (4) give the leader unqualified emotional commitment. Others (Willner, 1984) have added that charismatic leaders also must perform seemingly extraordinary feats and have outstanding speaking ability.

Sitting Bull and Red Cloud, for example, were charismatic leaders of the Sioux Indians. Their people followed them because they led by example and inspired personal loyalty. However, individuals were free to disagree, to refuse to participate in planned undertakings, and even

The power of a charismatic leader derives from the ruler's force of personality and ability to inspire passion and devotion among followers. President John F. Kennedy was the closest the United States has come to having a charismatic leader.

to leave and look for a group led by people with whom they were more likely to agree (Brown, 1971). This was not true in Russia under V. I. Lenin, in Germany under Adolf Hitler, or in Iran under the Ayatollah Ruholla Khomeini. These men all were charismatic rulers but also had the political authority necessary to enforce obedience or conformity to their demands.

Charismatic authorities and rulers emerge when people lose faith in their social institutions. Lenin led the Russian Revolution in the chaos left in the wake of World War I. Hitler rose to power in a Germany that had been defeated and humiliated in World War I and whose economy was shattered; inflation was so bad that money was almost worthless. Khomeini rose to power in a country in which rapid modernization had undercut traditional Islamic norms and values, in which great poverty and great wealth existed side by side, and in which fear of the previous leader's secret police left the people in constant fear for their personal safety.

The great challenge facing all charismatic rulers is to sustain their leadership after the crisis subsides and to create political institutions that will survive their death or retirement. Weber pointed out that if the program that the leader has implemented is to be sustained, the leader's charisma must be routinized in some form. For example, after Christ's death—and after it became apparent that his return to earth was not imminent—the apostles began to set up the rudiments of a religious organization with priestly offices.

## GOVERNMENT AND THE STATE

Governments vary according to the relationship that exists between the rulers and the ruled. In some societies, political power is shared among most or all adults. Isolated groups still exist in which the group is its own authority and decisions are made by a consensus among adults. In such societies, the concept of government is meaningless. However, in larger, more complex societies, government does exist.

In complex societies, the **state** is *the institutionalized way of organizing power within territorial limits.* The existence of the state shows that a high level of social and political development is present.

### Functions of the State

Although a preindustrial society can exist without an organized government, no modern industrial society can thrive without the functions the state performs: establishing laws and norms, providing social control, ensuring economic stability, setting goals, and protecting against outside threats.

*Establishing Laws and Norms*  The state establishes laws that formally specify what the society expects and what it prohibits. The laws are often a codification of specific norms; for example, one should not steal from or commit violent crimes against others. Establishing laws also means exacting penalties for violating the laws.

*Providing Social Control*  In addition to establishing laws, the state also has the power to enforce them. The police, courts, and various government agencies make sure that violators of those laws are punished. In the United States, the Internal Revenue Service seeks out tax evaders, the courts sentence criminals to prison, and the police attempt to maintain order.

*Ensuring Economic Stability*  In the modern world, no individual can provide entirely for his or her own needs. Large workforces must be mobilized to build roads, dig canals, and erect dams. Money must be minted, and standards of weights and measures must be set and checked; merchants must be protected from thieves and consumers from fraud. The state tries to ensure that a system for distributing goods and resources exists within the society.

*Setting Goals*  The state sets goals and provides a direction for the society. If a society is to limit the use of oil, for instance, the government must promote this as a goal. It must encourage conservation and the search for

In a democracy, groups must be free to try to change the laws.

alternative energy sources and must discourage (perhaps through taxation or rationing) the use of oil. How is the government able to accomplish these tasks? How can it bring about individual and organizational compliance? Obviously, it would be best if the government could rely on persuasion alone, but this of course is seldom enough. In the end, the government usually needs the power to compel compliance.

*Protecting Against Outside Threats* History leaves little doubt that the rise of the state was accompanied by the increased likelihood of war (Otterbein, 1970, 1973). As early as the fourteenth century, Ibn Khaldun (1958), an Islamic scholar, noted this connection and even attributed the rise of the state to the needs of sedentary farmers to protect themselves from raids by fierce nomads. His views were echoed by Ludwig Gumplowicz (1899): "States have never arisen except through the subjugation of one stock by another, or by several in alliance." In any event, it is clear that one of the tasks of maintaining a society is to protect it from outside threats, especially from hostile militaries. Hence, governments build and maintain armies.

Although there is widespread agreement that the functions just described are tasks that the state should and usually does perform, not all social scientists agree that the state emerged because of the need for those functions.

## TYPES OF STATES

Different types of states exist side by side and must deal with one another constantly in today's shrinking world. To comprehend their interrelationships, it is helpful to understand the structure of each main form of government: autocracy, totalitarianism, and democracy.

### Autocracy

In an **autocracy,** *the ultimate authority and rule of the government rest with one person, who is the chief source of laws and the major agent of social control.* For example, the pharaohs of ancient Egypt were autocrats. Contemporary examples of autocratic countries include Cuba, Vietnam, Laos, North Korea, Libya, Sudan, Eritrea, Tajikistan, Saudi Arabia, and Yemen.

In an autocracy, the loyalty and devotion of the people are required. To ensure that this requirement is met, dissent and criticism of the government and the person in power are prohibited. The government controls the media and can use terror to prevent or suppress dissent. For the most part, however, no great attempt is made to control the personal lives of the people. A strict boundary is set up between people's private lives and their public behavior. Individuals have a wide range of freedom in pursuing family concerns, and many other traditional elements of life. At the same time, virtually all present-day autocracies, even those professing to be communist, have witnessed exploitation of the poor by the rich and powerful.

### Totalitarianism

In a **totalitarian government,** *one group has virtually total control of the nation's social institutions.* Any other group is prevented from attaining power. Religious institutions, educational institutions, political institutions, and economic institutions all are managed directly or indirectly by the state. Typically, under totalitarian rule, several elements interact to concentrate political power.

1. *A single political party* controls the state apparatus. It is the only legal political party in the state. The party organization is itself controlled by one person or by a ruling clique.
2. *Terror* is implemented by an elaborate internal security system that intimidates the populace into conformity. It defines dissenters as enemies of the state and often chooses, arbitrarily, whole groups of people against whom it directs especially harsh oppression (for instance, the Jews in Nazi Germany or minority tribal groups in several recently created African states).
3. *Control of the media* (television, radio, newspapers, and journals) is in the hands of the state. Differing opinions are denied a forum. The media communicate only the official line of thinking to the people.
4. *Control over the military apparatus,* both the military personnel and the use of its weapons, is monopolized by those who control the political power of the totalitarian state.

5. *Control of the economy* is wielded by the government, which sets goals for the various industrial and economic sectors and determines both the prices and the supplies of goods.
6. *An elaborate ideology,* in which previous sociopolitical conditions are rejected, legitimizes the current state and provides more or less explicit instructions to citizens on how to conduct their daily lives. This ideology offers explanations for nearly every aspect of life, often in a simplistic and distorted way (Friedrich and Brzezinski, 1965).

### Democracy

Democracy has not always been regarded as the best form of government. People have often approved of the aims of democracy yet have argued that democracy was impossible to attain. Others argued that it was logically unsound. Today, however, there is hardly a government anywhere in the world that does not claim to have some sort of democratic authority. In the United States, we regard our political system as democratic, and the same claim is made by leaders in communist countries. The word *democracy* seems to have so many different meanings today that we face the problem of distinguishing democracy from other political systems.

*Democracy* comes from the Greek words *demos,* meaning "people," and *kratia,* meaning "authority." By democracy, then, the Greeks were referring to a system in which rule was by the people rather than by a few selected individuals. Because of the growth in population, industrialization, and specialization, it has become impossible for citizens to participate in politics today as they did in ancient Athens. Today, **democracy** refers to *a political system operating under the principles of constitutionalism, representative government, majority rule, civilian rule, and minority rights.*

**Constitutionalism** *means that government power is limited.* It is assumed that there is a higher law, which is superior to all other laws. The various agencies of the government can act only in specified legal ways. Individuals possess rights, such as freedom of speech, press, assembly, and religion, which the government cannot take away.

A basic feature of democracy is that it is rooted in **representative government,** which means that *the authority to govern is achieved through, and legitimized by, popular elections.* Every government officeholder has sought, in one way or another, the support of the **electorate** (*the citizens eligible to vote*) and has persuaded a majority of that group to grant its support (*through voting*). The elected official is entitled to hold office for a specified term and generally will be reelected as long as that body of voters is satisfied that the officeholder is adequately representing its interests.

Representative institutions can operate freely only if certain other conditions prevail. First, there must be what sociologist Edward Shils (1968) calls *civilian rule*—that is, every qualified citizen has the legal right to run for and hold an office of government. Such rights do not belong to any one class (say, of highly trained scholars, as in ancient China), caste, sect, religious group, ethnic group, or race. These rights, with certain exceptions, belong to every citizen. Further, there must be public confidence that such organized agencies as the police and the military will not intervene in, or change the outcome of, elections.

In addition, *majority rule* must be maintained. Because of the complexity of a modern democracy, it is not possible for the people to rule directly. One of the most important ways for people to participate in the political life of the country is to vote. For this to happen, people must be free to assemble, to express their views and seek to persuade others, to engage in political organizing, and to vote for whomever they wish.

Democracy also assumes that *minority rights* must be protected. The majority might not always act wisely, and it might be unjust. The minority abides by the laws as determined by the majority, but the minority must be free to try to change these laws.

Democratic societies contrast markedly with totalitarian societies. Ideally, democratic societies are open and culturally diverse, dissent is not viewed as disloyalty, there are two or more political parties, and terror and intimidation are not an overt part of the political scene.

The economic bases of democratic societies can vary considerably. Democracy can exist in a capitalistic country such as the United States and in a more socialistic one such as Sweden. However, it appears necessary for the country to have reached an advanced level of economic development before democracy can evolve. Such societies are most likely to have the sophisticated population and stability necessary for democracy (Lipset, 1960).

## FUNCTIONALIST AND CONFLICT THEORY VIEWS OF THE STATE

Functionalists and conflict theorists hold very different ideas about the function of the state. As our discussion of social stratification in Chapter 8, "Social Class in the United States," revealed, functionalist theorists view social stratification—and the state that maintains it—as necessary devices that recruit workers to perform the tasks needed to sustain society. Individual talents must be matched to jobs that need doing, and those with specialized talents must be given sufficiently satisfying rewards. Functionalists therefore maintain that the state emerged because society grew so large and complex that only a specialized, central institution (i.e., the state) could manage society's increasingly complicated and intertwined institutions (Davis, 1949; Service, 1975).

Marxists and other conflict theorists take a different view. They argue that certain groups were able to seize control of the means of production and distribution of commodities. In doing so, these groups were able to establish themselves as powerful ruling classes that dominated and exploited workers. Finally, the state emerged to allow the ruling classes to protect their institutionalized supremacy from the resentful and potentially rebellious lower classes. As Lenin (1949) explained, "The state is a special organization of force: it is an organization of violence for the suppression of some class."

There is evidence to support this view of the state's origins. The earliest legal codes of ancient states featured laws protecting the persons and properties of rulers, nobles, landholders, and wealthy merchants. The Code of Hammurabi of Babylon, dating to about 1750 BC, prescribed the death penalty for burglars and for anybody who harbored a fugitive slave. The code regulated wages, prices, and fees to be charged for services. It declared that a commoner must be fined six times as much for striking a noble or a landholder as for striking another commoner. It also condemned to death women who were proved by their husbands to be uneconomical in managing household resources (Durant, 1954).

Nevertheless, the functionalist view also has value. The state provides crucial organizational functions such as carrying out large-scale projects and undertaking long-range planning, without which complex society probably could not exist. Because it provides a sophisticated organizational structure, the state can—and does—fulfill many other important functions. In most modern societies, the state supports a public school system to provide a basic, uniform education for its members.

The health and well-being of its citizens also have become the concern of the state. In our own country, as in many others, the government provides some level of medical and financial support for its young, old, and disabled, and it sponsors scientific and medical research for the welfare of its people. Regulating industry and trade has become a function of the modern state, as has safeguarding the civil rights and liberties of its citizens. Certainly, one of the most important functions of any state is the protection of its people through armies, militias, and police forces.

When groups in a society become dissatisfied with their government, changes can take place. After such changes, the state might perform the same functions but do so in a different way and under different leadership.

## ● THE ECONOMY AND THE STATE

Philosophers have been writing about the relationship between the economy and the state for thousands of years. The Greek philosopher Plato, quoting Socrates, noted that both wealth and poverty are bad for society. Socrates suggested that a guard be placed at the gates of the city to keep wealth and poverty out. Wealth and poverty were not seen as two evils but as different sides to the same evil because the wealth of the rich, Plato believed, was the cause of the poverty of the poor. For Plato, this happens because the high consumption of the rich creates shortages for the poor.

For economist Adam Smith (1723–1790), poverty was due to the insufficient production of real wealth. The obvious solution to this problem was to increase the wealth of the community. "No society can surely be flourishing and happy, of which the far greater part of the members are poor and miserable" (quoted in Clark, 2002). These two views represent the starting points for the economic systems that came to be known as capitalism and socialism.

In its simplest terms, the economy is *the social institution that determines how a society produces, distributes, and consumes goods and services.* Money, goods, and services do not flow of their own accord. People work at particular jobs, manufacture particular products, distribute the fruits of their labor, purchase basic necessities and luxury items, and decide to save or spend their money. A very simple society may produce and distribute only food, water, and shelter. As a society becomes more complex and productive, the products produced and distributed become increasingly more elaborate. To be useful, all these goods and services must be distributed throughout the society. Every society must decide how much to be involved in the production and distribution of goods and services. In most economies, markets play the major role in determining what is produced, how, and for whom. Which means, is there a demand for something? Who is willing to produce it? How much are people willing to pay for it?

### Capitalism

In its classic form, **capitalism** is *an economic system based on private ownership of the means of production and in which resource allocation depends largely on market forces.* The government plays only a minor role in the marketplace, which works out its own problems through the forces of supply and demand.

Capitalism is built on two basic premises. The first, as Weber noted, is production "for the pursuit of profit and ever renewed profit." Capitalism entitles people to pursue their own self-interests, and this activity is desirable and eventually benefits society through the "invisible hand" of capitalism. For example, pharmaceutical companies may have no other goal in mind than profit when they develop new drugs. The fact that their products eventually benefit society is an indirect benefit brought about by this invisible hand.

The second basic premise is that the free market will determine what is produced and at what price. If people can profit from the production of a product, it will be produced. Adam Smith is regarded as the father of modern capitalism. He set forth his ideas in his book, *The Wealth of Nations* (1776), which is still used as a yardstick for analyzing economic systems in the Western world. According to Smith, capitalism has four features: private property, freedom of choice, freedom of competition, and freedom from government interference.

*Private Property* Smith believed that the ability to own private property acts as an incentive for people to be thrifty and industrious. These motivations, although selfish, will benefit society because those who own property will respect the property rights of others.

*Freedom of Choice* Along with the right to own property comes the right to do with it what one pleases as long as it does not harm society. Consequently, people are free to sell, rent, trade, give away, or retain whatever they possess.

*Freedom of Competition* Smith believed society would benefit most from a free market featuring unregulated competition for profits. Supply and demand would be the main factors determining the course of the economy.

*Freedom from Government Interference* Smith believed government should promote competition and free trade

---

### DAY-TO-DAY SOCIOLOGY

## Eat Your Fresh Fruit and Vegetables or Pay a Fine

*Sugar, rum, and tobacco are commodities which are nowhere necessaries of life, which are becoming objects of almost universal consumption, and which are therefore extremely proper subjects of taxation.*

—Adam Smith, *The Wealth of Nations*, 1776

An obesity epidemic seems to be sweeping through the United States. The medical community believes this is due to the increased consumption of sugar-sweetened beverages (soda and sports and energy drinks). These beverages cause increased body weight and poor nutrition when they replace more healthy beverages. They also put people at increased risk for obesity and diabetes.

Advertisers of sugared drinks target children and young people. A few examples show how successful they have been. By the mid-1990s, children began drinking more sugared beverages than milk. Also, in the past ten years, the average person's consumption of these drinks has increased 30 percent.

Sugared drinks have become more affordable compared to fresh fruits and vegetables, which also contributes to the rise in obesity in the United States. Even if one-quarter of the calories from sugared drinks were replaced by other food, it would translate into more than two pounds of body weight lost each year for the average person. Such a reduction could substantially reduce the risk of obesity and diabetes as well as heart disease and other conditions.

One controversial idea is to tax soda and other sweetened drinks. Maine and New York have proposed large taxes on sugared beverages, and similar discussions have begun in other states. The states like the idea because the taxes could generate large amounts of revenue. At the same time, they can show they are concerned about the excessive consumption of unhealthy foods.

Taxes on cigarettes have been very effective in reducing smoking, and it is expected that taxes could reduce sugared beverage consumption also. A study at Yale University suggests that for every 10 percent increase in price, consumption decreases by 7.8 percent. It is hoped the increased cost would encourage people to switch to more healthy drinks, thus curbing weight gain.

Some argue that government should not interfere in personal choices. Yet, about $79 billion is spent annually for overweight and obesity problems. Approximately half of the costs are paid by Medicare and Medicaid at taxpayers' expense. Obesity also costs society in terms of decreased work productivity, increased absenteeism, and poorer school performance.

It has also been pointed out that these taxes would affect the poor the most. But the poor are disproportionately affected by diet-related diseases and would derive the greatest benefit from reduced consumption. Americans consume about 250 to 300 more calories daily today than in the past, half of it coming from sugared beverages, which are not necessary for survival. In times of economic hardship for many states, taxes that both generate substantial revenue and promote health sound enticing.

**Sources:** Brownell, Kelly D., and Thomas R. Frieden. April 8, 2009. "Ounces of Prevention: The Public Policy Case for Taxes on Sugared Beverages" (www.nejm.org), accessed April 9, 2009; Vartanian, L. R., M. B. Schwartz, and K. D. Brownell. 2007. "Effects of Soft Drink Consumption on Nutrition and Health: A Systematic Review and Meta-Analysis." *American Journal of Public Health* 97:667–75.

and keep order in society but should not regulate business or commerce. The best thing the government can do for business, Smith said, is to leave it alone. *This view that government should stay out of business is referred to as* **laissez-faire capitalism.** The French words *laissez-faire* mean "allow to act."

In the United States, the government plays a vital role in the economy. Therefore, the U.S. system cannot be regarded as an example of pure capitalism. Rather, many have referred to our system as modified capitalism, also known as a mixed economy. (For a discussion about government intervention that would be considered controversial under laissez-faire capitalism, see "Day-to-Day Sociology: Eat Your Fresh Fruit and Vegetables or Pay a Fine.")

A **mixed economy** *combines free-enterprise capitalism with governmental regulation of business, industry, and social welfare programs.* Although private property rights are protected, the forces of supply and demand are not allowed to operate with total freedom. Resources are distributed through a combination of market and governmental forces.

Because there are few nationalized industries in this country (the Tennessee Valley Authority and Amtrak are two exceptions), the government uses its regulatory power to guard against private-industry abuse. Our government is also involved in such areas as antitrust violations, the environment, and minority employment. Ironically, this involvement may be even greater than it is in some of the more socialistic European countries.

Most countries have a mixed economy. Some countries, such as the United States and Taiwan, are closer to the free market end of the continuum, whereas China and Cuba are closer to the command economy or highly planned end. The absolute extremes of a complete command economy or complete capitalism do not exist, whether in Cuba or the United States. Command economy countries let consumers choose some of the goods they buy and allow private agricultural markets to some extent. In the United States, in addition to regulating economic activity by setting minimum wage levels, requiring safety standards for the workplace, and enacting antitrust laws and farm price supports, the federal government has also assumed partial or total control of privately owned businesses when their collapse would significantly affect the people as we have seen in the last few years.

### The Marxist Response to Capitalism

Smith believed that ordinary people would thrive under capitalism; not only would their needs for goods and services be met, but they also would benefit by being part of the marketplace. In contrast, Karl Marx was convinced that capitalism produces a small group of well-to-do individuals while the masses suffer under the tyranny of those who exploit them for profit.

Marx argued that capitalism causes people to be alienated from their labor and from themselves. Under capitalism, he said, the worker is not paid for part of the value of the goods produced. Instead, he said, this "surplus value" is taken by the capitalist as profit at the expense of the worker.

Workers also are alienated from their jobs. The worker feels no relationship to or pride in the product and merely works to obtain a paycheck and survive—a far cry from work being the joyous fulfillment of self that Marx believed it should be.

Marx believed that nineteenth-century capitalism contained the seeds of its own destruction. The main problem with capitalism, he contended, is that profits will decline as production expands. This, in turn, will force the industrialist to exploit the laborers and pay them less to continue to make a profit. As the workers are paid less or are fired, Marx said, they are less able to buy the goods being produced. This causes profits to fall even further, leading to bankruptcies, greater unemployment, and even a full-scale depression. After an increasingly severe series of depressions, Marx argued, the workers will rise up and take control of the state. They then will create a socialist form of government in which private property is abolished and turned over to the state. The workers will then control the means of production, and the exploitation of workers will end.

The reality of capitalism has not matched Marxist expectations. As we have seen, earlier capitalist economies have become much more mixed economies than the capitalist model developed by Smith. This has prevented some of Marx's prophesied outcomes. The level of impoverishment that Marx predicted for the workers has not taken place because labor unions have obtained higher wages and better working conditions for labor. Marx thought these changes could come about only through revolution. Labor-saving machinery has also led to higher profits without the predicted unemployment, and the production of goods to meet consumer demands has increased accordingly.

Marxists have offered a number of explanations for capitalism's continued success. Some have suggested that capitalism has survived because globalization has made it possible to sell excess goods to developing countries, maintaining high prices and profits. However, Marxists see this as only a temporary solution to the inevitable decline of capitalism. Eventually, the whole world will be industrialized, and the problems with capitalism will be revealed. They believe the movement toward socialism has not been avoided but only postponed.

In addition, critics of capitalism argue that true democracy is an impossible dream in a capitalist society. They claim that although in theory all members of a capitalist society have the same political rights, because the society is stratified, wealth, social esteem, and political power are unequally distributed (see Chapter 8).

Because of this, some critics contend, true democracy can be achieved only under socialism, to which our discussion now turns (Schumpeter, 1950).

## Socialism

**Socialism** is *an economic system in which the sources of production—including factories, raw materials, and transportation and communication systems—are collectively owned.* Socialism is an alternative to and a reaction against capitalism. Whereas capitalism views profit as the ultimate goal of economic activity, socialism is based on the belief that economic activity should be guided by public needs rather than private profit. Under an ideal socialist system, there would be public ownership of production and property, government control of the economy without a profit motive, and central planning.

*Public Ownership of Production and Property* The government owns the factories and apartment buildings. Housing and goods are then made available to everyone at a reasonable price.

*Government Control of the Economy Without a Profit Motive* Production and distribution are oriented toward output rather than profit. This orientation ensures that key industries run smoothly and that the public good is met. Individuals are heavily taxed to support a range of social-welfare programs that benefit every member of the society. Many socialist countries are described as having a cradle-to-grave welfare system.

*Central Planning* Instead of relying on the marketplace to determine prices, under socialism, prices for major goods and services are set by government agencies. Socialists believe that major economic, social, and political decisions should be made by elected representatives who control the economic system so that wealth and income are distributed as equally as possible. The belief is that everyone should have such essentials as food, housing, medical care, and education before some people can have luxury items such as cars and jewelry.

### The Capitalist View of Socialism

Capitalists view the centrally planned economies of socialist societies as inefficient and concentrating power in the hands of one group whose authority is based on party position. Among the problems they note are lack of incentive to increase production, waste of resources, overregulation and inflexibility, and corruption of power.

*No Incentive to Increase Production* Capitalists claim the main stimulus to investment is competition, the threat that if you do not improve your product or your production process, your rivals will, and they will take your market share. Under capitalism, it is necessary to cut costs and raise output to compete. Capitalists claim that if the producers of goods and services are immune from competition, they have few incentives to produce high-quality products.

*Waste of Resources* Socialist economies need to divert resources into planning rather than actually producing. In the former Soviet Union, *Gosplan*, the state planning organization, needed to calculate 12 million prices a year and plan the output of 24 million products.

*Overregulation and Inflexibility* Capitalists argue that socialist economies cannot be easily adjusted to take account of changing circumstances. They also note that if the state subsidizes essential goods and services and consumers do not pay full cost, nothing will prevent consumers from using more than they are entitled to and taking advantage of the system.

*Corruption of Power* Lenin argued that to consolidate power, socialists must use strong repressive measures against the old capitalist governments—in fact, build a dictatorship. As Lenin (1949) noted:

> The proletariat needs state power, a centralized organization of force, and organization of violence, both to crush the resistance of the exploiters and to lead the enormous mass of the population . . . in the work of organizing a socialist economy.

In China, North Korea, Cuba, and, more recently, in African and Southeast Asian countries, socialist revolutions have resulted in dictatorships. Members of the previous ruling classes have been executed, jailed, "reeducated," or exiled, and their properties have been seized and redistributed. In none of these societies has the dictatorship proved to be temporary, nor has the state gradually withered away, as Karl Marx and Friedrich Engels predicted it would after socialism was firmly established. Many Marxists contend that this will happen in the future, especially once capitalism has been defeated worldwide and socialist states no longer need to protect themselves against counterrevolutionary subversion and even direct military threats by capitalist nations. However, it is fair to observe that even the ancient Greeks knew that power corrupts and that those who have power are unlikely ever to give it up voluntarily.

### Democratic Socialism

**Democratic socialism** is *a convergence of capitalist and socialist economic theory in which the state assumes ownership of strategic industries and services but allows other enterprises to remain in private hands.* In Western Europe, democratic socialism has evolved as a political and economic system that attempts to preserve individual freedom in the context of social equality and a centrally planned economy.

With the parliamentary system of government present in many European countries, social democratic political parties have been able to win representation in the government. They have been able to enact their economic programs by being elected to office as opposed to producing a workers' uprising against capitalism. The social democrats have also attempted to appeal to middle-class workers and highly trained technicians as well as to industrial workers.

Under democratic socialism, the state assumes ownership of only strategic industries and services, such as airlines, railways, banks, television and radio stations, medical services, colleges, and important manufacturing enterprises. Certain enterprises can remain in private hands as long as government policies can ensure that they are responsive to the nation's common welfare. High tax rates prevent excessive profits and the concentration of wealth. In return, the population receives extensive welfare benefits such as free medical care, free college education, or subsidized housing.

Democratic socialism flourishes to varying degrees in the Scandinavian countries, in Great Britain, and in Israel. These countries all have a strong private (that is, capitalist) sector in their economies, but they also have extensive government programs to ensure the people's well-being. Those programs pertain to such things as national health service, government ownership of key industries, and the systematic tying of workers' pay to increases in the rate of inflation. Many observers believe that the American political economy has been moving in this direction. The social democratic movement is an example of the convergence of the capitalist and socialist economic theories, a trend that has been evident for some time. Capitalist systems have seen an ever-greater introduction of state planning and government programs, and socialist systems have seen the introduction of market forces and the profit motive. The growing economic interdependence of the world's nations will help continue this trend toward convergence.

## ● POLITICAL CHANGE

Political change can occur when there is a shift in the distribution of power among groups in a society. Political change can occur in a variety of ways, depending on the type of political structure the state has and the desire for change among the people. People can attempt to produce change through established channels within the government, or they can rise up against the political power structure with rebellion and revolution. Here we shall consider briefly three forms of political change: institutionalized change, rebellion, and revolution.

People can produce change through established channels or through rebellion and revolution.

### Institutionalized Political Change

In democracies, the way to change leaders is through elections. Usually, candidates representing different parties and interest groups must compete for a particular office at formally designated periods. There may also be laws that prevent a person from holding the same office for more than a given number of terms. If a plurality or a majority of the electorate is dissatisfied with a given officeholder, they can vote the incumbent out of office. Thus, the laws and traditions of a democracy ensure the orderly changeover of politicians and, usually, of parties in office.

In dictatorships and totalitarian societies, if a leader unexpectedly dies, is debilitated, or is deposed, a crisis of authority can occur. In dictatorships, illegal, violent means must often be used by an opposition to overthrow a leader or the government because no democratic means exist whereby a person or group can be legally voted out of power. Thus, we should not be surprised that revolutions and assassinations are most likely to occur in developing nations that have dictatorships. Established totalitarian societies, such as in China or the former Soviet Union, are more likely than dictatorships to have prescribed means by which a ruling committee decides who should fill a vacated position of leadership.

### Rebellions

**Rebellions** are *attempts—typically through armed force—to achieve rapid political change that is not possible within existing institutions.* Rebellions typically do not call into question the legitimacy of power but, rather, its uses. For example, consider Shays's Rebellion. Shortly after the American colonies won their independence from Britain, they were hit by an economic depression followed by raging inflation. Soon, in several states, paper money lost almost all its value. As the states began to pay off their war debts (which had been bought up by speculators), they were forced to increase the taxation of farmers, many of whom could not afford to pay those new taxes and consequently lost their land. Farmers began to band together to prevent courts from hearing debt cases, and state militias were called out to protect court hearings. Desperate farmers in the Connecticut River Valley armed themselves under Daniel Shays, an ex-officer of the Continental Army (Blum et al., 1981). This armed band was defeated by the Massachusetts militia, but its members eventually were pardoned and the debt laws were loosened somewhat (Parkes, 1968).

Shays's Rebellion did not intend to overthrow the courts or the legislature; rather, it was aimed at producing changes in their operation. Hence, it was a typical rebellion. (For a discussion of a current form of conflict designed to bring about change, see "Our Diverse World: Does Suicide Terrorism Make Sense?")

## Revolutions

*Revolution* is a powerful word. It evokes vivid images and strong emotions. It contains a mix of hope, excitement, and terror. It is small wonder that many great works of art, literature, and film have been inspired by revolution. In contrast to rebellions, **revolutions** are *attempts at rapidly and dramatically changing a society's*

---

### OUR DIVERSE WORLD: Does Suicide Terrorism Make Sense?

Why would someone become a suicide terrorist? How can we explain that someone would want to engage in an act that would surely kill him- or herself as well dozens or hundreds of innocent people? How can we possibly deter such fanaticism?

Robert Pape examined 188 suicide-terrorist attacks that took place between 1980 and 2001 throughout the world in places such as Lebanon, Israel, Sri Lanka, Chechnya, India, and Turkey. Pape found that far from being the irrational acts of desperate people, suicide terrorism was guided by identifiable strategic goals. He found that 95 percent of the incidents he studied were part of an organized political campaign. They were not unplanned acts of wanton cruelty. We may not understand the logic of the individual suicide bomber, but those that recruit and train them have a clear plan.

The terrorist organizations behind the attacks are trying to achieve specific political goals that include coercing the targeted government to change its policy, mobilizing additional recruits, and accumulating financial support. The attacks are meant to demonstrate that more and greater attacks are still to come. They are instruments of coercion, just as air power and dropping bombs is during a war.

Pape believes there are several principles to keep in mind when trying to understand suicide terrorists:

1. Suicide terrorism is strategic. The vast majority of the attacks are not isolated or random acts by individual fanatics. They are part of an organized campaign to achieve specific goals.
2. Suicide terrorism is designed to coerce democracies to make significant concessions. Pape found that every act of suicide terrorism since 1980 has been against a democratic form of government. In general, the goal has been to achieve specific territorial goals.
3. Suicide terrorism has increased because the organizers have found it works.
4. The leaders of terrorist organizations have been able to credit suicide attacks with significant progress toward their goals.
5. The gains are moderate and do not lead to complete victory. The power of suicide attacks is the threat of more attacks. The attacks do not compel nations to abandon compelling national interests.
6. The best way to reduce suicide terrorism is to increase security efforts.
7. Pape believes attacks will decrease if the attackers lose confidence in their ability to carry them out.

The main goal of suicide terrorism is to inflict enough pain that the interest in resisting the terrorist demands will weaken and the government will concede to the demands.

**Source:** Pape, Robert A. August 2003. "The Strategic Logic of Suicide Terrorism." *American Political Science Review* 97(3):1–19; Pape, Robert A. 2005. *Dying to Win: The Strategic Logic of Suicide Terrorism.* New York: Random House.

*previously existing structure.* Sociologists further distinguish between political and social revolutions.

*Political Revolutions* Relatively rapid transformations of state government structures without changes in social structure or stratification are known as **political revolutions** (Skocpol, 1979). The American War of Independence is a good example of a political revolution. The colonists were not seeking to change the structure of society or even necessarily to overthrow the ruling order. Their goal was to put a stop to the abuse of power by the British. After the war, they created a new form of government, but they did not attempt to change the fact that landowners and wealthy merchants held the reins of political power—just as they had before the shooting started. In the American Revolution, then, a lower class did not rise up against a ruling class. Rather, it was the American ruling class going to war to shake loose from inconvenient interference by the British ruling class. The initial result, therefore, was political change, not social change.

*Social Revolutions* In contrast, **social revolutions** are *rapid and basic transformations of a society's state and class structures.* They are accompanied and in part carried through by class-based revolts (Skocpol, 1979). Hence, they involve two simultaneous and interrelated processes: (1) the transformation of a society's system of social stratification, brought about by upheaval in the lower class(es), and (2) changes in the form of the state. Both processes must reinforce each other for a revolution to succeed. The French Revolution of the 1790s was a true social revolution. So were the Mexican Revolution of 1910, the Russian Revolution of 1917, the Chinese Revolution of 1949, and the Cuban Revolution of 1959, to name some of the most prominent social revolutions of the twentieth century. In all these revolutions, class struggle provided both the context and the driving force. The old ruling classes were stripped of political power and economic resources, wealth and property were redistributed, and state institutions were thoroughly reconstructed (Wolf, 1969).

Although the American Revolution did not arise from class struggle and did not result immediately in changes in the social structure, it did mark the beginning of a form of government that eventually modified the social stratification of eighteenth-century America.

## ● THE AMERICAN POLITICAL SYSTEM

The political system of the United States is unique in a number of ways, growing out of a strong commitment to a democratic political process and the influence of a capitalist economy. It has many distinctive features that are of particular interest to sociologists. In this section, we examine the role of the electorate and how influence is exerted on the political process.

### The Two-Party System

Few democracies have only two main political parties. Besides the United States, Australia, New Zealand, and Austria are further examples. Other democracies all have more than two major parties, thus providing proportional representation for a wide spectrum of divergent political views and interests. In most European democracies, if a political party receives 12 percent of the vote in an election, it is allocated 12 percent of the seats in the national legislature. Such a system ensures that minority parties are represented.

The American two-party political system, however, operates on a winner-take-all basis. Therefore, groups with differing political interests must face a lack of representation if their candidates lose. Conversely, candidates must attempt to gain the support of a broad spectrum of political interest groups because a candidate representing a narrow range of voters cannot win. This system forces accommodations between interest groups on the one hand and candidates and parties on the other.

Few, if any, individual interest groups (such as the National Organization for Women, the National Rifle Association, or the Conservative Caucus) represent the views of a majority of an electorate, whether local, state, or national. Hence, interest groups must find allies, hoping thereby to become part of a majority that can succeed in electing one or more candidates. Each interest group then hopes that the candidate(s) it has helped elect will represent its point of view. Interest groups must find a common ground with their allies in the party they have chosen to support. In doing so, they often have to compromise some strongly held principles. Hence, party platforms often tend to be composed of mild and noncontroversial issues, and party principles tend to adhere as closely as possible to the center of the American political spectrum.

Some have argued that the two-party system filters out extreme political views. A multiparty system allows extreme and sometimes destabilizing elements into the political system. The Nazi party gained credibility after entering a German coalition government when none of the major parties could gain a clear electoral victory.

When either party attempts to move away from the center to accommodate a very strong interest group with left- or right-wing views, the result generally is disaster at the polls. This happened to the Republican Party in 1964, when the politically conservative Barry Goldwater forces gained control of its organizational structure and led it to a landslide defeat. In that year's presidential election, the Democrats captured 61.1 percent of the national vote. Eight years later, the Democratic Party made the same mistake. It nominated George McGovern, a distinctly liberal candidate who, among other things, advocated a federally subsidized minimum income. The predictable landslide brought the Republicans and Richard Nixon 60.7 percent of the vote.

The candidates themselves have other problems. To gain support within their parties, they must somehow distinguish themselves from the other candidates. In other words, they must stake out identifiable positions. Yet, to win state and national elections, they must appeal to a broad political spectrum. To do this, they must soften the positions that first won them party support. Candidates thus often find themselves justly accused of double-talking and vagueness as they try to finesse their way through this built-in dilemma. This is most true of presidential candidates. It is no accident that once they have their party's nomination, some candidates might express themselves differently and far more cautiously than they did before. (For a discussion of how candidates distort each other's positions, see "Thinking About Social Issues: I Know It's Not True, but I'm Not Voting for Him Anyway.")

## Voting Behavior

In totalitarian societies, strong pressure is put on people to vote. Usually there is no contest between the candidates because there is no alternative to voting for the party slate. Dissent is not tolerated, and nearly everyone votes. In the United States, there is a constant progression of contests for political office and, in comparison with those of many other countries, the voter turnout is quite low. Since the 1920s, the turnout for presidential elections has ranged from about 50 percent to nearly

### THINKING ABOUT SOCIAL ISSUES: I Know It's Not True, But I'm Not Voting for Him Anyway

Any police officer will tell you that eyewitness accounts of events can differ dramatically. Not only can our memories of events be shaped by our interaction with other people, but it also turns out that they can be shaped by certain biases we might hold. Every day we encounter news reports. Most are accurate, but some turn out to be inaccurate. Which ones are we likely to believe? Which ones are we likely to remember?

For example, during the presidential election campaign of 2008, among the false rumors distributed was that Barack Obama (1) will *not* recite the Pledge of Allegiance nor will he show any reverence for the flag, (2) is a Muslim, and (3) his middle name is Mohammed.

John McCain was not spared either. False information was spread that McCain would (1) require workers to pay federal income tax on the value of their employer-provided health insurance, (2) support tax breaks for companies that ship jobs overseas, and (3) oppose stem cell research.

Many people continued to believe the rumors even when such information was shown to be false. They were so locked into a particular view that they rejected any contradictory information, no matter how well supported.

A couple of things seem to be happening here. People tend to remember things that logically go together even if they did not actually happen. Imagine that subjects are given a list of words that include, for example, *thermos*, *sandwich*, and *beverage*. They might later "remember" that a word such as *napkin* or *apple* was also on the list because they all relate to the topic "lunch."

It also turns out that people have mental maps of social life and the world in general. Once we have decided to either believe or not believe something, any later change or retraction will merely be filtered through our position on social life or politics.

This willingness to believe false information that fits our views can be particularly troubling in political campaigns. One side can make a false claim about the other side's candidate, allow it to echo through a news cycle, and then contritely retract it, knowing full well that receptive audiences are likely still to believe the initial lie.

We have to acknowledge that we have preconceived notions of people or events. We tend to believe what fits in with the social worlds that we are part of and reject that which does not fit our expectations. Accepting this would be the first step toward becoming more objective about the world around us.

**Sources:** Lewandowsky, Stephan, Werner G. K. Stritzke, Klaus Oberauer, and Michael Morales. March 2005. "Memory for Fact, Fiction, and Misinformation: The Iraq War 2003." *Psychological Science* 16(3):190–95; Begley, Sharon. February 4, 2005. "People Believe a 'Fact' That Fits Their Views Even If It's Clearly False." *Wall Street Journal*, p. B1; Robertson, Lori. March 18, 2008. "That Chain E-mail Your Friend Sent to You Is (Likely) Bogus. Seriously" (www.factcheck.org/specialreports/that_chain_e-mail_your_friend_sent_to.html), accessed July 1, 2009.

70 percent of registered voters. Even during the presidential election of 2008, only 66 percent of those eligible cast a vote. Still, this was the highest percentage of voters since 1908 (Marks, 2008).

Voting rates vary with the characteristics of the people. For example, those 45 and older and with a college education and a white-collar job have high rates of voter participation. Hispanics, the young, and the unemployed have some of the lowest voter participation rates. (See Figure 15-1 for voter participation by selected characteristics.) The age group with the highest proportion of voters is 60- to 74-year-olds, with more than 6 in 10 casting ballots. The lowest voting rates belong to 18- to 29-year-olds, with only 25 percent casting ballots.

The Democratic Party has tended to be the means through which the less privileged and the unprivileged have voted for politicians who they hoped would advance their interests. Since 1932, the Democratic Party has tended to receive most of its votes from the lower class, the working class, blacks, those of southern and eastern European descent, Hispanics, Catholics, and Jews. Thus, almost all the legislation that has been passed to aid these groups has been promoted by the Democrats and opposed by the Republicans. Democrats have sponsored legislation supporting unions, Social Security, unemployment compensation, disability insurance, antipoverty legislation, Medicare and Medicaid, civil rights, and consumer protection.

The Republican Party has tended to receive most of its votes from the upper-middle and lower-middle classes, Protestants, and farmers. Americans younger than 30 tend to vote the Democratic ticket; those older than 49 tend to vote for the Republican Party, even though the Democratic Party has been responsible for almost all the legislation to benefit older citizens. These voting patterns are, of course, generalizations and can change during any specific election.

Both parties tend to be most responsive to the needs of the best-organized groups with the largest sources of funds or blocks of votes. Thus, we would expect the Republicans to represent best the interests of large corporations and well-funded professional groups (such as the American Medical Association). The Democrats would logically be more closely aligned with the demands of unions.

Factors other than the social characteristics of voters and the traditional platforms of parties can affect the way people vote (Cummings and Wise, 1981). Indeed, the physical attributes, social characteristics, and personality of a candidate can prompt some people to vote against the candidate of the party they usually support. More important, the issues of the period can cause voters to vote against the party with which they usually identify. When people are frustrated by factors such as war, recession, inflation, and other international or national events, they often blame the incumbent president and the party he represents.

Since the 1960s, there have been efforts to increase the number of minority members who register to vote and to improve their voting rate. The success of Barack Obama has helped with this effort; minority groups are more likely to vote if they feel that the elections are relevant to their lives. As minority group members increase

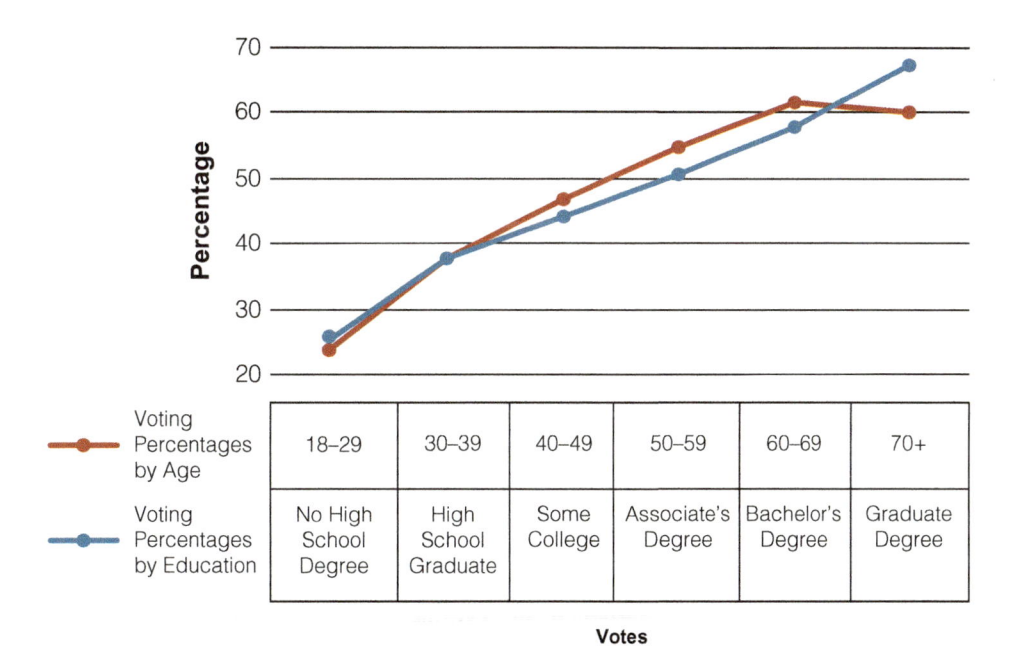

FIGURE 15-1  Who Votes? Voting by Age and Education, 2010

Source: U.S. Bureau of the Census. 2011. Current Population Survey, "Voting and Registration/Nov 2010" (http://smpbff1.dsd.census.gov/TheDataWeb_HotReport/servlet/HotReportEngineServlet?reportid=767b1387bea22b8d3e8486924a69adcd&emailname=essb@boc&filename=0328_nata.hrml), accessed September 4, 2012.

their voting rates, they also become successful in electing members of their groups.

Women also have been successful in increasing their representation in state legislatures. Figure 15-2 shows that the number of women holding such offices nearly tripled between 1975 and 2012.

Despite these advances, the members of Congress are still overwhelmingly white men older than 40. In 2011, the Senate had 17 women members and no African Americans; the other 83 members were men, including 2 Asian Americans and 2 Hispanics. In the House of Representatives, there were 74 women and 44 African Americans. (See Table 15-1 for a description of selected characteristics of members of Congress.)

## African Americans as a Political Force

In addition to Barack Obama being elected president, African Americans have been elected to statewide federal or administrative office in nineteen states. Illinois and Connecticut top the list of states with such black officeholders, each having elected African Americans to such posts five times. Illinois's record is most impressive with two U.S. senators, an attorney general, a secretary of state, and a comptroller, while in Connecticut, African Americans have served as state treasurer only. Georgia has elected African Americans to four statewide federal or administrative positions, and Colorado and Ohio have done so three times (Joint Center for Political and Economic Studies, 2009).

African Americans have increased their presence at the polls. In 1994, only 37 percent showed up at the polls to vote. By 2010, that number had increased to 43.5 percent (U.S. Bureau of the Census, 2011). Part of the explanation for this increase might be that African Americans always seem to receive attention before major political campaigns because each candidate attempts to convince them that he or she takes their interests to heart. Yet, African Americans have often been skeptical of those pre-election promises, sensing that once the election is won, their concerns will again be given low priority.

Although progress has occurred in a variety of areas in recent years, the economic gap between whites and African Americans remains large. The picture is somewhat more promising for education, but when one considers such areas as infant health, adult mortality rates, and income, there is still a great deal of room for progress. Statistics show an improvement in income levels of African American married-couple families, educational attainment and school enrollment, and home ownership among blacks during the previous decade.

**TABLE 15-1** Selected Characteristics of Members of the 112th Congress

|  | Senators | Representatives |
|---|---|---|
| **Sex** | | |
| Male | 83 | 361 |
| Female | 17 | 74 |
| **Race** | | |
| White | 96 | 350 |
| Black | 0 | 44 |
| Asian-Pacific Islander | 2 | 11 |
| Hispanic | 2 | 29 |
| Native American | 0 | 1 |
| **Age** | | |
| Average Age | 62.2 | 56.7 |
| Youngest | 39 | 29 |
| Oldest | 87 | 87 |

Source: Manning, Jennifer E. March 1, 2011. Congressional Research Service, "Membership of the 112th Congress: A Profile."

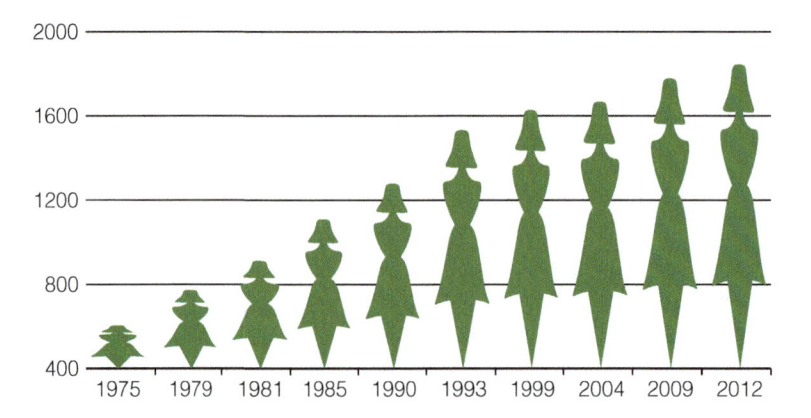

**FIGURE 15-2** Women Holding Office Within State Legislatures, 1975–2012

Source: https://docs.google.com/viewer?a=v&q=cache:Ysdsdtau0bgJ:www.cawp.rutgers.edu/fast_facts/levels_of_office/documents/elective.pdf+&hl=en&gl=us&pid=bl&srcid=ADGEESjcoxiJCbBTrqxaoc5AQr-mB1TVG8gwe52wvIgbWDY3HVGIIgCvCHrU9miFyH6eVoVmxEif6uyzkAaFQIEKLryRlzL8v4D5PRTQarK ZEtv2WMFL6sO1NsXLzVrCytU6fRULvlp9_&sig=AHIEtbRWf2C9LaQYDl50zAXi_thOBdEECw, accessed September 4, 2012.

There have been setbacks also, demonstrated by high African American unemployment, sharply increased divorce and separation rates, and a rise in family households maintained by black women.

A cursory look at income statistics makes blacks appear better off than they are. African American married couples have closed the gap with white incomes, but the proportionate share of married couples in the total black population is smaller than a decade ago.

African Americans also have fallen further behind whites in the accumulation of wealth. Part of this is because they are less likely to own property and therefore are less able to take advantage of rising real estate values. Regardless of income level, African Americans tend to have less wealth than whites. Even blacks who already own a home or eventually manage to purchase one must contend with the fact that most homes owned by blacks are in central cities, areas in which homes are less likely to appreciate than in the suburbs.

In the political arena, African Americans are not just another interest group. They experience deep-seated economic differences that make them skeptical of political promises. Those differences have made them cautious about supporting white Democrats, no matter how liberal the candidates' voting records are.

### Hispanics as a Political Force

There were more than 30 million Hispanics of voting age in November 2010. However, only 31.2 percent voted, compared with 48.6 percent of whites and 43.5 percent of African Americans (U.S. Bureau of the Census, 2011). The percentage of people voting is 15 to 20 percentage points higher during presidential election years.

Hispanic voters are crucial in many presidential elections. The main reason is the geographical distribution of the voters. Nearly half of all Hispanic voters are registered in six major states—New York, New Jersey, Florida, Illinois, Texas, and California—which together can give a presidential candidate most of the electoral votes needed to win the election.

Hispanic voters have traditionally been overwhelmingly Democratic. Hispanics tend to support Democratic candidates because of the support those candidates give to social programs that help the poor. The fact that Hispanics, like African Americans, have a harder time climbing the socioeconomic ladder keeps them in the Democratic camp.

With the Hispanic population expected to grow rapidly over the next few decades, Democrats and Republicans will be paying more attention to the needs of this group. It is also likely that Hispanics will be electing greater numbers of Hispanic candidates to political office.

There has been significant progress in electing African American and Hispanic political candidates. (See Figure 15-3, which shows the consistent rise in the number of African American and Hispanic elected officials.) In 1958, only 38 percent of the public would have voted for a black presidential candidate and 54 percent would have considered voting for a woman. By 2012, those percentages were 96 percent and 95 percent, respectively, and of course Barack Obama was elected president. Other groups would still have a difficult time getting elected. There is strong opposition to electing an atheist for president. Only 54 percent of the people would vote for this person. A Muslim presidential candidate would also have a difficult time getting elected, with 58 percent willing to cast their vote for this person. A gay or lesbian candidate would have a chance with only 68 percent of the public (Jones, Gallup Poll, 2012).

### The Role of the Media

The media have contributed to a radical transformation of election campaigns in the United States. This transformation involves changes in how political candidates communicate with the voters and in the information journalists provide about election campaigns.

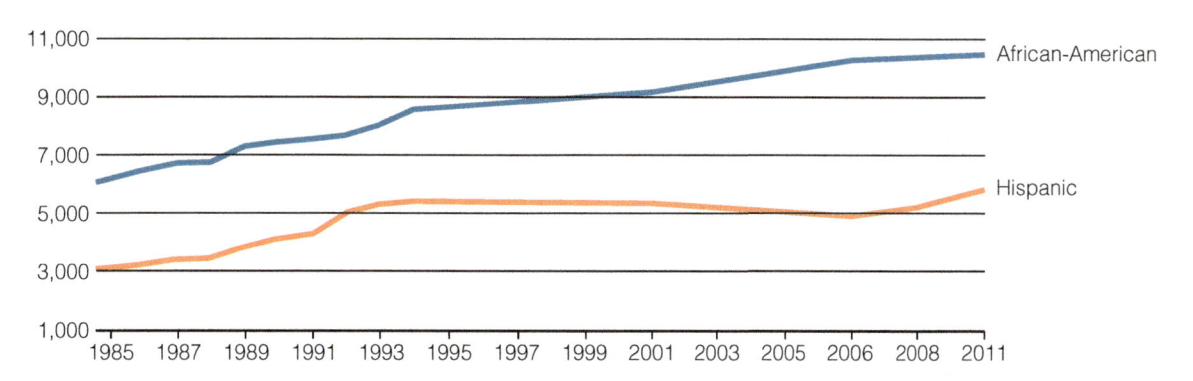

**FIGURE 15-3** African American and Hispanic Elected Officials, 1985–2011

Sources: Naleo Education Fund. "2011 Directory of Latino Elected Officials" (www.naleo.org/directory.html); Joint Center for Political and Economic Studies. National Roster of Black Elected Officials Fact Sheet (www.jointcenter.org/sites/default/files/upload/research/files/National%20Roster%20of%20Black%20Elected%20Officials%20Fact%20Sheet.pdf).

On one side, we have candidates who are trying to present self-serving and strategically designed images of themselves and their campaigns. On the other side are television and print journalists who believe that they should be detached, objective observers, motivated primarily by a desire to inform the U.S. public accurately.

Although the norm is often violated, there is no doubt that the journalistic perspective is markedly different from that of the candidate. Campaign coverage is far more apt to be critical and unfavorable toward a candidate than favorable. Journalists strive to reveal a candidate's flaws and weaknesses and to uncover tasty tidbits of hushed-up information.

Journalists exercise considerable political power in four important ways. First, the most obvious way involves deciding how much coverage to give a campaign and the candidate involved. A candidate who is ignored by the media has a difficult time becoming known to the public and acquiring important political resources such as money and volunteers. Such candidates have little chance of winning.

In the beginning stages of presidential nomination campaigns, for example, a candidate's goal is typically to do something that will generate news coverage and stimulate campaign contributions. These contributions can then be used for further campaigning, helping convince the press and the public that the candidate is credible and newsworthy.

Second, the media decide which of many possible interpretations to give to campaign events. Because an election is a complex and ambiguous phenomenon, different conclusions can be drawn about its meaning. Here the journalists help the public form specific impressions of the candidates.

For example, presidential candidates are always concerned with how the results of presidential primaries are interpreted. Candidates want to be seen as winners who are gaining momentum. At the same time, a candidate seen as the front-runner too early is open to being shot down later.

Third, the media exercise discretion in how favorably candidates are presented in the news. Although norms of objectivity and balance prevent most campaign coverage from including biased assertions, a more subtle and pervasive slant or theme to campaign coverage is possible and can be significant.

Finally, newspaper editors and publishers may officially endorse a candidate. This support can be particularly important to the candidate if the newspaper is one of the major national publications.

Politicians need the mass media to get the coverage they need to win. At the same time, they can very easily fall victim to the intense scrutiny that is likely to result.

## Special-Interest Groups

With the government spending so much money and regulating so many industries, special-interest groups constantly attempt to persuade the government to support them financially or through favorable regulatory practices. **Lobbying** refers to *attempts by special-interest groups to influence government policy.* Farmers lobby for agricultural subsidies, labor unions for higher minimum wages and laws favorable to union organizing and strike actions, corporate and big business interests for favorable legislation and less government control of their practices and power, the National Rifle Association to prevent the passage of legislation requiring the registration or licensing of firearms, consumer-protection groups for increased monitoring of corporate practices and product quality, the steel industry for legislation taxing or limiting imported steel, and so on.

*Lobbyists* The term *lobbyist* goes back to the presidency of Ulysses S. Grant. Grant's wife would not allow him to smoke his cigars in the White House. Grant would often spend time relaxing and smoking his cigars in the lobby of the nearby Willard Hotel. Those seeking a political favor from Grant became known as lobbyists.

Of all the pressures on Congress, none has received more publicity than the role of the Washington-based lobbyists and the groups they represent. The popular image of a lobbyist is an individual with unlimited funds trying to use devious methods to obtain favorable legislation. The role of today's lobbyist is far more complicated.

The federal government has tremendous power in many fields, and changes in federal policy can spell success or failure for special-interest groups. With the expansion of federal authority into new areas and the huge increase in federal spending, the corps of Washington lobbyists has grown markedly. The number of registered lobbyists swelled to 11,461 in 2012. The lobbyists spend more than $3 billion every year trying to get legislation passed. Some of the groups spending the most money on lobbying are the U.S. Chamber of Commerce, General Electric, the American Medical Association, and the American Association of Retired People (AARP). Each of these groups spends more than $15 million a year on lobbying (Congressional Budget Office, 2012).

Lobbyists usually are personable and extremely knowledgeable about every aspect of their interest group's concerns. They cultivate personal friendships with officials and representatives in all branches of the government, and they frequently have conversations with these government people, often in a social atmosphere such as over drinks or dinner.

The pressure brought by lobbyists usually has self-interest aims—that is, to win special privileges or financial benefits for the groups they represent. On some occasions, the goal can be somewhat more objective, as when the lobbyist is trying to further an ideological goal or to put forth a group's particular interpretation of what is in the national interest.

Certain liabilities are associated with lobbyists. The key problem is that they might lead Congress to make decisions that benefit the pressure group but might not serve the interests of the public. A group's influence might be based less on the arguments for their position than on the size of the membership, the amount of their financial resources, or the number of lobbyists and their astuteness.

Lobbyists might focus their attention not only on key members of a committee but also on the committee's professional staff. Such staffs can be extremely influential, particularly when the legislation involves highly technical matters about which the member of Congress might not be knowledgeable. Lobbyists also exert their influence through testimony at congressional hearings. Those hearings can give the lobbyist a propaganda forum as well as access to key lawmakers who could not have been contacted in any other way. The lobbyists can rehearse their statements before the hearing, ensure a large turnout from their constituency for the hearing, and can even give leading questions to friendly committee members so that certain points can be aired at the hearing.

Lobbyists do perform some important and indispensable functions such as helping inform Congress and the public about problems and issues that normally may not get much attention, thereby stimulating public debate and making known to Congress who would benefit and who would be hurt by specific legislation. Many lobbyists believe that their most important and useful role, both to the groups they represent and to the government, is the research and detailed information they supply. In fact, many members of the government find the data and suggestions they receive from lobbyists to be valuable in studying issues, making decisions, and even in voting on legislation.

*Political Action Committees* Special-interest groups called **political action committees (PACS)** are *organized to raise and spend money to elect and defeat political candidates, ballot initiatives, and other legislation.* Most PACs represent business, labor, or ideological interests. PACs have been around since 1944, when the first one was formed to raise money for the reelection of President Franklin D. Roosevelt. PACs contribute millions of dollars to congressional and senatorial political campaigns.

Several criticisms have been leveled at these special-interest groups. Among the most prominent is that they

Some people believe the ability of businesses to influence legislation through lobbying is dangerous.

represent neither the majority of the American people nor all social classes. Most PACs represent groups of affluent and well-educated individuals or large organizations. PACs have grown from 608 in 1974 to 4,600 in 2009. Many represent corporate interests and professional groups, but labor unions also use PACS (U.S. Bureau of the Census, 2010).

PACS are able to leverage the power of their group by combining the contributions from many members, which increases their influence tremendously. Only about 10 percent of the population is in a position to exert this kind of pressure on the government. Disadvantaged groups—those that most need the ear of the government—have no access to this type of political action (Clawson et al., 1998).

PACs also tend to favor incumbents. Two-thirds of all PAC contributions in recent elections have gone to incumbents. Challengers therefore end up being much more dependent on small donations from individuals or from the Democratic and Republican national committees. PACS ultimately might diminish the role of the individual voter.

In 2010, the Supreme Court decision in *Citizens United* ruled that PACs that did not make contributions to candidates, parties, or other PACs could accept unlimited contributions from individuals, unions, and corporations and could engage in unlimited political spending independently of the campaigns. These organizations became known as "Super PACs" and were able to have a large influence on political campaigns.

## SUMMARY

### What role does government play?
In modern, complex societies government is necessary, and the state is the institutionalized way of organizing power. In most societies, what laws are passed or not passed depends largely on which categories of people have power.

### What is politics?
Politics is the process by which power is distributed and decisions are made. As Weber defined it, power is the ability of a person or group to carry out its will, even in the face of resistance or opposition.

### How is power exercised?
Power is exercised in a broad spectrum of ways. At one pole is authority, power regarded as legitimate by those over whom it is exercised and who also accept the authority's legitimacy in imposing sanctions or even in using force if necessary. At the other extreme is coercion, power regarded as illegitimate by those over whom it is exerted.

### How else does authority and coercion differ?
Power based on authority is quite stable, and obedience to it is accepted as a social norm. Power based on coercion is unstable. It is based on fear; any opportunity to test it will be taken, and in the long run, it will fail.

### What are some of the functions of the state?
The state has a variety of functions. One is to establish laws that formally specify what is expected and what is prohibited in the society. A second function is to enforce those laws and make sure that violations are punished. In modern societies, the state must also try to ensure that a stable system of distribution and allocation of resources exists. The state also sets goals and provides a direction for society, usually through the power to compel compliance. Finally, the state must protect a society from outside threats, especially those of a military nature.

### What role does the economy play in politics?
Politics and economics are intricately linked, and the political form a state takes is tied to the type of economy. The economy is the social institution that determines how a society produces, distributes, and consumes goods and services.

### What is democracy?
Democracy refers to a political system operating under the principles of constitutionalism, representative government, civilian rule, majority rule, and minority rights.

### How do structural functionalists view the state and the economy?
Functionalists view social stratification—and the state that maintains it—as necessary for the recruitment of workers to perform the tasks required to sustain society.

### How does this differ from how Marxists and conflict theorists view the state and the economy?
Marxists and conflict theorists argue that the state emerged as a means of coordinating the use of force by means of which the ruling classes could protect their institutionalized supremacy from the resentful and potentially rebellious lower classes.

### What is unique about the U.S. political system?
Growing out of a strong commitment to a democratic political process and the influence of a capitalist economy, the political system in the United States is unique in that it has a winner-take-all, two-party system. In this system, political candidates are forced to gain the support of a broad spectrum of interest groups to be elected.

## Media Resources

**CourseMate for *Introduction to Sociology, Eleventh Edition***
Cengage Learning's Sociology CourseMate brings course concepts to life with interactive learning, study, and exam preparation tools that support the printed textbook. Access an integrated eBook, learning tools including glossaries, flashcards, quizzes, videos, and more in your Sociology CourseMate. Go to CengageBrain.com to register or purchase access.

# CHAPTER FIFTEEN STUDY GUIDE

## Key Concepts

*Match each of the following concepts with its definition, illustration, or explanation.*

a. Power
b. Legal-rational authority
c. Traditional authority
d. Charismatic authority
e. Coercion
f. State
g. Autocracy
h. Totalitarianism
i. Democracy
j. Constitutionalism
k. Laissez-faire capitalism
l. Mixed economy
m. Socialism
n. Capitalism
o. Democratic socialism
p. Rebellion
q. Revolution
r. Lobbying
s. PAC

\_\_\_\_ 1. Government in which most power rests with a single ruler
\_\_\_\_ 2. Government in which a single group controls all of the nation's social institutions
\_\_\_\_ 3. The use of power in a way that is regarded as illegitimate by those over whom it is exerted
\_\_\_\_ 4. A group formed to raise money and spend it on elections
\_\_\_\_ 5. A form of authority rooted in the assumption that the customs of the past legitimate the present
\_\_\_\_ 6. An economic system based on private ownership of the means of production and resource allocation through the market
\_\_\_\_ 7. The ability of people or groups to get their way, even in the face of resistance or opposition
\_\_\_\_ 8. A form of authority derived from a ruler's ability to inspire passion and devotion among followers
\_\_\_\_ 9. The attempt to influence government policymakers, especially legislators
\_\_\_\_ 10. The view that government should stay out of business affairs
\_\_\_\_ 11. An economic system in which the government owns the sources of production and sets production and distribution goals
\_\_\_\_ 12. A political system operating under the principles of constitutionalism, representative government, majority rule, civilian rule, and minority rights
\_\_\_\_ 13. A convergence of capitalist and socialist economic theory in which the state assumes ownership of strategic industries and services but allows other enterprises to remain in private hands
\_\_\_\_ 14. The institutionalized organization of power within a geographic territory
\_\_\_\_ 15. An attempt—typically through armed force—to achieve rapid political change not possible within existing institutions
\_\_\_\_ 16. A form of authority derived from the understanding that specific individuals have clearly defined rights and duties to uphold and implement rules and procedures impersonally
\_\_\_\_ 17. An attempt to change a society's previously existing structure rapidly and dramatically
\_\_\_\_ 18. The limiting of government power by a written system of laws
\_\_\_\_ 19. Capitalism combined with government regulation and social welfare

## Key Thinkers

*Match the thinkers with their main idea or contributions.*

a. Adam Smith
b. Thomas Jefferson
c. Karl Marx
d. Max Weber

\_\_\_\_ 1. Developed the sociological definitions of power and authority
\_\_\_\_ 2. Regarded as the father of modern capitalism; discussed many of the basic premises of this system
\_\_\_\_ 3. A severe critic of capitalism who argued that it was based on alienation and exploitation
\_\_\_\_ 4. Argued that a political system had to be taken out periodically, inspected, examined, and changed if necessary before passing it on to the next generation

## Central Idea Completions

*Fill in the appropriate concepts and descriptions for each of the following questions.*

1. Discuss how these three groups—white, black, and Hispanic—differ in regard to voter participation.

    a. Which of the three groups has the highest voter participation? _____

    b. Which of the three groups has the lowest voter participation? _____

    c. Explain why these groups differ in this regard: _____

2. How are age and education connected with voter participation in the United States?

3. Are rates of voter participation higher in totalitarian states or in democracies? Why?

4. What are the five functions of the state?

    a. _____
    b. _____
    c. _____
    d. _____
    e. _____

5. What are the four features of capitalism outlined by Adam Smith?

    a. _____
    b. _____
    c. _____
    d. _____

6. What are the six specific elements totalitarian governments use to concentrate political power?

    a. _____
    b. _____
    c. _____
    d. _____
    e. _____
    f. _____

## Critical Thought Exercises

1. Develop an essay, based on the materials presented in your text, on the extent to which Weber's three types of authority are represented in U.S. politics and government. How are these types of authority represented on your campus?

2. Look at some of the institutions of your life (school, family, workplace, and so on). In what ways do these perform functions similar to those of the state? How well do different models of government (democracy, aristocracy, and so on) fit with the way these institutions run?

3. How free is free enterprise? Look at all the products and services you pay for and note how the governments and authorities interfere in the market. For example, does the can of soda you buy have nutritional information on it? Did the manufacturer put it there voluntarily? Is there a McDonald's in the middle of your college campus? If not, is that because McDonald's didn't want to put one there, or did campus authorities restrict this type of enterprise? What might your world look like if there were no regulation of commerce? In what ways would true laissez-faire capitalism be an improvement?

## Internet Activities

1. Kim Jong-il, the recently deceased leader of North Korea, was widely recognized as a charismatic leader. He was unchallenged, ruled by decree, and inspired a personality cult. Millions of North Koreans died of famine while the armed forces and elite were provided for. Despite these facts, he was given blind obedience by the masses and was believed to have performed "miracles." Visit www.youtube.com/watch?v=mSLJYbhXCkE to witness the passing of Kim's funeral procession, paying particular attention to the "unorthodox" public displays of sorrow. Discuss the type of leadership Kim Jong-il displayed and explain the public response to his death. In what ways is this type of leadership beneficial to the masses and also not beneficial?

2. Visit www.politicalresources.net/. This site contains resources on political parties across the globe. Explore the site, and when you become familiar with how it is organized, visit three global locations and present a comparison of the activities of political parties within the countries you selected. In what ways are political party activities common across the three countries and in what ways is each nation's political parties' behavior unique to that country?

3. Visit www.fivethirtyeight.com/. The site has excellent data on U.S. political issues, often with easily understood graphs. Evaluate three of the articles presented and note whether you sense any bias on the blog. Describe the discussions that caused you to reach this conclusion.

## Answers to Key Concepts

1. g;  2. h;  3. e;  4. s;  5. c;  6. n;  7. a;  8. d;  9. r;  10. k;  11. m;  12. i;  13. o;  14. f;  15. p; 16. b;  17. q;  18. j;  19. l

## Answers to Key Thinkers

1. d;  2. a;  3. c;  4. b

# 16 Health and Aging

## The Experience of Illness
- Health Care in the United States
- Gender and Health

*Our Diverse World:* Women Live Longer than Men Throughout the World
- Race and Health
- Social Class and Health
- Age and Health

*Our Diverse World:* Why Isn't Life Expectancy in the United States Higher?
- Education and Health
- Women in Medicine

*Day-to-Day Sociology:* Marijuana: A Benign Drug or a Health Problem?

## Contemporary Health Care Issues
- Acquired Immunodeficiency Syndrome
- Health Insurance
- Preventing Illness

*How Sociologists Do It:* Can Your Friends Make You Fat?

## The Aging Population
*Thinking About Social Issues:* The Discovery of a Disease
- Composition of the Older Population
- Aging and the Sex Ratio
- Aging and Racial Minorities
- Aging and Marital Status
- Aging and Wealth

*Our Diverse World:* Stereotypes About the Elderly
- Global Aging
- Future Trends

*Our Diverse World:* Global Aging Quiz

## Summary

## LEARNING OBJECTIVES

After studying this chapter, you should be able to do the following:

- Know what sociologists mean by the sick role.
- Describe the basic characteristics of the U.S. health care system.
- Understand the link between demographic factors and health.
- Describe the three major models of illness prevention.
- Describe the basic demographic features of the older population in the United States.

In his book *The Youngest Science,* physician Lewis Thomas wrote about what it was like to be a medical intern at the Boston City Hospital in the pre-penicillin year of 1937. It was a time when medicine was cheap and very ineffective. If you were in a hospital, he said, it was going to do you good only because it offered you some warmth, food, shelter, and maybe the caring attention of a nurse. Doctors and medicine made little difference. Whether you survived or not depended on the natural progression of the disease itself.

Doctors still tried to figure out if the patient had a disease they could do something about. One disease that appeared at the time was Lobar pneumonia, which could be treated with an injection of rabbit antibodies. Patients in a diabetic coma could be given animal-extracted insulin and intravenous fluid. Acute heart failure patients could be saved by bleeding away a pint of blood from an arm vein, administering digitalis, and delivering oxygen by tent. If the patient had early signs of paralysis, the doctor might figure out that it was due to syphilis. The doctor would then give the patient mercury and arsenic, hopefully not in an overdose that would kill the patient. Beyond these sorts of things, there was not much a medical doctor could do.

In 1937, a doctor could learn nearly everything that was known about how to treat a patient. The distance medicine has traveled since then is almost unfathomable. We now have treatments for most of the tens of thousand of conditions that afflict human beings. We have more than four thousand medical and surgical procedures, and more than six thousand drugs. Such capabilities cannot guarantee everyone a long and healthy life, but they can make it possible for most.

Elias Zerhouni, the former director of the National Institutes of Health, calculated how many doctors, nurses, and staff were involved in the care of a typical hospital patient. In 1970, the number was 2.5. By 2000, it was more than 15. The number must be even larger today. Many people now have just a piece of patient care.

When people get sick, we have amazing physicians we can turn to. We have technologies that give us great hope, but little sense of how to make everything come together in a consistent way. Medicine has become considerably more expensive than in 1937, but many problems remain in making such a large system manageable and efficient (Gawande, 2011).

Medicine and health care issues are intertwined with our social, emotional, and cultural life. In this chapter, we examine these interactions as we explore health and illness.

## THE EXPERIENCE OF ILLNESS

Illness not only involves the body, but it also affects the individual's social relationships, self-image, and behavior. Being defined as "sick" has consequences that are independent of any physiological effects. Talcott Parsons (1951) suggested that to prevent the potentially disruptive consequences of illness on a group or society, a sick role exists. The **sick role** is *a shared set of cultural norms that legitimates deviant behavior caused by the illness and channels the individual into the health care system.*

According to Parsons, the sick role has four components. First, the sick person is excused from normal social responsibilities except to the extent that he or she is supposed to do whatever is necessary to get well. Second, the sick person is not held responsible for his or her condition and is not expected to recover by an act of will. Third, the sick person must recognize that being ill is undesirable and must want to recover. Finally, the sick person is obligated to seek medical care and cooperate with the advice of the designated experts, notably the physicians. In this sense, although sick people are not blamed for their illnesses, they must work toward regaining their health.

The sick role concept is based on the perspective that all roads lead to medical care, which tends to create a doctor-centered picture with the illness being viewed from outside the individual. Some (Schneider and Conrad, 1983; Strauss and Glaser, 1975) have suggested that the actual subjective experience of being sick should be examined more closely. These researchers suggest that we should focus more on individuals' perceptions of illness, their interactions with others, and the effects of the illness on the person's identity.

At the same time, society tends to define what should be considered a medical issue. Over the years, many issues have become medical issues that were not considered such before. **Medicalization** is *the process by which nonmedical problems become defined and treated as medical problems, usually in terms of illness or disorder.* Of late, such things as alcoholism, drug abuse, battering, gender confusion, obesity, anorexia and bulimia, and a host of reproductive issues from infertility to menopause have undergone medicalization (Conrad, 1992).

### Health Care in the United States

The United States has the most advanced health care resources in the world. We can scan a brain for tumors, reconnect nerve tissues and reattach severed limbs through microsurgery, and eliminate diseases such as poliomyelitis, which crippled a president.

We can do all this and much more; yet many consider our health care system wholly inadequate to meet the needs of all Americans. Critics maintain that the U.S. health care system is one that pays off only when the patient can pay.

The American health care system has been described as "acute, curative, [and] hospital based" (Knowles, 1977). This statement implies that our approach to medicine is organized around the cure or control of serious diseases and repairing physical injuries, rather than caring for the sick or preventing disease. The American medical-care system is highly technological, specialized, and increasingly centralized.

Medical-care workers include some of the highest-paid employees (physicians) in our nation and some of the lowest paid. About three-quarters of all medical workers are women, although the majority of doctors for the time being are men. Many of the workers are members of minority groups, and most come from lower-middle-class backgrounds. The majority of the physicians are white and upper-middle class.

The medical-care workforce can be pictured as a broad-based triangle with a small number of highly paid physicians and administrators at the top. Those people control the administration of medical-care services. As one moves toward the bottom of the triangle, there are increasing numbers of much lower-paid workers with little or no authority in the health care organization. This triangle is further layered with more than 300 licensed occupational categories of medical workers. There is practically no movement of workers from one category to another because each requires its own specialized training and qualifications (Conrad and Kern, 1986).

### Gender and Health

You probably are aware of the striking differences in life expectancy between men and women. The life expectancy of women in the United States has been continuously higher than that for men since records began being kept. In 1900, women could expect to live an average of two years longer than men. By the 1970s, life expectancy for women was seven years greater than for men. That number has since decreased to 4.9 years in 2010, when males lived an average of 76.2 years and women an average of 81.1 (National Center for Health Statistics, 2012).

Women have historically had high rates of death from complications related to pregnancy and childbirth. But improvements in prenatal and obstetric care have dramatically decreased the risk of pregnancy-related deaths.

Heart disease and cancer rank as the first and second leading causes of death for both men and women, accounting for nearly 50 percent of all deaths. Yet, women of all ages experience lower death rates. This difference is partly because men are three times more likely to die from unintentional injuries, homicide, accidents, or suicide than women. And programs to address these causes of death have achieved very slow progress during the past fifty years compared to programs for other causes of death. Studies also show that men are less likely to seek medical attention for health-related problems than are women, and when they do seek care, they are less likely to comply with medical treatments.

The leading cause of death among both women and men is heart disease. Women's higher levels of estrogen seem to provide protection against the incidence and severity of heart disease. In recent decades, the growth of estrogen replacement therapy for older women has also helped reduce deaths from heart disease.

Women suffer from illness and disability more frequently than do men, but their health problems are usually not as life threatening as those encountered by men. Women, of course, do suffer from most of the same diseases as men. The difference is at what point in life they encounter those diseases. For example, coronary heart disease is a leading cause of death for men and women, but the death rates for women are considerably lower than those for men until they reach age 80 (Blum, 2009).

Men appear to have lower life expectancies than women because of biological and sociological reasons. Males are at a biological disadvantage compared to females, as seen by the higher mortality rates from the prenatal and neonatal (newborn) stages onward. Although the percentages might vary from year to year, the chances of dying during the prenatal stage are approximately 12 percent greater among males than among females, and 130 percent greater during the newborn stage. Neonatal disorders common in male rather than in female babies include respiratory diseases, digestive diseases, certain circulatory disorders of the aorta and pulmonary artery, and bacterial infections. The male seems to be more vulnerable than the female even before being exposed to the different social roles and stress situations of later life.

A number of sociological factors also play an important role in the different life expectancies of men and women. Men are more likely to place themselves in dangerous situations both at work and during leisure activities. Therefore, it should be no surprise that accidents cause more than

## OUR DIVERSE WORLD

### Women Live Longer than Men Throughout the World

The widening of the sex differential in life expectancy has been a central feature of mortality trends in developed countries in the twentieth century. In 1900, in Europe and North America, women typically outlived men by two or three years. Today, the average gap between the sexes is roughly seven years, but it is twelve years in Russia because of the unusually high levels of male mortality.

This differential reflects the fact that in most nations, females have lower mortality than males in every age group and for most causes of death. Female life expectancy now exceeds 80 years in more than thirty countries and is approaching this level in many other nations. The gender differential usually is smaller in developing countries, commonly in the one- to three-year range, and is even reversed in a handful of African countries, such as South Africa and Swaziland, where cultural factors (such as low female social status and preference for male rather than female offspring) are thought to contribute to higher male than female life expectancy at birth (see Figure 16-1).

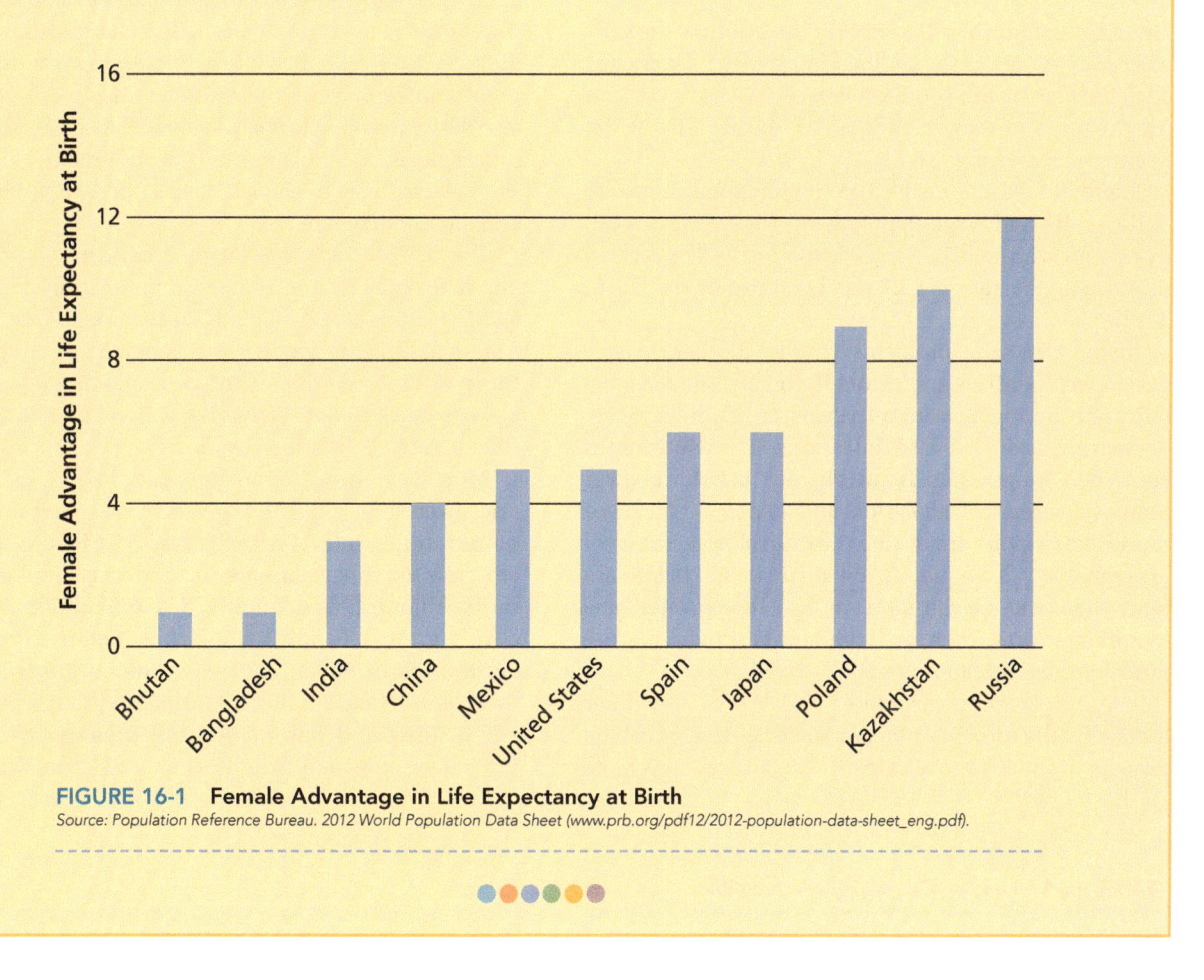

FIGURE 16-1 Female Advantage in Life Expectancy at Birth
Source: Population Reference Bureau. 2012 World Population Data Sheet (www.prb.org/pdf12/2012-population-data-sheet_eng.pdf).

---

three times as many deaths among younger males than among females. Men also are concentrated in some of the most dangerous jobs, such as structural steel workers, loggers, bank guards, coal miners, and state police. The rates of alcohol use, high-speed driving, and participation in violent sports are also much higher among men and contribute to the differences.

Although men have shorter life expectancies, women appear to be sick more often. Women have high rates of acute illnesses such as infectious and parasitic diseases, digestive problems, and respiratory conditions as well as chronic illnesses such as hypertension, arthritis, diabetes, and colitis. Some have suggested that women might not be sick more often but might just be more sensitive to bodily discomforts and more willing to report them to a doctor. (See "Our Diverse World: Women Live Longer than Men Throughout the World.")

Men are just as vulnerable to psychiatric problems as are women. A key difference, however, is that men, when emotionally disturbed, are likely to act out through drugs, liquor, and antisocial acts, whereas women display behaviors that show an internalization of their problems, such as depression or phobias.

The gap between male and female life expectancy during most of the twentieth century was attributed primarily to the fact that men smoked more than women. But in recent decades, the prevalence of smoking among women has increased while the prevalence among men has declined (Centers for Disease Control and Prevention [CDC], 2012).

## Race and Health

There are significant differences in the health of the various racial groups in the United States. Asian Americans have the best health profile, followed by whites. African Americans and Native Americans display the worst health data (see Table 16-1).

There are glaring disparities in childhood mortality of the white and black populations. The infant mortality rate for whites is 5.5 per 1,000 live births. For African Americans, the rate is 12.7 per 1,000 live births (CDC, 2012).

Many of the same problems that face mothers in less-developed countries are factors in the high infant mortality rate among African Americans. For example, black women giving birth are 2.5 times as likely as white mothers to be younger than 18, nearly one-third have fewer than twelve years of education, and 40 percent have not received prenatal care during the crucial first trimester of pregnancy. Consequently, low-birth-weight babies are more than twice as common among African Americans as among whites. There are high numbers of premature births among African Americans as well.

Life expectancies for whites and blacks also differ markedly. The African American male has the lowest life expectancy of any racial category. In 2010, a black male baby had a life expectancy of 71.4 years. For a white male baby, the figure was 76.4 years. The data for white and black females show 81.1 and 77.7 years, respectively (CDC, 2012).

Black health figures have changed in the past ten years. Homicide and HIV were both among the five leading causes of death until the mid-1990s. As deaths due to HIV disease declined, diabetes became one of the top five causes of death.

Native Americans have shown an improvement in overall health since 1950. Native Americans have the lowest cancer rates in the United States, and their mortality rates from heart disease are lower than those of the general population as well. In other areas, Native Americans fare much worse than other groups, however. Their mortality rates from diabetes are 2.3 times that of the general population. The complications from diabetes take a further toll by increasing the probability of kidney disease and blindness. Native Americans also suffer from high rates of venereal disease, hepatitis, tuberculosis, alcoholism, and alcohol-related diseases such as cirrhosis of the liver, gastrointestinal bleeding, and dietary deficiency.

The suicide rate for Native Americans is two-and-one-half times that of the white population and is the eighth leading cause of death for the population. Native American suicide victims are generally younger than those of other groups, with their suicide rate peaking between ages 15 and 39. For the general population, the suicide rate peaks after age 40.

Mortality rates for Asian and Pacific Islanders are lower than expected, given their social and economic status in the United States. The leading causes of death for this group are heart disease and cancers rather than injuries, suicide, or homicide, as in other ethnic groups. Studies show that Asians experience an immigrant advantage, which might contribute to their good health. International migrants are usually healthy and optimistic individuals who tend to eat better and take care of their overall health more

TABLE 16-1 Leading Causes of Death by Race

| White | Black | Asian or Pacific Islander |
|---|---|---|
| 1. Heart disease | 1. Heart disease | 1. Cancer |
| 2. Cancer | 2. Cancer | 2. Heart disease |
| 3. Lower respiratory disease | 3. Blood-vessel disease | 3. Blood-vessel disease |
| 4. Blood-vessel disease | 4. Accidents | 4. Accidents |
| 5. Accidents | 5. Diabetes | 5. Diabetes |
| 6. Alzheimer's disease | 6. Lower respiratory disease | 6. Pneumonia |
| 7. Diabetes | 7. Kidney disease | 7. Respiratory disease |
| 8. Pneumonia | 8. Homicide | 8. Kidney disease |

Sources: Centers for Disease Control and Prevention. 2012; National Vital Statistics. "Deaths: Leading Causes for 2008." Melonie Heron, 60(6), June 6 (www.cdc.gov/nchs/data/nvsr/nvsr60/nvsr60_06.pdf), accessed September 12, 2012.

conscientiously than nonimmigrants. Although the Asian population in the United States includes a high proportion of recent immigrants, the immigrant advantage alone does not fully account for the good health of Asian Americans (National Center for Health Statistics, 2012).

## Social Class and Health

It should surprise no one that poverty and the difficult life circumstances that accompany it, such as inadequate housing, malnutrition, stress, and violence, increase one's chance of getting sick. Recognition of the link between poverty and disease has existed for more than a century; however, today that link is viewed differently than in the past. Researchers in the past saw high disease rates among the poor as a result of moral weakness. In 1891, John Shaw Billings, the surgeon general of the United States at that time, after examining records from different hospitals, came to the surprising conclusion for that time that lower-income patients had higher death rates than those with more money. Billings decided that this reflected the moral failing of "a distinct class of people who are structurally and almost necessarily idle, ignorant, intemperate and more or less vicious, who are failures or the descendants of failures" (Billings, 1891).

Part of the reason the poor have higher death rates is because they have less access to high-quality medical care and good nutrition and are less likely to feel they have control over their life circumstances. They are also more likely to smoke, be overweight, and have low physical activity levels. Studies that have tried to account for these risk factors have shown that this is only part of the explanation, suggesting that the relationship between social class and health is more complex (National Center for Health Statistics, 2000).

Poverty contributes to disease and a shortened life span, both directly and indirectly. An estimated 25 million Americans do not have enough money to feed themselves adequately and, as a result, suffer from serious nutritional deficiencies that can lead to illness and death. Poverty also produces living conditions that encourage illness. Pneumonia, influenza, alcoholism, drug addiction, tuberculosis, whooping cough, and even rat bites are much more common in poor minority populations than among middle-class ones. Inadequate housing, heating, and sanitation all contribute to those acute medical problems, as does the U.S. fee-for-service system that links medical care to the ability to pay.

An example of how social class can account for race differences with respect to health issues can be seen by examining the health data of Asian Americans. Asian Americans have the highest levels of income, education, and employment of any racial or ethnic minority

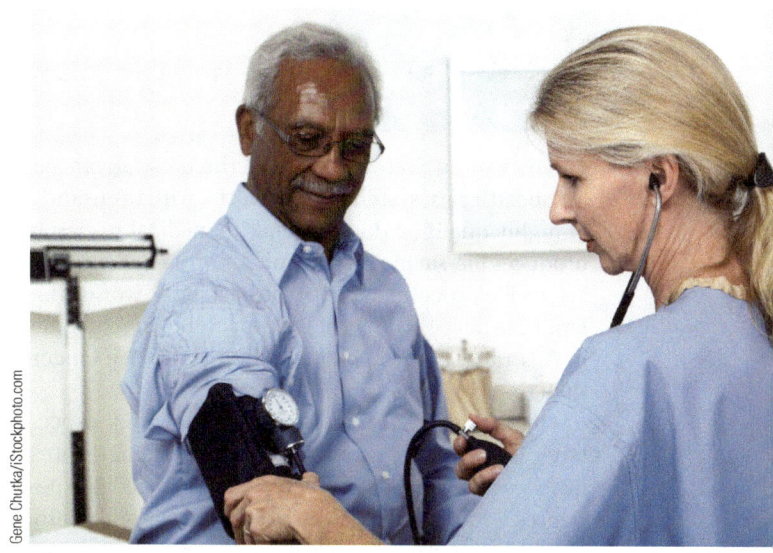

African American males have the shortest life expectancy of any racial group.

in the United States, often exceeding those of the general white population. At the same time, the lowest age-adjusted mortality rates in the United States are among Asian Americans. Even though heart disease is a leading cause of death for Asian Americans, their mortality from this disease is less than that for whites and for other minorities. Deaths from homicide and suicide are particularly low for Asian Americans. The infant mortality rates for Asian Americans range between 5 and 6 per 1,000, which are lower than those for whites. Although infant mortality rates are only one indicator of health within a group, they are nevertheless an important measure of the quality of life experienced by that population.

Studies of life expectancy show that on every measure, social class influences longevity. At age 65, white men in the highest-income families can expect to live 3.1 years longer than white men in the lowest-income families. (See "Our Diverse Society: Why Isn't Life Expectancy in the United States Higher?")

## Age and Health

As advances in medical science lengthen the life span of most Americans, the problem of medical care for the aged becomes more acute. In 1900, there were only 3.1 million Americans aged 65 or older, a group that constituted a mere 4 percent of the total population. Today, however, there are nearly 38 million Americans aged 65 or older, a full 13 percent of the total population. This change in the age structure of the American population has had important consequences for health and health care (U.S. Bureau of the Census, 2009).

## OUR DIVERSE WORLD

### Why Isn't Life Expectancy in the United States Higher?

The United States has one of the most advanced health care systems in the world, with thousands of highly qualified doctors, and its medical research produces life-saving breakthroughs on a regular basis. Yet, its life expectancy is not the best in the world. The Japanese have the longest life expectancy, and people in Australia, France, Spain, and Italy also live longer than they do in the United States.

One of the ways to understand why the U.S. life expectancy is not higher is to realize that the United States includes many groups who have vastly different life experiences. Some groups have life expectancies significantly above the average, others significantly below the average.

Studies have shown that the most educated Americans have significantly longer life expectancies than the least educated Americans. Surprisingly, that gap has been getting larger. Life expectancy for white women without a high school diploma is 73.5 years, compared with 83.9 years for white women with a college degree or more. For white men, the gap is even larger: 67.5 years for those without a high school degree compared with 80.4 for those with a college degree or better.

The steepest declines in life expectancy have been among white women without a high school degree. This group actually lost five years of life expectancy between 1990 and 2008, causing black women without a high school degree to have longer life expectancies than white women with the same education level. White men without a high school diploma lost three years of life during that period.

These drops in life expectancy, particularly for women, have put U.S. women in 41st place in 2010 on the world rankings, down from 14th place in 1985.

The reasons for the life expectancies of the least educated Americans becoming worse include an increase in prescription drug overdoses among young whites, higher rates of smoking among less educated white women, rising obesity, and a steady increase in the number of the least educated Americans who lack health insurance.

Additional reasons the average life expectancy in the United States is not higher include the following:

- In the United States, some groups, such as Native Americans, rural African Americans, and the inner-city poor, have extremely poor health that is more characteristic of a poor developing country rather than of a rich industrialized one.
- The United States is one of the leading countries for cancers related to tobacco use, especially lung cancer. Tobacco use also causes chronic lung disease.
- Americans exhibit a high coronary heart disease rate, which has dropped in recent years but remains high.
- High rates of HIV infection exist among young minority group members.
- The United States has fairly high levels of violence, especially homicides, when compared to other industrial countries.

*Source:* World Health Organization. April 2009. "World Development Report, 2009." Washington, D.C.; Olshansky et al. September 2012. "Differences in Life Expectancy Due to Race and Educational Differences Are Widening, and Many May Not Catch Up," *Health Affairs,* 31(9):1803–13; Tavernise, Sabrina. September 20, 2012. "Life Spans Shrink for Least-Educated Whites in the U.S." *The New York Times,* pp. A1, A3.

---

At the turn of the twentieth century, more Americans were killed by pneumonia, influenza, tuberculosis, infections of the digestive tract, and other microorganism diseases than by any other cause. By comparison, only 8 percent of the population died of heart disease and 4 percent of cancer. Today, this situation is completely reversed. Heart disease, cancer, chronic lower respiratory diseases, and stroke are now the most common causes of death. Those diseases are tied to the bodily deterioration that is a natural part of the aging process (CDC, 2012).

The result of these changes in health patterns is increased hospitalization for those older than 65. Only 1 percent of all Americans are institutionalized in medical facilities. However, of those 65 and older, 5 percent are institutionalized in convalescent homes, homes for the aged, hospitals, and mental hospitals.

### Education and Health

The age-adjusted death rate for those with less than a high school diploma or equivalent is 13.8 percent higher than the rate for those with a high school diploma or equivalent and 2.6 times the death rate for those with some college or collegiate degree (CDC, 2009 Report). We usually assume that the benefits of a college education include higher incomes and a richer intellectual life, but we should also add better health

## DAY-TO-DAY SOCIOLOGY

### Marijuana: A Benign Drug or a Health Problem?

In 2007, in Oakland, California, Oaksterdam University was founded to train students how to profit from selling marijuana. Their faculty includes many people working to legalize the drug. Marijuana was legalized in California in 1996 for medical reasons. Since then, many doctors have been writing prescriptions for the drug for people with borderline medical needs for it.

At the same time, the United Nations has found that marijuana is the most widely used illegal drug in the world. An estimated 146 million people use it each year. A substantial amount of marijuana is grown in the United States, and a large amount comes into the country from Mexico. The Mexican drug cartels get about 70 percent of their $38 billion profits from U.S. marijuana sales.

People are readily willing to admit that cocaine, methamphetamine, and heroin are addictive. Yet, many people believe marijuana is not addictive. It has been romanticized by musicians as diverse as Bob Dylan and Louis Armstrong. Some states have decriminalized usage and 18 states have legalized use of medical marijuana.

At normal levels, marijuana produces a relaxed euphoria. Taken in excess, a variety of unpleasant symptoms, ranging from anxiety and panic attacks to disorientation, can occur; and long-term use can increase the risk of developing schizophrenia or other psychoses. Smoking marijuana increases the risk of lung cancer, respiratory disease, and cardiovascular problems.

A variety of studies have found that between 2 and 3 percent of marijuana users become addicted within two years of first trying the drug. About 10 percent will become addicted at some point. According to a report by the Substance Abuse and Mental Health Service Administration, admissions for marijuana and hashish addictions are greater than for heroin, cocaine, and methamphetamine. A study of first-year college students at a mid-Atlantic university found one in ten reported problems related to marijuana use, such as concentration problems (40.1%) or putting themselves in physical danger (24.3%).

Marijuana addiction typically does not kill, wreck careers, or produce the sort of tragedies that make headlines. Memory of such events can keep a heroin user or alcoholic from relapsing. The less dramatic impact of marijuana addiction makes it easier to ignore; research shows that breaking from marijuana addiction can be very difficult.

People addicted to marijuana note that they are unable to move forward in their personal and professional lives while in a constant state of intoxication. Addiction has the potential to damage relationships and lead to arrest. Withdrawal symptoms include irritability, anger, nervousness, difficulty sleeping, and a variety of physical ailments.

If marijuana addiction were benign, thousands of people would not be seeking help each year. In recent years, the percentage of people in treatment who cited marijuana as their primary problem has more than doubled, from 7 percent in 1993 to 16 percent. Among illegal drugs, only opiates are more of a problem than marijuana.

The benign quality of marijuana has come into question. In recent years, studies have shown a link between marijuana use and the development of testicular cancer. Men who said they had ever smoked marijuana had more than twice the risk of aggressive testicular tumors, compared to men who did not smoke marijuana (Goodman, 2012).

Various studies have found that marijuana smoke contains 50–70 percent more carcinogens than tobacco smoke. Marijuana users are likely to inhale more deeply and hold their breath longer than tobacco smokers, which increases the lungs' exposure to carcinogenic smoke (National Institute of Drug Abuse, 2010).

*Sources:* Roffman, Roger A., and Robert S. Stephens (eds.). 2006. *Cannabis Dependence: Its Nature, Consequences, and Treatment.* Cambridge, UK: Cambridge University Press; United Nations Office of Drugs and Crime. 2004. "2004 World Drug Report." United Nations Publications Sales, No. E.04.XI.16; Caldeira, K. M., A. M. Arria, K. E. O'Grady, K. B. Vincent, and E. D. Wish. March 2008. "The Occurrence of Cannabis Use Disorders and Other Cannabis-Related Problems among First-Year College Students." *Addictive Behaviors* 33(3):397–411; Kershaw, Sarah, and Rebecca Cathcart. July 19, 2009. "Reefer Madness." *The New York Times*, St. 1, 2; National Institutes of Health. "DrugFacts: Marijuana, 2012" (www.drugabuse.gov/publications/drugfacts/marijuana), accessed September 8, 2012; Meier, M. H., et al., "Persistent Cannabis Users Show Neuropsychological Decline from Childhood to Midlife." Proceedings of the National Academy of Sciences, August 27, 2012; Zalesky, A., et al. June 4, 2012. "Source Effect of Long-Term Cannabis Use on Axonal Fibre Connectivity." *Brain.*

to the list. Not only are death rates for those with a college education considerably lower than for those with less education, particularly for those who have not completed high school, but those with a college education also are actually healthier and more active during those extra years they live. This is partly because college-educated people are less likely to smoke or take part in risky behavior.

Cigarette smoking by adults is strongly associated with educational attainment. Adults with less than a high school education were almost three times as likely to smoke as those with a bachelor's degree or more education. The percentage of high school students who smoke cigarettes increased in the early 1990s, but since 1997, the percentage of students who smoke has declined (Pastor, Makuc, and Xia, 2002). (For a related discussion, see "Day-to-Day Sociology: Marijuana: A Benign Drug or a Health Problem?")

### Women in Medicine

Women have always played a major role in U.S. health care and, by the late nineteenth century, the United States was a leader in the training of female doctors. In the twentieth century, the number of women in medical schools dropped markedly when a review of medical education closed many medical schools that had served them. There were fewer female physicians in Boston in 1950 than there had been in 1890.

The rise of the women's movement and affirmative action created an atmosphere more conducive to women becoming physicians. In 1960, only 5.8 percent of incoming medical students were female, but the proportion increased to 13.7 percent in 1971 following the passage of the Equal Opportunity Act. A good predictor of this trend continuing is that in 2011, 47 percent of new entrants into medical school were women. Women also represent 34 percent of all medical faculty compared to 26 percent in 1998 (American Association of Medical Colleges, 2012).

Female physicians earn less than male doctors. According to the American Medical Association, salaries of female family practitioners average 86 percent of that earned by men. The earnings ratio is higher for women who have been practicing medicine for twenty years or more.

Several factors account for these differences. The average woman in medical practice is younger and less experienced than the average man. She also earns less than a man partly because she works fewer hours, 55 a week compared with 61 for men. Doctors usually are not paid by the hour, but those in private practice are paid according to how many patients they see; fewer hours translate into fewer patients and less income.

Women doctors might work fewer hours than men for the same child-care-related reasons other women do. Two-thirds of practicing female physicians have children. Although they might have broken some traditional barriers at work, they still remain the primary family caretakers, being in charge of three times as much child care as men and more than twice the other household duties.

Studies also show that female doctors spend more time with each patient. Their per-visit fees are not necessarily smaller, but they see fewer patients per hour than men. Male physicians saw 117 patients per week compared with 97 for women.

This may explain why studies show that patients consider women physicians "more sensitive, more altruistic, and less egoistic" than men. Patients who see a female physician report a significantly higher total satisfaction level than those who see a male physician. These patients might be picking up on the fact that when men were asked why they entered medicine, they responded, "prestige and salary," whereas women said, "helping people." The interesting side effect of this differing approach to practicing medicine is that women physicians are far less likely to be sued for malpractice than are male physicians.

The type of practice a woman performs also determines her earnings. Female physicians are twice as likely as male physicians to be employed by a hospital, health maintenance organization (HMO), or group practice. Men are more likely to be in independent practices. Independent practitioners generally earn more money and tend to work longer hours.

Women also tend to be concentrated in lower-paying specialties such as internal medicine, pediatrics, and family practice. Women represent at least 60 percent of new residents in dermatology, medical genetics, and obstetrics. Specialists such as radiologists, surgeons, and cardiologists typically earn more than primary-care practitioners (American Association of Medical Colleges, 2012).

As the number of women entering the field continues to increase, we should see changes in the education of physicians and the manner in which medical care is carried out.

## ● CONTEMPORARY HEALTH CARE ISSUES

The American health care system is among the best in the world. The United States invests a large amount of social and economic resources into medical care. It has some of the world's finest physicians, hospitals, and medical schools. It is no longer plagued by infectious diseases and is in the forefront in developing medical and technological advances for the treatment of disease and illness. At the same time, there are many issues that the American health care system must address. In this section, we discuss a few of them.

## Acquired Immunodeficiency Syndrome

The Centers for Disease Control and Prevention (CDC) defines acquired immunodeficiency syndrome (AIDS) as a specific group of diseases or conditions that are indicative of severe immunosuppression related to infection with the human immunodeficiency virus (HIV). The CDC estimates that 1.2 million people are living with AIDS: 44 percent are black, 33 percent are white, 17 percent are Hispanic, 1 percent are Asian/Pacific Islander, and less than 1 percent are American Indian/Alaska Native. Newly developed antiretroviral drugs have produced a drop in the number of deaths from AIDS over the past ten years.

AIDS, a disease now known virtually everywhere in the world, was identified only in 1981, and the retrovirus that causes it was discovered only in 1983. Yet, this disease could transform the global future in ways no one imagined even a decade ago. For many developing countries, the impact of this disease could be as great as that of a major war unless a vaccine or low-priced cure can be developed soon.

AIDS is caused by the human immunodeficiency virus (HIV), which is a member of the retrovirus family. HIV gradually incapacitates the immune system by infecting at least two types of white blood cells. The depletion of white blood cells leaves the infected person vulnerable to a multitude of other infections and certain types of cancers. Those infections and cancers rarely occur, or produce only mild illness, in individuals with normally functioning immune systems. HIV also causes disease directly by damaging the central nervous system.

HIV is transmitted through sexual contact, piercing the skin with HIV-contaminated instruments, transfusion of contaminated blood products, and transplantation of contaminated tissue. An infected mother can transmit the virus to her child before, during, or shortly after giving birth. There is no evidence that HIV is transmitted by casual contact.

Nearly three-quarters of the people living with HIV/AIDS are male. According to the CDC, 61 percent of those males have been exposed through male-to-male sexual contact, and 36 percent through injection drug use.

AIDS is not as easily transmitted as is commonly assumed, perhaps because carriers are only intermittently infectious. Women are much more likely to contract the disease from infected men than are men from infected women. It takes as long as two years before half of the spouses or regular sex partners of infected people become infected.

On a national scale, the spread of AIDS is concentrated in a number of metropolitan areas. Areas hardest hit include Baton Rouge and New Orleans, Louisiana; Miami, Florida; Jackson, Mississippi; and Baltimore, Maryland (CDC, 2010).

Initially, the majority of U.S. AIDS victims were male homosexuals. AIDS, however, is not a gay disease; it can be transmitted between two people of any sex by the exchange of infected blood or semen.

In the United States, the impact of HIV and AIDS in the African American community has been devastating. Representing only an estimated 13 percent of the total U.S. population, blacks accounted for 44 percent of all HIV/AIDS cases diagnosed in 2009.

Today, the majority of AIDS deaths are black and Hispanic men. The disease has shifted from one that affects mainly white gay men to one whose victims are minority group members. Race, class, and gender issues affect how an illness is perceived and treated, and discrimination factors are not insignificant.

## Health Insurance

Most people pay for their health services through some form of insurance. Poor people, however, cannot afford premiums or the out-of-pocket expenses required before insurance coverage begins. They receive coverage through the government-sponsored Medicare and Medicaid programs.

Most of the money spent on medical care in the United States comes in the form of third-party payments as differentiated from direct or out-of-pocket payments. Third-party payments are those made through public or private insurance or charitable organizations. Essentially, insurance is a form of mass financing that ensures that medical-care providers will be paid and people will be able to obtain the medical care they need. Insurance involves collecting small amounts of money from a large number of people. That money is put into a pool, and when any of the insured people get sick, that pool pays for the medical services.

The United States has private and public insurance programs. Public insurance programs include Medicare and Medicaid, which are funded with tax money collected by federal, state, and local governments. The country also has two types of private insurance organizations: nonprofit, tax-exempt Blue Cross and Blue Shield plans, and for-profit commercial insurance companies. Blue Cross and Blue Shield emerged out of the Depression of the 1930s as a mechanism to pay medical bills to hospitals and physicians. People made regular monthly payments to the plan and, if they became sick, their hospital bills were paid directly by the insurance plan. Blue Cross and Blue Shield originally set the cost of insurance premiums by what was called community rating, giving everybody within a community the chance to purchase insurance at the same price. Commercial insurance companies appeared after World War II and based their prices on experience rating, which bases premiums on the statistical likelihood of the individual needing medical care. People more likely to need medical care are charged more than those less likely to need it. Eventually, to compete, Blue Cross and Blue Shield had to follow the path of the commercial insurers.

One unfortunate result of the use of experience ratings was that those who most needed insurance coverage, the old and the sick, were least able to afford or obtain it. Congress created Medicare and Medicaid in 1965 to help with this problem.

Medicare pays for medical care for people older than 65, and Medicaid pays for the care of those too poor to pay their own medical costs. Adults 18 to 24 years of age are most likely to lack coverage. Persons with incomes below or near the poverty level are almost four times as likely to have no health insurance coverage as those with incomes twice the poverty level or higher. Hispanic persons and non-Hispanic black persons are less likely to have health insurance than non-Hispanic white persons (Institute of Medicine, 2009).

Even with these forms of government insurance, many people believe that the way health care is delivered in the United States produces problems. Critics note that the United States is the only leading industrial nation that does not have an organized, centrally planned health care delivery system. Attempts to resolve this problem are often met with resistance from the medical establishment and the general public.

In addition, the care the poor receive is inferior to that received by the more affluent. Many doctors will not accept Medicare or Medicaid assignments, and most doctors will not practice in poor neighborhoods, with the result that the poor are relegated to overcrowded, demeaning clinics. Under conditions such as these, it is not a surprise that the poor generally wait longer before seeking medical care than do more affluent patients, and many poor people seek medical advice only when they are seriously ill and intervention is already too late.

At various times, a national health insurance program has been suggested. The American Medical Association is a leading opponent of such a plan. Doctors prefer a fee-for-service system of remuneration.

The fee-for-service approach has produced health care delivery problems such as the uneven geographic distribution of doctors and the overabundance of specialists. For example, only 8.3 percent of the nation's physicians are general practitioners, even though many more are needed to treat the total population, especially the poor and elderly. Part of the reason is that salaries for general practitioners are about half those of specialists such as radiologists or neurosurgeons (Medical Group Management Association, 2001).

Health maintenance organizations (HMOs) and other types of managed care have dramatically changed the face of health care in the United States in the past decades. HMOs try to control the cost of medical care by strictly regulating patients' access to doctors and treatments. This shift has meant lower insurance premiums for many individuals and employers. Many people, however, say those savings have come at too great a price. The HMOs have been accused of being more concerned with profits than with patients.

In 2010, President Obama signed the Affordable Care Act into law. The act puts comprehensive reforms into place that improve access to affordable health coverage for everyone and protect consumers from abusive insurance company practices. The new law makes it possible for all Americans to obtain health insurance and provides new ways to bring down costs and improve quality of care.

### Preventing Illness

Our cultural values cause us to approach health and medicine from a particular vantage point. We are conditioned to distrust nature and assume that aggressive medical procedures work better than other approaches. This situation has caused about one-quarter of all births in the United States to be by cesarean section. The rate of hysterectomy in the United States is twice that in England and three times that in France, and 60 percent of these hysterectomies are performed on women younger than 44. Our rate of coronary bypass operations is five times that of England. We tend to be a can-do society in which doctors emphasize the risk of doing nothing and minimize the risk of doing something. This approach is coupled with Americans' desire to be in perfect health. The result is that far more surgery is performed in the United States than in any other country.

Yet, our experience with heart disease has shown us that significant health benefits can be obtained from such adjustments as changing diets or engaging in healthier practices. At the moment, the best way to deal with the AIDS crisis is through prevention techniques that limit the spread of the illness.

If the health care system is to reorient its perspective from one that seeks cures to one that focuses on prevention, it must place greater emphasis on sociological issues. Illness and disease are socially as well as physiologically produced. During the past century, the medical system has devoted a great deal of effort to combating germs and viruses that cause specific illnesses, but we are starting to see that there are limitations to this viewpoint. We must now investigate environments, lifestyles, and social structures for the causes of disease with the same commitment we have shown to investigating germ theory.

At first glance, most of the factors that come to mind when thinking about preventing disease are little more than healthful habits. For example, if people adopt better diets, with more whole grains and less red meat, sugar, and salt; and if people stop smoking, exercise regularly, and keep their weight down, they will surely prevent illness.

Being overweight or obese are risk factors for a variety of chronic health conditions, including hypertension and diabetes mellitus. Despite public health efforts to encourage Americans to attain and maintain a healthy weight, it appears that much work remains to be done. The CDC (2009) uses a body mass index as a way to define which adults are overweight or obese. The body mass index (BMI) is determined by using weight and height to calculate a number. The BMI correlates with the amount of body fat a person has. A BMI between 25 and 29.9 is considered overweight. A BMI over 30 is considered obese. More than half of all American adults (54.7%) are overweight, and 1 in 5 (19.5%) are obese.

Obesity is most prevalent among middle-aged adults, black and Hispanic adults, and among peoples with less education and lower income. Among women, African American women have the highest percentage of overweight individuals (64.5%), followed by Hispanic women (56.8%), white women (43%), and Asian/Pacific Islander women (25.2%). In addition, black women are nearly twice as likely as white women and more than five times as likely as Asian/Pacific Islander women to be obese. There is little difference among white, Hispanic, and African American men, with more than 50 percent of each group being overweight. Only Asian/Pacific Islander men have significantly lower rates of being overweight (35.2%) than each of the other three groups.

Being overweight or obese is correlated with educational and social class levels. The prevalence of being overweight ranged from approximately 42 percent among women who had not finished high school to about 23 percent for women who had earned college degrees. In contrast, among men, the prevalence of being overweight remained high (and even increased) at all levels of education, dropping noticeably only among men who had attained college degrees. Even then, well over 50 percent of men were overweight.

Among men and women, rates of obesity were particularly high for those who had not graduated from high school and those who had earned a GED. Rates declined steadily with increasing education and were markedly lower among women who had earned a bachelor's degree or an advanced academic degree. Similarly among men, rates of obesity were highest among those with less education and lowest among those with college degrees, although the association was not as strong as for women (see Figure 16-2). (For more on this topic, see "How Sociologists Do It: Can Your Friends Make You Fat?")

We must also think of illness prevention as involving at least three levels: medical, behavioral, and structural. Medical prevention is directed at the individual's body, behavioral prevention is directed at changing behavior, and structural prevention is directed at changing the society or the environments within which people work and live.

We hear a great deal about trying to prevent disease on a behavioral level. Although this is an important level of prevention, we know little about how to change people's unhealthful habits. Education is not sufficient. Most people are aware of the health risks of smoking or not wearing seat belts, and yet roughly 30 percent of Americans smoke and 80 percent do not use seat belts regularly. Sometimes individual habits are responses to complex social situations, such as coping mechanisms to deal with stressful and alienating work environments. Behavioral approaches to prevention focus on the individual and place the burden of change on the individual.

The structural factors related to health care in the United States do not get as much attention as the medical and behavioral factors. We discussed the issues of gender, race, and class earlier in this chapter. Increasingly, people are realizing that health care and the prevention of illness take place on a number of levels.

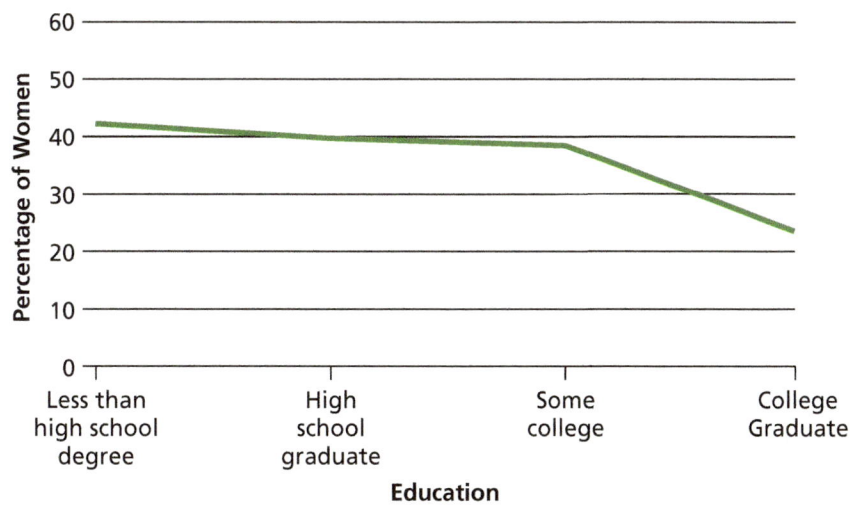

FIGURE 16-2 **Percentage of Women Overweight by Education**
Source: Ogden, Cynthia L., Molly M. Lamb, Margaret D. Carroll, and Katherine M. Flegal. December 2010. "Obesity and Socioeconomic Status in Adults: United States, 2005–2008." National Centers for Health Statistics, Data Brief.

## HOW SOCIOLOGISTS DO IT

### Can Your Friends Make You Fat?

Obesity has increased dramatically in the last thirty years. About 31 percent of the population is obese, which is defined as a body mass index of 30. Sixty-six percent of adults are overweight. People think it is due to a change in activity levels and overeating.

Can obesity spread from person to person, much like a virus? When one person gains weight, do his or her close friends gain weight also?

A study was done by doctors trained in medical sociology of 12,067 people who had been closely followed for 32 years. The investigators checked the subjects' weight gain against their friends, as well as spouses, siblings, and neighbors. The researchers found that if one person became obese, the odds of the friend becoming obese nearly tripled. The closer the friendship, the more likely the friends became fat together. The obesity relationship between close friends existed even when they were separated by hundreds of miles.

Family members had less influence than friends, and there was no relationship between a person becoming obese and his or her neighbor becoming obese. Among siblings, one sibling's chance of becoming obese increased by 40 percent if the other sibling became obese.

One way to explain these findings is that when a close friend becomes obese, obesity may not look so bad. It changes your idea of what is an acceptable body type as you look at the people around you. This may help to explain why Americans have become fatter in recent years. Every person who becomes fat may drag a friend with him or her. It can be thought of as a kind of social contagion that spreads through the social network.

The environment may play a large role in weight gain. As more and more people get fat, the social networks help obesity to spread. A side note from this study may be that if you want to avoid becoming fat, avoid having fat friends. Of course, most people do not want to lose a friend merely because he or she gains weight.

On average, the researchers found that a person who became obese gained 17 pounds. Those extra pounds were added onto the natural increases in weight that occur when people get older.

The study may help to explain why people have gotten so fat so fast. It appears that the social influence of our friends is very important even with something that seems unrelated like weight gain.

*Source:* Christakis, Nicholas A., and James H. Fowler. July 27, 2007. "The Spread of Obesity in a Large Social Network Over 32 Years." *New England Journal of Medicine* 357(4):370–79.

## THE AGING POPULATION

The United States is experiencing a major population shift. For the first time in our history, there are more people aged 65 and over in the population than there are teenagers. When the first U.S. census was taken in 1790, about 50,000 Americans were older than 65, representing 2 percent of the population of 2.5 million. One hundred years later, the over-65 population had grown to 2.4 million, or just under 4 percent of the population.

The life expectancy of a newborn white child in 1890 was only 50 years; it was less than 35 for African Americans. In just three decades, the average life expectancy shot up to 60 years for whites and 50 years for African Americans. At that point, the number of elderly people had more than doubled, to 6.7 million, about 5 percent of the total population.

By 1960, the elderly population more than doubled again. There were 16.7 million people over age 65, about 9 percent of the population. By 2010, the number had more than doubled again, to 40 million, representing 13 percent of the population (U.S. Bureau of the Census, March 2012).

By 2030, demographers estimate that one in five Americans will be age 65 or older, which is nearly four times the proportion of elderly one hundred years earlier, in 1930. The effects of this change will reverberate throughout the American economy and society (see Figure 16-3).

There are demographic and medical reasons for the growth in the older population. Many of the elderly were born when the birthrate was high or immigrated to the United States before World War II. The growth in the elderly population has also been influenced by the improvements in medical technology that have created a dramatic increase in life expectancy.

In the next twenty years, the growth in the elderly population is going to be influenced by the aging of the baby boom population, those people born between 1946 and 1964. This population bulge of 76 million people will place substantial demands on society as it tries to address their needs, whether for health care, housing, or leisure activities.

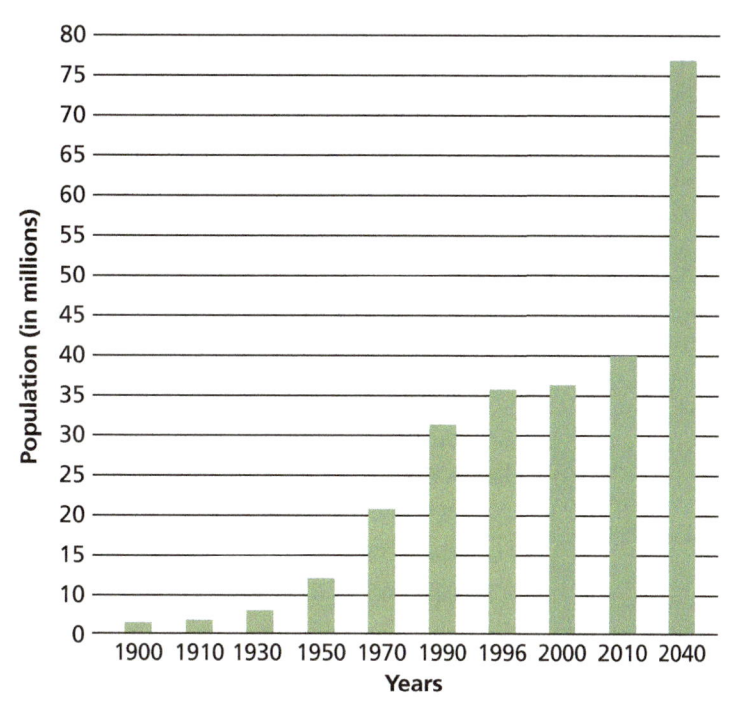

**FIGURE 16-3** United States Population Age 65 and Over, 1900–2040
Sources: U.S. Bureau of the Census. 1976. Historical Statistics of the United States: Colonial Times to 1970. Current Population Reports, pp. 25–1104, Table 2; U.S. Bureau of the Census. March 3, 2009. "Older Americans Month: May 2009." Facts for Feature (www.census.gov/Press-Release/www/releases/archives/facts_for_features_special_editions/013384.html), accessed August 13, 2009; Jacobsen, Linda A., et al. 2011. "America's Aging Population." Population Bulletin 66(1).

For the first time in our history, there are more people ages 65 and older in the population than there are teenagers.

Until the past fifty years, most gains in life expectancy came as the result of improved child mortality. The survival of larger proportions of infants and children to adulthood radically increased average life expectancy in the United States and many other countries over the past century. Now, gains are coming at the end of life as greater proportions of 65-year-olds are living into their 80s and 90s.

Historically, families have played a prominent role in the lives of elderly people. Is this likely to change? These changes raise a multitude of questions: How will these years of added life be spent? Will increased longevity lead to a greater role for the elderly in our society? What are the limits of life expectancy? (See "Thinking About Social Issues: The Discovery of a Disease.")

## Composition of the Older Population

At the dawn of the twentieth century, three demographic trends—high fertility, declining infant and child mortality, and high rates of international immigration—were acting together in the United States to keep the population young. All of this has changed, as we just noted.

People tend to think of the older population as a homogeneous group with common concerns and problems. This is not the case. We can divide the older population into three groups: the young-old, ages 65 to 74 (18.7 million); the middle-old, ages 75 to 84 (11.4 million); and

> ## THINKING ABOUT SOCIAL ISSUES
> ### The Discovery of a Disease
>
> Something had to be done with the woman. Frau August D. was 51 years old, socially prominent, and she was becoming an embarrassment to her family. She raged at her husband, accusing him of infidelity. She raged at her doctors, accusing them of rape. She wandered the city. She screamed in the streets. So, one day in 1906, in the city of Frankfurt in Germany, her family brought her to see a doctor, a promising 42-year-old Bavarian neuropathologist. Stout and balding, peering out at the world through thick spectacles, the doctor almost was a parody of the owlish academic. His name was Alois Alzheimer.
>
> Alzheimer's first thought was that the woman was suffering from "presenile dementia," a condition only recently defined, but one known to medical science from antiquity. The most recent studies available to Alzheimer suggested that the condition had something to do with a mysterious process by which the brain seemed to atrophy.
>
> Alzheimer studied the woman for nearly a year. At the end, she was screaming at him in a harsh alien voice. She knew this doctor. She knew what he wanted. He wanted what they all wanted. He wanted to cut her up.
>
> And, after she died, he did.
>
> He took tissue samples from her brain. He found them riven with sticky plaques and tangles containing a mysterious substance that would later be identified as a protein called beta-amyloid. The neurons around these areas were utterly destroyed, as if each one had been burned away.
>
> In 1907, Alzheimer published his paper. He argued that the woman had not been suffering from an unspecified condition known as premature senility. He maintained that she had been the victim of a specific disease. It sparked a great debate. Medical conferences coveted him as a speaker. In 1915, he died of heart failure, the same age as Frau D. was when her family had her brought to his clinic because she was screaming in the streets.
>
> *Source:* Excerpted from Pierce, Charles. 2000. *Hard to Forget: An Alzheimer's Story.* New York: Random House, pp. xvii–xix.

the old-old, ages 85 and older (3.8 million). The older population as a whole is, itself, getting older. Whereas, in 1960, about one-third of the older population was over 75, today nearly 50 percent of the older population is older than 75. In that same period of time, the old-old will have nearly tripled in numbers.

Elderly Americans are among the wealthiest and among the poorest in our nation. They come from a variety of racial and ethnic backgrounds. Some are employed full-time, whereas others require full-time care. Although general health has improved, many elderly suffer from poor health.

The older population in the twenty-first century will come to later life with different experiences than did older Americans in the last century—more women will have been divorced, more will have worked in the labor force, and more will be childless.

## Aging and the Sex Ratio

Women outnumber men at every age category among the elderly. Women at any age are less likely to die than men. Approximately 105 male babies are born for every 100 female babies, but higher male death rates cause the sex ratio to decline as age increases and, around age 35, females outnumber males in the United States.

The longer life spans of women will cause women to make up the majority of the older population in the future. The gap between the male and female life expectancy has been decreasing, however, and we will see a greater proportion of men in the older age categories going forward. At age 65, women can expect to live 2.7 years longer than a man aged 65. By 85, the gap between male and female life expectancy is only one year.

Most elderly women will outlive their spouses and are less likely to remarry than are men. They are more likely than older men to be poor, to live alone, or to need to enter a nursing home or be dependent on the care of others (Jacobsen, Lee, and Mather, 2011).

## Aging and Racial Minorities

Immigration will cause the United States to become a country where the majority of the people are racial and ethnic minorities by 2042. The population under 18 will reach that majority status earlier than the older population. About 80.1 percent of the people over 65 in the United States are white; 8.3 percent are African American; 7 percent Hispanic; and 3.4 percent Asian or Pacific Islander. "By 2050, the proportion of elderly who are non-Hispanic white is projected to drop to 58.5 percent as the growing minority populations move into old age" (U.S. Bureau of the Census, 2009 Population Estimates, National Population Projections, 2008) (see Figure 16-4).

The African American percentage of the over-65 group is lower than might be expected. That is, if African Americans make up 13 percent of the total population, then we would expect them to make up 13 percent of the older population instead of 8.3 percent. However, the difference is caused by the higher recent birthrate among African Americans as well as because life expectancy is lower among blacks than it is among whites (Jacobsen, Lee, and Mather, 2011). Life expectancy for white women is 4.1 years longer than for black women. For white men, it is 6.0 years longer than for black men. These differences narrow at older ages, however, so that the black-white difference in life expectancy at age 65 is about 1.7 years, and by age 85, it falls to zero. The death rates of elderly blacks fall below those of elderly whites at advanced ages. At ages 90 and older, black men and women have slightly more years of additional life expected than do their white counterparts (National Center for Health Statistics, 2009).

Why does the black-white life span gap disappear and then even reverse as the two races get older? One explanation for the crossover in mortality for black Americans is that blacks who are still alive at older ages are the hearty survivors of extraordinary mortality risks at younger ages. Because older whites were not exposed to the same mortality risks, it is believed they are more frail than blacks of the same age (Manton and Stallard, 1984). There is no real proof of this view, however.

## Aging and Marital Status

More than 95 percent of older Americans have been married. Marriage is important for older Americans for several reasons. The presence of a spouse provides a variety of resources in the household; married elderly are less likely to be poor, to enter a nursing home, or to be in poor health; spouses are the primary caregivers to their partners.

Women in the United States are more likely than men to outlive their spouse because they live longer. Seventy percent of women over age 75 are widows, whereas less than 20 percent of men that age are widowers. In addition to the longer life expectancy of women, this disparity also is caused by men's tendency to marry women younger than themselves. A widow will find few older men available to marry whereas a widower has many older women available to remarry. Consequently, it is more likely that those older people living alone or in institutions are women.

About 75 percent of men age 65 or older are currently married, compared with 44 percent of women. At the oldest ages, both men and women are less likely to have a surviving spouse, but the gender gap is even wider then. Although the likelihood of having experienced a divorce has increased over time, only a small percentage of Americans age 65 or older are divorced (Jacobsen, Lee, and Mather, 2011).

## Aging and Wealth

Special-interest groups lobby Washington politicians, claiming that millions of older adults could be thrown into poverty if there are cuts in Social Security. Meanwhile, cruise-ship companies, automobile manufacturers, and land developers present images of energetic, attractive, silver-haired couples with substantial amounts of money.

Which of these two images is correct? Both of these presentations are accurate depictions of some older Americans. The lobbyists are talking about the poorest 20 percent of households containing the elderly. These households have an average net worth of about $3,400. Luxury cruise lines are interested in the richest 20 percent, who are worth almost 90 times as much.

Some of the confusion over the economic condition of older Americans stems from the ways they use and save money. The median income of the average American household peaks when its primary working members are between the ages of 45 and 54. The incomes of households headed by people age 65 and older are two-fifths as high, on average. This would seem to indicate

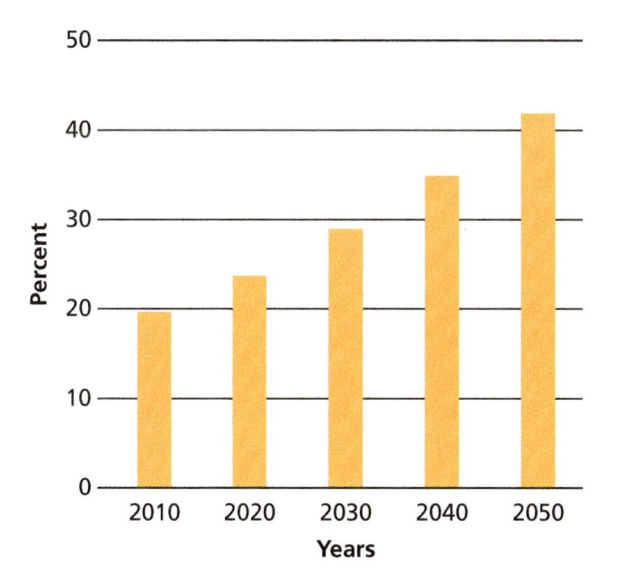

**FIGURE 16-4** Minority Elderly Americans, 2010 to 2050
*Source: U.S. Bureau of the Census, Population Division. August 14, 2008. "Percent Minority of the U.S. Population by Selected Age Groups: 2010 to 2050."*

that older Americans are significantly less well off than younger Americans.

Yet, lower incomes do not necessarily mean less spending power. Although the median income of households headed by people aged 65 and older is only about 40 percent that of households headed by 45- to 54-year-olds, the older group has a much smaller average household size, so its per-capita discretionary income is actually higher than it is among the younger group.

To get a more accurate picture, we should not focus on income but, instead, on the financial condition of older Americans—that is, net worth, or the market value of all assets minus all debts. This gives a more realistic picture of wealth than does income alone. For example, older adults can finance major purchases by cashing in on stocks, real estate, or other assets.

A Census Bureau survey found that median net worth rises with the age of the household, moving from $5,402 for households younger than 35, to $144,700 for those 45 to 54, to $170,128 for those over 65 (U.S. Bureau of the Census, 2012). Looking at net worth shows us that the oldest households control more wealth than do households in other age groups. Moreover, the share of wealth controlled by working-age Americans is eroding, whereas the share controlled by the elderly is increasing.

Three factors have caused the elderly to control a substantial and increasing portion of the nation's wealth. First, the share of households headed by the elderly has been increasing, thereby increasing the aggregate wealth of older Americans. Second, the stock market growth has benefited the affluent elderly who control a large portion of individual stock holdings. Finally, the escalation in home values in many states has boosted the net worth of the elderly because most older Americans own their own homes.

These facts paint a picture that is different from what is normally thought of as being the case. As with so much in sociology, the conclusion one draws appears to depend on which factors one considers to be most important in understanding the issue. (See "Our Diverse World: Stereotypes About the Elderly.")

## Global Aging

The numerical and proportional growth of older populations around the world is indicative of major achievements—decreased fertility rates, reductions in infant and maternal mortality, reductions in infectious and parasitic diseases, and improvements in nutrition and education—that have occurred, although unevenly, on a global scale.

The rapidly expanding numbers of older people represent a social phenomenon without historical precedent. The world's elderly population numbers half a billion people today and is expected to exceed 1 billion by the year 2020. In most countries, the elderly population is growing faster than the population as a whole. Almost half of the world's elderly live in China, India, the United States, and the countries of the former Soviet Union. China alone is home to more than 20 percent of the global total.

Increased longevity and the aging of the baby boom generation will mean that many people will find themselves caring for very old persons after they themselves have reached retirement age.

The oldest-old (85-plus) are the fastest-growing segment of the population in many countries worldwide. Unlike the elderly as a whole, the oldest-old today are more likely to reside in developed than developing countries, although this trend, too, is changing.

The percentage of the elderly population living alone varies widely among nations. In developed countries, percentages generally are high, ranging from 9 percent in Japan to 40 percent in Sweden. In developing countries, few elderly people live alone. In China, 3 percent of the elderly live alone; in South Korea, 2 percent; and in Pakistan, 1 percent. Living alone in those countries is often the result of having a spouse, siblings, and even children who have died.

To date, population aging has been a major issue mainly in the industrialized nations of Europe, Asia,

## OUR DIVERSE WORLD: Stereotypes About the Elderly

The elderly are the most heterogeneous of all age groups because we tend to age differently biologically, sociologically, and psychologically. Yet, people tend to have fairly specific stereotypes of the elderly and what it must be like to be old.

In fact, even the definition of who is "old" varies considerably. People 18 to 29 think someone is old at age 60. People who are 60 think someone is old at 72. Even among people age 75 or older, only 35 percent say they feel old.

In one study, people 18 to 64 were asked which issues were true for people over 65, including memory loss, depression, serious illness, loneliness, and several other experiences. Then, people over 65 were asked whether they had any of those experiences. Table 16-2 presents the results. As the table shows, there were vast differences between what problems people thought the elderly had and the reality of the elderly reporting having such problems.

We can understand that people would have distorted stereotypes about the elderly. Ageism has become common in our society and is one of the few areas in society in which using negative words such as *geezer* and *old coot* is not frowned upon. The belief that the elderly represent the least capable, least healthy, and least alert members of society has seeped into the thinking of many people. Ageism even exists among the elderly themselves. When elderly people believe that other older people are worse off than they are, it produces a higher level of life satisfaction. In a sense, holding the ageist stereotypes makes some elderly people feel better if they can say the stereotype does not apply to them. It is time for us all, no matter what our age, to rethink the distorted views we hold of the elderly.

**TABLE 16-2** The Elderly: Stereotypes vs. Reality

| Experience | % Ages 18–64 Expect | % Ages 65+ Reality | % Gap |
|---|---|---|---|
| Memory loss | 57 | 25 | 32 |
| Not able to drive | 45 | 14 | 31 |
| Serious illness | 42 | 21 | 21 |
| Not sexually active | 34 | 21 | 13 |
| Feeling sad or depressed | 29 | 20 | 9 |
| Not feeling needed | 29 | 9 | 20 |
| Loneliness | 29 | 17 | 12 |
| Trouble paying bills | 24 | 16 | 8 |
| Being a burden | 24 | 10 | 14 |

*Source:* Pew Research Center. June 29, 2009. "Growing Old in America: Expectations vs. Reality."

*Source:* Pew Research Center. June 29, 2009. "Growing Old in America: Expectations vs. Reality."

---

and North America. In at least thirty of these countries, 15 percent or more of the entire population is 60 and older. Those nations have experienced intense public debate over elder-related issues such as social security costs and health care provisions (see Figure 16-5). (To test your knowledge on this topic, see "Our Diverse World: Global Aging Quiz.")

### Future Trends

Population aging is already exerting major consequences and implications in all areas of day-to-day human life, and it will continue to do so. In the economic area, population aging will affect economic growth; savings, investment, and consumption; labor markets; pensions; taxation; and the transfers of wealth, property, and care from one generation to another. Population aging will continue to affect health and health care, family composition and living arrangements, housing, and

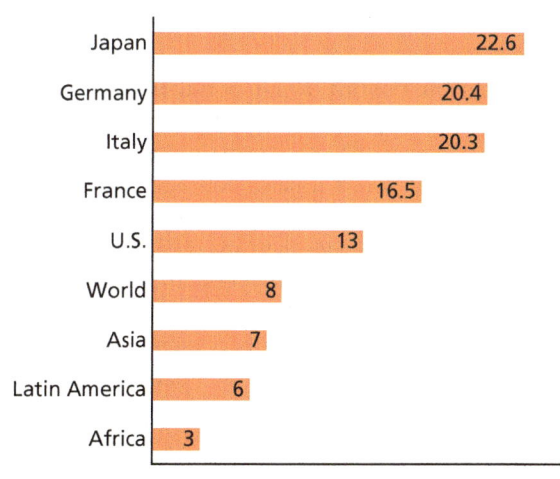

| Region | Percent |
|---|---|
| Japan | 22.6 |
| Germany | 20.4 |
| Italy | 20.3 |
| France | 16.5 |
| U.S. | 13 |
| World | 8 |
| Asia | 7 |
| Latin America | 6 |
| Africa | 3 |

**FIGURE 16-5** Older Populations in Selected Areas: Percent Age 65 and Older

*Sources:* Population Reference Bureau. 2012 World Population Data Sheet (www.prb.org/pdf12/2012-population-data-sheet_eng.pdf), accessed September 13, 2012; U.S. Bureau of the Census. 2009. International Population Reports, P95/09-1.

## OUR DIVERSE WORLD: Global Aging Quiz

1. *True or false?* By 2012, children under the age of 15 will still outnumber elderly people (aged 65 and over) in almost all nations of the world.
2. China has the world's largest total population (more than 1.2 billion people). Which country has the world's largest elderly (65+) population?

    **a.** Japan   **b.** Germany   **c.** China   **d.** Nigeria

3. *True or false?* More than half of the world's elderly today live in the industrialized nations of Europe, North America, and Japan.
4. Of the world's major countries, which had the highest percentage of elderly people in 2010?

    **a.** Sweden   **b.** Japan   **c.** Italy   **d.** France

5. *True or false?* Current demographic projections suggest that 35 percent of all people in the United States will be at least 65 years of age by the year 2050.
6. *True or false?* People 65 and over have the highest level of well-being in Denmark, Netherlands, Switzerland, and the United States.
7. *True or false?* In developing countries, older men are more likely than older women to be illiterate.

### Answers

1. **True.** Although the world's population is aging, children still outnumber the elderly in most areas; but in at least 16 countries, the elderly outnumber children. They include Sweden, Spain, Germany, Greece, Italy, and Japan.
2. **c.** China also has the largest elderly population, numbering nearly 178 million people over the age of 60 in 2010.
3. **False.** Although industrialized nations have higher percentages of elderly people than do most developing countries, 59 percent of the world's elderly now live in the developing countries of Africa, Asia, Latin America, the Caribbean, and Oceania.
4. **b.** Japan, with 22.6 percent of all people aged 65 or over. In Monaco, a small principality of about 32,000 people located on the Mediterranean, 22 percent of its residents were also aged 65 and over.
5. **False.** Although the United States will age rapidly when the baby boomers (people born between 1946 and 1964) begin to reach age 65 after the year 2010, the percentage of population aged 65 and over in the year 2050 is projected to be slightly more than 20 percent (compared with about 13% today).
6. **True.** These four countries have the highest level of well-being. These high levels are also seen in people aged 50 to 64 in these countries.
7. **False.** Older women are less likely to be literate. In China in 1990, for example, only 11 percent of women aged 60 and over could read and write, compared with half of men aged 60 and over.

*Source:* Population Reference Bureau. 2012 Population Data Sheet; Jacobsen, Linda A., et al. 2011. "America's Aging Population." *Population Bulletin* 66(1).

---

migration. In the political arena, population aging has already produced a powerful voice in developed countries because it can influence voting patterns and representation. Older voters usually read, watch the news, and educate themselves about the issues, and they vote in much higher percentages than any other age group.

The radical shift in the U.S. population age structure over the past one hundred years provides only one part of the story of the U.S. elderly population. Another remarkable aspect is the rapid growth in the number of elderly and the increasing number of Americans at the oldest ages, above ages 85 or 90.

As the baby boom population ages, its large numbers will cause cycles of relative growth and decline at each stage of life. The aging of the baby boom generation will push the median age, now at 34, to more than 38 by 2050.

In 2010, the oldest members of the baby boom generation turned 65, whereas the youngest were around 50. As the baby boom cohorts begin to reach age 65, the number of elderly people will rise dramatically.

In about 2030, the final phase of the elderly explosion caused by the baby boom will begin. At that point, the population aged 84 and older will be the only older age group still growing. It will increase from 2.7 million now to 8.6 million in 2030, to more than 16 million by 2050. To put it another way, today 1 in 100 people is 85 years old or older; in 2050, 1 in 20 people could be so old. People 85 and older could constitute close to one quarter of the older population by then.

Concerns associated with problems of the older population are exacerbated by the large excess of women over men in the older ages. Among the aged, women outnumber men 3 to 2. That imbalance increases to more than 2 to 1 for people 85 and over.

Increased longevity and the aging of the baby boom generation mean that many people will find themselves caring for very old people after they themselves have reached retirement age. Assuming that generations are separated by about twenty-five years, people 85, 90, or 95 would have children who are anywhere from 60 to 70 years old. It is estimated that every third person 60 to 64 years old will have a living elderly parent by 2015.

The rising cost of caring for the elderly will become a bigger problem. Without serious health care reform, the United States could spend 20 percent of its gross national product on health care.

Today, there are nearly twice as many people under 18 as people over 65. This is down from 1960 when there were four times as many people under 18 as over 65. As we go forward, this trend will continue, and the elderly will become an ever greater share of the population (U.S. Bureau of the Census, 2011).

There are few certainties about the future, but the demographic outlook seems relatively clear, at least for people already born. Careful consideration of the impact of our aging population can be an important tool in planning for the future.

## SUMMARY

### How are health and social issues related?

Medicine and health care issues are intertwined with social and cultural customs and reflect the society of which they are a part. Illness not only involves the body but also affects an individual's social relationships, self-image, and behavior.

### What is the sick role?

Talcott Parsons suggested that to prevent the potentially disruptive consequences of illness for a group or society, there exists a sick role, a shared set of cultural norms that legitimates deviant behavior caused by illness and channels the individual into the health care system.

### What is unique about health care in the United States?

The United States has the most advanced health care resources in the world. Critics maintain that the system pays off, though, only when the patient can pay.

The system has been described as acute, curative, and hospital based in that the focus is on curing or controlling serious diseases rather than on maintaining health.

### How does gender affect health?

Women seem to suffer from illness and disability more than men, but their health problems are usually not as life threatening. Some have suggested that women might not be sick more often but just more sensitive to bodily discomforts and more willing to report them to a doctor.

### How do male and female life expectancies differ?

Male infants are biologically more vulnerable than female infants in both the prenatal and neonatal stages. Male death rates exceed female death rates at all ages and for the leading causes of death such as heart disease, cancer, cerebrovascular diseases, accidents, and pneumonia. Sociologically, men are more likely to have dangerous jobs and more likely than women to place themselves in dangerous situations during both work and leisure.

### What differences are there in male and female health?

Men and women are equally vulnerable to psychiatric problems, but emotionally disturbed men are likely to act out through drugs, liquor, and antisocial acts, whereas women display behaviors such as depression or phobias that indicate an internalization of their problems.

### What differences are there in the health of different racial groups?

African Americans and Native Americans have the worst health profiles. On average, life expectancy for African American infants is about five years less than for white infants, for both males and females. Asian Americans have the best health profile, followed by whites.

### What connection is there between health and socioeconomic status?

The care the poor receive is inferior to that received by the more affluent.

Poverty contributes to disease and a shortened life span both directly and indirectly. Diseases such as tuberculosis, influenza, pneumonia, whooping cough, and alcoholism are more common among the poor than the middle class. Inadequate housing, heating, and sanitation all contribute to those medical problems.

### What role does insurance play in American health care?

The Affordable Health Care Act will change health care in the United States. Currently most Americans pay for their health services through some form of health insurance or third-party payments. Individuals—or their employers—pay premiums into a pool, which finances the medical care of those covered by the insurance. Other public insurance programs include Medicare, for those older than 65, and Medicaid for those who are below or near the poverty level.

▸ **Has there been a change in how health care is approached?**

Recently, there has been a shift among health care providers from the cure orientation to a prevention orientation. Prevention has three levels: medical prevention directed at the individual's body; behavioral prevention directed at changing the habits and behavior of individuals; and structural prevention directed at changing the social environments in which people work and live.

▸ **What is the size of the elderly population?**

For the first time in our history, there are more people aged 65 and older in the population than there are teenagers. The older population can be divided into the young-old (ages 65 to 74), the middle-old (ages 75 to 84), and the old-old (ages 85 and older). Fifty percent of the older population is older than 75.

▸ **What is the make-up of the elderly population?**

Because mortality rates are higher for men, there is a substantial sex ratio imbalance in the over-75 group that becomes more apparent as one goes up the age scale. Due to higher mortality rates among African Americans and other minorities, the older population is disproportionately white. However, that will reverse in future years.

## Media Resources

**CourseMate for *Introduction to Sociology*, Eleventh Edition**
Cengage Learning's Sociology CourseMate brings course concepts to life with interactive learning, study, and exam preparation tools that support the printed textbook. Access an integrated eBook, learning tools including glossaries, flashcards, quizzes, videos, and more in your Sociology CourseMate. Go to CengageBrain.com to register or purchase access.

# CHAPTER SIXTEEN STUDY GUIDE

## Key Concepts and Thinkers

*Match each of the following concepts with its definition, illustration, or explanation.*

a. Sick role
b. Medicalization
c. Talcott Parsons
d. Third-party payments
e. Experience ratings
f. WHO (World Health Organization)
g. Medicare
h. Medicaid
i. CDC (Centers for Disease Control)
j. Managed care
k. HMO (health maintenance organization)
l. Affordable Care Act
m. BMI

___ 1. A system in which the costs of an individual's health care are paid for by some form of public or private insurance or charitable organization
___ 2. A medical organization run by an insurance company, with doctors paid a salary and costs tightly controlled
___ 3. The process by which problems become defined and treated as medical problems, usually in terms of illness or disorder
___ 4. The U.S. government program to pay for medical costs of the poor
___ 5. A process used by health companies to distribute insurance coverage across different groups based on their likelihood of illness
___ 6. American sociologist who developed the concept of the sick role
___ 7. A system by which insurance companies set maximum fees they will pay to doctors
___ 8. A shared set of cultural norms that legitimates deviant behavior caused by illness and channels the individual into the health care system
___ 9. A part of the U.S. Public Health Service, this agency is charged with monitoring communicable ailments
___ 10. An international organization that monitors health issues
___ 11. The U.S. government program to pay for medical costs of the elderly
___ 12. Passed in 2010, it stresses cost-effective health care coverage for all
___ 13. A scale used to measure whether an individual is overweight or obese

## Central Idea Completions

*Fill in the appropriate concepts and descriptions for each of the following questions.*

1. What factors influence Asian and Pacific Islanders to have much lower mortality rates than would be expected?

   _____
   _____

2. Identify at least three factors that cause African Americans to have much higher mortality rates than those of other groups.

   a. _____
   b. _____
   c. _____

3. Most of us think of health in terms of becoming ill. What other factors could be used to evaluate the health of a population?

   a. _____

   b. _____

   c. _____

4. What changes in the AIDS population over the last 20 years might result in changes in how Americans view the seriousness of this disease?

   _____

   _____

5. What biological and sociological factors account for the differences in life expectancy between men and women?

   Biological: _____

   _____

   _____

   Sociological: _____

   _____

   _____

6. Identify at least two factors that result in U.S. life expectancy ranking in the lower half of the global scale of healthiest societies.

   a. _____

   b. _____

7. Identify three reasons why patients may respond more positively to female doctors than male doctors.

   a. _____

   b. _____

   c. _____

8. Elderly Americans represent a financial dichotomy: while many are counted among the poorest, many are also among the most financially secure. Discuss this phenomenon and explain some of the likely causes.

   a. _____

   b. _____

## ● Critical Thought Exercises

1. Consider the three levels of illness prevention—medical, behavioral, and structural. Discuss how each type might be related to public policy. How much does each type of intervention contribute to the differences in health among countries with similar levels of economic development?

2. Present a detailed discussion explaining the connection between education and health in the United States. How does the number of years of schooling affect health, and why does it? Consider this from the perspective of all three levels of illness prevention.

3. Discuss the changes in patterns of death among Americans since the turn of the twentieth century to the present. How have these changes affected the various social class and racial groups in America? Are these changes likely to continue in the next couple of decades?

4. What are some of the differences between the following three sub-classifications of the elderly: young-old; middle-old; old-old? Things to consider include health issues, stereotypes versus reality, health care needs now and in the future, and so on.

## Internet Activities

1. View the following YouTube video from the Centers for Disease Control and Prevention: http://www.youtube.com/playlist?list=PL9DE14257E1CF5CF4&feature=plcp. Summarize in your own words the CDC mission and how that mission is played out.

2. Go the American Association for Retired Persons' webpage on "Brain Games": www.aarp.org/health/brain-health/brain_games/. Take a look at the various games designed to keep the cognitive processes sharp in the elderly. Play a few brain games. Search for information on the Web as to whether there is merit to the claim of value of these games.

3. Take a look at the following timeline of women in medicine: www.ama-assn.org/resources/doc/wpc/wimtimeline.pdf. What about the list of female professionals, their contributions, and the timeline do you find interesting or historically enlightening?

## Answers to Key Concepts and Thinkers

1. d;  2. k;  3. b;  4. h;  5. e;  6. c;  7. j;  8. a;  9. i;  10. f;  11. g  12. l;  13. m

# Glossary

**Abstract ideals**   Aspects of a religion that focus on correct ways of thinking and behaving, rather than on a belief in supernatural forces, spirits, or beings.

**Achieved statuses**   Statuses obtained as a result of individual efforts.

**Acting crowd**   A group of people whose passions and tempers have been aroused by some focal event, who come to share a purpose, who feed off one another's arousal, and who often erupt into spontaneous acts of violence.

**Activity theory**   The view that satisfaction in later life is related to the level of activity the person engages in.

**Adaptation**   The process by which human beings adjust to the changes in their environment.

**Adult socialization**   The process by which adults learn new statuses and roles. Adult socialization continues throughout the adult years.

**Affiliation**   Meaningful interaction with others.

**Age-specific death rates**   The annual number of deaths per 1,000 people at specific ages.

**Agricultural societies**   Societies that use the plow in food production.

**Alienation**   The process by which people lose control over the social institutions they themselves invented.

**Analysis**   The process through which scientific data are organized so that comparisons can be made and conclusions drawn.

**Anglo conformity**   A form of assimilation that involves the renunciation of the ancestral culture in favor of Anglo-American behavior and values.

**Animism**   The belief in animate, personalized spirits or ghosts of ancestors that take an interest in, and actively work to influence, human affairs.

**Annihilation**   The deliberate practice of trying to exterminate a racial, religious, or ethnic group; also known as genocide.

**Anomie**   The feeling of some individuals that their culture no longer provides adequate guidelines for behavior; a condition of "normlessness" in which values and norms have little impact.

**Aristocracy**   The rule by a select few; a form of oligarchy.

**Ascribed statuses**   Statuses conferred on an individual at birth or on other occasions by circumstances beyond the individual's control.

**Assimilation**   The process whereby groups with different cultures come to have a common culture.

**Associations**   Purposefully created special-interest groups that have clearly defined goals and official ways of doing things.

**Attachment disorder**   A disorder in which children are unable to trust people and form relationships with others.

**Authoritarian leader**   A type of instrumental leader who makes decisions and gives orders.

**Authority**   Power regarded as legitimate by those over whom it is exercised, who also accept the authority's legitimacy in imposing sanctions or even in using force, if necessary.

**Autocracy**   A political system in which the ultimate authority rests with a single person.

**Bilateral descent system**   A descent system that traces kinship through both female and male family members.

**Bilocal residence**   Marital residence rules allowing a newly married couple to live with either the husband's or wife's family of origin.

**Blind investigator**   A researcher who does not know whether a specific subject belongs to the group of actual cases being investigated or to a comparison group. This is done to eliminate researcher bias.

**Bourgeoisie**   The term used by Karl Marx to describe the owners of the means of production and distribution in capitalist societies.

**Bureaucracy**   A formal, rationally organized social structure with clearly defined patterns of activity in which, ideally, every series of actions is fundamentally related to the organization's purpose.

**Capitalism**   An economic system based on private ownership of the means of production, in which resource allocation depends largely on market forces.

**Caste system**   A rigid form of social stratification based on ascribed characteristics that determines its members' prestige, occupation, residence, and social relationships.

**Casual crowd**   A crowd made up of a collection of people who, in the course of their private activities, happen to be in the same place at the same time.

**Charismatic authority**   The power that derives from a ruler's force of personality. It is the ability to inspire passion and devotion among followers.

**City**   A unit typically incorporated according to the laws of the state within which it is located.

**Class system**   A system of social stratification that contains several social classes and in which greater social mobility is permitted than in a caste or estate system.

**Closed society**   A society in which the various aspects of people's lives are determined at birth and remain fixed.

**Coalescence**   The second stage in the life cycle of a social movement, when groups begin to form around leaders, promote policies, and promulgate programs.

**Coercion**   A form of conflict in which one of the parties in a conflict is much stronger than the others and imposes its will because of that strength.

**Cognitive culture**   The thinking component of culture, consisting of shared beliefs and knowledge of what the world is like—what is real and what is not, what is important and what is trivial; one of the two categories of nonmaterial culture.

**Cohabitation**   Unmarried couples living together out of wedlock.

**Collective behavior**   Relatively spontaneous social actions that occur when people respond to unstructured and ambiguous situations.

**Collective conscience** A society's system of fundamental beliefs and values.

**Command economy** An economy in which the government makes all the decisions about production and consumption.

**Communism** The name commonly given to totalitarian socialist forms of government.

**Companionate marriage** Marriage based on romantic love.

**Competition** A form of conflict in which individuals or groups confine their conflict within agreed-upon rules.

**Concentric zone model** A theory of city development in which the central city is made up of a business district and, radiating from this district, zones of low-income, working-class, middle-class, and upper-class residential units.

**Conditioning** The molding of behavior through repeated experiences that link a desired reaction with a particular object or event.

**Conflict** The opposite of cooperation. People in conflict struggle against one another for some commonly prized object or value.

**Conflict approach** The view that the elite use their power to enact and enforce laws that support their own economic interests and go against the interests of the lower classes.

**Conflict theory** This label applies to any of a number of theories that assume society is in a constant state of social conflict, with only temporarily stable periods, and that social phenomena are the result of this conflict.

**Consensus approach** An approach to law that assumes laws are merely a formal version of the norms and values of the people.

**Conservative ideologies** Ideologies that try to preserve things as they are.

**Conservative social movement** A social movement that seeks to maintain society's current values.

**Constitutionalism** Government power that is limited by law.

**Contagion theory** The theory that states members of a crowd acquire a crowd mentality, lose their characteristic inhibitions, and become highly receptive to group sentiments.

**Context** The conditions under which an action takes place, including the physical setting or place, the social environment, and the other activities surrounding the action.

**Contractual cooperation** Cooperation in which each person's specific obligations are clearly spelled out.

**Conventional crowd** A crowd in which people's behavior conforms to some well-established set of cultural norms and in which people's gratification results from passive appreciation of an event.

**Convergence theory** Views collective behavior as the outcome of situations in which people with similar characteristics, attitudes, and needs are drawn together.

**Cooperative interaction** A form of social interaction in which people act together to promote common interests or achieve shared goals.

**Craze** A fad that is especially short-lived.

**Crime** Behavior that violates a society's legal code.

**Criminal justice system** Personnel and procedures for arrest, trial, and punishment to deal with violations of the law.

**Cross-sectional study** An examination of a population at a given point in time.

**Crowd** A temporary concentration of people who focus on some thing or event but who also are attuned to one another's behavior.

**Crude birthrate** The number of annual births per 1,000 population.

**Crude death rate** The annual number of deaths per 1,000 population.

**Cultural lag** A situation that develops when new patterns of behavior conflict with traditional values. Cultural lag may occur when technological change (material culture) is more rapid than are changes in norms and values (nonmaterial culture).

**Cultural relativism** The position that social scientists doing cross-cultural research should view and analyze behaviors and customs within the cultural context in which they occur.

**Cultural traits** Items of a culture such as tools, materials used, beliefs, values, and typical ways of doing things.

**Cultural transmission** The transmission of major portions of a society's knowledge, norms, values, and perspectives from one generation to the next. Cultural transmission is an intended function of education.

**Cultural universals** Forms or patterns for resolving the common, basic, human problems that are found in all cultures. Cultural universals include the division of labor, the incest taboo, marriage, the family, rites of passage, and ideology.

**Culture** All that human beings learn to do, to use, to produce, to know, and to believe as they grow to maturity and live out their lives in the social groups to which they belong.

**Culture shock** The reaction people may have when encountering cultural traditions different from their own.

**De facto segregation** Segregation of community or neighborhood schools that results from residential patterns in which minority groups often live in areas of a city where there are few whites or none at all.

**De jure segregation** Segregation that is an outgrowth of local laws that prohibit one racial group from attending school with another.

**Demise** The last stage in the life cycle of a social movement, when the movement comes to an end.

**Democracy** A political system operating under the principles of constitutionalism, representative government, majority rule, civilian rule, and minority rights.

**Democratic leader** A type of instrumental leader who attempts to encourage group members to reach a consensus.

**Democratic socialism** A political system that exhibits the dominant features of a democracy, but in which the control of the economy is vested in the government to a greater extent than under capitalism.

**Demographic transition theory** A theory that explains population dynamics in terms of four distinct stages from high fertility and high mortality to relatively low fertility and low mortality.

**Demography**  The study of the dynamics of human populations.

**Denomination**  A religious group that tends to draw its membership from a particular socially acceptable class or ethnic group, or at least to have its leadership positions dominated by members of such a group.

**Dependency ratio**  The number of people of nonworking age in a society for every 100 people of working age.

**Dependency theory**  A theory that proposes that the economic positions of rich and poor nations are linked and cannot be understood in isolation from each other. Global inequality is due to the exploitation of poor societies by the rich ones.

**Dependent variable**  A variable that changes in response to changes in the independent variable.

**Deviant behavior**  Behavior that fails to conform to the rules or norms of the group in which it occurs.

**Dictatorship**  A totalitarian government in which all power rests ultimately in one person, who generally heads the only recognized political party.

**Diffusion**  One of the two mechanisms responsible for cultural evolution. Diffusion is the movement of cultural traits from one culture to another.

**Directed cooperation**  Cooperation characterized by a joint effort under the control of people in authority.

**Discrimination**  Differential treatment, usually unequal and injurious, directed at individuals who are assumed to belong to a particular category or group.

**Diversion**  Steering youthful offenders away from the juvenile justice system to nonofficial social agencies.

**Double-blind investigator**  A researcher who does not know either the kind of subject being investigated or the hypothesis being tested.

**Dramaturgy**  The study of the roles people play to create a particular impression in others.

**Dyad**  A small group that contains only two members.

**Ecclesia**  A church that shares the same ethical system as the secular society and that has come to represent and promote the interests of the society at large.

**Economy**  An institution whose primary function is to determine the manner in which society produces, distributes, and consumes goods and services.

**Ecumenism**  The trend among many religions to draw together and project a sense of unity and common direction.

**Ego**  In Freudian theory, one of the three separately functioning parts of the self. The ego tries to mediate in the conflict between the id and the superego and to find socially acceptable ways for the id's drives to be expressed. This part of the self constantly evaluates social realities and looks for ways to adjust to them.

**Electorate**  Those citizens eligible to vote.

**Emergent norm theory**  A theory that notes that even though crowd members may have different motives for participating in collective behavior, they acquire common standards by observing and listening to one another.

**Emigration**  The movement of a population from an area.

**Empirical question**  A question that can be answered by observation and analysis of the world as it is known.

**Empiricism**  The view that generalizations are valid only if they rely on evidence that can be observed directly or verified through our senses.

**Endogamy**  Societal norms that limit the social categories from within which one can choose a marriage partner.

**Environmental determinism**  The belief that the environment dictates cultural patterns.

**Estate**  A segment of a society that has legally established rights and duties.

**Estate system**  A closed system of stratification in which social position is defined by law and membership is based primarily on inheritance. A very limited possibility of upward mobility exists.

**Ethnic group**  A group that has a distinct cultural tradition with which its members identify and which may or may not be recognized by others.

**Ethnocentrism**  The tendency to judge other cultures in terms of one's own customs and values.

**Ethnomethodology**  The study of the sets of rules or guidelines people use in their everyday living practices. This approach provides information about a society's unwritten rules for social behavior.

**Ethology**  The scientific study of animal behavior.

**Evolution**  The continuous change from a simpler condition to a more complex state.

**Exchange interaction**  An interaction involving one person doing something for another with the express purpose of receiving a reward or return.

**Exogamy**  Societal norms that require an individual to marry someone outside his or her culturally defined group.

**Experiment**  An investigation in which the variables being studied are controlled and the researcher obtains the results through precise observation and measurement.

**Expressive crowd**  A crowd that is drawn together by the promise of personal gratification for its members through active participation in activities and events.

**Expressive leadership**  A form of leadership in which a leader works to keep relations among group members harmonious and morale high.

**Expressive social movement**  A social movement that stresses personal feelings of satisfaction or well-being and that typically arises to fill some void or to distract people from some great dissatisfaction in their lives.

**Expulsion**  The process of forcing a group to leave the territory in which it resides.

**Extended families**  Families that include, in addition to nuclear family members, other relatives such as the parents' parents, the parents' siblings, and in-laws.

**External means of social control**  The ways in which others respond to a person's behavior that channel his or her behavior along culturally approved lines.

**External source of social change**  Changes within a society produced by events external to that society.

**Exurbs**  The fast-growing area located in a newer, second ring beyond the old suburbs.

**Fad** A transitory social change that has a very short life span marked by a rapid spread and an abrupt drop from popularity.

**Family of orientation** The nuclear family in which one is born and raised; also known as family of origin.

**Family of procreation** The family that is created by marriage.

**Fascism** A political-economic system characterized by totalitarian capitalism.

**Fashion** A transitory change in the standards of dress or manners in a given society.

**Fecundity** The physiological ability to have children.

**Felonies** Offenses punishable by a year or more in a state prison.

**Fertility** The actual number of births in the population.

**Fertility rate** The number of annual births per 1,000 women of childbearing age in a population.

**Folkways** Norms that permit a rather wide degree of individual interpretation as long as certain limits are not overstepped. Folkways change with time and vary from culture to culture.

**Forced acculturation** The situation that occurs when social change is imposed on weaker peoples by might or conquest.

**Forced migration** The expulsion of a group of people through direct action.

**Formal negative sanctions** Actions that express institutionalized disapproval of a person's behavior, such as expulsion, dismissal, or imprisonment. They are usually applied within the context of a society's formal organizations, including schools, corporations, and the legal system.

**Formal positive sanctions** Actions that express social approval of a person's behavior, such as public gatherings, rituals, or ceremonies.

**Formal sanctions** Sanctions that are applied in a public ritual, usually under the direct or indirect leadership of social authorities; examples: the award of a prize or the announcement of an expulsion.

**Fragmentation** The fourth stage in the life cycle of a social movement, when the movement gradually begins to fall apart.

**Functionalism (structural functionalism)** One of the major sociological perspectives, which assumes that society is a system of highly interrelated parts that operate (function) together rather harmoniously.

**Funnel effect** The situation in our criminal justice system whereby many crimes are committed, but few criminals seem to be punished.

**Game stage** According to George Herbert Mead, the stage in the development of the self when the child learns that there are rules that specify the proper and correct relationship among the players.

*Gemeinschaft* A community in which relationships are intimate, cooperative, and personal.

**Gender** The social, psychological, and cultural attributes of masculinity and femininity that are based on biological distinctions.

**Gender identity** The view of ourselves resulting from our sex.

**Gender-role socialization** The lifelong process whereby people learn the values, attitudes, and behavior considered appropriate to each sex by their culture.

**Generalized others** The viewpoints, attitudes, and expectations of society as a whole or of a general community of people that we are aware of and who are important to us.

**Genes** The set of inherited units of biological material with which each individual is born.

**Gentrification** A trend that involves wealthier, middle-class people moving to marginal urban areas, upgrading the neighborhood, and displacing some of the poor residents who become priced out of the available housing.

*Gesellschaft* A society in which relationships are impersonal and independent.

**Ghetto** A term originally used to refer to the segregated quarter of a city where the Jews in Europe were often forced to live. Today it is used to refer to any kind of segregated living environment.

**Globalization** The worldwide flow of goods, services, money, people, information, and culture. It leads to a greater interdependence and mutual awareness among the people of the world.

**Group** A collection of specific, identifiable people.

**Hidden curriculum** The social attitudes and values learned in school that prepare children to accept the requirements of adult life and to fit into the social, political, and economic statuses of adult life.

**Homogamy** The tendency to choose a spouse with a similar racial, religious, ethnic, educational, age, or socioeconomic background.

**Horizontal mobility** Movement that involves a change in status with no corresponding change in social class.

**Horticulture societies** Societies in which muscle power and handheld tools are used to cultivate gardens and fields.

**Hunting and food-gathering societies** Societies that survive by foraging for vegetable foods and small game, fishing, collecting shellfish, and hunting larger animals.

**Hypothesis** A testable statement about the relationship between two or more empirical variables.

**I** The portion of the self that wishes to have free expression, to be active, and to be spontaneous.

**Id** In Freudian theory, one of the three separately functioning parts of the self. The id consists of the unconscious drives or instincts that Freud believed every human being inherits.

**Ideal norms** Expectations of what people should do under perfect conditions. The norm that marriage will last "until death do us part" is an ideal norm in American society.

**Ideal type** A simplified, exaggerated model of reality used to illustrate a concept.

**Ideational culture** A term developed by Pitirim A. Sorokin to describe a culture in which spiritual concerns have the greatest value.

**Ideologies** Strongly held beliefs and values to which group members are firmly committed and which cement the social structure.

**Ideology**  A set of interrelated religious or secular beliefs, values, and norms justifying the pursuit of a given set of goals through a given set of means.

**Immigration**  The movement of a population into an area.

**Incest**  The term used to describe sexual relations within families. Most cultures have strict taboos against incest, which is often associated with strong feelings of horror and revulsion.

**Incest taboo**  A societal prohibition that forbids sexual intercourse among closely related individuals.

**Incipiency**  The first stage in the life cycle of a social movement, when large numbers of people perceive a problem without an existing solution.

**Independent variable**  A variable that changes for reasons that have nothing to do with the dependent variable.

**Industrial cities**  Cities established during or after the Industrial Revolution, characterized by large populations that work primarily in industrial or service-related jobs.

**Industrial societies**  Societies that use mechanical means of production instead of human or animal muscle power.

**Industrialism**  Consists of the use of mechanical means (machines and chemical processes) for the production of goods.

**Infant mortality rate**  The number of children who die within the first year of life per 1,000 live births.

**Informal negative sanctions**  Spontaneous displays of disapproval of a person's behavior. Disapproving treatment is directed toward the violator of a group norm.

**Informal positive sanctions**  Spontaneous actions such as smiles, pats on the back, handshakes, congratulations, and hugs, through which individuals express their approval of another's behavior.

**Informal sanctions**  Responses by others to an individual's behavior that arise spontaneously with little or no formal leadership.

**Innovation**  Any new practice or tool that becomes widely accepted in a society.

**Innovators**  Individuals who accept the culturally validated goal of success but find deviant ways of reaching it.

**Instincts**  Biologically inherited patterns of complex behavior.

**Institutionalization**  The third stage in the life cycle of a social movement, when the movement reaches its peak of strength and influence and becomes firmly established.

**Institutionalized prejudice and discrimination**  Complex societal arrangements that restrict the life chances and choices of a specifically defined group.

**Instrumental leadership**  A form of leadership in which a leader actively proposes tasks and plans to guide the group toward achieving its goals.

**Interactionist perspective**  An orientation that focuses on how individuals make sense of or interpret the social world in which they participate.

**Intergenerational mobility**  Changes in the social level of a family through two or more generations.

**Internal means of social control**  A group's moral code which becomes internalized and becomes part of each individual's personal code of conduct. It operates even in the absence of reactions by others.

**Internal migration**  The movement of a population within a nation's boundary lines.

**Internal sources of social change**  Those factors that originate within a specific society and that singly or in combination produce significant alterations in its social organization and structure.

**Interview**  A conversation between an investigator and a subject for the purpose of gathering information.

**Intragenerational mobility**  Social changes during the lifetime of one individual.

**Juvenile crime**  Refers to the breaking of criminal laws by individuals under the age of eighteen.

**Labeling theory**  A theory of deviance that assumes the social process by which an individual comes to be labeled a deviant contributes to causing more of the deviant behavior.

**Laissez-faire capitalism**  The view of capitalism that believes government should stay out of business.

**Laissez-faire leader**  A type of instrumental leader who is a leader in name or title only and does little actively to influence group affairs.

**Latent function**  One of two types of social functions identified by Robert Merton, referring to the unintended or not readily recognized consequences of a social process.

**Laws**  Formal rules adopted by a society's political authority.

**Leader**  Someone who occupies a central role or position of dominance and influence in a group.

**Legal code**  The formal body of rules adopted by a society's political authority.

**Legal-rational authority**  Authority that derives from the fact that specific individuals have clearly defined rights and duties to uphold and who implement rules and procedures.

**Liberal ideologies**  Ideologies that seek limited reforms that do not involve fundamental changes in the structure of society.

**Life expectancy**  The average number of years that a person born in a particular year can expect to live.

**Lobbying**  Attempts by special-interest groups to influence government policy.

**Longitudinal research**  A research approach in which a population is studied at several intervals over a relatively long period of time.

**Looking-glass self**  A theory developed by Charles Horton Cooley to explain how individuals develop a sense of self through interaction with others. The theory has three stages: (1) We imagine how our actions appear to others, (2) we imagine how other people judge these actions, and (3) we make some sort of self-judgment based on the presumed judgments of others.

**Magic**  Interaction with the supernatural. Magic does not involve the worship of a god or gods, but rather, it is an attempt to coerce spirits or control supernatural forces.

**Majority rule**   The right of people to assemble to express their views and seek to persuade others, to engage in political organizing, and to vote for whomever they wish.

**Mana**   A Melanesian/Polynesian concept of the supernatural that refers to a diffuse, nonpersonalized force that acts through anything that lives or moves.

**Manifest function**   One of two types of social functions identified by Robert Merton, referring to an intended and recognized consequence of a social process.

**Marital residence rules**   Rules that govern where a newly married couple settles down and lives.

**Marriage**   The socially recognized, legitimized, and supported union of individuals of opposite sexes.

**Mass**   A collection of people who, although physically dispersed, participate in some event either physically or with a common concern or interest.

**Mass hysteria**   A condition in which large numbers of people are overwhelmed with emotion and frenzied activity or become convinced that they have experienced something for which investigators can find no discernible evidence.

**Mass media**   Methods of communication, including television, radio, magazines, films, and newspapers, that have become some of society's most important agents of socialization.

**Master status**   One of the multiple statuses a person occupies that dominates the others in patterning that person's life.

**Material culture**   All the things human beings make and use, from small handheld tools to skyscrapers.

**Matriarchal family**   A family in which most family affairs are dominated by women.

**Matrilineal system**   A descent system that traces kinship through the females of the family.

**Matrilocal residence**   Marital residence rules that require a newly married couple to settle down near or within the wife's mother's household.

**Me**   The portion of the self that is made up of those things learned through the socialization process from the family, peers, school, and so on.

**Mechanically integrated society**   A type of society in which members have common goals and values and a deep and personal involvement with the community.

**Mechanisms of social control**   Processes used by all societies and social groups to influence or mold members' behavior to conform to group values and norms.

**Megalopolis**   Another term for Consolidated Metropolitan Statistical Area.

**Metropolitan area**   An area that has a large population nucleus, together with the adjacent communities that are economically and socially integrated into that nucleus.

**Middle-range theories**   Theories concerned with explaining specific issues or aspects of society instead of trying to explain how all of society operates.

**Migration**   The movement of populations from one geographical area to another.

**Millenarian movements**   Religious movements that prophesy the end of the world, the destruction of all evil people and their works, and the saving of the just.

**Minority**   A group of people who, because of physical or cultural characteristics, are singled out from others in the society in which they live for different and unequal treatment and who therefore regard themselves as objects of collective discrimination.

**Minority rights**   The minority has the right to try to change the laws of the majority.

**Mixed economy**   An economy that combines free-enterprise capitalism with government regulation of business, industry, and social-welfare programs.

**Modernization**   The complex set of changes that take place as a traditional society becomes an industrial society.

**Modernization theory**   Assumes a direct relationship between the extent of modernization in a society and the status and condition of the elderly.

**Monogamous marriage**   The form of marriage in which each person is allowed only one spouse at a time.

**Monotheism**   The belief in the existence of only one god.

**Moral code**   The symbolic system, made up of a culture's norms and values, in terms of which behavior takes on the quality of being good or bad, right or wrong.

**Moral order**   A society's shared view of right and wrong.

**Mores**   Strongly held norms that usually have a moral connotation and are based on the central values of the culture.

**Mortality**   The frequency of deaths in a population.

**Multiple marriage**   A form of marriage in which an individual may have more than one spouse (polygamy).

**Multiple nuclei model**   A theory of city development that emphasizes that different industries have different land-use and financial requirements, which determine where they establish themselves. As similar industries are established close to one another, the immediate neighborhood is strongly shaped by the nature of its typical industry, becoming one of a number of separate nuclei that together constitute the city.

**Negative sanctions**   Responses by others that discourage the individual from continuing or repeating the behavior.

**Neolocal residence**   Marital residence standards that allow a newly married couple to live virtually anywhere, even thousands of miles from their families of origin.

**Nonmaterial culture**   The totality of knowledge, beliefs, values, and rules for appropriate behavior that specifies how a people should interact and how they may solve their problems.

**Normal behavior**   Behavior that conforms to the rules or norms of the group in which it occurs.

**Norms**   Specific rules of behavior that are agreed upon and shared within a culture to prescribe limits of acceptable behavior.

**Nuclear family**   The most basic family form, made up of parents and their children, biological or adopted.

**Nurture**   The entire socialization experience.

**Oligarchy**   Rule by a few individuals who occupy the highest positions in an organization.

**Open-ended interview**   (See semistructured interview.)

**Open society**   A society that provides equal opportunity to everyone to compete for the role and status desired, regardless of race, religion, gender, or family history.

**Operational definition**  A definition of an abstract concept in terms of the observable features that describe the things being investigated.

**Opinion leaders**  Socially acknowledged experts to whom the public turns for advice.

**Organically integrated society**  A type of society in which social solidarity depends on the cooperation of individuals in many different positions who perform specialized tasks.

**Organized crime**  Structured associations of individuals or groups who come together for the purpose of obtaining gain, mostly from illegal activities.

**Panic**  An uncoordinated group flight from a perceived danger.

**Pantheon**  The hierarchy of deities in a religious belief system.

**Paradigm**  A basic model for explaining events that provides a framework for the questions that generate and guide research.

**Participant observation**  A research technique in which the investigator enters into a group's activities while, at the same time, studying the group's behavior.

**Pastoral societies**  Societies that rely on herding and the domestication and breeding of animals for food and clothing to satisfy most of the group's needs.

**Patriarchal family**  A family in which most family affairs are dominated by men.

**Patriarchal ideology**  The belief that men are superior to women and should control all important aspects of society.

**Patrilineal system**  A family system that traces kinship through the males of the family.

**Patrilocal residence**  Marital residence rules that require a newly married couple to settle down near or within the husband's father's household.

**Peers**  Individuals who are social equals.

**Personality**  The patterns of behavior and ways of thinking and feeling that are distinctive for each individual.

**Play stage**  According to George Herbert Mead, the stage in the development of the self when the child has acquired language and begins not only to imitate behavior, but also to formulate role expectation.

**Pluralism**  The development and coexistence of separate racial and ethnic group identities in a society in which no single subgroup dominates.

**Political action committees (PACs)**  Special-interest groups concerned with very specific issues who usually represent corporate, trade, or labor interests.

**Political revolutions**  Relatively rapid transformations of state or government structures that are not accompanied by changes in social structure or stratification.

**Politics**  The process by which power is distributed and decisions are made.

**Polyandrous family**  A polygamous family unit in which the central figure is female and the multiple spouses are male.

**Polygamous families**  Nuclear families linked together by multiple marriage bonds, with one central individual married to several spouses.

**Polygynous family**  A polygamous family unit in which the central person is male and the multiple spouses are female.

**Polytheism**  The belief in a number of gods.

**Positive checks**  Events, described by Thomas Robert Malthus, that limit reproduction, either by causing the deaths of individuals before they reach reproductive age or by causing the deaths of large numbers of people, thereby lowering the overall population; examples: famines, wars, and epidemics.

**Positive sanctions**  Responses by others that encourage the individual to continue acting in a certain way.

**Postindustrial societies**  Societies that depend on specialized knowledge to bring about continuing progress in technology.

**Power**  The ability of an individual or group to attain goals, control events, and maintain influence over others—even in the face of opposition.

**Power elite**  The group of people who control policymaking and the setting of priorities.

**Prayer**  A religious ritual that enables individuals to communicate with supernatural beings or forces.

**Preindustrial cities**  Cities established before the Industrial Revolution. Those cities were usually walled for protection, and power was typically shared between feudal lords and religious leaders.

**Prejudice**  An irrationally based negative, or occasionally positive, attitude toward certain groups and their members.

**Preparatory stage**  According to George Herbert Mead, the stage in the development of the self characterized by the child's imitating the behavior of others, which prepares the child for learning social-role expectations.

**Prestige**  The approval and respect an individual or group receives from other members of society.

**Preventive checks**  Practices, described by Thomas Robert Malthus, that limit reproduction; examples: contraception, prostitution, and other vices.

**Primary deviance**  A term used in labeling theory to refer to the original behavior that leads to the individual being labeled as deviant.

**Primary group**  A group characterized by intimate, face-to-face association and cooperation. Primary groups involve interaction among members who have an emotional investment in one another and who interact as total individuals rather than through specialized roles.

**Primary socialization**  The process by which children master the basic information and skills required of members of society.

**Profane**  All empirically observable things that are knowable through ordinary everyday experiences.

**Proletariat**  The label used by Karl Marx to describe the mass of people in society who have no resources to sell other than their labor.

**Propaganda**  Advertisements of a political nature seeking to mobilize public support behind one specific party, candidate, or point of view.

**Property crime**  An unlawful act that is committed with the intent of gaining property, but does not involve the use or threat of force against an individual.

**Psychoanalysis**  The form of therapy developed by Sigmund Freud for treating mental illness.

**Psychoanalytic theory**  A body of thought developed by Sigmund Freud that rests on two basic hypotheses: (1) Every human act has a psychological cause or basis, and (2) every person has an unconscious mind.

**Public opinion**  The beliefs held by a dispersed collectivity of individuals about a common concern, interest, focus, or activity.

**Race**  A category of people who are defined as similar because of a number of physical characteristics.

**Radical ideologies**  Ideologies that seek major structural changes in society.

**Random sample**  A sample selected purely on the basis of chance.

**Reactionary social movement**  A social movement that embraces the aims of the past and seeks to return the general society to yesterday's values.

**Real norms**  Norms that allow for differences in individual behavior. Real norms specify how people actually behave, not how they should behave under ideal circumstances.

**Rebellions**  Attempts to achieve rapid political change that is not possible within existing institutions.

**Rebels**  Individuals who reject both the goals of what to them is an unfair social order and the institutionalized means of achieving them. They propose alternative societal goals and institutions.

**Recidivism**  Repeated criminal behavior after punishment.

**Reference group**  A group or social category that an individual uses to help define beliefs, attitudes, and values, and to guide behavior.

**Reformulation**  The process in which traits passed from one culture to another are modified to fit better in their new context.

**Relative deprivation theory**  A theory that assumes social movements are the outgrowth of the feeling of relative deprivation among large numbers of people who believe they lack certain things they believe they are entitled to.

**Reliability**  The ability to repeat the findings of a research study.

**Religion**  A system of beliefs, practices, and philosophical values shared by a group of people that defines the sacred, helps explain life, and offers salvation from the problems of human existence.

**Religious taboo**  A sacred prohibition against touching, mentioning, or looking at certain objects, acts, or people.

**Replication**  Repetition of the same research procedure or experiment for the purpose of determining whether earlier results can be duplicated.

**Representative government**  The authority to govern is achieved through, and legitimized by, popular elections.

**Representative sample**  A sample that has the same distribution of characteristics as the larger population from which it is drawn.

**Researcher bias**  The tendency for researchers to select data that support their hypothesis and to ignore data that appear to contradict it.

**Research process**  A sequence of steps in the design and implementation of a research study, including defining the problem, reviewing previous research, determining the research design, defining the sample and collecting data, analyzing and interpreting the data, and preparing the final research report.

**Resocialization**  An important aspect of adult socialization that involves being exposed to ideas or values that conflict with what was learned in childhood.

**Resource-mobilization theory**  A theory that assumes social movements arise at certain times and not at others because some people know how to mobilize and channel popular discontent.

**Retreatists**  Individuals—such as drug addicts, alcoholics, drifters, and panhandlers—who have pulled back from society altogether and who do not pursue culturally legitimate goals.

**Revisionary social movement**  A social movement that seeks partial or slight changes within the existing order but does not threaten the order itself.

**Revitalization movements**  Powerful religious movements that stress a return to the religious values of the past. Those movements spring up when a society is under great stress or attack.

**Revolutionary social movement**  A social movement that seeks to overthrow all or nearly all of the existing social order and replace it with an order it considers to be more suitable.

**Revolutions**  Relatively rapid transformations that produce change in a society's power structure.

**Rites of passage**  Standardized rituals that mark the transition from one stage of life to another.

**Ritualists**  Individuals who deemphasize or reject the importance of success once they realize they will never achieve it and instead concentrate on following and enforcing rules more precisely than ever was intended.

**Rituals**  Patterns of behavior or practices related to the sacred.

**Role conflict**  The situation in which an individual who is occupying more than one status at the same time is unable to enact the roles of one status without violating those of another.

**Roles**  Culturally defined rules for proper behavior associated with every status.

**Role sets**  The roles attached to a single status.

**Role strain**  The stress that results from conflicting demands within a single role.

**Rumor**  Information that is shared informally and spreads quickly through a mass or a crowd.

**Sacred**  Things that are awe inspiring and knowable only through extraordinary experience. Sacred traits or objects symbolize important values.

**Sample**  The particular subset of a larger population that has been selected for study.

**Sampling**  A research technique in which a manageable number of subjects (a sample) is selected for study from a larger population.

**Sampling error**  The failure to select a representative sample.

**Sanctions** Rewards and penalties used to regulate an individual's behavior. All external means of control use sanctions.

**Sapir-Whorf hypothesis** A hypothesis that argues that the language a person uses determines his or her perception of reality.

**Science** A body of systematically arranged knowledge that shows the operation of general laws. The term also refers to the logical, systematic methods by which that knowledge is obtained.

**Scientific method** The approach to research that involves observation, experimentation, generalization, and verification.

**Secondary analysis** The process of making use of data that has been collected by others.

**Secondary deviance** A term used in labeling theory to refer to the deviant behavior that emerges as a result of a person being labeled as deviant.

**Secondary group** A group that is characterized by an impersonal, formal organization with specific goals. Secondary groups are larger and much less intimate than are primary groups, and the relationships among members are patterned mostly by statuses and roles rather than by personality characteristics.

**Sect** A small religious group that adheres strictly to religious doctrine involving unconventional beliefs or forms of worship.

**Sector model** A modified version of the concentric zone model, in which urban groups establish themselves along major transportation arteries around the central business district.

**Secularization** The process by which religious institutions are confined to ever-narrowing spheres of social influence, while people turn to secular sources for moral guidance in their everyday lives.

**Segregation** A form of subjugation that refers to the act, process, or state of being set apart.

**Selectivity** A process that defines some aspects of the world as important and others as unimportant. Selectivity is reflected in the vocabulary and grammar of language.

**Self** The personal identity of each individual that is separate from his or her social identity.

**Semistructured (open-ended) interview** An interview in which the investigator asks a list of questions but is free to vary them or make up new ones that become important during the interview.

**Sensate culture** A term developed by Pitirim A. Sorokin to describe a culture in which people are dedicated to self-expression and the gratification of their immediate physical needs.

**Sex** The physical and biological differences between men and women.

**Sick role** The sick role legitimates the deviant behavior caused by the illness and channels the individual into the health care system.

**Significant others** Those people who are most important in our development, such as parents, friends, and teachers.

**Signs** Objects or things that can represent other things because they share some important quality with them. A clenched fist, for example, can be a sign of anger because fists are used in physical arguments.

**Small group** A relative term that refers to the many kinds of social groups that actually meet together and contain few enough members so that all members know one another.

**Social action** Anything people are conscious of doing because of other people.

**Social aggregate** People who happen to be in the same place but share little else.

**Social attachments** The emotional bonds that infants form with others that are necessary for normal development. Social attachments are a basic need of human beings and all primates.

**Social change** Any modification in the social organization of a society in any of its social institutions or social roles.

**Social class** A category of people within a stratification system who share similar economic positions, similar lifestyles, and similar attitudes and behavior.

**Social Darwinism** The application of Charles Darwin's notion of "survival of the fittest" to society. Darwin believed those species of animals best adapted to the environment survived and prospered, while those poorly adapted died out.

**Social evaluation** The process of making qualitative judgments on the basis of individual characteristics or behaviors.

**Social function** A social process that contributes to the ongoing operation or maintenance of society.

**Social group** A number of people who have a common identity, some feeling of unity, and certain common goals and shared norms.

**Social identity** The statuses that define an individual. Social identity is determined by how others see us.

**Social inequality** The uneven distribution of privileges, material rewards, opportunities, power, prestige, and influence among individuals or groups.

**Social institutions** The ordered social relationships that grow out of the values, norms, statuses, and roles that organize those activities that fulfill society's fundamental needs.

**Social interaction** The interplay between the actions of one individual and those of one or more other people.

**Socialism** An economic system under which the government owns and controls the major means of production and distribution. Centralized planning is used to set production and distribution goals.

**Socialization** The long and complicated processes of social interactions through which a child learns the intellectual, physical, and social skills needed to function as a member of society.

**Social mobility** The movement of an individual or a group from one social status to another.

**Social movement** A form of collective behavior in which large numbers of people are organized or alerted to support and bring about, or to resist, social change.

**Social organization** The web of actual interactions among individuals and groups in society that defines their mutual rights and responsibilities and differs from society to society.

**Social revolutions**   Rapid and basic transformations of a society's state and class structures that are accompanied, and in part carried through, by class-based revolts.

**Social sciences**   All those disciplines that apply scientific methods to the study of human behavior. The social sciences include sociology, cultural anthropology, psychology, economics, history, and political science.

**Social solidarity**   People's commitment and conformity to a society's collective conscience.

**Social stratification**   The division of society into levels, steps, or positions that is perpetuated by the major institutions of society such as the economy, the family, religion, and education.

**Social structure**   The stable, patterned relationships that exist among social institutions within a society.

**Society**   A grouping of people who share the same territory and participate in a common culture.

**Sociobiology**   An approach that uses biological and evolutionary principles to explain the behavior of social beings.

**Sociological imagination**   The relationship between individual experiences and forces in the larger society that shape our actions.

**Sociology**   The scientific study of human society and social interactions.

**Specialization**   One of the two forms of adaptation. Specialization is developing ways of doing things that work extremely well in a particular environment or set of circumstances.

**Spontaneous cooperation**   The oldest and most common form of cooperation, which arises from the needs of a particular situation.

**State**   The institutionalized way of organizing power within territorial limits.

**Statement of association**   A proposition that changes in one thing are related to changes in another, but that one does not necessarily cause the other.

**Statement of causality**   A proposition that one thing brings about, influences, or changes something else.

**Statistical significance**   A mathematical statement about the probability that some event or relationship is not due to chance alone.

**Statuses**   The culturally and socially defined positions occupied by individuals throughout their lifetimes.

**Status inconsistency**   Situations in which people rank differently (higher or lower) on certain stratification characteristics than on others.

**Status offenses**   Behavior that is criminal only because the person involved is a minor.

**Stratified random sample**   A technique to make sure that all significant variables are represented in a sample in proportion to their numbers in the larger population.

**Structural conduciveness**   One of sociologist Neil Smelser's six conditions that shape the outcome of collective behavior. Structural conduciveness refers to the conditions within society that may promote or encourage collective behavior.

**Structural strain**   One of Smelser's six conditions that shape the outcome of collective behavior. Structural strain refers to the tension that develops when a group's ideals conflict with its everyday realities.

**Structured interview**   An interview with a predetermined set of questions that are followed precisely with each subject.

**Subculture**   The distinctive lifestyles, values, norms, and beliefs of certain segments of the population within a society.

**Subgroups**   Splinter groups within the larger group.

**Subjugation**   The subordination of one group and the assumption of a position of authority, power, and domination by the other.

**Suburbs**   Those territories that are part of a metropolitan statistical area but are outside the central city.

**Superego**   In Freudian theory, one of the three separately functioning parts of the self. The superego consists of society's norms and values, learned in the course of a person's socialization, that often conflict with the impulses of the id. The superego is the internal censor.

**Supernaturalism**   A belief system that postulates the existence of impersonal forces that can influence human events.

**Survey**   A research method in which a population or a sample is studied in order to reveal specific facts about it.

**Symbolic interactionism**   A theoretical approach that stresses the meanings people place on their own and one another's behavior.

**Symbols**   Objects that represent other things. Unlike signs, symbols need not share any of the qualities of whatever they represent.

**Taboo**   A sacred prohibition against touching, mentioning, or looking at certain objects, acts, or people.

**Techniques of neutralization**   A process that makes it possible to justify illegal or deviant behavior.

**Technological determinism**   The view that technological change has an important effect on a society and has an impact on its culture, social structure, and even its history.

**Theism**   A belief in divine beings—gods and goddesses—who shape human affairs.

**Theory of differential association**   A theory of juvenile delinquency based on the position that criminal behavior is learned in the context of intimate groups. People become criminals as a result of associating with others who engage in criminal activities.

**Threatened crowd**   A crowd that is in a state of alarm, believing itself to be in danger.

**Total institutions**   Environments, such as prisons or mental hospitals, in which the participants are physically and socially isolated from the outside world.

**Totalitarian capitalism**   A political-economic system under which the government retains control of the social institutions but allows the means of production and distribution to be owned and managed by private groups and individuals.

**Totalitarian government**   A government in which one group has virtually total control of the nation's social institutions.

**Totalitarian socialism**   In addition to almost total regulation of all social institutions, the government controls and owns all major means of production and distribution.

**Totem** An ordinary object, such as a plant or animal, which has become a sacred symbol to a particular group that not only reveres the totem but identifies with it.

**Tracking** The stratification of students by ability, social class, and various other categories.

**Tracks** The academic and social levels typically assigned to and followed by the children of different social classes.

**Traditional authority** Power that is rooted in the assumption that the customs of the past legitimize the present.

**Traditional cooperation** Cooperation that is tied to custom and is passed on from one generation to the next.

**Traditional ideology** An ideology that tries to preserve things as they are.

**Triad** A small group containing three members.

**Universal church** A church that includes all the members of a society within one united moral community.

**Urbanization** A process whereby a population becomes concentrated in a specific area because of migration patterns.

**Urban population** The inhabitants of an urbanized area and the inhabitants of incorporated or unincorporated areas with a population of 2,500 or more.

**Validity** The ability of a research study to test what it was designed to test.

**Value-added theory** A theory that attempts to explain whether collective behavior will occur and what direction it will take.

**Values** A culture's general orientations toward life—its notion of what is good and bad, what is desirable and undesirable.

**Variable** Anything that can change (vary).

**Vertical mobility** Movement up or down in the social hierarchy that results in a change in social class.

**Victimless crimes** Acts that violate those laws meant to enforce the moral code.

**Violent crime** Crime that involves the use of force or the threat of force against the individual.

**White-collar crime** Crime committed by individuals who, while occupying positions of social responsibility or high prestige, break the law in the course of their work for illegal, personal, or organizational gain.

# References

Albright, Joseph, and Marcia Kunstel. (1999, February 2). "Grim Odds for Chinese Girl Babies." *Atlanta Journal-Constitution*, p. 8A.

Amato, Paul, and Jacob Cheadle. (2005, February). "The Long Reach of Divorce: Divorce and Child Well-Being across Three Generations." *Journal of Marriage and Family* 67, pp. 191–206.

American Association of Retired People. (1994). *Images of Aging in America*. Washington, DC.

American Bar Association. (2012). "JD. and LL.B Degrees Awarded 1981–2011." Available at http://www.americanbar.org/content/dam/aba/administrative/legal_education_and_admissions_to_the_bar/statistics/jd_llb_degrees_awarded.authcheckdam.pdf.

American Institute of Philanthropy. (2012). "Charity Rating Guide & Watchdog Report." December 2011, http://www.charitywatch.org/.

American Psychological Association. (2006, March 22). "When the Boss Is a Woman." Available at http://www.psychologymatters.org/womanboss.html, accessed September 8, 2011.

AmeriStat. (2002, December). "The Gender Gap in U.S. Mortality." Population Reference Bureau.

___. (2002, August). "A Century of Progress in U.S. Infant and Child Survival." Population Reference Bureau.

___. (2002, August). "Higher Education Means Lower Mortality Rates." Population Reference Bureau.

___. (2001). "Solitary Living on the Rise in the United States." Population Reference Bureau.

___. (2001, November). "Regional Variations in the Traditional American Household Population." Population Reference Bureau.

___. (2000, February). "U.S. High School Dropouts: The Gender Gap." Population Reference Bureau.

Anderson, Craig A. (2010, January/February). "Violent Video Games and Other Media Violence (Part I)." *Pediatrics for Parents* 26(1 and 2), p. 2.

___. (2003). "Violent Video Games: Myths, Facts, and Unanswered Questions." Available at http://www.apa.org/science/psa/sb-andersonprt.html, accessed October 18, 2005.

Anderson, Craig A., and Karen E. Dill. (2000). "Video Games and Aggressive Thoughts, Feelings, and Behavior in the Laboratory and in Life." *Journal of Personality and Social Psychology* 78(4), pp. 772–790.

Anderson, Robert N. (2001). "Deaths: Leading Causes for 2000." *National Vital Statistics Reports* 50(16).

Anderson, Robert N., and Betty L. Smith. (2005, March 7). "Deaths: Leading Causes for 2002." *National Vital Statistics Reports* 53(17).

Annan, Kofi A. (2001). *We the Children: Meeting the Promises of the World Summit for Children*. Available at http://www.unicef.org, accessed February 15, 2002.

Ansolabehere, Stephen, and Shanto Iyengar. (1995). *Going Negative: How Political Advertisements Shrink and Polarize the Electorate*. New York: Free Press.

Argamon, Shlomo, Moshe Koppel, Jonathan Fine, and Anat Rachel Shimoni. (2003). "Gender, Genre, and Writing Style in Formal Written Texts." *Text* 23(3), pp. 321–346.

Aristotle. (1908). *The Politics and Economics of Aristotle*. In Edward English Walford and John Gillies (Trans.). London: G. Bell & Sons.

Armstrong, Karen. (1994). *A History of God*. New York: Ballantine Books.

Asch, Solomon. (1955). "Opinions and Social Press." *Scientific American* 193, pp. 31–35.

Associated Press. (2007, July 15). "India Tries to Stop Sex-Selective Abortions." *New York Times*. Available at http://www.nytimes.com/2007/07/15/world/asia/15india.html, accessed March 21, 2009.

Association of American Medical Colleges. (2012, March). "The Changing Gender Composition of U.S. Medical School Applicants and Matriculants," *Analysis in Brief* 12(1). Available at https://www.aamc.org/download/277026/data/aibvol12_no1.pdf.

___. (2002). *Medical School Graduation Questionnaire*.

A. T. Kearney, Inc. (2001, January/February). "Measuring Globalization." *Foreign Policy*.

Austen, Jane. (1813). *Pride and Prejudice*, reprint ed., 1998. Oxford: Oxford UP.

Austen-Smith, David, and Roland G. Fryer, Jr. (2005, May). "An Economic Analysis of 'Acting White.'" *Quarterly Journal of Economics* 120(2), pp. 551–583.

Autor, David. (2010, April). MIT Department of Economics and National Bureau of Economic Research, "The Polarization of Job Opportunities in the U.S. Labor Market: Implications for Employment and Earnings." Available at http://economics.mit.edu/files/5554.

Axell, Albert. (2002). *Kamikaze: Japan's Suicide Gods*. Boston: Longman.

Axtell, Roger E. (1998). *Gestures: The Do's and Taboos of Body Language around the World*. New York: John Wiley.

Bachman, Katy. (2007, September 26). "Nielsen's Nat'l PM TV Panel to Triple in Size." *Mediaweek*. Available at http://www.mediaweek.com/mw/news/recent_display.jsp?vnu_content_id=1003646903, accessed September 30, 2007.

Bahree, Megha. (2008, February 25). "Child Labor." *Forbes Magazine*. Available at http://www.forbes.com/forbes/2008/0225/072.html, accessed July 1, 2009.

Baker, Paul J., Louis E. Anderson, and Dean S. Dorn. (1993). *Social Problems: A Critical Thinking Approach*, 2nd ed. Belmont, CA: Wadsworth.

Baldus, David C., George Woodworth, and Charles A. Pulaski, Jr. (1990). *Equal Justice and the Death Penalty: A Legal and Empirical Study*. Boston: Northeastern University Press.

Bales, R. F. (1958). "Task Roles and Social Roles in Problem-Solving Groups." In E. E. Maccoby, T. M. Newcomb, and E. L. Hartley (Eds.), *Readings in Social Psychology*, 3rd ed. New York: Holt, Rinehart and Winston.

Bandura, Albert. (1969). *Principles of Behavior Modification*. New York: Holt, Rinehart and Winston.

Barnes, Jessica S., and Claudette E. Bennett. (2002, February). *The Asian Population: 2000*. U.S. Bureau of the Census.

Barnett, Cynthia. (2002). "The Measurement of White-Collar Crime Using Uniform Crime Reporting (UCR) Data." Federal Bureau of Investigation, Criminal Justice Information Services (CJIS) Division.

Barnett, Rosalind C., and Grace K. Baruch. (1987). "Social Roles, Gender, and Psychological Distress." In Rosalind C. Barnett, Lois Biener, and Grace K. Baruch (Eds.), *Gender and Stress*. New York: Free Press.

Barnett, R. C., and N. L. Marshall. (1991). "The Relationship between Women's Work and Family Roles and Their Subjective Well-Being and Psychological Distress." In M. Frankenhaeuser, V. Lundberg, and M. Chesney (Eds.), *Women, Work, and Health: Stress and Opportunities*. New York: Plenum.

Baron, Salo W. (1976). "European Jewry before and after Hitler." In Yisrael Gutman and Livia Rochkirchen (Eds.), *The Catastrophe of European Jewry*. Jerusalem: Yad Veshem.

Bartholomew, Robert E., and Erich Goode. (2000, May/June). "Mass Delusions and Hysterias: Highlights from the Past Millennium." *Skeptical Inquirer Magazine* 24(3), p. 20.

Baskin, Barbara H., and Karen Harris. (1980). *Books for the Gifted Child*. New York: R. R. Bowker.

Bateson, Mary Catherine. (1994). *Peripheral Visions*. New York: HarperCollins.

Bauman, Kurt J. (2001, August). "Home Schooling in the United States: Trends and Characteristics." U.S. Bureau of the Census, Working Paper Series No. 53.

Bazelon, Emily. (2008, March 9). "Hormones, Genes and the Corner Office," *New York Times*. Available at http://www.nytimes.com/2008/03/09/books/review/Bazelon-t.html?pagewanted=all&_r=0.

Bearak, Barry. (2008, September 5). "Destitute Swaziland, Leader Lives Royally." *New York Times*, p. A1. Available at http://http://www.nytimes.com/2008/09/06/world/africa/06king.html?pagewanted=all, accessed April 18, 2009.

Becker, Howard. (1963). *Outsiders: Studies in the Sociology of Deviance*. New York: Free Press.

Bedau, Hugo Adam. (Ed.). (1997). *The Death Penalty in America: Current Controversies*. New York: Oxford University Press.

Begley, Sharon. (2005, February 4). "People Believe a 'Fact' That Fits Their Views Even If It's Clearly False." *Wall Street Journal*, p. B1.

Bell, Nathan E. (2011, September). "Graduate Enrollment and Degrees: 2000 to 2010." Council of Graduate Schools. Available at http://www.cgsnet.org/ckfinder/userfiles/files/R_ED2010.pdf, accessed July 12, 2012.

Belsky, Jay. (2011, February). "Child Care and Its Impact on Young Children," *Encyclopedia on Early Childhood Development*, 2nd ed. rev. United Kingdom: Birkbeck University of London. Available at http://www.child-encyclopedia.com/documents/BelskyANGxp3-Child_care.pdf, accessed June 3, 2012.

Benedict, Ruth. (1961/1934). *Patterns of Culture*. Boston: Houghton Mifflin.

___. (1938). "Continuities and Discontinuities in Cultural Conditioning." *Psychiatry* 1, pp. 161–167.

Berbrier, Mitch. (2004, Winter). "Why Are There So Many 'Minorities'?" *Contexts* 3(1), pp. 38–44.

Berger, Peter. (1967). *The Sacred Canopy*. New York: Doubleday.

___. (1963). *Invitation to Sociology: A Humanistic Perspective*. New York: Doubleday.

Berger, Suzanne E. (1996). *Horizontal Woman*. Boston: Houghton Mifflin.

Bernard, L. L. (1924). *Instinct*. New York: Holt, Rinehart and Winston.

Berry, Brewton, and Henry L. Tischler. (1978). *Race and Ethnic Relations*, 4th ed. Boston: Houghton Mifflin.

Besharov, Douglas J. (1999, July 14). "Asking More from Matrimony." *New York Times*.

Best, Joel. (2002). "Monster Hype." *Education Next*, Washington, DC, Hoover Institution. Available at http://www.educationnext.org/20022/50.html#fig2, accessed January 9, 2006.

___. (2002, Summer). "Monster Hype: How a Few Isolated Tragedies—and Their Supposed Causes—Were Turned into a National 'Epidemic.'" *Education Next* 2, pp. 50–55.

___. (2001). *Damned Lies and Statistics*. Berkeley: University of California Press.

Bianchi, Suzanne, and Lynne M. Casper. (2000). "American Families." *Population Bulletin* 55(4). Population Reference Bureau.

Bierstadt, Robert. (1974). *The Social Order*, 4th ed. New York: McGraw-Hill.

Billings, John S. (1891, February). "Public Health and Municipal Government." *Annals of the American Academy of Political and Social Science* (Supplement).

Black, Dan, Gary Oates, Seth Sanders, and Lowell Taylor. (2000, May). "Demographics of the Gay and Lesbian Population of the United States: Evidence from Available Systematic Data Sources." *Demography* 37, pp. 139–154.

Blank, Jonah. (1992). *Arrow of the Blue-Skinned God*. Boston: Houghton Mifflin.

Blau, Peter M. (1964). *Exchange and Power in Social Life*. New York: John Wiley.

Blum, A., and Blum, N. (2009, September 6). "Coronary Artery Disease: Are Men and Women Created Equal?" *Gender Medicine* 6(3), pp. 410–418. Available at http://www.ncbi.nlm.nih.gov/pubmed/19850237.

Blum, John M., Edmund S. Morgan, Willie Lee Rose, Arthur Schlesinger, Jr., Kenneth M. Stamp, and C. Van Woodard. (1981). *The National Experience: A History of the United States*, 5th ed. New York: Harcourt Brace Jovanovich.

Bonczar, Thomas P. (2003, August). Bureau of Justice Statistics, Special Report, "Prevalence of Imprisonment in the U.S. Population, 1974–2001." NCJ 197976.

Borjas, George J. (1999). *Heaven's Door*. Princeton, NJ: Princeton University Press.

Bowles, Samuel, and Herbert Gintis. (1976). *Schooling in Capitalist America: Educational Reform and the Contradictions of Economic Life*. New York: Basic Books.

Bramlett, M. D., and W. D. Mosher. (2002). "Cohabitation, Marriage, Divorce, and Remarriage in the United States." *Vital Health Statistics* 23(22). National Center for Health Statistics.

Brehm, Sharon. (1992). *Intimate Relationships*, 2nd ed. New York: Random House.

Brenner, Joanna. (2012, March 29). Pew Internet: Social Networking. Available at http://pewinternet.org/Commentary/2012/March/Pew-Internet-Social-Networking-less-detail.aspx.

Brotz, Howard. (1966). *Negro Social and Political Thought 1850–1920*. New York: Basic Books.

Brown, Dee. (1971). *Bury My Heart at Wounded Knee*. New York: Holt, Rinehart and Winston.

Brownell, Kelly D., and Thomas R. Frieden, M.D. (2009, April 30). "Ounces of Prevention: The Public Policy Case for Taxes on Sugared Beverages." *New England Journal of Medicine*.

Buchmann, Claudia, and Thomas A. DiPrete. (2006, August). "The Growing Female Advantage in College Completion." *American Sociological Review* 71(4).

Bullough, Vern L. (1973). *The Subordinate Sex*. Chicago: University of Chicago Press.

Bumpass, Larry L. (1991). "What's Happening to the Family? Interactions between Demographic and Institutional Change." *Demography* 27(4), p. 485E.

Bumpass, Larry, and Lu Hsien-Hen. (2000). "Trends in Cohabitation and Implications for Children's Family Contexts in the U.S." *Population Studies* 54, pp. 29–41.

Bumpass, Larry, R. K. Raley, and J. A. Sweet. (1995). "The Changing Character of Stepfamilies: Implications of Cohabitation and Nonmarital Childbearing." *Demography* 32, pp. 425–436.

Bumpass, Larry L., James A. Sweet, and Andrew Cherlin. (1991, November). "The Role of Cohabitation in Declining Rates of Marriage." *Journal of Marriage and the Family* 53(4), pp. 913–927.

Bureau of the Census. (2005). Current Population Survey, 1960 to 2005 Annual Social and Economic Supplements.

___. (2005, October 11). Public-Use Microdata Sample (PUMS). Available at http://www.census.gov/main/www/pums.html, accessed November 8, 2005.

___. (2005, August 30). Historical Poverty Tables, Table 3. "Poverty Status of People, by Age, Race, and Hispanic Origin: 1959 to 2004." Housing and Household Economic Statistics Division. Available at http://www.census.gov/hhes/www/poverty/histpov/hstpov3.html, accessed August 31, 2005.

___. (2005, August 30). "Income Stable, Poverty Rate Increases, Percentage of Americans without Health Insurance Unchanged." Available at http://www.census.gov/Press-Release/www/releases/archives/income_wealth/005647.html, accessed January 9, 2006.

___. (2005, June 29). "Estimated Median Age at First Marriage, by Sex: 1890 to the Present." Available at http://www.census.gov/population/socdemo/hh-fam/ms2.pdf, accessed October 18, 2005.

___. (2005, June 9). "Hispanic Population Passes 40 Million, Census Bureau Reports." Available at http://www.census.gov/Press-Release/www/releases/archives/population/005164.html, accessed January 9, 2006.

___. (2005, January). Current Population Survey.

___. (2004, November). Current Population Survey.

___. (2004, March). International Data Base. Available at http://www.census.gov/ipc/www/idbnew.html, accessed January 9, 2006.

___. (2003). Current Population Survey, Annual Social and Economic Supplement.

___. (2002). *Statistical Abstract of the United States: 2001*, 121st ed. Washington, DC: U.S. Government Printing Office.

___. (2002, August). "Interracial Tables." Available at http://www.census.gov, accessed January 9, 2006.

___. (2001). *Statistical Abstract of the United States: 2000*, 120th ed. Washington, DC: U.S. Government Printing Office.

___. (2001, December). "Profile of the Foreign Born Population in the United States: 2000." Available at http://www.census.gov, accessed January 9, 2006.

___. (2001, October). Population Estimates Program. Washington, DC: U.S. Government Printing Office.

___. (2001, August). *The Black Population: 2000*. Census 2000 brief. Available at http://www.census.gov, accessed January 9, 2006.

___. (2001, May). *The Hispanic Population: 2000*. Census 2000 brief. Available at http://www.census.gov, accessed January 9, 2006.

___. (2000). *Statistical Abstract of the United States: 1999*, 119th ed. Washington, DC: U.S. Government Printing Office.

___. (2000, March). "The Foreign Born Population in the United States, March 2000." Available at http://www.census.gov, accessed January 9, 2006.

___. (1999). "Resident Population of the United States: Middle Series Projections. 2015–2030 by Age and Sex." Available at http://www.census.gov/population/projections/nation/nas/npas1530.txt, accessed February 12, 2001.

___. (1999, October). "Region of Birth a Key Indicator of Well-Being for America's Foreign-Born Population, Census Bureau Reports." Press release. Available at http://www.census.gov, accessed November 3, 2000.

___. (1999, March). "Educational Attainment in the United States," pp. 205–228.

___. (1998, March). *World Population Profile: 1998*. Washington, DC: U.S. Government Printing Office.

___. (1997). *Statistical Abstract of the United States: 1997*, 117th ed. Washington, DC: U.S. Government Printing Office.

___. (1995, April). "Housing of American Indians on Reservations." Available at http://www.census.gov, accessed March 10, 1998.

___. (1981). *Statistical Abstract of the United States: 1980*. Washington, DC: U.S. Government Printing Office.

___. (1976). *Historical Statistics of the United States: Colonial Times to 1970*. Washington, DC: U.S. Government Printing Office.

Bureau of Justice Statistics. (2005). "Nation's Prison and Jail Population Grew by 932 Inmates per Week, Number of Female Inmates Reached More Than 100,000." Available at http://www.ojp.usdoj.gov/bjs, accessed July 21, 2005.

___. (2002). "Capital Punishment Statistics 2001, Summary Findings." U.S. Department of Justice.

___. (2002, September). "The Nation's Two Crime Measures." U.S. Department of Justice press release.

___. (2002, June). "Two-Thirds of Former State Prisoners Rearrested for Serious New Crimes." U.S. Department of Justice press release.

___. (1997). *Criminal Victimizations in the United States, 1996*. U.S. Department of Justice.

Bureau of Labor Statistics. (2005). U.S. Department of Labor, *Occupational Outlook Handbook, 2004–05 Edition*, Childcare Workers. Available at http://www.bls.gov/oco/ocos170.htm, accessed August 09, 2005.

Bures, Frank. (2008, June). "A Mind Dismembered: In Search of the Magical Penis Thieves." *Harpers Magazine*. Available at http://harpers.org/archive/2008/06/0082063, accessed July 6, 2009.

Buss, David M. (1994). *The Evolution of Desire*. New York: Basic Books.

Buttner, E. H., and M. McEnally. (1996). "The Interactive Effect of Influence Tactic. Applicant Gender and Type of Job on Hiring Recommendations." *Sex Roles* 34, pp. 581–591.

Bylinsky, Gene. (1988, July 19). "Technology in the Year 2000." *Fortune*, pp. 92–98.

Caldeira, K. M., Arria, A. M., O'Grady, K. E., Vincent, K. B., and Wish, E. D. (2008, March). "The Occurrence of Cannabis Use Disorders and Other Cannabis-Related Problems among First-Year College Students." *Addictive Behaviors* 33(3), pp. 397–411.

Canetti, Elias. (1978/1960). *Crowds and Power*. New York: Seabury Press.

Capizzano, Jeffrey, and Regan Main. (2005, March 31). "Many Young Children Spend Long Hours in Child Care." Washington, DC: Urban Institute.

Cardozo Law Innocence Project. Available at http://www.cardozo.yu.edu/innocence_project/, accessed January 9, 2006.

Carli, L. L. (2001). "Gender and Social Influence." *Journal of Social Issues* 57(4), pp. 725–737.

Carli, L. L., S. J. LaFleur, and C. C. Loeber. (1995). "Nonverbal Behavior, Gender, and Influence." *Journal of Personality and Social Psychology* 68(6), pp. 1030–1041.

Carlson, Darren K. (2001, February 14). "Over Half of Americans Believe in Love at First Sight." Princeton, NJ: Gallup News Service.

Casper, Lynne M., and Philip Cohen. (2000, May). "How Does POSSLQ Measure Up? Historical Estimates of Cohabitation." *Demography* 37(2), pp. 237–245.

Cassidy, Tina. (1999, September 28). "Job Complaint Recalls 'Racial Charade.' " *Boston Globe*, pp. B1, B5.

"Caste." (2002). *The Columbia Encyclopedia*, 7th ed.

Cato Institute. (2000, December 12). "China's One-Child Policy." *Cato Daily Dispatch*.

CBS News. (2005, April 20). "Faithful See Image of Virgin Mary." Available at http://www.cbsnews.com/stories/2005/04/20/national/main689630.shtml, accessed January 9, 2006.

CBS News Poll. (2005, July 13–14). "Poll: More Concerned about Terror." Available at http://www.cbsnews.com/stories/2005/07/15/opinion/polls/main709488_page2.shtml?CMP=ILC-Search Stories, accessed January 9, 2006.

Centers for Disease Control and Prevention. (2012, June 6). National Vital Statistics, "Deaths: Leading Causes for 2008," *Melonie Heron* 60(6). Available at http://www.cdc.gov/nchs/data/nvsr/nvsr60/nvsr60_06.pdf.

___. (2012, July). "Death in the United States, 2010," *NCHS Brief* No. 99. Available at http://www.cdc.gov/nchs/data/databriefs/db99.htm#Fig5.

___. (2011, January 14). "Health Disparities and Inequalities Report—United States, 2011," *Morbidity and Mortality Weekly Report* 60. Available at http://sharing.govdelivery.com/bulletins/GD/USCDC-11D2BC.

___. (2010). *HIV Surveillance Report* 22. Available at cdc.gov/hiv/surveillance/resources/reports/2010report, accessed September 13, 2012.

___. (2009, May 28). "Defining Overweight and Obesity." Available at http://www.cdc.gov/obesity/defining.html, accessed August 14, 2009.

___. (2005). "Life Expectancy Hits Record High, Gender Gap Narrows." Available at http://www.cdc.gov/nchs/pressroom/05facts/lifeexpectancy.htm, accessed January 9, 2006.

Centerwall, B. (1992). "Television and Violence: The Scale of the Problem and Where to Go from Here." *Journal of the American Medical Association* 267, pp. 3059–3061.

Central Intelligence Agency. (2012). *The World Factbook*. Available at https://www.cia.gov/library/publications/the-world-factbook/fields/2103.html, accessed August 30, 2012.

___. (2012, August 10). *The World Factbook*. Available at https://www.cia.gov/library/publications/the-world-factbook/fields/2102.html.

___. (2009). "Life Expectancy at Birth." *The World Factbook*. Available at https://www.cia.gov/library/publications/the-world-factbook/fields/2102.html, accessed May 15, 2009.

___. (2009). *The 2008 World Factbook*. Available at https://www.cia.gov/library/publications/the-world-factbook/fields/2127.html, accessed May 15, 2009.

___. (2005, July). "Rank Order—Population." Available at http://www.cia.gov/cia/publications/factbook/rankorder/2119rank.html, accessed January 9, 2006.

Chafe, Zoe. (2007, November 8). "Child Labor Harms Many Young Lives." Worldwatch Institute. Available at http://www.worldwatch.org/node/5479, accessed July 1, 2009.

Chaiken, Jan M., and Marcia R. Chaiken. (1982). *Varieties of Criminal Behavior*. Santa Monica, CA: RAND Corporation.

Chambliss, William J. (1973). "Elites and the Creation of Criminal Law." In William J. Chambliss (Ed.), *Sociological Readings in the Conflict Perspective*. Reading, MA: Addison-Wesley.

Chandler, David L. (1992, November 2). "Polling: The Methods behind the Madness." *Boston Globe*, pp. 35, 37.

Cherlin, Andrew J. (2009). *The Marriage-Go-Round*. New York: Alfred A. Knopf.

___. (1992). *Marriage, Divorce, and Remarriage*. Cambridge, MA: Harvard University Press.

Cherlin, Andrew J., and Frank F. Furstenberg, Jr. (1994, January 1). "Stepfamilies in the United States: A Reconsideration." *Annual Review of Sociology* 20.

China Daily. (2008, April 25). "Penis Theft Panic Hits City." Available at http://www.chinadaily.net/world/2008-04-25/content_6644724.htm, accessed July 6, 2009.

Cholo, Ana Beatriz. (2007, March 6). "Homeschoolers Find University Doors Open." Associated Press.

Chomsky, Noam. (1975). *Language and Mind*. New York: Harcourt Brace Jovanovich.

Chorover, Stephan L. (1979). *From Genesis to Genocide: The Meaning of Human Nature and the Power of Behavior Control*. Cambridge, MA: MIT Press.

Christakis, Nicholas A., and James H. Fowler. (2007, July 27). "The Spread of Obesity in a Large Social Network Over 32 Years." *New England Journal of Medicine* 357(4), pp. 370–379.

Clark, Charles M. A. (2002, June). "Wealth and Poverty: On the Social Creation of Scarcity." *Journal of Economic Issues* 36(2).

Clark, Robert D., and Elizabeth Hatfield. (1989). "Gender Differences in Receptivity to Sexual Offers." *Journal of Psychology and Human Sexuality* 2, pp. 39–55.

Clarke-Stewart, Alison, and Virginia D. Allhusan. (2005). *What We Know about Childcare*. Cambridge, MA: Harvard University Press.

Cleland, John G., and Jerome K. Van Ginneken. (1988). "Maternal Education and Child Survival in Developing Countries: The Search for Pathways of Influence." *Social Science and Medicine* 27(12), pp. 357–368.

Cloninger, Dale. O., and Roberto Marchesini. (2001). "Execution and Deterrence: A Quasi-Controlled Group Experiment." *Applied Economics* 35(5), pp. 569–576.

CNN News. (2000, May 17). "British Soccer Fans, Danish Police Clash."

Cohany, Sharon R., and Emy Sok. (2007, February). "Trends in Labor Force Participation of Married Mothers of Infants." *Monthly Labor Review* 130(2). Available at http://www.bls.gov/opub/mlr/2007/02/art2exc.htm, accessed, May 7, 2008.

Cohn, D'Vera. (2001, August 22). "Count of Gay Couples up 300%; 2000 Census Ranks DC, Arlington, Alexandria among Top Locales." *Washington Post*, p. A3.

Colapinto, John. (2000). *As Nature Made Him*. New York: HarperCollins.

Coleman, James S. (1977). *Parents, Teachers, and Children*. San Francisco: San Francisco Institute for Contemporary Studies.

___. (1966). *Equality of Educational Opportunity*. Washington, DC: U.S. Government Printing Office.

Collins, Randall. (1979). *The Credential Society: An Historical Sociology of Education and Stratification*. New York: Academic Press.

___. (1975). *Conflict Sociology: Toward an Explanatory Science*. New York: Academic Press.

Comte, Auguste. (1968/1851). *System of Positive Policy*, Vol. 1 (John Henry Bridges, Trans.). New York: Burt Franklin.

Conrad, Peter. (1992). "Medicalization and Social Control." *Annual Review of Sociology* 18, pp. 209–232.

Conrad, Peter, and Rochelle Kern. (Eds.). (1986). *The Sociology of Health and Illness*, 2nd ed. New York: St. Martin's Press.

Cook, Phillip, and Jens Ludwig. (1997). "Weighing the 'Burden of Acting White': Are There Race Differences in Attitudes towards Education?" *Journal of Public Policy and Analysis* 16, pp. 256–278.

Cooley, C. H. (1909). *Social Organization*. New York: Scribner's.

Cooper, Edith Fairman. (2005). "Missing and Exploited Children: Overview and Policy Concerns." CRS Report for Congress, April 29, 2005, Washington, DC: Congressional Research Service, Library of Congress.

Copeland, Libby. (2011). "America's Next Top Sociologist," *Slate*. Available at http://www.slate.com/articles/double_x/doublex/2011/09/americas_next_top_sociologist.html.

Corcoran, Paul E. (1994). "Presidential Concession Speeches: The Rhetoric of Defeat." *Political Communication* 11, pp. 109–131.

Cornell, Vincent J. (2001). "Islam." *World Book Online Americas Edition*. Available at http://www.aolsvc.worldbook.aol.com/wbol/wbPage/na/ar/co/282380, accessed May 23, 2003.

"Corruption in the Higher Education System of Kazahkistan." (2002, June). KIMEP Times No. 7. Kazakhstan Institute of Management Economics and Strategic Research.

Cose, Ellis. (1997). "Census and the Complex Issue of Race." *Commentary* 34(6), pp. 9–13.

Coser, L. A. (1977). *Masters of Sociological Thought*, 2nd ed. New York: Harcourt Brace Jovanovich.

___. (1967). *Continuities in the Study of Social Conflict*. New York: Free Press.

___. (1956). *The Functions of Social Conflict*. Glencoe, IL: Free Press.

"Covenant Marriage." (1999, May 7). *Issues and Controversies on File*, pp. 187–189.

Crano, William D., and Joel Aronoff. (1978, August). "A Cross-Cultural Study of Expressive and Instrumental Role Complementarity in the Family." *American Sociological Review* 43, pp. 463–471.

Crichton, Judy. (1998). *America 1900*. New York: Henry Holt.

Crossen, Cynthia. (2007, November 5). "For a Time in the '50s, a Huckster Fanned Fears of Ad 'Hypnosis.'" *Wall Street Journal*, p. B1.

___. (2003, March 5). "Even Women Didn't Want to Give Women the Vote." *Wall Street Journal*, p. B1.

___. (1994). *Tainted Truth: The Manipulation of Fact in America*. New York: Simon & Schuster.

Cummings, Milton C., and David Wise. (1981). *Democracy under Pressure: An Introduction to the American Political System*, 4th ed. New York: Harcourt Brace Jovanovich.

Current Population Survey. (2002). *Annual Demographic Survey*, March Supplement, 1960–2002. Washington, DC: U.S. Bureau of the Census.

___. (2001). *Annual Demographic Survey*, March Supplement, U.S. Bureau of the Census.

Curtiss, Susan. (1977). *Genie: A Psycholinguistic Study of a Modern-Day Wild Child*. New York: Academic Press.

Cuzzort, R. P., and E. W. King. (1980). *Twentieth Century Social Thought*, 3rd ed. New York: Holt, Rinehart and Winston.

Dahrendorf, R. (1959). *Class and Conflict in Industrial Society*. Stanford, CA: Stanford University Press.

___. (1958, September). "Out of Utopia: Toward a Reorientation of Sociological Analysis." *American Journal of Sociology* 64, pp. 158–164.

Daly, M., and M. Wilson. (1988). *Homicide*. Hawthorne, NY: Aldine de Gruyter.

D'Andrade, Roy G. (1966). "Sex Differences and Cultural Institutions." In Eleanor Emmons Maccoby (Ed.), *The Development of Sex Differences*. Stanford, CA: Stanford University Press.

Darwin, Charles. (1964/1859). *On the Origin of Species*. Cambridge, MA: Harvard University Press.

Davis, F. James. (1979). *Understanding Minority-Dominant Relations*. Arlington Heights, IL: AHM Publishing.

Davis, Kingsley. (1949). *Human Society*. New York: Macmillan.

___. (1940). "Extreme Social Isolation of a Child." *American Journal of Sociology* 45, pp. 554–565.

Davis, Kingsley, and W. E. Moore. (1945). "Some Principles of Stratification." *American Sociological Review* 10, pp. 242–249.

Deacon, Terrence W. (1997). *The Symbolic Species*. New York: Norton.

DeLisi, Matt, Anna Kosloski, Molly Sween, Emily Hachmeister, Matt Moore, and Alan Drury. (2010, August). "Murder by Numbers: Monetary Costs Imposed by a Sample of Homicide Offenders," *Journal of Forensic Psychiatry and Psychology* 21(4), pp. 501–513.

DeNavas-Walt, Carmen, and Robert Cleveland. (2002, September). *Money Income in the United States, 2001*. U.S. Bureau of the Census.

DeNavas-Walt, Carmen, Bernadette D. Proctor, and Cheryl Hill Lee. (2005). *Income, Poverty, and Health Insurance Coverage in the United States: 2004*. Current Population Reports, P60-229. Washington, DC: U.S. Bureau of the Census.

DeNavas-Walt, Carmen, Bernadette D. Proctor, and Jessica C. Smith. (2011). "U.S. Census Bureau, Current Population Reports, P60-239," *Income, Poverty, and Health Insurance Coverage in the United States: 2010*. U.S. Government Printing Office, Washington, DC, 2011.

Dershowitz, Alan M. (1996). *Reasonable Doubts: The O.J. Simpson Case and the Criminal Justice System*. New York: Simon & Schuster.

Dezhbakhsh, Hashem, Paul H. Rubin, and Joanna M. Shepherd. (2002, January). "Does Capital Punishment Have a Deterrent Effect? New Evidence from Post-moratorium Panel Data." Department of Economics, Emory University.

Diamond, Milton, and H. Keith Sigmundson. (1997, March). "Sex Reassignment at Birth: A Long Term Review and Clinical Implications." *Archives of Pediatric and Adolescent Medicine* 151, pp. 298–304.

Dinan, T. G. (1996). "Serotonin: Current Understanding and the Way Forward." *International Clinical Psychopharmacology* 11(1) Supplement, pp. 19–21.

Ditton, Paula M., and Doris Wilson. (1999). *Truth in Sentencing in State Prisons*. U.S. Department of Justice, Bureau of Justice Statistics.

Domhoff, G. William. (1983). *Who Rules America Now?* Englewood Cliffs, NJ: Prentice Hall.

Dowd, Maureen. (2002, April 10). "The Baby Bust." *New York Times*, p. 27A.

Dreifus, Claudia. (2005, May 31). "Declaring with Clarity, When Gender Is Ambiguous." *New York Times*, p. 2F.

Droege, Kristen. (1995, Winter). "Child Care: An Educational Perspective." *MIJCF, Jobs and Capital* 4, pp. 1–8.

DuBois, W. E. B. (1968). *Autobiography: A Soliloquy on Viewing My Life from the Last Decade of Its First Century*. New York: International Publishers.

Dugger, Celia W. (2001, May 6). "Modern Asia's Anomaly: The Girls Who Don't Get Born." *New York Times*, p. 4.

Duhart, Detis T. (2000). *Urban, Suburban, and Rural Victimization, 1993–98*. U.S. Department of Justice, Bureau of Justice Statistics.

Durant, Will. (1954). "Our Oriental Heritage." *The Story of Civilization*, Vol. 1. New York: Simon & Schuster.

Durkheim, Émile. (1961/1915). *The Elementary Forms of Religious Life*. New York: Collier Books.

___. (1960/1893). *The Division of Labor in Society* (G. Simpson, Trans.). New York: Free Press.

___. (1958/1895). *The Rules of Sociological Method*. Glencoe, IL: Free Press.

Dwork, Deborah, and Robert Jan van Pelt. (2002). *Holocaust: A History*. New York: Norton.

Eberstadt, Nicholas. (1992, January 20). "America's Infant Mortality Problem: Parents." *Wall Street Journal*, p. A14.

Eck, Diana L. (2001). *A New Religious America: How a "Christian Country" Has Become the World's Most Religiously Diverse Nation*. New York: HarperCollins.

Edmondson, Brad. (1996, October). "How to Spot a Bogus Poll." *American Demographics* 18(10), pp. 10, 12–15.

Edsall, Thomas B. (2012, March 12). "The Reproduction of Privilege," *New York Times*. Available at http://campaignstops.blogs.nytimes.com/.

Edwards, D. H., and E. A. Kravitz. (1997). "Serotonin. Social Status and Aggression." *Current Opinions in Neurobiology* 7(6), pp. 812–819. Review.

Eggdonation.com. The Egg Donor Program. Available at http://www.eggdonation.com/SampleDonor/Donor1.asp?DonorID=1026, accessed January 9, 2006.

Ehrlich, Isaac. (1975). "The Deterrent Effect of Capital Punishment: A Question of Life and Death." *American Economic Review* 65(3), pp. 397–417.

___. (1974). *The End of Affluence.* New York: Simon & Schuster.
Ekman, Paul, William V. Friesen, and John Bear. (1984, May). "The International Language of Gestures." *Psychology Today*, pp. 64–69.
El-Badry, Samia. (1994, January). "Understanding Islam in America." *American Demographics*, pp. 10–11.
Elkind, David. (1981). *The Hurried Child.* Reading, MA: Addison-Wesley.
Ellen, Elizabeth Fried. (2002, August). "Identifying and Treating Suicidal College Students." *Psychiatric Times* 19(8).
Elliott, Diana B., Simmons, Tavia, and Jamie M. Lewis. (2010). "Evaluation of the Marital Events Items on the ACS." Available at www.census.gov/hhes/socdemo/marriage/data/acs/index.html.
Ellis, Joseph J. (1997). *American Sphinx: The Character of Thomas Jefferson.* New York: Knopf.
Elshtain, Jean Bethke. (2001). *Jane Addams and the Dream of American Democracy.* New York: Basic Books.
Ember, Carol R., and Melvin Ember. (1981). *Anthropology*, 3rd ed. Englewood Cliffs, NJ: Prentice Hall.
Embree, Edwin R. (1967/1939). *Indians of the Americas.* Boston: Houghton Mifflin.
Enard, Wolfgang, et al. (2002, April 12). "Intra- and Interspecific Variation in Primate Gene Expression Patterns." *Science* 296, pp. 340–344.
Encyclopaedia Britannica. (2006). "Muslims." *Book of the Year.*
Engels, Friedrich. (1973/1845). *The Condition of the Working Class in England in 1844.* Moscow: Progress.
___. (1942/1884). *The Origin of the Family, Private Property and the State.* New York: International Publishers.
Erikson, Erik H. (1968). *Identity, Youth and Crisis.* New York: Norton.
___. (1964). *Childhood and Society.* New York: Norton.
Erikson, Kai T. (1966). *Wayward Puritans: A Study in the Sociology of Deviance.* New York: John Wiley.
Exline, Christopher H., Gary L. Peters, and Robert P. Larkin. (1982). *The City: Patterns and Processes in the Urban Ecosystem.* Boulder, CO: Westview Press.
Federal Communications Commission. (2005). "The V-Chip: Putting Restrictions on What Your Children Watch." Available at http://www.fcc.gov/cgb/consumerfacts/vchip.html, accessed October 18, 2005.
Feldman, M. D., and J. M. Feldman. (1998). *Stranger Than Fiction: When Our Minds Betray Us.* Washington, DC: American Psychiatric Press.
Festinger, Leon, Henry W. Rieken, and Stanley Schacter. (1956). *When Prophesy Fails.* New York: Harper Torchbooks.
Fields, Jason. (2004). "America's Families and Living Arrangements: 2003." Current Population Reports, pp. 20–553.
Fields, Jason, and Lynne M. Casper. (2001, June). "America's Families and Living Arrangements." Current Population Reports, P20–537. Washington, DC: U.S. Bureau of the Census.
Fine, Gary Alan. (1993). "Ten Lies of Ethnography: Moral Dilemmas of Field Research." *Journal of Contemporary Ethnography*, 22: 267–294.
Fine, Mark A. (1994, May). "An Examination and Evaluation of Recent Changes in Divorce Laws in Five Western Countries: The Crucial Role of Values." *Journal of Marriage and the Family* 56, pp. 249–263.
Fishbein, Diana. (2001). *Biological Perspectives in Criminology.* Belmont, CA: Wadsworth/Thomson Learning.
Fisher, Helen. (1994, October 16). " 'Wilson,' They Said, 'You're All Wet!' " *New York Times Book Review*, pp. 15–17.
___. (1992). *Anatomy of Love.* New York: Ballantine Books.
Fisher, Robert. (1984). *Let the People Decide: Neighborhood Organizing in America.* Boston: G. K. Hall.

Fishman, Charles. (2007, December 19). "Message in a Bottle." *Fast Company.* Available at http://www.fastcompany.com/magazine/117/features-message-in-a-bottle.html, accessed June 7, 2009.
Ford, Clellan S. (1970). "Some Primitive Societies." In Georgene H. Seward and Robert C. Williamson (Eds.), *Sex Roles in Changing Society.* New York: Random House.
Fortes, M., R. W. Steel, and P. Ady. (1947). "Ashanti Survey, 1945–46: An Experiment in Social Research." *Geographical Journal* 110, pp. 149–179.
Fouts, Roger. (1997). *Next of Kin: What Chimpanzees Have Taught Me about Who We Are.* New York: William Morrow.
Frankfort, H. (1956). *The Birth of Civilization in the Near East.* Garden City, NY: Doubleday/Anchor.
Freud, Sigmund. (1930). "Civilization and Its Discontents." *Standard Edition of the Complete Psychological Works of Sigmund Freud,* Vol. 29. London: Hogarth Press.
___. (1928). *The Future of an Illusion.* New York: Horace Liveright and the Institute of Psycho-Analysis.
___. (1923). "The Ego and the Id." *Standard Edition of the Complete Psychological Works of Sigmund Freud,* Vol. 19. London: Hogarth Press.
___. (1920). "Beyond the Pleasure Principle." *Standard Edition of the Complete Psychological Works of Sigmund Freud,* Vol. 14. London: Hogarth Press.
___. (1918). *Totem and Taboo.* New York: Moffat, Yard.
Fried, Morton. (1967). *The Evolution of Political Society.* New York: Random House.
Friedrich, Carl J., and Zbigniew Brezinski. (1965). *Totalitarian Dictatorship and Autocracy,* Vol. 2. Cambridge, MA: Harvard University Press.
Fryer, Roland G., Jr., and Paul Torelli. (2005, May 1). "An Empirical Analysis of 'Acting White.' " Harvard University Society of Fellows and NBER. Available at http://post.economics.harvard.edu/faculty/fryer/papers/fryer_torelli.pdf, accessed January 9, 2006.
Galinsky, Ellen, Caroline Howe, Susan Koutos, and Marybeth Shinn. (1994). *The Study of Children in Family Child Care and Relative Care: Highlights and Findings.* New York: Families and Work Institute.
Gallup, George, Jr., and Jim Castelli. (1987). *The American Catholic People: Their Beliefs, Practices, and Values.* Garden City, NY: Doubleday.
Gallup Poll. (2002, June 18). "Fewer Blacks Say Anti-White Sentiment Is Widespread in Black Community."
___. (2001, February 14). "Over Half of Americans Believe in Love at First Sight."
___. (2000, December 26). "When It Comes to Having Children, Americans Still Prefer Boys."
___. (2000, August 24–27). "Religion."
___. (1972). *The Gallup Poll: Public Opinion 1934–1971.* New York: Random House.
Gandhi, M. K. (1962). *All Religions Are True.* Ed. and published by Anand T. Hingorani. Bombay, India: Bharatiya Vidya Bhavan.
Gans, Herbert J. (1979, May). "Deception and Disclosure in the Field." *The Nation* 17, pp. 507–512.
___. (1962). *The Urban Villagers.* New York: Free Press.
Gardner, Howard. (1978). *Developmental Psychology.* Boston: Little, Brown.
___. (1972). "Studies of the Routine Grounds of Everyday Activities." In David Snow (Ed.), *Studies in Social Interaction.* New York: Free Press.
Garfinkel, Harold. (1972). "Conditions of Successful Degradation Ceremonies." In J. Manis and B. Meltzer (Eds.), *Symbolic Interactionism*, pp. 201–208. New York: Allyn & Bacon.
___. (1967). *Studies in Ethnomethodology.* Englewood Cliffs, NJ: Prentice Hall.

Gawande, Atul. (2012, March). "How Do We Heal Medicine?" *Ted Talk*. Available at http://www.ted.com/talks/atul_gawande_how_do_we_heal_medicine.html.

Gayflor, Vabah. (2009). "The Challenges Faced by Adolescent Girls in Liberia." In *The State of The World's Children 2009*. Population Reference Bureau.

Geertz, Clifford. (1998). "The World in Pieces: Culture and Politics at the End of the Century." *Focaal: Tijdschrift voor Antropolgie* 32, pp. 91–117.

___. (1973). *The Interpretation of Cultures*. New York: Basic Books.

Gehrke, Robert. (2001, January 8). "Utah Case Tests American Indian Law." Associated Press.

Gelles, Richard J. (1996). *The Book of David*. New York: Basic Books.

Gelles, Richard J., and J. R. Conte. (1990, November). "Domestic Violence and Sexual Abuse of Children: A Review of Research in the Eighties." *Journal of Marriage and the Family* 52, pp. 1045–1058.

Gelles, Richard J., and Murray A. Straus. (1988). *Intimate Violence*. New York: Simon & Schuster.

Gerth, Hans, and C. Wright Mills. (1953). *Character and Social Structure*. New York: Harcourt Brace.

Gewertz, Deborah. (1981). "A Historical Reconsideration of Female Dominance among the Chambri of Papua New Guinea." *American Ethnologist* 8(1), pp. 94–106.

Gilbert, Geoffrey. (2005). *World Population: A Reference Handbook*, 2nd ed. Santa Barbara, CA: ABC-CLIO.

Gilligan, Carol. (1982). *In a Different Voice*. Cambridge, MA: Harvard University Press.

Ginsberg, Morris. (1958). "Social Change." *British Journal of Sociology* 9(3), pp. 205–229.

Glazer, Nathan, and Daniel P. Moynihan. (Eds.). (1975). *Ethnicity: Theory and Experience*. Cambridge, MA: Harvard University Press.

Goble, Paul. (2009, June 16). "Half of Russian University Students 'Regularly' Bribe Instructors, Experts Say," *Georgian Daily*. Available at http://georgiandaily.com/index.php?option=com_content&task=view&id=12221&Itemid=78, accessed April 27, 2012.

Goffman, E. (1971). *Relations in Public*. New York: Basic Books.

___. (1963). *Behavior in Public Places*. New York: Free Press.

___. (1961). *Asylums: Essays on the Social Situation of Mental Patients and Other Inmates*. Chicago: Aldine.

___. (1959). *The Presentation of Self in Everyday Life*. Garden City, NY: Doubleday.

Goldhagen, Daniel. (2002). *A Moral Reckoning: The Role of the Catholic Church in the Holocaust and Its Unfulfilled Duty of Repair*. New York: Knopf.

___. (1996). *Hitler's Willing Executioners*. New York: Knopf.

Goldin, Claudia, Lawrence F. Katz, and Ilyana Kuziemko. (2006, Fall). "The Homecoming of American College Gap Women: The Reversal of the College Gender." *Journal of Economic Perspectives* 20(4), pp. 133–156.

Goldman, Ari L. (1991). *The Search for God at Harvard*. New York: Times Books.

Goliber, Thomas J. (1997, December). "Population and Reproductive Health in Sub-Saharan Africa." *Population Bulletin* 52(4). Population Reference Bureau.

Good, Kenneth. (1991). *Into the Heart*. New York: Simon & Schuster.

Goode, W. J. (1963). *World Revolution and Family Patterns*. New York: Free Press.

___. (1960, August 25). "A Theory of Role Strain." *American Sociological Review*, pp. 902–914.

Goodman, Brenda. (2012). "Smoking Marijuana Tied to Testicular Cancer." Available at http://www.webmd.com/cancer/news/20120910/marijuana-tied-to-testicular-cancer.

Gordon, Milton M. (1975/1961). "Assimilation in America: Theory and Reality." In Norman R. Yetman and C. Hoy Steele (Eds.), *Majority and Minority: The Dynamics of Racial and Ethnic Relations*. Boston: Allyn & Bacon.

___. (1964). *Assimilation in American Life*. New York: Oxford University Press.

Goudreau, Jenna. (2011). "The Best and Worst Countries for Women." Available at http://www.forbes.com/sites/jennagoudreau/2011/11/01/the-best-and-worst-countries-for-women/.

Gould, H. (1971). *Caste and Class: A Comparative View*. Reading, MA: Addison-Wesley, pp. 1–24.

Gould, Stephen Jay. (1976, May). "The View of Life: Biological Potential versus Biological Determinism." *Natural History Magazine* 85, pp. 34–41.

Graeff, F. G., F. S. Guimaraes, T. G. De Andrade, and J. F. Deakin. (1996). "Role of 5HT in Stress, Anxiety, and Depression." *Pharmocology and Biochemistry of Behavior* 54(1), pp. 129–141.

Granovetter, Mark. (1983). "The Strength of Weak Ties: A Network Theory Revisited." *Sociological Theory* 1, pp. 201–233.

Greenberg, David. (1999, June 15). "White Weddings: The Incredible Staying Power of the Laws against Interracial Marriage." *Slate*. Available at http://slate.msn.com/id/30352/, accessed December 9, 1999.

Greenberg, J. (1980). "Ape Talk: More than Pigeon English?" *Science News* 117(19), pp. 298–300.

Grieco, Elizabeth M., Yesenia D. Acosta, G. Patricia de la Cruz, Christine Gambino, Thomas Gryn, Luke J. Larsen, Edward N. Trevelyan, and Nathan P. Walters. (2012, May). "The Foreign-Born Population in the United States: 2010." *American Community Survey Reports*. Available at http://www.census.gov/prod/2012pubs/acs-19.pdf, accessed July 30, 2012.

Gross, Neil, and Solon Simmons. (2009). "The Religiosity of American College and University Professors." *Sociology of Religion* 70(2), pp. 101–129.

___. (2007, September 24) "The Social and Political Views of American Professors." Working Paper. Available at http://74.125.95.132/search?q=cache:kjC0tnrrzb0J:www.wjh.harvard.edu/~ngross/lounsbery_9-25.pdf+Neil+Gross+Solon+Simmons+professors&cd=1&hl=en&ct=clnk&gl=us, accessed May 19, 2009.

Grossman, Cathy Lynn. (2007, February 6). "Are Women Suited for Politics? Americans Are Deeply Divided." *USA Today*. Available at http://www.bls.gov/opub/mlr/2007/02/art2exc.htm, accessed May 15, 2008.

Guillen, Mauro F. (2001). "Is Globalization Civilizing, Destructive or Feeble? A Critique of Five Key Debates in the Social Science Literature." *Annual Review of Sociology* 27, pp. 235–260.

Gumplowicz, Ludwig. (1899). *The Outlines of Sociology*. Philadelphia: American Academy of Political and Social Sciences.

Gurr, Ted Robert. (1970). *Why Men Rebel*. Princeton, NJ: Princeton University Press.

Hall, Edward T. (1981). *The Silent Language*. New York: Doubleday.

___. (1974). *Handbook for Proxemic Analysis*. Washington, DC: Society for the Anthropology of Visual Communication.

___. (1969). *The Hidden Dimension*. Garden City, NY: Doubleday.

Hall, Edward T., and Mildred Reed Hall. (1990). *Hidden Differences*. New York: Anchor Books.

Hamilton, Brady E., Joyce A. Martin, and Stephanie J. Ventura. (2011, November 17). "Births: Preliminary Data for 2010." *National Vital Statistics Reports* 60(2). Available at http://www.cdc.gov/nchs/data/nvsr/nvsr60/nvsr60_02.pdf, accessed August 10, 2012.

Hammer, Heather, David Finkelhor, and Andrea J. Sedlak. (2002, October). "Runaway/Thrownaway Children: National Estimates and Characteristics." Washington, DC: U.S. Department of Justice, Office of Justice Programs.

Hammons, Christopher W. (2001). "School @ Home." *Education Next*, pp. 1–10.

Hampden-Turner, Charles, and Fons Trompenaars. (2000). *Building Cross-Cultural Competence*. New Haven, CT: Yale University Press.

Hare, Paul A. (1976). *Handbook of Small Group Research*, 2nd ed. New York: Free Press.

Harlow, Harry F. (1959, June). "Love in Infant Monkeys." *Scientific American*, pp. 68–74.

Harlow, Harry F., and M. Harlow. (1962). "The Heterosexual Affectional System in Monkeys." *American Psychologist* 17, pp. 1–9.

Harris, C. D., and E. L. Ullman. (1945). "The Nature of Cities." *Annals of the American Academy of Political and Social Science* 242, p. 12.

Harris, Marvin. (1975). *Culture, People, and Nature: An Introduction to General Anthropology*, 2nd ed. New York: Crowell.

___. (1966). "The Cultural Ecology of India's Sacred Cattle." *Current Anthropology* 7, pp. 51–63.

*Harvard Law Review*. (1993, June). "Notes." 106, pp. 1905–1925.

Hatfield, E. (1988). "Passionate and Companionate Love." In R. J. Steinberg and M. L. Barnes (Eds.), *The Psychology of Love*. New Haven, CT: Yale University Press.

Haub, Carl. (2011). "2011 World Population Data Sheet." Available at http://www.prb.org/pdf11/2011population-data-sheet_eng.pdf, accessed July 17, 2012.

Haub, Carl, and Toshiko Kaneda. (2012). "2012 World Population Data Sheet, July 2012." Available at http://www.prb.org/pdf12/2012-population-data-sheet_eng.pdf, accessed July 19, 2012.

Hausmann, Ricardo, Laura D. Tyson, and Saadia Zahidi. (2011). "The Global Gender Gap Report 2011." World Economic Forum Geneva, Switzerland 2011. Available at http://www3.weforum.org/docs/WEF_GenderGap_Report_2011.pdf.

Havemann, Ernest. (1967, October). "Computers: Their Scope Today." *Playboy Magazine*.

Hawes, Alex. (1995). "Machiavellian Monkeys and Shakespearean Apes: The Question of Primate Language." *Zoogoer* 24(6), pp. 1–10.

Hawkins, Sanford A., and Robert Hastie. (1990). "Hindsight: Biased Judgments of Past Events after the Outcomes Are Known." *Psychological Bulletin* 107(3), pp. 311–327.

Hawley, Amos. (1981). *Urban Society*, 2nd ed. New York: John Wiley.

Heath, Chip, and Dan Heath. (2007). *Made to Stick: Why Some Ideas Survive and Others Die*. New York: Random House.

Heer, David M. (1980). "Intermarriage." *Harvard Encyclopedia of American Ethnic Groups*. Cambridge, MA: Harvard University Press, pp. 513–521.

Held, D., A. McGrew, D. Goldblatt, and J. Perraton. (1999). *Global Transformations*. Stanford, CA: Stanford University Press.

Henig, Robin Marantz. (1997). *The People's Health: A Memoir of Public Health and Its Evolution at Harvard*. Washington, DC: Joseph Henry Press.

Hernandez, Peggy. (1989, July 26). "Firemen Who Claimed to Be Black Lose Appeal." *Boston Globe*, p. 13.

___. (1988, November 7). "Many Chances to Dispute Malones Firefighters' Minority Status Unchallenged for 10 Years." *Boston Globe*, p. 1.

Heron, Melonie. (2007, November 20). U.S. Centers for Disease Control and Prevention, "Deaths: Leading Causes for 2004." *National Vital Statistics Reports* 56(5). Available at www.cdc.gov/nchs/data/nvsr/nvsr56/nvsr56_05.pdf, accessed June 25, 2009.

Heron, M. P., Hoyert, D. L., Murphy, S. L., Xu, J. Q., Kochanek, K. D., and Tejada-Vera, B. (2009). "Deaths: Final Data for 2006." *National Vital Statistics Reports* 57(14). Hyattsville, MD: National Center for Health Statistics.

Heron, Melonie, Donna L. Hoyert, Sherry L. Murphy, Jiaquan Xu, Kenneth D. Kochanek, and Betzaida Tejada-Vera. (2008). "Women in U.S. Academic Medicine Statistics and Medical School Benchmarking 2007–2008." Available at http://www.aamc.org/members/wim/statistics/stats08/start.htm, accessed June 21, 2009.

Herring, Susan. (2000, Winter). "Gender Differences in CMC: Findings and Implications." *CPSR Newsletter* 18(1).

Herring, S. C., and J. C. Paolillo. (2006). "Gender and Genre Variation in Weblogs." *Journal of Sociolinguistics*, 10(4), 439–459.

Hesketh, Therese, Li Lu, and Zhu Wei Xing. (2005, September). "The Effect of China's One-Child Family Policy after 25 Years." *New England Journal of Medicine* 353(11), pp. 1171–1176.

Hesketh, Therese, and Zhu, Wei Xing. (1997). "The One Child Policy: The Good, the Bad, and the Ugly." *British Medical Journal* 314(7095).

Hetherington, E. Mavis. (2002, April 8). "Marriage and Divorce American Style." *American Prospect*.

Hetherington, E. Mavis, and John Kelly. (2002). *For Better or for Worse: Divorce*

Himes, Christine L. (2001, December). "The Elderly American." *Population Bulletin* 56(4). Population Reference Bureau.

Hingson, Ralph W. (1998, January/February). "College-Age Drinking Problems." *Public Health Reports*, p. 113.

Hirschi, Travis. (1969). *Causes of Delinquency*. Berkeley: University of California Press.

Hirschi, Travis, and Michael Gottfredson. (1993). "Commentary: Testing the General Theory of Crime." *Journal of Research in Crime and Delinquency* 30, pp. 47–54.

Hirst, P., and G. Thompson. (1996). *Globalization in Question*. London: Polity.

Hochschild, Arlie Russell. (1997). *Time Bind: When Work Becomes Home and Home Becomes Work*. New York: Henry Holt.

___. (1997, April 20). "There's No Place Like Work." *New York Times Magazine*, pp. 51–55, 81, 84.

Hodges, Jim. (2012, March 22). "Cover Story: U.S. Army's Human Terrain Experts May Help Defuse Future Conflicts." *Defense News*. Available at http://www.defensenews.com/article/20120322/C4ISR02/303220015/Cover-Story-U-S-Army-8217-s-Human-Terrain-Experts-May-Help-Defuse-Future-Conflicts, accessed May 1, 2012.

Hoebel, E. Adamson. (1960). *The Cheyennes: Indians of the Great Plains*. New York: Holt, Rinehart and Winston.

Hoecker-Drysdale, Susan. (1992). *Harriet Martineau: First Woman Sociologist*.

Hoefer, Michael, Nancy Rytina, and Bryan Baker. (2012, March). "Estimates of the Unauthorized Immigrant Population Residing in the United States: January 2011." Department of Homeland Security.

Holden, Constance. (1997, March). "Changing Sex Is Hard to Do." *Science* 275(5307), p. 1745.

Holmes, Steven A. (2000, March 11). "New Policy on Census Says Those Listed as White and Minority Will Be Counted as Minority." *New York Times*.

Holt, Jim. (2004, November 7). "The Other National Conversation." *New York Times Magazine*, section 6, p. 17.

Homans, G. C. (1950). *The Human Group*. New York: Harcourt Brace.

"Homelessness." (2000, January 21). *Issues and Controversies on File*.

Homer-Dixon, Thomas. (2000). *The Ingenuity Gap*. New York: Knopf.

Hooton, E. A. (1939a). *Crime and the Man*. Cambridge: Harvard University Press.

___. (1939b). *Twilight of Man*. New York: G. P. Putnam's Sons.

___. (1939c). *The American Criminal*. Cambridge: Harvard University Press.

Horgan, John. (2005, August 12). "In Defense of Common Sense." *New York Times*. Available at http://www.census.gov/hhes/www/poverty/histpov/hstpov3.html, accessed May 13, 2009.

Horn, Wade F. (1998). *Father Facts*, 3rd ed. Gaithersburg, MD: National Fatherhood Initiative.

Howes, C., and C. Hamilton. (1991). "Child Care for Young Children." In B. Spodek (Ed.), *Handbook of Research in Early Childhood Education*. New York: Macmillan.

Hoyt, H. (1943). "The Structure of American Cities in the Post-War Era." *American Journal of Sociology* 48, pp. 475–492.

Hudson, Valerie M., and Andrea M. den Boer. (2004). *Bare Branches: The Security Implications of Asia's Surplus Male Population*. Cambridge, MA: MIT Press.

Huntington, Samuel P. (1996). *The Clash of Civilizations and the Remaking of World Order*. New York: Simon & Schuster.

___. (1968). *Political Order in Changing Societies*. New Haven, CT: Yale University Press.

Hvistendahl, Mara. (2011). "Unnatural Selection: Choosing Boys Over Girls and the Consequences of a World Full of Men." New York, Public Affairs.

Hyland, N. (2006). "Detracking in the Social Studies: A Path to a More Democratic Education?" *Theory into Practice* 45(1), pp. 64–71. Available at doi:10.1207/s15430421tip4501_9.

Iceland, John. (2005, August). "Why Concentrated Poverty Fell in the United States in the 1990s." Population Reference Bureau. Available at http://www.prb.org/Template.cfm?Section=PRB&template=/ContentManagement/ContentDisplay.cfm&ContentID=12769, accessed September 12, 2005.

Immerwahr, John, and Jean Johnson. (1996). "Incomplete Assignment. America's Views on Standards: An Assessment by Public Agenda." *Progress of Education Reform* 8.

"Income Gap." (1999, December 17). *Issues and Controversies on File* 4(3), pp. 489–496.

Inkeles, Alex, and David H. Smith. (1974). *Becoming Modern: Individual Change in Six Developing Countries*. Cambridge, MA: Harvard University Press.

Institute of Medicine Committee on the Consequences of Uninsurance. (2001). *Coverage Matters: Insurance and Health Care*. Washington, DC: National Academies Press.

Institute of Medicine of the National Academies. (2010). "America's Uninsured Crisis: Consequences for Health and Health Care." Available at http://www.iom.edu/Reports/2009/Americas-Uninsured-Crisis-Consequences-for-Health-and-Health-Care.aspx.

International Centre for Prison Studies. (2012). "World Prison Brief." Available at http://www.prisonstudies.org/info/worldbrief/, accessed June 25, 2012.

International Conference on Child Labor. (1998). Oslo. October 27–30, 1997. Geneva: International Labour Office.

Itard, J. (1932). *The Wild Boy of Aveyron*. In G. Humphrey and M. Humphrey (Trans.). New York: Appleton-Century-Crofts.

Jablon, Robert. (2000, May 19). "Study: Women's Response to Stress May Lead to Healthier Lives." Associated Press.

Jacob, H. (1988). *Silent Revolution: The Transformation of Divorce Law in the United States*. Chicago: University of Chicago Press.

Jacobs, Garry, Robert Macfarlane, and N. Asokan. (1997, November 15). "Comprehensive Theory of Social Development." Napa, CA: International Center for Peace and Development.

Jacobs, J. (1961). *The Death and Life of American Cities*. New York: Vintage.

Jacobsen Linda A., et al. (2011). "America's Aging Population." *Population Bulletin* 66(1). Available at http://www.prb.org/pdf11/aging-in-america.pdf.

Jacoby, Tamar. (2000, July–August). "In Asian America." *Commentary* 110(1), pp. 58–62.

Jaffe, S., and J. S. Hyde. (2000). "Gender Differences in Moral Orientation." *Psychological Bulletin* 126, pp. 703–726.

Jain, Anita. (2005). "Is Arranged Marriage Really Any Worse than Craigslist?" *New York Magazine*. Available at http://www.nymag.com/nymetro/news/culture/features/11621, accessed February 19, 2009.

Jamison, Kay Redfield. (1995). *An Unquiet Mind*. New York: Knopf.

Janis, I., Dwight W. Chapman, John P. Gillin, and John P. Spiegel. (1964). "The Problem of Panic." In Duane P. Schultz (Ed.), *Panic Behavior*. New York: Random House.

Jarvik, L. F., V. Klodin, and S. Matsuyama. (1973). "Human Aggression and the Extra Y Chromosome: Fact or Fantasy." *American Psychologist* 28, pp. 674–682.

Jencks, Christopher. (1994). *The Homeless*. Cambridge, MA: Harvard University Press.

Johnson, George. (1995, June 6). "Chimp Talk Debate: Is It Really Language?" *New York Times*, p. C1.

Johnson, Jeffrey. G. Patricia Cohen, Elizabeth M. Smailes, Stephanie Kasen, and Judith S. Brook (2002). "Television Viewing and Aggressive Behavior during Adolescence and Adulthood." *Science* 295, pp. 2468–2471.

Johnson, Julie. (1988, August 11). "President Signs Law to Redress Wartime Wrong." *New York Times*, p. A16.

Johnson, Tallese, and Jane Dye. (2005). "Indicators of Marriage and Fertility in the United States from the American Community Survey: 2000 to 2003." Available at http://www.census.gov/population/www/socdemo/fertility/mar-fert-slides.html, accessed January 9, 2006.

Johnston, L. D., P. M. O'Malley, J. G. Bachman, and J. E. Schulenberg. (2011). "Monitoring the Future National Survey Results on Adolescent Drug Use: Overview of Key Findings, 2010." Ann Arbor: Institute for Social Research, University of Michigan.

Joint Center for Political and Economic Studies. (2012). "National Roster of Black Elected Officials Fact Sheet." Available at http://www.jointcenter.org/sites/default/files/upload/research/files/National%20Roster%20of%20Black%20Elected%20Officials%20Fact%20Sheet.pdf.

Jones, C. G., and J. H. Lawton (Eds.). (1995). *Linking Species and Ecosystems*. New York: Chapman and Hall.

Jones, Jeffrey M. (2012, September 5). "Atheists, Muslims See Most Bias as Presidential Candidates." Gallup Poll. Available at http://www.gallup.com/poll/155285/atheists-muslims-bias-presidential-candidates.aspx, accessed June 12, 2012.

___. (2012, June 21). "Confidence in Local Police Drops to 10-Year Low." Gallup Poll. Available at http://www.gallup.com/poll/19783/confidence-local-police-drops-10year-low.aspx.

___. (2011, September 12). "Record-High 86% Approve of Black-White Marriages." Gallup Poll. Available at http://www.gallup.com/poll/149390/Record-High-Approve-Black-White-Marriages.aspx, accessed September 12, 2011.

___. (2002, August 16). "Public Divided on Benefits of Living Together before Marriage." Gallup News Service.

Jones, Sydney, and Susannah Fox. (2009, January 28) "Generations Online 2009," Pew Internet and American Life Project. Available at http://www.pewinternet.org/pdfs/PIP_Generations_2009.pdf, accessed May 17, 2009.

Jones-Correa, Michael. (2012). *Contested Ground: Immigration in the United States*. Washington, DC: Migration Policy Institute, July 2012.

Josephy, Alvin M., Jr. (1994). *500 Nations: An Illustrated History of North American Indians*. New York: Knopf.

Kahneman, Daniel, Alan B. Krueger, David Schkade, Norbert Schwarz, and Arthur A. Stone. (2006, May). "Would You Be Happier If You Were Richer? A Focusing Illusion." *CEPS Working Paper* No. 125.

Kallen, Horace M. (1956). *Cultural Pluralism and the American Idea: An Essay in Social Philosophy*. Philadelphia, PA: University of Pennsylvania Press.

Katz, Elihu. (1957). "The Two-Step Flow of Communication: An Up-to-Date Report on an Hypothesis." *Public Opinion Quarterly* 21, pp. 61–78.

Kelling, George L., and Catherine M. Coles. (1996). *Fixing Broken Windows: Restoring Order and Reducing Crime in Our Communities*. New York: Martin Kessler Books (Free Press).

Kennedy, John F. (1961). "Introduction." In William Brandon (Ed.), *The American Heritage Book of Indians*. New York: Dell.

Kennedy, Randall. (2003). *Interracial Intimacies: Sex, Marriage, Identity and Adoption*. New York: Pantheon Books.

Kent, Mary, and Robert Lalasz. (2006). "Speaking English in the United States." Population Reference Bureau. Available at http://prb.org/Articles/2006/IntheNewsSpeakingEnglishintheUnitedStates.aspx, accessed February 10, 2009.

Kent, Mary M., and Mark Mather. (2002, December). "What Drives U.S. Population Growth?" *Population Bulletin* 57(4). Population Reference Bureau.

Kent, Mary Mederios. (2007). "Immigration and America's Black Population." Population Reference Bureau. Available at http://www.prb.org/Publications/PopulationBulletins/2007/blackimmigration.aspx?p=1, accessed June 19, 2009.

Kent, Mary Mederios, and Carl Haub. (2005). "The Demographic Divide: What It Is and Why It Matters." Available at http://prb.org/Articles/2005/TheDemographicDivideWhatItIsandWhyItMatters.aspx, accessed March 10, 2009.

Kershaw, Sarah, and Rebecca Cathcart. (2009, July 19). "Reefer Madness," *New York Times*, St 1, 2.

Khaldun, Ibn. (1958). *The Muqaddimah*. Bolligen Series 43. Princeton, NJ: Princeton University Press.

Kilker, Ernest Evans. (1993). "Black and White in America: The Culture and Politics of Racial Classification." *International Journal of Politics, Culture and Society* 7(2), pp. 229–257.

Kinsella, Kevin, and Victoria Velkoff. (2001, November). *An Aging World: 2001*. U.S. Bureau of the Census.

Kishkovsky, Sophia. (2011, July 15). "Russia Enacts Law Opposing Abortion," *New York Times*. Available at http://www.nytimes.com/2011/07/15/world/europe/15iht-russia15.html.

Klaus, Patsy, and Callie M. Rennison. (2002). *Age Patterns in Violent Victimization, 1976–2000*. Washington, DC: U.S. Department of Justice, Bureau of Justice Statistics.

Klick, Jonathan, and Alexander Tabarrok. (2005, April). "Using Terror Alert Levels to Estimate the Effect of Police on Crime." *Journal of Law and Economics* 48.

Koenig, Frederick. (1985). *Rumor in the Marketplace: The Social Psychology of Commercial Heresy*. Dover, MA: Auburn House.

Kohlberg, Lawrence. (1969). "Stage and Sequence: The Cognitive-Developmental Approach to Socialization." In David A. Goslin (Ed.), *Handbook of Socialization Theory and Research*. Chicago: Rand McNally.

___. (1967). "Moral and Religious Education in the Public Schools: A Developmental View." In T. Sizer (Ed.), *Religion and Public Education*. Boston: Houghton Mifflin.

Kohut, Andrew, Richard Wike, Juliana Menasce Horowitz, Jacob Poushter, Cathy Barker. (2012, February 29). "American Exceptionalism Subsides: The American-Western European Values Gap," Updated. Pew Research Center, Pew Global Attitudes Project.

Kolata, Gina. (2007, July 25). "Study Says Obesity Can Be Contagious." *New York Times*.

Kopel, David B. (1995). "Massaging the Medium: Analyzing and Responding to Media Violence without Harming the First Amendment." *Kansas Journal of Law and Public Policy* 4, p. 17.

Kosmin, Barry A., and Seymour P. Lachman. (1993). *One Nation under God*. New York: Harmony.

Kozol, Jonathan. (2007, September 10) "Why I Am Fasting: An Explanation to My Friends." *Huffington Post*. Available at http://www.huffingtonpost.com/jonathan-kozol/why-i-am-fasting-an-expl_b_63622.html, accessed May 15, 2009.

Krause, Michael. (1966). *Immigration: The American Mosaic*. New York: Van Nostrand-Reinhold.

Kreider, Rose M., and Jason M. Fields. (2005). "Number, Timing and Duration of Marriages and Divorces, 2001." Current Population Reports, P70–80. Washington, DC: U.S. Bureau of the Census.

Kristof, Nicholas, and Sheryl Wudunn. (1994). *China Wakes*. New York: Random House.

Kroeber, Alfred A. (1963/1923). *Anthropology: Culture Patterns and Processes*. New York: Harcourt, Brace & World.

Kumar, Meira. (2009). "A Dalit Leader is the New Lok Sabha Speaker," NCHRO. Available at http://www.nchro.org/index.php?option=com_content&view=article&id=6863:meira-kumar-a-dalit-leader-is-the-new-lok-sabha-speaker&catid=5:dalitsatribals&Itemid=14.

Kwintessential. (2012). Available at http://www.kwintessential.co.uk/resources/culture-tests.html, accessed June2, 2012.

Lacayo, Richard. (1985, June. 10). "Blood in the Stands." *Time*, p. 35.

LaFraniere, Sharon. (2009, April 11). "Chinese Bias for Baby Boys Creates a Gap of 32 Million." *New York Times*. Available at http://www.nytimes.com/2009/04/11/world/asia/11china.html?ref=global-home&pagewanted=print, accessed April 19, 2009.

Laing, Aislinn. (2011, November 20). "Twelfth Wife of Swaziland Kicked Out of Palace over 'Affair' with Justice Minister." *The Telegraph*. Available at http://www.telegraph.co.uk/news/worldnews/africaandindianocean/swaziland/8902417/Twelfth-wife-of-Swaziland-kicked-out-of-palace-over-affair-with-justice-minister.html, accessed April 27, 2012.

Lakshmanan, Indira A. R. (1997, August 26). "Love No Match for Custom: India's Young Favor Arranged Unions." *Boston Globe*, pp. A1, A10.

Lampman, Jane. (2006, February 6). "Megachurches' Way of Worship Is on the Rise." Christian Science Monitor. Available at http://www.csmonitor.com/2006/0206/p13s01-lire.html, accessed July 16, 2009.

___. (2002, October 10). "Charting America's Religious Landscape." *Christian Science Monitor*. Available at http://www.csmonitor.com/2002/1010/p12s01-lire.html, accessed December 18, 2002.

Landes, Richard. (1998). *While God Tarried: Disappointed Millennialism and the Making of the Modern West*. New York: Houghton Mifflin.

Lantz, Herman R. (1982, Spring). "Romantic Love in the Pre-Modern Period: A Sociological Commentary." *Journal of Social History*, pp. 349–370.

Larson, Jan. (1992, July). "Understanding Stepfamilies." *American Demographics*, pp. 36–40.

Laslett, P. (1965). *The World We Have Lost: England before the Industrial Age*. New York: Scribner's.

Laslett, P., and Phillip W. Cummings. (1967). "History of Political Philosophy." In Paul Edwards (Ed.), *The Encyclopedia of Philosophy*, Vols. 5, 6. New York: Macmillan.

Lattimore, Pamela K., and Cynthia A. Nahabedian. (2000). "The Nature of Homicide: Trends and Changes." *Sourcebook of Criminal Justice Statistics: 1999.* U.S. Department of Justice. Washington, DC: U.S. Government Printing Office.

Lauzen, Martha M. (2011). "It's a Man's (Celluloid) World: On-Screen Representations of Female Characters in the Top 100 Films of 2011." Center for the Study of Women in Television and Film, San Diego, CA 9218. Available at http://womenintvfilm.sdsu.edu/files/2011_Its_a_Mans_World_Exec_Summ.pdf, accessed August 10, 2012.

Lazarsfeld, Paul F., Bernard Berelson, and Hazel Gaudet. (1968). *The People's Choice*, 3rd ed. New York: Columbia University Press.

Leakey, Richard E., and Roger Lewin. (1977). *Origins*. New York: Dutton.

Le Bon, Gustave. (1960/1895). *The Crowd: A Study of the Popular Mind*. New York: Viking.

Leland, Elizabeth. (2002, June 30). "Born White, Raised Black." *Charlotte Observer*, p. 7.

Lemert, Edwin. (1972). *Human Deviance, Social Problems and Social Control*, 2nd ed. Englewood Cliffs, NJ: Prentice Hall.

Lenhart, Amanda. (2009, January 14). "Adults and Social Network Websites." Pew Internet and American Life Project. Available at http://www.pewinternet.org/Reports/2009/Adults-and-Social-Network-Websites.aspx, accessed June 22, 2009.

Lenin, Vladimir I. (1949/1917). *The State and Revolution*. Moscow: Progress.

Lenski, Gerhard. (1966). *Power and Privilege: A Theory of Social Stratification*. New York: McGraw-Hill.

Lenski, Gerhard, and Jean Lenski. (1982). *Human Societies*, 4th ed. New York: McGraw-Hill.

Leonard, Ira M., and R. D. Parmet. (1972). *American Nativism: 1830–1860*. New York: Van Nostrand-Reinhold.

Leslie, Gerald R. (1979). *The Family in Social Context*, 4th ed. New York: Oxford University Press.

Lester, Toby. (2002, February). "Oh, Gods!" *Atlantic Monthly*, pp. 37–45.

Levin, Jack. (1993). *Sociological Snapshots*. Newbury Park, CA: Pine Forge Press.

Levine, Robert. (1993, February). "Is Love a Luxury?" *American Demographics*, pp. 27, 29.

Lewandowsky, Stephan, Werner G. K. Stritzke, Klaus Oberauer, and Michael Morales. (2005, March). "Memory for Fact, Fiction, and Misinformation: The Iraq War 2003." *Psychological Science* 16(3).

Lewin, Kurt. (1948). *Resolving Social Conflicts*. New York: Harper.

Lewis, Bernard. (2002). *What Went Wrong: Western Impact and Middle Eastern Response*. New York: Oxford University Press.

Lewis, David Levering. (2000). *W. E. B. DuBois: The Fight for Equality and the American Century, 1919–1963*. New York: Henry Holt.

___. (1993). *W. E. B. DuBois: Biography of a Race/1868–1919*. New York: Henry Holt.

Liens, G. E. (1992). *The Ecosystem Approach: Use and Abuse, Excellence in Ecology,* Vol. 3. Oldendorf'Luhe Ecology Institute.

Lindesmith, Alfred R., and Anselm L. Strauss. (1956). *Social Psychology*. New York: Holt, Rinehart and Winston.

Ling Li. (2007)."Bottled Water Consumption Jumps." *Vital Sign 2007–2008*. Worldwatch Institute.

Linton, R. (1936). *The Study of Man*. New York: Appleton-Century-Crofts.

Lipset, Seymour Martin. (1996). *American Exceptionalism: A Double-Edged Sword*. New York: Norton.

___. (1960). *Political Man*. Garden City, NY: Doubleday.

Lipset, Seymour Martin, and Gabriel Salman Lenz. (2000). "Corruption, Culture and Markets." In Lawrence E. Harrison and Samuel P. Huntington (Eds.), *Culture Matters*. New York: Basic Books, pp. 112–124.

Lipsey, R. G., and P. D. Steiner. (1972). *Economics*. New York: Harper & Row.

Loyd, Janice. (2009, January 5). "Number of Home-Schooled Children on the Rise." *USA Today*. Available at http://www.usatoday.com/news/education/2009-01-04-homeschooling_N.htm, accessed June 28, 2009.

Lombroso-Ferrero, Gina. (1972). *Criminal Man*, reprint ed. Montclair, NJ: Patterson Smith.

Longman, Phillip. (2004). *The Empty Cradle: How Falling Birthrates Threaten World Prosperity and What to Do about It*. New York: Basic Books.

Lown, Bernard. (1996). *The Lost Art of Healing*. Boston: Houghton Mifflin.

Lutz, Wolfgang. (1994). *The Future Population of the World: What Can We Assume Today?* London: Earthscan.

Lynch, K. (1960). *The Image of the City*. Cambridge, MA: MIT Press.

Mackenzie, Doris Layton. (2000). "Sentencing and Corrections in the 21st Century: Setting the Stage for the Future." Washington, DC: U.S. Department of Justice.

MacMurray, V. D., and P. H. Cunningham. (1973). "Mormons and Gentiles." In Donald E. Gelfand and Russell D. Lee (Eds.), *Ethnic Conflicts and Power: A Cross-National Perspective*. New York: John Wiley.

Madsen, William. (1973). *The Mexican-Americans of South Texas*, 2nd ed. New York: Holt, Rinehart and Winston.

Maines, David R., and Monica J. Hardesty. (1987, September). "Temporality and Gender: Young Adults' Career and Family Plans." *Social Forces* 66(1), pp. 102–120.

Malinowski, Bronislaw. (1954). *Magic, Science and Religion*. New York: Free Press.

___. (1922). *Argonauts of the Western Pacific*. New York: Dutton.

Manning, Jennifer E. (2011, March 1). "Membership of the 112th Congress: A Profile." Congressional Research Service.

Manning, Wendy D., and Daniel T. Lichter. (1996). "Parental Cohabitation and Children's Economic Well-Being." *Journal of Marriage and the Family* 58, pp. 998–1010.

Martin, Joyce A., Brady E. Hamilton, Stephanie J. Ventura, Michelle J. K. Osterman, Sharon Kirmeyer, T. J. Mathews, and Elizabeth C. Wilson. (2011, November 3). "Births: Final Data for 2009, National Center for Health Statistics." *National Vital Statistics Reports* 60(1). Available at http://www.cdc.gov/nchs/data/nvsr/nvsr60/nvsr60_01.pdf, accessed August 15, 2012.

Martin, Philip, and Elizabeth Midgley. (1994, September). "Immigration to the United States: Journey to an Uncertain Destination." *Population Bulletin* 49(2). Population Reference Bureau.

Marx, Karl. (1968). "Wage, Labour and Capital." In Karl Marx and Friedrich Engels, *Selected Works in One Volume*. New York: International Publishers.

___. (1967/1867a). *Capital: A Critique of Political Economy*. Friedrich Engels (Ed.). New York: New World.

___. (1967/1867b). *Das Kapital*, Vols. 1–3. Friedrich Engels (Ed.). New York: International Publishers.

___. (1959/1847). *Class and Class Conflict in Industrial Society*. Stanford, CA: Stanford University Press.

___. (1906). *The Process of Capitalist Production*. The Modern Library. New York.

Marx, Karl, and Friedrich Engels. (1961/1848). "The Communist Manifesto." In Arthur P. Mendel (Ed.), *Essential Works of Marxism*. New York: Bantam Books.

Mather, Mark. (2009, February). "Children in Immigrant Families Chart New Path." Population Reference Bureau.

Mathews, T. J., and Marian F. MacDorman. (2012, May 10). Centers for Disease Control and Prevention, "Infant Mortality Statistics from the 2008 Period Linked Birth/Infant Death Data Set," *National Vital Statistics Reports* 60(5). Available at http://www.cdc.gov/nchs/data/nvsr/nvsr60/nvsr60_05.pdf.

Matschiner, M., and S. K. Murnen. (1999). "Hyperfemininity and Influence." *Psychology of Women Quarterly* 23, pp. 631–642.

Mauss, Armand I. (1975). *Social Problems of Social Movements.* Philadelphia: Lippincott.

Maybury, Kelly. (2002, January 22). "I Do? Marriage in Uncertain Times." Gallup Poll.

McCown, Karen Stone, Joshua M. Freedman, and Marsha C. Rideout. (1998). *The Emotional Intelligence Curriculum.* San Mateo, CA: Six Seconds Press.

McFalls, Joseph A. (1991, October). "Population: A Lively Introduction." *Population Bulletin* 46(2). Population Reference Bureau.

McKinnon, Jesse, and Karen Humes. (2000, September). *The Black Population in the United States.* Washington, DC: U.S. Bureau of the Census.

McLuhan, M. (1964). *Understanding Media.* London: Routledge.

McNeill, William H. (1976). *Plagues and People.* New York: Anchor/Doubleday.

Mead, George H. (1934). *Mind, Self, and Society.* C. W. Morris (Ed.). Chicago: University of Chicago Press.

Mead, Margaret. (1943). "Our Educational Emphases in Primitive Perspectives." *American Journal of Sociology* 48, 633–639.

___. (1935). *Sex and Temperament in Three Primitive Societies.* New York: William Morrow.

Meadows, Donelle H., Dennis L. Meadows, Jorgan Randers, and William Behrens III. (1972). *The Limits of Growth: A Report of the Club of Rome's Project on the Predicament of Mankind.* New York: Universe Books.

Mears, Ashley. (2011). *Pricing Beauty: The Making of a Fashion Model.* Berkeley: University of California Press.

___. (2011, September 15). "Poor Models. Seriously." *New York Times*, p. A35.

Mediascope. (1996). *National Television Violence Study.* Findings available at http://www.mediascope.org, accessed September 14, 1998.

Mednick, Sarnoff A. (1977). "A Biosocial Theory of the Learning of Law-Abiding Behavior." In Sarnoff A. Mednick and Karl O. Christiansen (Eds.), *Biosocial Bases of Criminal Behavior.* New York: Gardner Press.

Mednick, Sarnoff A., Terrie E. Moffitt, and Susan A. Stacks. (Eds.). (1987). *The Causes of Crime: New Biological Approaches.* Cambridge, England: Cambridge University Press.

Meier, M. H., A. Caspi, A. Ambler, H. Harrington, R. Houts, R. S. Keefe, K. McDonald, A. Ward, R. Poulton, and T. E. Moffitt. (2012, August 27). "Persistent Cannabis Users Show Neuropsychological Decline from Childhood to Midlife." Proceedings of the National Academy of Sciences. Melbourne Neuropsychiatry Centre, The University of Melbourne and Melbourne Health, Melbourne, 3053, Australia.

Mendes, Elizabeth, and Kyley McGeeney. (2012, August 16). "In U.S., Majority Overweight or Obese in All 50 States." Gallup Poll. Available at http://www.gallup.com/poll/156707/Majority-Overweight-Obese-States.aspx.

Merton, Robert K. (1969/1949). *Social Theory and Social Structure.* New York: Free Press.

___. (1938). "Social Structure and Anomie." *American Sociological Review* 3, pp. 672–682.

Michels, R. (1966/1911). *Political Parties* (Eden Paul and Adar Paul, Trans.). New York: Free Press.

Miller, Thomas A. W., and Geoffrey D. Feinberg. (2002, March/April). "Culture Clash: Personal Values Are Shaping Our Times." *Public Perspective*, pp. 6–9.

Mills, C. Wright. (1959). *The Sociological Imagination.* New York: Oxford University Press.

Miniño, Arialdi M., Sherry L. Murphy, and Jiaquan Xu. (2011, December 7). "Deaths: Final Data for 2008." *National Vital Statistics Reports* 59(10). Available at http://www.cdc.gov/nchs/data/nvsr/nvsr59/nvsr59_10.pdf.

Mishel, Lawrence, and Gared Bernstein. (1995). *The State of Working America, 1994–95.* Washington, DC: Economic Policy Institute Service, M. E. Sharpe.

Mocan, Naci, and Kaj Gittings. (2001). "Pardons, Executions and Homicide." Working Paper 8639, National Bureau of Economic Research.

Monaghan, Peter. (1993, November 11). "Free After 6 Months: Sociologist Who Refused to Testify Is Released." *Chronicle of Higher Education.*

Money, John, and Paul Tucker. (1975). *Sexual Signatures: On Being a Man or Woman.* Boston: Little, Brown.

Montagu, Ashley. (Ed.). (1973). *Man and Aggression*, 2nd ed. London: Oxford University Press.

___. (1964a). *The Concept of Race.* New York: Collier Books.

___. (1964b). *Man's Most Dangerous Myth: The Fallacy of Race.* New York: Meridian.

Morris, Desmond. (1970). *The Human Zoo.* New York: McGraw-Hill.

Moskos, Charles C., and John Sibley Butler. (1996). *All That We Can Be: Black Leadership and Racial Integration in the Army.* New York: Basic Books.

Mthethwa, Thulani. (2004, September 6). "Swazi Schools Closed for Wedding." *BBC News.* Available at http://news.bbc.co.uk/2/hi/africa/3420775.stm, accessed July 20, 2005.

Murdock, George P. (1949). *Social Structure.* New York: Macmillan.

___. (1937). "Comparative Data on the Division of Labor by Sex." *Social Forces* 15(4), pp. 551–553.

Murray, Charles. (2008). *Real Education.* New York: Random House.

___. (1994, December). "What to Do about Welfare." *Commentary* 98(6), pp. 26–34.

Murray, Christopher J. L., and Alan D. Lopez (Eds.). (1996). *The Global Burden of Disease.* Cambridge, MA: Harvard School of Public Health.

Murray, David W., Joel Schwartz, and S. Robert Lichter. (2001). *It Ain't Necessarily So: How Media Make and Unmake the Scientific Picture of Reality.* New York: Rowman and Littlefield.

Myrdal, Gunnar. (1969). *Objectivity in Social Research.* New York: Pantheon Books.

Nakao, Keoko, and Judith Treas. (1993). *General Social Surveys, 1972–1991: Cumulative Codebook.* Chicago: National Opinion Research Center, pp. 827–835.

___. (1990). "Occupational Prestige in the United States Revisited: Twenty-five Years of Stability and Change." Paper presented at the annual meeting of the American Sociological Association, Washington, DC.

Nash, Jonathan G., and Roger-Mark De Souza. (2002). *Making the Link: Population, Health, Environment.* Population Reference Bureau.

National Association of Latino Elected and Appointed Officials (NALEO). (2012). "2011 Directory of Latino Elected Officials." Education Fund. Available at http://www.naleo.org/directory.html.

___. (2009). "National Directory of Latino Elected Officials." Washington, DC. Available at http://www.naleo.org/, accessed June 25, 2009.

National Center for Education Statistics. (1990). "The Nation's Report Card: 1988." Available at http://nces.ed.gov/nationsreportcard/about/.

National Center for Health Statistics. (2012). "National Marriage and Divorce Rate Trends, Provisional Number of Marriages and Marriage Rate: United States, 2000–2010." Available at http://www.cdc.gov/nchs/nvss/marriage_divorce_tables.htm, accessed August 15, 2012.

___. (2012, June 6). "Deaths: Leading Causes for 2008," *National Vital Statistics Reports* 60(6). Available at http://www.cdc.gov/nchs/data/nvsr/nvsr60/nvsr60_06.pdf, accessed June 6, 2012.

___. (2002). "The Condition of Education 2001." U.S. Department of Education.

___. (2002). "Health, United States. 2001."

___. (2002, June 26). "Births, Marriages, Divorces, and Deaths: Provisional Data for October 2001." *National Vital Statistics Reports* 50(11).

___. (2001). "Health, United States. 2000, with Adolescent Chartbook."

___. (2001). "Urban and Rural Health Chartbook."

___. (2000). "Indicators of School Crime and Safety, 2000." U.S. Department of Education.

National Commission on Excellence in Education. (1998). *The NICHD Study of Early Child Care*.

___. (1988). *American Education: Making it Work*. Washington DC: U.S. Government Printing Office.

___. (1983). *A Nation at Risk*. Washington, DC: U.S. Government Printing Office.

National Education Goals Panel. (1998). "1998 Key Findings." Available at http://govinfo.library.unt.edu/negp/page7-3.htm, accessed January 16, 2000.

National Institute of Child Health and Human Development (NICHD). (2001). "Results of NICHD Study of Early Child Care."

National Institutes of Health. (2012). "DrugFacts: Marijuana, 2012." Available at http://www.drugabuse.gov/publications/drugfacts/marijuana, accessed September 8, 2012.

___. (2012). Health Resources and Services Administration, "Women's Health USA 2011." Available at http://mchb.hrsa.gov/whusa11/more/sitemap.html, accessed August 1, 2012.

___. (2009). Health Resources and Services Administration, "Women's Health, 2008." Available at http://mchb.hrsa.gov/whusa08/index.html, accessed June 11, 2009.

___. (2001). Health Resources and Services Administration, "Women's Health."

___. (1997, April 3). "Results of NICHD Study of Early Child Care Reported at Society for Research in Child Development Meeting." *NIH News Release*.

National Research Council. (2009). "Strengthening Forensic Science in the United States: A Path Forward." Washington, DC: The National Academies Press. Available at http://www.nap.edu/catalog.php?record_id=12589#toc, accessed June 25, 2012.

Neilsen Wire. (2011). "U.S. Homes Add Even More TV Sets in 2010." Available at http://blog.nielsen.com/nielsenwire/consumer/u-s-homes-add-even-more-tv-sets-in-2010/, accessed August 10, 2012.

___. (2010, August 27). "Number of U.S. TV Households Climbs by One Million for 2010–11 TV Season." Available at http://blog.nielsen.com/nielsenwire/media_entertainment/number-of-u-s-tv-households-climbs-by-one-million-for-2010-11-tv-season/, accessed August 10, 2012.

Newport, Frank. (2012, August 8). "Americans Still Prefer Male Bosses; Many Have No Preference." Gallup Poll. Available at http://www.gallup.com/poll/149360/americans-prefer-male-bosses-no-preference.aspx, accessed August 10, 2012.

___. (2012, May 8). "Half of Americans Support Legal Gay Marriage." Gallup Poll. Available at http://www.gallup.com/poll/154529/Half-Americans-Support-Legal-Gay-Marriage.aspx, accessed August 10, 2012.

___. (2010, May 21). "In U.S., Increasing Number Have No Religious Identity." Gallup Poll. Available at http://www.gallup.com/poll/128276/Increasing-Number-No-Religious-Identity.aspx, accessed August 27, 2012.

NICEF. (2010). "The Children Left Behind: A League Table of Inequality in Child Well-Being in the World's Rich Countries." *Innocenti Report Card 9*, UNICEF Innocenti Research Centre, Florence.

Niebuhr, Gustav. (2000, October 31). "Marriage Issue Splits Jews, Poll Finds." *New York Times*.

Nisbett, Richard E., and Dov Cohen. (1996). *Culture of Honor: The Psychology of Violence in the South*. Boulder, CO: Westview Press.

Nock, Albert Jay. (1996). *Jefferson*. New York: John Day Company.

Nolan, Patrick, and Gerhard E. Lenski. (1999). *Human Societies: An Introduction to Macrosociology*. New York: McGraw-Hill.

NORC. GSS 1972–2002 Cumulative Datafile. Available at http://sda.berkeley.edu:7507/quicktables/quickoptions.do, accessed January 9, 2006.

NORC Public Affairs. (2004). "America's Protestant Majority Is Fading, University of Chicago Research Shows." Available at http://www-[KF2] news.uchicago.edu/releases/04/040720.protestant.shtml, accessed July 20, 2004.

Novit-Evans, Bette, and Ashton Wesley Welch. (1983). "Racial and Ethnic Definition as Reflections of Public Policy." *Journal of American Studies* 17(3), pp. 417–435.

Office of International Religious Freedom. (2012). *The International Religious Freedom Report for 2011*, Washington, DC: Department of State. Available at http://www.state.gov/j/drl/rls/irf/religiousfreedom/index.htm#wrapper, accessed August 29, 2012.

Ogburn, William F. (1964). *On Culture and Social Change*. Chicago: University of Chicago Press.

Ogunwole, Stella U. (2002 February). *The American Indian and Alaska Native Population: 2000*. Washington, DC: U.S. Bureau of the Census.

O'Hare, William P. (1996, September). "A New Look at Poverty in America." *Population Bulletin* 5(2). Population Reference Bureau.

O'Hare, William P., and William H. Frey. (1992, September). "Booming, Suburban, and Black America." *American Demographics*, pp. 30–38.

Oinas-Kukkonen, Harri. (2008). "Network Analysis and Crowds of People as Sources of New Organisational Knowledge." In A. Koohang, et al. (Eds.), *Knowledge Management: Theoretical Foundation*. Santa Rosa, CA: Informing Science Press, pp. 173–189.

Oliver, Melvin L., and Thomas M. Shapiro. (1997). *Black Wealth/White Wealth*. New York: Routledge.

Olshansky, S. Jay, et al. (2012, September). "Differences in Life Expectancy due to Race and Educational Differences Are Widening, and Many May Not Catch Up," *Health Affairs* 31(9).

Onishi, Norimitsu. (2007, July 16). "Japan Learns Dreaded Task of Jury Duty." *New York Times*, p. 1.

Organization for Economic Co-operation and Development. (1997). INES Project, International Indicators Project. *The Condition of Education 1997* (Indicator 23).

Otterbein, Keith. (1973). "The Anthropology of War." In John J. Honigmann (Ed.), *Handbook of Social and Cultural Anthropology*. Chicago: Rand McNally.

___. (1970). *The Evolution of War*. New Haven, CT: Human Relations Area Files.

Oxana, Malaya. (2005). "The Ukrainian Dog Girl." Available at http://www.feralchildren.com/en/showchild.php?ch=oxana, accessed October 18, 2005.

Palen, John J. (1995). *The Suburbs*. New York: McGraw-Hill.

Palfrey, John, Dena T. Sacco, Danah Boyd, Laura DeBonis. (2009, January 14). "Enhancing Child Safety and Online Technologies: Final Report of the Internet Safety Technical Task Force to the Multi-State Working Group on Social Networking of State Attorneys General of the United States." Berkman Center for Internet & Society at Harvard University. Available at http://cyber.law.harvard.edu/pubrelease/isttf/, accessed June 27, 2009.

Pankhurst, Alula. (1999). " 'Caste' in Africa: The Evidence from South-Western Ethiopia Reconsidered." Edinburgh, Scotland: Edinburgh University Press.

Pape, Robert A. (2003, August). "The Strategic Logic of Suicide Terrorism." *American Political Science Review* 97(3), pp. 1–19.

Park, R., E. Burgess, and R. McKenzie. (Eds.). (1925). *The City*. Chicago: University of Chicago Press.

Parker, R. N. (1995). *Alcohol and Homicide: A Deadly Combination of Two American Traditions*. Albany, NY: State University Press.

Parkes, Henry Bamford. (1968). *The United States of America: A History*, 3rd ed. New York: Knopf.

Parrillo, Vincent N. (1997). *Strangers to These Shores*, 5th ed. Boston: Allyn & Bacon.

Parsons, Talcott. (1971). *The System of Modern Societies*. Englewood Cliffs, NJ: Prentice Hall.

___. (1966). *Societies: Evolutionary and Comparative Perspectives*. Englewood Cliffs, NJ: Prentice Hall.

___. (1954). *Essays in Sociological Theory*, rev. ed. New York: Free Press.

___. (1951). *The Social System*. New York: Free Press.

___. (1937). *The Structure of Social Action: A Study in Social Theory with Special Reference to a Group of Recent European Writers*. New York: McGraw-Hill.

Parsons, Talcott, and Robert F. Bales. (1955). *Family Socialization and Interaction Process*. New York: Free Press.

Passel, Jeffrey, Gretchen Livingston and D'Vera Cohn. (2012, May 17). Pew Research Center, "Explaining Why Minority Births Now Outnumber White Births." Available at http://www.pewsocialtrends.org/2012/05/17/explaining-why-minority-births-now-outnumber-white-births/, accessed July 25, 2012.

Patterson, Orlando. (1997). *The Ordeal of Integration*. Washington, DC: Civitas Counterpoint.

Patterson, T. E. (2002). *The Vanishing Voter*. New York: Knopf.

___. (2002, August 25). "Disappearing Act." *Boston Globe*, pp. D1, D2.

Pavlov, I. P. (1927). *Conditioned Reflexes* (G. V. Anrep, Trans.). New York: Oxford University Press.

Perkins, Wesley H., Michael P. Haines, and Richard Rice. (2005). "Misperceiving the College Drinking Norm and Related Problems: A Nationwide Study of Exposure to Prevention Information, Perceived Norms and Alcohol Misuse." *Journal of Studies of Alcohol* 66(4), pp. 470–478.

Perkins, Wesley H., William DeJong William, and Jeff Linkenbach. (2001). "Estimated Blood Alcohol Levels Reached by 'Binge' and 'Nonbinge' Drinkers." *Psychology of Addictive Behaviors* 15(4), pp. 317–320, 319.

Petersen, William. (1975). "On the Subnations of Western Europe." In Nathan Glazer and Daniel P. Moynihan (Eds.), *Ethnicity: Theory and Experience*. Cambridge, MA: Harvard University Press.

Pettigrew, Thomas F., and Robert C. Green. (1975). "School Desegregation in Large Cities: A Critique of the Coleman White Flight Thesis." *Harvard Educational Review* 46(1), pp. 1–53.

Pew Forum on Religious and Public Life. (2008, June). "Faith in Flux: Changes in Religious Affiliation in the U.S." Available at http://religions.pewforum.org/, accessed June 28, 2009.

___. (2008, June). U.S. Religious Landscape Survey, "Religious Beliefs and Practices: Diverse and Politically Relevant." Washington, DC. Available at http://religions.pewforum.org/, accessed April 29, 2009.

Pew Global Attitudes Project. (2005, June 23). "American Character Gets Mixed Reviews." Available at http://pewglobal.org/reports/display.php?ReportID=247, accessed January 9, 2006.

Pew Hispanic Center. (2006). "Hispanic Attitudes toward Learning English." Available at http://pewhispanic.org/files/factsheets/20.pdf, accessed June 3, 2009.

Pew Hispanic Center/Kaiser Family Foundation. (2003, August 7–October 15). "National Survey of Latinos: Education." Available at http://pewhispanic.org/reports/report.php?ReportID=25, accessed August 25, 2009

Pew Research Center. (2012). "A Gender Reversal on Career Aspirations." Available at http://www.pewsocialtrends.org/files/2012/04/Women-in-the-Workplace.pdf, accessed April 23, 2012.

___. (2010, November 18). "The Decline of Marriage and Rise of New Families." *Social and Demographic Trends Project*, p. 62. Available at http://www.pewsocialtrends.org/files/2010/11/pew-social-trends-2010-families.pdf, accessed August 23, 2012.

___. (2009, June 29). "Growing Old in America: Expectations vs. Reality." Pew Research Center. Available at http://www.pewsocialtrends.org/2009/06/29/growing-old-in-america-expectations-vs-reality/.

Pew Research Center of People and the Press. (2005, July 26). "Views of Muslim-Americans Hold Steady after London Bombings." Available at http://people-press.org/reports/display.php3?ReportID= 252, accessed January 9, 2006.

Pew Research Center's Religion and Public Life. (2011, January). "The Future of the Muslim Population." Available at http://www.pewforum.org/The-Future-of-the-Global-Muslim-Population.aspx, accessed August 27, 2012.

Piaget, J., and B. Inhelder. (1969). *The Psychology of the Child*. New York: Basic Books.

Pierce, Charles. (2000). *Hard to Forget: An Alzheimer's Story*. New York: Random House.

Pinker, Steven. (2002). *The Blank Slate*. New York: Viking.

___. (1994). *The Language Instinct*. New York: Harper Perennial.

Pinker, Susan. (2008). *The Sexual Paradox*. New York: Scribner.

Plucker, J. A. (Ed.). (2003). "Human Intelligence: Historical Influences, Current Controversies, Teaching Resources." Available at http://www.indiana.edu/~intell, accessed July 12, 2005.

Popenoe, David. (2005). "The Top Ten Myths of Marriage." National Marriage Project. Available at http://marriage.rutgers.edu/Publications/Print/Print%20Myths%20of%20Marriageaccessed October 18, 2005.

___. (1994). "The Evolution of Marriage and the Problem of Stepfamilies." In A. Booth and J. Dunn (Eds.), *Stepfamilies: Who Benefits? Who Does Not?* Hillsdale, NJ: Lawrence Erlbaum, pp. 3–27.

Popenoe, David, and Barbara Defoe Whitehead. (2009, February). "The State of Our Unions 2008, Updates of Social Indicators: Tables and Graphs." The National Marriage Project. Available at http://, accessed July 29, 2009.

Popenoe, David, and Barbara Dafoe Whitehead. (2004, June). "The State of Our Unions 2004." The National Marriage Project. Available at http://marriage.rutgers.edu/Publications/SOOU/SOOU2004.pdf, accessed October 14, 2005.

"Population and the Environment." (1998, July 17). *CQ Researcher* 8(26).

Population Reference Bureau. (2012). "The World's Women and Girls, 2011 Data Sheet." Available at http://www.prb.org/pdf11/world-women-girls-2011-data-sheet.pdf, accessed June 25, 2012.

___. (2009). *2008 World Population Data Sheet*.

___. (2008, September). "World Population Highlights." *Population Bulletin* 63(3).

___. (2006, August). "Elderly White Men Afflicted by High Suicide Rates." Washington, DC. Available at http://www.prb.org/Articles/2006/ElderlyWhiteMenAfflictedbyHighSuicideRates.aspx, accessed February 6, 2009.

___. (2005a). *Human Development Report 2005.*

___. (2005b). *2005 World Population Data Sheet.*

___. (2004, March). *Transitions in World Population.* 59(1).

___. (2002). *Human Population: Fundamentals of Growth Environmental Relationships.*

___. (1995, April). "Women, Children, and AIDS." *Population Today* 23(4).

Power, Carla, and Sudip Mazumdar. (2000, July 3). "Caste Struggle." *Newsweek International*, p. 30.

Prejean, Helen. (1993). *Dead Man Walking.* New York: Random House.

Prentice, Deborah A. (2008). "Mobilizing and Weakening Peer Influence as Mechanisms for Changing Behavior." In M. J. Prinstein and K. A. Dodge (Eds.), *Understanding Peer Influence in Children and Adolescents,* pp. 161–180. New York: Guilford.

Proctor, Bernadette D., and Joseph Dalaker. (2001, September). *Poverty in the United States: 2001.* Washington, DC: U.S. Bureau of the Census.

Programme for International Student Assessment. (2007). "PISA 2006 Science Competencies for Tomorrow's World." Available at http://www.pisa.oecd.org/document/2/0,3343,en_32252351_32236191_39718850_1, accessed January 25, 2009.

Propp, K. M. (1995). "An Experimental Examination of Biological Sex as a Status Cue in Decision-Making Groups and Its Influence on Information Use." *Small Group Research* 26, pp. 451–474.

Provine, Robert A. (2000). *Laughter: A Scientific Investigation.* New York: Penguin Books.

Pulliam, Sarah. (2008, September 19). "Magazines with Women Pastors on Cover Pulled from Bookstores." *Christianity Today.* Available at http://blog.christianitytoday.com/ctpolitics/2008/09/magazines_with.html, accessed July 27, 2009.

Putka, Gary. (1984, April 13). "As Jewish Population Falls in U.S., Leaders Seek to Reverse Trend." *Wall Street Journal*, pp. 1, 10.

Putnam, Robert D. (2000). *Bowling Alone.* New York: Simon & Schuster.

___. (1996, Winter). "The Strange Disappearance of Civic America." *American Prospect* 24.

Putnam, Robert D., and David E. Campbell. (2010). *American Grace.* New York: Simon and Schuster, pp. 121–129.

Pyeritz, R., C. Madansky, H. Schreier, L. Miller, and J. Beckwith. (1977). "The XYY Male: The Making of a Myth." In Ann Arbor Science for the People (Ed.), *Biology as a Social Weapon.* Ann Arbor, MI: Burgess Publishing.

Radelet, Michael L., Hugo Adam Bedau, and Constance E. Putnam. (1992). *In Spite of Innocence.* Boston: Northeastern University Press.

Radford, Benjamin. (2006, September 20). "Predator Panic: A Closer Look." *Skeptical Inquirer.* Available at http://csicop.org/si/2006-05/panic.html, accessed March 17, 2009.

Raines, Howell. (1988, June 17). "British Government Devising Plan to Curb Violence by Soccer Fans." *New York Times*, p. A1.

Rainwater, Lee, and T. M. Smeeding. (1995). "Doing Poorly: The Real Income of Children in Comparative Perspective." Working Paper No. 127. Luxembourg Income Study. Maxwell School of Citizenship and Public Affairs. New York: Syracuse University.

Ramo, Joshua Cooper. (1996, December 16). "Finding God on the Web." *Time*, pp. 60–64, 66–67.

Ravitch, Diane. (2010, November). "Dictating to the Schools: A Look at the Effects of the Bush and Obama Administrations on Schools." *Virginia Journal of Education.* Available at http://www.veanea.org/home/907.htm.

___. (2000). *Left Back: A Century of Failed School Reforms.* New York: Simon & Schuster.

Reich, Robert. (1991). *The Work of Nations: Preparing Ourselves for 21st Century Capitalism.* New York: Knopf.

Reid, Sue Titus. (1991). *Crime and Criminology*, 6th ed. Fort Worth, TX: Harcourt Brace Jovanovich.

Reiman, Jeffrey H. (1990). *The Rich Get Richer and the Poor Get Prison: Ideology, Class, and Criminal Justice*, 3rd. ed. New York: John Wiley.

"Religious Congregations and Membership in the United States: 2000." (2002). Cincinnati, OH: Glenmary Research Center.

"Religious Freedom Abroad." (2000, January 21). *Issues and Controversies on File*, pp. 17–24.

Renka, Russell D. (2010, February 22). "The Good, the Bad, and the Ugly of Public Opinion Polls." Available at http://cstl-cla.semo.edu/renka/Renka_papers/polls.htm.

Rennison, Callie M. (2002). "Criminal Victimization 2001: Changes 2000–2001 with Trends 1993–2001." Washington, DC: U.S. Department of Justice, Bureau of Justice Statistics.

___. (2001). "Criminal Victimization 2000: Changes 1999–2000 with Trends 1993–2000." Washington, DC: U.S. Department of Justice, Bureau of Justice Statistics.

Reuters News Service. (1997, July 11). "Television Industry Submits a Voluntary System of Parental Guidelines for Rating Television Programming to the FCC for Review."

Reverby, Susan M. (2011, January). " 'Normal Exposure' and Inoculation Syphilis: A PHS 'Tuskegee' Doctor in Guatemala, 1946–48." *Journal of Policy History.* Available at http://www.scribd.com/doc/38583804/%E2%80%9CNormal-Exposure%E2%80%9D-and-Inoculation-Syphilis-A-PHS-%E2%80%9CTuskegee%E2%80%9D-Doctor-in-Guatemala-1946-48.

Rheingold, Howard. (1999, January). "Look Who's Talking." *Wired*, pp. 35–39.

Rhymer, Russ. (1993). *Genie: An Abused Child's Flight from Silence.* New York: Basic Books.

Richerson, Peter J. and Robert Boyd. (2004). *Not by Genes Alone: How Culture Transformed Human Evolution.* Chicago, IL: University of Chicago Press.

Richmond, Riva. (2009, July 9). "Does Social Networking Breed Social Division?" *New York Times.* Available at http://gadgetwise.blogs.nytimes.com/2009/07/09/does-social-networking-breed-social-division/, accessed July 1, 2009.

Ricks, Thomas E. (1997). *Making the Corps.* New York: Scribner's.

Riesman, D. (1950). *The Lonely Crowd: A Study of the Changing American Character.* New Haven, CT: Yale University Press (in collaboration with Nathan Glazer and Reuel Denney).

Riley, Glenda. (1991). *Divorce: An American Tradition.* New York: Oxford University Press.

Riley, Nancy E. (1997, May). "Gender, Power, and Population Change." *Population Bulletin* 52(1). Population Reference Bureau.

___. (1996, February). "China's 'Missing Girls': Prospects and Policy." *Population Today* 24(2), pp. 4–5.

Rob, A. K. Ubaidur. (1988). "Community Characteristics, Leaders, Fertility and Contraception in Bangladesh." *Asia-Pacific Population Journal* 3(2), pp. 55–72.

Robers, Simone, Jijun Zhang, Jennifer Truman, and Thomas D. Snyder. (2012). "Indicators of School Crime and Safety: 2011" (NCES 2012-002/NCJ 236021). National Center for Education Statistics, U.S. Department of Education, and Bureau of Justice Statistics, Office of Justice Programs, U.S. Department of Justice, Washington, DC. Available at http://bjs.ojp.usdoj.gov/content/pub/pdf/iscs11.pdf.

Robertson, Lori. (2008, March 18). "That Chain E-mail Your Friend Sent to You Is (Likely) Bogus. Seriously." Factcheck.org. Available at http://www.factcheck.org/specialreports/that_chain_e-mail_your_friend_sent_to.html, accessed July 1, 2009.

Robinson, Jacob. (1976). "The Holocaust." In Yisrael Gutman and Livia Rothkirchen (Eds.), *The Catastrophe of European Jewry*. Jerusalem: Yad Veshem.

Roffman, Roger A., and Robert S. Stephens (Eds.). (2006). *Cannabis Dependence: Its Nature, Consequences, and Treatment*. Cambridge, UK: Cambridge University Press.

Rohde, David. (2007, October 5). "Army Enlists Anthropology in War Zones." *New York Times*, pp. A1, A14.

Roos, Patricia. (1995, August). "Occupational Feminization, Occupational Decline?" Paper presented at the annual meeting of the American Sociological Association.

Rosaldo, Michelle Zimbalist. (1974). "Woman, Culture and Society: A Theoretical Overview." In Michelle Zimbalist Rosaldo and Louise Lamphere (Eds.), *Woman, Culture and Society*. Stanford, CA: Stanford University Press.

Rose, Elaina. (2004, March). "Education and Hypergamy in Marriage Markets." Paper No. 353330, Department of Economics, University of Washington, Seattle.

Rosenthal, Elizabeth. (1998, November 1). "For One Child Policy: China Rethinks Iron Hand." *New York Times*, pp. 1, 20.

Rosenthal, R., and L. Jacobson. (1966). "Teachers' Expectancies: Determinants of Pupils' I.Q. Gain." *Psychological Reports* 18, pp. 115–118.

Rosenzweig, Jane. (1999, July/August). "Can TV Improve Us?" *American Prospect* 45.

Roth, Philip. (2000). *The Human Stain*. New York: Houghton Mifflin.

___. (1998). *American Pastoral*. New York: Vintage Books.

Rowland, Christopher. (2003, March 10). "R.I. Club's Exits at Issue in Fire Probe." *Boston Globe*, p. A1.

Rubin, Zick. (1973). *Liking and Loving*. New York: Holt, Rinehart and Winston.

___. (1970). "Measurement of Romantic Love." *Journal of Personality and Social Psychology* 16(2), pp. 265–273.

Rubinstein, Moshe. (1975). *Patterns of Problem Solving*. Englewood Cliffs, NJ: Prentice Hall.

Rumbaut, Rubén G. (2005, June 30–July 1). "A Language Graveyard? Immigration, Generation, and Linguistic Acculturation in the United States." Paper presented to the International Conference on The Integration of Immigrants: Language and Educational Achievement, Social Science Research Center, Berlin, Germany.

Rybczynski, Witold. (1983). *Taming the Tiger: The Struggle to Control Technology*. New York: Viking Press.

Saad, Lydia. (2012, June 21). "Most Americans Believe Crime in U.S. Is Worsening." Gallup Poll. Available at http://www.gallup.com/poll/150464/Americans-Believe-Crime-Worsening.aspx, accessed October 31, 2011.

___. (2008, August 11). "By Age 24, Marriage Wins Out." Gallup Poll. Available at http://www.gallup.com/poll/109402/Age-24-Marriage-Wins.aspx?version=print, accessed September 20, 2008.

Sabol, William J. (1999, May). *Crime Control and Common Sense Assumptions Underlying the Expansion of the Prison Population*. Washington, DC: Urban Institute.

Sahlins, Marshall D., and Elman R. Service. (Eds.). (1960). *Evolution and Culture*. Ann Arbor: University of Michigan Press.

Sailer, Steve. (1997, July 14). "Is Love Colorblind?" *National Review*.

"Same-Sex Partnerships." (2000, February 18). *Issues and Controversies on File*, pp. 49–56.

Samovar, Larry A., Richard Porter, and Nemi C. Jain. (1981). *Understanding Intercultural Communication*. Belmont, CA: Wadsworth.

Sanderson, Warren, and Sergei Scherbov. (2008). "Rethinking Age and Aging." *Population Bulletin* 63(4). Population Reference Bureau.

Sapir, Edward. (1961). *Culture, Language and Personality*. Berkeley and Los Angeles: University of California Press.

Sappenfield, Mark. (2002, March 29). "Mounting Evidence Links TV Viewing to Violence." *Christian Science Monitor*.

Sawhill, Isabel V. (2012, July 9). "Are We Headed toward a Permanently Divided Society?" *CCF Briefs*, No.47. Available at http://www.brookings.edu/research/papers/2012/03/30-divided-society-sawhill.

Scarce, Rik. (1994, July). "(No) Trial (But) Tribulations." *Journal of Contemporary Ethnography* 23(2), pp. 123–149.

Schneider, Joseph W., and Peter Conrad. (1983). *Having Epilepsy: The Experience and Control of Illness*. Philadelphia: Temple University Press.

Schoenborn, Charlotte A., Patricia Adams, and Patricia Barnes. (2002, September). "Body Weight Status of Adults: United States, 1997–1998. Advance Data from Vital Health and Statistics." Washington, DC: U.S. Centers for Disease Control and Prevention.

Schumpeter, Joseph A. (1950). *Capitalism, Socialism and Democracy*, 3rd ed. New York: Harper Torchbooks.

Schur, Edwin M., and Hugo A. Bedau. (1974). *Victimless Crimes: Two Sides of a Controversy*. Englewood Cliffs, NJ: Prentice Hall.

Schützwohl, Achim, Amrei Fuchs, William F. McKibbin, and Todd K. Shackelford. (2009). "How Willing Are You to Accept Sexual Requests from Slightly Unattractive to Exceptionally Attractive Imagined Requestors?" *Human Nature* 20, pp. 282–293.

Schwarz, Frederic. (1997, February/March). "Women Who Smoke and the Men Who Arrest Them." *American Heritage*, pp. 108–111.

Scott, Elizabeth. (1990). "Rational Decision Making about Marriage and Divorce." *Virginia Law Review* 76, pp. 9–94.

Sharp, Lauriston. (1952). "Steel Axes for Stone-Age Australians." *Human Organization* 11, pp. 17–22.

Shattuck, R. (1980). *The Forbidden Experiment*. New York: Farrar, Straus & Giroux.

Shaw, Clifford R., and Henry D. McKay. (1942). *Juvenile Delinquency and Urban Areas*. Chicago: University of Chicago Press.

___. (1931). "Social Factors in Juvenile Delinquency." In *National Committee on Law Observance and Law Enforcement, Report on the Causes of Crime*, Vol. 2. Washington, DC: U.S. Government Printing Office.

Sheldon, W. H., E. M. Hartl, and E. McDermott. (1949). *The Varieties of Delinquent Youth*. New York: Harper.

Sheldon, W. H., and S. S. Stevens. (1942). *The Varieties of Temperament*. New York: Harper.

Sheldon, W. H., and W. B. Tucker. (1940). *The Varieties of Human Physique*. New York: Harper.

Sheler, Jeffrey L. (1997, December 15). "Dark Prophecies." *U.S. News and World Report*, pp. 62–63, 64, 68–71.

Shelton, Deborah L. (1999, March 2). "Scientists Study Why Women Get Certain Diseases More Readily than Men Do." *Los Angeles Times*.

Shepardson, Mary. (1963). *Navajo Ways in Government*. Manasha, WI: American Anthropological Association.

Shermer, Michael. (2000). *How We Believe: The Search for God in an Age of Science*. San Francisco: W. H. Freeman.

Shibutani, Tamotsu. (1966). *Improvised News: A Sociological Study of Rumor*. Indianapolis, IN: Bobbs-Merrill.

Shils, Edward. (1971/1960). "Mass Society and Its Culture." In Bernard Rosenberg and David Manning (Eds.), *Mass Culture Revisited*. New York: Van Nostrand-Reinhold.

___. (1968). *Political Development in the New States*. The Hague: Mouton.

Simmel, Georg. (1957). "Fashion." *American Journal of Sociology* 62, pp. 541–588.

___. (1955). *Conflict: The Web of Group Affiliations*. Glencoe, Illinois: The Free Press.

___. (1950). *The Sociology of Georg Simmel*. Kurt Wolff (Ed.). New York: Free Press.

Simon, Julian L., and Herman Kahn (Eds.). (1984). *The Resourceful Earth: A Response to Global 2000*. New York: Basil Blackwell.

Simpson, George E., and Milton Yinger. (1972). *Racial and Cultural Minorities: An Analysis of Prejudice and Discrimination*, 4th ed. New York: Harper & Row.

Sjoberg, Gideon. (1956). *Preindustrial City: Past and Present*. New York: Free Press.

Skocpol, Theda. (1979). *States and Social Revolutions*. New York: Cambridge University Press.

Smeeding. T., L. Rainwater, and G. Burtless. (2001, May). "United States Poverty in Cross-National Context." Prepared for the IRP Conference Volume: "Understanding Poverty in America: Progress and Problems."

Smelser, Neil J. (1971). "Mechanisms of Change and Adjustment to Change." In George Dalton (Ed.), *Economic Development and Social Change*. Garden City, NY: Natural History Press.

___. (1962). *Theory of Collective Behavior*. New York: Free Press.

Smith, Aaron. (2011, November 15). "Why Americans Use Social Media." Available at http://pewinternet.org/Reports/2011/Why-Americans-Use-Social-Media/Main-report.aspx.

Smith, Adam. (1976/1776). *An Inquiry into the Nature and Causes of the Wealth of Nations*. Oxford: Oxford University Press.

Smock, Pamela J. (2000). "Cohabitation in the United States: An Appraisal of Research Themes: Findings and Implications." *Annual Review of Sociology* 26.

Sniffen, Michael J. (1999, December 6). "Most Female Crime Is Simple Assault." Associated Press.

Social Security Administration. (2005, May 6). "Most Popular Baby Names, 1880–2004." Available at http://www.ssa.gov/OACT/babynames/, accessed January 9, 2006.

Solecki, Ralph. (1971). *Shanidar: The First Flower People*. New York: Knopf.

Solomon, Amy L., Vera Kachnowski, and Avi Bhati. (2005, March 31). "Does Parole Work? Analyzing the Impact of Postprison Supervision on Rearrest Outcomes." Washington, DC: Urban Institute.

Solomon, Robert C. (2002, March). "Reasons for Love." *Journal for the Theory of Social Behaviour* 32(1), pp. 1–28.

Solomon-Fears, Carmen. (2008, November 20). "Nonmarital Childbearing: Trends, Reasons, and Public Policy Interventions." Congressional Research Service. Available at http://www.fas.org/sgp/crs/misc/RL34756.pdf, accessed August 15, 2012.

Sommers, Christina Hoff. (2000). *The War against Boys: How Misguided Feminism Is Harming Our Young Men*. New York: Simon & Schuster.

___. (1994). *Who Stole Feminism?* New York: Simon & Schuster.

Sourcebook of Criminal Justice Statistics 2003, Table 4.19. Available at http://www.albany.edu/sourcebook/pdf/t419.pdf, accessed January 9, 2006.

Southall, E. Woodrow Eckard, and Mark Nagel. (2012). "Adjusted Graduation Gap Report: NCAA Division-I Football." NCAA College Sport Research Institute at the University of North Carolina. Available at https://docs.google.com/viewer?a=v&q=cache:8cJPgT3f0yIJ:chronicle.com/blogs/ticker/files/2012/09/AGG-Report.pdf+&hl=en&gl=us&pid=bl&srcid=ADGEESgFlxC7udChVsG3yL2g20F6ZfGk0vTexOZXH67P_Em96YkEehT6iAZdW30Oo1x8_-t8koZAi7rSOhTNjv_blMRDeBJelGuQe4nmWPTsTf4xINh6AfM7cneT4LvP3oU0S6GFqVSy&sig=AHIEtbS85weHrtswG0tp5IP6phFvJqrozA—'

Spar, Debora, and Cate Reavis. (2004). *The Business of Life*. Boston: Harvard Business School Publishing.

Spencer, Herbert. (1864). *Principles of Biology*, Vol. 1, p. 444.

Spitz, Rene A. (1945). "Hospitalism: An Inquiry into the Genesis of Psychiatric Conditions in Early Childhood." In Anna Freud, et al. (Eds.), *The Psychoanalytic Study of the Child*. New York: International University Press.

Sprecher, S., and S. Metts. (1989). "Development of the 'Romantic Beliefs Scale' and Examination of the Effects of Gender and Gender-Role Orientation." *Journal of Personal and Social Relationships* 6, pp. 387–411.

Squire, Peverill. (1988, Spring). "Why the 1936 Literary Digest Poll Failed." *Public Opinion Quarterly* 52, pp. 125–133.

Stark, Rodney, and William Sims Bainbridge. (1985). *The Future of Religion: Secularization, Revival, and Cult Formation*. Berkeley: University of California Press.

Stark, Rodney, and Charles Y. Glock. (1968). *American Piety: The Nature of Religious Commitments*. Berkeley and Los Angeles: University of California Press.

Starr, Paul. (1992). "Social Categories and Claims in the Liberal State." *Social Research* 59(2), pp. 263–296.

___. (1982). *The Social Transformation of American Medicine*. New York: Basic Books.

*State of Maryland v. Bryan Rose.* (2008). In the Circuit Court for Baltimore County. Case No. K06-545. Available at http://www.baltimoresun.com/media/acrobat/2007-10/33446162.pdf.

*Statistical Abstract of the United States: 2004–2005*, 124th ed. (2005). U.S. Bureau of the Census, Washington, DC: U.S. Government Printing Office.

Stockman, Farah. (2009, February 12). "Anthropologist's War Death Reverberates." *Boston Globe*.

Stone, Brad. (2009, January 13). "Report Calls Online Threats to Children Overblown." *New York Times*. Available at http://www.nytimes.com/2009/01/14/technology/internet/14cyberweb.html?ref=internet, accessed February 19 2009.

Stouffer, Samuel A. (Ed.). (1950). *The American Soldier*. Princeton, NJ: Princeton University Press.

Strauss, Anselm, and Barney Glaser. (1975). *Chronic Illness and the Quality of Life*. St. Louis: Mosby.

Surowiecki, James. (2004). *The Wisdom of Crowds*. New York: Doubleday.

Susskind, Ron. (1998). *A Hope in the Unseen*. New York: Broadway Books.

Sutherland, Edwin H. (1961). *White Collar Crime*. New York: Holt, Rinehart and Winston.

___. (1940). "White Collar Criminality." *American Sociological Review* 40, pp. 1–12.

___. (1924). *Criminology*. New York: Lippincott.

Sutherland, Edwin H., and D. R. Cressey. (1978). *Principles of Criminology*, 10th ed. Chicago: Lippincott.

Suttles, G. (1972). *The Social Construction of Communities*. Chicago: University of Chicago Press.

___. (1968). *The Social Order of the Slum*. Chicago: University of Chicago Press.

Swarthmore College Peace Collection. "Introduction to an Exhibit of Photographs of Jane Addams, Her Family, and Hull-House." Available at http://www.swarthmore.edu, accessed March 7, 2002.

"Swazi King Marries Eleventh Wife." (2005, May 30). BBC News. Available at http://news.bbc.co.uk/2/hi/africa/4592961.stm, accessed July 24, 2005.

Sykes, Gresham, and David Matza. (1957). "Techniques of Neutralization: A Theory of Delinquency." *American Sociological Review* 22(6), pp. 664–670.

Szymanski, Albert T., and Ted George Goertzel. (1979). *Sociology: Class, Consciousness, and Contradictions*. New York: Van Nostrand-Reinhold.

Tabuchi, Hiroko. (2012, May 29). "Young and Global Need Not Apply in Japan." *New York Times*, Venture Japan, "Japanese Business Culture." Available at http://www.venturejapan.com/japanese-business-culture.htm, accessed June 3, 2012.

Tannen, Deborah. (1998). *The Argument Culture: Moving from Debate to Dialogue*. New York: Random House.

___. (1994). *Talking from 9 to 5*. New York: William Morrow.

___. (1990). *You Just Don't Understand: Women and Men in Conversation*. New York: William Morrow.

Tavernise, Sabrina. (2012, September 20). "Life Spans Shrink for Least-Educated Whites in the U.S." *New York Times*. Available at http://www.nytimes.com/2012/09/21/us/life-expectancy-for-less-educated-whites-in-us-is-shrinking.html?nl=todaysheadlines&emc=edit_th_20120921&moc.semityn.www&pagewanted=all.

Taylor, I., P. Walton, and J. Young. (1973). *The New Criminology*. London: Routledge & Kegan Paul.

Taylor, O. L. (1990). *Cross-Cultural Communication: An Essential Dimension of Effective Education*, rev. ed. Washington, DC: Mid-Atlantic Equity Center.

Taylor, Paul, Mark Hugo Lopez, Jessica Hamar Martínez, and Gabriel Velasco. (2012, April 4). Pew Hispanic Research Center, "When Labels Don't Fit: Hispanics and Their Views of Identity." Available at http://www.pewhispanic.org/2012/04/04/when-labels-dont-fit-hispanics-and-their-views-of-identity/.

Teigen, K. H. (1986). "Old Truths or Fresh Insight? A Study of Students' Evaluation of Proverbs." *British Journal of Social Psychology* 25, pp. 43–50.

Tejada-Vera, B., and P. D. Sutton. (2009). "Births, Marriages, Divorces, and Deaths: Provisional Data for August 2008." *National Vital Statistics Reports* 57(15). Hyattsville, MD: National Center for Health Statistics.

Terrace, Herbert S., L. A. Petitto, R. J. Sanders, and T. G. Bever. (1979). "Can an Ape Create a Sentence?" *Science* 206, pp. 891–902.

Thernstrom, Stephen, and Abigail Thernstrom. (1997). *America in Black and White: One Nation Indivisible*. New York: Simon & Schuster.

Thomas, Karen. (1994, April 6). "Learning at Home: Education outside School Gains Respect." *USA Today*, p. 5D.

Thomas, W. I. (1928). *The Child in America*. New York: Knopf.

Thumma, Scott, and Dave Travis. (2007). *Beyond Megachurch Myths: What We Can Learn from America's Largest Churches*. San Francisco: Jossey-Bass.

Tierney, John J., Jr. (2000, August). "The World of Child Labor." *The World and I* 15, p. 54.

Tiger, Lionel, and Robin Fox. (1971). *The Imperial Animal*. New York: Holt, Rinehart and Winston.

Tischler, Henry L. (1994, March). "The Message behind the Image: A Comparison of Rock Music Videos with Country Music Videos." Paper presented at the Eastern Sociological Society Meetings.

Tönnies, Ferdinand. (1963). *Community and Society*. New York: Harper & Row. (Originally published in German as *Gemeinschaft und Gesellschaft* in 1887.)

Toynbee, Arnold. (1946). *A Study of History*. New York and London: Oxford University Press.

"Tracking." (2000, September 15). *Issues and Controversies on File*. Facts on File.

Trickey, H. (2006, August 24). "No Child Left out of the Dodgeball Game?" CNN.com. Available at http://www.cnn.com/2006/HEALTH/08/20/PE.NCLB/index.html, accessed June 12, 2009.

Tsertishvili, Tamar. (2010, August 8). "A Love Story." Available at http://tamartsertishvili.blogspot.com/2010/08/yarima-and-ken-good.html#!/2010/08/yarima-and-ken-good.html, accessed March 5, 2012.

Tulshyan, Ruchika. (2010, March 29). "Match Dot Mom and Dad." Available at http://www.forbes.com/2010/03/29/arranged-marriage-love-tradition-forbes-woman-well-being-family.html, accessed April 22, 2012.

Turkle, Sherry. (2011). *Alone Together: Why We Expect More from Technology and Less from Each Other*. New York: Basic Books.

Turnbull, Mark. (2012, February 14). "America's Big Wealth Gap: Is It Good, Bad, or Irrelevant?" *Christian Science Monitor*. Available at http://www.csmonitor.com/USA/Politics/2012/0214/America-s-big-wealth-gap-Is-it-good-bad-or-irrelevant, accessed July 9, 2012.

Turner, James C., and Adrienne Keller. (2011, October 29–November 2). "Leading Causes of Mortality Among American College Students at 4-Year Institutions." Paper presented at the American Public Health Association annual meeting, Washington, DC.

Tylor, E. (1958/1871). *Primitive Culture: Researches into the Development of Mythology, Philosophy, Religion, Art and Custom*, Vol. 1. London: John Murray.

UNAIDS. (2011). Data Tables. Geneva, Switzerland. Available at http://www.unaids.org/en/media/unaids/contentassets/documents/unaidspublication/2011/JC2225_UNAIDS_datatables_en.pdf1, accessed June 9, 2012.

___. (2009). "2008 Report on the Global AIDS Epidemic." Available at http://www.unaids.org/en/KnowledgeCentre/HIVData/GlobalReport/2008/2008Global_report.asp, accessed May 18, 2009.

UND Women's Center News. (2003, January). Available at http://www.und.edu/dept/womenctr/newsletters/January2004.pdf, accessed February 12, 2004.

UNESCO. (2004). *EFA Global Monitoring Report 2005: Education for All, the Quality Imperative*, Statistical Annex, Table 2. Paris: UNESCO.

Uniform Crime Reports. (2005). "Untangling the Statistics: Numbers Don't Lie—But They Can Deceive." Available at http://www.whitehousedrugpolicy.gov/publications/whos_in_prison_for_marij/untangling_the_stats.pdf, accessed August 31, 2005.

Uniform Crime Reports; FBI Supplement. (2005). "Homicide Reports 1976–2002."

United Jewish Communities. (2002). "U.S. Jewish Population Fairly Stable over Decade, According to Results of National Jewish Population Survey 2000–01." Press release, October 8, 2002. Available at http://www.ujc.org, accessed June 12, 2003.

United Nations. (2009). "2007/2008 Human Development Report, Indicators." Available at http://hdrstats.undp.org/indicators/, accessed May 18, 2009. Population Division.

___. (2008). "World Urbanization Prospects: The 2007 Revision." Division of the Department of Economic and Social Affairs of the United Nations Secretariat.

___. (2005, June). "World Population Prospects: The 2004 Revision." *Population Newsletter* 79.

___. (2002). *UN Human Development Report: Deepening Democracy in a Fragmented World*. New York: United Nations Publications.

___. (2002, April). "Population Aging: Facts and Figures." Second World Assembly on Ageing. New York: United Nations Publications.

___. (2002, March–May). *UN Chronicle* 39(1). New York: United Nations Publications.

___. (1948). *Convention on the Prevention and Punishment of the Crime of Genocide*. General Assembly resolution, December 9, 1948.

United Nations Children's Fund (UNICEF). (2012, February). "The State of the World's Children 2012." New York.

___. (2008, December). "The State of the World's Children 2009." New York.

___. (2008, January). "Child Marriage and the Law." New York.

___. (2007). "Child Poverty in Rich Countries, 2007." *Innocenti Report Card* No. 7. Florence, Italy: UNICEF Innocenti Research Centre.

___. (2005). "Child Poverty in Rich Countries, 2005." *Innocenti Report Card* No. 6. Florence, Italy: UNICEF Innocenti Research Centre.

___. (2005). "Factsheet: Early Marriage." Unicef.org. Available at http://www.unicef.org/protection/files/earlymarriage.pdf, accessed January 9, 2006.

United Nations Office of Drugs and Crime. (2004). "2004 World Drug Report." United Nations Publications Sales, No. E. 04.XI.16.

United Nations, Population Division. (2011, April). "World Population Prospects: The 2010 Revision." Available at http://esa.un.org/unpd/wpp/Excel-Data/fertility.htm, accessed July 17, 2012.

___. (2005). "World Marriage Patterns 2000." Department of Economic and Social Affairs.

U.S. Census Bureau, Current Population Reports, P60–239, *Income, Poverty, and Health Insurance Coverage in the United States: 2010*, U.S. Government Printing Office, Washington, D.C., 2011.

U.S. Bureau of the Census. (2012). "Educational Attainment in the United States: 2011." Available at http://www.census.gov/hhes/socdemo/education/data/cps/2011/tables.html, accessed August 31, 2012.

___. (2012). "U.S. Census of Population: 1960," Vol. I, Part 1; J. K. Folger and C. B. Nam, "Education of the American Population" (1960 Census Monograph); Current Population Reports, Series P-20, various years; and Current Population Survey (CPS), March 1970–March 2011.

___. (2012, June). International Population Reports, P95/09-1, 2009.

___. (2012, June 18). "Changes in Household Net Worth from 2005 to 2010," Alfred Gottschalck and Marina Vornovytskyy. Available at http://blogs.census.gov/2012/06/18/changes-in-household-net-worth-from-2005-to-2010/, accessed June 18, 2012, September 12, 2012.

___. (2012, May 17). Newsroom. Census Bureau Reports, "Most Children Younger Than Age 1 are Minorities." Available at http://www.census.gov/newsroom/releases/archives/population/cb12-90.html, accessed July 25, 2012.

___. (2012, April). Current Population Survey, "1960 to 2011 Annual Social and Economic Supplements."

___. (2012, February). Current Population Reports, "What It's Worth: Field of Training and Economic Status in 2009," accessed August 30, 2012.

___. (2011). 2010 Census. Summary File 1, Table PCT 11. Available at http://factfinder2.census.gov/faces/tableservices/jsf/pages/productview.xhtml?pid=DEC_10_SF1_QTP10&prodType=table, accessed July 25, 2012.

___. (2011). Current Population Reports, P60-239, "Income, Poverty, and Health Insurance Coverage in the United States: 2010." Washington, DC: U.S. Government Printing Office.

___. (2011). Current Population Survey. "Voting and Registration/Nov 2010." Available at http://smpbff1.dsd.census.gov/TheDataWeb_HotReport/servlet/HotReportEngineServlet?reportid=767b1387bea22b8d3e8486924a69adcd&emailname=essb@boc&filename=0328_nata.hrml, accessed September 4, 2012.

___. (2011). "Educational Attainment in the United States: 2011." Available at http://www.census.gov/hhes/socdemo/education/.

___. (2011, December 10). *Statistical Abstract of the United States: 2012*, 131st ed., Table 1334. Washington, DC. Available at http://www.census.gov/compendia/statab/.

___. (2011, May). "Age and Sex Composition, 2010." Available at http://www.census.gov/prod/cen2010/briefs/c2010br-03.pdf, accessed September 12, 2012.

___. (2011, May). 2010 Summary File 1; The Hispanic Population 2010, Census Brief. Available at http://www.hacu.net/images/hacu/OPAI/2012_Virtual_Binder/2010%20census%20brief%20-%20hispanic%20population.pdf.

___. (2009). "2008 Annual Social and Economic Supplement." Available at http://www.census.gov/population/www/socdemo/hh-fam/cps2008.html, accessed May 19 2009. Current Population Survey.

___. (2009). "2007 and 2008 Annual Social and Economic Supplements." Current Population Survey.

___. (2009). "Information and Communications: Internet Publishing and Broadcasting and Internet Usage." *Statistical Abstract of the United States, 2009*. Available at http://www.census.gov/compendia/statab/cats/information_communications/internet_publishing_and_broadcasting_and_internet_usage.html, accessed May 21, 2009.

___. (2009). "World Population: 1950–2050." International Database. http://www.census.gov/ipc/www/idb/worldpopinfo.php, accessed June 7, 2009.

___. (2009). "World Vital Events per Time Unit: 2009." Available at http://www.census.gov/cgi-bin/ipc/pcwe, accessed June 7, 2009.

___. (2009, April 6). "Decennial Census of Population, 1940 to 2000."

___. (2009, March 3). Facts for Feature, "Older Americans Month: May 2009." Available at http://www.census.gov/Press-Release/www/releases/archives/facts_for_features_special_editions/013384.html, accessed August 13, 2009.

___. (2009, January). Current Population Survey.

___. (2009, January 13). "2008 Annual Social and Economic Supplement." Available at http://www.ssa.gov/OACT/babynames/http://www.ssa.gov/OACT/babynames/, accessed February 15 2009.

___. (2008). "Characteristic Percentage Voting 2006." Internet release date: July 1, 2008. Current Population Survey, November 2006.

___. (2008). *Statistical Abstract of the United States: 2009*, 128th ed. Washington, DC: U.S. Government Printing Office.

___. (2008). *Statistical Abstract of the United States: 2008*. Washington, DC: U.S. Government Printing Office.

___. (2008, August 14). "Percent Minority of the U.S. Population by Selected Age Groups: 2010 to 2050." Population Division.

___. (2008, July 14). "Births, Marriages, Divorces, and Deaths: Provisional Data for 2007." *National Vital Statistics Reports* 56(21).

___. (2007). "Annual Social and Economic Supplement." Population Division, Current Population Survey, 2004. Available at http://www.census.gov/population/socdemo/hhfam/cps2007, accessed July 15 2009.

___. (2007). Current Population Reports. Series P20–537.

___. (2007). "Population Estimates, July 1, 2006."

___. (2007). "Poverty Status of People, by Age, Race, and Hispanic Origin: 1959 to 2006." Housing and Household Economic Statistics Division. Available at http://www.census.gov/hhes/www/poverty/histpov/hstpov3.html, accessed May 13, 2009.

___. (2007, December 27). "Educational Attainment in the United States: 2007." Current Population Reports.

___. (2005). "African-American History Month: February 2005." Available at http://www.census.gov/Press-Release/www/releases/archives/facts_for_features_special_editions/003721.html, accessed January 9, 2006.

___. (2005). "2004 American Community Survey." Available at http://www.census.gov/acs/www/, accessed January 9, 2006.

___. (2005). *Statistical Abstract of the United States: 2004–2005*, p. 254. Washington, DC: U.S. Government Printing Office.

___. (2005). "Unmarried and Single Americans Week September 18–24, 2005." Available at http://www.census.gov/Press-Release/www/releases/archives/facts_for_features_special_editions/005384.html, accessed September 12, 2005.

___. (2004). "50th Anniversary of 'Wonderful World of Color' TV." Available at http://www.census.gov/Press-Release/www/releases/archives/facts_for_features/001702.html, accessed October 18, 2005.

U.S. Census Bureau News. (2005). "Income Stable, Poverty Rate Increases, Percentage of Americans without Health Insurance Unchanged." Available at http://www.census.gov/Press-Release/www/releases/archives/income_wealth/005647.html, accessed August 30, 2005.

U.S. Centers for Disease Control and Prevention. (2009). "Births, Marriages, Divorces, and Deaths: Provisional Data for October 2008." *National Vital Statistics Reports* 57(17).

___. (2009, April). "Deaths: Final Data for 2006." *National Vital Statistics Reports* 57(14). Available at http://www.cdc.gov/nchs/data/nvsr/nvsr54/nvsr54_13.pdf, accessed July 10, 2009.

U.S. Department of Agriculture. (2004). Economic Research Service, Rural Development Research Report 100. Washington, DC.

U.S. Department of Education. (2011, September). *Digest of Education Statistics 2012*; Nathan E. Bell, *CGS/GRE Survey of Graduate Enrollment and Degrees,* Graduate Enrollment and Degrees: 2000 to 2010.

___. (2009). *Indicators of School Crime and Safety, 2008.*

___. (2008). *Digest of Education Statistics 2007*. Washington, DC: NCES.

U.S. Department of Education, National Center for Education Statistics. (2012). *The Condition of Education 2012* (NCES 2012-045), Indicator 33.

___. (2012). *The Condition of Education 2011* (NCES 2011-045), Indicator 6.

___. (2012). "Fast Facts." Available at http://nces.ed.gov/fastfacts/display.asp?id=372, accessed August 31, 2012.

U.S. Department of Health and Human Services, Administration on Aging. (2004). "A Profile of Older Americans: 2004. Racial and Ethnic Composition." Available at http://www.aoa.gov/prof/Statistics/profile/2004/7.asp, accessed September 30, 2005.

U.S. Department of Homeland Security. (2012, May). Annual Flow Reports, "Refugees and Asylees: 2011." Available at http://www.dhs.gov/xlibrary/assets/statistics/publications/ois_rfa_fr_2011.pdf.

___. (2009, March). Annual Flow Report.

U.S. Department of Justice. (2004, November). Bulletin NCJ 206627. Capital Punishment, 2003, p. 9, Table 9.

U.S. Department of Justice, Bureau of Justice Statistics. (2012). FBI, Uniform Crime Reports; FBI Supplement. "Homicide Reports 1976–2007." Available at http://www.icpsr.umich.edu/icpsrweb/ICPSR/studies/24801.

___. (2011, December). "Capital Punishment, 2010." Bulletin NCJ 236510. Washington, DC: U.S. Department of Justice. Available at http://bjs.ojp.usdoj.gov/content/pub/pdf/cp10st.pdf, accessed July 5, 2012.

___. (2011, December). "Capital Punishment, 2010—Statistical Tables." Bulletin NCJ 236510. Available at http://www.bjs.gov/content/pub/pdf/cp10st.pdf, accessed June 25, 2012.

___. (2011, September). "Criminal Victimization, 2010." Bulletin NCJ 235508. Available at http://bjs.ojp.usdoj.gov/content/pub/pdf/cv10.pdf, accessed June 21, 2012.

___. (2011, May 5). "Time Served in Prison, by Offense and Release Type, Sex and Race." Available at http://bjs.ojp.usdoj.gov/index.cfm?ty=pbdetail&iid=2045, accessed June 21, 2012.

U.S. Department of Justice, Federal Bureau of Investigation. (2012). Uniform Crime Reports, "Crime—National or State Level Data with One Variable." Available at http://www.ucrdatatool.gov/Search/Crime/State/RunCrimeTrendsInOneVar.cfm, accessed June 21, 2012.

___. (2011). Uniform Crime Reports, "Crime in the United States, 2010." Available at http://www.fbi.gov/about-us/cjis/ucr/crime-in-the-u.s/2010/crime-in-the-u.s.-2010/tables/10tbl01.xls, accessed June 22, 2012.

___. (2011). "Crime in the United States, 2009, Table 38." Available at http://www2.fbi.gov/ucr/cius2009/data/table_38.html, accessed June 22, 2012.

___. (2011, September). Uniform Crime Reports, "Crime in the United States, 2010." Available at http://www.fbi.gov/about-us/cjis/ucr/crime-in-the-u.s/2010/crime-in-the-u.s.-2010/clearancetopic.pdf.

___. (2005). "Crime in the United States, 2004." Available at http://www.fbi.gov/ucr/cius_04/offenses_cleared/index.html, accessed October 28, 2005.

___. (2005). "Crime in the United States, 2004, Table 38." Available at http://www.fbi.gov/ucr/cius_04/documents/04tbl38a.xls, accessed January 9, 2006.

___. (2005). *Uniform Crime Reports: Supplementary Homicide Reports, 1976–2002.*

U.S. Department of Justice, Office of Justice Programs. (2012, June). "Violent Crime against the Elderly Reported by Law Enforcement in Michigan, 2005–2009." Available at http://bjs.gov/content/pub/pdf/vcerlem0509.pdf, accessed June 23, 2012.

___. (1996). *The Nature of Homicide: Trends and Changes.* Washington, DC: U.S. Government Printing Office.

U.S. Department of Labor. (2011, December). "Women in the Labor Force: A Databook." Available at http://www.bls.gov/cps/wlf-databook-2011.pdf, accessed August 8, 2012.

___. (2005). "Women in the Labor Force in 2004." Available at http://www.dol.gov/wb/factsheets/Qf-laborforce-04.htm, accessed November 18, 2005.

U.S. Department of Labor, Bureau of Labor Statistics. (2012). "Women in the Labor Force: A Databook." Available at http://www.bls.gov/cps/wlf-intro-2011.pdf.

___. (2012, April 19). "College Enrollment and Work Activity of 2011 High School Graduates." Available at http://www.bls.gov/news.release/hsgec.nr0.htm.

___. (2009). "Employment and Earnings, 2008 Annual Averages." Available at http://govguru.com/monthly-labor-review.

___. (2009). "Employment Status of Women and Men in 2008." Available at http://www.dol.gov/wb/factsheets/Qf-ESWM08.htm, accessed June 3, 2009.

___. (2009). *Occupational Outlook Handbook, 2008–2009*. Available at http://www.bls.gov/oco/oco2003.htm, accessed July 15, 2009.

___. (2007, November). "Employment and Earnings, 2008 Annual Averages and the Monthly Labor Review."

U.S. Department of Labor, Women's Bureau. (2011). "Women in the Labor Force in 2010." Available at http://www.dol.gov/wb/factsheets/Qf-laborforce-10.htm, accessed August 8, 2012.

___. (2009). "Women in the Labor Force in 2008." Available at http://www.dol.gov/wb/factsheets/Qf-laborforce-08.htm, accessed June 11, 2009.

U.S. Department of State. (2009). "The International Religious Freedom Report for 2008." Washington, DC: Office of International Religious Freedom.

___. (2002, October 7). "The International Religious Freedom Report for 2002." Washington, DC: Office of International Religious Freedom.

U.S. Federal Register. (2011, January 20). Vol. 76, No. 13, pp. 3637–3638, Washington, DC: U.S. Government Printing Office.

U.S. Office of Refugee Resettlement. (1995). "Annual Report to Congress."

Van de Kaa, Dirk J. (1987, March). "Europe's Second Demographic Transition." *Population Bulletin*. Population Reference Bureau.

van den Haag, Ernest. (1991). *Punishing Criminals: Concerning a Very Old and Painful Question.* New York: University Press of America.

___. (1986). "The Ultimate Punishment: A Defense." *Harvard Law Review*, pp. 124–128.

___. (1981). "Punishment as a Device for Controlling the Crime Rate." *Rutgers Law Review*, pp. 706, 719.

Van Lawick-Goodall, Jane. (1971). *In the Shadow of Man.* Boston: Houghton Mifflin.

Vartanian, L. R., M. B. Schwartz, K. D. Brownell. (2007). "Effects of Soft Drink Consumption on Nutrition and Health: A Systematic Review and Meta-Analysis." *American Journal of Public Health* 97, pp. 667–675.

Ventura, S. J. (2009). "Changing Patterns of Nonmarital Childbearing in the United States." NCHS data brief, no 18. Hyattsville, MD: National Center for Health Statistics. Available at http://www.cdc.gov/nchs/data/databriefs/db18.htm, accessed May 9, 2009.

Ventura, Stephanie J., and Christine A. Bachrach. (2000, October 18). "Nonmarital Childbearing in the United States, 1940–99." National Center for Health Statistics, *National Vital Statistics Reports* 48(16) (Revised). Available at http://www.cdc.gov/nchs/data/nvsr/nvsr48/nvs48_16.pdf, accessed August 15, 2012.

Verba, Sidney. (1972). *Small Groups and Political Behavior: A Study of Leadership.* Princeton, NJ: Princeton University Press.

Wade, Nicholas. (2007, October 19). "Neanderthals Had Important Speech Gene, DNA Evidence Shows." *New York Times.* Available at http://www.nytimes.com/2007/10/19/science/19speech-web.html?pagewanted=1&_r=1&ref=world.

Walker, T. B., and L. D. Elrod. (1993). "Family Law in the Fifty States: An Overview." *Family Law Quarterly* 26, pp. 319–421.

Wang, Wendy. (2012, February 16). Pew Research Center, "The Rise of Intermarriage: Rates, Characteristics Vary by Race and Gender." Available at http://www.pewsocialtrends.org/2012/02/16/the-rise-of-intermarriage/?src=prc-headline, accessed June 30, 2012.

Watson, J. B. (1925). *Behavior.* New York: Norton.

Weber, Max. (1968/1922). *Economy and Society* (Ephraim Fischoff, Trans.). New York: Bedminster Press.

___. (1958/1921). *The City.* New York: Collier.

___. (1957). *The Theory of Social and Economic Organization.* New York: Free Press.

___. (1956). "Some Consequences of Bureaucratization." In J. P. Mayer (Trans.), *Max Weber and German Politics*, 2nd ed. New York: Free Press.

___. (1930/1920). *The Protestant Ethic and the Spirit of Capitalism* (Talcott Parsons, Trans.). New York: Scribner's.

Wechsler, Henry, Jae Eun Lee, Meichun Kuo, Mark Seibring, Toben F. Nelson, and Hang Lee. (2002). "Trends in College Binge Drinking during a Period of Increased Prevention Efforts: Findings from 4 Harvard School of Public Health College Alcohol Study Surveys: 1993–2001." *Journal of American College Health* 50(5).

Weeks, John R. (1994). *Population*, 5th ed. Belmont, CA: Wadsworth.

West, Heather C., William J. Sabol, and Sarah J. Greenman. (2011, December). U.S. Department of Justice, Bureau of Justice Statistics, "Capital Punishment, 2010—Statistical Tables." NCJ 236510. Available at http://www.bjs.gov/content/pub/pdf/cp10st.pdf, accessed June 25, 2012.

___. (2010, December; rev. 2011, October 27). U.S. Department of Justice, Bureau of Justice Statistics, "Prisoners in 2009." NCJ 231675, revised. Available at http://bjs.ojp.usdoj.gov/content/pub/pdf/p09.pdf.

Westoff, Charles F., and Elise F. Jones. (1977, September/October). "The Secularization of Catholic Birth Control Practice." *Family Planning Perspectives* 9, pp. 96–101.

Whitehead, Barbara Dafoe, and David Popenoe. (2002). "Why Men Won't Commit: Exploring Young Men's Attitudes about Sex, Dating, and Marriage." *The State of Our Unions: The Social Health of Marriage in America.* Piscataway, NJ: Rutgers.

Whorf, B. (1956). *Language, Thought, and Reality.* Cambridge, MA: MIT Press.

Whyte, W. H. (1988). *City: Rediscovering the Center.* New York: Doubleday.

___. (1956). *The Organization Man.* New York: Simon & Schuster.

Whyte, William Foote. (1943). *Street Corner Society.* Chicago: University of Chicago Press.

Willner, Ruth Ann. (1984). *The Spellbinders: Charismatic Political Leadership.* New Haven, CT: Yale University Press.

Wilson, Barbara Foley, and S. C. Clarke. (1992, June). "Remarriages: A Demographic Profile." *Journal of Family Issues* 13, pp. 123–141.

Wilson, Doris, James, and Paula M. Ditton. (1999, January). "Truth in Sentencing in State Prisons." U.S. Department of Justice, Office of Justice Programs, Bureau of Justice Statistics. Special Report. NCJ 1700–32.

Wilson, Edmund. (1969). *The Dead Sea Scrolls 1947–1969*, rev. ed. London: W. H. Allen.

Wilson, Edward O. (1994). *Naturalist.* Washington, DC: Island Press.

___. (1979). *Sociobiology*, 2nd ed. Cambridge, MA: Belknap.

___. (1978). *On Human Nature.* Cambridge, MA: Harvard University Press.

___. (1975). *Sociobiology: The New Synthesis.* Cambridge, MA: Harvard University Press.

Wilson, G. Willow. (2005, May 29). "The Comfort of Strangers." *New York Times*, p. 22.

Wilson, James Q. (1996, March). "Against Homosexual Marriage." *Commentary.*

Wilson, James Q., and Richard J. Herrnstein. (1985). *Crime and Human Nature.* New York: Simon & Schuster.

Wilson, Woodrow. (1915, May 10). "Americanism." Speech given at Convention Hall, Philadelphia.

Winner, Ellen. (1996). *Gifted Children: Myths and Realities.* New York: Basic Books.

Wirth, Louis. (1944, March). "Race and Public Policy." *Scientific Monthly* 58(4), p. 303.

___. (1938). "Urbanism as a Way of Life." *American Journal of Sociology* 64, pp. 1–24.

Wiseman, Paul. (2002, November 19). "China Thrown off Balance as Boys Outnumber Girls." *USA Today.*

Woessman, Ludger. (2001, Summer). "Why Students in Some Countries Do Better." *Education Next.*

World Bank. (2012). "Birth Rate, Crude (per 1,000 People)." Available at http://data.worldbank.org/indicator/SP.DYN.CBRT.IN, accessed July 23, 2012.

___. (2009). "World Development Indicators 2008." Available at http://hdrstats.undp.org/indicators/24.html, accessed May 18, 2009.

___. (1984). "Measuring the Value of Children." In *World Development Report 1984.* New York: Oxford University Press.

World Health Organization. (2011, October). Fact Sheets, "Child Health." Available at http://www.who.int/topics/child_health/factsheets/en/index.html.

___. (2009, April). "World Development Report 2009." Washington, DC.

Worldwatch Institute. (2007). International Labour Organization, "The End of Child Labour: Within Reach," Geneva, 2006, pp. 112–113.

*Yearbook of Immigration Statistics: 2004.* (2005). U.S. Department of Homeland Security.

Yin, Sandra. (2007, November). "Gender Disparities in Health and Mortality." Population Reference Bureau. Available at http://prb.org/Articles/2007/genderdisparities.aspx, accessed June 7, 2009.

Yinger, J. Milton. (1970). *The Scientific Study of Religion*. New York: Macmillan.

Zalesky A., N. Solowij, M. Yücel, D. I. Lubman, M. Takagi, I. H. Harding, V. Lorenzetti, R. Wang, K. Searle, C. Pantelis, and M. Seal. (2012, June 4). "Source Effect of Long-Term Cannabis Use on Axonal Fibre Connectivity." *Brain*.

Zaslow, Martha J., and Kathryn Tout. (2002, April 8). "Child-Care Quality Matters." *American Prospect* 13(7).

Zeitlin, I. M. (1981). *Social Condition of Humanity*. New York: Oxford University Press.

Zimmer, Michael. (2010, January). "But the Data Is Already Public: On the Ethics of Research in Facebook." *Ethics Information Technology*, Special Issue on Human Subjects. Springer. Available at http://collections.lib.uwm.edu/cipr/image/592.pdf.

Zimring, Franklin E., and Gordon Hawkins. (1997). *Crime Is Not the Problem: Lethal Violence in America*. New York: Oxford University Press.

Zoll, Rachel. (2002, October 9). "U.S. Jewish Population Declining." Information about the 2000–2001 National Jewish Population Survey. Available at http://www.ujc.org/content-display.html?ArticleID 60346, accessed February 21, 2003.

# Index

Page numbers with "t" denote tables; those with "f" denote figures; and those with "b" denote boxes.

## A

AARP (American Association of Retired Persons), 372, 401
Abortion, 220
Abstract ideals, in religion, 312
Academic skills, education for, 336–337
Acculturation, 249
Achieved status, 110, 259
ACS (American Community Survey), 48
ACT (American College Test), 349
"Acting white," 88b
Adams, Donovan James, 164b
Adams, John, 241
Adaptation
　culture and, 61–64
　religious consequences, 315
Addams, Jane, 16–17
Addams, John, 16
Addiction, 7
Adolescent gender-role socialization, 269
Adult socialization, 91–94
Advertising, subliminal, 37
Affordable Care Act of 2010, 388
African Americans. *See also* Racial and ethnic minorities
　in contemporary US, 245–246
　as crime victims, 165
　disenfranchisement of, 17
　elected officials, 371f
　HIV/AIDS impact, 387
　infant mortality, 192
　life expectancy, 382, 393
　peer pressure, 88b
　political impact, 370–371
　poverty rates, 185f, 188
　unwed mothers, 279
AFT (American Federation of Teachers), 354
Age. *See also* Gender stratification; Health
　attitude toward, 94
　crime victimization, 165
　global, 221
　income distribution, 184
　living arrangements, 292
　poverty, 190f
　voting behavior, 369f
Aggravated assault, 158f, 159f
Aggression
　gender differences, 262
　serotonin levels, 148
　television and video games, 89–91
Agnosticism, 308, 321b. *See also* Religion
AIDS education, 87. *See also* HIV/AIDS
Alcoholism, 7, 90b, 130b
Alienation, 316
*All Religions Are True* (Gandhi), 307
Altruism, 75–77
Altruistic suicide, 14
Al-Wahhab, Muhammad Ibn Abd, 317b
Alzheimer, Alois, 392b
Alzheimer's disease, 263, 392b
American Anthropological Association, 67b, 77

American Association of Retired Persons (AARP), 372, 401
American Association of University Women, 277
American Bar Association, 294
American College Test (ACT), 349
American Community Survey (ACS), 48
*American Education: Making It Work* (1988), 337
*American Educator* magazine, 354
American Federation of Teachers (AFT), 354
American Heart Association, 338
American Institute of Philanthropy, 2
American Medical Association (AMA), 194, 372, 386, 388
*American Pastoral* (Roth), 132
American sign language (ASI), 63
American Sociological Association, 5b, 25
American Sociological Society, 335
American War of Independence, 367
Amnesty International USA, 169b
Amway Corporation, 117
Ancestor spirits, in religion, 311
Anglo-conformity, assimilation as, 236, 245
Animals, culture of, 63–64
Animal studies, 262–263
Animism, in religion, 311
Annihilation of racial and ethnic minorities, 239–240
Anomic suicide, 14
Anomie theory, 150, 315
Apocalyptic beliefs, 318
Aristotle, 259
Armstrong, Karen, 309b
Ascribed status, 110, 259
ASI (American sign language), 63
Asian Americans. *See also* Racial and ethnic minorities
　description, 249–251
　health, 383
　life expectancy, 382–383
　poverty rates, 185f
Assault, 158f, 159f
Assimilation, 236–237, 249
Association, statement of, 31
Association of Internet Researchers, 42b
Associations, 131–132
Atheism, 308, 309b, 317b, 321b. *See also* Religion
Atlas of Global Inequality, 226
Attachment disorder, 79
Austen, Jane, 282
Authority
　disdain for, 57
　political, 357–358
　religious legitimizing of, 314
Autocratic government, 359
Ayala, Don, 67b

## B

Baby stealing, 2
Baha'is (religion), 317b
Bainbridge, William, 310
Bales, Robert, 267
Barron's Profile of American Colleges, 338b
Batista, Fulgencio, 248

435

Behavior. *See also* Deviant behavior and social control
  collective, 114–118
  female gender role, 137
  group control of, 127–128
  instinctive, 75
  nonverbal, 105b, 106–107
  voting, 368–370
Behavioral theories of crime, 149
Beliefs, in religion, 308, 321b
Bell, Joshua, 101
Belsky, Jay, 86b
Benedict, Ruth, 59, 92
Berger, Peter, 114
Berger, Suzanne, 103
Berkshire Hathaway, 181
Beverly Hills Supper Club fire of 1977 (KY), 118
Bhatia, Michael, 67b
Bhutto, Benazir, 284
Bias, 37
Bierstadt, Robert, 50
Bilateral system, 282
Bilingual education, 345
Billings, John Shaw, 383
Bilocal residence rules, 283–284
Binge drinking, 27–28
Biocriminology, 148
Biological theories of deviance, 147–148
Biology, 74–82
  culture and, 51
  deprivation and development, 77–79
  human development, 80–82
  nature *versus* nurture, 75
  self concept, 79–80
  sociobiology, 75–77
Birth control, 210, 220, 326
Birthrates, 221b
Blank, Jonah, 51
Blau, Peter, 107
Blind investigators, 37
Blogs, 25, 270b, 277
Blue Cross and Blue Shield health insurance, 387
Blumenbach, Johann, 229
Blumer, Herbert, 118f
Bond, Charles, 105b
"Born criminal," 147
Boucicault, Dion, 232
Boundary definition, of groups, 126
Bourgeoisie, in capitalist society, 195
Breast-feeding, 217
Broken-windows theory of crime, 155b
*Brown v. Board of Education of Topeka, Kansas* (1954), 344
Brutality, police, 165
Buffett, Warren, 181
Bureaucracy, 16, 134–136
Bureau of Labor Statistics, 273b
Burgess, Ernest W., 16
Burglary, 158f, 159f
Burt, Cyril, 43b
Bush, George W., 61, 337
Busing, school, 345
Butler, Nicholas Murray, 17

## C

Calvinism, 314. *See also* Religion
Capitalism, 314–315, 361–364
Capital punishment, 169b–171b
Careers in sociology, 5b
Carson, Christopher ("Kit"), 239
Caste system of stratification, 205
Castro, Fidel, 248
Catholicism, 324–326
Causality, statement of, 31
Center for Education Reform, 354
Centers for Disease Control and Prevention (CDC), 389, 401
Central planning, in socialism, 364
Chambliss, William, 156
Charismatic authority, 357–358
Cherlin, Andrew, 279
Chicago school of sociologists, 16, 132
*Childhood and Society* (Erikson), 83
Childless couples, 291
Children. *See also* Education; Socialization and development
  abused, 292
  birthrates, 243f
  child labor laws, 17, 336
  custody laws, 296
  deprivation, 78–79
  in developing countries, 214–215
  gender-role socialization, 268–269
  infant mortality, 192, 208, 214t, 217
  poverty, 191b
  schools caring for, 339
  tax incentives to have, 222b
  unwed mothers, 279
Chinese Communist Party (CCP), 317b
Chinese revolution of 1949, 367
Choice, freedom of, in capitalism, 362
Choice, individual, 67–69
Christianity, 307, 309b, 317b. *See also* Religion
Clark University, 115
Class. *See* Social class
Class conflict, 12
Code of Hammurabi, 361
Cognitive development, 81
Cohabitation, 289–290, 298
Cohen, Dov, 113b
Coleman, James S., 344–345
Collective behavior, 114–118
Collective conscience, 133
Collective identity, 313
College Sport Research Institute, University of North Carolina-Southall, 27
College students, suicide of, 15b
Columbus, Christopher, 318–319
Common sense, in sociological perspective, 4–6
Communism, 314
Companionate marriage, 288
Competition
  freedom of, in capitalism, 362
  for jobs, 5b
  as social interaction, 108–109
  social stratification and, 194
Comte, Auguste, 9–10, 260
Conflict approach to laws, 156
Conflict theory
  education, 340–343
  gender stratification, 267–268
  Marx and, 13
  overview, 18–21
  political and economic systems, 360–361
  religion, 315–316
  role, 113–114
  social class, 195–197, 196t
  social interaction, 108

Consensus approach to law, 155
Constitutionalism, 360
Contemporary sociology, 20–21
Context of social interaction, 102–103
Contraception, 220
Control theory of crime, 151–152
Convictions, wrongful, 160b
Cooley, Charles Horton, 82, 125
Cooperation, 108
Cornerville study (Whyte), 33
Correctional officials, 166t
Corruption, 56, 135
Coser, Lewis, 107
Cost-of-living adjustment, 187
Countries of Particular Concern (CPCs), 317b
*Cours de Philosophie Positive* (Comte), 10–11
Courts, 166
Credentialized society, 342–343
Cressey, Donald R., 152
Crichton, Michael, 313
Crime in U.S.
   juvenile, 159–161
   property, 161–162
   social class and, 192–193
   statistics, 156–159
   victimless, 164–165
   victims of, 165
   violent, 161
   white-collar, 162–164
Criminal justice in U.S., 165–174
   courts, 166
   funnel effect, 172–173
   police, 165–166
   prisons, 166–172
   truth in sentencing, 173–174
*Crisis, The* (National Association for the Advancement of Colored People), 17
Critical theory, 21
Cross-cultural social interaction, 104b
Cross-sectional studies, 31–32
Crowds, 118f
Crowdsourcing, 136b
Crude birthrate, 208
Crude death rate, 208
Cuban Americans, 248–249. *See also* Hispanic Americans
Cuban revolution of 1959, 367
Cults, 91, 319b, 377
Cultural anthropology, 6
Cultural relativism, 52
Cultural transmission theory of crime, 152–153
Culture, 49–72. *See also* Racial and ethnic minorities; Social interaction
   adaptation and, 61–64
   components of, 54–59
   concept of, 50–53
   deprivation and development, 77–79
   distance zones, 101
   education to transmit, 336
   human development, 80–82
   individual choice and, 67–69
   job success and, 93b
   nature *versus* nurture, 75
   self concept, 79–80
   sociobiology, 75–77
   subcultures, 64–65
   symbolic nature of, 59–61
   universals in, 65–67

*Culture of Honor: The Psychology of Violence in the South* (Nisbett and Cohen), 113b
Culture shock, 51–52

### D

Dahrendorf, Ralf, 197
Dalai Lama, 317b
Dalit population (India), 205
Darnell Army Medical Center, Fort Hood, TX, 2
Darwin, Charles, 11, 75, 195
Data
   analysis of, 37–41
   collection of, 35–37
Davis, Kingsley, 194
Day care, 86b
*Dead Poets Society, The* (film), 25
Dead Sociologists' Society, 25
Death penalty, 169b–171b
De facto segregation, 344–345
De jure segregation, 344
Delegation of power, 134
Democratic government, 360
Democratic Party, U.S., 367–369
Democratic socialism, 364–365
Demographic transition theory of population growth, 211–212
Demography, 5b, 207–208
Denominations, religious, 318, 323
Dependency ratio, in population growth, 211
Dependent variables, 31
"Depressive pessimists," in groups, 128b
Deprivation and development, 77–79
Design, research, 31–35
Designated driver, 90b
Deterrent, capital punishment as, 169b
Development. *See* Socialization and development
Deviant behavior and social control, 142–179
   biological theories of, 147–148
   crime in U.S., 156–165
      juvenile, 159–161
      property, 161–162
      statistics, 156–159
      victimless, 164–165
      victims of, 165
      violent, 161
      white-collar, 162–164
   criminal justice in U.S., 165–174
      courts, 166
      funnel effect, 172–173
      police, 165–166
      prisons, 166–172
      truth in sentencing, 173–174
   defining, 143–145
   external means of control, 146–147
   internal means of control, 145
   law, 154–156
   psychological theories of, 148–150
   sociological theories of, 150–154
   subcultures, 65
DeWitt, John L., 250
Diamond, Milton, 76b
Differential association, theory of, 152–153
Diffusion, of cultural change, 62
Disclosure, 33, 43
Discrimination, 235–236, 270–273
Disease, gender and, 264t
Distance zones, 101
Diversion, 161

Division of labor, 65, 134
*Division of Labor in Society, The* (Durkheim), 14, 150
Divorce
  child custody, 296
  data on, 287b
  individualism and, 57
  laws on, 294–296
  rate of, 291f, 293–294
DNA evidence, 160b
Domestic abuse, 267f, 292–293
Donohue, John J., 170b
Doomsday beliefs, 318
Double-blind investigators, 37
Dowd, Maureen, 290b
Dramaturgy, 19, 105–106
Dropouts, high-school, 346
Drunk driving, 90b
Du Bois, W. E. B., 17–18
Durkheim, Émile, 13–14, 18, 34, 133, 144, 145f, 150, 307, 313–314, 333
Dyads, 129

## E

Early socialization, 84–91
  family, 85
  peer groups, 87–89
  school, 85–87
  television, movies, and video games, 89–91
Earned Income Tax Credit, 186
Ecclesia, in religion, 316–317
Eckard, E. Woodrow, 27
Economics, 7–8. *See also* Political and economic systems
Ectomorphic body types, 148
Ecumenism, 323
Education, 334–354
  conflict theory, 340–343
  early socialization and, 85–87
  functionist perspective, 335–340
  gender and, 272b
  health and, 385–386
  obesity and, 389f
  religious ties to, 321b
  U.S. issues
    English as second language, 345–346
    gifted students, 349–351
    high-school dropouts, 346
    home schooling, 347–349
    school violence, 346–347
    standardized testing, 349
    unequal access, 344–345
  voting behavior and, 369f
  women and, 290b
Educational Testing Service (ETS), 337, 349
*Ego*, in self, 83
Egoistic suicide, 14–15b
Einstein, Albert, 339
Ekman, Paul, 122
El-Badry, Samia, 328
*Elementary Forms of the Religious Life, The* (Durkheim), 14, 307
Elkind, David, 89
Ellison, Larry, 181
E-mail, symbols in, 60b
Emerson, Ralph Waldo, 319b
Emoticons, 60b
Emotion, in religion, 308
Empirical questions, 28
Empiricism, 6, 9

Employment. *See also* Education
  gender stratification in, 270–273
  job hunting, 132b
  lifelong, 93b
  qualifications for, 135
Endogamy, rule of, 283
Endomorphic body types, 148
Engels, Friedrich, 195, 268, 364
English as second language (ESL), 345–346
*Equality of Educational Opportunity* (Coleman), 344–345
Equal Opportunity Act, 386
Erikson, Erik, 82–84, 269
Erikson, Kai T., 118
ESL (English as second language), 345–346
Estate system of stratification, 205–206
Ethics, 42–44, 67b
Ethnic subcultures, 64–65. *See also* Racial and ethnic minorities
Ethnocentrism, 52
Ethnomethodology, 19, 103–105
Ethology, 262–263
Evaluation research, 5b
Exchange, as social interaction, 107
Executions, 170f
Exogamy, rule of, 283
Experiments, 33, 36t
Expressive leadership, 127
Expulsion of racial and ethnic minorities, 239
Extended families, 282
External social control, 146–147
Eye contact, 107

## F

Facebook, 20b, 42b, 114b, 132b, 181
Fads, 115–116
Family. *See also* Gender stratification; Marriage
  child custody laws, 296
  cultural universal of, 66
  divorce, 293–296
  early socialization, 85
  household size changes, 291–292
  income distribution, 183f, 184f
  marriage rate changes, 287–291
  overview, 280–282
  remarriage and stepfamilies, 297–298
  as social institution, 136–137
  violence, 292–293
  women in labor force, 292
Family Law Act of 1969 (CA), 294
Fashions, 115–116
Fashion Week, 34b
Fecundity, 208
Federal Bureau of Investigation (FBI), 34, 47, 157–158, 179
Felony crime, 157
Fertility
  breast-feeding and, 217
  decisions related to, 216t
  explanation of, 207–208
  gender preferences and, 218
Festinger, Leon, 33
Fingerprint identification, 160b
Fisher, Helen, 30
Folkways, 55–56, 233
Food stamps, 186
*Forbes* magazine, 181
Forced migration, 239
Forensic science, 160b
Formal, logical thought stage of development, 81

Formal sanctions, 146–147
Formal structure, of associations, 131
Fort Hood, TX, 2
Fox, Robin, 263
*Frankenstein* (Shelley), 62b
Franklin, Benjamin, 236, 336
Franklin and Marshall College, 115
Frank W. Ballou High School (Washington, DC), 88b
French revolution of 1790, 9, 367
Freud, Sigmund, 82–83, 149, 313
Functionalist perspective
　on education, 335–340
　on gender stratification, 267
　overview, 14, 18, 20
　on political and economic systems, 360–361
　on religion, 312–315
　on social class, 194–195, 196t
Fundamentalist religious views, 323–324
Funnel effect, in criminal justice, 172–173
Furstenburg, Frank, 298

### G

Gallup, George, 37
Gallup poll, 48, 90b, 162
Game stage, of self development, 83
Gandhi, Mohandas, 205, 307
Gangs, 347
Gans, Herbert, 33, 42
Gapminder, 226
Gardner, Allen, 63
Gardner, Beatrix, 63
Garfinkel, Harold, 19, 103–105
Gates, Bill, 181, 343b
Gay and lesbian marriage, 300–301
Gemeinschaft and gesellschaft, 132–133
Gender identity
　family impact, 85
　female, 137
　nature *versus* nurture, 76b
　self concept, 82
Gender preferences, fertility and, 218, 219b
Gender stratification, 258–277
　aging and, 392
　conflict theory viewpoint, 267–268
　employment, 270–273
　equality of sexes, 259–267
　　biological views, 262–265
　　historical views, 259–260
　　religious views, 260–262
　　sex perceptions, 265–266
　　sociological view, 266–267
　functionalist viewpoint, 267
　gender-role socialization, 268–269
　health, 380–382
General Electric Co., 372
Generalized others, 83
General Motors, 126
General Social Surveys, 323
Genetics, 74, 229–230
Genocide, 239
Geographic subcultures, 65
Ghost Dance (Native American), 315
Ghosts, in religion, 311
Gifted students, 349–351
Gilder, Jeannette, 261b
Gillie, Oliver, 43b
Globalization, 181

Global stratification, 202–226. *See also* Social class
　caste system, 205
　class system, 206
　diversity, 213–222
　　HIV/AIDS, 215–216
　　infant and child health, 214–215
　　population growth, 216–222
　　world health trends, 213–214
　estate system, 205–206
　Japan *versus* Nigeria, 204b
　population dynamics, 206–210
　population theories, 210–212
Goal setting, in groups, 127
Goffman, Erving, 19, 91, 105
Goldwater, Barry, 367
Good, Kenneth, 68b–69b
Gottfredson, Michael R., 151
Gould, Stephen Jay, 77
Government, 358–360, 362–363
Granovetter, Mark, 132b
Grant, Ulysses S., 372
Groups. *See also* Deviant behavior and social control
　cohesion, 13–14
　functions, 126–128
　nature, 124–126
　reference, 129–134
　religious unity, 313
　sociology focus, 3
　special-interest, 372–374
　subcultures, 64
Groupthink, 136b
Guatemala syphilis studies, 43b
Gumplowicz, Ludwig, 359
Gutenberg, Johannes, 62b
Gypsies, 240

### H

Halfway houses, 167
Hall, Edward T., 102, 106
Harlow, Harry F., 79
Harris, Marvin, 283, 315
Harrison, Benjamin, 238
Hartford Institute for Religion Research, 333
Harvard University, 17–18, 27–28, 42b, 64, 74, 77, 90b, 115, 257, 321b, 345
Health, 378–401. *See also* Age
　aging population and, 390
　　composition, 391–392
　　future trends, 395–397
　　gender, 392
　　global, 394–395
　　marital status, 393
　　racial minorities, 392–393
　　wealth, 393–394
　children in developing countries, 214–215
　health insurance, 387–388
　HIV/AIDS, 215–216, 387
　illness experience, 379–386
　　age and, 383–384
　　education and, 385–386
　　gender and, 380–382
　　race and, 382–383
　　social class and, 383
　　in U.S., 380
　　women in medicine, 386
　preventing illness, 388–390
　trends in, 213–214

Herrnstein, Richard, 149–150
Hidden curriculum, in schools, 340
Higher education, 272b, 321b. *See also* Education
Higher Education Research Institute, University of California at Los Angeles (UCLA), 56
High-school dropouts, 346
Hinduism, 307. *See also* Religion
Hirschi, Travis, 151
Hispanic Americans. *See also* Racial and ethnic minorities
 elected officials, 371f
 overview, 246–249
 political impact, 371
 poverty rates, 185f
History, 8
Hitler, Adolf, 239, 358
HIV/AIDS, 203, 215–216, 387
Holt, John, 348
Homelessness, 143f
Home schooling, 347–349
Homicide rates, 156–157
Homogamy, 284–285
Honor-killings, 72
Hooten, E. A., 147
Hoover Commission Report, 237
Household size, 280, 291–292
Housing subsidies, 186
Hsuan, Fu, 260
Hull House (Chicago, IL), 16–17
Human development, 80–82, 84t
Human Genome Project, 74
Human Terrain program, 67b
*Human Zoo, The* (Morris), 263
Humphreys, Laud, 43b
Hypotheses, research, 30–31
Hysteria, mass, 117–118

# I

*I*, in self, 83
Ibn Khaldun, 359
*Id*, in self, 83
Ideal norms, 56
Ideal type bureaucracy (Weber), 134
Ideology, in culture, 66–67
Illegal immigration, 243–244
Illiteracy, 341b
Immigration
 minority proportion in U.S., 392–393
 population declines and, 222b
 to U.S., 240–244
Immigration Reform and Control Act, 241f, 244
Immunizations, 209
Impact assessments, 5b
Impartiality, in bureaucracy, 135
*Imperial Animal, The* (Tiger and Fox), 263
Incest taboo, 66, 281
Income distribution
 African American, 246, 371
 aging, 393–394
 education, 343f, 347f
 racial, 186t
 rich and poor gap, 189b
 social class, 183–184
Independent variables, 31
Indiana Women's Prison, 172
Individualism
 crime as choice, 149–150

culture and, 67–69
divorce rate and, 57
in US *versus* Europe, 58b
Industrial Revolution, 9
Infant mortality
 diseases, 214–215
 global, 208–209, 214t
 high fertility and, 217
 Native American, 252
 social class, 192
Infectious diseases, 214
Informal sanctions, 146–147
Informal structure, of associations, 131–132
Informed consent, 43
Innovation, education for, 339
Innovations, cultural change from, 61
Innovators, in strain theory of crime, 151
Instinctive behavior, 75
Institutionalized political change, 365
Institutional prejudice and discrimination, 236
Institutions, 136–137
Instrumental leadership, 127
Integration, school, 344–345
Interactionist perspective, 19–20b
Interference, government, 362–363
Internal migration, 210
Internal social control, 145
International Court of Justice, 240
Internet
 education, 63
 ethics, 42b
 social interaction, 20b
 symbols, 60b
Interracial marriage, 231–233, 285
Interview schedules, 32
Intimate distance, 101
Iron law of oligarchy, 135
Islam, 309b, 317b, 327–329
Itard, Jean-Marc, 78

# J

Jacobson, L., 342
Jain, Anita, 53b
Jamison, Kay Redfield, 111–112
Jefferson, Thomas, 241, 260, 356
Jehovah's Witnesses, 317b
Jenkins, Florence Foster, 143
"Jerks," in groups, 128b
Job hunting, 5b, 132b, 339–340
John Hopkins University Center for Talented Youth, 351
John Jay College of Criminal Justice, 74
John Paul II (Pope), 326
Johns Hopkins Hospital, 76b
John XXIII (Pope), 325
Joint custody of children, 296
Journalism, sociology *versus*, 9b
*Journal of the American Medical Association (JAMA)*, 27
Judaism, 309b, 326–327
Judges, in courts, 166t
Juvenile crime, 152, 159–161

# K

Kaczynski, Ted, 62b
Kallen, Horace, 237
Kamikaze attacks (altruistic suicide, Japan), 14

Kennedy, John F., 358f
Kharijite Muslims, 328
Khomeini, Ayatollah Ruholla, 358
Kim, Jong-il, 377
Kinesics, 106
King, Martin Luther, Jr., 17
Klyamkin, Igor, 56
Koga, Kenta, 93b
Kohlberg, Lawrence, 81
Kozol, Jonathan, 338
Kroeber, Alfred A., 115
Kruger, Scott, 88
Kumar, Meira, 205

## L

Labeling theory of crime, 153–154
Labor
    child labor laws, 17, 336
    division of, 65–66, 134
    women in workforce, 292
Lag, cultural, 63
Laissez-faire capitalism, 363
Landon, Alfred E., 36–37
Language
    animal, 63–64
    culture and, 58–59
    gender usage differences, 270b
    origin of, 57–58
Larceny, 159f
Large groups, 131–132
Lasswell, Thomas, 64
Latent functions, 18
Laughing, 109b
Laws. *See also* Deviant behavior and social control
    child labor laws, 17, 336
    emergence of, 154–156
    racial and ethnic minorities defined by, 230
    state establishment of, 358
Lazarus, Emma, 241
Leaders, 126–127, 273b
Learning disabilities, 350
Left-Handers International, 125
Legal code, 154
Legal-rational authority, 357
Lemert, Edwin, 153, 161
Lenin, V. I., 358, 361, 364
Lenski, Gerhard, 238
Lesbian marriage, 300–301
Letter bombs, 62b
Lewontin, Richard, 77
Liars, 105b
*Lie to Me* (TV show), 122
Life expectancy, 208–209, 380, 381b, 384b
Lifelong employment, 93b
Limbaugh, Rush, 117
Lincoln, Abraham, 16, 209
LinkedIn, 20b, 132b
Linton, Ralph, 29–30
Linton, Robert, 110
Lipset, Seymour Martin, 57
*Literary Digest* magazine, 36–37
Liz Claiborne, Inc., 117
Lobbying, 372–373
Lombroso, Cesare, 147
*London Sunday Times,* 43b
*Lonely Crowd, The* (Riesman), 89

Longitudinal research, 32
Looking-glass self, 82
Loving, Mildred, 285
Loving, Richard P., 285
Lower class, 183
Lower-middle class, 182–183
Loyd, Paula, 67b

## M

Magic, in religion, 310
Magistrates, in courts, 166t
Malaya, Oxana, 78
Malinowski, Bronislaw, 281, 310
Malnutrition, 214–215
Malone, Paul, 228
Malone, Philip, 228
Malthus, Thomas, 210, 214f
Mana (Melanesian/Polynesian concept), 311
Manifest functions, 18
Mann, Horace, 342
Manners, 55
Marginal groups, 64
Margin of error, 39b
Marijuana use, 164b, 385b
Marital residence rules, 283–284
Market research, 5b
Marriage, 278–305. *See also* Family
    aging population and, 393
    arranged, 53b
    child custody laws, 296
    cultural universal of, 66
    defining, 282–287
    divorce, 293–296
    early, 217
    family life, 280–282
    future of, 301
    gay and lesbian, 300–301
    household size changes, 291–292
    interracial, 231–232
    length of, 7b
    marriage rate changes, 287–291
    patterns in, 4
    remarriage and stepfamilies, 297–298
    single-parent families *versus,* 299–300
    single population *versus,* 298–299
    violence, 292–293
    women in labor force, 292
Marshall, Thurgood, 285
Martineau, Harriet, 10–11
Marx, Karl, 12–13, 16, 195, 210–211, 315–316, 363–364
Massachusetts Institute of Technology (MIT), 88
Mass hysteria, 117–118
Mass media, 89
Master status, 110
Material culture, 54
Matriarchal families, 282
Matrilineal system, 282, 327
Matrilocal residence rules, 283–284
Matza, David, 152
McCain, John, 368
McCord, Linda Fay, 229
McCowan, Karen Stone, 350
McDonald's Corp., 117
McFate, Montgomery, 67b
McGovern, George, 368
McKay, Henry, 152

McMoon, Cosme, 143
*Me,* in self, 83
Mead, George Herbert, 19, 82–83
Mead, Margaret, 77, 301, 335
Mean, calculating, 38
Means-tested government assistance, 188
Mears, Ashley, 34b–35b
Mechanically integrated society, 133
Media, 359, 371–372
Median, calculating, 38–39
Mediascope, 90
Medicaid, 186, 387–388
Medicalization, 379
Medicare, 387–388
Medicine. *See* Health
Mednick, Sarnoff, 148
Melvin, Edith, 261b
Mendel, Gregor, 339
Mercy Medical Center (Springfield, MA), 309
Merton, Robert K., 18, 21, 134, 150, 235
Mesomorphic body types, 148
Mexican Americans, 248. *See also* Hispanic Americans
Mexican revolution of 1910, 367
Michels, Robert, 135
Microsoft Corporation, 181
Middle-middle class, 182
Middle-range theories, 21
Migration, 209–210, 239
Milgram, Stanley, 43b
Military, sociology work for, 67b
Millenarian movements, 318–319
Mills, C. Wright, 4
Misdemeanor crime, 157
Mixed economy, 363
Mobility, social, 205–206
Mode, calculating, 39
Money, John, 76b
Monogamous marriage, 283
Monotheism, 312
Moore, Wilbert, 194
Moral code, 143
Moral development, 81
Mores, 55, 233
Morgan, Thomas Jefferson, 238
Mormon Church, 313, 319b. *See also* Religion
Morris, Desmond, 263
Mortality, 208
Moskos, Peter, 74
Mother Theresa, 196
Motorola Nippon, 93b
Motor vehicle theft, 158f, 159f
Movies, early socialization and, 89–91
Mswati III of Swaziland, 52
Multiple marriages, 283
Multiracial ancestry, 230–231
Murdock, George P., 280–281
Murray, Charles, 343b
Muslim population, 327t. *See also* Religion

## N

Nagel, Mark, 27
Naqshabandiya order of Sufi Islam, 319b
National Association for the Advancement of Colored People (NAACP), 17, 228
National Center for Education Statistics, 354
National Center for Health Statistics, 34
National Center for Missing and Abducted Children, 2
National Center for Missing and Exploited Children (NCMEC), 2
National Collegiate Athletic Association (NCAA), 27
National Commission on Excellence in Education, 336
National Crime Victimization Survey (NCVS), 47, 158
National Education Association, 341
National Geographic, 69b
National Incident-Based Reporting System (NIBRS), in Uniform Crime Reports (UCR), Federal Bureau of Investigation (FBI), 157–158
National Institute of Justice, 171
National Institutes of Health (NIH), 263, 379
National Opinion Research Center, 323
*Nation at Risk, A* (1983), 336–337
Nation's Report Card (1988), 337
Native Americans, 237, 251–252, 382
Nature *versus* nurture, 75
Negative reference groups, 129
Negative sanctions, 146–147
Neolocal residence rules, 283–284
Neo-Marxism, 21
Network of Concerned Anthropologists, 67b
Neutralization, techniques of, 152
*Newer Ideals of Peace* (Addams), 17
*New York Times,* 201, 257, 290b
New York University, 34b
*1984* (Orwell), 129
Nisbett, Richard E., 113b
Nixon, Richard, 368
Nobel Prize, 17, 41, 196
No Child Left Behind Act of 2002, 337–338, 346, 350, 354
No-fault divorce laws, 294–296
Nonmaterial culture, 54–55
Nonverbal behavior, 105b, 106–107
Norms
   group, 124, 130b
   law and, 155
   marriage, 283
   overview, 55–56
   social interaction in, 103
   state establishment of, 358
Nuclear family, 282
Nurture, nature *versus,* 75

## O

Oaksterdam University, 385b
Obama, Barack, 88b, 110, 233, 338, 368–371, 388
*Obedience to Authority* (Milgram), 43b
Obesity epidemic, 362b, 388–389, 390b
Objectivity, research, 41–42
Observation of participants, 33–34b, 36t
Occupational subcultures, 65
*Octoroon, The* (Boucicault), 232
Ogburn, William F., 63
Older Americans Act, 190
Oligarchy, 135
*On the Origin of Species* (Darwin), 75
Open-ended interviews, 32
Operational definitions, 29
Operational stage of development, 81
Opinion leaders, 117
Oracle Corporation, 181
Organically integrated societies, 133, 150
Organization of the Islamic Conference, 317b
Organizations, 137
Origin, families of, 283

Orshansky, Mollie, 186
Orwell, George, 129
"Other directed" (peer pressure), 89
Overpopulation, 211
Oxford University (UK), 16

# P

Pacifism, 17
PACs (political action committees), 373
Panic, 117–118
Pape, Robert, 366b
Paradigms, 18
Parcells, Bill, 313
Parenthood, 92
Park, Robert E., 16
Paroling authorities, 166t
Parsons, Talcott, 18, 267, 379
Participant observations, 33–34b, 36t
Patriarchal families, 282
Patriarchal ideology, 261
Patrilineal system, 282
Patrilocal residence rules, 283–284
Patterson, Francine, 64
Pavlov, Ivan, 75
PBS (Public Broadcasting System), 201
Peer groups, socialization and, 87–89
Pentecostalists, 319b
Personal contacts, in job hunting, 132b
Personal distance, 101
Personality, 74
Personality cult, 377
Petraeus, David, 67b
Pew Hispanic Center, 247b
Pew Research Center for the People & the Press, 38b
Peyote, in Native American religion, 308
Phipps, Susie Guillory, 230
Piaget, Jean, 81
Pinker, Steven, 64
Plato, 170b, 361
Play stage, of self development, 83
Pliny the Younger, 319b
Pluralism, for racial and ethnic minorities, 237–238
Pluralism Project, 333
Poe, Edgar Allan, 62b
Police, 158f, 165–166
Political action committees (PACs), 373
Political and economic systems, 355–377
    economy, 361–365
    functionalist and conflict theory perspectives, 360–361
    government, 358–360
    political change, 365–367
    power and authority, 357–358
    in U.S., 367–374
        African American impact, 370–371
        Hispanic impact, 371
        media role, 371–372
        special-interest groups, 372–374
        two-party system, 367–368
        voting behavior, 368–370
Political revolutions, 367
Political science, 8
Political subcultures, 65
Polling, 36–38b, 47–48
Polyandrous families, 282
Polygamous families, 282
Polygyny, 283

Polytheism, in religion, 312
Population
    declining, 221b–222b
    dynamics of, 206–210
    growth of, 216–222
    Japan *versus* Nigeria, 204b
    Muslim, 327t
    theories of, 210–212
Positive checks on population, 210
Positive reference groups, 129
Positive sanctions, 146–147
Postmodernism approach, 21
Poststructuralism approach, 21
Poverty, 185–191
    changing face of, 190–191
    counting the poor, 186–187
    feminization of, 185–186
    global, 203f
    government assistance, 188–190
    health issues, 383
    myths about, 187–188
Power, 134, 357–358
Prayer, religious, 308
Prejudice, 234–236
Prejudiced discriminators, 235
Prejudiced nondiscriminators, 235
Preparatory stage, of self development, 83
Preventive checks on population, 210
*Pride and Prejudice* (Austen), 282
Primary deviance, 154
Primary groups, 125–127
Primary socialization, 91
*Principles of Biology* (Spencer), 11
*Principles of Ethics, The* (Spencer), 11
*Principles of Psychology* (Spencer), 11
*Principles of Sociology* (Spencer), 11
Prisons, 166–172
    capital punishment, 169b–171b
    goals of, 167
    incarceration rates, 168b
    shortage of, 171–172
    women in, 172
Private property, in capitalism, 362
Private sphere, 135
Probable cause, 74
Problem definition, in research, 28–29
Procreation, families of, 283
Procter & Gamble, Inc., 116–117
Profane things, 307. *See also* Religion
Programme for International Student Assessment (PISA), 338
Proletariat, in capitalist society, 195
Propaganda, 117
Property, private, in capitalism, 362
Property crime, 157, 158f, 159f, 161–162
Prophets, in religion, 308
Prosecutors, in courts, 166t
*Protestant Ethic and Spirit of Capitalism, The* (Weber), 16, 314
Protestantism, 319b, 324
Psychoanalytic theory of crime, 149
Psychological theories of deviance, 148–150
Psychology, 6–7
Public distance, 101
Public opinion, 5b, 117
Public property ownership, in socialism, 364
Public sphere, 135
Puerto Ricans, 248. *See also* Hispanic Americans
Puritanism, 319b

## Q

Quality-of-life issues, 155b
Questionnaires, 32
Quinney, Richard, 156

## R

Racial and ethnic minorities, 227–257
   African Americans, 245–246
   aging population and, 392–393
   annihilation, 239–240
   Asian Americans, 249–251
   assimilation, 236–237
   discrimination, 235
   elderly proportion, 393f
   ethnic group concept, 233
   expulsion, 239
   genetic definitions, 229–230
   health, 382–383
   high-school dropout rates, 346
   Hispanics, 246–249
   immigration to U.S., 240–244
   institutional prejudice and discrimination, 236
   legal definitions, 230
   minority group concept, 233–234
   Native Americans, 251–252
   pluralism, 237–238
   prejudice, 234–235
   segregation, 239
   social definitions, 230–233
   subjugation, 238
   U.S. ethnic composition today, 244–252
   voting behavior, 369–370
   white Anglo-Saxon Protestants (WASPs), 245
Rape/sexual assault, 158f, 159f
Reagan, Ronald, 240, 250, 294
*Real Education* (Murray), 343b
Real norms, 56
Reardon, Sean, 338b
Rebellions, for political change, 365–366
Rebels, in strain theory of crime, 151
Recidivism, 161
Reciprocal power, of social networks, 132b
Red Cloud (Native American), 357
Reference groups, 129–134
Reformulation, 62
Regional culture, 113b
Rehabilitation of criminals, 167
Reimer, David, 76b
Reliability, of studies, 41
Religion, 306–333
   abstract ideals, 312
   animism, 311
   cults, 91
   denominations, 318
   ecclesia, 316–317
   ecumenism, 323
   elements of, 308–310
   magic, 310
   marriage and, 286
   Marx *versus* Weber on, 16
   millenarian movements, 318–319
   monotheism, 312
   sects, 318
   secularism, 321–322
   sexual equality, 260–262
   sociological perspective, 312–316
      conflict theory, 315–316
      functionalist, 312–315
   subcultures, 65
   supernaturalism, 311
   theism, 312
   universal church, 316
   in U.S.
      Catholicism, 324–326
      diversity of, 319–321
      Islam, 327–329
      Judaism, 326–327
      overview, 323–324
      Protestantism, 324
      social aspects, 329
Remarriage, 297–298
Reporting research, 41
Representative government, 360
Representative samples, 35
Republican Party, U.S., 367
Researcher bias, 37
Research methods, 26–48
   ethical issues in, 42–44
   objectivity in, 41–42
   process, 28–41
      problem definition, 28–29
      review of previous research, 29–30
      hypotheses development, 30–31
      research design, 31–35
      sample definition and data collection, 35–37
      data analysis and conclusions, 37–41
      reporting, 41
"Reservations" affirmative action (India), 205
Resocialization, 91, 93
Retreatists, in strain theory of crime, 151
Reverse assimilation, 237
Revitalization movements, in religion, 315
Revolutions, 366–367
Ridener, Larry, 25
Riesman, David, 89
Rites of passage, 66
Ritualists, in strain theory of crime, 151
Rituals, religious, 307–308
Robbery, 158f, 159f
Rockford Female Seminary (Illinois), 16
Rodin, Auguste, 55
Roentgen, Wilhelm, 41
Role conflict, 113–114
Role playing, 114
Roles, 110–112, 379
Role sets, 112
Role strain, 112–113
Romney, Mitt, 356
Roosevelt, Franklin D., 36, 250, 373
Roosevelt, Theodore, 250
Roper organization polls, 90b
Rosenthal, R., 342
Rotating question order, 39b
Roth, Philip, 132
Rubin, Zick, 29
Rubinstein, Moshe, 52
Rules and regulations, 134–135
*Rules of the Sociological Method* (Durkheim), 13
Rumors, 116–117
Russian revolution of 1917, 367

## S

Sacred things, 307. *See also* Religion
Saffin, John, 314
Saint-Simon, Henri, 9
Same-sex marriage, 300–301
Sampling, 35–37, 39b
Sanctions, 146–147
Sapir-Whorf hypothesis, 58
Savage-Rumbaugh, Sue, 64
Scholastic Aptitude Test (SAT), 349
School of Public Health, Harvard University, 27–28
Schools. *See* Education
Scientific method, 6
Secondary analysis, 33–34, 36t
Secondary deviance, 154
Secondary groups, 125–127
Sects, 318
Secularism, 321–322
Segregation, 239, 344
Selectivity of culture, 58
Self concept, 79–80
Semistructured interviews, 32
Sensorimotor stage of development, 81
Seppuku (altruistic suicide, Japan), 14
Serotonin, 148, 263
Settlement houses, 16
Sex change, surgical, 76b
Sex Offender Registry, FBI, 179
Sexual equality, 259–267
   biological views, 262–265
   historical views, 259–260
   religious views, 260–262
   sex perceptions, 265–266
   sociological view, 266–267
Shakespeare, William, 28
"Shallow cover," in observations, 34b
Shanidar Cave, Iraq, 310
Shaw, Clifford, 152
Shays, Daniel, 366
Shays's Rebellion, 365–366
Sheldon, William H., 147–148
Shelley, Mary, 62b
Shiite Muslims, 328
Shils, Edward, 360
Sick role, 379
Significant others, 83
Simmel, Georg, 115–116, 129
Simple assault, 158f
Single-parent families, 299–300
Single population, marriage *versus*, 298–299
Sitting Bull (Native American), 357
"Slackers," in groups, 128b
Slavery, 17, 314
Small groups, 129–130
Smith, Adam, 203, 361–362
Smith, Goldwin, 261b
Smith, Joseph, 313
Smoking, 192b, 265b, 362b, 386
Social action, 102
Social aggregate, 124
Social attachments, 79
Social class, 180–201. *See also* Global stratification
   caste and estate system *versus*, 206
   conflict theory of, 195–197
   education to equalize, 342
   functionalist theory of, 194–195
   health, 383
   marriage and, 286
   poverty, 185–191
      changing face of, 190–191
      counting the poor, 186–187
      feminization of, 185–186
      government assistance, 188–190
      myths about, 187–188
   religious differences, 324, 326
   stratification consequences, 192–194
   structure of, 181–184
   subcultures, 65
Social control, 13, 358. *See also* Deviant behavior and social control
Social Darwinism, 11–12, 195
Social distance, 101
Social functions, 18
Social groups and organizations, 123–141
   bureaucracy, 134–136
   group functions, 126–128
   group nature, 124–126
   reference groups, 129–134
   social institutions, 136–137
   social organizations, 137
Social identity, 80
Social interaction, 100–122
   collective behavior, 114–118
   competition, 108–109
   conflict, 108
   context, 102–103
   cooperation, 108
   dramaturgy, 105–106
   ethnomethodology, 103–105
   exchange, 107
   nonverbal behavior, 106–107
   norms, 103
   role conflict, 113–114
   role playing, 114
   roles, 110–112
   role sets, 112
   role strain, 112–113
   statuses, 109–110
Socialism, 364. *See also* Political and economic systems
Socialization and development, 73–99
   adult, 91–94
   aging, 94
   biology and culture, 74–82
      deprivation and development, 77–79
      human development, 80–82
      nature *versus* nurture, 75
      self concept, 79–80
      sociobiology, 75–77
   crime and, 149
   early, 84–91
      family, 85
      peer groups, 87–89
      school, 85–87
      television, movies, and video games, 89–91
   education for, 335–336, 340
   gender-role, 268–269
   longevity and, 265b
   theories of, 82–84
Social psychology, 6–7
Social revolutions, 367
Social science, sociology as, 6–9
Social Security Administration, 186

Social solidarity, 133
*Social System, The* (Parsons), 18
*Social Theory and Social Structure* (Merton), 18
Social work, 8
Society for the Advancement of Women's Health Research, 263
Sociobiology, 75–77, 262
*Sociobiology: The New Synthesis* (Wilson), 75
Sociobiology Study Group, 77
Sociological Images blog, 277
Sociological imagination, 4
Sociological perspective, 1–25
    development of, 9–18
        Comte, August, 9–10
        Durkheim, Émile, 13–14
        Martineau, Harriet, 10–11
        Marx, Karl, 12–13
        Spencer, Herbert, 11–12
        in United States, 16–18
        Weber, Max, 14–16
    point of view, 3–9
        common sense, 4–6
        social science, 6–9
        sociological imagination, 4
    theoretical, 18–21
Sociological theories of deviance, 150–154
Socrates, 361
Solecki, Ralph, 310
Souder, Susan, M., 160b
*Souls of Black Folk, The* (Du Bois), 17
Special-interest groups, politics and, 372–374
Spencer, Herbert, 11–12, 335
Spitz, Rene, 79
Splinter groups, 129
Standardized testing, in education, 349
Stanford University, 43b
Stanford University Education Program for Gifted Youth, 351
Star, Ellen Gates, 16
Stark, Rodney, 310
State functions, 358–359. *See also* Political and economic systems
Statement of association, 31
Statement of causality, 31
Statistics, crime, 156–159
Statuses
    formally defined, 134
    gender stratification, 259
    social interaction, 109–110
    in socialization, 79
Stealing babies, 2
Stebbins, Mark Linton, 228, 256
Stepfamilies, 297–298
Stevens, John Paul, 170b
Stradivari, Antonio, 101
Strain theory of crime, 150–151
Stratification, 181, 192–194. *See also* Gender stratification
*Street Corner Society* (Whyte), 33
Stress response, 264–265
Structural functionalism, 18
*Structure of Social Action, The* (Parsons), 18
Study of Early Child Care and Youth Development, 86b
*Study of Sociology, The* (Spencer), 11
Subcultures, 64–65, 233
Subjugation of racial and ethnic minorities, 238
Subliminal advertising, 37
Substance Abuse and Mental Health Service Administration, 385b
Sufi Muslims, 317b, 319b
Suicide
    college students, 15b
    Durkheim on, 34
    Native American, 252, 382
    social control and, 13
    terrorism by, 366b
Sunni Muslims, 317b, 328
*Superego,* in self, 83
Supernaturalism, in religion, 311
Super PACs (political action committees), 374
Superstition, 310
*Suppression of the African Slave-Trade to the United States, The* (Du Bois), 17
Surveys
    advantages and disadvantages, 36t
    bogus, 38b
    in research design, 31–32
    sociology and, 5b
Survival of the fittest (Darwin), 11
Sutherland, Edwin H., 152, 162, 163
Suveges, Nicole, 67b
Sykes, Gresham, 152
Symbolic interactionist perspective, 19–20
Symbolic nature of culture, 59–64

## T

Tables, reading, 40b
Taboos, 66, 281, 311
Taliban, 319b
Tannen, Deborah, 270b, 276–277
Task assignment, in groups, 127
Taxes on sugared beverages, 362b
Tearoom Trade Observation (Humphreys), 43b
Techniques of neutralization, 152
Technology, cultural change from, 62b
Television, early socialization and, 89–91
Terrorism, suicide, 14, 366b
Texting, 114b
Theft, 158f, 159f
Theism, 312
Theoretical sociological perspectives, 18–21
*Theory and Practice of Society in America* (Martineau), 10
Thomas, Lewis, 379
Thomas, W. I., 16, 82
Thompson, Warren, 211
Thoreau, Henry David, 62b
Tiger, Lionel, 263
Tinker, Grant, 90b
Tischler, Henry L., 178, 305, 333
Tocqueville, Alexis de, 101, 168b
*Today's Education* journal, 341
Tönnies, Ferdinand, 132–133
Total fertility rate (TFR), 208, 221b
Total institutions, 91
Totalitarian government, 359–360
Totemism, in religion, 313–314. *See also* Religion
Toynbee Hall (UK), 16
Tracking systems, in education, 341–342
Traditional authority, 357
Trajan (Roman Emperor), 319b
Translation, problems in, 59
Triads, 129
Truman, Harry S., 240
Truth in sentencing, 173–174
Tulshyan, Ruchika, 53b
Tuskegee syphilis studies, 43b
Twain, Mark, 106b
*Twelfth Night* (Shakespeare), 28

*Twin studies* (Burt), 43b
Twitter, 20b
Two-party political system, in U.S., 367–368
Tylor, Edward, 50

## U

UNICEF, 217
Uniform Crime Reports (UCR), Federal Bureau of Investigation (FBI), 47, 157–158
United Nations, 239–240, 385b
United States Navy, 136b
Universal church, 316
University of Bordeaux (France), 13
University of California at Los Angeles (UCLA), 56
University of California at Santa Cruz, 226
University of Chicago, 152
University of Kansas, 76b
University of Nevada, 63
University of North Carolina-Southall, 27
University of Wisconsin, 115
Unprejudiced discriminators, 235
Unprejudiced nondiscriminators, 235
Unwed mothers, 279
Upper class, 181
Upper-middle class, 182
*Urban Villagers, The* (Gans), 33
U.S. Armed Forces, 141
U.S. Bureau of Justice Statistics (BJS), 164b
U.S. Bureau of Labor Statistics, 5b
U.S. Census, 25, 32, 34, 48, 183, 187, 189b, 231t, 247b
U.S. Chamber of Commerce, 372
U.S. Department of Education, 345
U.S. Department of Labor, 34
U.S. Immigration and Nationality Act of 1965, 243b
U.S. Public Health Service, 43b, 115
U.S. Supreme Court, 232, 285, 344, 373
*USA Today*, 47

## V

Validity, of studies, 39
Values
    in collective conscience, 133
    law and, 155
    overview, 56–57
    U.S. *versus* European, 58b
Variables, 30–31
V-chip device, for television, 90
Vicary, James, 37
Victimless crime in U.S., 164–165
Victims of crime in U.S., 165
Video games, early socialization and, 89–91
Violence
    family, 292–293
    mass media and, 89
    school, 346–347
Violent crime, 156, 158f, 159f, 161, 163b
Voting behavior in U.S., 368–370

## W

*Walden* (Thoreau), 62b
War, 67b, 359
Ward, Lester Frank, 335
*War of the Worlds* (Wells), 117
War on Poverty, 187

Washington, George, 241
WASPs (White Anglo-Saxon Protestants), 245
Water, clean, 203, 209, 214. *See also* Global stratification
Watson, John B., 75
Weak ties (informal structure), 132b
*Wealth of Nations, The* (Smith), 362
Weber, Max, 14–16, 18, 41, 102, 134, 195–196, 314–315, 357–358, 361
Welles, Orson, 117
Wells, H. G., 117
Wendell, Barrett, 237
White, Ralph Lee, 228
White, Walter, 228
White Anglo-Saxon Protestants (WASPs), 245
White-collar crime in U.S., 162–164
*Why Children Fail* (Holt), 348
Whyte, William Foote, 33
Williams, Jason, 101
Williams, Robin, 25
Wilson, Edward O., 75–77
Wilson, James Q., 149–150
Wilson, Woodrow, 236
Winfrey, Oprah, 117
Winner, Ellen, 350
Winsten, Jay, 90b
Wirth, Louis, 233, 234b
Witchcraft trials, 118
Wolfers, 170b
Women. *See also* Gender stratification; Global stratification
    aging, 392
    education of, 220
    Jewish rabbis, 326
    in labor force, 292
    life expectancy, 380–382, 392
    Martineau on treatment of, 11
    in medicine, 386
    poverty and, 185–186
    in prisons, 172
    in state legislatures, 370f
Women's suffrage, 261b
Work and Family Equity Index, 305
*World Christian Encyclopedia*, 319b
World Economic Forum, 266
World Health Organization (WHO), 213, 215, 217
World Values Survey, 72
Worldviews, religion and, 314
World Wide Web, 63
Wovoda (Native American), 315
Writing, gender differences in, 270b
Wrongful convictions, 160b

## X

XYY chromosome theory, 148

## Y

Yale University, 93b
*You Just Don't Understand* (Tannen), 270b, 276
*Youngest Science, The* (Thomas), 379
YouTube, 305

## Z

Zerhouni, Elias, 379
Zimbardo, Philip, 43b
Zoroastrians, 317b
Zuckerberg, Mark, 181

# Practice Tests

## CHAPTER 1   THE SOCIOLOGICAL PERSPECTIVE

### Sociology as a Point of View

1. **T F** According to Tischler, the social science most closely related to sociology is cultural anthropology.
2. **T F** A sociologist studying steroid use among high-school students would first want to know whether such behavior was actually increasing or decreasing before formulating an explanation of why it was occurring.
3. **T F** The term "the sociological imagination" refers to the use of sociology to solve complex social issues.
4. **T F** The physical sciences (physics, chemistry, and so on) use the scientific method. Social sciences use unscientific, social methods.
5. **T F** To say that an argument is not empirical is to say that it is not based on evidence.
6. Fear and anxiety of infant abduction in the United States have reached epidemic proportions. In almost three decades of research, of 108 million babies born, there were a total of _____ attempts to steal infants.
    a. approximately 5,800
    b. approximately 7,500
    c. less than 300
    d. more than 100,000
7. Although sociology and psychology have similarities, they differ in that psychology usually emphasizes
    a. human society.
    b. individual motivation and behavior.
    c. the differences among different cultures.
    d. the operations of government.
8. According to Tischler, people confuse sociological theorems with
    a. attempts to influence social policy.
    b. attempts to justify research.
    c. psychology.
    d. philosophy.
9. Who developed the concept of the sociological imagination?
    a. Auguste Comte
    b. Émile Durkheim
    c. Wright Mills
    d. W. E. B. DuBois
    e. Karl Marx
10. Systematically arranged knowledge demonstrating the operation of general laws is
    a. general laws production.
    b. science.
    c. philosophy.
    d. problem solving.
11. Which of the following is *not* a component of the scientific method?
    a. experimentation
    b. generalization
    c. verification
    d. personalization
    e. observation
12. The main focus of sociology is
    a. the individual.
    b. the group.
    c. humanity as a whole.
    d. all of the above.

13. What is the main difference between sociology and social work?
    a. Sociology uses theory and social work does not.
    b. Social work overlaps with psychology whereas sociology does not.
    c. Social workers help people solve their problems while sociologists try to understand why the problems exist.
    d. There really is no difference between sociology and social work.

## ● The Development of Sociology

14. T F The term *sociology* dates back to the philosophers of ancient Greece.
15. T F Sociology emerged as a separate field of study during the Renaissance.
16. T F Teaching sociology at the university level typically requires a PhD.
17. T F Comte thought sociology would move society toward perfection.
18. T F The doctrine of social Darwinism calls on the government to intervene actively to help the less fortunate in society.
19. T F Most people who enter the field of sociology become field researchers.
20. T F Karl Marx thought the most important fact about determining a person's social class was the lifestyle a person chose.
21. T F Max Weber and Karl Marx were in close agreement in their explanations of the role of religion in society.
22. T F Durkheim's research on suicide relied on several in-depth studies of individuals who had killed themselves.
23. T F Max Weber maintained that a system of beliefs he called "the Protestant ethic" enabled the development of capitalism in places where the ethic existed.
24. T F Whereas Marx believed that socialism and communism would ultimately bring an end to capitalistic exploitation, Weber believed bureaucracy would characterize both socialist and capitalist societies.
25. T F Weber predicted that religion would lead to a gradual turning away from science as the main mode of accounting for what was occurring in the real world.
26. T F Suicides resulting from a sense of feeling disconnected from society's values are termed *anomic suicides*.
27. T F Latent functions are the unintended or not readily recognized consequences of social processes.
28. T F The application of sociological knowledge to the corporate world is fairly limited.
29. T F Functionalists view society as a system of highly interrelated parts that generally operate together harmoniously.
30. Sociology is typically classified as a _____ science.
    a. physical
    b. research
    c. social
    d. sociocultural
31. According to Durkheim, the most important factor in suicide rates was
    a. biochemical imbalances in the brain.
    b. depression.
    c. social solidarity.
    d. economic well-being.

32. The process by which generalizations can be validated is
    a. research.
    b. pragmatism.
    c. speculation.
    d. empiricism.

33. Which of the following correctly reflects the ideas of Auguste Comte?
    a. He preferred "armchair philosophy" to empirical data.
    b. He believed all societies move through fixed stages of development toward perfection.
    c. He thought suicide was an aspect of the most advanced stage of society.
    d. He separated the analysis of society into latent and manifest functions.
    e. He predicted that the proletariat would rise to overthrow the capitalists.

34. Which of the following is true of the work of Harriet Martineau?
    a. She translated into English Comte's major work.
    b. Her major research involved observing day-to-day life in the United States.
    c. She compared social stratification systems in Europe with those in America.
    d. These are all true of the work of Harriet Martineau.

35. Durkheim identified three types of suicide. These are based on
    a. the personality of the individual who might commit suicide.
    b. the particular method (gun, pills, and so on) of committing suicide.
    c. social solidarity.
    d. the age of the individual.

36. A man in a sinking life raft jumps overboard into shark-infested waters in order to lighten the load and provide his comrades with a greater chance of survival. The man is demonstrating which type of suicide?
    a. anomic
    b. egoistic
    c. altruistic
    d. humanitarian

37. Durkheim would say that Kamikaze pilots and suicide bombers were products of a society
    a. with a very high level of social solidarity.
    b. with a very low level of social solidarity.
    c. utterly lacking in social solidarity.
    d. with an ideology of evil and hate.

38. Many people who commit suicide in the United States today are depressed and isolated from others. They represent which type of suicide?
    a. anomic
    b. egoistic
    c. altruistic

39. Which aspect of U.S. society would Durkheim see as contributing to egoistic suicide?
    a. the high level of patriotism
    b. the high standard of living
    c. the emphasis on individualism and self-reliance
    d. the gap between rich and poor

40. Which of the following represents Marx's view of the law?
    a. The law is impartial and serves the interests of the entire society.
    b. The law is created by the wealthy, and it protects their interests from the rest of society.
    c. The law is the main tool the poor can use to protect themselves from exploitation by the wealthy.
    d. The law represents the ideas of the most intelligent and most fair-minded people in the society.

41. According to Marx, all history is the history of
    a. class struggle.
    b. technological progress.
    c. intellectual progress.
    d. religious morality.

42. According to Max Weber, what was the relation between Protestantism and capitalism?
    a. The ideas of religion created an atmosphere in which capitalism could flourish.
    b. Capitalists promoted Protestantism as a justifying ideology for their economic pursuits.
    c. Protestant ideas served as a restraint on the excesses of capitalism.
    d. Protestant leaders accepted large donations from successful capitalists.
    e. The rise of capitalism and Protestantism in the same historical period was mostly a coincidence.

43. Statistically speaking, which couple is likely to have the longest marriage?
    a. the Madisons from Indiana
    b. the Bensons from Maryland
    c. the Coles form Florida
    d. the Lodges from Nevada

44. Sociology and psychology overlap in the field of
    a. social psychology.
    b. social anthropology.
    c. sociobiology.
    d. synecology.

## Theoretical Perspectives

45. T F A paradigm is an overall framework that shapes the questions a sociologist is likely to ask about a social issue.

46. T F Symbolic interactionists usually base their studies on data from national surveys.

47. T F Sociologists who are influenced by Marx's ideas are most likely to emphasize the conflict theory paradigm.

48. A sociologist studied higher education in the United States to find out how well universities trained students to do important and necessary work that benefited the society. This sociologist was working mainly from
    a. the functionalist paradigm.
    b. the conflict theory paradigm.
    c. the social interactionist paradigm.

49. A sociologist studied higher education to see how students tried in subtle ways to influence teachers into giving less work and better grades. This sociologist was working mainly from
    a. the functionalist paradigm.
    b. the conflict theory paradigm.
    c. the symbolic interactionist paradigm.

50. An auto executive once proclaimed, "What's good for the country is good for General Motors and vice versa." This notion that GM and the entire United States have identical interests is most likely to be supported by sociologists using
    a. the functionalist paradigm.
    b. the conflict theory paradigm.
    c. the symbolic interactionist paradigm.

51. Egoistic suicide comes from
    a. overinvolvement with others.
    b. a general uncertainty from norm confusion.
    c. overall feelings of depression resulting from economic setbacks.
    d. low group solidarity and underinvolvement with others.

## Essay Question

52. How might each of the main sociological paradigms be used to look at the sociology class you are now in? What aspects of the class would each paradigm be most interested in, and how would each go about exploring those questions?

# CHAPTER 2   DOING SOCIOLOGY: RESEARCH METHODS

## The Research Process

1. **T F** An empirical question is a question posed in such a way that it can be studied through observation.
2. **T F** A testable statement about the relationship between two or more empirical variables is known as a hypothesis.
3. **T F** The research problems of greatest interest to sociologists are most easily and accurately studied by means of controlled experiments.
4. **T F** The mean is the number that occurs most often in a dataset.
5. **T F** The phenomenon studied by the researchers and the one they are trying to explain is called the independent variable.
6. **T F** A self-fulfilling prophecy is produced when a researcher who is strongly inclined toward a particular point of view communicates that attitude to the research subjects so that their responses end up consistent with the initial point of view.
7. **T F** A structured interview is a form of research conversation in which the interviewer follows the questionnaire rigidly.
8. **T F** Validity is the extent to which a study tests what it was intended to test.
9. **T F** The reason forecasters inaccurately predicted the winner of the 1936 Landon-Roosevelt presidential election was that their sample was too small.
10. **T F** Secondary analysis is useful for collecting historical and longitudinal data.
11. **T F** The number of nonrespondents in a study has the capacity to skew the results.
12. **T F** Surveys in which participation is mandatory produce more accurate results than their counterparts in which participation is voluntary.
13. **T F** Sociologists try to provide answers to two general questions: "Why did it happen?" and "Under what circumstances is it likely to happen again?"
14. **T F** The first step in the research process is to develop one or more hypotheses.
15. **T F** A specific statement about an abstract concept in terms of the observable features that describe the thing being studied is called a hypothesis.
16. **T F** An operational definition is not required if the term being studied is widely understood.
17. **T F** In setting up tables, it is often all right to omit the headings for rows and columns.
18. **T F** An empirical question can be answered by observing the world as it is known.
19. **T F** A statement of association says that one thing causes another.
20. **T F** A variable that changes another variable is a dependent variable.
21. **T F** Survey research is not as effective as interviewing or participant observation for allowing the researcher to understand the feelings and attitudes of the respondents.
22. **T F** Participant observation studies are usually less objective and less replicable than are studies that use survey research.
23. **T F** The three measures of central tendency always give identical results.
24. **T F** The median is preferable to the mean when there are one or two scores that are far away from the other scores of the sample.

25. Your score on the SAT the second time you took it was 80 points higher than the first time you took it. This difference is relevant for what aspect of the SAT?
    a. validity
    b. reliability
    c. central tendency
    d. double-blind investigation

26. A subset of the population that exhibits, in correct proportion, the significant characteristics of the population as a whole is known as a
    a. dependent variable.
    b. nonrandom sample.
    c. representative sample.
    d. cross-sectional sample.

27. The figure that falls in the exact middle of a ranked series of scores is the
    a. median.
    b. mean.
    c. mode.
    d. average.

28. A longitudinal study is research
    a. aimed at predicting the future.
    b. investigating a population over a period of time.
    c. examining a population at a given point in time.
    d. conducted without the participants' knowledge.

29. Which of the following steps in the research process must come last?
    a. developing hypotheses
    b. reviewing previous research
    c. defining the problem
    d. determining research design
    e. analyzing the data and drawing a conclusion

30. Which of the following is a measure of central tendency that is commonly referred to as the average?
    a. median
    b. mean
    c. mode
    d. meridian

31. Which of the following research methods is the most subjective?
    a. surveys
    b. controlled experiments
    c. participant observation
    d. secondary analysis

32. Which of the following is a statement of causality?
    a. Rural areas have fewer services than do urban areas.
    b. This sociology course is difficult.
    c. Poverty produces low self-esteem.
    d. Mean income in New York is higher than in Florida.

33. Researchers ride with police in unmarked police cars to collect data on drug dealers the police encounter. They are using the _____ method of research.
    a. longitudinal survey
    b. participant observation
    c. laboratory experiment
    d. semistructured interview

34. Consultants for political candidates often use polls to judge how their candidates' positions are being perceived by the public. This method of research is closest to
    a. survey.
    b. interview.
    c. experiment.
    d. participant observation.

35. Consultants for political candidates sometimes use focus groups to judge how their candidates' positions are being perceived by the public. In a focus group, a small number of people—a dozen or so—are questioned and encouraged to talk freely about specific issues. This method of research is closest to
    a. survey.
    b. interview.
    c. experiment.
    d. participant observation.

36. What element of a statistical table should tell you the topic the data in the table are all about?
    a. footnotes
    b. column headings
    c. headnotes
    d. title
    e. source

37. Some critics of IQ tests claim that lower-class people do poorly not because they are less intelligent but because they have not been exposed to the kinds of cultural things that the tests ask about. The critics are claiming that as a test of intelligence, IQ tests are not
    a. valid.
    b. reliable.
    c. blind.
    d. double-blind.

38. Using the following quiz scores, 2, 4, 6, 7, 7, 10, calculate the mode.
    a. 4
    b. 7
    c. 5
    d. 6
    e. 8

39. Bill Gates walks into your classroom. What statistic about the income of the population in the classroom has been most affected?
    a. the mean
    b. the median
    c. the mode
    d. the reliability

40. A researcher publishes findings that several other researchers doing similar research cannot replicate. The first researcher's data therefore were not
    a. valid.
    b. reliable.
    c. scientific.
    d. objective.

41. People are selected from a group in such a way that every person has the same chance of being selected. The people who are selected make up what type of sample?
    a. representative
    b. cross-sectional
    c. random
    d. unintentional
    e. stratified

42. In a study of a student population at a university, researchers selected a preset number of students from each major so that these reflected the same proportion of students in that major at the university. Their technique for selecting students was
    a. secondary analysis.
    b. random sample.
    c. stratified random sample.
    d. double-blind investigation.

## Objectivity and Ethical Issues in Sociological Research

43. T F Complete objectivity in research may be impossible to achieve, but it is still a reasonable goal to strive for.

44. T F A researcher decides to investigate binge-drinking. The researcher can be relatively certain that the definition of "binge-drinking" is standard from location to location.

45. T F Social scientists regard deception of research participants as a necessary evil in most social research.

46. T F One advantage of social research is that it frequently benefits the research subjects directly and immediately.

47. T F Most subjects of sociological research belong to groups with little or no power.

## Essay Questions

48. Drawing on your reading, discuss the techniques you would use to determine whether an opinion poll is bogus.

49. Discuss three important criteria that a person could use to determine whether the results of a research study were accurate.

50. Generate a hypothesis about a current social issue. Identify the dependent and independent variables and indicate whether the hypothesis is a statement of causality or association.

# CHAPTER 3   CULTURE

## The Concept of Culture

1. T F  Human beings, like most other species, pass a wide variety of behavioral patterns from one generation to the next through their genes.
2. T F  Social scientists have identified some societies that, for all practical purposes, do not have a culture.
3. T F  Genetically determined behavior is innate; culturally determined behavior is learned or acquired.
4. T F  As of the writing this edition (2012), scientists still had not discovered any genes related to language.
5. T F  The culture of a group is generally shared by all group members rather than being a matter of individual preference.
6. If we based our study of another society on the principle of cultural relativism, we would
   a. pay close attention to how people treated their relatives.
   b. try to see whether that society was superior to others we had studied.
   c. try to understand that society on its own terms and withhold moral judgment.
   d. look at the art, music, and other high culture to see whether it was as good as ours.
7. The behavior of most nonhuman animal species is largely determined by
   a. norms.
   b. culture.
   c. instinct.
   d. mores.
8. Anthropologist Kenneth Good, seeing a Yanomoma woman about to be raped—a commonplace event—says, "I stood there, my heart pounding," uncertain what to do. This response is an example of
   a. cultural lag.
   b. culture shock.
   c. instinct.
   d. material culture.
9. Some Chinese journalists visiting the United States for the first time were horrified to be served a steak—meat in one huge slab just sitting on a plate rather than cut up into small pieces and mixed with vegetables. They concluded that Americans were barbaric in their food preferences. Their reaction is an example of
   a. ethnocentrism.
   b. cultural innovation.
   c. cultural relativism.
   d. reformulation.
10. An example of nonmaterial American culture is
    a. patriotism.
    b. Big Macs.
    c. MacBook computers.
    d. Mack trucks.

## Components of Culture

11. T F  Ideal norms are expectations of what people should do under perfect conditions.
12. T F  The Sapir-Whorf hypothesis suggests language affects how humans perceive the world around them.
13. T F  Real norms are strongly held rules of behavior that have a genuine moral connotation.
14. T F  Culture is transmitted to children by DNA and other genetic material.
15. T F  The same value in a culture can produce both positive and negative behaviors.
16. Folkways and mores are both examples of
    a. norms.
    b. values.
    c. beliefs.
    d. ethnocentrism.
    e. material culture.

17. The view that "time is money" is an example of
    a. the material aspect of culture.
    b. a universal shared by all cultures.
    c. cultural relativism.
    d. a nonmaterial aspect of culture.

18. The gift of a clock is a sign of respect in the United States; in China, it is considered bad luck. This difference in ideas about clocks shows that the two countries differ in their
    a. nonmaterial culture.
    b. material culture.
    c. ethnocentrism.
    d. cultural lag.

19. It is considered polite to send regrets when you are invited to a party but cannot attend. Failure to do so, however, is not considered a serious moral lapse. This rule about RSVPs is the type of norm called
    a. an Emily Post modernism.
    b. a taboo.
    c. a folkway.
    d. a subculture.

20. In American culture, such things as freedom, individualism, and equal opportunity are deemed to be highly desirable. In sociological terms, these concepts are
    a. values.
    b. folkways.
    c. cultural universals.
    d. norms.

21. The Hopi use the same word for everything that flies except birds (planes, flying insects, and so on). According to the Sapir-Whorf hypothesis, this means that
    a. the Hopi language is generally inferior to English.
    b. Hopi are less likely to perceive differences between planes and helicopters than are most Americans.
    c. Hopi are afraid of flying.
    d. the Hopi have much to teach us about flying.

22. The study of U.S. and Saudi values showed that these two cultures differed in the importance they place on all the following *except*
    a. religion.
    b. modesty.
    c. freedom.
    d. individualism.

23. The American Anthropological Association termed the Human Terrain Program
    a. insightful.
    b. a new beginning in understanding cultures.
    c. violates the group's code of ethics.
    d. dangerous and reckless.

## The Symbolic Nature of Culture

24. T F A gesture such as a "thumbs up" is a symbol.
25. T F Within a culture, there is a wide variation among people as to what most symbols mean.
26. T F Some simple gestures, such as a nod of the head to mean "yes," are universal and always have the same meaning in all cultures and societies.
27. T F Tischler's overall conclusion regarding culture is that all aspects of culture—nonmaterial and material—are symbolic.

## Culture and Adaptation

28. T F Cultural adaptation is a response to changes in the environment.
29. T F Cultures that are isolated are unlikely to experience cultural diffusion.
30. T F The fewer cultural items in a society's inventory, the less likely it is to welcome cultural innovation.
31. T F Chimpanzees cannot speak, but they can learn simple sign language.
32. T F Although chimpanzees can recognize icons on a computer, they cannot make new combinations of these icons to express their desires.

33. Humans are superior to other animals in the area of
    a. surviving in natural environments.
    b. using symbols.
    c. physical strength.
    d. physical agility.
    e. self-preservation.

34. Many hallucinogenic drugs that were once used for religious ceremonies by indigenous peoples are now used and abused as recreational drugs. This is an example of
    a. ethnocentrism.
    b. cultural relativism.
    c. cultural reformulation.
    d. cultural lag.

35. Although cell phones have been around for more than a decade, the norms regarding how, when, and where they should be used are still in dispute. The failure of society to formulate norms in response to technological change is an example of
    a. innovation.
    b. diffusion.
    c. cultural lag.
    d. ethnocentrism.

## Subcultures

36. T F Subcultures are the ways of life of smaller groups within the larger society.

37. T F An immigrant group that retains many of the ways of the old country constitutes a deviant subculture.

38. T F Subcultures always represent a threat to the dominant culture's major values.

39. Novelist F. Scott Fitzgerald once commented that "the rich really are different from you and me." He was implying that the rich are
    a. ethnocentric.
    b. a deviant subculture.
    c. a product of innovation.
    d. a social class subculture.

40. Capital-murder cases are among the most divisive types of court cases. Why do prosecutors and defense attorneys typically question potential jurors in a very in-depth manner?
    a. to assess cultural backgrounds of potential jurors
    b. to rule out cultural lag
    c. to assess the possibility of membership in certain subcultures and the effect it might have on the decision-making process
    d. to assess the beliefs and adherence to norms, values, and mores of potential jurors

41. _____ percent of French citizens believe that religion is a very important factor in their lives.
    a. 25
    b. 60
    c. 13
    d. 77

## Universals of Culture

42. T F About 20 percent of cultures throughout the world do not have an incest taboo.

43. T F Although all cultures have marriage as an institution, the specific form of marriage and the rules of marriage vary greatly.

44. T F Chimpanzees and other primates are generally more cooperative and less self-sufficient than humans.

45. T F Even the simplest human societies, such as the Yanomami, have a division of labor that is much more complex than that of groups of the higher-order primates (apes and chimpanzees).

46. According to Tischler, the most important function of the incest taboo is that it
    a. prevents immoral behavior.
    b. protects against genetic mutation and disease caused by inbreeding.
    c. expands social bonds outside the family, creating larger, stronger groups.
    d. prevents conflict and jealousy within the nuclear family.

47. Confirmations, weddings, and funerals are examples of
    a. division of labor.
    b. rites of passage.
    c. cultural diffusion.
    d. cultural innovation.

48. The belief that America is the "land of opportunity for all" is part of the American
    a. ideology.
    b. division of labor.
    c. social class subculture.
    d. material culture.

49. The conflict between being a researcher and being a human being is basically the conflict between
    a. cultural innovation and cultural lag.
    b. cultural relativism and ethnocentrism.
    c. norms and values.
    d. mores and folkways.

### Essay Questions

50. Tischler says that humans inevitably feel somewhat dissatisfied, no matter to which group they belong. What is the logic of his argument? Suppose someone (perhaps even you) says, "If I could be a _____ [or "If I could be a member of the _____ group], I would never feel dissatisfied." Why might that person's prediction turn out to be wrong?

51. Think of some subculture of which you are or have been a member—a high-school clique, a student subculture at your college, a group at work. How do the norms and values of this subculture differ from those of the dominant institution in which the subculture exists? How does the group try to reinforce its cultural ideas when these are in conflict with those of the institution?

52. A German immigrant to the United States is approached by an organization that assists the needy. According to the text, what is the likelihood that the immigrant will become involved in supporting the goals of the organization?

## CHAPTER 4   SOCIALIZATION AND DEVELOPMENT

### Becoming a Person: Biology and Culture

1. **T F** Most of our bodily characteristics, such as how tall we will be, are completely controlled by our genes and uninfluenced by factors in the environment.

2. **T F** Sociobiologists such as Edward O. Wilson believe that human behavior can be understood as continuing attempts to ensure the transmission of one's genes to a new generation.

3. **T F** The concept of instinct is more important than culture for biological explanations of behavior.

4. **T F** Behavioral psychologists such as John B. Watson believed that almost all human behavior was controlled by principles of conditioning.

5. **T F** Although human babies raised in isolation suffer serious and permanent social deficiencies, young monkeys raised without mothers show no lasting negative effects of this deprivation.

6. **T F** Attachment disorder results in children being unable to form relationships.

7. **T F** Overall, the fields of academic sociology and anthropology readily accepted the new ideas offered by theorists in sociobiology.

8. Among the great apes (chimps, gorillas), mothers have sole responsibility for rearing the young. But among humans, "it takes a village to raise a child." The principal reason for the difference is that
   a. humans are physically weaker than apes.
   b. human babies have a much longer period of dependency.
   c. humans must follow laws regarding child neglect.
   d. humans aren't as good at childrearing so it takes more of them to do it.

9. Behavioral psychologists such as John B. Watson, who emphasize "conditioning," believe that most human behavior is controlled by
   a. biochemical factors.
   b. genetic makeup.
   c. forces outside the individual.
   d. God.

10. The debate about nature versus nurture is basically a debate about the relative importance of
    a. the theories of Darwin and E. O. Wilson.
    b. biology and culture.
    c. personality and nourishment.
    d. peer group and school.

11. Stephen Jay Gould's critique of sociobiology maintained that human behavior was
    a. almost entirely controlled by genes.
    b. a product of evolution just like physical characteristics.
    c. completely unaffected by genes.
    d. limited by genetics but shaped by culture.

12. Harlow studied monkeys raised in isolation, with no contact with other monkeys. When these monkeys were placed among other monkeys, they
    a. eventually learned to interact like the others.
    b. sought out a single other monkey to play with.
    c. could interact only to mate with monkeys of the opposite sex.
    d. never learned to interact normally with other monkeys.

13. Children raised with little contact with others often suffer a psychological condition known as
    a. attachment disorder.
    b. identity crisis.
    c. id.
    d. operant conditioning.

14. **T F** The case of David Reimer clearly supports the ideas of John Money that gender identity is almost entirely under the influence of social forces and that biogenetic makeup counts for very little.

15. Stephen J. Gould was a leading critic of
    a. moral development.
    b. Freudian psychology.
    c. socialization.
    d. sociobiology.

16. The ideas of sociobiology are inspired mostly by the theories of
    a. Mead.
    b. Cooley.
    c. Darwin.
    d. Freud.

## ● Theories of Development

17. **T F** According to Piaget, the stage when the infant relies on touch and the manipulation of objects for information about the world is called the operational stage.

18. **T F** Moral order is a society's shared view of right and wrong.

19. **T F** According to Kohlberg, concepts such as good and bad, right and wrong, once established, carry the same meaning for us throughout our lives.

20. **T F** Piaget was more interested in the development of logical thought than in the development of moral reasoning.

21. **T F** In Kohlberg's theory, the highest stage of moral development is an orientation toward the law exactly as it is written.

22. **T F** According to Kohlberg's research, once a person has achieved a higher level of moral development, he or she never regresses to a lower level of moral reasoning.

23. **T F** In Freudian theory, the part of the psyche of which we are most conscious is the superego.

24. **T F** In Freudian theory, the ego is the repository of the thoughts and feelings that we are not aware of.

25. **T F** According to Mead, the portion of the self that is made up of everything learned through the socialization process is called the "me."

26. **T F** In Mead's theory, the term *significant other* refers to romantic partners but not parents or friends.

27. **T F** The terms *personal identity* and *social identity* mean essentially the same thing.

28. In Mead's terms, a significant other is
    a. a romantic partner.
    b. a representative of society in general.
    c. anyone important to a person's development.
    d. the "looking-glass self."

29. In Mead's theory, the play of children was
    a. an important source of the formation of the self.
    b. a temporary release from the demands of forming a self.
    c. a time when children often acquired the wrong kinds of self.
    d. irrelevant to the development of self.

30. Which of the following is *not* one of the components of the looking-glass self, according to Cooley?
    a. our imagination of what we must really be like
    b. our imagination of how our actions appear to others
    c. our imagination of how others judge our actions
    d. a self-judgment in reaction to the imagined judgments of others

31. Each individual's social identity consists of
    a. all the behaviors he or she has learned to imitate.
    b. his or her changing yet enduring view of himself or herself.
    c. the sum total of all the statuses he or she occupies.
    d. all the ways other people view him or her.

32. Social identity, unlike personal identity, does not involve a person's
    a. ideas.
    b. occupation.
    c. sex.
    d. statuses.

33. Students were asked why it was wrong to plagiarize a paper. Which response represents the *lowest* level on Kohlberg's scale of moral reasoning?
    a. because I'd be looked down on by my family
    b. because the teacher would fail me for the course if I got caught
    c. because it wouldn't be fair to others who wrote their own papers
    d. because I want my grade to represent my own abilities

34. Which of the following theorists coined the term *identity crisis*?
    a. Pavlov
    b. Freud
    c. Erikson
    d. Mead

35. Erikson thought identity crisis occurred most frequently among
    a. children.
    b. adolescents.
    c. middle-aged people.
    d. the elderly.

## Early Socialization in American Society

36. **T F** Young people who feel ignored by their parents seem to be more vulnerable to peer pressure.

37. **T F** The majority of American families fit the traditional model, with the husband as wage earner outside the home and the wife as homemaker.

38. T F According to research cited by Tischler, peer groups of black high-school students encourage their members to try to get better grades than whites.

39. T F In most communities in the United States, schools are havens isolated from the conflicting values of the wider society.

40. Tischler writes that "as the authority of the family diminishes under the pressures of social change, peer groups move into the vacuum and substitute their own morality for that of the parents." The youth who changes his or her morality in this way is undergoing a process of
    a. attachment disorder.
    b. cognitive development.
    c. resocialization.
    d. preparatory stage.

41. Preschool-age children of parents who work are most frequently cared for by
    a. day-care centers.
    b. relatives.
    c. babysitters, nannies, or other nonrelatives.
    d. nobody; they stay home alone.

42. The studies on day care cited by Tischler show that when kids go on to elementary school, compared with those who were raised mostly at home, those who attended day-care centers were
    a. more disobedient.
    b. more cooperative.
    c. more athletic.
    d. more depressed.

43. The process of social interaction that teaches a child the intellectual, physical, and social skills needed to function as a member of society is called
    a. identification.
    b. social adjustment.
    c. socialization.
    d. social conditioning.

44. Studies cited by Tischler on the effects of television support the idea that
    a. violent TV shows are responsible for much real violence.
    b. watching nonviolent shows such as *Sesame Street* and *Barney* makes children less violent.
    c. regardless of the content of the shows they watch, children who watch more television are more likely to be violent than are children who watch less TV.
    d. there is no relationship between television and real-life acts of violence.

## ● Adult Socialization

45. T F Resocialization often involves exposure to people with ideas and values different from those a person was raised with.

46. T F Marriage is an important event in adult socialization.

47. T F Total institutions are environments in which people are physically and socially isolated from the outside world.

48. T F Studies of mating strategies show that men and women are looking for the same qualities in their romantic and sexual partners.

49. The structure of total institutions often makes them very effective places for
    a. cognitive development.
    b. evolutionary psychology.
    c. attachment disorder.
    d. resocialization.

50. An older man was required to retire. Then his wife died; his children were grown and no longer needed his parenting. These facts were most important for his
    a. social identity.
    b. moral reasoning.
    c. cognitive development.
    d. superego.

51. The results of using television to promote the concept of the designated driver illustrate the idea that television is
    a. an agent of socialization.
    b. a way to sell more beer.
    c. an ineffective way of trying to reduce drunk driving.
    d. a major cause of violence in society.

● **Essay Questions**

52. Total institutions often attempt to resocialize adults. In what ways do the resocialization practices of these institutions resemble the ordinary practices of childhood socialization?

53. This chapter outlines different theories to explain moral behavior. In these theories, some forces that produce moral behavior are unconscious or out of the person's usual awareness; other forces for moral behavior are more visible to the person. How does each kind of influence—visible or invisible—work, and which is more important?

54. Adam worked for eight years as a logger in a mill in Oregon. He has decided to apply for a job in the world of advertising. Adam shows up for his initial interview heavily bearded and in jeans, work boots, and a lumberjack-style shirt. As Adam's career counselor, what types of resocialization might he be in need of?

# CHAPTER 5    SOCIAL INTERACTION

● **Understanding Social Interaction**

1. **T F** Human behavior is not random. It is patterned and, for the most part, predictable.
2. **T F** In his "dramaturgical" approach to sociology, Goffman was interested in how frequently people of different cultures attended the theater.
3. **T F** Until the norms of social distance are broken, most people are unaware that these rules exist even though they abide by them.
4. **T F** When students acted like boarders in their own homes, family members knew them quite well, so the family members were rarely upset by the students' behavior.
5. **T F** All people in a social setting are participants in the sense that their behavior sends a message, even if they remain absolutely silent.
6. To an observer, two people engaging in a fistfight on the street would mean something entirely different from the same two people fighting in a boxing ring. This illustrates how the meaning of social interaction is dependent on
   a. personalities.
   b. competition.
   c. status.
   d. context.
7. In a large lecture class, it is generally expected that students will raise their hands and be called upon before speaking. This illustrates the operation of
   a. bureaucracy.
   b. norms.
   c. roles.
   d. statuses.
8. The process of two or more people taking each other's actions into account is called
   a. social accounting.
   b. social organization.
   c. social interaction.
   d. social action.
9. A white teacher is talking to an African American child. The child looks away. From the child's perspective, according to the text, looking away is a sign of
   a. shyness.
   b. embarrassment.
   c. respect.
   d. disrespect.

● **Types of Social Interaction**

10. **T F** When people do something for each other with the express purpose of receiving something in return, they are engaged in exchange.

11. **T F** A gesture, such as waving hello or the thumbs up to indicate approval, mean the same thing in nearly all cultures the world over.

12. **T F** Coercion is a form of conflict in which one of the parties is much stronger than the other and can impose its will on the weaker party.

13. **T F** Competition is a form of conflict within agreed-upon rules.

14. **T F** A successful society is one in which conflict has been eliminated.

15. **T F** Relations within a small group are based on either cooperation or conflict, never both.

16. Staring, smiling, nodding one's head, and using one's hands while talking are all examples of
    a. nonverbal communication.
    b. instinctive behavior.
    c. cooperation.
    d. ethnomethodology.

17. "I'm certainly not going to watch Marge's kids this afternoon. She didn't even bring them to Jason's birthday party last week." The person speaking is thinking of her relation to Marge in terms of
    a. competition.
    b. conflict.
    c. cooperation.
    d. exchange.

18. The main difference between cooperation and exchange is that
    a. cooperation is based upon shared goals, exchange on individual goals.
    b. cooperative relationships are voluntarily; exchange relationships are not.
    c. cooperation has no material awards; exchange always has material rewards.
    d. exchange relationships are almost always successful and long-lived; cooperative relationships are short and rarely successful.

19. Using power in a way that is not considered legitimate by those on whom it is used is called
    a. conflict.
    b. competition.
    c. coercion.
    d. comportment.

20. A sociologist was interested in impostors and how they managed to create the impression that they were doctors or lawyers or pilots or other things they were not. This sociologist's perspective is that of
    a. dramaturgy.
    b. conflict.
    c. exchange.
    d. competition.

## ● Elements of Social Interaction

21. **T F** Socially defined positions, such as teacher, student, athlete, and daughter, are called statuses.

22. **T F** The term *master status* refers to one of the multiple statuses a person occupies that seems to dominate the others in patterning his or her life.

23. **T F** The status of mayor is basically defined by the person who occupies that position.

24. **T F** Sociologists would say that a person is playing a role even when that person feels entirely comfortable and natural and sincerely believes in what he or she is doing.

25. Some students feel pressured by the need to keep up good grades, hold down a job, and maintain a social life with friends. These pressures on a student constitute
    a. role conflict.
    b. role competition.
    c. role exchange.
    d. role strain.

26. College professors are usually expected to engage in at least three types of activities: teaching effectively, conducting research and publishing the results, and devoting time to institutional governance and civic activities. Taken together, these activities constitute a
    a. role conflict.
    b. role set.
    c. role reversal.
    d. master status.

27. Which of these multiple statuses would *most likely* function as a master status?
    a. president of the United States
    b. husband
    c. father
    d. none; all are equal statuses

28. Which of the following is an ascribed status?
    a. male
    b. employee
    c. student
    d. shortstop

29. The relationship between roles and statuses is that
    a. a status may have a number of roles attached to it.
    b. a role may include many statuses.
    c. not all statuses have roles attached.
    d. statuses are dynamic, whereas roles are not.

30. Marissa, a police officer, begins to suspect that her teenage son is involved in selling drugs. She is torn between responding as a parent and responding as a police officer. This is an example of
    a. role conflict.
    b. status conflict.
    c. role strain.
    d. role playing.

31. Some colleges have a policy of legacy admissions, by which the children of alumni are admitted even though their grades and exam scores might be lower than those of other applicants who were rejected. The policy is based on
    a. conflict.
    b. exchange.
    c. achieved status.
    d. ascribed status.

32. A primary school made sure that a child whose father was a teacher in the school was not assigned to his father's class. The policy was designed to avoid
    a. role conflict.
    b. role strain.
    c. achieved status.
    d. exchange.

## Essay Questions

33. Discuss the influence of television on American social interaction.
34. Discuss the ways in which the context of the college classroom structures at least three types of social interaction.
35. Drawing on jnformation on the Web, create an essay in which you describe the ways blacks and whites in American society offend each other without realizing it.
36. Tischler says that a role is "a collection of rights and obligations." Choose some role you play and detail the rights and obligations it includes.
37. Some people claim that television detracts from social interaction and the general level of citizen involvement. But among young people today, some television time has given way to time spent online. Is sitting in front of a computer screen different from sitting in front of a TV screen? How does the Internet affect social interaction?

# CHAPTER 6  SOCIAL GROUPS AND ORGANIZATIONS

## The Nature of Groups

1. T F Because they share a particular characteristic common only to them, left-handed people constitute what sociologists would classify as a social group.
2. T F Social groups and social aggregates both cease to exist when members are apart from one another.
3. T F An aggregate is a group of people with close emotional ties and long-term relationships.
4. T F In primary groups, people's identities are determined more on personal characteristics; in secondary groups, identities are more a matter of what people do.

5. **T F** Cooley called primary groups the nursery of human nature because they are effective only during early childhood.

6. **T F** A secondary group may have specific goals, but it is impersonal and offers less intimacy among the members.

7. **T F** Primary groups are more likely than are secondary groups to rely on formal procedures and sanctions as a means of social control.

8. Which of the following would most likely be a secondary group?
   a. a family
   b. a juvenile gang
   c. a college club
   d. a high-school clique

9. Which of the following is *not* a fundamental characteristic of social groups?
   a. Members have a common identity.
   b. Members do not have to follow any norms of behavior.
   c. There is some feeling of unity.
   d. There are common goals and shared norms.

10. Strangers waiting in line to buy movie tickets constitute what sociologists call a
    a. social group.
    b. social aggregate.
    c. clique.
    d. social category.

11. The most important characteristic of primary groups that is missing in secondary groups is
    a. intimacy.
    b. interaction.
    c. small size.
    d. shared expectations.

## ● Functions of Groups

12. **T F** According to Tischler, a group can usually get along quite well without expressive leadership.

13. **T F** In nearly all groups, the function of instrumental leadership and the function of expressive leadership are fulfilled by the same person.

14. **T F** Instrumental leadership is not necessary in an acapella singing group.

15. **T F** A reference group is a purposefully created special-interest group that has clearly defined goals and official ways of doing things.

16. Tischler says that by performing tasks, members can also increase their commitment to one another and to the group. This function of increasing commitment would be classified as
    a. instrumental.
    b. expressive.
    c. primary.
    d. goal-oriented.

17. Some adults fear that youths might choose "gangsta" rap artists as role models, dressing and talking like them, displaying the attitudes promoted in rap music, and even asking themselves, "What would Jay-Z do?" as a guide to behavior. Such youths would be using the rappers as
    a. a primary group.
    b. a secondary group.
    c. a reference group.
    d. an aggregate.

18. "Why do I have to go to Uncle Mike's party? I barely know him, and I don't really like him," says a youngster. "Because you're a member of this family," answers his parent. What essential function of groups is the parent invoking in her reasoning?
    a. defining boundaries
    b. choosing leaders
    c. setting goals
    d. assigning tasks

19. During jury deliberations, two members get into an angry argument about a piece of evidence. The jury foreman says, "I think tempers are a bit short; let's take a ten-minute break, cool off, and discuss this calmly." The foreman is exhibiting
    a. task leadership.
    b. instrumental leadership.
    c. expressive leadership.
    d. legal leadership.

## Reference Groups

20. T F  The smallest group is a dyad.
21. T F  New members are more threatening to small groups than to large groups.
22. T F  Large groups are more likely to rely on formal procedures than are small groups.
23. T F  Decades of research show that people in small groups are more likely to have specific and clearly spelled-out tasks than are people in large groups.
24. T F  In Tönnies's view, we can expect to find more individualism and less of a cooperative spirit in *gemeinschaft* than in *gesellschaft.*
25. T F  Mechanical solidarity is more characteristic of small, preliterate societies than is organic solidarity.
26. T F  Durkheim felt that the trouble with modern society was that it had no collective conscience.
27. Three students always study together. This group is a
    a. trio.
    b. triage.
    c. dyad.
    d. triad.
28. In *gemeinschaft and gesellschaft*, Tönnies was most concerned with the change from
    a. capitalism to socialism.
    b. democracy to dictatorship.
    c. dyads to triads.
    d. rural society to urban society.
29. A good example of a *gemeinschaft*-like society in the United States is
    a. the Goth subculture.
    b. the Republican Party.
    c. the March of Dimes.
    d. the Amish.
30. The core beliefs and values shared by the members of a society, in Durkheim's terms, is the
    a. *gesellschaft.*
    b. collective unconscious.
    c. collective conscience.
    d. guilty conscience.
31. When people interact on the basis of many specialized positions, their coordination, in Durkheim's terms, is an example of
    a. mechanical integration.
    b. organic solidarity.
    c. racial solidarity.
    d. social disharmony.
32. Durkheim thought that the collective conscience did *not* appear in
    a. simple, preliterate societies.
    b. agricultural societies.
    c. industrial societies.
    d. postindustrial societies.
    e. none of these; he thought it appeared in all societies.

## Bureaucracy

33. T F  Many people tend to think that only government agencies are bureaucracies, but most large private corporations are bureaucracies as well.
34. T F  According to Weber's model of bureaucracy, bureaucrats (people who work in a bureaucracy) are likely to have special training and competency for the office they hold.
35. T F  In bureaucracies that resemble Weber's ideal type, workers are especially likely to give special favorable treatment to their friends and family members but not to strangers.
36. T F  In the view of Robert Michels, democracy is nearly impossible to maintain in a large organization.
37. T F  The more like an oligarchy an organization becomes, the closer the leaders are to the ordinary members of the organization.
38. T F  Michels felt that the iron law of oligarchy could be avoided by using personality tests to get the right kinds of leaders.

39. **T F** In the ideal-type bureaucracy as outlined by Weber, workers frequently bend or break the rules to make their work more interesting.

40. **T F** For something to be a social institution, it must have a specially designated building or group of buildings.

41. Which of the following is *not* an essential aspect of bureaucracy as outlined in Weber's ideal type?
    a. extensive division of labor
    b. personalized service
    c. impartiality toward everyone seeking service
    d. hierarchy of authority

42. One advantage of the division of labor in a bureaucracy is that it allows for
    a. a high level of specialization and expertise.
    b. the ability to get around official regulations.
    c. a friendly atmosphere.
    d. personalized service.

43. "Boy, did I get the runaround in the administration building. They kept sending me from one office to the next till I finally found someone who knew how to solve my problem." The problem referred to here is caused by what aspect of bureaucracy?
    a. division of labor
    b. hierarchy of authority
    c. impartiality
    d. efficiency

44. After an African American church was destroyed by arson, the reverend said, "They didn't burn down the church. They burned down the building." His remark illustrates the concept of the church as
    a. an institution.
    b. an aggregate.
    c. a *gesellschaft*.
    d. an oligarchy.

45. According to Tischler, the large increase in the percentage of women working outside the home led to a change in
    a. *gemeinschaft*.
    b. social aggregates.
    c. social organization.
    d. oligarchy.

● **Essay Questions**

46. Distinguish between primary and secondary groups and weigh the relative advantage of each. Describe at least two circumstances or contexts in which primary groups are more necessary, advantageous, or effective than secondary groups.

47. Responding to an announcement on the website Meetup, a group of strangers get together to form a writers' group. Discuss from a sociological point of view the six tasks these people must successfully accomplish to create a viable group.

48. Describe how a relatively informal group such as a family might look if you were to impose bureaucratic structures on it. Pay special attention how the group fulfills the functions mentioned in the text.

49. What groups serve as reference groups for you? Describe the ways in which these are reference groups and the nature of your relationship with them.

# CHAPTER 7   DEVIANT BEHAVIOR AND SOCIAL CONTROL

● **Defining Normal and Deviant Behavior**

1. **T F** Sociologically speaking, behavior is classified as normal or deviant only with reference to the group in which it occurs.

2. **T F** Durkheim maintained that a society without any deviant behavior is both desirable and possible.

3. **T F** In a large society, different groups can have differences of opinion as to which acts are deviant and which are not.

4. **T F** Although we usually think of deviance as bad for society, Durkheim thought deviance served important positive functions for society.

## ● Mechanisms of Social Control

5. **T F** Internal mechanisms of control are more effective and cost-efficient than are external mechanisms of control.

6. **T F** The term *sanctions* refers only to negative punishments, not to positive rewards.

7. Which of the following is an *internal* means of social control?
   a. ridicule
   b. guilt
   c. imprisonment
   d. exclusion from the group

8. At a mostly black high school, the principal instituted a program in which students who got all A's were given $100 at an awards ceremony. The sanctions the program uses are
   a. formal positive.
   b. formal negative.
   c. informal positive.
   d. informal negative.

9. At the awards assemblies for "A" students, other students often jeered and called the recipients names such as "nerd" and "Whitey." The sanctions used by the students were
   a. formal positive.
   b. formal negative.
   c. informal positive.
   d. informal negative.

## ● Theories of Crime and Deviance

10. **T F** Lombroso's theory of atavism is based on the idea that criminal behavior is learned and that people become criminals because others teach them the wrong kinds of behavior.

11. **T F** Shaw and McKay's study of Chicago neighborhoods showed that as the ethnicity of an area changed, its crime rate changed dramatically, and an area high in crime in one period might be low in crime a decade later.

12. **T F** Neutralization theory emphasizes the idea that the law should be neutral on matters of race and social class.

13. "Sure, I lifted his wallet. Anybody who gets drunk and flashes a lot of money around is just asking to get ripped off." This is an example of which technique of neutralization?
    a. denial of responsibility
    b. appeal to a higher principle
    c. denial of injury
    d. denial of the victim

14. Which of the following theories is less concerned with the causes of norm violations than with the way others react to the deviance?
    a. psychoanalytic theory
    b. anomie theory
    c. cultural transmission theory
    d. labeling theory
    e. control theory

15. Freudian theory sees a source of crime in the failure to develop a proper superego. The superego is
    a. an internal mechanism of control.
    b. an external mechanism of control.
    c. a mechanism of control that is sometimes internal, sometimes external.
    d. not a mechanism of control at all.

16. _____ theory emphasizes the social bond between the individual and the society as the most important factor in determining whether someone will become criminal.
    a. Psychoanalytic
    b. Anomie
    c. Cultural transmission
    d. Labeling
    e. Control

17. A youth wants to achieve the sort of affluent lifestyle he sees on television. Unable to find a well-paying job, he turns to drug dealing to pay for it. In Merton's typology, he would be classified as a(n)
    a. retreatist.
    b. rebel.
    c. innovator.
    d. conformist.
    e. danger to society.

18. Sykes and Matza's neutralization theory emphasizes which of the following as a factor in crime?
    a. thought processes
    b. biological and genetic makeup
    c. punishment and deterrence
    d. attachment to society

19. Durkheim saw anomie as a condition of
    a. weak law enforcement.
    b. overemphasis on the welfare of the group.
    c. normlessness.
    d. dependency.

20. Lombroso and Sheldon, each in his own way, attempted to explain deviant behavior as rooted in
    a. psychological orientation.
    b. early childhood experiences.
    c. anatomical characteristics.
    d. differential association.

21. "People commit crimes because they see the rewards as outweighing the risks, and because it is more profitable than anything someone with their abilities might do." This statement is most consistent with which type of theory?
    a. biological
    b. behavioral
    c. psychoanalytic
    d. rational choice

22. Fourteen-year-old Janet is arrested for shoplifting. Even though the charges are later dropped, Janet's teachers designate her as a troublemaker and someone not to be trusted. Because the teachers make school an unwelcoming place for her, Janet begins to skip school frequently and to get into fights when she is there. Lemert and others would see her truancy and violence as examples of
    a. primary deviance.
    b. secondary deviance.
    c. recidivism.
    d. anomie.

23. The research of Shaw and McKay, which linked crime to certain types of urban neighborhoods, provided the foundation for _____ theories of deviance.
    a. control
    b. labeling
    c. genetic transmission
    d. cultural transmission

## ● The Importance of Law

24. Conflict theory maintains that laws have the effect of protecting
    a. law-abiding people from harm.
    b. poor people against exploitation by the wealthy and powerful.
    c. the wealthy and powerful from losing their position of dominance.
    d. the entire society from harmful conflicts.

25. "The law is merely a formal and enforceable statement of widely accepted norms and values." The idea expressed here is most consistent with
    a. control theory.
    b. labeling theory.
    c. conflict theory.
    d. consensus theory.

## ● Crime in the United States

26. T F Because armed robbery involves the taking of property, it is classified as a property crime.

27. T F The FBI's Uniform Crime Reports is based on a nationwide, door-to-door survey of crime victims.

28. T F The major reason the Uniform Crime Reports gives an inaccurate count of the number of crimes in the United States is that victims of crime frequently do not call the police.

29. T F The United States violent crime rate in 2011 reached the highest level since the Bureau of Justice Statistics started measuring it in 1973.

30. T F Violent crimes are committed far more frequently than property crimes and far outnumber property crimes in both the UCR and the NCVS.

31. T F The UCR's measure of serious crime does not include white-collar crimes such as bribery and fraud.

32. T F The amount lost to white-collar crime far exceeds the amount lost to street crimes such as burglary and robbery.

33. T F The United States has much higher rates of murder than do other countries with similar economic and political systems (Western European countries, Canada, and so on).

34. T F The most frequent reason given by victims for not reporting crime to the authorities is the belief that the crime was not important enough.

35. T F Execution in the United States is declining in frequency, primarily due to cost factors.

36. T F A major difference between adult and juvenile crime is that juveniles are much more likely to commit offenses in groups.

37. T F Females are much more likely to be victims of serious crimes than are males.

38. Which of the following crimes is most likely to be reported to the police?
    a. robbery
    b. rape
    c. motor vehicle theft
    d. burglary

39. Relatively minor crimes that are usually punishable by a fine or less than a year's confinement are called
    a. civil offenses.
    b. misdemeanors.
    c. larcenies.
    d. felonies.

40. A status offense is an act that is
    a. against the law if committed by a juvenile but not if committed by an adult.
    b. committed by a criminal in to achieve higher status.
    c. committed by a person of high status.
    d. committed by a person of low status.

41. Violation of laws meant to enforce the moral code, such as public drunkenness, prostitution, gambling, and possession of illegal drugs, are called _____ crimes.
    a. moral
    b. victimless
    c. organized
    d. status

## ● Criminal Justice in the United States

42. T F Unlike policing in countries with a national police force, policing in the United States is controlled largely at the local level.

43. T F Truth-in-sentencing laws are an attempt to reduce prison overcrowding.

44. T F Women who commit crimes are more likely to commit property crimes as opposed to violent crimes.

45. T F About half of all inmates on death row come from middle-class backgrounds.

46. The increased police presence during terror alerts had the greatest impact on which crime?
    a. murder
    b. rape
    c. motor vehicle theft
    d. fraud

47. Steering offenders away from the justice system and into social agencies is known as
    a. diversion.
    b. labeling.
    c. recidivism.
    d. rehabilitation.

48. We can get an idea of whether prisons deter criminals who are sent there by looking at rates of
    a. diversion.
    b. recidivism.
    c. incarceration.
    d. capital punishment.

49. Which of the following countries executes more people than does the United States?
    a. Japan
    b. France
    c. Mexico
    d. India
    e. none of these

50. The process in which a large number of crimes committed results in only a small number of criminals going to prison is called the _____ effect.
    a. diversion
    b. deterrence
    c. funnel
    d. recidivism

51. At which stage in the funnel effect does the greatest number of crimes disappear?
    a. victims not reporting crimes to the police
    b. police not arresting the criminal
    c. arrested criminals not being convicted
    d. judges giving convicted criminals sentences other than prison

52. Which of the following countries imprisons a greater proportion of its population than does the United States?
    a. Russia
    b. France
    c. Italy
    d. Spain
    e. none of these; the United States imprisons more than all of them.

53. In Lowell, Massachusetts, the broken-windows theory inspired the city to take which action in some hot spots to reduce crime?
    a. increase the number of drug raids
    b. increase the number of police on patrol
    c. search all homes for drugs and weapons
    d. clean the streets and fix the streetlights

● Essay Questions

54. Since the 1960s, America has seen a succession of youth cultures, each of which has been self-consciously deviant from mainstream attitudes and values. Evaluate the functional and dysfunctional aspects of deviant youth culture.

55. Using the rules and regulations on your own campus as examples, compare and contrast the *conflict* and *consensus* theories of law.

56. Describe Merton's five modes of individual adaptation to the discrepancy between cultural goals and institutionalized means.

## CHAPTER 8   SOCIAL CLASS IN THE UNITED STATES

● The American Class Structure

1. T F Social class is a fact of all industrialized societies, even in socialist and communist countries.

2. T F In the last quarter-century, the income gap between the rich and the poor in the United States has generally increased.

3. **T F** Members of a social class usually have roughly similar lifestyles.

4. **T F** Historically, the members of the upper class in the United States have been predominantly Jewish.

5. **T F** In the United States, members of the upper class, despite their great wealth, often live, work, and associate socially with members of the upper-middle and lower-middle classes.

6. **T F** According to the U.S. Census Report (2010), the richest one-fifth of Americans was responsible for 70 percent of the country's total earned income.

7. **T F** Lower-class people typically share the American desire for advancement and achievement found in the other social classes.

8. **T F** Because most Americans work for a living, well over 50 percent of the population is usually considered working class.

9. **T F** Great Britain, with its long history of nobility and aristocracy, has a much greater gap between the rich and the rest of society than does the United States.

10. Nearly all sociologists agree that there are _____ social classes in the United States.
    a. three
    b. four
    c. five
    d. six
    e. none of these; sociologists do not agree on a number of social classes.

11. About what proportion of the total amount of income in the United States goes to the poorest one-fifth of the population?
    a. 20 percent
    b. 35 percent
    c. 3 percent
    d. 75 percent

12. The upper class of the United States is approximately _____ of the population.
    a. 1–3 percent
    b. 7–10 percent
    c. 3–5 percent
    d. < 1 percent

13. The term for stocks, real estate, and other owned property not including the money a person earns at work is called
    a. income.
    b. wealth.
    c. prosperity.
    d. risk.

14. Managers and other people who work in the professional and technical fields are members of the _____ class.
    a. upper
    b. upper-middle
    c. lower-middle
    d. working

15. Which data would you look at if you wanted to show a larger degree of inequality between the rich and the rest of the population?
    a. distribution of income
    b. distribution of wealth
    c. neither of these; they show the same degree of inequality.

16. In which of these countries is income inequality—the difference between rich and poor—greatest?
    a. Sweden
    b. Germany
    c. France
    d. Italy
    e. United States

17. Bob is a carpenter. He holds a high-school diploma, owns a modest home, but struggles to give his family the things they need and want. He considers himself politically conservative and moderately religious. According to your text, what social class does he belong to?
    a. upper-middle
    b. lower-middle
    c. lower
    d. working

## Poverty

18. **T F** Basically, poverty refers to a condition in which people maintain a standard of living that includes the basic necessities but none of the extras and luxuries.

19. **T F** Most government agencies define a family as poor if its income is less than half of the average American family income.

20. **T F** More than one American in ten was below the poverty line in 2011.

21. **T F** The poverty index includes only cash income and does not take into account noncash benefits such as Medicaid and food stamps.

22. **T F** Mothers who are divorced are usually economically worse off than mothers who were never married.

23. **T F** Poor people who work are often not eligible for the assistance that jobless poor people are eligible for.

24. **T F** Since the 1970s, the percentage of the elderly living in poverty has decreased while the percentage of children living in poverty has increased.

25. **T F** The minimum wage in the United States is set so that a single parent of two children who works full time at a minimum-wage job will have an income above the poverty index.

26. Since the passage of the Welfare Reform Act in 1996, the number of families receiving aid has
    a. increased greatly.
    b. decreased greatly.
    c. remained about the same.

27. Because of the way the government computes the official poverty index—the income level below which a family is defined as poor—this index can be affected by a change in the cost of
    a. food.
    b. clothing.
    c. shelter.
    d. fuel.
    e. all of these.

28. Of the following nations, the highest rate of poverty among children is found in
    a. the United States.
    b. Germany.
    c. the United Kingdom.
    d. Sweden.

29. According to the Bureau of the Census, in 2011 a family of four was living in poverty if their income was below
    a. $18,200.
    b. $22,350.
    c. $28,200.
    d. $32,000.

30. Poverty is more frequent
    a. among men than among women.
    b. among whites than among minorities.
    c. among the elderly than among middle-aged people.
    d. in rural than in urban areas.
    e. in the Northeast than in the South.

31. The feminization of poverty refers to
    a. the trend in which females represent an increasing proportion of the poor.
    b. a federal program eliminating aid to unwed mothers.
    c. an increase of feminist politicians involved in aid programs.
    d. the vast overrepresentation of women among social workers who work with the poor.

## Government Assistance Programs

32. **T F** The amount of government benefits going to the middle class (Social Security, Medicare, and so on) is greater than the amount going to the poor in welfare programs.

33. **T F** The amount of government money going to female-headed families in poverty is less than one-tenth of the amount going to Social Security payments to the elderly.

34. T F **Government assistance** programs are generous enough that they provide a comfortable lifestyle for most of the poor.

35. If a government program is means tested, it is
    a. a beta version, still being tested.
    b. available to all people regardless of income.
    c. available only to people who can prove that they earn below a certain income.
    d. available to people with an average income.

36. U.S. government programs to combat poverty have been most successful among
    a. children.
    b. working-age adults.
    c. the elderly.
    d. women.

## Consequences of Social Stratification

37. T F Even when a middle-class person and a poor person commit similar crimes, the poor person is more likely to be arrested.

38. T F Because of the right to a free lawyer, a poor person charged with a crime is no more likely to be convicted than is a middle-class person.

39. T F Rich-poor differences in health do not begin until late childhood. Babies of the poor are no more likely to die in the first year than are babies of the middle or upper classes.

40. T F Prison systems are heavily populated by the poor.

41. T F The crimes committed by the poor do much more financial harm to society than do the crimes of those in the middle and upper classes.

42. In which social class are parents most likely to treat sons and daughters similarly?
    a. lower class
    b. lower middle class
    c. middle class
    d. none of these; there are no real differences among social classes when it comes to childrearing.

## Why Does Social Inequality Exist?

43. T F The functionalist explanation maintains that inequality exists and persists because it benefits the society as a whole.

44. T F Kingsley Davis and Wilbert Moore contend that social stratification exists in most industrialized nations except the United States.

45. T F According to functionalist theory, a society must pay some jobs more to attract people to important positions.

46. T F According to functionalist theory, low-paying jobs are low-paying because they are not as necessary to the society as better-paying jobs.

47. T F Conflict theorists argue that inequality persists because those with wealth and power use that wealth and power to protect and further their own interests, even at the expense of the interests of others in the society.

48. T F In Marx's terms, the bourgeoisie comprises the owners of the means of production.

49. T F Max Weber, unlike Marx, argued that status and power could be separated from economic position.

50. Conflict theories of inequality are often based on the ideas of
    a. Barack Obama.
    b. Kingsley Davis and Wilbert Moore.
    c. Karl Marx.
    d. Émile Durkheim.

51. According to Marx, what was the relation between the owners of the means of production (factories, land, and so on) and those who worked for them?
    a. Owners, if they were wise, made sure that workers were satisfied with their jobs.
    b. Owners usually took workers' interests into consideration.
    c. Owners and workers cooperated because the success of a business meant increased income for everyone in it.
    d. Owners exploited workers and paid them as little as possible.

## Essay Questions

52. Which factors that determine a person's social class does the person have control over; which factors does the person have no control over? Which factors are more important in affecting the class position of the individual? Which are more important in affecting the overall degree of inequality in the society?

53. In its proportion of poor people and the consequences of being poor, how does the United States compare with other advanced countries (Western Europe, Japan, Australia, Canada)? What accounts for these differences in poverty?

54. Some people attribute poverty to the personal qualities and lifestyle choices of poor people. In this view, people are poor because they are lazy or unintelligent or because they have chosen to drop out of school, to have children at an early age, to take drugs, and so on. How valid is this individual-based explanation of poverty?

# CHAPTER 9   GLOBAL STRATIFICATION

## Stratification Systems

1. T F More than half the people in the world live on less than $2 a day.
2. T F In most countries, the gap between rich and poor has gotten narrower over the past half century.
3. T F In all societies, the basis of social stratification is income.
4. T F Social class systems are more open than social caste systems.
5. T F The caste system in India is just as rigid today as it was 100 years ago.
6. The chances of a woman in sub-Saharan Africa dying during childbirth in her lifetime is approximately 1 in _____.
    a. 10
    b. 50
    c. 5
    d. 20
7. In India, as in most societies with a caste system, the justification of the caste system is based mostly on
    a. legal principles.
    b. religious ideas.
    c. economic necessity.
    d. the teachings of a charismatic leader.
8. T F The principle of untouchability and the untouchable caste (Dalits) is currently illegal in India.
9. Despite being made illegal nearly 60 years ago, the caste system in India still persists, especially in
    a. business.
    b. poor neighborhoods of large cities.
    c. rural areas.
    d. government.
    e. none of these; untouchability has been nearly eliminated everywhere.

10. **T F** Estate systems are usually more rigid than the caste systems.

11. The system of stratification prevalent in Europe during the Middle Ages was
    a. a caste system.
    b. a class system.
    c. an estate system.
    d. a religious hierarchy.
    e. a classless society.

12. A person with great ambition would fit best in
    a. a caste system.
    b. a class system.
    c. an estate system.
    d. a religious hierarchy.
    e. a classless society.

13. **T F** In medieval Europe, the clergy was not considered an estate.

14. **T F** During the Middle Ages, merchants were not actually a formally recognized estate.

15. **T F** The estate system of the Middle Ages provided some limited social mobility.

## Theories of Population

16. **T F** There is no caste system whatsoever in Great Britain.

17. **T F** Africa is currently the continent that is displaying the fastest population growth.

18. **T F** The physiological capacity to bear human offspring is known as fecundity.

19. **T F** A child born in Afghanistan has a 25 percent chance of dying before his/her fifth birthday.

20. **T F** Malthus embraced the theories of utopian socialists.

21. An example of a positive check is
    a. contraception.
    b. famine.
    c. celibacy.

22. **T F** Marxists point to industrialization as the core problem that effects population growth.

23. Dependency ratio is the number of non-working-age individuals in comparison to every _____ of working age.
    a. 1,000
    b. 10
    c. 10,000
    d. 100

24. India is expected to surpass China as the most populous country on earth in approximately
    a. 2030.
    b. 2050.
    c. 2040.
    d. 2020.

25. **T F** According to statistics, the world population has doubled since 1960.

26. Even with the diversity throughout the world, what factors are common to a significant percentage of the population in most developing nations?
    a. poor sanitation
    b. no access to modern health services
    c. inadequate housing
    d. many children not attending school
    e. all of the above

27. **T F** In the past century, the worldwide average life expectancy has at least doubled.

28. **T F** Because of great advances in medicine and communication, the majority of people in the world now have access to professional heath care.

29. Approximately what percent of the population in China today may be expected to reach age 70?
    a. less than 10 percent
    b. about 25 percent
    c. 60 percent
    d. 90 percent

30. T F Because of the existence of antibiotics, death rates in developing nations are similar to those in advanced nations.

31. In developing nations, the most frequent cause of death for children under age five is
    a. child abuse.
    b. malnutrition.
    c. predatory animals.
    d. infectious diseases.
    e. accidents.

32. In developed countries (Europe, the United States, and so on), infectious diseases such as measles and diarrhea cause about what percentage of all childhood death?
    a. 1 percent
    b. 25 percent
    c. 40 percent
    d. more than 50 percent

33. T F Most childhood deaths in developing countries could be prevented by sanitary drinking water and inexpensive medicines.

34. T F In rich countries and poor, an increase in the mother's education is associated with decrease in child mortality.

35. In which area of the world are HIV/AIDS rates highest (that is, the largest percentage of the population is infected)?
    a. East Asia
    b. North America
    c. Russia and Eastern Europe
    d. sub-Saharan Africa
    e. the Caribbean

36. T F Despite the increase of HIV/AIDS worldwide, in wealthier countries the number of AIDS deaths has declined since the 1990s.

37. In countries outside Europe and North America, the most common source of HIV/AIDS transmission is
    a. intravenous drug use.
    b. homosexual sex.
    c. heterosexual sex.
    d. blood transfusion.

38. Roughly how many people are there currently on planet Earth?
    a. 200 million
    b. 1–2 billion
    c. 7 billion
    d. 10–12 billion

39. The population of the earth did not begin to grow until about _____ years ago.
    a. 1 million
    b. 50,000
    c. 10,000
    d. 1,000
    e. 400

40. T F The increase in population on earth is accounted for mostly by countries in the developing world.

41. T F There is no connection between the age at which a woman marries and the number of children she has. Fertility is determined by economic and health factors.

42. In countries with little modern contraception (for instance, Pakistan, Bangladesh, sub-Saharan Africa), which factor does the most to lower fertility?
    a. abortion
    b. abstinence
    c. breast feeding
    d. herbal contraceptive potions

43. Worldwide, the most widely used form of birth control is
    a. condoms.
    b. the pill.
    c. abstinence.
    d. abortion.
    e. sterilization.

44. If you wanted to reduce fertility in a country, you would do best to increase the years of education of
    a. women.
    b. men.
    c. either one; there is no difference.

45. Which of the following is a successful government policy that has reduced fertility?
    a. raising the legal minimum age for marriage
    b. banning abortion
    c. strengthening the role of religion
    d. requiring teenagers to take vows of abstinence until marriage

46. T F In the developing world, birthrates are generally higher in cities than in rural areas.

47. T F By the year 2050, the number of elderly people will exceed the number of younger persons worldwide.

48. T F Because girls have a higher survival rate than boys, parents in the developing world prefer to have girls rather than boys.

49. T F In countries such as India, where the parents of a girl must pay a high dowry, birthrates for girls will be unnaturally low.

50. T F Throughout all countries, people with the greatest income usually have more children than those with lower income, largely because wealthier people can support additional family members.

51. T F In poor regions such as sub-Saharan Africa, one of the more acute problems is that of taking care of the elderly.

## Essay Questions

52. Aside from the obvious fact that poor people cannot afford to buy material goods, what are some of the differences between life in poor, underdeveloped countries and life in countries such as the United States? Consider the nature of basic areas of life such as family, work, education, health, and so on.

53. What factors contribute to the gap between rich countries and poor countries? What factors might help reduce that gap?

# CHAPTER 10   RACIAL AND ETHNIC MINORITIES

## The Concept of Race

1. T F Racial classifications are not simple or obvious; a person considered white in one society might be categorized as nonwhite in another.

2. T F Biologists today rely on outward physical traits such as skin color, hair texture, and nose type in making biological classifications of human beings.

3. T F Nearly half of the U.S. population now classifies themselves as multiracial.

4. T F Most interracial couples list their children as multiracial on census forms.

5. T F The number of interracial marriages in the United States has increased greatly in the past three decades.

6. T F Despite changes in race relations and laws over the past fifty years, a majority of Americans still say they disapprove of interracial marriage.

7. The term *race* refers to a category of people who are defined as similar because they
    a. have a unique and distinctive genetic makeup.
    b. share a number of physical characteristics.
    c. exhibit similar behaviors.
    d. express comparable attitudes.

8. Legal definitions of race in the United States usually were based on the person's
    a. physical characteristics.
    b. ancestry.
    c. educational achievement.
    d. lifestyle.

9. According to the *social* definition of race, a woman is black if
   a. she has certain physical characteristics: darker-than-average skin, flat nose, and so on.
   b. she has at least one grandparent who is black.
   c. she has African genetic markers in her DNA.
   d. she and others think she is black.

10. The U.S. census counts a person as African American if he or she
    a. has at least one African American parent (one-half black).
    b. has at least one African American grandparent (one-fourth black).
    c. has at least one African American great-grandparent (one-sixteenth black).
    d. defines himself or herself as African American regardless of ancestry.

11. In what year were laws prohibiting interracial marriage declared unconstitutional by the Supreme Court?
    a. shortly after the founding of the United States
    b. shortly after the Civil War
    c. shortly after World War II
    d. in the 1960s
    e. none of these; the Court has never declared anti-interracial-marriage laws unconstitutional.

12. About what percentage of married couples in the United States are interracial?
    a. 5 percent
    b. 25 percent
    c. 50 percent
    d. 75 percent

13. The text's example of Philip and Paul Malone was used to illustrate
    a. the failure of multiculturalism.
    b. a humorous look at the problems of birth records.
    c. the difficulties in accurately assigning race.
    d. that racial appearance and racial definition generally match closely.

14. In the United States, people with one white parent and one black parent have usually thought of themselves as
    a. multiracial.
    b. white.
    c. black.
    d. about half of them think of themselves as white, half as black.

## The Concept of Ethnic Group

15. T F  To qualify as an ethnic group, members must be unified politically.

16. According to Tischler, the distinguishing features of an ethnic group are usually
    a. cultural.
    b. physical.
    c. economic.
    d. political.

## The Concept of Minority

17. Women can be regarded as a minority group in American society because
    a. there are fewer women than there are men.
    b. they belong to a variety of ethnic and racial groups.
    c. they are discriminated against because of their sex.
    d. none of these; they cannot be regarded as a minority by any definition.

## Problems in Race and Ethnic Relations

18. T F  Prejudice is a negative attitude that is nevertheless rational.

19. T F  Prejudice serves some positive functions for the people who hold that prejudice.

20. When psychologists say that prejudice allows for "projection," they mean that the prejudiced person
    a. sees in some other group what are really his or her own faults.
    b. speaks loudly about his or her opinions of other groups.
    c. prefers movies about the groups he or she dislikes.
    d. feels guilty about his or her own hatreds.

21. The difference between prejudice and discrimination is a difference between
    a. attitudes and behavior.
    b. positive and negative.
    c. socially unacceptable and socially acceptable.
    d. rational and irrational.

22. T F Prejudice is more likely to develop between groups that are competing against each other for scarce resources.

23. T F According to Tischler, people with prejudiced attitudes may not engage in discriminatory behavior.

24. Social arrangements that restrict a group's life chances, even though there is no apparent hatred, prejudice, or stereotyping, are called
    a. racism.
    b. socialism.
    c. institutionalized discrimination.
    d. projection.

25. A major difference between prejudice and discrimination is that
    a. prejudice involves ethnic groups; discrimination involves race.
    b. prejudice involves thoughts; discrimination involves actions.
    c. prejudice involves economics; discrimination involves culture.
    d. prejudice can be reversed; discrimination is permanent.

26. Many white shopkeepers in U.S. southern towns during the 1950s and 1960s depended upon African American customers for a large part of their business, so although they welcomed blacks as customers, they considered them social inferiors. These merchants are examples of
    a. unprejudiced nondiscriminators.
    b. unprejudiced discriminators.
    c. prejudiced nondiscriminators.
    d. prejudiced discriminators.

27. Sonia is a member of a minority group. She is unable to obtain a well-paying, secure job not because of outright racism but, rather, because, like many others of her minority group, she attended a less-than-adequate school and lacks connections. Sonia is a victim of
    a. subtle and unrecognized personal prejudice.
    b. unfortunate accidental discrimination.
    c. institutionalized prejudice and discrimination.
    d. bad luck that has nothing to do with her being a minority.

28. An individual who feels very uncomfortable when friends tell a racist joke and yet does not speak out of fear of being ridiculed would be classified by Merton as a(n)
    a. unprejudiced nondiscriminator.
    b. unprejudiced discriminator.
    c. prejudiced nondiscriminator.
    d. prejudiced discriminator.

● Patterns of Racial and Ethnic Relations

29. Canada's maintaining two official languages, English and French, is an example of
    a. pluralism.
    b. assimilation.
    c. Anglo conformity.
    d. subjugation.

30. During the nineteenth century, Native Americans were pushed off land desired by white settlers and onto small and distant reservations. This exemplifies all of the following *except*
    a. forced migration.
    b. segregation.
    c. assimilation.
    d. expulsion.

31. A 1946 government report on Native Americans called for "their complete integration into the mass of the population." The report was advocating
    a. assimilation.
    b. pluralism.
    c. forced migration.
    d. segregation.

32. "Not the elimination of differences but the perfection and conservation of differences" is a statement favoring
    a. pluralism.
    b. assimilation.
    c. Anglo conformity.
    d. subjugation.

33. Genocide, a practice banned by international law, is another term for
    a. forced migration.
    b. segregation.
    c. subjugation.
    d. annihilation.

34. Most of the nineteenth-century immigrants to America had distinctive subcultures, with their own unique language, style of dress, norms, and values. The children of these immigrants, however, rapidly learned to speak English and to adopt mainstream American cultural styles. The children's behavior is an example of
    a. segregation.
    b. pluralism.
    c. assimilation.
    d. subjugation.

35. The development of ethnic neighborhoods such as Chinatowns and Little Italy is an example of voluntary
    a. segregation.
    b. Anglo conformity.
    c. subjugation.
    d. assimilation.

36. The policy of the Nazis toward the Jews during the Holocaust of the 1940s was one of
    a. pluralism.
    b. Anglo conformity.
    c. annihilation.
    d. subjugation.

37. To advance his career in radio, Jaime Fernandez learns to speak English without a trace of a Spanish accent and changes his name to Jim Fox. This is an example of
    a. subjugation.
    b. Anglo conformity.
    c. Chicano conformity.
    d. annihilation.

## ● Racial and Ethnic Immigration to the United States

38. T F Even during its more restrictive periods, the United States has had one of the most open immigration policies in the world.

39. T F Hispanics are now the largest ethnic group in the United States.

40. The immigration quotas of the 1920s were designed to limit immigration from countries such as
    a. Ireland and England.
    b. Germany and France.
    c. Italy and Poland.
    d. Mexico and Cuba.

41. About what percentage of the U.S. population was born outside the United States?
    a. 1 percent
    b. 10 percent
    c. 25 percent
    d. 50 percent

42. The largest proportion of Hispanics in the United States come from
    a. Cuba.
    b. the Dominican Republic.
    c. Mexico.
    d. Argentina.

43. The largest number of *illegal* immigrants to the United States come from which country?
    a. Cuba
    b. the Dominican Republic
    c. Mexico
    d. China

44. Which Hispanic group in the United States has the highest average income?
    a. Mexicans
    b. Cubans
    c. Dominicans
    d. Puerto Ricans

45. According to Tischler, compared with other Hispanics, Cuban Americans
    a. are less educated.
    b. have a lower average income.
    c. are more likely to resist assimilation.
    d. are more likely to vote Democratic.

46. The racial or ethnic group with the largest percentage growth in recent years is
    a. Hispanics.
    b. Asians.
    c. Africans.
    d. Canadians.

47. The earliest large wave of immigrants from China worked mostly in
    a. restaurants.
    b. laundries.
    c. opium dens.
    d. railroad construction.

48. Which ethnic or regional group has the highest average levels of education?
    a. African Americans
    b. Italian Americans
    c. Asian Americans
    d. Hispanics

49. Italian Americans today are most likely to trace their ancestors back to
    a. the old migration.
    b. the new migration.
    c. the modern migration.

50. In recent years, some states have proposed laws making English the official state language and eliminating or restricting bilingual education. The aim of these measures is
    a. annihilation.
    b. Anglo conformity.
    c. exploitation.
    d. pluralism.
    e. segregation.

51. The largest proportion of immigrants to the United States today is accounted for by people from
    a. Europe.
    b. Africa.
    c. Asia.
    d. Latin America.
    e. Antarctica.

52. T F More than half of all Native Americans live on or near reservations administered by the U.S. government.

53. *Old migration* to the United States consisted of people from
    a. the ancient civilizations of the Mediterranean.
    b. northern Europe who came prior to 1880.
    c. eastern Europe who came after 1880.
    d. age 50 upward.

54. The largest number of African Americans is found in what part of the United States?
    a. Northeast
    b. Midwest
    c. South
    d. Mountain states
    e. West Coast

55. According to Tischler, a major reason for the increasing income gap between African Americans and whites is
    a. racism.
    b. the decline of unionized jobs.
    c. the influx of African immigrants.
    d. the increase in female-headed families among African Americans.

56. African Americans make up approximately _____ percent of the total population of the United States.
    a. 2.5
    b. 13.5
    c. 24.6
    d. 43.3

## Essay Question

57. Choose a specific group, and explain why it qualifies as a race, ethnic group, or minority.

# CHAPTER 11   GENDER STRATIFICATION

## Are the Sexes Separate and Unequal?

1. **T F** In movies, there are about as many female characters as male characters.
2. **T F** Traditional Chinese society had much greater equality between the sexes than did traditional European society.
3. **T F** Critics of sociobiology are likely to emphasize learned behavior and socialization.
4. **T F** The biblical story of creation has been used as a theological justification for patriarchal ideology.
5. **T F** Pioneering sociologist Auguste Comte differed from most other thinkers of his time (the early 1800s) in that he saw women as intellectually equal to men.
6. **T F** Although societies might treat the sexes unequally, diseases such as diabetes and multiple sclerosis affect men and women with equal frequency.
7. **T F** Jeanette Gilder used the analogy of being John F. Kennedy's mother as an illustration of the right of women to vote.
8. **T F** Unlike modern Western society, many simpler, preliterate societies have no division of labor by sex. Men and women do pretty much the same things.
9. The branch of science that tries to establish the genetic bases of human behavior is called
    a. biogenetics.
    b. ethology.
    c. ethnography.
    d. sociobiology.
10. Ethology is the study of
    a. women named Ethel.
    b. ethical behavior.
    c. ethanol use among primates.
    d. animal behavior.
    e. ethnic groups.
11. In Clark and Hatfield's study of pick-ups, which group was most likely to agree to have sex with an attractive stranger of the opposite sex?
    a. men
    b. women
    c. no difference
12. In Clark and Hatfield's study of pick-ups, which group was most likely to agree to a date with an attractive stranger of the opposite sex?
    a. men
    b. women
    c. no difference
13. A patriarchal ideology is
    a. the study of powerful males in past societies.
    b. the belief that there are differences in the social behavior of men and women.
    c. an attempt to find a genetic basis for human behavior.
    d. the belief that men are superior to women and should control all aspects of society.
14. Among mathematicians, men far outnumber women. A university president explained this difference by saying that men are "naturally" better at math than women. His statement implies that mathematical ability is a matter of
    a. gender.
    b. sex.
    c. socialization.
    d. unequal opportunity.

15. Gender is best understood as an _____ status.
    a. achieved
    b. ascribed
    c. ideal
    d. irrelevant

16. In stressful situations
    a. males and females react with similar intensity.
    b. males and females react with similar behaviors, although women are more intense.
    c. men react more slowly.
    d. women react more slowly.

17. Critics of sociobiology assert that
    a. even among animals and insects, much gender-related behavior is learned, not innate.
    b. sex differences have no biological basis.
    c. gender differences do not exist among nonhuman primates.
    d. it is not valid to generalize from animal to human behavior.

18. T F Differences in hormones in men's and women's brains might explain why aging women suffer cognitive decline, for instance, Alzheimer's disease, at much higher incidence than men.

19. Research suggests that "befriending" behavior among females might be caused by the way their bodies process oxytocin. This research supports the _____ explanation of male-female difference.
    a. cultural
    b. social
    c. socialization
    d. biological

## What Produces Gender Inequality?

20. Functionalists argue that the family functions best when
    a. the father focuses on things outside the home while the mother focuses on relationships within the family.
    b. gender roles are more equal.
    c. men share more equally in the internal life of the family.
    d. the father deals with sons' emotions while the mother deals with daughters' emotions.

21. The person who takes the instrumental role in the family usually
    a. accompanies family sing-alongs.
    b. nurtures children.
    c. earns money.
    d. listens carefully to family members' emotional needs.

22. The person who takes the expressive role in the family usually
    a. earns money for the family's expenses.
    b. deals with interpersonal conflicts.
    c. pays the American Express bill.
    d. does the cleaning and makes household repairs.

23. Conflict theorists argue that gender inequality is _____ based.
    a. biologically
    b. functionally
    c. economically
    d. psychologically

24. Conflict theorists see the traditional male-female relationship as one of
    a. cooperation.
    b. exploitation.
    c. mutuality.
    d. passive-aggressive behavior.

25. Karl Marx's sometime collaborator, Friedrich Engels, argued that the basis for inequality between the sexes was
    a. biology.
    b. capitalism.
    c. socialism.
    d. communism.

## Gender-Role Socialization

26. Women today, says Tischler, are encouraged to pursue careers before, during, and after marriage. The increased presence of women in the labor force, according to this theory, arises from changes in
    a. the economy.
    b. socialization.
    c. education.
    d. hormones.

27. Women are _____ times more likely to develop chronic fatigue syndrome than men.
    a. 4
    b. 3
    c. 5
    d. 6

28. Glaucoma is found to be developed
    a. in greater frequency among women.
    b. in greater frequency among men.
    c. approximately equally among women and men.

29. Social psychologist Erik Erikson argued that in Western society it is more difficult for girls than for boys to
    a. achieve a positive identity.
    b. learn to moderate their innate aggression.
    c. learn to be nurturing.
    d. develop behaviors to attract a suitable mate.

## Gender Inequality and Work

30. T F Even when women and men have similar job titles and do equivalent work, women on average receive lower pay.

31. T F Despite recent changes, less than half the women in the United States have jobs in the paid labor force.

32. T F Women are now just as likely as men to win career-enhancing promotions.

33. Which gender earns the greatest share of undergraduate degrees?
    a. females
    b. male
    c. no difference

34. According to Deborah Tannen, men, more so than women, use language to
    a. make jokes with their buddies.
    b. come off as better than others.
    c. blow off steam.
    d. convey information.

35. According to Deborah Tannen, women, more so than men, use language to
    a. convey information.
    b. connect with other people.
    c. solve problems.
    d. complain.

36. According to Deborah Tannen, when someone brings up a personal problem in conversation, men are more likely to offer _____, whereas women are more likely to offer _____.
    a. money; ideas
    b. cynicism; sincerity
    c. bafflement; knowledge
    d. solutions; sympathy

37. T F Because Internet users need not disclose their true identities, men and women have similar styles when they contribute to discussions on the Internet.

38. T F Recent research now indicates that a majority of women and men now favor female bosses over male bosses in the workplace.

39. T F Although the proportion of female executives has increased, most of these women executives work for large corporations rather than as entrepreneurs who have started their own small businesses.

## Essay Questions

40. Present at least two arguments in support of the view that American society is biased against assertive women.

41. Discuss at least three problems posed by traditional gender-role socialization for women and men.

42. Describe the major patterns of job discrimination, which are related to gender, in the United States.

# CHAPTER 12   MARRIAGE AND CHANGING FAMILY ARRANGEMENTS

## The Nature of Family Life

1. **T F** Murdock's study of 250 societies found that in a small number of these societies—about six out of the 250—the family does not exist.

2. **T F** Every known human society has an incest taboo.

3. **T F** The relationships covered by the incest taboo vary from one society to another.

4. **T F** The nuclear family is found in all societies regardless of degree of industrialization.

5. **T F** Polyandry is a very rare form of family structure, existing in only a few societies.

6. **T F** Most of the world's societies have bilateral systems of descent.

7. **T F** Patriarchal family systems are far more common than matriarchal systems.

8. Patriarchy and matriarchy refer to
    a. whether a child takes its name from its mother or father.
    b. whether a couple lives near the husband's family or the wife's.
    c. the distribution of power between men and women in the family.
    d. whether the father or mother has primary responsibility for socializing children.

9. The family structure that consists of a married couple and their children is called the
    a. normal family.           c. matrilineal family.
    b. nuclear family.          d. extended family.

10. A society that traces descent through the mother's side of the family would be characterized as a _____ system.
    a. matriarchal              c. matrilocal
    b. matrilineal              d. polygynous

11. Dwayne and Katrina are a married couple. They have two children and live with Katrina's brother, his wife, and their three children as well as with Katrina's mother. This is an example of a(n) _____ family.
    a. extended                 c. nuclear
    b. polygamous               d. matriarchal

## Defining Marriage

12. **T F** Almost all societies allow for divorce.

13. **T F** Marriage for love has been the most common type of marriage throughout the history of the human species.

14. **T F** Many European nations have had laws restricting marriage between people of different religions; the United States has not.

15. **T F** Societies that see marriage as an economic and political institution are less likely to emphasize romantic love.

16. **T F** The most common type of interracial marriage in the United States involves a marriage between a white person and an African American.

17. According to anthropologist Marvin Harris, the major factor restraining the amount of polygamy in most cultures of the world is
    a. social disapproval.
    b. the likely increase in marital discord with more than one spouse.
    c. the high cost of maintaining more than one spouse.
    d. religious prohibitions against more than one marriage at a time.

18. One negative consequence of marriage based on romantic love is that it
    a. is less likely to produce children.
    b. is less likely to make for a strong husband-wife relationship.
    c. weakens ties to families of origin.
    d. is less suited to socializing children.

19. Homogamy refers to
    a. gay marriage.
    b. marriage in which family functions are shared equally between spouses.
    c. marriage based on love.
    d. marriage between people of similar social backgrounds.

20. In the United States today, the most common form of residence for newly married couples is
    a. patrilocal.
    b. matrilocal.
    c. bilocal.
    d. neolocal.
    e. uptown local.

21. The incest taboo ensures a certain degree of
    a. homogamy.
    b. exogamy.
    c. endogamy.
    d. monogamy.

22. Which of the following is a feature of marriage in all societies?
    a. Marriage is a public, socially approved relationship.
    b. The husband and wife are expected to love one another.
    c. The husband and wife must support one another emotionally.
    d. The husband and wife must socialize the children.

23. When did the Supreme Court rule that state laws banning interracial marriage were unconstitutional?
    a. shortly after the founding of the country
    b. shortly after the Civil War
    c. during the immigration waves of the late 1890s
    d. during the 1960s
    e. none of these; such laws have not been declared unconstitutional.

24. "Stick to your own kind," sings a character in *West Side Story*. She is advocating
    a. exogamy.
    b. polyandry.
    c. matriarchy.
    d. homogamy.

25. Since the 1950s, the age at which Americans first marry
    a. has been getting younger.
    b. has been getting older.
    c. has remained relatively unchanged.
    d. has fluctuated randomly with no discernible pattern.

26. Marriage in the United States has become less homogamous on all the following variables *except*
    a. race.
    b. religion.
    c. age.
    d. none of these; all exhibit a decreasing degree of homogamy.

27. The family in which a person is raised is his or her
    a. nuclear family.
    b. extended family.
    c. family of orientation.
    d. family of procreation.
    e. family of aggravation.

## ● The Transformation of the Family

28. T F The United States has the highest divorce rate in the world.
29. T F A twentieth-century leader of a modernized country had an arranged marriage.
30. T F Arizona's interracial marriage ban was reversed in 1867.
31. T F The percentage of births to unmarried mothers is higher in the United States than in any other industrialized country.
32. T F Nearly all studies of cohabitation show that couples who live together before marriage are less likely to get divorced than are couples who do not try living together before marriage.
33. T F Households in nineteenth-century America were much more likely than those today to include people not related to the family.
34. T F In divorce cases in the United States today, legal custody is given to the father about as often as it is given to the mother.
35. T F A large majority of divorced persons remarry.
36. T F Joint custody laws require children to spend equal time living with each of the divorced parents.
37. T F Children who are victims of abuse are more likely to be abusive as adults than are children who have not already experienced family violence.
38. T F The presence of children lowers the probability of remarriage for women but not for men.
39. T F Stepparents are more likely to abuse children than are biological parents.
40. T F The majority of gay and lesbian households include children.

41. Americans believe that marriage should be permanent, yet the United States has a high divorce rate. According to Andrew Cherlin, the value on permanent marriage is undercut by another aspect of American culture. What is it?
    a. the decline in religion
    b. hypocrisy and immorality
    c. violence
    d. the belief in the individual pursuit of happiness

42. Cohabitation is most common among those who
    a. are less religious than their peers.
    b. have been divorced.
    c. those who have experienced parental divorce.
    d. all of the above.

43. A term for marriage based on romantic love is
    a. homogamy.
    b. alternate marriage.
    c. companionate marriage.
    d. compassionate marriage.
    e. compatible marriage.

44. According to Tischler, World War II affected marriage in the United States because it
    a. led to the baby boom.
    b. led to an economic boom.
    c. moved women into the paid labor force.
    d. militarized the country.

45. Which of the following is a feature of the transformation of the American family in the six decades following the end of World War II?
    a. Average family size increased.
    b. The family reverted to a more extended family structure.
    c. The family decreased in importance in socializing children.
    d. Family roles became less equal.

46. Which of the following variables is *not* related to family violence?
    a. social class
    b. religion
    c. number of children
    d. unemployment

47. The largest differences in divorce rates in the United States are between
    a. upper-income and low-income.
    b. suburban and urban.
    c. Protestant and Catholic.
    d. Republican and Democrat.

48. No-fault divorce laws have
    a. made it more difficult to get a divorce.
    b. made it easier to get a divorce.
    c. had no effect on divorce.
    d. made it easier for men, but more difficult for women, to get a divorce.

49. T F Because divorce allows people to leave unhappy marriages, Americans in the twenty-first century who stay married report greater happiness in their marriages than did Americans of earlier generations.

50. T F Cohabitation refers to a situation in which a newly married couple settles down in a home that is near neither the bride's family nor the groom's family.

● Essay Questions

51. Family ranks high on the list of things Americans say they value. Yet, the divorce rate in the United States is much greater than it was in earlier generations and is much higher than in other advanced countries. How can you explain the value of family and the prevalence of divorce?

52. What functions does the family fulfill? How are these related to the way the society is organized? How have social changes in the United States over the past three centuries changed the functions that the family fulfills? How might these functions be fulfilled by institutions other than the family in other societies?

53. Some societies welcome romantic love. In other societies, it is seen as an aberration. What kinds of societies downplay romantic love?

# CHAPTER 13   RELIGION

● The Nature of Religion

1. T F According to Durkheim, the profane consists of objects that people are prohibited from touching, looking at, or even mentioning.

2. **T F** Unlike marriage, which is a universal institution, religion exists in many, but far from all, societies. About 15 percent of societies that we know about do not have any religion.

3. **T F** The concept "God" has had a fixed, unchanging meaning throughout human history.

4. Émile Durkheim observed that all religions, regardless of their particular doctrines, divide the universe into two mutually exclusive categories:
   a. the natural and the unnatural.
   b. the good and the bad.
   c. the ugly and the beautiful.
   d. the sacred and the profane.

5. A group of teenagers met each month at the full moon to worship Satan, curse God, and place voodoo hexes on their teachers. Durkheim would classify this as an example of
   a. the sacred.
   b. the profane.
   c. evil.
   d. rationality.

6. Compared with sacred things, things that are profane are more likely to be
   a. solid.
   b. large.
   c. useful.
   d. expensive.

7. Standardized behaviors or practices such as receiving holy communion, singing hymns, praying while bowing toward Mecca, and the bar mitzvah ceremony are examples of
   a. totems.
   b. magic.
   c. rituals.
   d. shamanism.

8. Which of the following, according to Durkheim, is an element in all religions?
   a. priests
   b. a holy book
   c. ritual
   d. a house of worship

9. All religions
   a. rely on the teachings of a sacred book.
   b. use magic in their rituals.
   c. promote social equality.
   d. demand some public, shared participation.
   e. include all of the above elements.

10. The use of peyote and other drugs in some religions is part of what more universal quality of religion?
    a. the violation of dominant norms and laws
    b. separation of people from one another by emphasizing individual experience
    c. creation of a special and heightened emotional state
    d. availability only to those with enough money

● Magic

11. One difference between magic and religion is that
    a. magic relies on the supernatural; religion is more rational.
    b. magic is usually used to benefit the individual; religion benefits the group.
    c. magic uses special objects and words; religion does not.
    d. magic is rational; religion is supernatural.

12. **T F** According to Stark and Bainbridge, Christianity has always been opposed to magic.

13. According to Stark and Bainbridge, the church's saints and shrines for specialized miracles, such as healing at Lourdes, are really a form of
    a. magic.
    b. the profane.
    c. totemism.
    d. Protestantism.
    e. rationalism.

14. **T F** According to Stark and Bainbridge, established churches do not want to get rid of all magic; they merely want to monopolize it for themselves.

## Major Types of Religions

15. **T F** Only three world religions are known to be monotheistic: Judaism and its two offshoots, Christianity and Islam.

16. An ordinary object that has become a sacred symbol for a group or clan is said to be its
    a. mana.
    b. mama.
    c. mojo.
    d. totem.

17. **T F** The earliest archaeological evidence of religious practice has been found in northern Europe.

18. In *The Exorcist*, a priest heals a girl by forcing a demon to leave her body. This idea behind exorcism is closest to
    a. monotheism.
    b. polytheism.
    c. animism.
    d. totemism.

19. The largest of the world's religions, in terms of the size of its membership, is
    a. Christianity.
    b. Islam.
    c. Hinduism.
    d. Confucianism.

20. Touching is important in many religions (for instance, in blessings and spiritual healing) because of a powerful but invisible force that can be transferred by touch. This force is generally known as
    a. mana.
    b. charisma.
    c. the force.
    d. kavorka.

21. "The carcasses of every beast which divideth the hoof, and is not cloven-footed, nor cheweth the cud, are unclean unto you: every one that toucheth them shall be unclean." (Leviticus 11:26) In other words, a person who touches an unclean animal, such as a pig, becomes himself unclean. This statement combines the concepts of
    a. ritual and prayer.
    b. cleanliness and godliness.
    c. taboo and mana.
    d. animism and theism.

22. Animism involves a belief in the power of
    a. animals.
    b. vegetables.
    c. minerals.
    d. spirits.

23. Most theistic societies practice
    a. monotheism.
    b. totemism.
    c. polytheism.
    d. ecumenism.

24. Ghosts are spirits of the dead that can haunt or help people. Belief in ghosts is a form of
    a. rationalism.
    b. animism.
    c. naturalism.
    d. theism.

## A Sociological Approach to Religion

25. Sigmund Freud emphasized which function of religion?
    a. helping individuals deal with guilt and anxiety
    b. bringing about group cohesion
    c. bringing spiritual enlightenment to individuals
    d. getting individuals to accept inequality in the society at large

26. **T F** Durkheim thought that even atheistic societies would develop rituals that served the same functions as religious rituals.

27. **T F** According to Durkheim, when people worship supernatural entities such as God, they are really worshiping their own society.

28. For Durkheim, the most important function of religion was to
    a. provide comfort for the individual.
    b. foster social cohesion.
    c. suppress social revolt.
    d. prevent suicide.

29. According to Max Weber, which of the following belief systems fostered a worldview that promoted the development of capitalism?
    a. Judaism
    b. Catholicism
    c. Confucianism
    d. Calvinism

30. According to Weber, which of the following provided the background in which European capitalism could flourish?
    a. Jewish values that emphasized education and learning
    b. Catholic values that emphasized glorifying God with expensive churches, clothes, art, and so on
    c. Protestant values that emphasized hard work and self-denial
    d. pagan values that emphasized innovation and problem solving

31. Which of the following is *not* a function common to most religions?
    a. emotional integration and the reduction of personal anxiety
    b. providing charity for the poor
    c. legitimizing arrangements in the secular society
    d. establishing a worldview that helps explain the purpose of life

32. According to Marvin Harris, the Hindu taboo against eating beef is most important because it
    a. promotes public health because many cattle in India carry disease.
    b. promotes public health by eliminating a source of cholesterol.
    c. allows for more economic sources of food and fuel.
    d. promotes social cohesion.
    e. gives support to the animal-rights movement.

33. The Ghost Dance of the Plains Indians is an example of a
    a. millenarian movement.
    b. revitalization movement.
    c. religious sect.
    d. universal church.

34. The best-selling *Left Behind* books and movie suggest that we are near the end of time, when the faithful will be taken to Heaven and the rest left behind to a dismal fate. The ideas behind *Left Behind* are an example of
    a. millenarianism.
    b. revitalization.
    c. denominationalism.
    d. totemism.

35. T F Alienation refers to the process by which people lose control over the social institutions they themselves invented.

36. What did Karl Marx refer to as "the opiate of the masses"?
    a. capitalism
    b. communism
    c. religion
    d. opium

37. From Marx's perspective, one of the chief dysfunctions of religion was that it
    a. discouraged people from taking action against injustice.
    b. alienated people from their society.
    c. took people away from the work they should be doing in the capitalist system.
    d. allowed priests to abuse their power for personal ends.

## Organization of Religious Life

38. The animistic beliefs and rituals of a Native American tribe, in which all members of the tribe participate, are an example of a(n)
    a. ecclesia.
    b. universal church.
    c. sect.
    d. denomination.

39. The Church of England, or Anglican Church, is the official church of that country, and its titular head is the king or queen of England. This would make the Anglican Church a(n)
    a. ecclesia.
    b. universal church.
    c. sect.
    d. denomination.

40. Which of the following types of religious organization is most likely to reject ideas of the dominant society?
    a. ecclesia
    b. universal church
    c. sect
    d. denomination

41. The Puritans who left England for America because they felt they were being persecuted are an example of a(n)
    a. ecclesia.
    b. universal church.
    c. sect.
    d. denomination.

42. Once established in Massachusetts, the Puritans became dominant, and their religious ideas became the basis for politics and society. The Puritans of Massachusetts thus constituted a(n)
    a. ecclesia.
    b. universal church.
    c. sect.
    d. denomination.

## Aspects of American Religion

43. T F  A majority of Americans believe that the Bible is the word of God.

44. T F  The United States has a much broader diversity of denominations than do European countries.

45. During the 1960s, many religious organizations and many different churches came together to work for civil rights. This cooperation is an example of
    a. secularism.
    b. ecumenism.
    c. denominationalism.
    d. an ecclesia.

46. T F  Unlike in Europe, ecumenism has flourished in the United States because the boundaries between denominations here are less rigid and more fluid.

## Major Religions in the United States

47. Which of these religious categories has the most followers in the United States today?
    a. Catholicism
    b. Protestantism
    c. Judaism
    d. Buddhism
    e. Atheism

48. Which type of Protestant denomination has been gaining members more rapidly?
    a. Conservative (Mormons, evangelicals)
    b. Liberal (Episcopalians, Presbyterians)
    c. Neither; in recent years they have both lost membership.

49. Vatican II, the ecumenical council called by Pope John XXIII in the 1960s, generally had the effect of making church doctrine more
    a. conservative.
    b. liberal.
    c. traditional.
    d. theological.

50. T F  Although the Catholic Church officially condemns artificial means of birth control, most American Catholics do not support this ban.

51. T F  Because of the great diversity of religions in the United States, religious affiliation is a very poor predictor of political attitudes.

52. T F  Mega churches appeal mostly to older people and have a membership whose average age is well above the national average.

53. People born into which faith are most likely to change their religion?
    a. conservative Protestant
    b. liberal Protestant
    c. Catholic
    d. Judaism
    e. Islam

## Essay Questions

54. List and define the basic elements of religion. Which element do you see as most important?

55. Choose a religion with which you are familiar and show how it fulfills the four major functions of religion.

# CHAPTER 14   EDUCATION

## Education: A Functionalist View

1. **T F** The functionalist perspective stresses the role of schools in perpetuating class differences from generation to generation.

2. **T F** Preliterate societies that have no schools do not fulfill the basic functions of education.

3. **T F** Many preliterate societies do not distinguish between education and socialization.

4. **T F** The American educational system helps slow the entry of young adults into the labor market.

5. **T F** Even in colonial times, patriots such as Ben Franklin saw schools as a way of Americanizing immigrants.

6. **T F** Lester Frank Ward believed the main purpose of education was to equalize society.

7. To say that schools are places of socialization means that schools
    a. provide a place for children to interact with friends.
    b. are anticapitalist and run by the government.
    c. inculcate the ways of the society in children.
    d. should focus on basic academic skills.

8. Which of the following is a latent function of education?
    a. reducing unemployment rates by keeping youths out of the labor market
    b. teaching basic academic skills
    c. transmitting cultural knowledge
    d. generating innovation

9. To say that child care is a "latent" function of schools means that child care
    a. does not come until later in life.
    b. is necessary for children whose parents work late.
    c. is not an officially stated goal of schools.
    d. is unimportant.

10. Opponents of bilingual education argue that it
    a. is too expensive.
    b. hurts immigrant children by not giving them necessary language skills.
    c. favors nonnative children at the expense of native-born children.
    d. confuses children as to which language is the best.

11. What, according to Lester Frank Ward, is the main source of inequality in society?
    a. the difference in intellectual abilities of those at the top and bottom of society
    b. the differences in the way rich and poor families socialize their children
    c. the unequal distribution of knowledge
    d. the poor academic skills of teachers in the inner cities
    e. all of the above

12. The basic message of the report titled *A Nation at Risk* was that U.S. schools
    a. were the best in the world but would soon face competition from Asian countries.
    b. had held steady in their performance.
    c. were improving but not rapidly enough.
    d. were so bad they put the future of the country in danger.

13. Major assessments of U.S. students have found them most lacking in ability in
    a. math.
    b. English composition.
    c. English grammar.
    d. geography.
    e. athletics.

14. The single most important element in the phenomenon of continuing innovation in American society is the
    a. work done by garage and basement hobbyists.
    b. continuous effort to recruit foreign geniuses.
    c. performance of high-caliber academic and research universities.
    d. increased attention to standardized testing in science education.

15. The "back to basics" movement in education emphasized rote learning, for instance, memorizing the multiplication table. The effect of this kind of education was that
    a. students were left unprepared to deal with more complex problems.
    b. students were better behaved.
    c. students got bored quickly.
    d. students came to dislike their teachers.

16. According to some studies, what proportion of fourth graders in the United States can read proficiently?
    a. less than 33 percent
    b. 50 percent
    c. 75 percent
    d. 90 percent

17. In 2006, four years after No Child Left Behind was passed, 15-year-olds in the United States, compared with those in 57 other countries
    a. did not score in the top 20 in math, science, or literacy.
    b. were in the top 10 in literacy but only 15th in science and math.
    c. did well in science and math but not in literacy.
    d. were in the top 10 in math, science, and literacy.

## ● The Conflict Theory View

18. T F To the conflict theorist, the function of school is to help each student develop his or her particular talents and abilities so as to live a more fulfilling life.

19. T F To the conflict theorist, the function of school is to produce the kind of people the system needs.

20. T F According to conflict theorists, what is important about obtaining a degree from Harvard or Yale or some other elite college is that it is a guarantee that a person has received quality training.

21. T F According to conflict theorists, what is important about obtaining a degree from Harvard or Yale or some other elite college is that it is means the person probably will not rock the boat.

22. T F According to conflict theorists, the hidden curriculum of schooling subtly promotes creativity and imagination and downplays rote learning.

23. In the view of conflict theory, the most important thing that schools actually do is
    a. provide child care.
    b. provide skills so that children will become successful.
    c. provide employment for teachers.
    d. preserve the existing class system.

24. The "hidden curriculum," according to conflict theory, includes
    a. advanced placement classes for the lucky few.
    b. after-school programs.
    c. getting students to accept society as it is.
    d. secret clubs like the Dead Poets' Society.

25. "Tracking" in schools, according to conflict theory, is a way of
    a. winning track meets.
    b. keeping track of problem students.
    c. increasing racial and class inequality.
    d. adjusting teaching to students' abilities.
    e. finding talented but underprivileged students.

26. Rosenthal and Jacobson found that student performance is substantially affected by the
    a. location of their school.
    b. occupational status of their parents.
    c. level of their innate intelligence.
    d. expectations of their teachers.

27. Which of the following is an aspect of the credentialized society?
    a. Credentials are a sign of competence.
    b. For employers, socialization matters more than does competence.
    c. The United States is moving away from an emphasis on credentials.
    d. Credentials are a way for less-privileged people to get ahead.

28. Ted Brown, an elementary school teacher for the past fifteen years, has returned to school to get a master's degree. The schooling will not make him a better teacher, but it will get him a raise. This is an example of
    a. functional illiteracy.
    b. the hidden curriculum.
    c. de facto segregation.
    d. the credentialized society.

## Issues in American Education

29. T F Cross-district busing of schoolchildren was a direct outgrowth of the *Coleman Report* of 1966.

30. T F The United States adopted the concept of mass public education only after it had been accepted in Europe.

31. T F The *Coleman Report* of 1966 found that minority students perform better when they go to school with others like them in predominantly minority schools.

32. T F The high-school dropout rate has been increasing gradually since 1980.

33. T F The *Coleman Report* and much subsequent research have shown that the more money a school district spends, the greater the increase in the achievement of its students.

34. T F Teachers often associate giftedness with children who come from prominent families, who have traveled widely, and who have extensive cultural advantages.

35. T F Standardized tests can accurately measure intelligence and abilities, especially among younger children.

36. T F The level of violence in schools has been decreasing since the mid-1990s.

37. T F In the past 20 years, the difference between males and females on the math part of the SAT has all but disappeared.

38. The Supreme Court decision declaring school segregation policies unconstitutional was handed down
    a. in the early days of the republic.
    b. shortly after the Civil War.
    c. shortly after World War II.
    d. in the 1950s.
    e. none of these; the Court has never declared segregation unconstitutional.

39. In the South, state policy required blacks and whites to attend separate schools. This system was known as
    a. de facto segregation.
    b. de jure segregation.
    c. de minimis segregation.
    d. de gustibus segregation.

40. The major cause of de facto segregation is
    a. the hidden curriculum.
    b. No Child Left Behind.
    c. de jure segregation.
    d. residential segregation.
    e. tracking.

41. What is the relationship between education and median income?
    a. Each higher level of education attained brings higher median income.
    b. Level of education attained has little effect on median income.
    c. Although obtaining a high-school diploma increases median income, going to college results in little additional earnings.
    d. A four-year college degree increases median income, but postgraduate degrees add little in median incomes.

42. From 1977 to 2011, total enrollment in two- and four-year colleges _____.
    a. increased by about 50%
    b. decreased
    c. more than doubled
    d. stayed the same

43. Which of the following is *not* a social consequence of dropping out of high school?
    a. increased crime
    b. decreased tax revenues
    c. increased intergenerational mobility
    d. reduced political participation

44. T F Nearly 20 percent of all students in grades 9–12 reported they had carried a weapon at least once during the previous month.

45. The *Coleman Report* of 1966 concluded that
    a. the quality of school experiences for black and white students had become approximately equal.
    b. the only way to improve the quality of school experiences for blacks was to spend more money on their schools.
    c. schools play a less important role in student academic achievement than once thought.
    d. academic success is most powerfully influenced by individual merit rather than by social factors.

46. Dropping out of high school affects not only those who leave school, but also society in general because dropouts
    a. pay less in taxes because of their lower earnings.
    b. are less likely to vote.
    c. have poorer health.
    d. increase the demand for social services.
    e. all of the above.

47. In its famous *Brown v. Board of Education* decision in 1954, the United States Supreme Court banned
    a. de jure segregation
    b. busing.
    c. standardized testing.
    d. tracking.

48. Over the past 75 years, the percentage of Americans completing high school has
    a. declined slightly.
    b. risen dramatically.
    c. risen slightly.
    d. remained about the same.

49. **T F** The early home schoolers in the 1950s and 60s objected to the permissiveness and liberalism of public schools.

50. **T F** One problem with home schooling today is that it deprives children of the chance to participate in extracurricular activities.

## Essay Questions

51. Compare and contrast the functionalist and conflict theory views of education as socialization.

52. Discuss the issue of gender bias in the classroom. Based upon your reading of this chapter, what is your position on this issue?

53. Discuss the pro and con issues connected with the phenomenon of home schooling.

# CHAPTER 15   POLITICAL AND ECONOMIC SYSTEMS

## Politics, Power, and Authority

1. **T F** The founders of the United States favored a strong central government.

2. **T F** Power that is regarded as illegitimate by those over whom it is exercised is called coercion.

3. **T F** Power based on fear is the most stable and enduring form of power.

4. **T F** Legitimacy refers to the condition in which people accept the idea that the allocation of power is as it should be.

5. **T F** Thomas Jefferson believed that, once established, a political system should be changed only in unusually dire circumstances.

6. **T F** Even a small tribe, in which decisions are made by group consensus, as long as it exercises control over its members, constitutes a state.

7. **T F** In most societies, what laws are passed or not passed depends to a large extent on which categories of people have power.

8. According to the text, the concession speech of the loser in a presidential election is important because it
   a. avoids long, drawn-out recounting of ballots.
   b. makes the loser look like a nice guy in case he or she decides to run again.
   c. affirms the legitimacy of the winner and of the political system as a whole.
   d. provides a sense of closure for all citizens.

9. Regardless of his personal abilities or popularity, Charles, Prince of Wales, will become King of England when his mother, Elizabeth II, abdicates the throne or dies. This is an example of _____ authority.
   a. rational-legal
   b. appointive
   c. traditional
   d. charismatic

10. Establishing durable institutions of government is a problem most likely to arise for rule based on
    a. charismatic authority.
    b. traditional authority.
    c. rational-legal authority.

11. Although Ronald Reagan was considered by many to have a magnetic personality and to be a symbol of the conservative movement in America, as president, he was nevertheless limited by the Constitution and the system of checks and balances established there. Thus, in Weber's terms, he is best thought of as a _____ authority.
    a. legal-rational
    b. representative
    c. traditional
    d. charismatic

12. _____ is the ability to carry out one's will, even in the face of opposition.
    a. Power
    b. Politics
    c. Force
    d. Authority

## Government and the State

13. Which of the following is *not* one of the basic functions of the state?
    a. establishing laws and norms
    b. ensuring economic stability
    c. protecting against outside threats
    d. socializing the young

14. The institutionalized way of organizing power within territorial limits is known as
    a. politics.
    b. authority.
    c. the state.
    d. the market.

## The Economy and the State

15. T F  In theory, "the invisible hand of the market" means that when all people act in their own self-interest, the result is of benefit to the society as a whole.

16. T F  In socialist societies, luxury goods tend to be expensive, whereas necessities are kept affordable.

17. According to Adam Smith, everyone in a society benefits most from
    a. competition among producers.
    b. centralized government planning.
    c. local government control of economic processes.
    d. democratic decision making in the workplace.

18. Which of the following is characteristic of mixed economies?
    a. All economic decisions are made by central planners.
    b. Private property is virtually abolished.
    c. The government intervenes to prevent industry abuses.
    d. Government planners decide on the best mix of public goods, such as education, and private consumer goods such as televisions.

19. Marx predicted that, as production expands in capitalist economies
    a. profits will decline and wages will fall, leading to revolution.
    b. wages will increase and profits will decline, leading to bankruptcy.
    c. profits and wages will both increase, leading to inflation.
    d. profits and wages will be become irrelevant as a decent standard of living for all is attained.

20. In many localities in the United States, primary and secondary schools are public; that is, they are provided by and ultimately run by the government. This form of education is an example of
    a. capitalism.
    b. laissez-faire.
    c. socialism.
    d. democracy.

21. Which of the following is one of the basic premises behind capitalism?
    a. Producers attempt to serve the best interests of society as a whole.
    b. Consumers attempt to serve the best interests of society as a whole.
    c. Democratically elected leaders decide what should be produced to meet the needs of the society.
    d. Free markets decide what is produced and for what price.

22. Marx thought there was an inherent conflict between the interests of capitalists and the interests of
    a. owners.
    b. consumers.
    c. professionals.
    d. workers.

23. _____ is a mechanism for determining the supply, demand, and price of goods and services through consumer choice.
    a. Command economy
    b. Legal-rational authority
    c. The market
    d. Representative government

24. The Federal Communications Commission is supposed to regulate broadcasting in the public interest. One FCC commissioner, implying that such regulation was not necessary, said, "The public interest is what interests the public." His statement is most in keeping with
    a. a mixed economy.
    b. laissez-faire capitalism.
    c. socialism.
    d. state capitalism.

25. In politics, people who speak most favorably about the virtues of the market are most likely to support
    a. grocery shopping.
    b. socialism.
    c. laissez-faire capitalism.
    d. a command economy.

26. The U.S. economic system is best described as
    a. laissez-faire capitalism.
    b. democratic socialism.
    c. a mixed economy.
    d. a command economy.

27. Which of the following is a basic principle of socialism?
    a. Everyone should have the essentials before some can have luxury items.
    b. Major decisions should be made by professional economists appointed by the government leaders.
    c. The government should give the greatest economic rewards to those who contribute the most to the society.
    d. Individual freedom cannot exist without economic freedom.

## Types of States

28. T F Democracy has always been regarded as the best form of government.

29. T F There are both capitalist and socialist states that are totalitarian in nature.

30. T F Representative government is a form of government in which a select few have all the power.

31. T F According to Edward Shils, civilian rule means that no member of the military may hold public office.

32. T F Democracy can exist only in capitalist societies.

33. Under democratic socialism
    a. private ownership of means of production is abolished.
    b. taxes are kept low.
    c. the state assumes ownership of strategic industries.
    d. little effort is made to expand social welfare programs or redistribute income.

34. _____ refers to the principle of limited government.
    a. The invisible hand
    b. Autocracy
    c. Constitutionalism
    d. Democracy

35. A major difference between an autocratic government and a totalitarian government is that under an autocratic government
    a. there is less repression of political dissent.
    b. people have more freedom in nonpolitical areas.
    c. there is greater economic planning.
    d. there is more influence by the military.

36. Minimum-wage laws require employers to pay workers at least a certain amount even if workers are willing to work for less. The government's setting wages violates the principle of
    a. the market.
    b. democracy.
    c. totalitarianism.
    d. socialism.

37. Which of the following is *not* a characteristic of democratic political systems?
    a. majority rule
    b. civilian rule
    c. direct government by the people
    d. constitutionalism

38. Socialists argue that
    a. true democracy is impossible in a capitalist society.
    b. with appropriate modifications, capitalism can be democratic.
    c. democracy is just an illusion in any form of society.
    d. socialism and democracy are in theory incompatible.

## Functionalist and Conflict Theory Views of the State

39. T F Functionalists maintain that the state emerged to manage and stabilize an increasingly complex society.

40. T F Historical evidence indicates that the earliest legal codes were enacted to protect the property of the wealthy.

41. Conflict theorists see which of the following as the key to the origin of the state?
    a. the need to coordinate increasingly large and complex societies
    b. the desire of populations to control their own destiny
    c. the nature of human nature, in which some will always dominate others
    d. the elite attempting to take control over surplus production

42. One main difference between the functionalist theory and the conflict theory regarding the origins of the state is that conflict theory emphasizes the idea that
    a. the state arose to deal fairly with conflicts.
    b. the state arose to make life better for all citizens.
    c. the state arose to protect the interests of the powerful.
    d. the state arose to reduce inequalities that could lead to conflict.

## Political Change

43. When the Chinese Communists under Mao Zedong took power in China in 1949, they sought to institute an entirely new way of life for their people, transforming politics, economics, and culture. This was an example of a
    a. rebellion.
    b. political revolution.
    c. political disorder.
    d. social revolution.

44. One thing that democracies have that dictatorships tend to lack is a mechanism for
    a. dealing with dissent.
    b. protection against foreign powers.
    c. implementing laws and official policies.
    d. institutionalized political change.

45. The American Revolution is an example of a
    a. rebellion that produced social change.
    b. rebellion that produced political change.
    c. revolution that produced social change.
    d. revolution that produced political change.

46. Rebellions differ from revolutions in that they
    a. do not use force.
    b. question the uses of power but not its legitimacy.
    c. are rarely successful.
    d. attack social but not political issues.

## The American Political System

47. T F Lobbyists are people paid by special-interest groups to attempt to influence government policy.

48. T F The percentage of women voting is substantially lower than the percentage of men, thus accounting for the small percentage of women elected to high public office.

49. **T F** Journalists and the media report the news on politics but have no real political influence on their own.

50. **T F** Men in the United States are more involved in politics, so they have higher rates of voting than do women.

51. Which of the following groups has the highest rate of voting?
    a. 18- to 20-year-olds
    b. the unemployed
    c. college graduates
    d. those with high-school diplomas but no college education

52. Which of the following statements concerning the political influence of Hispanics is correct?
    a. They have virtually no influence because of their very low numbers in the electorate.
    b. They have rather little influence because many do not vote.
    c. They have disproportionate influence because of their concentration in states with many electoral votes.
    d. They have disproportionate influence because of their large numbers and their relatively high rate of voter registration.

53. Data on members of Congress indicate that they are disproportionately
    a. Hispanic.
    b. young (under 50).
    c. male.
    d. intelligent.

## Essay Questions

54. Winston Churchill said, "Democracy is the worst form of government except for all those others that have been tried." What's wrong with democracy? What's wrong with those other forms of government?

55. Airlines, railways, banks, television and radio stations, medical services, colleges, and important manufacturing enterprises are some of the industries run by the state in various democratic socialist governments. What are the advantages and disadvantages of having the state run each of these?

# CHAPTER 16   HEALTH AND AGING

## The Experience of Illness

1. **T F** Being sick has consequences that are independent of any physiological effects.
2. **T F** In his book *The Youngest Science*, Lewis Thomas wrote about medical care prior to the advent of penicillin.
3. **T F** The United States has achieved the lowest infant mortality rate in the world.
4. **T F** By 2050, baby boomers are expected to cause the median age of the U.S. population to increase by 25 percent.
5. **T F** In most of the wealthier, industrialized countries, population is likely to decrease rather than increase.
6. **T F** It is only in the past 75 years that women in the United States have begun to experience a longer life span than their male counterparts.
7. **T F** The medical care system in the United States displays a great deal of mobility among the different types of health care providers.
8. **T F** In terms of the disparity in life span between the genders, men can be said to be at a distinct biological disadvantage.

9. Differences in life expectancy among countries are most affected by the rate of
   a. infant mortality.
   b. adult disease.
   c. abortion.
   d. heart attacks.
   e. AIDS.

10. According to statistics, which gender has higher rates of acute and chronic diseases and illnesses?
    a. males
    b. females
    c. there is no statistical difference.

11. According to 2012 CDC data, infant mortality rates of African Americans exceed those of their white counterparts by
    a. over 100 percent.
    b. approximately 50 percent.
    c. approximately 25 percent.
    d. approximately 15 percent.

12. A well-known example of an "age-specific death rate" is
    a. war.
    b. famine.
    c. the crude mortality rate.
    d. the infant mortality rate.

13. An African American baby born in 2010 could be expected to have a life expectancy _____ years shorter than a white female baby born in the same year.
    a. 10
    b. 6
    c. 14
    d. 3

14. One of the reasons women have, on average, longer life spans than men is that
    a. they are less likely to die young from non-medical causes such as accidents and homicides.
    b. parents take better care of girls than boys.
    c. they spend more of their income on preventive medical care.
    d. women get diseases at an earlier age and thus build up their immunity.

15. One hundred years ago, the poor health of low-income people was attributed to
    a. moral weakness.
    b. lack of medical care.
    c. pollution and other environmental factors in poor neighborhoods.
    d. hazardous work conditions.
    e. poor mental health.

16. The lowest cancer rates in the United States are found within which population?
    a. Caucasian
    b. Native American
    c. Asian American
    d. Hispanic

17. The infant mortality rate refers to
    a. babies who are born dead.
    b. miscarriages before the ninth month of pregnancy.
    c. babies who die within the first year of life.
    d. babies who die before reaching their tenth birthday.

18. Which continent has the highest rate of population growth?
    a. South America
    b. Europe
    c. Africa
    d. Asia
    e. North America

19. Women are more likely than men to experience psychological problems in the form of
    a. obsessive-compulsive disorder.
    b. drug abuse.
    c. depression.
    d. schizophrenia.

20. T F  The leading cause of death among African Americans is cancer.

21. T F  John Shaw Billings, U.S. Surgeon General in the late nineteenth century, attributed the high mortality rate of certain ethnic groups to their moral shortcomings.

22. According to the American Medical Association (AMA), a female doctor employed as a family practitioner can expect to receive approximately _____ percent of the salary of her male counterpart.
    a. 77
    b. 91
    c. 68
    d. 85

23. Patients who are attended to by female physicians typically
    a. report that they have divulged more information and have been more forthcoming.
    b. report a higher level of satisfaction than those attended to by a male physician.
    c. switch to a male physician within three to five years.
    d. live in more urban areas.

24. According to the Substance Abuse and Mental Health Service Administration, which of the following accounts for the highest proportion of admissions for addiction?
    a. methamphetamine
    b. cocaine
    c. heroin
    d. marijuana and hashish

25. Medicare reimbursement is _____ based, while Medicaid is _____ based.
    a. income | age
    b. age | income

26. Of every 100 practicing doctors in the United States, only about _____ are general practitioners.
    a. 15
    b. 12
    c. 8
    d. 4

27. Taken as a whole, approximately _____ percent of the U.S. population is classified as overweight while _____ percent is classified as obese.
    a. 45 | 30
    b. 70 | 15
    c. 30 | 20
    d. 55 | 20

28. The model of illness prevention stated in the text is composed of three distinct components. They are
    a. medical | cognitive | behavioral
    b. medical | deterrent | systemic
    c. medical | societal | behavioral
    d. medical | behavioral | structural

29. It is estimated that by the year _____, 20 percent of the U.S. population will be aged 65 or older.
    a. 2020
    b. 2040
    c. 2030
    d. 2050

30. Alzheimer's disease—before its naming and diagnostic criteria—was originally classified as a type of _____ of unknown and unspecified origin.
    a. paranoia
    b. senility
    c. reaction to trauma
    d. hysterical reaction

31. At the turn of the twentieth century, three demographic trends helped to account for the relative youngness of the U.S. population. Which of the following is *not* one of those factors?
    a. high fertility rates
    b. high rates of tobacco-related deaths among adults
    c. immigration
    d. declining mortality rates among infants and children

32. The "old-old"—those aged 85 years or older—represent approximately what percentage of the elderly population (aged 65 and older)?
    a. 11
    b. 7
    c. 16
    d. 3

33. In classifying the American elderly in terms of financial position, they are seen as
    a. among the poorest and most prosperous.
    b. generally among the poorest.
    c. generally among the wealthiest.
    d. impossible to classify due to their diversity.

34. Birthrates in the United States are not equal among males and females. One gender outnumbers the other by a ratio of 105:100. To which gender is the higher proportion accorded?
    a. female
    b. male

35. According to Knowles (1977), the American health care system can be aptly described as "acute, curative, and _____."
    a. preventative
    b. hospital-based
    c. inherently costly
    d. essentially for the wealthy

36. The American medical system, in terms of workers, can be described as a triangle, whereby categories of workers are divided according to income, status, training, and so on. The classification system can be looked at vertically and/or horizontally in terms of its over 300 categories. Movement from one category to another is
    a. almost nonexistent.
    b. commonplace.
    c. almost always vertical.

37. T F Sociologists have found that, as is the case with many transmittable illnesses, obesity can often be "contagious."

38. T F Holding stereotypical attitudes toward the elderly, such as "Old people are incapable," is known as "gerentophobia."

39. T F The world's elderly population is expected to meet or exceed one-half billon by the year 2020.

40. The percentage of elderly people who live alone is lowest by far in which region of the world?
    a. North America
    b. Western Europe
    c. Asia
    d. Scandinavia

41. As far as politics are concerned, the elderly can be described as which of the following in terms of voter turnout?
    a. ambivalent
    b. "no shows" at the polls
    c. highly motivated
    d. disenfranchised

42. In present-day America, the ratio of those under the age of 18 to those over the age of 65 is approximately:
    a. 2:1
    b. 1:1
    c. 5:1
    d. 3:1

43. T F The set of norms whose end product is the transfer of the unwell individual into the health care system is known as the sick roll.

44. T F People aged 60 who were surveyed and asked the question, "Who is old?" responded on average by stating "those aged 72 or above."

45. T F Investments in the stock market is one of the chief factors contributing to the higher-than-average income of many elderly individuals.

46. T F Schneider, Conrad, Strauss, and Glaser suggest that the experience of being ill be looked at in a more objective and nonpersonal manner.

47. **T F** As of late, the study of such disorders/illnesses as obesity, drug abuse, and alcoholism has begun to move away from "medicalization."

48. **T F** The American health care system represents a dichotomy. While many view it as the most advanced in the world, it is still viewed by a majority as wholly inadequate.

49. **T F** Among whites, African Americans, and Asian Americans, only among the second group is homicide one of the eight leading causes of death.

50. **T F** In the United States, more than 10 percent of the elderly population is "institutionalized" (includes homes for the aged, convalescent homes, mental-health facilities, and so on).

## Essay Questions

51. We usually think of disease prevention as resting with the individual. Some people make healthy choices; others don't. But structural prevention suggests that forces beyond the control of the individual are also important. Explain how these larger forces affect the health of individuals and of the society generally.

52. Describe an occasion when you were sick, showing how your behavior was shaped not just by your own reactions but by the demands of the sick role. What aspects of this role fit with what you wanted to do, and what aspects of the role were in conflict with your own preferences? How did others try to get you to conform to the demands of the sick role?

53. Write an essay describing how demographic factors affect a person's health. How do sex, race, age, social class, and so on figure in the probability of various health outcomes (disease, recovery, death, and so on)?

54. Many sociological factors affect life expectancy. Discuss two of the following: (1) occupation; (2) leisure activities; (3) friendships and acquaintances.

55. If you were put in charge of a nusing home, what types of services would you make available? What types of issues would you need to plan for among the residents? What changes would you expect to make in services in the coming decades?

# Practice Test Answers

## CHAPTER 1  THE SOCIOLOGICAL PERSPECTIVE

| | | | | | | | |
|---|---|---|---|---|---|---|---|
| 1. T | 8. a | 15. F | 22. F | 29. T | 36. c | 43. a | 50. a |
| 2. T | 9. c | 16. T | 23. T | 30. c | 37. a | 44. a | 51. d |
| 3. F | 10. b | 17. T | 24. T | 31. c | 38. b | 45. T | |
| 4. F | 11. d | 18. F | 25. F | 32. d | 39. c | 46. F | |
| 5. T | 12. b | 19. F | 26. T | 33. b | 40. b | 47. T | |
| 6. c | 13. c | 20. F | 27. T | 34. d | 41. a | 48. a | |
| 7. b | 14. F | 21. F | 28. F | 35. c | 42. a | 49. c | |

## CHAPTER 2  DOING SOCIOLOGY: RESEARCH METHODS

| | | | | | | | |
|---|---|---|---|---|---|---|---|
| 1. T | 7. T | 13. T | 19. F | 25. b | 31. c | 37. a | 43. T |
| 2. T | 8. T | 14. F | 20. F | 26. c | 32. c | 38. b | 44. F |
| 3. F | 9. F | 15. F | 21. T | 27. a | 33. b | 39. a | 45. F |
| 4. F | 10. T | 16. F | 22. T | 28. b | 34. a | 40. b | 46. F |
| 5. F | 11. T | 17. F | 23. F | 29. e | 35. b | 41. c | 47. T |
| 6. T | 12. T | 18. T | 24. T | 30. b | 36. d | 42. c | |

## CHAPTER 3  CULTURE

| | | | | | | | |
|---|---|---|---|---|---|---|---|
| 1. F | 8. b | 15. T | 22. a | 29. T | 36. T | 43. T | |
| 2. F | 9. a | 16. a | 23. c | 30. T | 37. F | 44. F | |
| 3. T | 10. a | 17. d | 24. T | 31. T | 38. F | 45. T | |
| 4. F | 11. T | 18. a | 25. F | 32. F | 39. d | 46. c | |
| 5. T | 12. T | 19. c | 26. F | 33. b | 40. d | 47. b | |
| 6. c | 13. F | 20. a | 27. T | 34. c | 41. c | 48. a | |
| 7. c | 14. F | 21. b | 28. T | 35. c | 42. F | 49. c | |

## CHAPTER 4  SOCIALIZATION AND DEVELOPMENT

| | | | | | | | |
|---|---|---|---|---|---|---|---|
| 1. F | 8. b | 15. d | 22. F | 29. a | 36. T | 43. c | 50. a |
| 2. T | 9. c | 16. c | 23. F | 30. a | 37. F | 44. c | 51. a |
| 3. T | 10. b | 17. T | 24. F | 31. c | 38. F | 45. T | |
| 4. T | 11. d | 18. T | 25. T | 32. a | 39. F | 46. T | |
| 5. F | 12. d | 19. T | 26. F | 33. b | 40. c | 47. T | |
| 6. T | 13. a | 20. F | 27. F | 34. c | 41. b | 48. F | |
| 7. F | 14. F | 21. F | 28. c | 35. b | 42. a | 49. d | |

## CHAPTER 5  SOCIAL INTERACTION

| | | | | | | | |
|---|---|---|---|---|---|---|---|
| 1. T | 5. T | 9. c | 13. T | 17. d | 21. T | 25. b | 29. a |
| 2. F | 6. d | 10. T | 14. F | 18. a | 22. F | 26. b | 30. a |
| 3. T | 7. b | 11. F | 15. F | 19. a | 23. T | 27. a | 31. d |
| 4. F | 8. c | 12. T | 16. a | 20. a | 24. F | 28. a | 32. a |

## CHAPTER 6  SOCIAL GROUPS AND ORGANIZATIONS

| | | | | | | | |
|---|---|---|---|---|---|---|---|
| 1. F | 7. F | 13. F | 19. c | 25. T | 31. b | 37. F | 43. a |
| 2. F | 8. c | 14. F | 20. T | 26. F | 32. e | 38. F | 44. a |
| 3. F | 9. b | 15. F | 21. T | 27. d | 33. T | 39. F | 45. c |
| 4. T | 10. b | 16. b | 22. T | 28. d | 34. T | 40. F | |
| 5. F | 11. a | 17. c | 23. F | 29. d | 35. F | 41. b | |
| 6. T | 12. F | 18. a | 24. F | 30. c | 36. T | 42. a | |

## CHAPTER 7   DEVIANT BEHAVIOR AND SOCIAL CONTROL

| | | | | | | | |
|---|---|---|---|---|---|---|---|
| 1. T | 8. a | 15. a | 22. b | 29. F | 36. T | 43. F | 50. c |
| 2. F | 9. d | 16. e | 23. d | 30. F | 37. F | 44. T | 51. a |
| 3. T | 10. F | 17. c | 24. c | 31. T | 38. c | 45. F | 52. e |
| 4. T | 11. F | 18. a | 25. d | 32. T | 39. b | 46. c | 53. d |
| 5. T | 12. F | 19. c | 26. F | 33. T | 40. a | 47. a | |
| 6. F | 13. d | 20. c | 27. F | 34. T | 41. b | 48. b | |
| 7. b | 14. d | 21. d | 28. T | 35. T | 42. T | 49. e | |

## CHAPTER 8   SOCIAL CLASS IN THE UNITED STATES

| | | | | | | | |
|---|---|---|---|---|---|---|---|
| 1. T | 8. F | 15. b | 22. F | 29. b | 36. c | 43. T | 50. c |
| 2. T | 9. F | 16. e | 23. T | 30. d | 37. T | 44. F | 51. d |
| 3. T | 10. e | 17. b | 24. T | 31. a | 38. F | 45. T | |
| 4. F | 11. c | 18. F | 25. F | 32. T | 39. F | 46. T | |
| 5. F | 12. a | 19. F | 26. b | 33. T | 40. T | 47. T | |
| 6. T | 13. b | 20. T | 27. a | 34. F | 41. F | 48. T | |
| 7. T | 14. b | 21. T | 28. a | 35. c | 42. c | 49. T | |

## CHAPTER 9   GLOBAL STRATIFICATION

| | | | | | | | |
|---|---|---|---|---|---|---|---|
| 1. T | 8. T | 15. T | 22. T | 29. c | 36. T | 43. d | 50. F |
| 2. F | 9. c | 16. F | 23. d | 30. F | 37. c | 44. a | 51. F |
| 3. F | 10. F | 17. T | 24. a | 31. d | 38. c | 45. a | |
| 4. T | 11. c | 18. T | 25. T | 32. a | 39. c | 46. F | |
| 5. F | 12. b | 19. T | 26. e | 33. T | 40. T | 47. T | |
| 6. d | 13. F | 20. F | 27. T | 34. T | 41. F | 48. F | |
| 7. b | 14. T | 21. b | 28. F | 35. d | 42. c | 49. T | |

## CHAPTER 10  RACIAL AND ETHNIC MINORITIES

| | | | | | | | |
|---|---|---|---|---|---|---|---|
| 1. T | 8. b | 15. F | 22. T | 29. a | 36. c | 43. c | 50. b |
| 2. F | 9. d | 16. a | 23. T | 30. c | 37. b | 44. b | 51. d |
| 3. F | 10. d | 17. c | 24. c | 31. a | 38. T | 45. c | 52. T |
| 4. F | 11. d | 18. F | 25. b | 32. a | 39. T | 46. b | 53. b |
| 5. T | 12. a | 19. T | 26. c | 33. d | 40. c | 47. d | 54. c |
| 6. F | 13. c | 20. a | 27. c | 34. c | 41. b | 48. c | 55. d |
| 7. b | 14. c | 21. a | 28. b | 35. a | 42. c | 49. b | 56. b |

## CHAPTER 11  GENDER STRATIFICATION

| | | | | | | | |
|---|---|---|---|---|---|---|---|
| 1. F | 6. F | 11. a | 16. d | 21. c | 26. b | 31. F | 36. d |
| 2. F | 7. F | 12. c | 17. d | 22. b | 27. a | 32. F | 37. F |
| 3. T | 8. F | 13. d | 18. T | 23. c | 28. b | 33. a | 38. F |
| 4. T | 9. d | 14. b | 19. d | 24. b | 29. a | 34. d | 39. F |
| 5. F | 10. d | 15. a | 20. a | 25. b | 30. T | 35. b | |

## CHAPTER 12  MARRIAGE AND CHANGING FAMILY ARRANGEMENTS

| | | | | | | | |
|---|---|---|---|---|---|---|---|
| 1. F | 8. c | 15. T | 22. a | 29. T | 36. F | 43. c | 49. F |
| 2. T | 9. b | 16. F | 23. d | 30. F | 37. T | 44. c | 50. F |
| 3. T | 10. b | 17. c | 24. d | 31. F | 38. T | 45. c | |
| 4. T | 11. a | 18. c | 25. b | 32. F | 39. T | 46. b | |
| 5. T | 12. T | 19. d | 26. c | 33. T | 40. F | 47. a | |
| 6. F | 13. F | 20. d | 27. c | 34. F | 41. d | 48. b | |
| 7. T | 14. T | 21. b | 28. T | 35. T | 42. d | | |

## CHAPTER 13   RELIGION

| | | | | | | | |
|---|---|---|---|---|---|---|---|
| 1. F | 8. c | 15. T | 22. d | 29. d | 36. c | 43. T | 50. T |
| 2. F | 9. d | 16. d | 23. c | 30. c | 37. a | 44. T | 51. F |
| 3. F | 10. c | 17. T | 24. b | 31. b | 38. b | 45. b | 52. F |
| 4. d | 11. b | 18. c | 25. a | 32. c | 39. a | 46. T | 53. c |
| 5. a | 12. F | 19. a | 26. T | 33. b | 40. c | 47. b | |
| 6. c | 13. a | 20. a | 27. T | 34. a | 41. c | 48. a | |
| 7. c | 14. T | 21. c | 28. b | 35. T | 42. a | 49. b | |

## CHAPTER 14   EDUCATION

| | | | | | | | |
|---|---|---|---|---|---|---|---|
| 1. F | 8. a | 15. a | 22. F | 29. T | 36. T | 43. c | 49. F |
| 2. F | 9. c | 16. a | 23. d | 30. F | 37. F | 44. T | 50. F |
| 3. T | 10. b | 17. a | 24. c | 31. F | 38. d | 45. c | |
| 4. T | 11. c | 18. F | 25. c | 32. F | 39. b | 46. e | |
| 5. T | 12. d | 19. T | 26. d | 33. F | 40. d | 47. a | |
| 6. T | 13. a | 20. F | 27. b | 34. T | 41. a | 48. b | |
| 7. c | 14. c | 21. T | 28. d | 35. F | 42. a | | |

## CHAPTER 15   POLITICAL AND ECONOMIC SYSTEMS

| | | | | | | | |
|---|---|---|---|---|---|---|---|
| 1. F | 8. c | 15. T | 22. d | 29. F | 36. a | 43. d | 50. F |
| 2. T | 9. c | 16. T | 23. c | 30. F | 37. c | 44. d | 51. c |
| 3. F | 10. a | 17. a | 24. b | 31. F | 38. a | 45. d | 52. c |
| 4. T | 11. a | 18. c | 25. c | 32. F | 39. T | 46. b | 53. c |
| 5. F | 12. a | 19. a | 26. c | 33. c | 40. T | 47. T | |
| 6. F | 13. d | 20. c | 27. a | 34. c | 41. d | 48. F | |
| 7. T | 14. c | 21. d | 28. F | 35. b | 42. c | 49. F | |

# CHAPTER 16   HEALTH AND AGING

| | | | | | | | |
|---|---|---|---|---|---|---|---|
| 1. T | 8. T | 15. a | 22. d | 29. c | 36. a | 43. T | 49. T |
| 2. T | 9. a | 16. b | 23. b | 30. b | 37. T | 44. T | 50. F |
| 3. F | 10. c | 17. c | 24. d | 31. b | 38. F | 45. T | |
| 4. T | 11. a | 18. c | 25. b | 32. a | 39. F | 46. F | |
| 5. T | 12. d | 19. c | 26. c | 33. a | 40. c | 47. F | |
| 6. F | 13. a | 20. F | 27. d | 34. b | 41. c | 48. T | |
| 7. F | 14. a | 21. T | 28. d | 35. b | 42. a | | |